High-speed heterostructure devices

From device concepts to circuit modeling

Fuelled by rapid growth in the communications industry, compound heterostructures and related high-speed semiconductor devices are spearheading the drive toward smaller, faster and lower-power electronics.

High-speed heterostructure devices is a textbook on modern high-speed semiconductor devices intended for both graduate students and practising engineers. This book is concerned with the underlying physics of heterostructures as well as practical analytical techniques for modeling and simulating these devices. Emphasis is placed on heterostructure devices of the present and of the immediate future such as the MODFET, HBT and RTD. The principles of operation of other devices such as the Bloch Oscillator, RITD, Gunn diode, quantum cascade laser and SOI and LD MOSFETs are also introduced.

Initially developed for a graduate course taught at The Ohio State University, the book comes with a complete set of homework problems and a web link to homework solutions and MATLAB programs supporting the lecture material.

Patrick Roblin is an Associate Professor in the Department of Electrical Engineering at The Ohio State University.

Hans Rohdin is a member of the technical staff at Agilent Laboratories in Palo Alto, California.

High-speed heterostructure devices

From device concepts to circuit modeling

Patrick Roblin and Hans Rohdin

CAMBRIDGE
UNIVERSITY PRESS

PUBLISHED BY THE PRESS SYNDICATE OF THE UNIVERSITY OF CAMBRIDGE
The Pitt Building, Trumpington Street, Cambridge, United Kingdom

CAMBRIDGE UNIVERSITY PRESS
The Edinburgh Building, Cambridge CB2 2RU, UK
40 West 20th Street, New York, NY 10011-4211, USA
477 Williamstown Road, Port Melbourne, VIC 3207, Australia
Ruiz de Alarcón 13, 28014, Madrid, Spain
Dock House, The Waterfront, Cape Town 8001, South Africa

http://www.cambridge.org

First published 2002

Printed in the United Kingdom at the University Press, Cambridge

Typeface Times 10.5/14pt. *System* LaTeX 2_ε [DBD]

A catalogue record of this book is available from the British Library

Library of Congress Cataloguing-in-Publication data

Roblin, Patrick, 1958–
High-speed heterostructure devices / Patrick Roblin and Hans Rohdin.
p. cm.
Includes bibliographical references and index.
ISBN 0 521 78152 3 hardback
1. Semiconductors. 2. Very high speed integrated circuits. 3. Heterostructures.
4. Transistors. 5. Low voltage integrated circuits. I. Rohdin, Hans, 1954– II. Title.
TK7871.85.R56 2001
621.38154′2–dc21 00-066719

ISBN 0 521 78152 3 hardback

To Gloria, Sébastien and Sophie
To the memory of Daga and Göte Rohdin

Contents

Preface

High-speed heterostructure devices is a textbook on modern high-speed semiconductor devices intended for both graduate students and practising engineers. This book is concerned with the physics and processes involved in the devices' operation as well as some of the most recent techniques for modeling and simulating these devices. Emphasis is placed on the heterostructure devices of the immediate future: namely the MODFET, HBT and RTD. The principle of operation of other devices such as the Bloch oscillator, RITD, Gunn diode, quantum cascade laser and SOI and LD MOSFETs is also introduced.

This text was initially developed for a graduate course taught at The Ohio State University and comes with a complete set of homework problems. MATLAB* programs are also available for supporting the lecture material. They can be used to regenerate a number of the pictures in the book and to assist the reader with some of the homework assignments.

This book should also prove useful to researchers and engineers, as it presents research material which is disseminated throughout the research literature and has never before been presented together in a book.

This text starts with two chapters reviewing the semiclassical theory of heterostructure devices. Five chapters are dedicated to presenting a realistic picture of heterostructures, introducing quantum devices and developing practical tools for analyzing quantum transport in these devices in the presence of scattering, and at high frequencies. One chapter is focused on the Boltzmann equation and its application to the derivation of moment equations for high-field transport. Five chapters are dedicated to reviewing the modeling of long- and short-channel FETs, including charge control, DC and high-frequency characteristics and the electrothermal modeling of FETs. This is followed by four chapters providing advanced DC and microwave modeling techniques, including a detailed analysis of parasitics in these devices. Finally the book concludes with two chapters dedicated to HBTs. A number of the chapters also provide practical design examples.

* MATLAB is a registered trademark of the MathWorks, Inc.

Required background

This text is intended for graduate students who have been introduced to semiconductor devices by a textbook of the level of Streetman's *Solid State Electronic Devices*. A more advanced introduction to quantum mechanics, thermodynamics, band structure, phonons and devices is not assumed, as it is not realistic to request that the reader be familiar with all these theories. Our strategy is therefore to start from an undergraduate level and construct a more advanced theory of electronic heterostructure devices on this basis. However, there are a few concepts that we have not derived. The Boltzmann, Fermi–Dirac and Bose–Einstein distributions are postulated without a derivation from more fundamental principles. The results of the harmonic oscillator are also presented without a derivation. It is hoped that those graduate students not familiar with these topics will be motivated to take additional courses in classical thermodynamics, quantum theory, and semiconductor theory to enhance their understanding of those topics. But again this book is sufficiently self-contained that this is not a requirement.

Outline for the reader

Chapter 1 gives an overview of the device concepts introduced in all the chapters and motivates the need for these studies. This chapter also introduces MBE technology and its application to the growth of materials (alloys, pseudomorphic, modulation doped) for new device structures. The chapter concludes with a review of the cubic crystal structure and its reciprocal lattice.

Chapter 2 introduces the concept of heterostructures. Both gradually varying semiconductors (alloys) and abrupt heterojunctions are analyzed using the Anderson band-diagram model, and dipole correction effects are considered. The generalized low-field transport equations which apply to heterostructures are reviewed including the drift-diffusion and thermionic-diffusion models. A phenomenological model of ballistic electron launching is also presented. The principle of the heterojunction bipolar transistor is then introduced.

Chapter 3 presents a rigorous introduction to the concept of spatially-varying band structure using the generalized Wannier picture. Following a derivation of the Bloch theorem, the Wannier functions are introduced as the Fourier coefficients of the Bloch states and the Wannier recurrence equation is derived. For abrupt heterojunctions, the matrix elements of the heterojunction Hamiltonians are derived in the limit of the maximally transparent heterojunction. A multi-band density of states based on the impulse response is also introduced for the spatial identification of quantum resonances. The principal advantages of using the Wannier picture lie in: (1) its inherent capability to account rigorously for both the spatial variation of the band structure and its periodicity in $\mathbf{k}$ space, and (2) its representation in terms of difference equations which are easily amenable to numerical solution.

Chapter 4 presents some fundamental one-dimensional quantum devices realizable using semiconductor heterostructures. The first class of devices discussed is that involving an accelerated electron in a band structure subjected to an applied electric field. Topics covered include the Wannier ladder and the Zener resonant tunneling effect, the Houston state and the acceleration theorem and finally wave-packets and squeezed states. The second class of devices considered is quantum wells. Topics covered include rectangular and triangular wells and the formation of a two-dimensional electron gas and subbands. The third class of devices considered is resonant tunneling and resonant interband tunneling diodes. Finally the fourth class of devices discussed focuses on superlattices, including the formation of minibands, wave-function localization in random superlattices and the fractal spectrum in Fibonacci superlattices.

Chapter 5 introduces the major scattering processes which limit the performance of quantum devices. Both elastic and inelastic scattering are considered. First the spectrum of lattice vibrations is presented and a semiclassical phonon model is introduced. The general form of the electron–phonon interaction Hamiltonian is then derived and the specific matrix elements for polar, acoustic, and intervalley phonon scattering processes evaluated. Next interface roughness scattering is analyzed using a model of uncorrelated terraces with a Gaussian distribution in size, and alloy scattering is analyzed using the virtual-crystal model. The chapter finishes with a discussion of electron–electron scattering.

Chapter 6 presents a realistic treatment of the impact of scattering upon tunneling-based devices using a direct three-dimensional ensemble-average solution of the Schrödinger equation. The importance of a three-dimensional analysis is first demonstrated. Next the scattering-assisted tunneling theory is shown to lead to a system of coupled Wannier recurrence equations enforcing current conservation. The formalism is then generalized to handle multiple sequential scattering processes and the Pauli exclusion effect with the introduction of the self-energy and the impulse response. Results for various resonant tunneling diodes (RTDs) are then presented for each scattering process, both individually and combined.

Chapter 7 studies tunneling in the presence of a time-varying interaction potential. The problem of an accelerated electron in a band subjected to both uniform DC and AC fields is solved exactly. A general rigorous analysis in terms of Fourier series is then given. The importance of self-consistently solving the Poisson and Schrödinger equations for calculating the current is demonstrated. Calculated small- and large-signal device impedances are presented for RTDs and an equivalent circuit is developed for their microwave simulation. The chapter concludes by studying how infrared radiation is coupled to ballistic quantum transport, and presents the principles of operation and recent results for the quantum cascade mid-infrared (10 μm) laser.

Chapter 8 covers the problem of the calculation of the 2DEG concentration in both gated and ungated MODFET capacitors. The self-consistent solution of the Schrödinger and Poisson equations is discussed, and an approximate analytic solution based on the triangular well approximation is presented. The control of the 2DEG by a Schottky barrier, its high-frequency response and the MODFET capacitance are then modeled or analyzed. The chapter finishes with the modeling of the Schottky-barrier gate under forward bias.

Chapter 9 introduces simple transport models applicable to the MODFET and HBT. Transport in the electron gas is discussed using the Boltzmann equation. Approximate solutions are obtained for both small and large electric fields using the assumption of a drifted and heated Maxwell–Boltzmann distribution, and they are used to derive a generalized drift-diffusion current equation and its associated energy balance equation. These equations are solved to obtain the velocity–field relation in bulk silicon and GaAs, to analyze the Gunn effect, and to discuss transient and stationary overshoot in short-channel MOSFETs and MODFETs.

Chapter 10 is concerned with the I–V modeling of the MOSFET/MODFET. The I–V characteristic MOSFET/MODFET is studied using a simple charge control model and transport model. Emphasis is placed on studying short-channel effects and velocity saturation, and the threshold for their occurence. A discussion of the two-dimensional field effects and their impact on the drain conductance is presented. For this analysis the Grebene–Ghandhi model, the channel opening model, and a full two-dimensional solution are compared.

Chapter 11 develops and solves the long-channel MOSFET/MODFET wave-equation. An optimal non-quasi-static equivalent circuit and a large-signal model approximating the large-signal MODFET wave-equations are then presented. The large-signal state equations are shown to conserve charge, and a charge-based representation suited for a circuit simulator is presented.

In Chapter 12 the velocity-saturated MODFET wave-equation is developed and solved for the short-channel MODFET. An optimal non-quasi-static equivalent circuit is presented and compared with the exact solutions. The long- and short-channel model topologies are also compared. Finally a charge-based large-signal model is presented for the short-channel MODFET.

Chapter 13 is concerned with the table-based electrothermal modeling of FETs for use in microwave circuit simulation. This chapter covers various topics such as device physics and model topology, measurement and characterization, parameter extraction and data presentation algorithms, and finally circuit design and simulation. The FET model topology introduced in Chapter 11 is augmented to account for the low-frequency dispersions associated with self-heating and the parasitic bipolar transistor. The need for and application of isothermal and pulsed DC and RF measurement techniques are reviewed. These concepts and modeling techniques are illustrated with examples from two major technologies: SOI for low-power RF CMOS and LDMOS for high-power linear amplification. However, the material presented is general enough that the techniques discussed can be applied to other devices.

Chapter 14 develops an accurate analytical model for the DC characteristics of MODFETs designed in industry for ultimate performance in high-speed communication and instrument applications. A thorough motivation for the approximate treatment is given. An overview of materials issues and evolution follows. The high doping and short gates used for these devices require a refined treatment of the charge control and transport, respectively. A quasi-two-dimensional model that includes mixed gate and drain charge control in regions internal and external to the gate is developed. I–V characteristics and the internal field distribution are obtained. These allow prediction of basic breakdown characteristics, which, in turn, affect the reliability of the devices, as is discussed.

Chapter 15 continues the analysis of cutting-edge MODFETs, but switches gear from DC to AC performance. The equivalent circuit is developed based on the theory in Chapter 11, and includes some important effects which occur in a velocity-saturated MODFET. These are studied with classical electrostatic approaches. In addition to standard capacitances, interesting effects are induced by transit delays in the device. The output conductance, a notoriously elusive parameter, is analyzed and predicted. The chapter concludes with an almost-complete extrinsic equivalent circuit. What is left for later is the inclusion of the distributed gate metalization resistance. First, an important topic that requires its own chapter has to be covered.

Chapter 16 focuses on an effect that requires a rather deep and different detour into semiconductor physics. The effect is the interfacial gate resistance which is of significant importance for device performance and scaling. In its purest form it is also of interest in the context of Schottky-barrier formation, a topic that has inspired a plethora of models, several of which are reviewed. Theories for dispersion and tunneling at the gate–semiconductor interface are developed. These require, in addition to familiar device and circuit analyses, a quantum mechanical treatment of a rather complex nature. Bardeen's powerful view of tunneling is reviewed. The overlapping metal and semiconductor wave-functions are derived and motivated, respectively. The tunneling resistance is then derived. The various Schottky-barrier models can be accommodated by the model to produce theoretical values for the interfacial gate resistance. These are compared with the typical range of experimental values. After a summary and discussion of the results, the final extrinsic equivalent circuit for the velocity-saturated MODFET is arrived at.

After a brief overview of some high-frequency measurement issues, Chapter 17 uses the analytical physics-based MODFET equivalent circuit to predict and optimize the gain and noise. Two fundamental power gains and their cut-off frequencies are reviewed, as are three commonly used FET noise models. A general thermal noise model that accommodates the full extrinsic equivalent circuit is formulated and exercised. Some process and manufacturability issues affecting performance, yield, cost, and reliability are discussed. A very brief discussion on reverse modeling concludes this chapter on high-performance MODFETs.

Chapter 18 focuses on the modeling of the heterojuction bipolar transistor (HBT). Compact models for HBTs are developed with the intention of providing tractable equations for predicting the DC, small-signal AC, and large-signal properties of high-frequency and high-power devices. The models are connected to fundamental theory by appealing to results from a microscopic theory of transport based on a direct solution of the Boltzmann transport equation.

Chapter 19, our last chapter, gives the reader an in-depth look at examples of the device physics issues that must be faced in realizing the HBT devices described theoretically in Chapter 18. It covers the application of arsenide and phosphide compound semiconductor material systems to HBTs in detail. The main device-design problems for high-speed HBTs, and their interaction with fabrication, are described. An example of the problems posed by practical III–V surfaces is provided by an examination of the emitter–base saddle-point effect in AlGaAs/GaAs HBTs. The effect of material choice on the important area of thermal properties is described. Finally, this chapter examines long-term device degradation, using beryllium diffusion as an example to study the defect chemistry behind the problem.

Recommendations for the instructor

This book is best suited for a semester course. By focusing on device concepts rather than mathematical derivations during the lecture it is possible to cover one chapter a week. The mastery of the mathematical techniques presented is then acquired by the students when they complete the homework problems. These homework problems indeed usually motivate a careful reading of the derivations presented in each chapter.

New graduate students are generally sufficiently prepared by conventional undergraduate textbooks/courses to take this graduate course. An exception, however, is the concept of Brillouin zone and **k** space which is not often well mastered if it has been covered at all. To address this problem a review of the cubic crystal structure and its reciprocal lattice is included in Chapter 1. The concept of **k** space is also heuristically introduced in Chapter 1 before being rigorously derived in Section 3.2.2 for one dimension and in Section 3.2.4 for three dimensions using the translation operator.

We have found it to be of critical importance to provide the students with simple MATLAB* programs implementing the techniques presented. These MATLAB* programs serve multiple purposes. First they allow many of the figures in the text to be regenerated. The students can then vary the parameters and do simple experiments. Sometimes these tools are also used in exercises to verify the validity of analytic calculations. This is particularly important for the quantum calculations which can be quite abstract until the students start reproducing the results themselves. This literally

* MATLAB is a registered trademark of the MathWorks, Inc.

brings this material to life. A special web site is available from Cambridge University Press or from http://eewww.eng.ohio-state.edu/~roblin/cupbook for downloading these MATLAB† programs. We will keep adding new problems and programs to support this text. A correction set for most of the homework problems can also be downloaded by instructors from the same web site.

For the MATLAB product information, please contact

The MathWorks Inc.
3 Apple Hill Drive,
Natick, MA, 01760-2098 USA
Tel: 508-647-7000
Fax: 508-647-7101
E-mail: info@mathworks.com
Web: www.mathworks.com

† MATLAB is a registered trademark of the MathWorks, Inc.

Acknowledgements

Acknowledgements by Patrick Roblin

The writing of a graduate textbook on *high-speed heterostructure devices* has been a project that I conceived early in my research and teaching career with the motivation of presenting fundamental device concepts as well as addressing challenging modeling issues faced by researchers and engineers in this field. The realization of such a book would not have been possible, however, without the help of many researchers, and I would like to acknowledge them here. First, this book is the result of the cooperative writing of many experts in their fields. My coauthor contributed four key chapters on the state-of-the-art design and modeling of MODFETs, in addition to contributing an insightful and critical review of the overall manuscript. My PhD student, Dr Siraj Akhtar, now at Texas Instruments, cowrote with me a chapter on electrothermal modeling of FETs. Finally we invited two experts in HBT to contribute to this book. Prof. David Pulfrey of the University of British Columbia contributed a chapter on the modeling of high-frequency HBTs, and Dr Nick Moll of Agilent Laboratories contributed a chapter on the practical and theoretical know-how required for building high-performance HBTs.

The chapters that I contributed are for the most part based on original research papers published in the literature or on research conducted with my MS and PhD students at OSU. Therefore I would like in particular to thank my PhD students, Dr Young Min Kim, Dr Sung Choon Kang, Prof. Wan Rone Liou, Dr Chih Ju Hung and Dr Siraj Akhtar for their key contributions. My group was also enhanced by the important contributions of several postdoctoral researchers, Dr Paul Sotirelis, the late Prof. Gene Cao, and Dr Dae Kwan Kim. The contributions of many of these researchers would not have been possible without the research funding support provided by US government agencies (NSF and NEMO project) and US industry, principally Cray, Allied Signal, Texas Instruments and Lucent Technologies.

I am particularly indebted to the Electrical Engineering department at The Ohio State University for the support they provided for me throughout the years. Indeed this book could not have been written without the friendly and intellectually stimulating environment provided by my colleagues at OSU: Professors Steve Bibyk, Furrukh Khan, George Valco, Mohammed Ismail, Betty Lise Anderson, Steve Ringel, Bob

Sacks, Roberto Rojas, Len Brillson, and Paul Berger as they successively joined OSU. Likewise the development of this manuscript greatly benefited from the insightful comments, questions and evaluations of the graduate students who took my electrical engineering graduate course from which this manuscript was developed. Their input inspired the revision of many chapters and homework problems and the development of a number of supportive MATLAB* programs. I am also thankful to Julie Kasick for contributing many of the illustrations. Finally I owe so much to the Torrini and Roblin families for their support and encouragement throughout the years.

Acknowledgements by Hans Rohdin

I would like to acknowledge my colleagues at Agilent Laboratories (formerly part of Hewlett-Packard Laboratories) without whom I would have learned and experienced so much less. I feel particular gratitude towards Avelina Nagy, a close coworker and friend of 13 years, for her superb processing work and wisdom in all matters of life. Knowing that the limitations of the fabricated devices and circuits are set by physics alone makes the design, measurement, and analysis so much easier and more rewarding. Of course, a necessary condition for this to be the case is high-quality epitaxial material, and for this I thank Virginia Robbins, Alice Fisher-Colbrie, Dan Mars, and Midori Kanemura. Another special thank-you goes to Nick Moll for continually sharing his deep physical insight. In particular, I would like to thank him for letting me use his work on drain delays prior to publication. Of the many other present and former colleagues at Hewlett-Packard and Agilent Labs who have contributed greatly I have to, in the present context, 'single' out Chung-yi Su, Arlene Wakita, Judith Seeger, Alex Bratkovski, Rolf Jaeger, Chris Madden, and Greg Lee.

Joint acknowledgements

The authors both completed their doctoral work at Washington University in St Louis where they shared the same advisor and many professors.

The interactions with the professors and students at WU were stimulating on both an intellectual and a personal level, in a way that neither of us had experienced before. We are greatly indebted to our advisor Prof. Marcel Muller for his mentoring, guidance, and amazing physical insights. The friendship with him and his wife Ester has enriched our lives. We are also deeply grateful to the late Prof. Fred Rosembaum for his outstanding series of courses on microwave devices and circuits and for involving us in his exciting ONR research project 'Semiconductor Millimeter Wavelength Electronics'. We would also like to thank Profs. C. M. Wolfe and D. L. Rode for outstanding and stimulating courses in semiconductor physics.

* MATLAB is a registered trademark of the MathWorks, Inc.

Finally we would like to express our gratitude to our editors Dr Phil Meyler and Eric Willner for their support and guidance in this book project. It has been a great pleasure working with them. We are also indebted to Maureen Storey for her thoughtful and detailed editing of the entire manuscript, to Jane Williams for the jacket design, to St.John Hoskyns and Jane Williams for the excellent book layout and to Lucille Murby for guiding the production of this book to completion. The accuracy of the presentation owes a lot to the cheerful professionalism of the staff of Cambridge University Press. Any short-comings that remain are our own responsibility. Heterostructure device physics has fascinated the authors for more than two decades. We hope in turn that this book will be useful and inspiring to students and researchers alike.

Columbus P. R.
Palo Alto H. R.
July 2001

List of abbreviations

1SS	single-sequential scattering
2DEG	two-dimensional electron gas
2DHG	two-dimensional hole gas
3DEG	three-dimensional electron gas
AMPS	advanced mobile phone service
AUDM	advanced unified defect model
bcc	body-centered cubic
BJT	bipolar junction transistor
BTE	Boltzmann transport equation
CAD	computer-aided design
CDMA	code division multiple access
DBS	direct broadcast satellite
DDE	drift-diffusion equation
DHBT	double-heterojunction bipolar transistor
DIGS	disorder-induced gap state
DUT	device under test
FATFET	long-gate FET
fcc	face-centered cubic
FDMA	frequency division multiple access
FET	field-effect transistor
GCA	gradual-channel approximation
HBT	heterojunction bipolar transistor
HEMT	high-electron-mobility transistor
HFET	heterostructure field-effect transistor
HTOL	high-temperature operating lifetime
IC	integrated circuit
IF	intermediate frequency
JWKB	Jeffreys–Wentzel–Kramers–Brillouin
LGLTB	linearly-graded low-temperature buffer
LNA	low-noise amplifier
LRM	line, reflect, match
MBE	molecular beam epitaxy

MESFET	metal–semiconductor field-effect transistor
MIGS	metal-induced gap state
MISFET	metal–insulator–semiconductor field-effect transistor
MMIC	monolithic microwave integrated circuits
MOCVD	metal organic chemical vapor deposition
MODFET	modulation-doped field-effect transistor
MSS	multiple-sequential scattering
MSSCAT	multiple-sequential scattering-assisted tunneling
MTTF	mean time to failure
NDC	negative differential conductivity
NDR	negative differential resistivity
NEGF	non-equilibrium Green's function
NWA	network analyzer
OEIC	optoelectronic integral circuits
OMVPE	organometallic vapor phase epitaxy
PA	power amplifier
PAE	power-added efficiency
PDC	positive differential conductivity
PE	Poisson's equation
PECVD	plasma-enhanced chemical vapor deposition
PHEMT	pseudomorphic high-electron-mobility transistor
PHS	Pucel, Haus and Statz
POP	polar-optical phonon
RF	radio frequency
RFOL	RF operating lifetime
RIE	reactive ion etch
RITD	resonant interband tunneling diode
RTD	resonant tunneling diode
SBGFET	Schottky-barrier-gate field-effect transistor
SDHT	selectively doped heterojunction transistor
SEM	scanning electron microscopy
SHBT	single-heterojunction bipolar transistor
SII	screened ionized impurity
SOI	silicon on insulator
SOLT	short, open, load, thru
SPICE	simulation program for integrated circuit emphasis
SWE	Schrödinger wave-equation
TEGFET	two-dimensional electron gas field-effect transistor
TPS	tensor product spline
VCO	voltage controlled oscillator
WKB	Wentzel–Kramers–Brillouin

Introduction

It is the trend in the silicon and compound microelectronic technology to continuously develop semiconductor circuits which are faster, smaller, and consume less power for a similar level of integration. This has been recently fueled in part by the rapid growth of digital wireless communication, which relies on both low-power high-speed digital and high-frequency analog electronics. As part of this trend, microwave, RF and IF analog and digital circuits are being integrated in 'mixed-signal' circuits for wireless applications. Both silicon and compound state-of-the-art integrated circuits presently rely on high-speed state-of-the-art submicron devices. However, research in microelectronic technology is always expanding its frontier; new heterostructure semiconductor materials and devices are continuously being developed or improved in a process often referred to as bandgap engineering. These heterostructure devices, in particular, and high-speed devices, in general, constitute the subject of this book. In this book we take the readers on a journey providing them with an understanding of both fundamental and advanced device-physics concepts as well as introducing them to the development of realistic device models which can be used for the design, simulation and modeling of high-speed electronics.

The journey in this book takes the reader from the fundamental physical processes taking place in heterostructures to the practical issues involved in designing high-performance heterostructure devices.

Ever shrinking high-speed devices

It is a basic requirement that high-speed devices must be small. Reducing the device reduces the transit-time and the capacitances in devices. The operating voltage is also reduced, and this helps with the reduction of the power dissipation. There are a few exceptions, i.e., devices which do not rely on the transit-time principle, but essentially this principle holds so far for the field-effect and bipolar transistors which are the engine of today's microelectronics. The shrinking of the device is occuring both horizontally, as defined by lithography, and vertically, as defined by growth and processing techniques. For example MOSFETs and MODFETs of 0.085 μm or 850 Å gate length are becoming very common. Even more striking are the modern growth techniques which have made possible the vertical growth of new semiconductor devices with unprecedented control. One of the most versatile growth techniques available for research is molecular beam epitaxy (MBE) which permits one to deposit one atomic layer at a time while abruptly or gradually changing the semiconductor material and doping type. The capability of MBE growth techniques will be reviewed in Chapter 1.

Quantum effects

In Chapter 2 we will explore the semiclassical modeling of heterostructures by reviewing how the bulk and junction theory has been extended to deal with them. However, as the device size keeps shrinking, quantum effects clearly become important and must be considered. This occurs when the device dimension compares with the mean free path of the electrons. In fact many fundamental questions are raised when dealing with very small devices. Traditionally semiconductors are theoretically introduced as crystals which are by definition periodic structures repeating indefinitely. But how can a band structure now be rigorously defined in spatially-varying semiconductors? To address this question we will introduce in Chapter 3 a special quantum picture, the generalized Wannier representation, which will describe the formation of the bands at the lattice level. In fact we shall see that it takes typically about ten lattice parameters for the band structure to be well defined away from an interface or surface. The generalized Wannier representation will also permit us to discuss in Chapter 4 transport problems such as the Bloch oscillations which have long both fascinated and challenged device physicists.

Quantum devices

Quantum devices are devices which directly exploit quantum effects. Various types of quantum devices have been conceived, including quantum wells, superlattices and resonant tunneling diodes (RTDs). Superlattices are periodic heterostructures forming a synthesized one-dimensional crystal. Superlattices can be used, for example, to generate the elusive Bloch oscillations. The study of random superlattices will permit us to gain insight into the conductor–insulator transition which takes place when the superlattice periodicity is destroyed.

With a couple of periods of a superlattice we can form a double-barrier potential system, which is transparent to electrons when the barrier separation corresponds to a multiple of half their wavelength. This effect is the basis for the RTD which exhibits a negative differential up to terahertz. The RTD, which is the fastest active semiconductor diode available so far, is a very important test device, as it is based on a quantum effect and yet operates at room temperature. It also finds applications in high-speed digital and microwave circuits.

Finally, one of the key quantum effects we shall study is the creation of a two-dimensional electron gas (2DEG) by quantum confinement. The 2DEG is of particular importance as it is used as the channel of the fastest FET developed: the MODFET. Chapter 8 will therefore be dedicated to studying how we can control this 2DEG with a gate voltage at both DC and high frequency.

From quantum transport to Boltzmann equation

To the first order, transport in quantum devices is typically ballistic. That is the electrons travel with out being scattered. However, even in quantum devices the ballistic transport approximation is not realistic, and scattering processes must be accounted for. How do we solve the Schrödinger equation in the presence of phase-breaking scattering processes? We shall address this subject in Chapter 6 and develop a realistic theory bridging the gap between the ideal ballistic transport model and the semiclassical Boltzmann transport theory. As we shall see, the electron wave-functions are effectively attenuated in their propagation as they spawn new scattered waves through various possible scattering processes. The exploitation of quantum effects in quantum devices is therefore only possible when the spatial variation device structure is smaller than the mean free path. Indeed, it is only when the electron wave-function has a well-defined phase that interferences, which are a requirement of quantum wave effects, can effectively take place.

Ballistic transport versus drift-diffusion transport

Even when quantum effects are negligible we will find it necessary to identify whether or not ballistic transport or/and drift-diffusion is taking place. These are indeed the two fundamentally different regimes of transport which can both take place, sometimes simultaneously, inside a device.

Consider the simple pastoral scene of a lake with a waterfall on one side and a small creek on the other. There is clearly a continuous flow of water from the waterfall, through the lake and into the creek. However, if the lake is very wide, the water drift might not even be perceptible to a fisherman on its bank fishing for trout. But a fly-fisherman fishing in the creek will see his dry fly quickly drift away and will need to recast his line often. If a red dye or some kind of liquid trout food is poured into the lake at one spot we expect it to slowly diffuse and spread throughout the lake. A drift-diffusion model therefore applies well to the lake area. On the other hand, no dye is expected to be able to diffuse up the waterfall into its upper reservoir. Clearly the waterfall is operating in a ballistic mode. The creek, however, might be operating in some kind of mixed mode with faster water at the center moving in a near-ballistic way and pockets of slower water near its bank trapping some of the dye.

In the ballistic regime, drift not only dominates over diffusion but also no thermalization is expected in that region. The particles acquire mostly kinetic energy from the potential (gravitational in the case of the waterfall, electrostatic in the case of electrons). However, a release of thermal energy is still expected due to the Joule effect. In ballistic transport it is in the collector (the lake in our example) that the particle kinetic energy is converted into thermal energy as is evidenced by the increased agitation or random motion of the particles in the lake region surrounding the waterfall.

For ballistic transport in electron devices we shall find that the current is also space-charge-limited: that is the charge distribution of the electrons actually screens the applied potential accelerating the electrons. This self-consistent process profoundly alters the current flow and effectively shapes the I–V characteristic of ballistic devices.

In the course of our study of high-speed devices we shall find that ballistic and space-charge-limited transport takes place not only in quantum devices (Chapters 4 and 7), but also in the drain region of FETs (Chapters 10 and 14) and in the base region of heterojunction bipolar transistors (HBTs) (Chapters 2 and 18) with an abrupt emitter–base junction.

Importance of a microscopic study

We have discussed above some of the key physical processes we must address when dealing with heterostructure devices with dimensions below 0.1 μm. The very nature of the junctions is of critical importance in heterostructure devices. New modes of transport such as ballistic and space-charge limited transport can be expected in these devices. For even smaller devices, quantum transport is fully required for their analysis. These effects justify our detailed study in the first half of this book of the various microscopy processes taking place in devices. Then, equipped with this physical understanding we will be ready to complete our journey in the second half of this book on the detailed analysis of the operating principles, modeling, and design of high-speed heterostructure transistors.

Big Science, Little Science by Sidney Harris.

1 Heterostructure materials

It is as easy to count atomies, as to resolve the propositions of a lover.

As You Like It, WILLIAM SHAKESPEARE

1.1 Introduction

Modern growth technologies have made possible the growth of new semiconductor devices with unprecedented control on the atomic level. In this chapter we shall briefly introduce the molecular beam epitaxy (MBE) growth technique and discuss its application to the growth of materials (alloys, pseudomorphic, modulation doped) for new device structures. The chapter will conclude with a review of the cubic crystal structure and its reciprocal lattice, as these concepts are used extensively in Chapters 2 and 3.

1.2 MBE technology

One of the most versatile growth techniques available for research is the MBE. In this growth technique a semiconductor substrate is placed in a high-vacuum chamber (see Figure 1.1). Different components such as Ga, Al, As, In, P, and Si are heated in separate closed cylindrical cells. These components escape through an opening in the cylindrical cell and form a molecular beam. These beams are directed toward the substrate. A shutter positioned in front of each cell is used to select the desired molecular beams. By selecting a low temperature for the substrate growth and a slow growth rate (a few micrometers per hour), it is possible to grow high-quality crystals, while making abrupt changes in doping and crystal composition.

This growth technique can be used to grow semiconductor alloys such as $Al_xGa_{1-x}As$, $In_xGa_{1-x}As$, $In_xAl_{1-x}As$, and $Si_{1-x}Ge_x$, where x, the mole fraction, specifies the composition of the alloy. For example in $Al_xGa_{1-x}As$, Al and Ga are randomly distributed over the same Ga lattice site of the GaAs crystal and x gives the fraction of Ga sites occupied by Al.

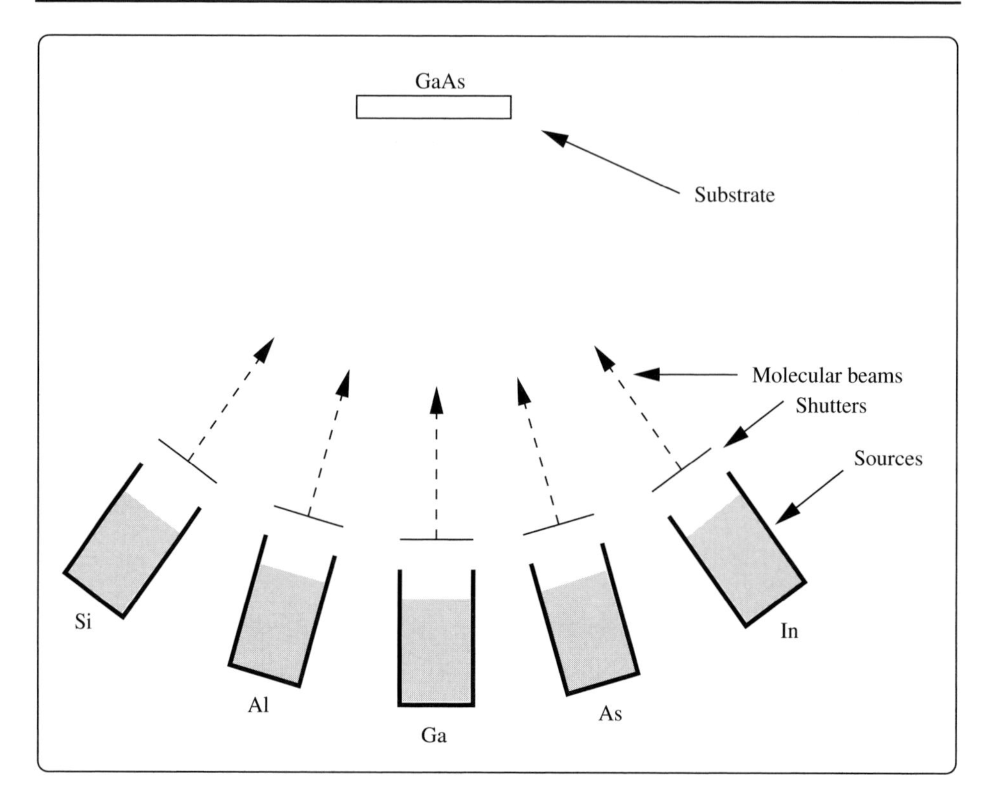

Fig. 1.1. Simplified diagram of an MBE growth chamber.

1.2.1 Lattice-matched systems

The epitaxial growth of one semiconductor on top of another requires that they have a similar lattice parameter so as to minimize the number of defects in the epitaxial layer. As can be seen in Table 1.1, Ge, GaAs, and AlAs have nearly the same lattice parameters. The lattice mismatch measured as $\Delta a/a$ is a fraction of 1%, and these semiconductors can be grown epitaxially on top of each other with extremely small concentrations of defects. This is therefore also the case of the alloy $Al_xGa_{1-x}As$. Complex heterostructures making use of the bandgap variation between the GaAs and AlAs bandgaps can then be grown on a binary GaAs substrate as is illustrated in Figure 1.2.

The $Al_xGa_{1-x}As$ and Ge system is, however, an exceptional case. For example GaAs, InAs, and their alloy $In_xGa_{1-x}As$ can be seen in Figure 1.2 to have quite different lattice parameters leading to a lattice mismatch of up to 7%. In such a case the growth of lattice-matched semiconductor alloys can be achieved by selecting the mole fraction such that both semiconductor alloys have the same lattice parameter. From Figure 1.2 one can verify that $In_xGa_{1-x}As$ and $In_xAl_{1-x}As$ can be grown epitaxially

Table 1.1. *Material parameters for Ge, GaAs and AlAs.*

	$a(\bar{A})$	$\frac{\Delta a}{a_{GaAs}}$	E_g	χ (affinity)
Ge	5.6461	1.3×10^{-3}	0.663	4.13
GaAs	5.6533	0	1.424	4.07
AlAs	5.6614	1.2×10^{-3}	2.16	3.5

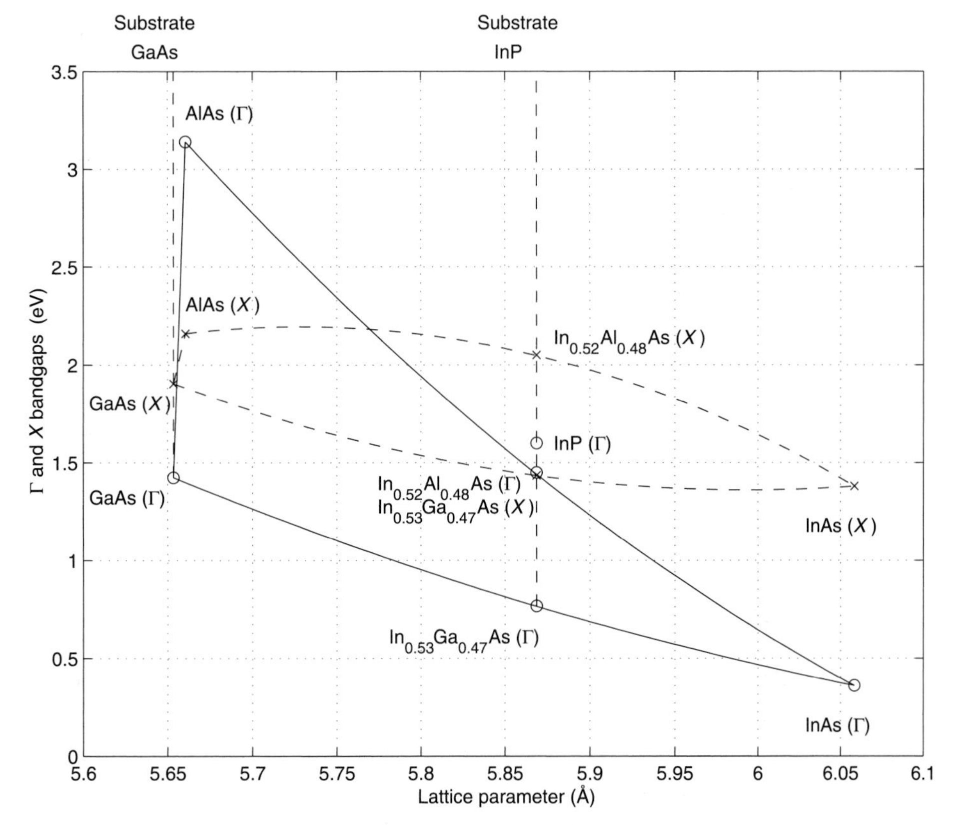

Fig. 1.2. Direct (optical) Γ (plain line) and indirect X (dashed line) bandgaps (see Figure 2.1 for a definition) of the alloys of the semiconductor binaries GaAs, AlAs, and InAs, plotted versus their lattice parameters for all mole fractions x.

on an InP substrate when using the In mole fractions $x = 0.53$ and $x = 0.52$, respectively.

1.2.2 Pseudomorphic materials

The MBE growth technique also permits one to grow lattice-mismatched alloys if the mismatch is only a few percent and the layer is thin. The epitaxial layer grown will

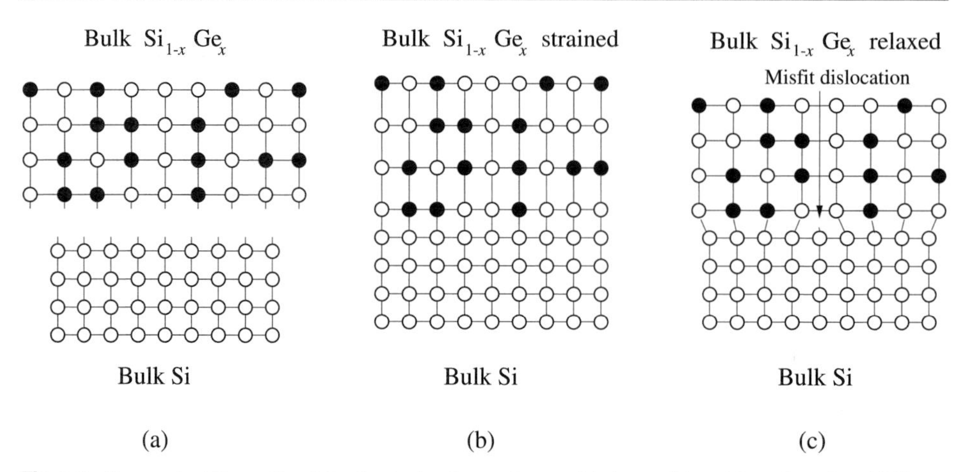

Fig. 1.3. Conceptual formation (a) of a strained pseudomorphic layer (b) and its relaxation (c) for a large thickness.

then assume the lattice constant of the substrate semiconductor on which it is grown. The resulting epitaxial layer is therefore subject to a strain (compression if its natural lattice parameter is larger than that of the substrate or tension if it is smaller) which modifies its physical properties. Such a layer is called pseudomorphic as it assumes a new crystal structure. One can verify, for example, that if the lattice parameter of the epitaxial layer parallel to the interface is reduced, then the transversal lattice parameter is increased as is shown in Figure 1.3(b) in which $Si_{1-x}Ge_x$ is grown on Si.

A pseudomorphic material can only be formed if the epitaxial film thickness is smaller than a critical thickness h_c. This critical thickness corresponds to the thickness at which it becomes energetically more favorable for the epitaxial layer to generate dislocations than to maintain the lattice strain [2]. Figure 1.4 shows the critical thickness for growing $Si_{1-x}Ge_x$ on Si as a function of the Ge mole fraction x. Note that thicker defect-free $Si_{1-x}Ge_x$ films can be grown at a lower growth temperature [3], but these films are metastable and require appropriate subsequent thermal treatments [4].

Other examples of pseudomorphic material systems are $In_xGa_{1-x}As$ and $In_xAl_{1-x}As$ on GaAs and InP substrates and $Al_xGa_{1-x}N$ on GaN. For example $In_{0.25}Ga_{0.75}As$ can be grown for a thickness up to 124 Å on GaAs. For pseudomorphic films of thickness larger than the critical thickness h_c, the film relaxes to its original unstrained bulk lattice, and dislocations are formed, usually rendering the material unusable for making devices.

Pseudomorphic layers can present much improved physical properties. Consider the light- and heavy-hole band structures in Figure 1.5 [1]. In the absence of strain the heavy-hole band is populated due to its higher density of states, and the holes typically exhibit an effective mass much larger than that of electrons. The presence of strain breaks the degeneracy of the light- and heavy-hole bands (see Figure 1.5)

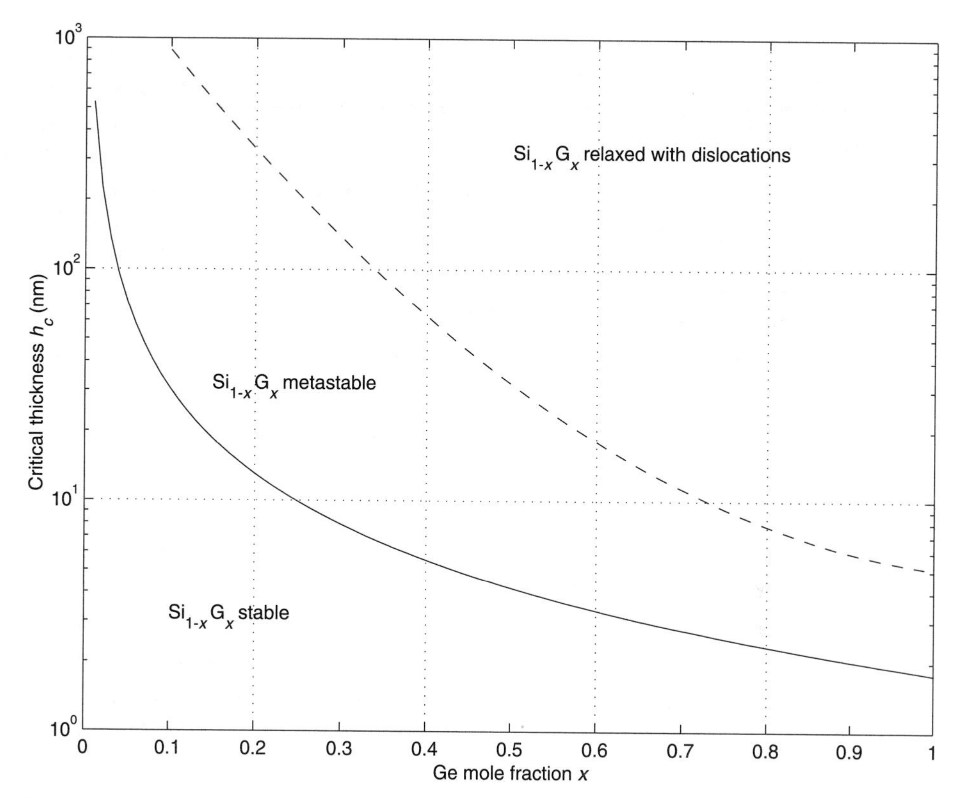

Fig. 1.4. Approximate critical thickness for strained pseudomorphic (full line) and metastable (dashed line) layer in $Si_{1-x}Ge_x$.

at the top of the valence band. When the light-hole band is raised, the light-hole band is preferentially populated, and a high hole mobility results, due to the smaller effective mass of the light holes. In $Si_{1-x}Ge_x$ the heavy-hole mass can even become smaller than the light-hole mass. Pseudomorphic field-effect transistors (FETs) are therefore potentially important for the generation of high-speed complementary logic with p-channel FETs of improved performance.

Other interesting physics can also occur in pseudomorphic materials. For example the large strain present at the interface of $Al_xGa_{1-x}N$ and GaN can, via the piezoelectric effect, enhance the electron charge density.

1.2.3 The materials game and bandgap engineering

Since each semiconductor alloy has a different bandgap (see Figure 1.2) novel heterostructure semiconductor devices can be created using either lattice-matched or pseudomorphic semiconductor epitaxial layers. The most important examples include the semiconductor laser diodes, the heterostructure field-effect transistors (HFETs), the heterojunction bipolar transistors (HBTs), and the resonant tunneling

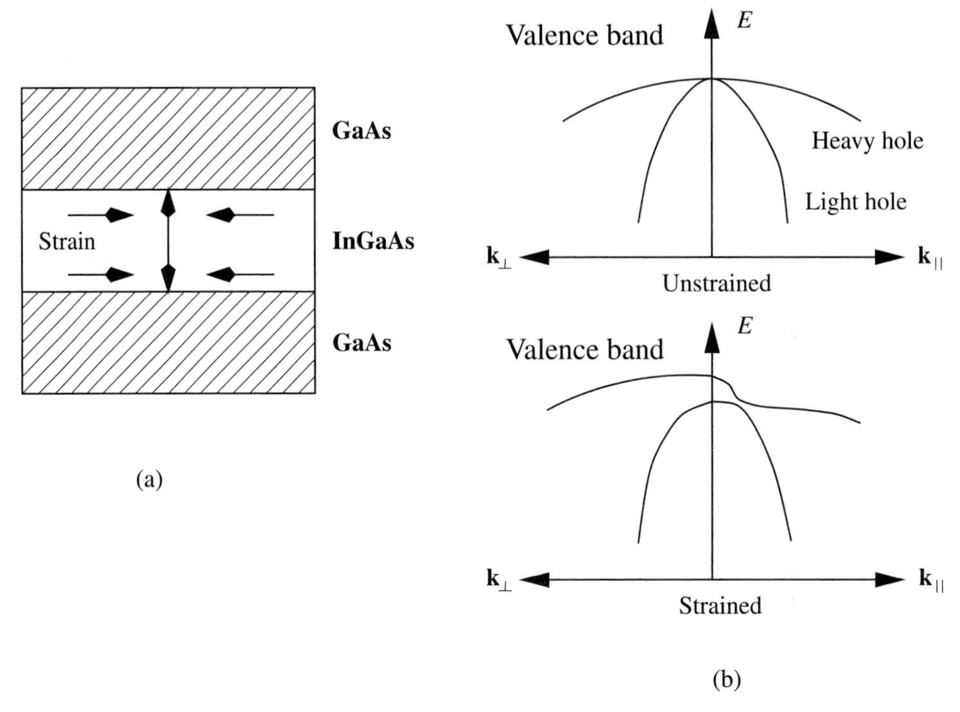

Fig. 1.5. Effect of strain on the band structure of an InGaAs layer grown on the (100) surface of GaAs. The arrows in (a) indicate the direction of the lateral compressive strain and the resulting tensile perpendicular strain. The heavy-hole band structure shown in (b) has shifted upward for transverse wave-vectors and downward for longitudinal wave-vectors resulting in a smaller longitudinal effective mass.

diodes (RTDs). Furthermore, since modern growth techniques can be used to grow semiconductor structures with dimensions as small as a few lattice parameters, new devices making use of the quantum properties of electrons are possible. Among these are quantum wells (applications include laser diodes, HFETs, light switches and so on), superlattices, and RTDs.

Since so many types of semiconductor materials can be grown epitaxially together, it is natural that researchers continuously investigate new materials for the improvement of devices such as HEMTs (High-Electron-Mobility Transistors), HBTs, and RTDs as well as means for further improving the growth techniques. The $Al_xGa_{1-x}As$ system was one of the first material systems used to fabricate HEMTs, HBTs, and RTDs due to its advantageous flexibility when growing lattice-matched structures on the readily available GaAs substrate.

The $In_xGa_{1-x}As/In_xAl_{1-x}As$ system has been subsequently investigated on both InP and GaAs substrates. As we shall see one of its advantages is that it provides a higher electron concentration and is not afflicted by DX (deep trap) center problems for In mole fractions below 60%.

The $Si_{1-x}Ge_x$ system also permits the fabrication of HEMTs, HBTs, RTDs and RITDs (Resonant Interband Tunneling Diodes). It has the advantage of being more easily integrated with other silicon processes. The reader is referred to [5] for a review of its performance achievements and future potentials.

The $Al_xGa_{1-x}N/Al_xAl_{1-x}N$ system has recently attracted a lot of interest. These materials are based on the binaries InN, GaN, and AlN which have large direct (Γ) bandgaps of 1.9, 3.4 and 6.2 eV, respectively. These wide-bandgap materials find application in the creation of green, blue, and violet lasers. This material system is also being investigated for use in high-power microwave electronic devices. High-temperature electronics is also being pursued with this material system and others such as SiGeC. A more complete review of materials and their impact on HBT and HEMT performance is given in Chapters 14 and 19.

1.2.4 Limitations and applications of modern growth techniques

MBE provides the means to grow high-quality materials, with an excellent control of material composition and of epitaxial layer thickness. MBE materials suffer, however, from morphological defects which can affect the smoothness of the semiconductor wafer and the yields. In addition, MBE is an expensive growth technique and other growth techniques such as metal organic chemical vapor deposition (MOCVD) are usually used for high-volume production. However, both MBE and MOCVD have been used in a production environment for the growth of the low-cost laser diodes found in the compact disk recorder and for discrete high-speed HEMT and HBT devices.

MBE and MOCVD technologies are presently the only viable approaches to developing semiconductor transistors which operate at millimeter wavelengths. However, the use of MBE for the production of large-scale digital integrated circuits is limited by the integrated circuit yield. MBE and MOCVD technologies are therefore finding mostly low-scale integrated circuit applications, in digital and analog circuits such as pre-scalers and A/D converters, and in RF (RFIC) and microwave (MMIC) front-end integrated circuits for wireless applications. Another emerging area in which MBE and MOCVD technology will have an important impact is the field of OptoElectronics Integrated Circuits (OEIC). Fiber optics present unrivaled potential for high-speed communication with gigabit bandwidths. Presently this potential is exploited only in expensive communication systems. The development of low-cost OEICs and their integration with the present Si technology will multiply the use of optical local area networks. Applications include the optical wiring of computers, cars, airplanes and so on.

1.3 Crystal and reciprocal lattices

Before discussing heterostructure physics and devices, let us briefly review in this section crystals and reciprocal lattices.

1.3.1 Crystals and lattices

The semiconductor materials used for making devices are in a crystalline state of matter. So a brief review of some of the techniques used for the characterization of crystals is in order.

Crystal structures are defined in terms of a lattice and a basis. The semiconductors we shall consider in this book have either a zinc-blende and or a diamond crystal structure which is realized with a face-centered cubic (fcc) lattice and a basis consisting of two atoms.

The three-dimensional lattice consists of all the points generated by the lattice vectors

$$\mathbf{R} = n_1\mathbf{a}_1 + n_2\mathbf{a}_2 + n_3\mathbf{a}_3, \tag{1.1}$$

where n_1, n_2, and n_3 are integers and where a_1, a_2 and a_3 are the lattice translation vectors.

For a fcc lattice the lattice vector $\mathbf{R}$ is given in the orthonormal coordinates of the Bravais cell (a cube whose side is the lattice parameter a, see Figure 1.6(a) by

$$\mathbf{R} = \frac{a}{2}\begin{bmatrix} 0 & 1 & 1 \\ 1 & 0 & 1 \\ 1 & 1 & 0 \end{bmatrix}\begin{bmatrix} n_1 \\ n_2 \\ n_3 \end{bmatrix}, \tag{1.2}$$

where the lattice translation vectors $\mathbf{a}_1$, $\mathbf{a}_2$ and $\mathbf{a}_3$ for the fcc crystal are defined in Figure 1.6(a).

The fact that fcc or body-centered (bcc) lattices are referred to as cubic lattices (Figure 1.6) is not due to the cubic appearance of their Bravais cell but rather to the fact that an fcc or bcc crystal is left invariant under the 48 symmetries of the O_h (or m3m) group [6] (see Problem 1.2). These 48 symmetries consist of the identity E, inversion I and the rotations of angles $2\pi/2$, $2\pi/3$ and $2\pi/4$ (respectively denoted: C_2, C_3 and C_4) and of their products.

The zinc-blende (or sphalerite) crystal structure of conventional III–V semiconductors is obtained by selecting a basis with two atoms as shown in Figure 1.7. such that, for example, for GaAs, the crystal consists of an fcc lattice of Ga and an fcc lattice of As separated by the vector $a/4(\hat{\mathbf{x}} + \hat{\mathbf{y}} + \hat{\mathbf{z}})$. For C, Si, and Ge crystals the same atoms are used for the basis, and the structure is referred to as the diamond structure.

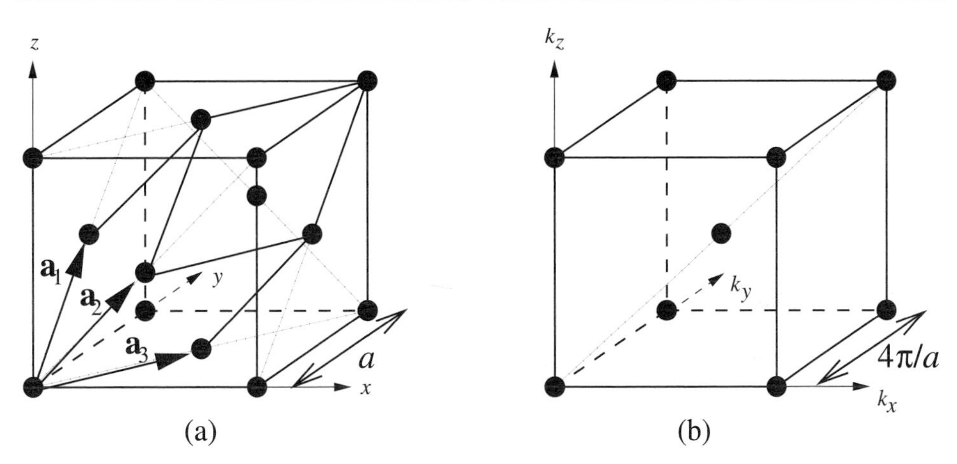

Fig. 1.6. (a) Bravais cell for a face-centered cubic crystal in (direct) space and (b) Bravais cell of the associated body-centered cubic crystal in reciprocal (indirect) space.

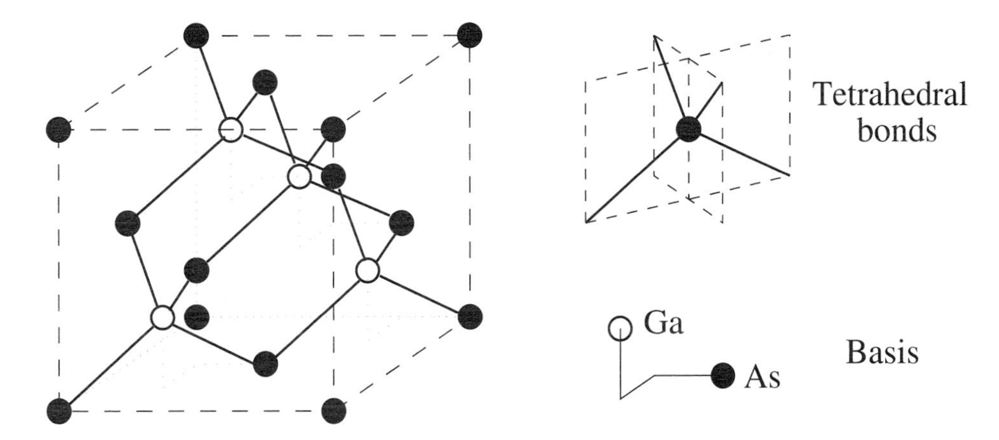

Fig. 1.7. Zinc-blende crystal, its two-atom basis, and the tetrahedral bond structure.

It is interesting to note that the zinc-blende structure does not have an inversion symmetry (I) but is still invariant under the 24 symmetry operations of the T_d subgroup of O_h ($O_h = T_d + I \times T_d$). However, the inversion symmetry is recovered in the band structure due to the time-reversal symmetry of the Schrödinger equation when spin degeneracy is neglected.

1.3.2 The reciprocal lattice

A lattice is a periodic structure in three dimensions. As a consequence, any local physical property of the crystal $f(\mathbf{r})$ is invariant under a translation by a lattice vector $\mathbf{R}$:

$$f(\mathbf{r}) = f(\mathbf{r} + \mathbf{R}). \tag{1.3}$$

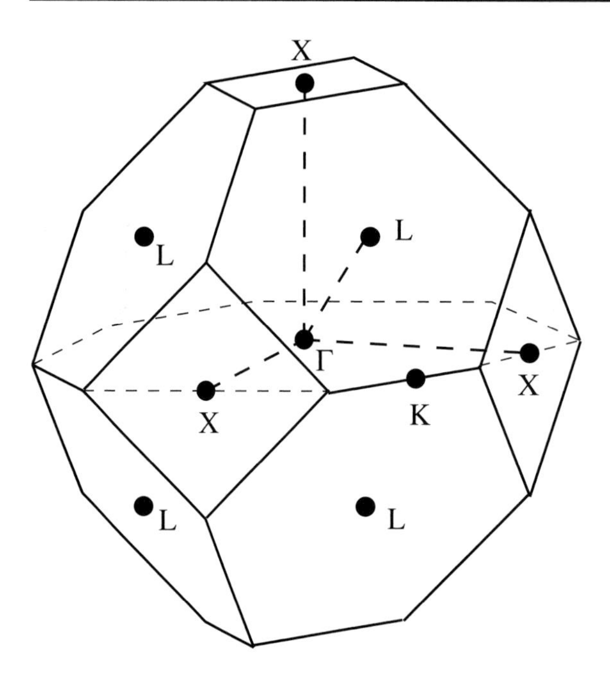

Fig. 1.8. Brillouin zone for the zinc-blende reciprocal lattice.

As $f(\mathbf{r})$ is periodic we can expand it using a Fourier series.

$$f(\mathbf{r}) = \sum_{\mathbf{K}} A_{\mathbf{K}} \exp(i\mathbf{K} \cdot \mathbf{r}),$$

where $\mathbf{K}$ is the so-called reciprocal-lattice vector and $A_{\mathbf{K}}$ the Fourier coefficient.

Now using the translation property we have

$$f(\mathbf{r} + \mathbf{R}) = \sum_{\mathbf{K}} A_{\mathbf{K}} \exp[i\mathbf{K} \cdot (\mathbf{r} + \mathbf{R})] = f(\mathbf{r}).$$

Therefore we must have $\exp(i\mathbf{K} \cdot \mathbf{R}) = 1$ such that the reciprocal-lattice vectors verify

$$\mathbf{K} \cdot \mathbf{R} = n \times 2\pi$$

with n an integer.

For the fcc lattice the reciprocal-lattice vectors $\mathbf{K}$ satisfying this relation are given in the orthonormal coordinates of the Bravais cell by

$$\mathbf{K} = \frac{2\pi}{a} \begin{bmatrix} -1 & 1 & 1 \\ 1 & -1 & 1 \\ 1 & 1 & -1 \end{bmatrix} \begin{bmatrix} m_1 \\ m_2 \\ m_3 \end{bmatrix}, \tag{1.4}$$

where m_1, m_2, and m_3 are integers.

An inspection of $\mathbf{K}$ for the fcc lattice (in direct space) reveals that the reciprocal lattice (in indirect space) is a bcc lattice (see Figure 1.6). The Brillouin zone (the

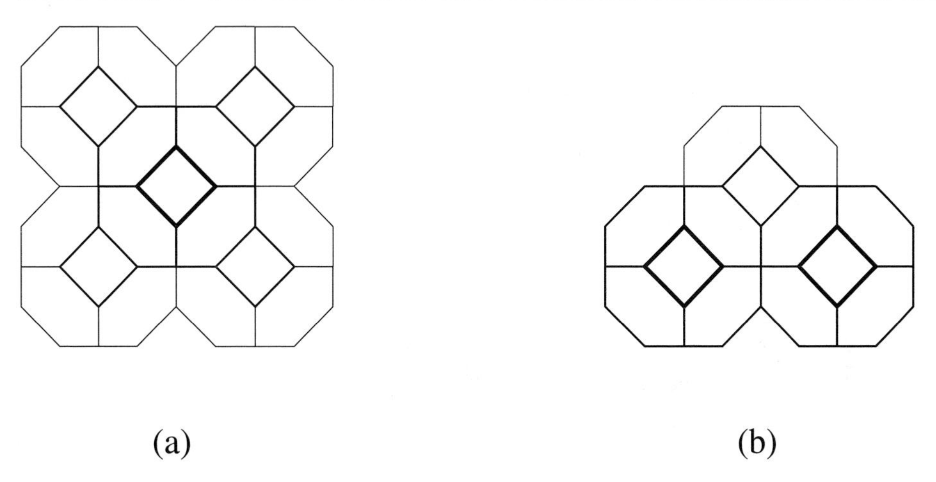

Fig. 1.9. Top (a) and side (b) views of the partial reconstruction of the bcc reciprocal Bravais cell using five Brillouin zones.

Wigner–Seitz cell in reciprocal space) is defined by the median boundary between nearest neighbors in the bcc reciprocal lattice. The three-dimensional Brillouin zone of a zinc-blende crystal is shown in Figure 1.8. Figure 1.9 shows how five Brillouin zones can be packed together to reconstruct the bottom half of the bcc Bravais cell.

1.3.3 Application to band structures

Let us briefly review how the semiconductor band structure is formed and how the concept of the reciprocal lattice can be applied to the band structure.

We have seen that the diamond and zinc-blende structures effectively originate from the tetrahedral nature of the bonds formed between the atoms. For example, for C, Si, and Ge the last shell is occupied by four electrons in the s^2p^2 electronic configuration. By bonding with four other atoms, each atom reaches a stable rare-gas configuration by effectively surrounding itself with eight electrons with two electrons per bond. In this process the tetrahedral bonding structure can be verified to result from a preliminary hybridization of the four valence electrons in a hybrid state of configuration s^1p^3 denoted:

$$|\psi_{hybrid}\rangle = |\psi_s\rangle \pm |\psi_{p,x}\rangle \pm |\psi_{p,y}\rangle \pm |\psi_{p,z}\rangle.$$

This hybridization generates four distinct hybrids, which establish the tetrahedral structure shown on Figure 1.7. The bonding of two hybrids generates an energy level associated with the valence band, and the anti-bonding (broken bond) generates a high energy level associated with the conduction band. Therefore in a system with N atoms, the bonding of the hybrids will form the $4N$ states constituting the valence band, and the anti-bonding of the hybrids will form the $4N$ states constituting the conduction band. At 0 K the states of the valence band are all filled while the states of the

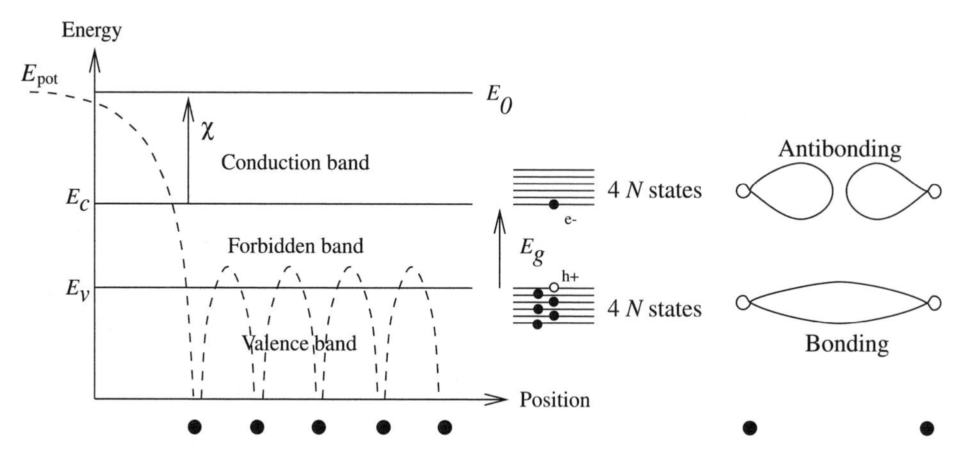

Fig. 1.10. One-dimensional band diagram of a uniform semiconductor. The average crystal potential generated near the surface of that one-dimensional crystal is shown using a dashed line.

conduction band are all empty. As we increase the temperature, the thermal agitation breaks some of the bonds and introduces holes (broken bond) at the top of the valence band and electrons at the bottom of the conduction band.

Clearly semiconductors are complex many-body systems. A conventional approach is to study the motion of a single electron in the average potential defined by all the other electrons and nucleus. As expected this average crystal potential $E_{pot}(\mathbf{r})$ is invariant under a translation by a lattice vector $\mathbf{R}$

$$E_{pot}(\mathbf{r}) = E_{pot}(\mathbf{r} + \mathbf{R}). \tag{1.5}$$

The periodic potential resulting for the average potential of five atoms near the surface of a one-dimensional crystal is shown in Figure 1.10 (dashed line). One sees how no barriers impede the motion of the electron in the conduction band. Similarly the electrons at the top of the valence band can easily move by tunneling through the potential barriers. In a crystal the propagation (diffraction) of the electron wave-function throughout the entire crystal is actually facilitated by the crystal periodicity (lattice), while, as we will find in Chapter 4 (Figure 4.19) when studying random superlattices, the electron wave-function is localized in a random structure.

Actually our discussion could be generalized to any type of wave. These waves could be the electron wave-functions, the vibration waves of the atoms in the crystal structure itself, or even electromagnetic waves as in the case of the X-ray photons used to analyze a crystal structure. A wave $f(\mathbf{r}, t)$ can generally be expressed in the form

$$f(\mathbf{r}, t) = A_{\mathbf{k}} \exp[i\omega(\mathbf{k})t - \mathbf{k} \cdot \mathbf{r}],$$

where $\omega(\mathbf{k})$ is the wave frequency and $\mathbf{k}$ the wave-vector. For crystal vibrations the $\omega(\mathbf{k})$ dispersion relation defines the vibration spectra possible in that crystal. We shall discuss these further in Chapter 5 and also introduce their particle embodiment in

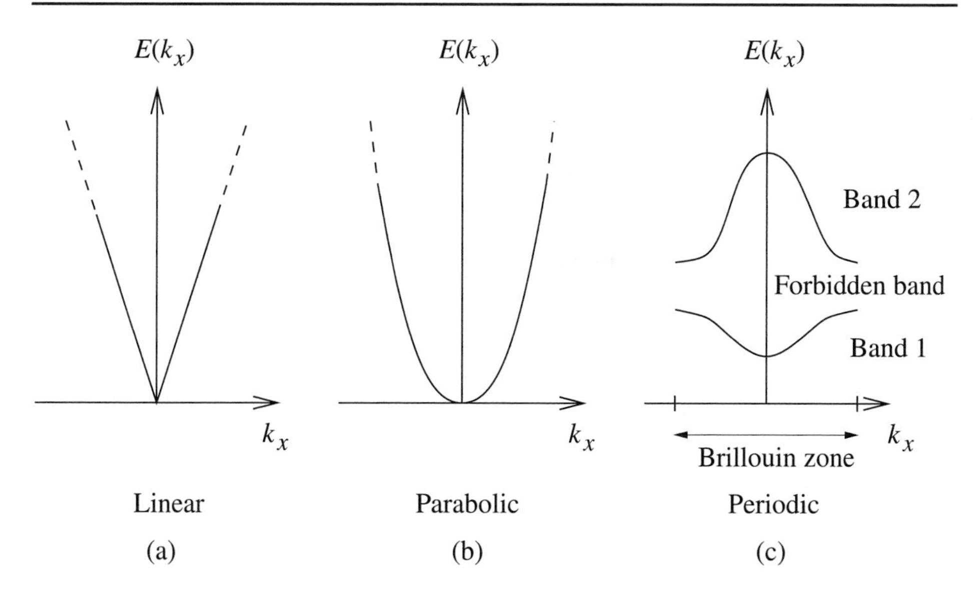

Fig. 1.11. Energy dispersion for: (a) a photon, (b) a free electron, and (c) a crystal electron.

phonons with energy $E = \hbar\omega(\mathbf{k})$. For electrons the $\omega(\mathbf{k})$ dispersion relation defines the electron band structure $\mathcal{E}(\mathbf{k}) = \hbar\omega(\mathbf{k})$.

We shall demonstrate in Chapter 3 that waves propagating in a crystal undergo (Bragg) reflections such that $\mathcal{E}(\mathbf{k}) = \hbar\omega(\mathbf{k})$ becomes a periodic function in **k** space in the reciprocal lattice. This means that the band structure $\mathcal{E}(\mathbf{k}) = \hbar\omega(\mathbf{k})$ is invariant under a translation of a reciprocal-lattice vector **K**:

$$\mathcal{E}(\mathbf{k}) = \mathcal{E}(\mathbf{k} + \mathbf{K}). \tag{1.6}$$

In Figure 1.11 we compare the one-dimensional energy dispersion for a photon, a free electron, and an electron in a crystal. For the photon we have a linear dispersion

$$E = \hbar\omega = \hbar c|k_x|,$$

where c is the speed of light and k_x is the photon wave-vector. For the free electron we have a parabolic dispersion

$$E = \frac{1}{2}mv_x^2 = \frac{p_x^2}{2m} = \frac{\hbar^2 k_x^2}{2m},$$

where m is the mass, v_x is the velocity, $p_x = mv_x = \hbar k_x$ is the momentum, and k_x is the wave-vector of the free electron.

For the crystal band structure a more complex periodic dispersion is shown in Figure 1.11 in the one-dimensional Brillouin zone with various bands of energy and energy gaps. The creation of such an energy spectrum can be intuited by considering the case of the empty lattice. An empty lattice consists of a free electron system in which we insist in treating the constant reference potential of the free electron

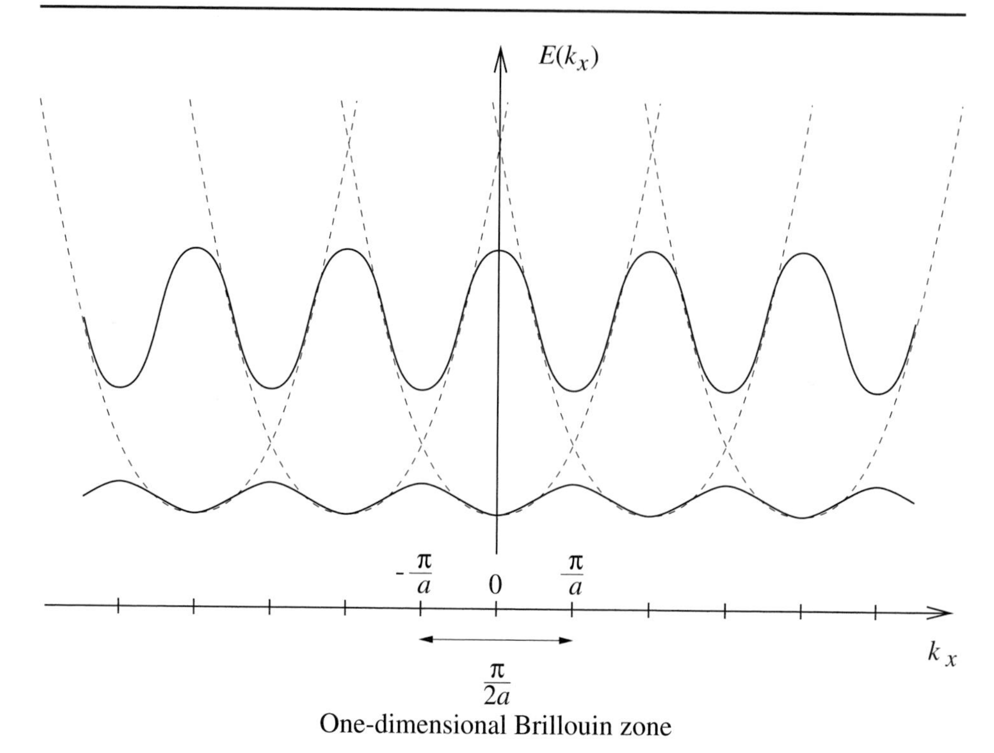

Fig. 1.12. Formation of the band structure for a 1D empty lattice (dashed line) and 1D crystal (plain line).

as a periodic potential along x. The associated periodic energy dispersion can then be obtained from the construction shown in Figure 1.12 using the parabolic energy dispersion (dashed line) of the free electron. In a real crystal the electron wave-function is perturbed by the crystal potential leading to the formation of bands and energy gaps (full line) in the original periodic band structure of the empty lattice.

The band structures obtained for various semiconductor materials along the Γ to X direction in the Brillouin zone are shown in Figure 1.13. These band structures were calculated using the empirical pseudopotential method of [7] and [8]. The wave-functions of the three-dimensional empty (see Problem 1.1) fcc lattice are used as a basis in the empirical pseudopotential method for calculating the three-dimensional band structures.

1.4 Conclusion

In this chapter we have introduced the new epitaxial materials (alloys, pseudomorphic, modulation doped) which can be fabricated with modern growth techniques. The various epitaxial films realized with these materials are the building blocks of modern heterostructure devices. They provide a seemingly infinitely rich

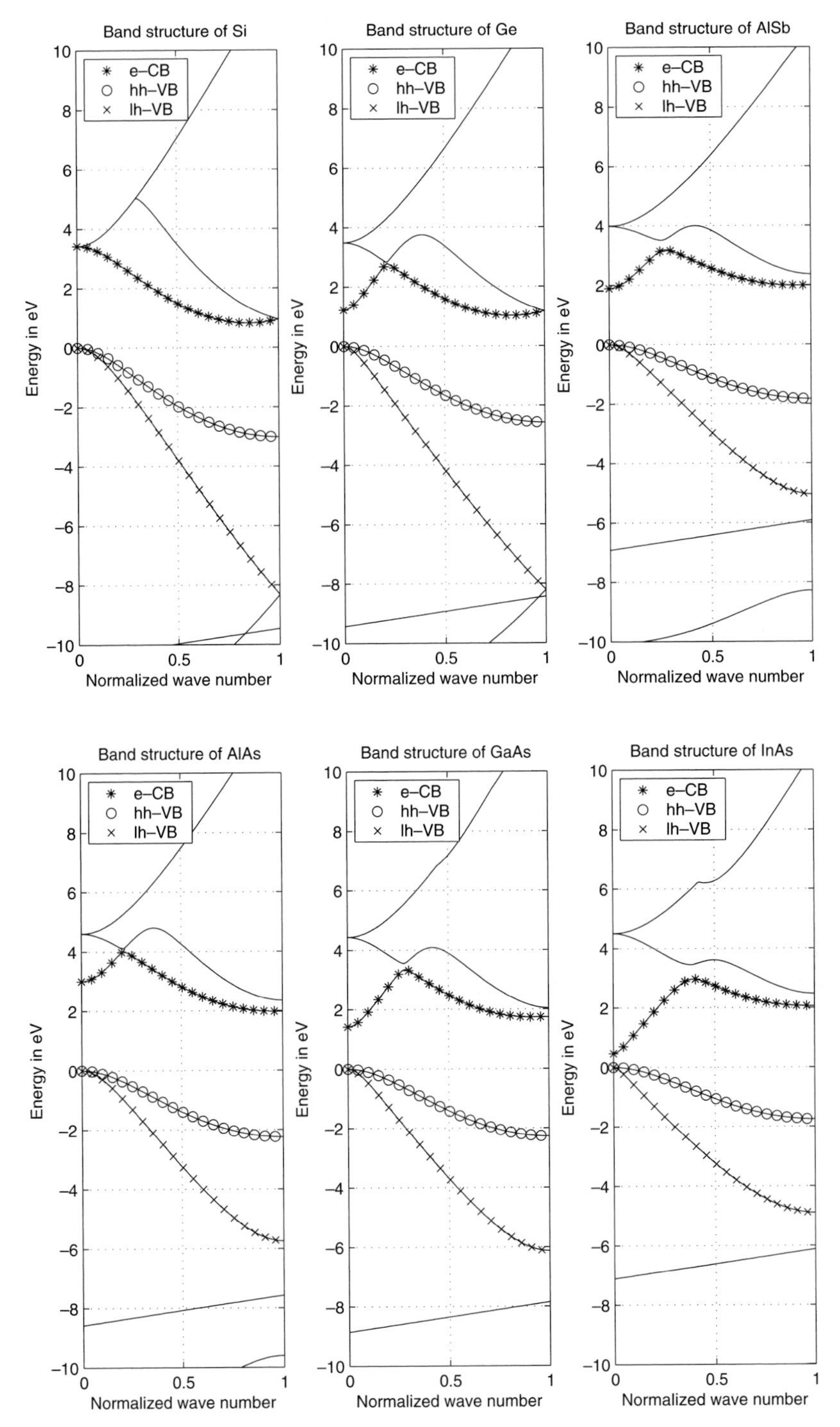

Fig. 1.13. Band structure of Si, Ge, AlSb, AlAs, GaAs, and InAs along the Γ to X direction, obtained using the empirical pseudopotential method of Cohen and Bergstresser [7]. The band structures were generated using the code of Dr. Paul von Allmen [8].

realm of material/doping combinations, stimulating the imagination of the device researcher.

Having introduced the players and reviewed some basic properties of crystals, we will now proceed in the next chapters with the development of a semiclassical and quantum theory of heterostructures, which will permit us to describe their electronic properties and study their device applications.

1.5 Bibliography

1.5.1 Recommended reading

W. A. Harrison, *Electronic Structure and Property of Solids*, Freeman, San Francisco, 1980.

C. M. Wolfe, N. Holonyak Jr, Gregory E. Stillman, *Physical Properties of Semiconductors*, Prentice Hall, Englewood Cliffs, 1989.

1.5.2 References

[1] P. Ruden, M. Shur, D. K. Arch, R. R. Daniels, D. E. Grider, and T. E. Nohava, 'Quantum-well p-channel AlGaAs/InGaAs/GaAs heterostructure insulated-gate field-effects transistors,' *IEEE Transactions on Electron Devices*, Vol. 36, No. 11, p. 2371, November 1989.

[2] J. W. Matthews and A. E. Blakeslee, 'Defects in epitaxial multilayers. I. Misfit Dislocations,' *Journal of Crystal Growth*, Vol. 27, p. 118, 1974.

[3] J. C. Bean, L. C. Feldman, A. T. Fiory, S. Nakahara and I. K. Robinson, 'Ge_xSi_{1-x}/Si strained-layer superlattice grown by molecular beam epitaxy,' *Journal of Vacuum Science Technology*, Vol. 2, p. 436, 1984.

[4] D. C. Houghton, 'Strain relaxation kinetics in $Si_{1-x}Ge_x/Si$ heterostructures,' *Journal of Applied Physics*, Vol. 70, p. 2136, 1991.

[5] D. J. Paul, 'Silicon–germanium strained layer materials in microelectronics,' *Advanced Materials*, Vol. 11, No. 3, pp. 191–204, 1999.

[6] Heine Volker, *Group Theory in Quantum Mechanics*, Pergamon Press, London, 1960.

[7] M. L. Cohen and T. K. Bergstresser, 'Band structures and pseudopotential form factors for fourteen semiconductors of the diamond and zinc-blende structures,' *Physical Review*, Vol. 141, No. 2, p. 789, 1963.

[8] W. Andreoni and R. Car, 'Similarity of (Ga, Al, As) alloys and ultrathin heterostructures: electronic properties from the empirical pseudopotential method,' *Physical Review*, Vol. B 21, No. 8, p. 3334, 1980.

1.6 Problems

1.1 (a) Derive an expression for the wave-functions of the empty lattice for a zinc-blende crystal.
(b) Plot the first four bands of the empty-lattice band structure from Γ to X.

Table 1.2. *Organization in five classes of the 24 symmetries of the T_d group.*

E	xyz							
$3C_4^2$	$\bar{x}\bar{y}z$	$x\bar{y}\bar{z}$	$\bar{x}y\bar{z}$					
$6C_4 \times I$	$y\bar{x}\bar{z}$	$\bar{y}x\bar{z}$	$\bar{x}z\bar{y}$	$\bar{x}\bar{z}y$	$\bar{z}\bar{y}x$	$z\bar{y}\bar{x}$		
$6C_2 \times I$	$\bar{y}\bar{x}z$	$\bar{z}y\bar{x}$	$x\bar{z}\bar{y}$	yxz	zyx	xzy		
$8C_3$	zxy	yzx	$z\bar{x}\bar{y}$	$\bar{y}\bar{z}x$	$\bar{z}\bar{x}y$	$\bar{y}z\bar{x}$	$\bar{z}x\bar{y}$	$y\bar{z}\bar{x}$

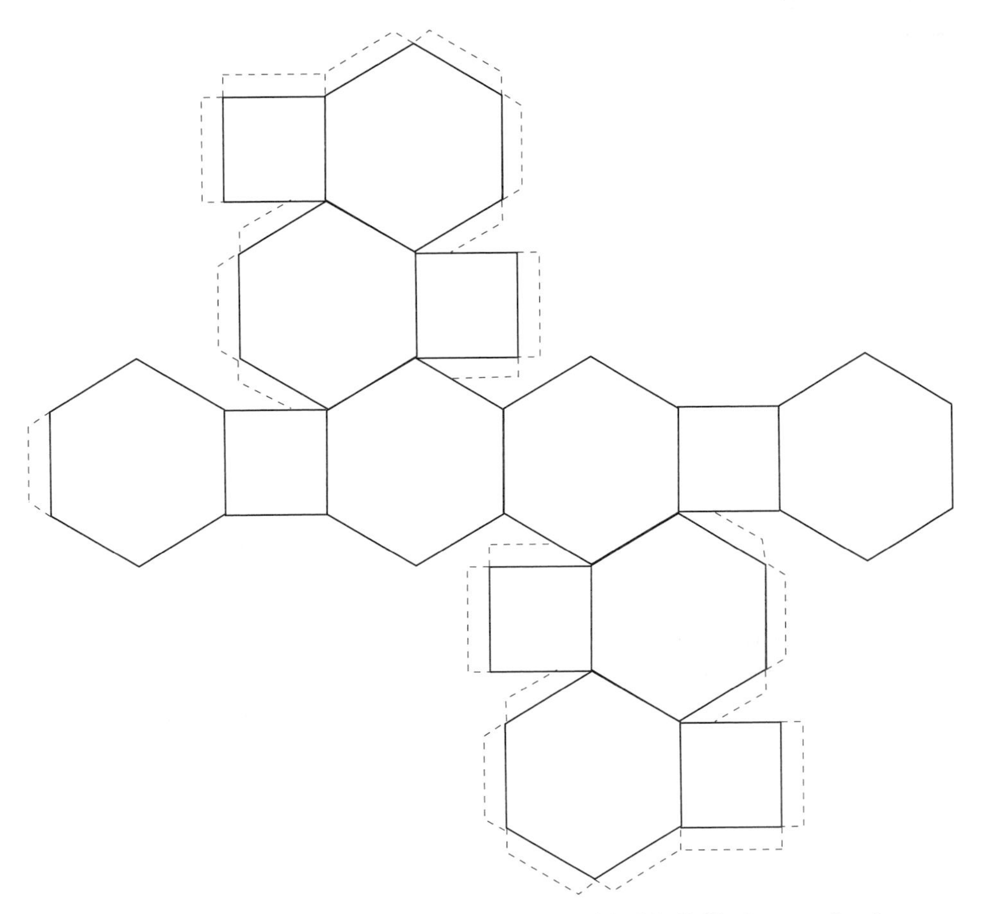

Fig. 1.14. Pattern to be copied and cut out for building a model of the Brillouin zone of an fcc crystal (in real space).

1.2 The 24 symmetries of the T_d group which can be applied to the zinc-blende crystal are shown in Table 1.2. The notation used is such that for a rotation of $2\pi/4$ around the z axis, the notation $S_i(y, \bar{x}, z)$ represents the transformation $S_i f(x, y, z) = f(x', y', z') = f(y, -x, z)$, where $f(x, y, z)$ is a physical property of the crystal.

(a) For each of the five symmetries of the second column, write the associated transformation matrix $[S_i(n, m)]$

$$\begin{bmatrix} x' \\ y' \\ z' \end{bmatrix} = \begin{bmatrix} S_{11} & S_{12} & S_{13} \\ S_{21} & S_{22} & S_{23} \\ S_{31} & S_{32} & S_{33} \end{bmatrix} \begin{bmatrix} x \\ y \\ z \end{bmatrix}.$$

(b) The 24 symmetries of T_d are organized in Table 1.2 in five classes of representative symmetries. A class is a group of symmetries satisfying the following property: for any symmetry S_i of T_d, and any symmetry T_j of the class T, the symmetry $S_i^{-1} T_j S_i$ is also a member of the class T.

Verify that this property indeed holds for the symmetries T_j of the class $3C_4^2$ when using the symmetries S_i of the 2nd column of Table 1.2.

1.3 Build several Brillouin zones of an fcc crystal using the cutout shown in Figure 1.14 and verify (in class if each student builds one) that they assemble into a bcc crystal.

2 Semiclassical theory of heterostructures

In Nature's infinite book of secrecy
A little can I read.

Anthony and Cleopatra, WILLIAM SHAKESPEARE

2.1 Introduction

We start our study of heterostructure devices with a review of the semiclassical theory of semiconductor heterostructures. Our goals in this chapter are to introduce semiconductor heterostructures and to present their application in semiclassical devices. For this purpose we first introduce various tools, such as band-diagram theory and transport models. We then discuss the modeling of the p–n heterojunction and its application to the high-performance heterojunction bipolar transistor (HBT).

2.2 Spatially-varying semiconductors

Semiconductor heterostructures involve semiconductors which vary with position. We will mostly be concerned with layered heterostructures grown by molecular beam epitaxy (MBE) which vary along a one-dimensional axis. We will call this axis the superlattice axis. In the direction perpendicular to the superlattice axis the lattice is uniform. The variation of the semiconductor material along the superlattice axis can be either abrupt or gradual.

An example of an abrupt heterostructure or heterojunction is Ge/GaAs. As indicated in Table 2.1, Ge and GaAs have nearly the same lattice parameters and can therefore be grown on top of one another with a very small number of defects*.

* Although the Ge/GaAs structure was the first heterostructure to be experimentally studied (Anderson 1960 [13]) it is actually one of the most tricky heterostructures to grow. Ga and As are doping impurities for Ge, and Ge is a doping impurity for GaAs. To avoid contamination, the Ge and GaAs epitaxial layers can be grown in separate MBE chambers. The sample is moved from one MBE chamber to the other without breaking the vacuum. Quasi-ideal Ge/GaAs heterojunctions (ideality factor of 1) have been grown in this way [11]. However, under thermal cycling (annealing) the heterojunction degrades. A very thin (10 Å) Si layer at the Ge/GaAs interface can then be used to stabilize the heterojunction. No such problems are encountered with AlAs/GaAs heterojunctions.

Table 2.1. *Bandgap, lattice parameter and electron affinity of several semiconductors (from [5]).*

	a (Å)	E_g (eV)	χ (eV)
Ge	5.6461	0.663	4.13
GaAs	5.6533	1.424	4.07
AlAs	5.6614	2.16	3.5
Si	5.451	1.11	4.01
GaP	5.431	2.25	4.3
InP	5.8687	1.34	4.35
$In_{0.53}Ga_{0.47}As$	5.8687	0.77	4
$In_{0.52}Al_{0.48}As$	5.8687	1.49	
AlSb	6.136	1.6	3.65
GaSb	6.095	0.68	4.06
InAs	6.058	0.36	4.9

2.2.1 Semiconductor alloys

Gradual variation of the semiconductor material can be achieved with semiconductor alloys[†] such as $Al_mGa_{1-m}As$. In the AlGaAs alloy, the atoms Al and Ga are randomly distributed over the same lattice sites, say Ga in the GaAs fcc lattice. The Al mole fraction, m, gives the fraction of Ga sites occupied by Al and varies between 0 and 1. Because Al and Ga both have an s^2p electronic structure (see Streetman [1]) we can expect the alloy $Al_mGa_{1-m}As$ to behave like a semiconductor crystal despite the random nature of its structure. AlGaAs can therefore be seen as a novel semiconductor with properties intermediate between the semiconductors GaAs and AlAs. This is the so-called virtual-crystal approximation. We will see in Chapter 5 that the local departures of the alloy from this virtual crystal can be treated as crystal defects which are at the origin of a scattering process called alloy scattering. Alloy scattering is usually small and the virtual-crystal approximation is quite accurate.

The band structure of an alloy can be obtained to first order using a simple averaging rule (Vegards' law). For example, the conduction band of $Al_mGa_{1-m}As$ can be

[†] The first alloy made by man was bronze. Its introduction in prehistoric times signaled the end of the 'stone age' (8000 BC) and the beginning of the 'bronze age' (1000 BC). Bronze is an alloy of tin and copper. It provides an improved material strength over copper or tin alone and its introduction is considered to be an important step in the evolution of mankind. In some sense we could say that with the use of semiconductor alloys we have entered the 'microbronze age' of semiconductor technology.

approximated by

$$\mathcal{E}_{c,AlGaAs}(\mathbf{k}, m) = m\ \mathcal{E}_{c,AlAs}(\mathbf{k})\ +\ (1-m)\ \mathcal{E}_{c,GaAs}(\mathbf{k}).$$

Similarly the valence band can be approximated by

$$\mathcal{E}_{v,AlGaAs}(\mathbf{k}, m) = m\ \mathcal{E}_{v,AlAs}(\mathbf{k})\ +\ (1-m)\ \mathcal{E}_{v,GaAs}(\mathbf{k}).$$

In the absence of experimental data or an accurate theory, other physical properties such as the dielectric constant and density of states can be estimated using this average rule. For example, to first order the dielectric constant can be estimated by

$$\epsilon_{AlGaAs}(m) = m\ \epsilon_{AlAs}\ +\ (1-m)\ \epsilon_{GaAs}.$$

Experimental data are available, however, for various mole fractions for the most common alloys. The physical parameters of AlGaAs, InAlAs, InGaAs, and InP are given in Tables 2.3 and 2.10 (see chapter appendix) for the mole fractions specified and can be extrapolated for other values.

To study the dependence of the band structure upon the mole fraction m we have plotted in Figure 2.1 the GaAs and AlAs band structure and the variation with the Al mole fraction m of the AlGaAs conduction band $\mathcal{E}_c(\mathbf{k}, \mathbf{m})$ at the points Γ, X and L, and of the valence band $\mathcal{E}_v(\mathbf{k})$ at the point Γ in the Brillouin zone. Note that in Figure 2.1, GaAs has a direct bandgap and AlAs an indirect bandgap. Note that the alloy $Al_mGa_{1-m}As$ will switch from a direct to an indirect bandgap when m increases beyond 0.4.

In graded heterostructures, the mole fraction m varies along the superlattice axis (say x) and the band structure varies spatially:

$$\mathcal{E}_c(\mathbf{k}, x) = \mathcal{E}_c(\mathbf{k}, m(x)),$$
$$\mathcal{E}_v(\mathbf{k}, x) = \mathcal{E}_v(\mathbf{k}, m(x)).$$

The concept of a spatially-varying band structure requires clarification since the band structure is an eigenvalue derived for infinite crystals. A rigorous quantum model providing a theoretical justification for this semiclassical picture will be given in the next chapter.

For GaAs/$Al_mGa_{1-m}As$ it has been found that the variation of the conduction and valence bands can be obtained from the variation of the bandgap using the so-called 68/32% rule:

$$\begin{aligned}\Delta E_{c,AlGaAs}(m) &= \mathcal{E}_{c,Al_mGa_{1-m}As} - \mathcal{E}_{c,GaAs}\\ &= 0.68(E_{g,Al_mGa_{1-m}As} - E_{g,GaAs})\\ &= 0.68\Delta E_g(m),\end{aligned}$$

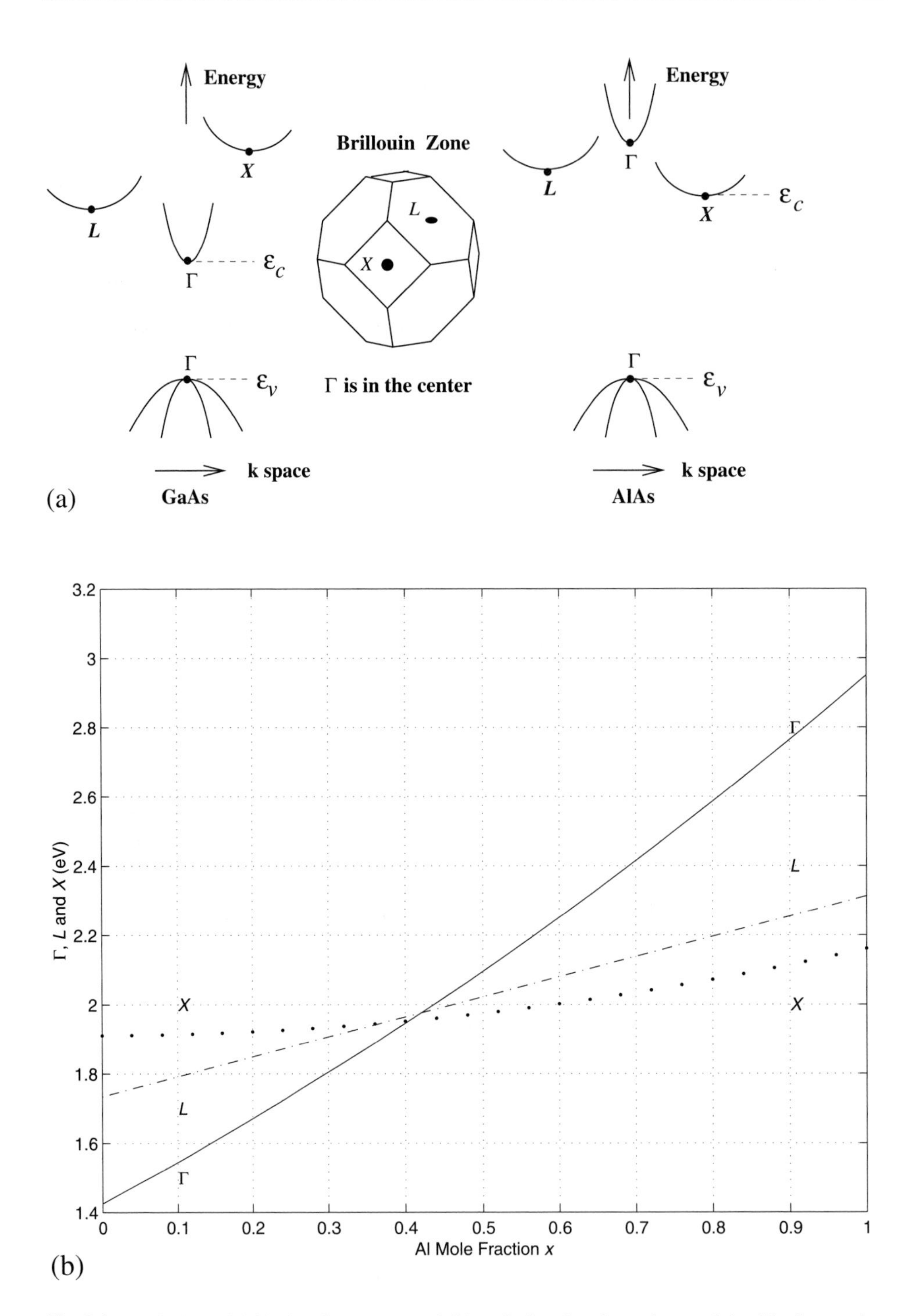

Fig. 2.1. (a) GaAs and AlAs band structure and (b) mole fraction dependence of the $Al_mGa_{1-m}As$ conduction band at the points Γ, X and L relative to the top of the valence band (Γ).

$$\Delta E_{v,AlGaAs}(m) = \mathcal{E}_{v,Al_mGa_{1-m}As} - \mathcal{E}_{v,GaAs}$$
$$= -0.32(E_{g,Al_mGa_{1-m}As} - E_{g,GaAs})$$
$$= -0.32\Delta E_g(m).$$

Using such a rule we can estimate the conduction and valence-band discontinuities between two different AlGaAs alloys:

$$\mathcal{E}_{c,Al_{m'}Ga_{1-m'}As} - \mathcal{E}_{c,Al_mGa_{1-m}As} = \Delta E_{c,AlGaAs}(m') - \Delta E_{c,AlGaAs}(m),$$
$$\mathcal{E}_{v,Al_{m'}Ga_{1-m'}As} - \mathcal{E}_{v,Al_mGa_{1-m}As} = \mathcal{E}_{c,Al_{m'}Ga_{1-m'}As} - \mathcal{E}_{c,Al_mGa_{1-m}As} - (\mathcal{E}_{g,Al_{m'}Ga_{1-m'}As} - \mathcal{E}_{g,Al_mGa_{1-m}As}).$$

For $In_mGa_{1-m}As$ and $In_mAl_{1-m}As$ (see Table 2.10 in chapter appendix) 73/27% and 62/38% rules, respectively, are used. To estimate the conduction and valence discontinuities between an AlGaAs and InGaAs alloy we could use GaAs as a reference:

$$\mathcal{E}_{c,In_{m'}Ga_{1-m'}As} - \mathcal{E}_{c,Al_mGa_{1-m}As} = \Delta E_{c,InGaAs}(m') - \Delta E_{c,AlGaAs}(m),$$
$$\mathcal{E}_{v,In_{m'}Ga_{1-m'}As} - \mathcal{E}_{v,Al_mGa_{1-m}As} = \mathcal{E}_{c,In_{m'}Ga_{1-m'}As} - \mathcal{E}_{c,Al_mGa_{1-m}As} - (\mathcal{E}_{g,In_{m'}Ga_{1-m'}As} - \mathcal{E}_{g,Al_mGa_{1-m}As}).$$

The band structure discontinuities between InAlAs and InGaAs are calculated similarly using InAs as a reference. Such estimates should be verified against the most recent experimental data since the use of a third material as a reference (e.g., GaAs and InAs in the previous examples) means that we are indirectly using the affinity rule despite its potential inaccuracy.

Within the effective-mass approximation the conduction band can be approximated by

$$\mathcal{E}_c(\mathbf{k}, x) \simeq \frac{\hbar^2 k^2}{2m_n^*(x)} \simeq \frac{\hbar^2 k^2}{2m_n^*(m(x))},$$

where the dependence of the effective mass m_n^* upon the mole fraction m can be obtained from Table 2.3 or from a linear extrapolation using the data of Table 2.10.

2.2.2 Modulation doping

Pure semiconductors grown by MBE have an unintentional (p or n) doping typically smaller than 10^{14} cm^{-3}. The semiconductor heterostructures grown by MBE can also be doped during growth. Silicon is a commonly used dopant. Silicon (s^2p^2) is an amphoteric dopant in III–V compounds as it can be either a donor or an acceptor. For example, in GaAs (and AlGaAs), a Si impurity is a donor when it occupies a Ga (s^2p) site and an acceptor when it occupies an As (s^2p^3) site. AlGaAs is usually grown in an overpressure of As such that Si preferentially occupies vacant Ga sites and is therefore used as an n dopant. The highest doping concentration that can be achieved with MBE growth is about 5×10^{18} cm^{-3}. As for the semiconductor material itself, the doping can be varied through the MBE growth. Such a technique is referred to as modulation doping. Abrupt variation of doping is called pulse doping. Spike doping, which involves growing a monoatomic layer of silicon (10^{12} cm^{-2}), is also sometimes used. Naturally, the silicon atoms are expected to diffuse. However, the diffusion can be limited to a few atomic layers (5–10) for low MBE growth temperature.

In a semiconductor such as GaAs the silicon impurities are well modeled by the conventional hydrogenic model using the effective mass m^* and dielectric constant ϵ of the harboring semiconductor. The ionization energy of the electron of a hydrogenic donor impurity is given by its ground state E_d:

$$E_d = E_c - \frac{q^4 m_n^*}{2(4\pi\epsilon)^2\hbar^2} = E_c - 13.6\left(\frac{\epsilon_0}{\epsilon}\right)^2 \frac{m_n^*}{m_0}\ \text{eV},$$

where ϵ_0 is the vacuum dielectric constant and m_0 the rest mass of a free electron.

The population of such a donor level is then given by Fermi–Dirac statistics using a spin degeneracy factor of 2:

$$N_D^+ = \frac{N_D}{1 + 2\exp\left(\dfrac{E_F - E_d}{k_B T}\right)}, \tag{2.1}$$

where E_F is the Fermi energy. As the donor concentration N_D is increased the donor hydrogenic potentials overlap (the Bohr radius is of the order of 100 Å) and the donor energy levels form a band of energies. In the silicon crystal, due to the random nature of the donor impurity distribution, the electron mobility in this band is that of an insulator. As the donor concentration is increased above a critical concentration of about 10^{17} cm^{-3}, an insulator–metal transition (of the combined Mott and Anderson type) takes place and the impurity band becomes metallic (conductor).

In GaAs, due to the very small effective mass, the donor energy level is very close to the conduction-band edge and will merge with the conduction band for high donor concentration. This can be explained by the screening of the donor potential. Under a degenerate doping condition a large number of electrons is present in the conduction band. The electron distribution adjusts itself to screen the Coulombic potential of the donor impurity [14]. The Coulombic potential of the impurity $q^2/(4\pi\epsilon r)$ is replaced by $q^2\exp(-\lambda r)/(4\pi\epsilon r)$, where λ is the Debye length. As a result of the shorter range of the impurity potential, the donor energy level moves toward the conduction-band edge. The donor ionization energy can be expressed by the following empirical relation [15]:

$$E_d(N_D) = E_d(0)\left[1 - \left(\frac{N_D}{N_{crit}}\right)^{1/3}\right],$$

where $N_{crit} \simeq 2 \times 10^{17}$ cm^{-3} is the critical donor ionization for which the donor ionization energy vanishes. For such high concentrations the donors (or acceptors) are simply assumed to be all ionized independently of the Fermi energy,

$$N_D^+ = N_D.$$

Under degenerate doping conditions a large number of electrons is present in the conduction band. Electron–electron interactions take place and the band structure is modified. A reduction of the bandgap is observed experimentally. An empirical formula for the dependence of the bandgap upon the doping is given by Casey and Stern [16] for bulk GaAs:

$$E_g(n, p) = E_g(0, 0) - 1.6 \times 10^{-8}(p^{1/3} + n^{1/3}) \text{ eV} \tag{2.2}$$

with p and n in cm^{-3} units. A reduction of 16 meV occurs for a donor concentration of about 10^{18} cm^{-3}.

In AlGaAs the behavior of silicon doping is more complicated. Each silicon donor impurity seems to be associated with three different states: an ionized donor level, a donor with one electron in the shallow hydrogenic level (H), and a donor with an electron in a deep trap level (the so-called DX center). The following model can then be derived [2] assuming a degeneracy factor of 2 for both the H and DX levels:

$$N_{HD}^+ = N_D\left[\frac{2\exp\left(\dfrac{E_F - E_{d,H}}{k_BT}\right)}{1 + 2\exp\left(\dfrac{E_F - E_{d,H}}{k_BT}\right) + 2\exp\left(\dfrac{E_F - E_{d,DX}}{k_BT}\right)}\right],$$

$$N_{DX}^+ = N_D\left[\frac{2\exp\left(\dfrac{E_F - E_{d,DX}}{k_BT}\right)}{1 + 2\exp\left(\dfrac{E_F - E_{d,H}}{k_BT}\right) + 2\exp\left(\dfrac{E_F - E_{d,DX}}{k_BT}\right)}\right].$$

The deep donor level dominates in equilibrium since due to its lower energy it is preferentially populated. The donor ionization energy given in Table 2.3 for the AlGaAs versus Al mole fraction m is that of the deep donor. However, the capture and emission time of this deep donor level is very large (sometimes on the order of seconds) due to an unusually large energy barrier on the order 0.3 eV. Consequently in transient or AC situations the deep trap does not have time to respond (its population remains frozen) and only the population of the shallow level is updated.

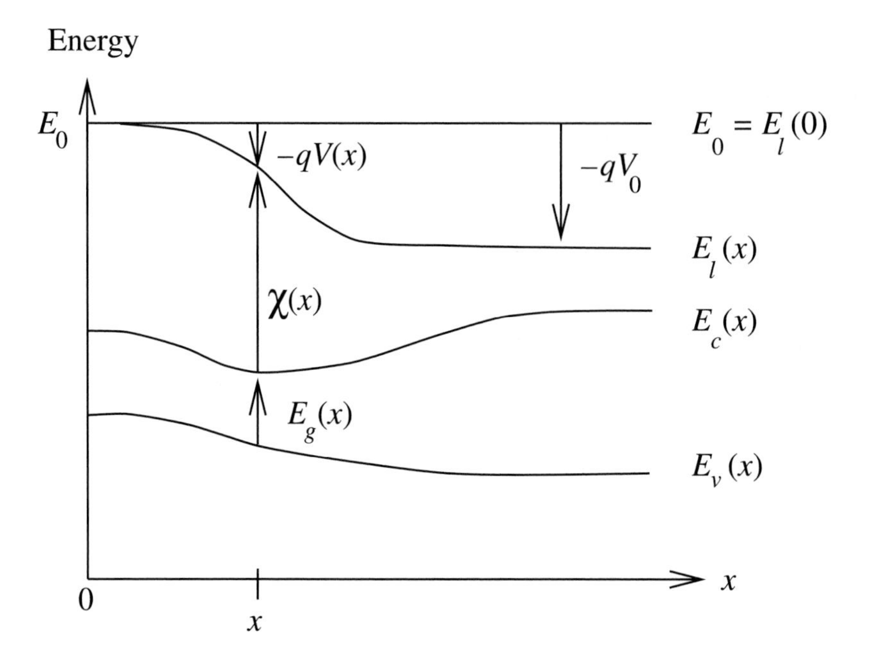

Fig. 2.2. Band diagram of an arbitrary heterostructure.

2.3 The Anderson band-diagram model

Equilibrium band diagrams provide a graphical representation which is traditionally used as a starting point in the analysis of semiconductor devices. The analysis of heterostructures requires a special extension of band-diagram techniques as the semiconductor is now allowed to vary with position.

First let us consider the one-dimensional band diagram of a uniform semiconductor in equilibrium shown in Figure 1.10. The atomic potentials of the crystal atoms are represented by dashed lines. The vacuum level, E_0, is the energy defined at the crystal surface as the energy required to free an electron from the crystal. The electron affinity, $\chi(x)$, is the energy separation between the conduction-band edge $\mathcal{E}_c$ and the vacuum level.

Consider now the general band diagram of a heterostructure shown in Figure 2.2. We assume initially that no tunneling is taking place and that no dipoles and surface states are present in the heterostructures. The conduction-band edge $E_c(x)$ and the valence band edge $E_v(x)$ now vary spatially according to

$$E_c(x) = \mathcal{E}_c(x) - qV(x) = E_0 - qV(x) - \chi(x),$$
$$E_v(x) = \mathcal{E}_v(x) - qV(x) = E_0 - qV(x) - \chi(x) - E_g(x),$$

where $V(x)$ is the built-in electrostatic potential in the heterostructure, $E_l(x)$ is the

Table 2.2. *Valence-band offsets in eV calculated by Ruan and Ching [6] with ($|\Delta E_v|$) and without ($|\Delta E_{v0}|$) the dipole correction δE_v.*

Material	$\|\Delta E_{v0}\|$	δE_v	$\|\Delta E_v\|$
Ge/Si	0.32	0.10	0.22
Ge/AlAs	1.22	0.30	0.92
Ge/GaAs	0.70	0.19	0.51
Ge/InAs	0.51	0.13	0.38
Si/GaAs	0.38	0.16	0.22
Si/InAs	0.19	0.09	0.10
Si/InP	0.58	0.22	0.36
GaAs/AlAs	0.52	0.16	0.36
GaAs/InAs	0.19	0.03	0.16
GaAs/InP	0.20	0.07	0.13

local vacuum level at the position x, E_0 is the local vacuum level $E_l(x)$ at $x = 0$, $\chi(x)$ is the spatially-varying electron affinity, $\mathcal{E}_c(x) = \mathcal{E}_c(\mathbf{k} = 0, x)$ is the spatially-varying bottom edge of the conduction band, and $\mathcal{E}_v(x) = \mathcal{E}_v(\mathbf{k} = 0, x)$ is the spatially-varying top edge of the valence band.

The electron and hole populations are given by the usual Fermi–Dirac statistics (see Marshak and Vliet [4] for a demonstration):

$$n = N_c(x) F_{1/2}\left[\frac{E_F - E_c(x)}{k_B T}\right],$$

$$p = N_v(x) F_{1/2}\left[\frac{E_v(x) - E_F}{k_B T}\right],$$

where the density of states N_c and N_v now vary spatially (see also Table 2.3):

$$N_c(x) = 2\left(\frac{2\pi m_{de}(x) k_B T}{h^2}\right)^{3/2},$$

$$N_v(x) = 2\left(\frac{2\pi m_{dh}(x) k_B T}{h^2}\right)^{3/2},$$

with m_{de} and m_{dh} the density-of-states electron and hole effective-masses, respectively:

$$m_{de} = N_\alpha^{2/3}\, m_{t\alpha}^{*\,2/3}\, m_{l\alpha}^{*\,1/3},$$

$$m_{dh} = \left(m_{lh}^{*\,3/2} + m_{hh}^{*\,3/2}\right)^{2/3}.$$

m_{lh} and m_{hh} are the light- and heavy-hole masses, $m_{t\alpha}$ and $m_{l\alpha}$ are the transverse and longitudinal masses at the conduction-band minimum α, and N_α is the number of minima of the conduction: $N_\alpha = 1, 3$ or 4 for $\alpha = \Gamma$, X and L, respectively. For direct bandgap materials ($\alpha = \Gamma$) we have $m_{de} = m^*_{t\Gamma} = m^*_{l\Gamma} = m^*_n$.

The Fermi–Dirac integral $F_{1/2}(\eta)$

$$F_{1/2}(\eta) = \frac{2}{\pi^{1/2}} \int_0^\infty \frac{x^{1/2}}{1 + \exp(x - \eta)} dx$$

is tabulated in Appendix B of [5]. For non-degenerate doping the equilibrium concentration reduces to the simpler Boltzmann expression:

$$n = N_c(x) \exp\left[\frac{E_F - E_c(x)}{k_B T}\right], \tag{2.3}$$

$$p = N_v(x) \exp\left[\frac{E_v(x) - E_F}{k_B T}\right]. \tag{2.4}$$

The charge concentration ρ in a heterostructure is then given by

$$\rho(x) = q(\, p(x) - n(x) + N_D^+(x) - N_A^-(x)\,).$$

The electrostatic potential $V(x)$ is obtained from the Gauss equation

$$\frac{dD(x)}{dx} = \rho(x),$$

where $D(x) = \epsilon(x)F(x)$ is the displacement field and $F(x)$ the electric field. The Gauss equation for a spatially-varying dielectric constant is then

$$\frac{d}{dx}[\epsilon(x)F(x)] = \rho(x)$$

The electrostatic potential is given by

$$F(x) = -\frac{dV(x)}{dx}$$

and the Poisson equation in a spatially-varying heterostructure is given by

$$\epsilon(x)\frac{d^2V(x)}{dx^2} + \frac{dV(x)}{dx}\frac{d\epsilon(x)}{dx} = -\rho(x).$$

The solution of this Poisson equation is sufficient to calculate the semiclassical band diagram of a heterostructure. The equilibrium band diagram is obtained by assuming that the Fermi level is constant over the entire heterostructure.

In Problem 2.2 an exact solution is developed for this general one-dimensional Poisson equation for the useful case in which the charge distribution is assumed to be uniform in each atomic layer (lattice parameter wide).

2.4 The abrupt heterojunction case

The Anderson band-diagram model for heterostructures was presented above for the general case of a spatially-varying band structure. Let us now consider the simpler and more common case of an ideal abrupt heterojunction. The heterojunction is located at $x = 0$. The semiconductor on the left-hand side ($x < 0$) is labeled 1. The semiconductor on the right-hand side ($x > 0$) is labeled 2.

The conduction-band edge $\mathcal{E}_c(x)$ is discontinuous and can be written

$$\mathcal{E}_c(x) = \mathcal{E}_{c,1} + \mathrm{u}(x)(\mathcal{E}_{c,2} - \mathcal{E}_{c,1}), \tag{2.5}$$

where $\mathrm{u}(x)$ is the step function. The effective density of states $N_c(x)$ is also discontinuous and can also be written

$$N_c(x) = N_{c,1} + \mathrm{u}(x)(N_{c,2} - N_{c,1}). \tag{2.6}$$

Similar expressions can also be written for the valence band edge $\mathcal{E}_v$ and the effective density of states N_v.

The field boundary conditions at $x = 0$ are

$$D(0^-) = D(0^+),$$

$$\epsilon_1 \frac{dV(0^-)}{dx} = \epsilon_2 \frac{dV(0^+)}{dx}.$$

Here we are assuming that no surface states are present at the junctions. In lattice-matched heterostructures such as GaAs/AlGaAs, surface states are usually negligible. However, the presence of surface states certainly depends on the material and growth parameters and each case must be considered separately. When the left-hand layer is a strained material, such as AlGaN pseudomorphically grown on GaN, the piezoelectric effect must be taken into account by using the boundary condition

$$D(0^-) + P(0^-) = D(0^+),$$

where P is the piezoelectric field (polarization) induced by the tensile strain in the strained layer [3].

In the absence of an interface dipole layer the electrostatic potential is continuous

$$V(0^-) = V(0^+).$$

The equilibrium band diagram can now be constructed by setting the Fermi level constant across the heterostructure. The electrostatic potential $V(x)$ cannot be expressed in closed form (a transcendental equation is obtained) and a numerical solution is required. It is, however, possible to calculate the built-in electrostatic potential which appears across the heterostructure if we assume a non-degenerate statistic. Using Equation (2.3) we can express the Fermi level E_F as

$$E_F = \mathcal{E}_c(x) - qV(x) + k_B T \ln n(x) - k_B T \ln N_c(x). \tag{2.7}$$

In equilibrium the Fermi level is constant and we have

$$0 = \frac{dE_F}{dx} = \frac{d\mathcal{E}_c}{dx} - q\frac{dV}{dx} + k_B T\frac{d\ln(n)}{dx} - k_B T\frac{d\ln(N_c)}{dx}.$$

Integrating from $-\infty$ to ∞ we have

$$0 = [\mathcal{E}_c(\infty) - \mathcal{E}_c(-\infty)] - q[V(\infty) - V(-\infty)] + k_B T \ln\left[\frac{n(\infty)}{n(-\infty)}\right] - k_B T \ln\left[\frac{N_c(\infty)}{N_c(-\infty)}\right].$$

The built-in potential V_0 is then

$$qV_0 = q[V(\infty) - V(-\infty)] = (\mathcal{E}_{c,2} - \mathcal{E}_{c,1}) + k_B T \ln\frac{n_2}{n_1} - k_B T \ln\frac{N_{c,2}}{N_{c,1}} \tag{2.8}$$

if we define $n_1 = n(-\infty)$ and $n_2 = n(\infty)$. Using $\mathcal{E}_c = E_0 - \chi$ we can also express V_0 in terms of the affinity χ:

$$qV_0 = (\chi_1 - \chi_2) + k_B T \ln\frac{n_2}{n_1} - k_B T \ln\frac{N_{c,2}}{N_{c,1}}. \tag{2.9}$$

The same result can also be obtained from a graphical analysis, as will be shown with an example below.

As usual, the electrostatic potential V_0 cannot be measured. A measurable voltage always appears between two different Fermi levels, and here the Fermi level is constant (zero voltage). The physical reason why the variation of the electrostatic potential $V(x)$ cannot be measured stems from the fact that its variation is canceled by the variation of the chemical potential $\mu(x)$. Indeed the Fermi energy E_F used here is a thermodynamic quantity called an electrochemical potential (often labeled ξ):

$$E_F = \xi = \mu(x) - qV(x),$$

where the chemical potential $\mu(x)$ is easily identified in Equation (2.7).

An example of the graphical construction of the band diagram of a p-Ge–n-GaAs heterojunction is shown in Figure 2.3(a) and (b). Figure 2.3(a) shows the equilibrium band diagram before the two semiconductors are joined and Figure 2.3(b) shows the equilibrium band diagram after the two semiconductors have been joined. For this heterostructure we have

$$\mathcal{E}_{c,2} - \mathcal{E}_{c,1} = \chi_1 - \chi_2 = 0.06 \text{ eV},$$
$$\mathcal{E}_{v,2} - \mathcal{E}_{v,1} = \mathcal{E}_{c,2} - \mathcal{E}_{c,1} - (E_{g,2} - E_{g,1}) = 0.06 - 0.772 = -0.712 \text{ eV}.$$

An alternative model based on the alignment of the intrinsic energy level was proposed for the construction of heterostructure band diagrams, but unlike the Anderson theory, it was found not to be self-consistent. However, the use of the affinity in the

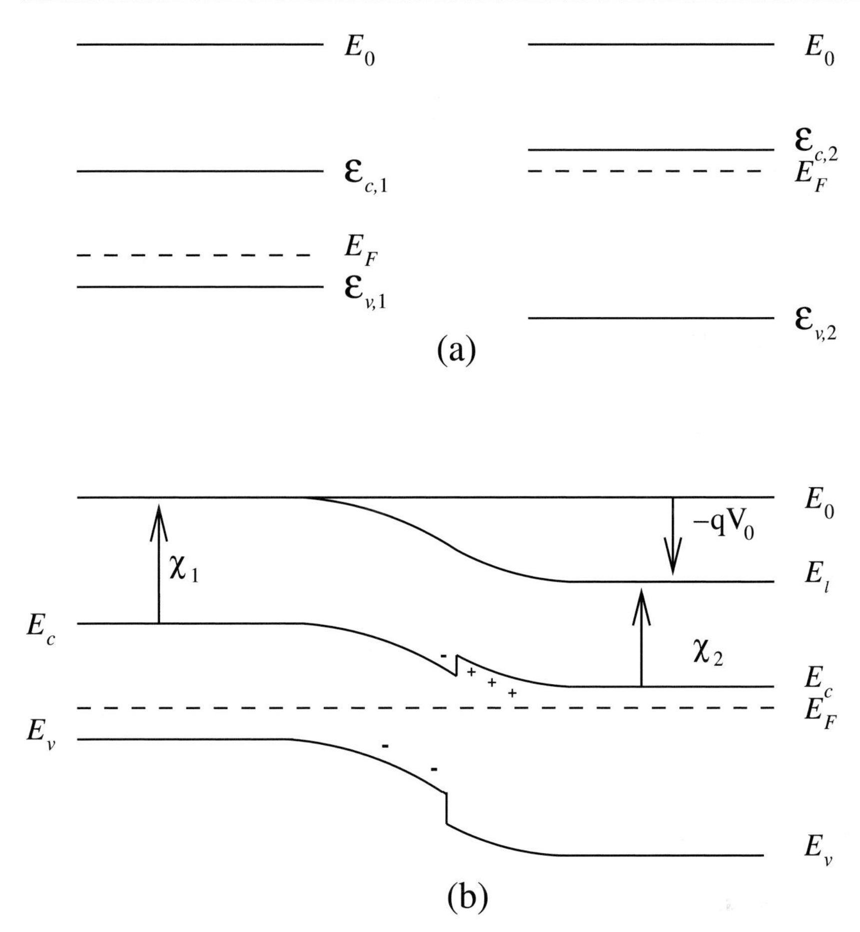

Fig. 2.3. Band diagram of a Ge/GaAs heterojunction: (a) equilibrium band diagram before the two semiconductors are joined. (b) equilibrium band diagram once the two semiconductors have been joined.

Anderson model is not very practical, because the affinities of semiconductors are not accurately known. A tabulation of electron affinity is given in Table 2.1 (from Wolfe [5]). The electron affinities are large numbers (about 4 eV), and the calculation of conduction and valence-band discontinuities (which are typically on the order of fractions of an electron volt) is therefore not very accurate. Furthermore, dipoles arise in heterostructures which offset the effective conduction and valence-band discontinuities. Thus the use of electron affinities alone is not usually sufficiently accurate for arbitrary heterostructures. In practice, instead of using the affinity, the conduction and valence discontinuities are measured experimentally for each type of heterostructure.

The treatment used here to analyze the heterojunctions is semiclassical. We shall see that in some heterostructures a large electric field can arise at the interface of the heterojunction, and the potential barrier created by the conduction or valence band

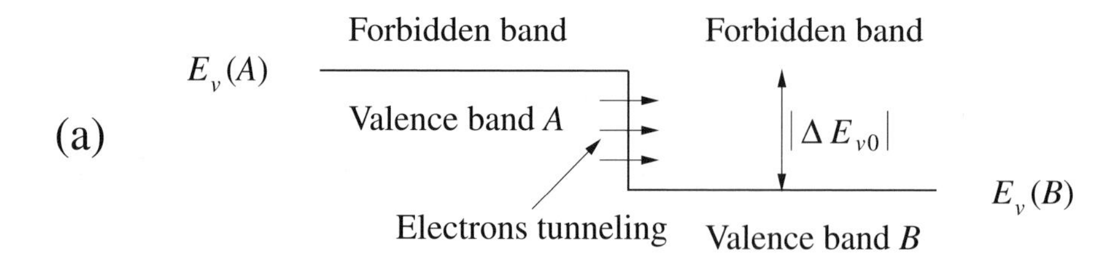

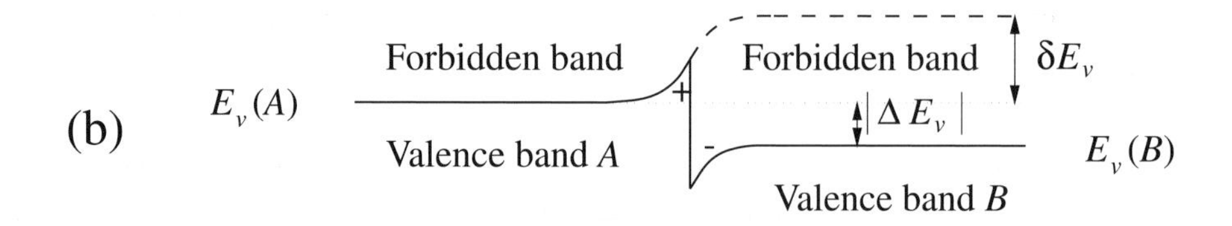

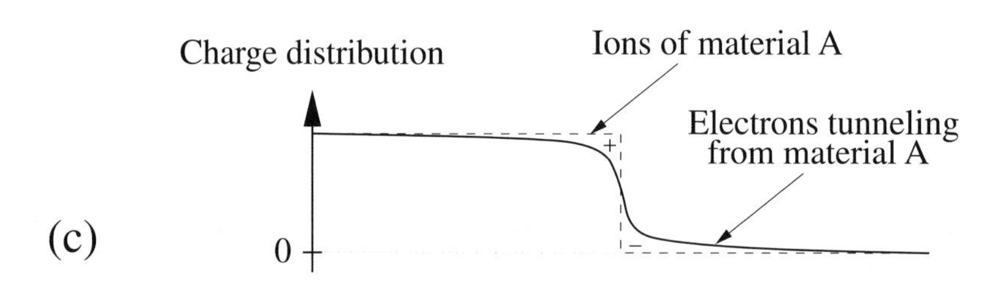

Fig. 2.4. Band diagram of a heterojunction showing the valence band edge (a) before and (b) after the two semiconductors are joined. The charge distribution supported by the electrons tunneling from material A to material B and the ions of material A which contributed these electrons are shown in (c).

edges and the electric field can introduce a narrow triangular quantum well. In such instances a quantum treatment of the heterostructure becomes necessary.

An additional quantum effect which explains the presence of a dipole at the heterojunction interface is the penetration by tunneling of the valence-band electrons from the material with the higher valence band into the forbidden band of the material with the lower valence band (see Figure 2.4(a) and (b)). The resulting charge transfer leads to the formation of an effective dipole (see Figure 2.4(c)) which induces an electrostatic potential barrier δE_v partially hindering in turn the tunneling of electrons. This dipole typically extends over a region of about 10 Å wide. The effect of the resulting electrostatic potential is to reduce the effective band discontinuity by δE_v:

$$|\Delta E_v| = |\Delta E_{v0}| - \delta E_v$$

with

$$\Delta E_{v0} = \mathcal{E}_{v,A} - \mathcal{E}_{v,B} = -(\chi_A + E_{g,A}) + (\chi_B + E_{g,B}).$$

The valence-band offsets with and without the dipole correction δE_v calculated by Ruan and Ching [6] are given in Table 2.2.

2.5 Drift-diffusion transport model for heterostructures

The calculation of the I–V characteristic of a heterostructure requires the use of a transport model appropriate for heterostructures. A semiclassical theory of transport using the Boltzmann equation permits us (see Chapter 7) to derive the following drift-diffusion current equations for a non-degenerate heterostructure under low-field conditions (small departure from equilibrium):

$$J = J_n + J_p, \tag{2.10}$$

$$J_n = \mu_n(x)n(x)\frac{d\xi_n}{dx}, \tag{2.11}$$

$$J_p = \mu_p(x)p(x)\frac{d\xi_p}{dx}, \tag{2.12}$$

where J is the total current, J_n the electron current, J_p the hole current, and where ξ_n and ξ_p are the electrochemical potentials of the electron and hole gases, respectively. The use of two different electrochemical potentials for electrons and holes implies that the electron and hole gases are *not* in equilibrium together but are each in local equilibrium. For non-degenerate conditions the electron and hole concentrations are each given by their own Boltzmann distribution

$$n = N_c(x)\exp\left[\frac{\xi_n - E_c(x)}{k_BT}\right], \tag{2.13}$$

$$p = N_v(x)\exp\left[\frac{E_v - \xi_p(x)}{k_BT}\right]. \tag{2.14}$$

Note that the product $n \times p$ is not equal to n_i^2 when electrons and holes are *not* in equilibrium with one another ($\xi_n \neq \xi_p$). The current Equations (2.11) and (2.12) and the Boltzmann Equation (2.13) and (2.14) are actually the same as in the case of the uniform semiconductor except that now the material parameters vary abruptly at the interface.

Substituting ξ_n obtained from Equation (2.13) into Equation (2.11) we obtain

$$\begin{aligned} J_n = & -q\mu_n(x)n(x)\frac{dV}{dx} + \mu_n(x)k_BT\frac{dn(x)}{dx} \\ & + \mu_n(x)n(x)\frac{d\mathcal{E}_c}{dx} - \mu_n(x)k_BT\frac{n}{N_c}\frac{dN_c}{dx}. \end{aligned} \tag{2.15}$$

The first two terms are easily recognized to be the drift and the diffusion (with the diffusion constant $qD_n = k_B T\mu_n$). The third term is a quasi-drift current arising from the variation of the conduction-band edge, since the latter is equivalent to an electrostatic potential step accelerating or decelerating carriers. The fourth current term is a quasi-diffusion term arising from the variation of the effective density of states.

The continuity of the current J_n is regulated by the conservation of particles, which is enforced by the usual steady-state continuity equations:

$$\frac{dJ_n}{dx} = q[R(x) - G(x)], \tag{2.16}$$

$$\frac{dJ_p}{dx} = -q[R(x) - G(x)], \tag{2.17}$$

where $R(x)$ and $G(x)$ are, respectively, the recombination and generation rates of carriers. For direct (valence- to conduction-band) recombination we have

$$R(x) - G(x) = r(T)[n(x)p(x) - n_i(x)^2]$$

with $r(T)$ a temperature-dependent proportionality constant. We can rewrite $n(x) = n_0(x) + \delta n(x)$ and $p(x) = p_0(x) + \delta p(x)$ in terms of their equilibrium values, n_0 and p_0 and excess concentrations δn and δp. If space-charge neutrality is enforced we must have $\delta n(x) = \delta p(x)$ and the recombination and generation terms reduce to

$$R(x) - G(x) = \frac{\delta n}{\tau_n} = \frac{\delta p}{\tau_p},$$

where $\tau_n = \tau_p = [r(T)(n_0 + p_0)]^{-1}$ are the electron and hole recombination lifetimes.

Consider now the case of an abrupt heterojunction. Let us first check the continuity of the current J_n at the junction ($x = 0$). For this purpose we integrate the continuity Equation (2.16) across the heterojunction

$$J_n(0^+) - J_n(0^-) = \int_{0^-}^{0^+} \frac{dJ_n(x)}{dx}\, dx = \int_{0^-}^{0^+} q(R(x) - G(x))\, dx = 0. \tag{2.18}$$

The right-hand term vanishes if we assume that no impulse function is contained in the recombination and generation term and therefore in n and p; an assumption which will be verified to be self-consistent. Note that an electron current discontinuity would be observed if there were interface states at the junction contributing to the recombination and generation of carriers. Therefore in the absence of surface states, the electron current J_n and similarly the hole current J_p are continuous at the interface. However, their derivatives are not continuous at the interface.

Let us verify whether the electrochemical potentials ξ_n and ξ_p are continuous at the heterojunction:

$$J_n = \mu_n(x)n(x)\frac{d\xi_n}{dx} = \mu_n(x)N_c(x)\exp\left[\frac{\xi_n - E_c(x)}{k_B T}\right]\frac{d\xi_n}{dx}. \tag{2.19}$$

Separating the variables and integrating from 0^- to 0^+ we have

$$\int_{\xi_n(0^-)}^{\xi_n(0^+)} \exp\left[\frac{\xi_n}{k_B T}\right] d\xi_n = \int_{0^-}^{0^+} \frac{J_n}{\mu_n(x) N_c(x)} \exp\left[\frac{E_c(x)}{k_B T}\right] dx = 0. \tag{2.20}$$

This demonstrates that the electrochemical potential $\xi_n(x)$ (and similarly $\xi_p(x)$) is continuous at $x = 0$, namely $\xi_n(0^-) = \xi_n(0^+)$ since the integration from 0^- to 0^+ of an integrand with only step discontinuities and no impulses vanishes. Note that in the derivation of the continuity of the quasi-Fermi levels $\xi_n(x)$ and $\xi_p(x)$ we assumed that there was no recombination singularity at the interface and that the drift-diffusion transport model was applicable to an abrupt heterostructure. As we shall see in Section 2.7 this is a reasonable assumption for small discontinuities.

The derivatives of $\xi_n(x)$ and $\xi_p(x)$ will in general be discontinuous at $x = 0$ in a heterojunction since we have

$$J_n(0^-) = \mu_1 n(0^-) \frac{d\xi_n(0^-)}{dx} = \mu_2 n(0^+) \frac{d\xi_n(0^+)}{dx} = J_n(0^+). \tag{2.21}$$

Using the continuity of the electrochemical potentials we can infer from the Boltzmann Equations (2.13) and (2.14) that the electron and hole concentrations n and p must be discontinuous at $x = 0$:

$$\frac{n(0^+)}{n(0^-)} = \frac{N_{c,2}}{N_{c,1}} \exp\left[\frac{\mathcal{E}_{c,1} - \mathcal{E}_{c,2}}{k_B T}\right]. \tag{2.22}$$

Let us verify now that the impulses introduced in the current Equation (2.15) by the discontinuities of the conduction band $\mathcal{E}_c(x)$ and the effective carrier concentration $N_c(x)$ are canceled by the discontinuity of the electron concentration n. For this purpose let us evaluate the term $d\xi_n/dx$. Using Equations (2.5) and (2.6) and writing $n(x) = n_1(x) + \mathrm{u}(x)(n_2(x) - n_1(x))$ the following distribution identities are obtained for each component of $d\xi_n/dx$:

$$\frac{d(\mathcal{E}_c - qV)}{dx} = (\mathcal{E}_{c,2} - \mathcal{E}_{c,1})\delta(x) - q\frac{dV}{dx},$$

$$k_B T \frac{d\ln(n)}{dx} = k_B T\left[\frac{dn_1}{dx} + \left(\frac{dn_2}{dx} - \frac{dn_1}{dx}\right)\mathrm{u}(x)\right] + k_B T \ln\left[\frac{n(0^+)}{n(0^-)}\right]\delta(x),$$

$$-k_B T \frac{d\ln(N_c)}{dx} = -k_B T \ln\left(\frac{N_{c,2}}{N_{c,1}}\right)\delta(x).$$

Summing all these terms and using Equation (2.22) we easily verify that all the impulses disappear in $d\xi_n/dx$ and that Equation (2.15) gives the drift-diffusion equations for the left-hand (1) and right-hand (2) sides.

2.6 $I-V$ characteristics of p–n heterojunctions

An introduction to the semiclassical theory of heterostructures would not be complete without a discussion of the $I-V$ characteristic of the heterostructure version of the p–n junction diode. The p–n junction is a fundamental semiclassical device particularly since its principle of operation is of critical importance in the analysis of the bipolar transistor. Of particular interest is the impact of the conduction and valence-band discontinuities on the $I-V$ characteristic.

The analysis of the ideal p–n heterojunction diode within the drift-diffusion transport model is similar to the analysis of the ideal p–n homojunction diode, the only difference being that the material parameters are different on each side of the junction. The detailed derivation is therefore left as an exercise to the reader. Here we only outline the procedure given in [5]. The depletion region is assumed to extend from $-x_P$ to x_N. Solving the Poisson equation across the depletion region we have

$$x_P^2 = \frac{2\epsilon_N \epsilon_P N_D (V_0 - V)}{q N_A(\epsilon_P N_A + \epsilon_N N_D)},$$

$$x_N^2 = \frac{2\epsilon_N \epsilon_P N_A (V_0 - V)}{q N_D(\epsilon_P N_A + \epsilon_N N_D)},$$

where N_A and N_D are the dopings of the p and n sides respectively. The applied voltage V is positive on the p side and negative on the n side. The built-in voltage V_0 was derived in the previous section to be (now 1 is p and 2 is n)

$$q V_0 = q(V_N - V_P) = (\mathcal{E}_{c,N} - \mathcal{E}_{c,P}) + k_B T \ln \frac{n_N}{n_P} - k_B T \ln \frac{N_{c,N}}{N_{c,P}}. \tag{2.23}$$

The total depletion width is simply $W = x_P + x_N$.

If we neglect the recombination and generation of carriers in the depletion region, the current is given by (see, for example, Streetman [1])

$$J = q\left(\frac{D_{p,N}}{L_{p,N}} \Delta p_N + \frac{D_{n,P}}{L_{n,P}} \Delta n_P\right), \tag{2.24}$$

where Δn_P and Δp_N are the excess minority-carrier concentrations, $L_{n,P}$ and $L_{p,N}$ the diffusion lengths, and $D_{n,P}$ and $D_{p,N}$ the diffusion constants, in the p and n sides, respectively. The only difference with the homojunction diode theory is that care must be taken to specify whether the physical quantities D and L are measured on the p side or the n side of the junction.

To calculate the diode current we must now evaluate the excess minority-carrier concentrations Δn_P and Δp_N on the p and n sides of the depletion region. In equilibrium we have established a relation (Equation (2.23)) between the built-in potential and the equilibrium carrier concentration which we can rewrite

$$q V_0' = k_B T \ln \frac{n_N}{n_P} \tag{2.25}$$

or equivalently

$$\frac{n_N}{n_P} = \exp\frac{qV_0'}{k_B T} \tag{2.26}$$

where the quantity V_0' is defined as

$$qV_0' = qV_0 - (\mathcal{E}_{c,N} - \mathcal{E}_{c,P}) + k_B T \ln\frac{N_{c,N}}{N_{c,P}}. \tag{2.27}$$

Under non-equilibrium conditions we can assume that the same expression can be used to calculate the excess carriers Δn_P and Δp_N. (It can be shown that this procedure is equivalent to assuming that the total (diffusion and drift) diode current is negligible compared to the drift or diffusion component of the diode current)

$$\frac{n(x_N)}{n(-x_P)} = \frac{n_N + \Delta n_N}{n_P + \Delta n_P} = \exp\left[\frac{q(V_0' - V)}{k_B T}\right]. \tag{2.28}$$

Using $n = n_i^2/p$, a similar expression results for the hole excess carrier

$$\frac{p(-x_P)}{p(x_N)} = \frac{p_P + \Delta p_P}{p_N + \Delta p_N} = \frac{n_{i,N}^2}{n_{i,P}^2}\exp\left[\frac{q(V_0' - V)}{k_B T}\right]. \tag{2.29}$$

We assume that space-charge neutrality is enforced locally outside the space-charge region so that we have $\Delta n_P = \Delta p_P$ and $\Delta n_N = \Delta p_N$. We can now solve the resulting system of equations to obtain the excess electron concentration on the p side of the space-charge region:

$$\Delta n_P = n_P \left\{\frac{\exp\left(\frac{qV}{k_B T}\right) - 1}{1 - \frac{n_{i,N}^2}{n_{i,P}^2}\exp\left[\frac{2q(V - V_0')}{k_B T}\right]}\right\} \times \left[1 + \frac{n_{i,N}^2}{n_N^2}\exp\left(\frac{qV}{k_B T}\right)\right]. \tag{2.30}$$

Similarly the excess hole concentration on the n side of the space-charge region is given by

$$\Delta p_N = p_N \left\{\frac{\exp\left(\frac{qV}{k_B T}\right) - 1}{1 - \frac{n_{i,N}^2}{n_{i,P}^2}\exp\left[\frac{2q(V - V_0')}{k_B T}\right]}\right\} \times \left[1 + \frac{n_{i,P}^2}{p_P^2}\exp\left(\frac{qV}{k_B T}\right)\right]. \tag{2.31}$$

The total diode current is then obtained by substituting the excess carriers Δn_P and Δp_N derived into Equation (2.24)

$$J = q \left\{\frac{\exp\left(\frac{qV}{k_B T}\right) - 1}{1 - \frac{n_{i,N}^2}{n_{i,P}^2}\exp\left[\frac{2q(V - V_0')}{k_B T}\right]}\right\} \tag{2.32}$$

$$\times \left\{ \frac{D_{p,N} p_N}{L_{p,N}} \left[1 + \frac{n_{i,P}^2}{p_P^2} \exp\left(\frac{qV}{k_B T}\right) \right] + \frac{D_{n,P} n_P}{L_{n,P}} \left[1 + \frac{n_{i,N}^2}{n_N^2} \exp\left(\frac{qV}{k_B T}\right) \right] \right\}.$$

From Equation (2.32) we see that the current is infinite when the term $1 - (n_{i,N}^2/n_{i,P}^2) \exp\left[2q(V - V_0')/k_B T\right]$ is zero. In the regular p–n homojunction diode this occurs when V is equal to V_0. The contact potential V_0 is therefore the limiting forward voltage. In a p–n heterojunction diode the limiting forward voltage has been changed by the discontinuities of the bandgaps and effective carrier densities to

$$\begin{aligned} qV_0' - k_B T \ln \frac{n_{i,N}}{n_{i,P}} &= qV_0 - (\mathcal{E}_{c,N} - \mathcal{E}_{c,P}) + k_B T \ln \frac{N_{c,N}}{N_{c,P}} - k_B T \ln \frac{n_{i,N}}{n_{i,P}} \\ &= qV_0 - (\mathcal{E}_{i,N} - \mathcal{E}_{i,P}), \end{aligned}$$

where $\mathcal{E}_i$ is the intrinsic level which is defined as the Fermi level for which $n = p = n_i$.

Under low injection ($V \ll V_0$) Equation (2.32) reduces to the simple diode equation

$$J = q \left(\frac{D_{p,N}}{L_{p,N}} p_N + \frac{D_{n,P}}{L_{n,P}} n_P \right) \times \left[\exp\left(\frac{qV}{k_B T}\right) - 1 \right]. \tag{2.33}$$

For the p–n heterojunction of Figure 2.3(b) where the p side is the narrow-bandgap semiconductor we see that the injection of electrons will dominate because of the reduced potential barrier seen by the electrons compared to the large potential barrier seen by the holes. *In general the discontinuity in the bandgap favors the injection of majority carriers from the larger bandgap material.* Note that this injection is nearly independent of the doping concentration. This is in contrast with the homojunction which favors the injection of majority carriers from the strongly doped semiconductor. A demonstration of this effect (using Equation (2.33)) and its application to the bipolar heterojunction transistor is given in Section 2.9.

2.7 The thermionic model of heterojunctions

The diffusion theory developed above for the p–n junction performs well for graded heterojunctions. For abrupt p–n heterojunctions the diffusion theory works well when the band structure discontinuities are small compared with the built-in potential. Under forward bias the built-in potential is reduced by the applied voltage, and thermionic emission over the discontinuities becomes a rate-limiting process (see the band diagram of Figure 2.5). This situation is even more dramatic in n–n heterojunctions for which the built-in potential is comparable to the band structure discontinuities. In such instances the current at the heterojunction becomes limited by the thermionic emission of electrons over the heterojunction barriers. We therefore need to develop a new model for the heterojunction I–V characteristic which accounts for this thermionic emission effect in addition to drift diffusion of excess carriers. Our approach follows the analysis given in [9] for a p–n heterojunction.

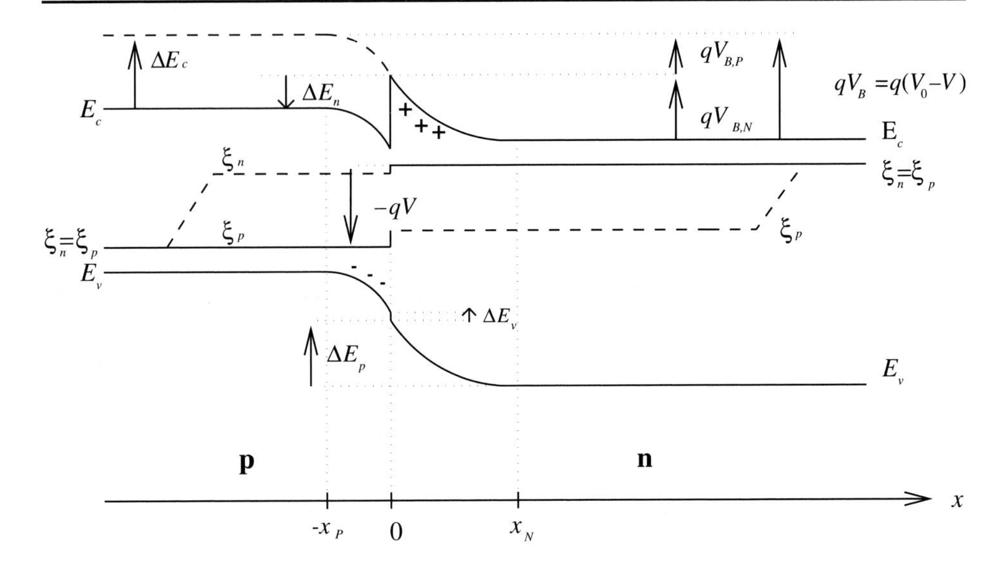

Fig. 2.5. Band diagram of a p–n heterojunction.

An analysis of thermionic emission including tunneling will be given in Chapter 3. When tunneling is neglected the classical thermionic emission model predicts a diode current at the heterojunction given by (see Chapter 8 for a derivation)

$$J_n(0) = A_n^* T^2 \left\{ \exp\left[\frac{\xi_n(0^+) - E_c(0^+)}{k_B T}\right] - \exp\left[\frac{\xi_n(0^-) - E_c(0^+)}{k_B T}\right]\right\}.$$

This result holds for the case shown in Figure 2.5 in which we have $E_c(0^+) > E_c(0^-)$. For simplicity we have assumed that the variation of the effective mass across the junction is negligible. The constant A_n^* is the Richardson constant given by

$$A_n^* = \frac{4\pi m_n^* q k_B^2}{\hbar^3}.$$

Note that in the thermionic model the quasi-Fermi level of the electron ξ_n will no longer be continuous at the heterojunction (see Figure 2.5) in contrast with the diffusion model where it was continuous. Let us rewrite this equation in terms of the electron concentration:

$$n(0^-) = N_c \exp\left[\frac{\xi_n(0^-) - E_c(0^-)}{k_B T}\right], \tag{2.34}$$

$$n(0^+) = N_c \exp\left[\frac{\xi_n(0^+) - E_c(0^+)}{k_B T}\right]. \tag{2.35}$$

We have introduced $N_c = N_{c,N} = N_{c,P}$ since we have assumed that both semiconductors have the same effective mass. The current at the junction can therefore be rewritten

$$J_n(0) = q v_{n,x}\left[n(0^+) - n(0^-)\exp\left(-\frac{\Delta E_c}{k_B T}\right)\right], \tag{2.36}$$

with $\Delta E_c = E_c(0^+) - E_c(0^-)$ the discontinuity of the conduction band, and with v_n given by

$$v_{n,x} = \frac{A_n^* T^2}{q N_c} = \left(\frac{k_B T}{2 m_n^* \pi} \right)^{1/2}.$$

$v_{n,x}$ can be identified to be the average longitudinal velocity $\langle v_x \rangle$ of the electrons moving toward the heterojunction ($v_x > 0$). $v_{n,x}$ is similar to the thermal velocity v_{th} introduced in Chapter 8.

The quasi-Fermi energies ξ_n can be assumed to be constant (see Figure 2.5) through the depletion regions on both sides of the heterojunction. This is equivalent to assuming that, in the depletion region, the total current (diffusion compensated by drift) is quite small compared to the large drift and diffusion components. The variation of the quasi-Fermi level is therefore negligible and the electrons are nearly in equilibrium with themselves on each side of the depletion region (even though they are not in equilibrium with the holes: $\xi_n \neq \xi_p$). However, the discontinuity of the conduction band prevents the electron gas on the n and p sides of the depletion from being in mutual equilibrium and the quasi-Fermi level ξ_n is discontinuous at the heterojunction itself. Assuming that this near-equilibrium assumption is valid on both sides of the depletion region we can write

$$n(0^-) = n(-x_P) \exp\left(\frac{q V_{B,P}}{k_B T} \right), \tag{2.37}$$

$$n(0^+) = n(x_N) \exp\left(\frac{-q V_{B,N}}{k_B T} \right), \tag{2.38}$$

using the non-equilibrium potential drops on the n and p sides:

$$V_{B,N} = (V_0 - V) \times \frac{N_A \epsilon_P}{\epsilon_P N_A + \epsilon_N N_D}, \tag{2.39}$$

$$V_{B,P} = (V_0 - V) \times \frac{N_D \epsilon_N}{\epsilon_P N_A + \epsilon_N N_D}. \tag{2.40}$$

Under low injection the hole concentration on the n side is small and the electron concentration is approximately the equilibrium value.

$$n(x_N) = n_N = N_D = n_P \exp\left(\frac{q V_0 - \Delta E_c}{k_B T} \right), \tag{2.41}$$

where n_N and n_P are the equilibrium values on the n and p sides of the depletion region, respectively. Let us now calculate the excess minority-carrier concentration $n(-x_P) - n_P$ on the p side. These minority carriers support a diffusion/recombination current at $-x_P$:

$$J_n(-x_P) = \frac{q D_{n,P}}{L_{n,P}} \left[n(-x_P) - n_P \right]. \tag{2.42}$$

In the absence of recombination in the depletion region the electron current J_n is constant through the depletion region. Equating Equations (2.42) and (2.36) and using Equation (2.37), (2.38), and (2.41), we obtain the excess minority-carrier concentration:

$$n(-x_P) - n_P = \frac{n_P}{R_n} \times \left[\exp\left(\frac{qV}{k_B T}\right) - 1\right], \tag{2.43}$$

where R_n is given by

$$R_n = 1 + \frac{D_{n,P}}{L_{n,P} v_{n,x}} \exp\left(\frac{-\Delta E_n}{k_B T}\right)$$

with $\Delta E_n = qV_{B,P} - \Delta E_c$ (see Figure 2.5) the difference between the electrostatic barrier on the p side and the conduction band discontinuity. A similar equation can be derived for the holes. The diode current J is therefore given by the summation of electron current $J_n(-x_P)$ (see Equation (2.42)) with the hole current $J_p(x_N)$:

$$J = q\left(\frac{D_{p,N}}{L_{p,N} R_p} p_N + \frac{D_{n,P}}{L_{n,P} R_n} n_P\right) \times \left[\exp\left(\frac{qV}{k_B T}\right) - 1\right], \tag{2.44}$$

where R_p is given by

$$R_p = 1 + \frac{D_{p,N}}{L_{p,N} v_{p,x}} \exp\left(\frac{-\Delta E_p}{k_B T}\right),$$

with $\Delta E_p = qV_{B,N}$ (see Figure 2.5). The p–n heterojunction diode current (Equation (2.44)) obtained for the thermionic emission model is the same as the current obtained for the diffusion model (Equation (2.33)) except for the new prefactors R_n and R_p.

Under small forward-bias voltages, when the diode built-in potentials are large enough that $qV_{B,P} - \Delta E_c$ and $qV_{B,N}$ are large (positive) compared to $k_B T$, the thermionic emission model reduces to the diffusion model ($R_p = R_n = 1$).

For large forward-bias voltages, ΔE_n becomes negative ($qV_{B,P} < \Delta E_c$), and the thermionic emission becomes dominant. The I–V characteristic of the diode then reduces to that of a Schottky barrier

$$J = J_s \left[\exp\left(\frac{qV}{k_B T}\right) - 1\right] \tag{2.45}$$

with J_s given by

$$J_s = qn_P \left(\frac{k_B T}{2\pi m_n^*}\right)^{1/2} \exp\left(\frac{-V_{D,N}}{k_B T}\right), \tag{2.46}$$

where $V_{D,N} = -\Delta E_n$ is the diffusive potential (positive).

The thermionic model derived here assumed that ΔE_c and ΔE_v as defined in Figure 2.5 are both positive. For arbitrary signs of ΔE_c and ΔE_v one can verify that the following generalized ΔE_p and ΔE_n should be used to handle the four possible sign combinations:

$$\Delta E_p = qV_{B,N} + \min(0, \Delta E_v),$$
$$\Delta E_n = qV_{B,P} - \max(0, \Delta E_c).$$

The reader is referred to [12] for further examples of p–n and n–n heterojunction band diagrams and I–V characteristics.

2.8 Ballistic launching

We have seen in the previous section that for large forward bias the built-in potential in a p–n heterojunction is effectively reduced by the applied voltage and thermionic emission over the band structure discontinuities becomes the rate-limiting process. Consider the n–p junction shown in Figure 2.6(a). The current voltage characteristic of the diode is then of the form of Equation (2.45).

Such an abrupt heterojunction can serve as a launching ramp to inject high-energy quasi-ballistic electrons into the p side. Figure 2.6(a) shows the n(emitter)–p(base) heterostructure of a bipolar transistor studied by Pelouard *et al.* [10] using a Monte-Carlo simulator. The electron velocity distribution obtained near the heterojunction is sketched in Figure 2.6(b). In the base (p) region, two electron populations can be distinguished in Figure 2.6(b): one corresponding to relaxed velocity (or thermalized) electrons (n_R) and the other centered around 10^8 cm/s corresponding to quasi-ballistic electrons (n_{QB}). The electron population with a negative velocity around -10^8 cm/s capable of retrodiffusing into the emitter is negligible. This is the required condition for thermionic emission. The total diode current is still given by Equation (2.45) but now includes two components: a diffusion current and a quasi-ballistic current (after [10])

$$J = J_R(x) + J_{QB} = J_s \left[\exp\left(\frac{qV}{k_B T}\right) - 1 \right],$$

$$J_R(x) = qD_{n,P} \frac{dn_R}{dx},$$

$$J_{QB} = J_s \exp\left(\frac{qV}{k_B T}\right) \exp\left(\frac{-x}{\lambda}\right),$$

where $D_{n,P}$ is the diffusion constant for the relaxed electron population and λ is the mean free path of the quasi-ballistic electrons. The quasi-ballistic electrons do not change the total electron current and have a small impact on the potential dropped across the diode. However, if such a heterojunction is used as the emitter–base diode of a bipolar transistor, the fraction of electrons recombining in the base will decrease.

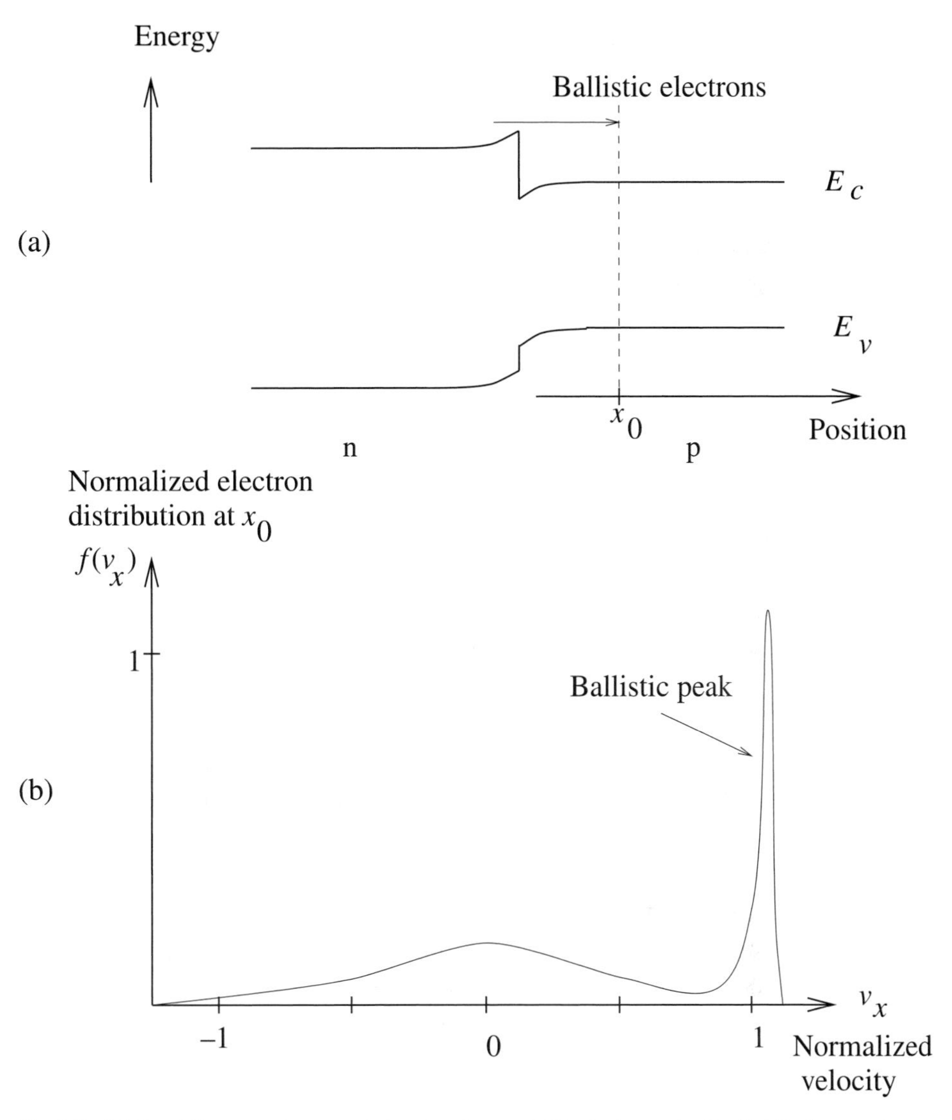

Fig. 2.6. (a) Band diagram of an n–p heterojunction and (b) electron velocity distribution. v_x is normalized by 10^8 cm/s.

Indeed the recombination current in the base J_{BR} involves only the thermalized electron population

$$J_{BR} = q \int_{BASE} \frac{n_R - n_B}{\tau_{n,B}} dx \tag{2.47}$$

with $\tau_{n,B}$ the lifetime of electrons in the base (p) and n_B the equilibrium electron concentration in the base (p). These formulas are used in Problem 2.3 to demonstrate that the ballistic component of the emitter current increases the current gain of an HBT. Pelouard *et al.* [10] quotes a factor 3 increase in β for a 0.1 μm InGaAs HBT due to the quasi-ballistic transport in the base (see Problem 2.3). Unfortunately three-dimensional geometrical effects in HBTs seem to reduce this effect. To date the only devices to successfully benefit from ballistic transport are quantum devices such as the resonant tunneling diode (RTD) (see Chapter 3).

2.9 The HBT

The p–n heterojunction diode finds an important application as the emitter–base diode of a bipolar transistor. Such a transistor is called an HBT. The idea of the HBT was proposed by W. Shockley [7] and was later developed by H. Kroemer [8]. The advantage provided by using a heterojunction diode for the emitter junction is evidenced by the ratio of the electron to hole emitter current. Consider an n–p–n transistor (E–B–C). According to our previous drift-diffusion analysis we have

$$\begin{aligned}\frac{I_{nE}}{I_{pE}} &= \frac{D_{n,B}}{D_{p,E}}\frac{L_{p,E}}{L_{n,B}}\frac{n_B}{p_E} = \frac{D_{n,B}}{D_{p,E}}\frac{L_{p,E}}{L_{n,B}}\frac{N_{D,E}}{N_{A,B}}\frac{n_{i,B}^2}{n_{i,E}^2} \\ &= \frac{D_{n,B}}{D_{p,E}}\frac{L_{p,E}}{L_{n,B}}\frac{N_{D,E}}{N_{A,B}}\frac{N_{c,E}N_{v,E}}{N_{c,B}N_{v,B}}\exp\left(\frac{E_{g,E}-E_{g,B}}{k_BT}\right).\end{aligned}$$

The use of a wider bandgap for the emitter (E) region and a smaller bandgap for the base (B) region permits us to obtain very large I_{nE}/I_{pE} ratios, and therefore to preferentially inject electrons rather than holes. This can be used to improve the performance of a bipolar transistor. The resulting band diagram of such a heterojunction bipolar transistor is shown in Figure 2.7.

Consider the transistor current gain α which is given by

$$\alpha = \frac{I_C}{I_E} = \frac{I_C}{I_{nE}}\frac{I_{nE}}{I_{nE}+I_{pE}} = B\,\gamma,$$

where B is the base transport factor and γ the emitter injection efficiency. Clearly the emitter injection efficiency approaches unity as the ratio I_{nE}/I_{pE} is increased. In practice the recombination current in the depletion region of the emitter becomes the limiting factor. The base transport factor is given by the usual ratio

$$B = \frac{\tau_n}{\tau_{TR}},$$

where τ_n is the electron recombination time in the base and τ_{TR} is the transit time of the electron across the base from emitter to collector. Since MBE growth is used, the base can be made as short as desired to minimize the transit time. However, a base width which is too small would increase the lateral base resistance and decrease the high-frequency performance of the HBT.

The current gain can be made very large (close to unity) and very large $\beta = I_C/I_B = \alpha/(1-\alpha)$ can theoretically be obtained. βs as high 1600 have been reported.

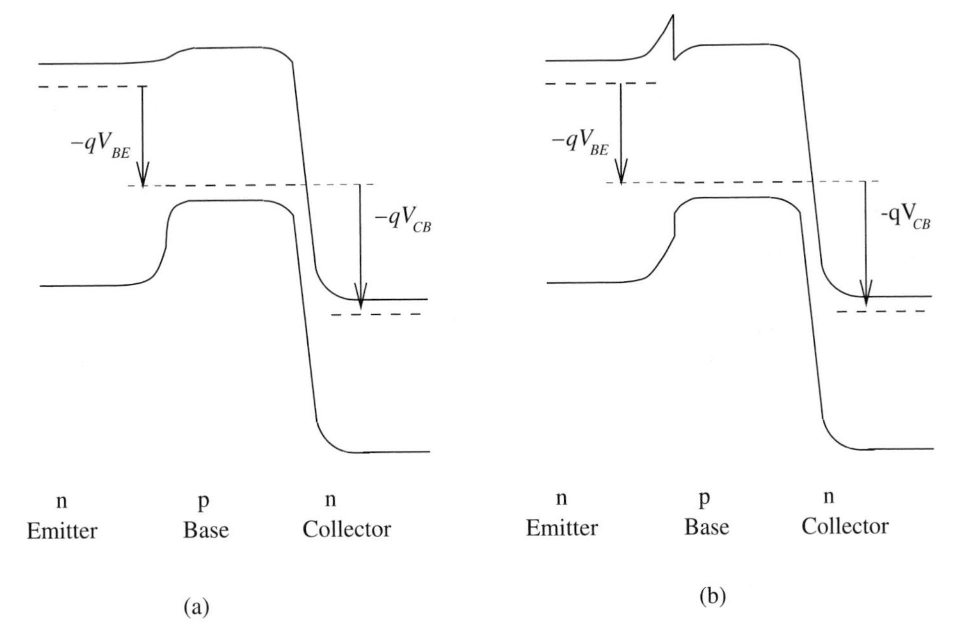

Fig. 2.7. Band diagram of an HBT with (a) a graded emitter–base heterojunction and (b) an abrupt emitter–base heterojunction.

The presence of the heterojunction also permits one to increase the doping in the base to lower the base resistance without seriously decreasing the emitter efficiency. Similarly, a smaller emitter doping can be used to reduce the junction capacitance. An improved high-frequency performance results from these modifications. The small base–emitter capacitance and its weak voltage variation permits ones to design HBT power amplifiers with an extremely high linearity (IP_3 of 40 dBm) which greatly reduces intermodulation distortion. Linearity is particularly important in cellular phone applications, where a single power amplifier is used at each base-station for all the frequency channels.

Abrupt heterojunctions feature a spike in the conduction band which can impede the injection of electrons into the base (see Figure 2.7(b)). A thermionic model is then required to predict the performance of the emitter–base diode [9]. (The diffusion analysis applies best to graded heterojunctions such as the smooth heterojunction shown in Figure 2.7(a).) An accurate analysis of the HBT also requires the inclusion of the tunneling current [9]. Tunneling through the heterojunction spike can improve the performance of the HBT by reducing the effective barrier height of the spike. Shur [9] quotes a factor of 10 increase in β due to tunneling.

The discontinuity of the conduction band can also be smoothed out by grading the composition of the alloy from the emitter to the base. This grading typically takes place over a distance of 50–200 Å. A factor of 6 increase in β has been observed experimentally [9]. Note that grading the heterojunction can also be used to design a better launching ramp (avoiding tunneling) for high-energy ballistic electrons (see previous section).

The reader is referred to Chapters 18 and 19 for a deeper discussion of the device physics, modeling techniques, and device characteristics of HBTs.

2.10 Conclusion

In this chapter we have presented a semiclassical theory of heterostructures which relies on simple band-diagram and transport models. Note that the thermionic and drift-diffusion transport equations introduced in this chapter will be derived in Chapters 4, 8, and 9. Also these low-field transport models will be extended in Chapter 9 to the case of high-electric-field conditions to account for self-heating effects. Finally ballistic transport in HBTs will be further studied in Chapter 18 using a direct solution of the Boltzmann equation.

The semiclassical transport picture adopted in this chapter is only applicable to relatively large and slowly-varying heterostructures. For small or rapidly-varying heterostructures the wave nature of the electron plays a dominant role, and a quantum band and transport picture is required. Chapters 3–7 are dedicated to the development of such a quantum picture and its application to practical and high-performance quantum devices.

Appendix: Semiconductor parameter tables

The material parameters presented in Tables 2.3 and 2.10 are based on [17,18,19,20, 21,22]. The temperature is 300 K unless otherwise specified. For a definition of the various bandgaps refer to Figure 2.8.

2.11 Bibliography

2.11.1 Recommended reading

S. Tiwari, *Compound Semiconductor Transistors: Physics & Technology* EDS Press Books, New York, 1993.

J. Singh, *Physics of Semiconductors and their Heterostructures* McGraw-Hill, New York, 1993.

C. M. Wolfe, N. Holonyak Jr, Gregory E. Stillman, *Physical Properties of Semiconductors*, Prentice Hall, Englewood Cliffs, New Jersey, 1989.

M. Shur, *GaAs Devices and Circuits*, Plenum, New York, 1987.

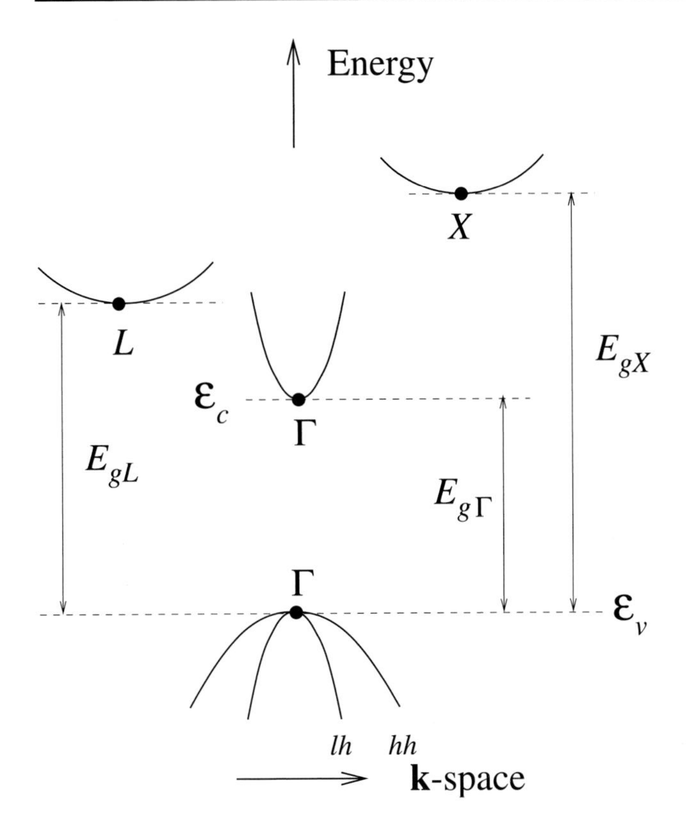

Fig. 2.8. Band structure for the definition of the various direct and indirect bandgaps.

Table 2.3. *Parameters of* $Al_mGa_{1-m}As$.

$Al_mGa_{1-m}As$	Values or formula	m range	Temp. range
N_i (m^{-3})	$1.79 \times 10^{12} m^{-3}$ (experimental)	0	300 K
N_c (m^{-3})	$4.35\ (6.988) \times 10^{23}$	0 (0.3)	300 K
N_v (m^{-3})	7×10^{24}	0	300 K
ϵ/ϵ_0	$13.18 - 3.12m$	from 0 to 1	300 K
$\epsilon_\infty/\epsilon_0$	$10.89 - 2.73m$	from 0 to 1	300 K
m_n^*/m_0 (Γ)	$0.067 + 0.083m$	from 0 to 1	300 K
m_{lh}^*/m_0	$0.087 + 0.063m$	from 0 to 1	300 K
m_{hh}^*/m_0	$0.62 + 0.14m$	from 0 to 1	300 K
E_g (eV)	$1.424 + 1.594m + (1.310m - 0.127)m(m - 1)$	from 0 to 1	300 K
	1.52	0	0 K
E_{gX} (eV)	$1.991 + 0.005m + 0.245m^2$	from 0 to 1	300 K
E_{gL} (eV)	$1.734 + 0.574m + 0.055m^2$	from 0 to 1	300 K
a (Å)	$5.6533 + 0.008\,09m$	from 0 to 1	300 K
E_d (eV)	0.004, 0.016, and 0.05	0, 0.2 and 0.3	300 K
SBH: Al (eV)	1.1 and 1.3	0.3 and 0.5	300 K

Table 2.4. *300 K material parameters for the interpolation of the material parameters of the $Si_{1-x}Ge_x$, $Al_xGa_{1-x}As$, $In_xGa_{1-x}As$, and $In_xAl_{1-x}As$ material systems.*

300 K values	Si	Ge	GaAs	AlAs	InAs	$In_{0.53}Ga_{0.47}As$	$In_{0.52}Al_{0.48}As$	InP
a (Å)	5.4309	5.6461	5.6533	5.6614	6.0584	5.8687	5.8687	5.8687
Bandgap	indirect	indirect	direct	indirect	direct	direct	direct	direct
$E_{g\Gamma}$ (eV)	4.08	0.89	1.424	3.018	0.356	0.77	1.49	1.34
E_{gL} (eV)	1.87	0.663	1.734	2.308	1.08	1.323	1.78	1.93
E_{gX} (eV)	1.124	0.96	1.911	2.161	1.37	1.438	2.048	2.19
χ (eV)	4.05	4.13	4.07	3.5	4.9	4		4.38
m_l^*/m_0	0.9163	1.59	0.067	0.15 (Γ)	0.024	0.043	0.075	0.075
m_t^*/m_0	0.1905	0.0823						
m_{lh}^*/m_0	0.153	0.043	0.087	0.15	0.025	0.053	0.096	0.12
m_{hh}^*/m_0	0.537	0.284	0.62	0.76	0.37	0.56	0.041	0.56
μ_n (cm^2/(V s))	1450	3900	8000	400	30 000			5000
μ_p (cm^2/(V s))	370	1800	400	100	480			180
ϵ/ϵ_0	11.9	16.2	13.18	10.06	15.15	13.94	12.46	12.61
$\epsilon_\infty/\epsilon_0$	11.9	16.2	10.89	8.16	12.25	11.61	9.84	9.61
ω_{LO} (meV)	63	37	35	48	30		39.5	43
ρ (g/cm^3)	2.329	5.3235	5.3165	3.729	5.667	5.504	4.90	4.791
SBH:Al (eV)	0.72 (n)	0.48 (n)	0.7–0.8			0.53	0.8	
SBH:Au (eV)	0.8 (n)	0.59 (n)	0.95		0.47		0.64	0.52

2.11.2 References

[1] B. G. Streetman, *Solid State Electronic Devices*, Prentice Hall, Englewood Cliffs, New Jersey, 1990.

[2] N. Chand, T. Henderson, J. Klem, W. T. Masselink, R. Fischer, Y. Chang, and H. Morkoç, 'Comprehensive analysis of Si-doped $Al_xGa_{1-x}As$ ($x = 0$ to 1): Theory and experiments', *Physical Review B*, Vol. 30, pp. 4481–4492, 1984.

[3] A. Bykhovski, B. Gelmont, and M. S. Shur, 'The influence of the strain-induced electric field on the charge distribution in BaN–AlN–GaN structure', *Journal of Applied Physics*, Vol. 74, p. 6734, 1993.

[4] A. H. Marshak and K. M. Van Vliet, 'Electrical current in solids with position-dependent band structure', *Solid State Electronics*, Vol. 21, pp. 417–427, 1978.

[5] C. Wolfe, N. Holonyak, and G. Stillman, *Physical Properties of Semiconductors, Prentice Hall*, Englewood Cliffs, New Jersey, 1989.

[6] Ying-Chao Ruan and W. Y. Ching, 'An effective dipole theory for band lineups in semiconductor heterojunctions', *Journal of Applied Physics*, Vol. 61, No. 7, pp. 2885–2897, 1987.

[7] W. Shockley, US Patent No. 2 569 347, issued 1951.

[8] H. Kroemer, 'Theory of a wide-gap emitter for transistors', *Proceedings of the IRE*, Vol. 45, pp. 1535–1537, 1957.

[9] M. Shur, *GaAs Devices and Circuits*, Plenum Press, New York, 1986.

[10] J.-L. Pelouard, P. Hesto, and R. Castagne, 'Modeling of the InP/InGaAs Ballistic HJBT', *Proceedings of the IEEE/Cornell Conference on Advanced Concepts in High Speed Semiconductor Devices and Circuits*, July 29, 30, & 31, Ithaca, New York, 1985.

[11] M. S. Unlu, S. Strite, G. B. Gao, K. Adomi, and H. Morkoç, 'Electrical characteristics of p^+-Ge/(N-GaAs and N-AlGaAs) junctions and their applications to Ge base transistors', *Applied Physics Letters*, Vol. 56, No. 9, pp. 842–844, February 1990.

[12] K. Horio and H. Yania, 'Numerical modeling of heterojunctions including the thermionic emission mechanism at the heterojunction interface', *IEEE Transactions on Electron Devices*, Vol. 37, No. 4, pp. 1093–1098, April 1990.

[13] R. L. Anderson, 'Experiments on Ge–GaAs heterojunctions', *Solid State Electronics*, Vol. 5, p. 341, 1962.

[14] D. K. Ferry, *Semiconductors*, Macmillan Publishing Company, New York, 1991.

[15] J. Singh, *Physics of Semiconductors and their Heterostructures*, McGraw-Hill, New York, 1993.

[16] H. C. Casey, Jr and F. Stern, *Journal of Applied Physics*, Vol. 47, p. 631, 1976.

[17] A. Katz, Ed, *Indium Phosphide and Related Materials: Processing, Technology, and Devices*, Artech House, Boston, 1992.

[18] S. Adachi, *Physical Properties of III–V Semiconductor Compounds: InP, InAs, GaAs, GaP, InGaAs, & InGaAsP*, John Wiley & Sons, Incorporated, New York, 1992.

[19] S. Adachi, *Properties of Aluminium Gallium Arsenide*, EMIS Datareviews Series No. 7, INSPEC, London, 1993.

[20] P. Bhattacharya, *Properties of Lattice-Matched & Strained Indium Gallium Arsenide*, EMIS Datareviews Series No. 8, INSPEC, London, 1993.

[21] E. Rosencher and B. Vinter, *Optoelectronique*, Thomson-CSF, Masson, Paris, 1998.

[22] S. M. Sze, *Physics of Semiconductor Devices*, 2nd Edition, John Wiley & Sons, Wiley Interscience, New York, 1981.

2.12 Problems

2.1 p-GaAs–n-AlGaAs heterojunction: Consider a p-GaAs–n-$Al_{0.3}Ga_{0.7}As$ heterojunction at 300 K. The acceptor doping on the p side is $N_A = 10^{16}$ cm^{-3}. The donor doping on the n side is $N_D = 10^{16}$ cm^{-3}. Assume that all donor and acceptor impurities are ionized and that the doping is non-degenerate. The temperature is 300 K.

(a) Calculate the conduction and valence-band discontinuities. Use the 68/32% rule and the AlGaAs material parameters given in Table 2.3.

(b) Sketch the band diagram.

(c) Calculate the built-in potential V_0. Rank by order of importance the various terms contributing to V_0.

2.2 An efficient one-dimensional Poisson solver: We wish to solve the Poisson equation in a heterostructure. The heterostructure is divided into series of atomic layers i of width a_i and centered at the position x_i. The dielectric constant ϵ_i is uniform in each atomic layer i.

(a) We first assume that the charge distribution $\rho(x)$ and dielectric constant ϵ_i are uniform in each atomic layer i:

$$\rho(x) = \sum_i \rho_i \left[\mathrm{u}\left(x - x_i + \frac{a_i}{2}\right) - \mathrm{u}\left(x - x_i - \frac{a_i}{2}\right) \right],$$

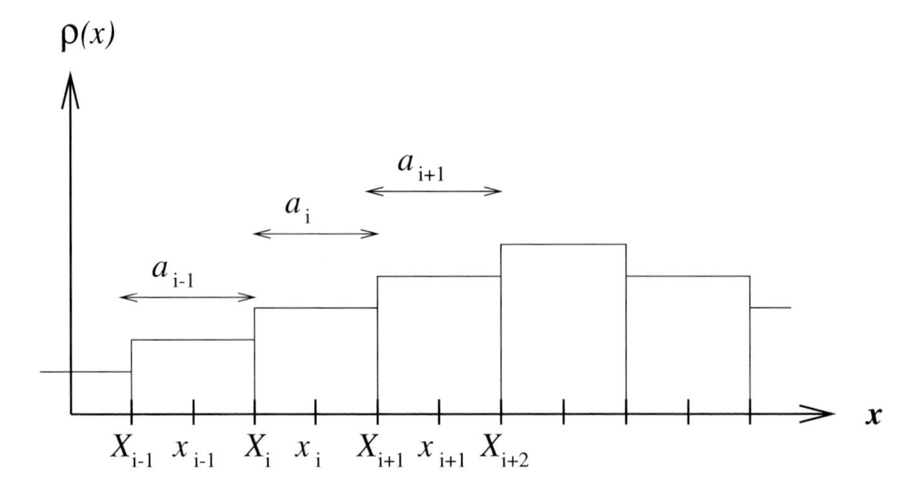

Fig. 2.9. Charge distribution $\rho(x)$ for the step approximation.

where ρ_i is the value of $\rho(x)$ in the interval i (see Figure 2.9). We assume initially that the charge distribution ρ_i is known.

Verify that the following relations are obtained if we integrate the Poisson equation exactly from the site 1 to the site n:

$$\epsilon_n F(X_n^+) = \epsilon_1 F(X_1^+) + \sum_{j=1}^{n-1} \rho_j a_j,$$

$$V(X_n) = V(X_1) - \sum_{j=1}^{n-1} F(X_j^+) a_j - \frac{1}{2} \sum_{j=1}^{n-1} \frac{\rho_j a_j^2}{\epsilon_j}$$

with $X_i = x_i - a_i/2$.

(b) Assume now that the charge distribution in each atomic layer is given by an impulse function (see Figure 2.10):

$$\rho(x) = \sum_i \rho_i \, a_i \, \delta(x - x_i).$$

Compare the field and potential distribution. Do the expressions given in (a) still hold? Assume that ρ_i is known and is the same as in (a).

(c) The charge distribution $\rho(x, V)$ is now assumed to depend on the voltage $V(x)$ as well as position x. The charge distribution is assumed to be $\rho_i = \rho(x_i, V(x_i))$ for both the uniform charge distribution (a) and the impulse distribution (b). Verify that a close-form solution can easily be obtained with method (b) but not with method (a).

Note: This method of integrating a non-linear Poisson equation using the initial conditions $F(X_i^+)$ and $V(X_i)$ gives results quite similar to the Runge-Kutta algorithm. This new algorithm, however, is much simpler and therefore faster than the Runge-Kutta algorithm. Note that the solution obtained is exact. However, the uniform charge distribution (a) is easier to justify than the impulse distribution (b).

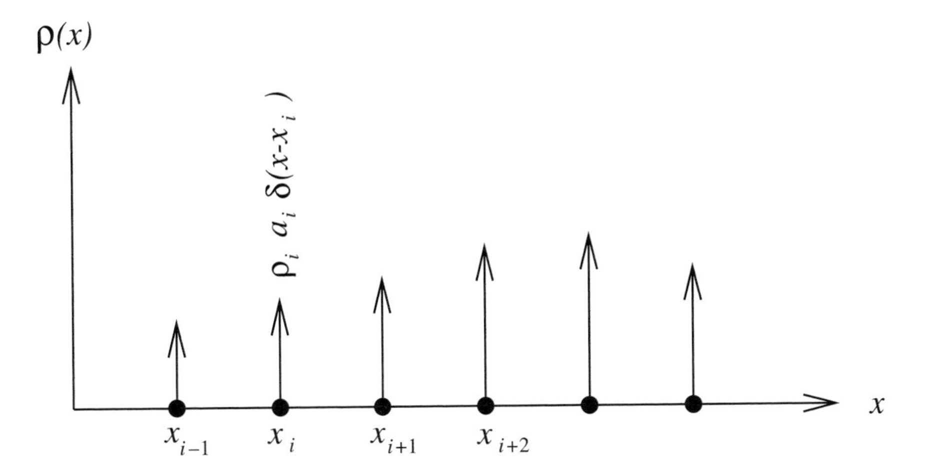

Fig. 2.10. Charge distribution $\rho(x)$ for the impulse approximation.

2.3 Ballistic electrons in the HBT (after Pelouard *et al.* [8]): Consider a n–p–n HBT of base width W. We wish to study the impact of the quasi-ballistic electrons launched at the emitter–base heterojunction upon the performance of the HBT.

The total emitter current which flows in the base includes two components: a diffusion current and a quasi-ballistic current:

$$J_E = J_R(x) + J_{QB}(x) = J_s \left[\exp\left(\frac{qV_{BE}}{k_B T}\right) - 1\right],$$

$$J_R(x) = qD_n \frac{dn_R}{dx},$$

$$J_{QB}(x) = J_s \exp\left(\frac{qV_{BE}}{k_B T}\right) \exp\left(\frac{-x}{\lambda}\right),$$

where D_n is the diffusion constant for the relaxed electron population and λ is the mean free path of the quasi-ballistic electrons. Note that the ballistic current J_{QB} in the base is maximum at $x = 0$, which is the location of the emitter–base heterojunction. The x axis is assumed to be oriented from the emitter to the collector. Therefore J_s must be negative for this n–p–n transistor.

(a) Express the thermalized electron concentration $n_R(x)$ at position x in the base in terms of the electron concentration $n_R(W)$.

(b) Calculate the equilibrium ($V_{BE} = V_{BC} = 0$) electron concentration $n_{R0}(x)$ at the position x in the base in terms of $n_{R0}(W)$.

(c) Calculate the base current J_B resulting from the recombination of electrons with holes in the base

$$J_B = q \int_0^W \frac{n_R(x) - n_{R0}(x)}{\tau_n} dx \tag{2.48}$$

with τ_n the lifetime of electrons in the base. Note that the excess of electrons at the edge of the collector–base depletion region is controlled by the collector–base diode

$$\Delta n_P = n_R(W) - n_{R0}(W) = n_P \left[\exp\left(\frac{qV_{BC}}{k_B T}\right) - 1 \right],$$

assuming low injection.

(d) Verify that the common base current gain is

$$\alpha = 1 - \frac{W^2}{2L_n^2} \left\{ 1 + \frac{2\lambda}{W} \exp\left(\frac{-W}{\lambda}\right) + 2\left(\frac{\lambda}{W}\right)^2 \left[\exp\left(\frac{-W}{\lambda}\right) - 1 \right] \right\},$$

where $L_n^2 = D_n \tau_n$ is the diffusion length of the electrons in the base. Note that for $\lambda = 0$ the current gain α reduces to the classical bipolar expression for the base transport $B = \mathrm{sech}(W/L_n) \simeq 1 - W^2/2L_n^2$. The emitter efficiency γ is assumed to be 1 ($\alpha = B$) which is quite reasonable for an HBT.

(e) Plot the ratio of the common emitter current gains $\beta(\lambda)/\beta(\lambda = 0)$ for a base width W varying from 0.6 to 0.1 μm and $\lambda = 0.19$ μm. Assume that the gate length is much smaller than the diffusion length so that we can use the approximation $\beta = \alpha/(1-\alpha) \simeq 1/(1-\alpha)$.

3 Quantum theory of heterostructures

. . . I think I can safely say that nobody understands quantum mechanics.

The Character of Physical Law, RICHARD FEYNMAN

3.1 Introduction

Modern technology has made possible the growth of thin crystalline epitaxial layers of different semiconductors. These epitaxial layers can be as small as a few lattice parameters. For small heterostructures (100 Å or less) a quantum treatment of heterostructures becomes necessary. In this chapter we will attempt to build a quantum picture of heterostructures. Note that the conventional quantum picture of a semiconductor crystal cannot be applied to rapidly varying semiconductor heterostructures since crystals are defined as periodic structures extending up to infinity. New theoretical techniques are thus required to describe these 'spatially-varying crystals'.

Our quantum picture will be based upon an envelope model. An envelope model focuses on calculating the relative distribution of the wave-function from atomic cell to atomic cell rather than on the detailed distribution of the electron wave-function in each atomic cell.

The particular envelope picture we shall use is the so-called generalized Wannier picture. The generalized Wannier picture is capable of handling both the concept of band structure and the concept of its variation in space in a rigorous fashion. This model will therefore permit us to understand the impact of the interface upon the band structure in a heterojunction.

Other envelope pictures have been developed such as the Ben Daniel Duke Hamiltonian (effective-mass model [10], see also [11]), the $\mathbf{k} \cdot \mathbf{p}$ envelope model ([11]), and the tight-binding model ([8]). The major advantage of the generalized Wannier picture is that like the tight-binding model, it can handle the full band structure (the very accurate $\mathbf{k} \cdot \mathbf{p}$ formalism, introduced in Chapter 16, is limited to the bottom of the conduction band and the top of the valance band). The generalized Wannier picture for the conduction band is, however, a much simpler formalism than the tight-binding and $\mathbf{k} \cdot \mathbf{p}$ envelope models. Finally, a critical advantage of the Wannier picture is that the envelope function (electron wave-function) is obtained from the solution of a difference equation. Analytic solutions are therefore available for simple systems and numerical solutions are readily obtained for more complicated systems.

We first introduce the Wannier picture for homojunctions (Section 3.2) before extending it to the case of heterojunctions (Section 3.3). Both the one-dimensional approximation and the general three-dimensional picture are presented. A major topic also discussed in this chapter is the definition of electron current in spatially-varying crystals. The application of the quantum theory of heterostructures presented here to quantum devices will be covered in the next chapter.

3.2 Band structures, Bloch functions and Wannier functions

In this section we shall review the quantum properties of an electron in a uniform one-dimensional crystal.

3.2.1 The Schrödinger equation

In classical mechanics the total energy of an electron is the sum of its kinetic energy K and of its potential energy E_{pot}:

$$E = K + E_{pot} = \frac{p^2}{2m_0} + E_{pot}(x),$$

where p is the electron momentum, and m_0 the electron mass. In quantum mechanics the momentum p is replaced by the operator p_{op}

$$p \rightarrow p_{op} = -i\hbar\frac{\partial}{\partial x},$$

so that the Hamiltonian $K + E_{pot}$ is now an operator H^0

$$H^0(x) = -\frac{\hbar^2}{2m_0}\frac{\partial^2}{\partial x^2} + E_{pot}(x). \tag{3.1}$$

Similarly the energy is replaced by the operator

$$E \rightarrow i\hbar\frac{\partial}{\partial t}.$$

The Schrödinger equation is then obtained by letting these operators operate on the wave-function $\varphi(x, t)$

$$H^0(x)\varphi(x, t) = i\hbar\frac{\partial}{\partial t}\varphi(x, t).$$

$|\varphi(x, t)|^2$ is postulated to be the probability of the presence of the electron at position x at time t.

In the case of a time-independent Hamiltonian, the Schrödinger equation admits, using separation of variables, an eigenstate solution of the form

$$\varphi(x,t) = \exp\left(\frac{-iEt}{\hbar}\right)\varphi(x,E),$$

where $\varphi(x,E)$ is a solution of the eigenstate equation

$$H^0(x)\varphi(x,E) = E\varphi(x,E)$$

with the energy E a real eigenvalue because H^0 is Hermitian. We will find it convenient to rewrite this result using the bra-ket notation

$$H^0|E\rangle = E|E\rangle,$$

where $|E\rangle$ is the eigenvector of H in an arbitrary representation. In the position representation, the eigenvector is then

$$\langle x|E\rangle = \varphi(x,E).$$

3.2.2 Electron in a periodic potential

In a crystal the potential energy E_{pot} is a periodic function of position

$$E_{pot}(x) = E_{pot}(x+a),$$

where the period a is the lattice parameter. Let us analyze the properties of the electron wave-function in a periodic potential. For this purpose we shall introduce the translation operator.

$$T(na) = \exp\left[\frac{i(p_{op}na)}{\hbar}\right] = \exp\left[na\frac{\partial}{\partial x}\right]$$

As indicated by its name the translation operator $T(na)$ operated on a test function $\varphi(x)$ translates it by a distance na

$$\begin{aligned} T(na)\varphi(x) &= \exp\left(na\frac{\partial}{\partial x}\right)\varphi(x) \\ &= \left[1 + \frac{1}{1!}\left(na\frac{\partial}{\partial x}\right) + \frac{1}{2!}\left(na\frac{\partial}{\partial x}\right)^2 + \frac{1}{3!}\left(na\frac{\partial}{\partial x}\right)^3 + \cdots\right]\varphi(x) \\ &= \varphi(x+na). \end{aligned}$$

With the use of a test function $\varphi(x)$ (not shown here) we can easily verify that the translation operator $T(na)$ commutes with H^0

$$T(na)H^0(x) = T(na)\left[-\frac{\hbar^2}{2m_0}\frac{\partial^2}{\partial x^2} + E_{pot}(x)\right]$$

$$= \left[-\frac{\hbar^2}{2m_0} \frac{\partial^2}{\partial x^2} + E_{pot}(x + na) \right] T(na)$$
$$= H^0(x + na)\, T(na)$$
$$= H^0(x) T(na).$$

This commutation property will permit us to classify the eigenvectors of the crystal Hamiltonian with the eigenvectors of the translation operator. For this purpose we first need to find the eigenvalues of the translation operator. Consider

$$T(na)\, \varphi(x) = \lambda_n \varphi(x) = \varphi(x + na),$$

where λ_n and $\varphi(x)$ are the eigenvalues and eigenvectors respectively. The complex conjugate of this expression is

$$\lambda_n^* \, \varphi^*(x) = \varphi^*(x + na).$$

Multiplying the last two expressions together and integrating over space, we obtain

$$\int_{-\infty}^{\infty} |\lambda_n|^2 |\varphi(x)|^2 \, dx = \int_{-\infty}^{\infty} |\varphi(x + na)|^2 \, dx,$$

which implies that $|\lambda_n|^2 = 1$. Let us introduce the variable k defined from λ_n using

$$\lambda_n = \exp(ikna)$$

Note that k and $k + m(2\pi/a)$ are equivalent since they are associated with the same eigenvalue λ_n. As a consequence of the commutation property we have

$$T\, H^0 \,|E\rangle = T\, E\, |E\rangle,$$
$$H^0\, T\, |E\rangle = E\, T\, |E\rangle.$$

This indicates that if $|E\rangle$ is an eigenvector of H^0 then $T|E\rangle$ is also an eigenvector of H^0 with the same eigenvalue E. If there is only one state $|E\rangle$ for each energy E (non-degenerate case) then we must have

$$T(na)\, |E\rangle \propto |E\rangle.$$

Since this is an eigenvalue equation, the proportionality constant is an eigenvalue λ_n:

$$T(na)\, |E\rangle = \lambda_n \, |E\rangle = \exp(ikan)|E\rangle.$$

Consequently, $|E\rangle$ is an eigenvector of both $T(na)$ and H^0. This result still holds in the degenerate case (several states $|E\rangle$ for a single energy E) by selecting the eigenvectors $|E\rangle$ which simultaneously diagonalize the translation operator $T(na)$ and the Hamiltonian H^0 (see Ziman, pp. 17–19, [8] for a proof). In practice the eigenvectors of H^0 are always degenerate and we need to establish a system for labeling the different states $|E\rangle$. Since the translation operator $T(na)$ and the crystal Hamiltonian H^0 admit the same eigenvectors we can use the eigenvalue k to label the states $|E\rangle$ using $|E(k)\rangle$ or even simply $|k\rangle$. The eigenwave-function $|k\rangle$ of an electron in a crystal is a solution of the eigenvalue equation

$$H^0\,|k\rangle = \mathcal{E}(k)\,|k\rangle,$$

where we use $E = \mathcal{E}(k)$. The dispersion characteristics $\mathcal{E}(k)$ between the energy and the wave-vector k define the so-called band structure.

Using $\varphi(k, x) = \langle x|k\rangle$, we can simply write the same equation in the position representation:

$$H^0(x)\varphi(k, x) = \mathcal{E}(k)\varphi(k, x).$$

This is called the Bloch equation and $\varphi(k, x)$ is called a Bloch function. Note that Bloch functions are eigenvectors and therefore verify the orthogonality property

$$\langle k'|k\rangle = \int_{-\infty}^{\infty} \varphi^*(k', x)\varphi(k, x)\,dx = \delta(k' - k).$$

We can now very easily verify that the translation of a Bloch state by one lattice parameter a gives

$$\varphi(k, x + a) = T(a)\varphi(k, x) = \exp(jka)\varphi(k, x).$$

This is the so-called Bloch theorem which identifies k as a wave-vector. Note that since k and $k + m(2\pi/a)$ are equivalent, both the band structure $\mathcal{E}(k)$ and the wave-function $\varphi(x, k)$ are periodic functions of k with periodicity $2\pi/a$.

Properties of the band structure $\mathcal{E}(k)$

Let us state some of the properties of the band structure $\mathcal{E}(k)$.

From time-reversal considerations one can show that $\mathcal{E}(k)$ is an even function of k:

$$\mathcal{E}(-k) = \mathcal{E}(k).$$

The even and periodic properties of the band structure result in the slope of the band structure $\mathcal{E}(k)$ being zero at $k = \pm\pi/a$:

$$\frac{d\mathcal{E}(k = \pm\pi/a)}{dk} = 0.$$

The Bloch equation admits eigenvectors for energies $\mathcal{E}(k)$ located in intervals which are called bands. The ranges or bands of energy for which the Bloch equation admits no eigenvector solutions are called forbidden bands. In the case of the one-dimensional periodic potentials studied here, one can label successive bands of energies in the direction of increasing energy in such a way that each band contains the entire spectrum of eigenvalues k without energy overlap between successive bands. Bands and wave-functions can thus be labeled by successive integers, which we call band indices b: $\mathcal{E}_b(k)$, $\varphi_b(k, x)$. In a three-dimensional crystal, however, overlaps of bands do take place (for a discussion of the three-dimensional case refer to Wannier's original paper [7]). This is illustrated in Figure 1.13 of Chapter 1 which shows the various bands $\mathcal{E}_b(\mathbf{k})$ obtained along the Γ to X direction in the Brillouin zone for various semiconductor materials.

In the remaining analysis we shall focus our interest on a single band, for example, the conduction band or valance band. In this one-band approximation we can drop the band index b.

Since $\mathcal{E}(k)$ and $\varphi(k, x)$ are periodic functions of k we can express them in terms of a Fourier series

$$\varphi(k, x) = \left(\frac{a}{2\pi}\right)^{1/2} \sum_{m=-\infty}^{\infty} w(m, x) \exp(ikma),$$

$$\mathcal{E}(k) = \sum_{m=-\infty}^{\infty} \mathcal{E}_m \exp(ikma)$$

where $w(m, x)$ and $\mathcal{E}_m$ are the Fourier coefficients. $w(m, x)$ is called a Wannier function after Wannier who first studied them [7].

3.2.3 Wannier functions

In the remaining work we shall adopt the Wannier functions as a basis. The main motivation as we shall see is that Wannier functions are spatially localized in contrast with Bloch functions which extend over the entire crystal. Such a localized picture will provide a better basis for spatially-varying crystals.

First let us derive some of the properties of the Wannier functions. From their very definition the Wannier functions are the Fourier coefficients of the Bloch waves:

$$w(m, x) = \left(\frac{a}{2\pi}\right)^{1/2} \int_{-\frac{\pi}{a}}^{\frac{\pi}{a}} \exp(-ikma)\varphi(k, x)\, dk.$$

In the empty-lattice case ($E_{pot} = 0$), the Bloch function $\varphi(k, x)$ is a plane wave:

$$\varphi(k, x) = \frac{1}{(2\pi)^{1/2}} \exp(ikx),$$

where we limit the wave-vector k to the first Brillouin zone:

$$-\frac{\pi}{a} \leq k \leq \frac{\pi}{a},$$

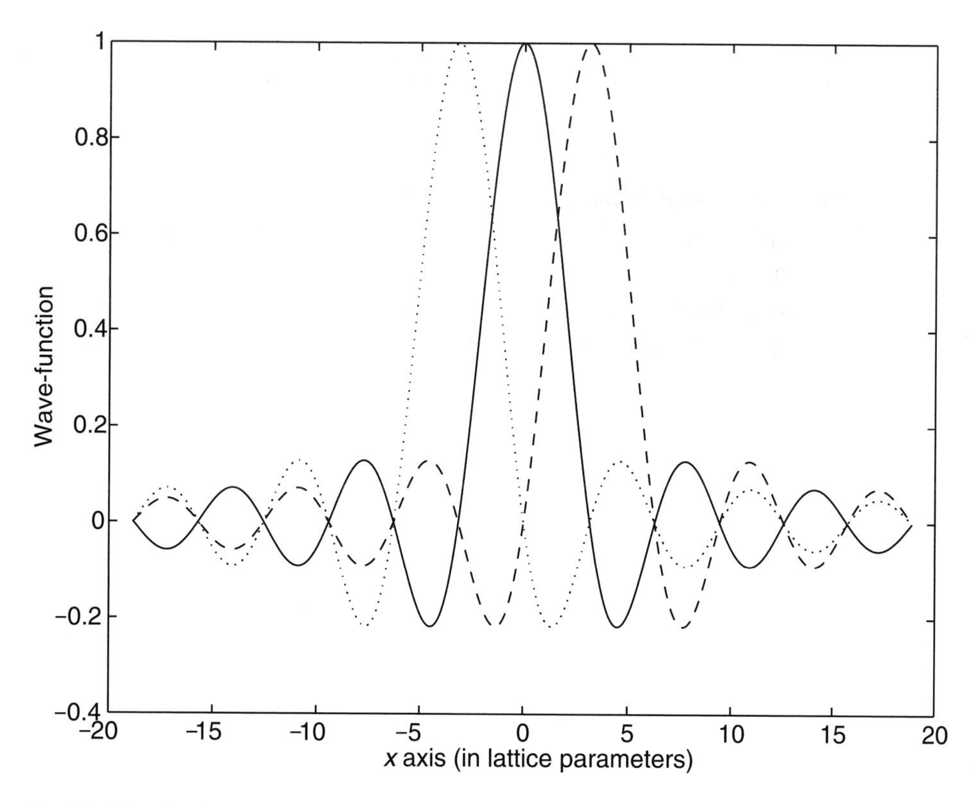

Fig. 3.1. Wannier functions of the empty lattice for adjacent sites.

for the first band of the empty lattice. One can then readily calculate the Wannier function for the empty lattice

$$w(m, x) = \frac{\sqrt{a}}{2\pi} \int_{-\frac{\pi}{a}}^{\frac{\pi}{a}} \exp(ik(x - ma))\, dk,$$

$$w(m, x) = \frac{1}{\sqrt{a}} \frac{\exp\left[i\frac{\pi}{a}(x - ma)\right] - \exp\left[-i\frac{\pi}{a}(x - ma)\right]}{2i\dfrac{\pi}{a}(x - ma)}$$

$$= \frac{1}{\sqrt{a}} \operatorname{sinc}\left[\frac{\pi}{a}(x - ma)\right].$$

As can be seen in Figure 3.1, the Wannier function $w(m, x)$ of the empty lattice appears to be an electron state localized around the lattice site m. The Wannier function of the empty lattice is a singular example, but it illustrates some of the properties of the Wannier functions.

Let us now list some of the properties of the Wannier functions $w(m, x) = \langle x|m\rangle$

- Translation invariance:
  The Wannier function is invariant under a lattice translation
  $$w(m, x) = w(x - ma).$$
- Normalization and orthogonality:
  Wannier functions are normalized and orthogonal (see Problem 3.1)
  $$\langle n|m\rangle = \int_{-\infty}^{\infty} w^*(n, x) w(m, x)\, dx = \delta_{nm}.$$
- Hamiltonian matrix elements:
  The matrix elements of the lattice Hamiltonian in the Wannier representation are the Fourier coefficients of the band structure (see Problem 3.1 for a proof)
  $$H_{nm} = \langle n|H^0|m\rangle = \int_{-\infty}^{\infty} w^*(n, x) H^0 w(m, x)\, dx = \mathcal{E}_{m-n}.$$
- Completeness:
  The Wannier functions $w_q(m, x)$ form a complete basis. In the one-band approximation only one band is considered.
- Convergence:
  The Bloch functions are defined up to an arbitrary phase. The Wannier functions are therefore not uniquely defined. Kohn [1] has shown that for a 'best' choice of the Bloch function phase, the Wannier function vanishes exponentially for large $x - na$.

Proof of translation invariance

Let us now demonstrate the translation invariance property of the Wannier functions: Consider a Bloch function translated by a distance na. Let us expand it in terms of Wannier functions

$$\varphi(k, x - na) = \left(\frac{a}{2\pi}\right)^{1/2} \sum_m \exp(ikma)\, w(m, x - na).$$

Using the Bloch theorem we can write

$$\varphi(k, x - na) = \left(\frac{a}{2\pi}\right)^{1/2} \sum_m \exp(-ikna)\, \exp(ikma)\, w(m, x).$$

Let us change the index of summation $m - n = m'$:

$$\varphi(k, x - na) = \left(\frac{a}{2\pi}\right)^{1/2} \sum_{m'} \exp(ikm'a)\, w(m' + n, x).$$

As a result we have

$$w(m, x - na) = w(m + n, x).$$

This justifies using the following notation for the Wannier functions

$$w(n, x) = w(0, x - na) = w(x - na).$$

The Wannier picture

We shall now select the Wannier functions as a basis and derive the Hamiltonian in this representation. We consider the lattice Hamiltonian H^0 with a superimposed potential energy $-qV(x)$ resulting from an electrostatic potential $V(x)$:

$$\begin{aligned} H &= \frac{p_{op}^2}{2m_0} + E_{pot}(x) - qV(x) \\ &= H^0 - qV(x). \end{aligned}$$

Within the one-band approximation the Wannier functions form a complete basis with which we can expand the electron wave-function φ:

$$\varphi(x, t) = \sum_{n=-\infty}^{\infty} f(n, t)\, w(x - na)$$

or using bra-ket notation

$$|\varphi\rangle = \sum_{n=-\infty}^{\infty} f(n, t)|n\rangle,$$

where $|n\rangle$ is the Wannier state centered around the site n and $f(n, t)$ is called an envelope function. $|f(n, t)|^2$ gives the probability of the presence of the electron at time t at the site n:

$$H|\varphi\rangle = i\hbar \frac{d}{dt}|\varphi\rangle$$

$$H \sum_{n=-\infty}^{\infty} f(n, t)|n\rangle = i\hbar \frac{d}{dt} \sum_{n=-\infty}^{\infty} f(n, t)|n\rangle.$$

Multiplying the last equation by $\langle m|$, we obtain the Wannier recurrence equation

$$\sum_{n=-\infty}^{\infty} \langle m|H|n\rangle f(n, t) = i\hbar \frac{d}{dt} \sum_{n=-\infty}^{\infty} f(n, t)\langle m|n\rangle$$

$$\sum_{n=-\infty}^{\infty} H_{mn} f(n, t) = i\hbar \frac{d}{dt} f(m, t)$$

with

$$H_{mn} = H_{mn}^0 - qV_{mn} = \mathcal{E}_{n-m} - qV_{mn}$$

and with

$$V_{mn} = \langle m|V(x)|n\rangle = \int_{-\infty}^{\infty} w^*(m,x)V(x)\,w(n,x)\,dx.$$

For an electrostatic potential $V(x)$ slowly varying in space, V_{mn} reduces to

$$V_{mn} \simeq V(ma)\delta_{mn} = V_m.$$

Note that if the electrostatic potential is varying too abruptly in space, the one-band approximation breaks down.

Eigenstate solution

When the Hamiltonian is not time-varying, it is easy to verify that the eigenstate solutions are of the form

$$f(n,t) = \exp\left(-i\frac{Et}{\hbar}\right) f(n,E)$$

with $f(n,E)$ given by the stationary Wannier recurrence equation

$$\sum_{n=-\infty}^{\infty} H_{mn}\, f(n,E) = E\, f(m,E).$$

The tight-binding band

Let us consider the simple but important case of a tight-binding band

$$\begin{aligned}\mathcal{E}(k) &= A - A\cos(ka)\\ &= A - \frac{A}{2}\exp(ika) - \frac{A}{2}\exp(-ika).\end{aligned}$$

A section of the resulting Hamiltonian matrix is:

$$[H_{mn}] = \begin{bmatrix} \ddots & \ddots & \ddots & 0 & & & \\ 0 & -A/2 & A-qV_{-1} & -A/2 & 0 & & \\ & 0 & -A/2 & A-qV_0 & -A/2 & 0 & \\ & & 0 & -A/2 & A-qV_1 & -A/2 & 0 \\ & & & 0 & \ddots & \ddots & \ddots \end{bmatrix}.$$

The Hamiltonian matrix takes the form of a band matrix. For the tight-binding band structure this band matrix is tridiagonal.

The flat-band case

Consider the case of a band structure represented by N Fourier coefficients

$$\mathcal{E}(k) = \sum_{m=-N}^{N} \mathcal{E}_m \exp(-ikma).$$

We are interested in obtaining the stationary solutions of the Wannier recurrence equation for the case in which no potential is applied $V = 0$ (flat band). The Wannier recurrence equation gives the following eigenvalue problem:

$$\left.\begin{aligned} &\sum_{m=n-N}^{n+N} f(m, E)\mathcal{E}_{m-n} = Ef(n, E), \\ &\sum_{m=-N}^{N} f(n+m, E)\mathcal{E}_m = Ef(n, E), \end{aligned}\right\} \quad (3.2)$$

where E is the electron energy, $\mathcal{E}_{m-n}$ are the Fourier coefficients of the band structure and m is a dummy integer. This difference equation admits $2N$ solutions which we shall label with the index j running from 1 to $2N$. The general solution of the Wannier difference equation is therefore a linear superposition of these:

$$f(n, E) = \sum_{j=1}^{2N} \lambda_j f_j(n, E).$$

In the flat-band case, $f_j(n, E)$ is of the type

$$f_j(n, E) = [r_j(E)]^n.$$

Substituting r^n in the Wannier difference equation (3.2) we have

$$r^n \times \left[\sum_{m=-N}^{N} r^m \mathcal{E}_m - E = 0\right].$$

The $r_j(E)$ terms are therefore the roots of the $2N$ order polynomial obtained by multiplying by r^N

$$\sum_{m=0}^{2N} r^m (\mathcal{E}_{m-N} - E\,\delta_{m\,N}) = 0. \quad (3.3)$$

Note that due to the even property of the band structure we have $\mathcal{E}_m = \mathcal{E}_{-m}$. For such polynomials one can demonstrate [6] that if r_j is a root of Equation (3.3) then $1/r_j$ is also a root. Let us replace r_j by $\exp(ik_j a)$ in this polynomial:

$$\sum_{m=-N}^{N} \exp(ik_j ma)\mathcal{E}_m = E.$$

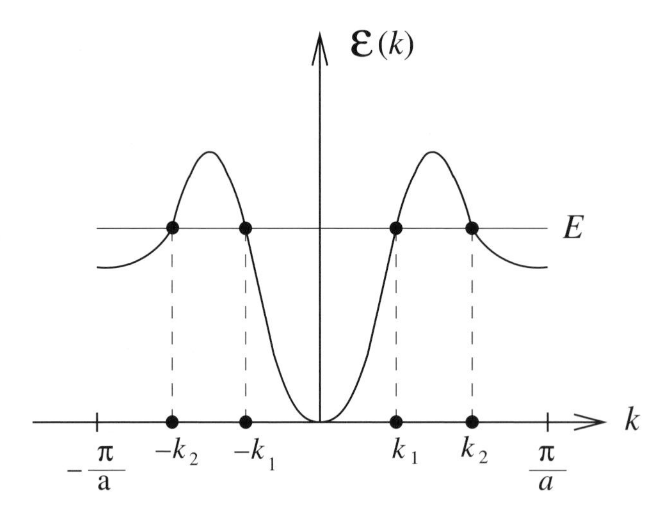

Fig. 3.2. Graphical solution of $\mathcal{E}(k) = E$. Only the four propagating wave-vectors k are shown (k real).

The wave-vectors k_j are seen to be the solution of

$$\mathcal{E}(k_j) = E.$$

Figure 3.2 shows a graphical solution which can be used to determine the real wave-vectors k. For a flat band the Wannier envelope functions $f_j(n, E) = \exp[jk_j(E)na]$ are simply the Fourier coefficients of the expansion of a Bloch function $|k_j\rangle$ in terms of Wannier functions. $f_j(n, E)$ is therefore a Bloch state which we can write $f(n, k_j)$. Complex wave-vectors k_j are obtained by extending the band structure $\mathcal{E}(k)$ to the complex plane k using analytic continuation. These complex wave-vectors correspond to damped or evanescent waves which do not propagate. These waves will be excited if we introduce a discontinuity of the potential energy (for example, heterostructure). As an example in Figure 3.3, we have plotted as a function of energy E, the amplitude of the 20 resulting propagating and evanescent waves for the GaAs conduction-band structure shown in Figure 3.4 and represented using $N = 10$. Note the disappearance of two evanescent waves for energies above X when the system switches from two to four propagating waves.

3.2.4 Three-dimensional crystal

The one-dimensional theory introduced above also applies to the three-dimensional crystal (see [4]). As we shall see the wave-vector k is the projection of the three-dimensional wave-vector $\mathbf{k}$ along a desired crystal axis (typically the superlattice axis).

Consider a three-dimensional crystal. The crystal potential $E_{pot}(\mathbf{r})$ is invariant under a translation by a lattice vector $\mathbf{R}$:

$$E_{pot}(\mathbf{r}) = E_{pot}(\mathbf{r} + \mathbf{R}). \tag{3.4}$$

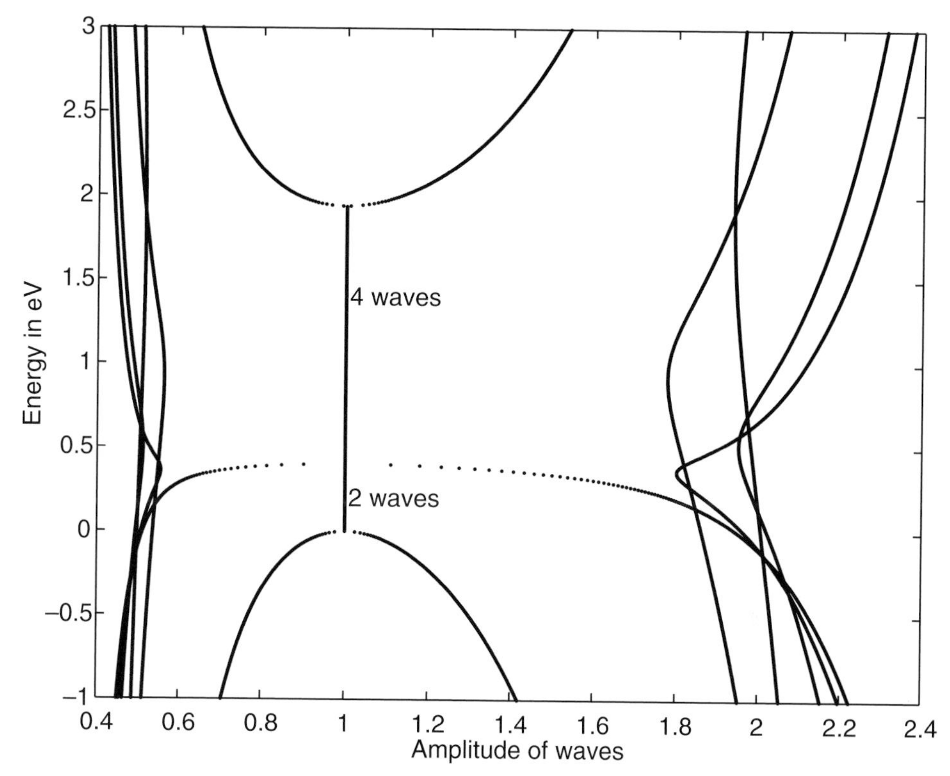

Fig. 3.3. Variation of the amplitude $|r(E)| = |\exp(jk(E)a)|$ versus energy E for both the propagating roots ($|r(E)| = 1$, $k(E)$ real) and the evanescent roots ($|r(E)| <> 1$, $k(E)$ complex) for $\mathbf{k}_\perp = 0$.

For a face-centered cubic (fcc) crystal the lattice vector R is given in the orthonormal coordinates of the Bravais cell (see Figure 1.6) by

$$\mathbf{R} = n_1\mathbf{a}_1 + n_2\mathbf{a}_2 + n_3\mathbf{a}_3 = \frac{a}{2}\begin{bmatrix} 0 & 1 & 1 \\ 1 & 0 & 1 \\ 1 & 1 & 0 \end{bmatrix}\begin{bmatrix} n_1 \\ n_2 \\ n_3 \end{bmatrix}, \tag{3.5}$$

where a is the lattice parameter and n_1, n_2 and n_3 are integers.

For a three-dimensional lattice the translation operator is now

$$T(\mathbf{R}) = \exp\left(\frac{i\mathbf{p}_{op}\cdot\mathbf{R}}{\hbar}\right) = \exp\left(\mathbf{R}\cdot\frac{\partial}{\partial\mathbf{r}}\right), \tag{3.6}$$

where $\mathbf{p}_{op}$ is the momentum operator. Its eigenvalue, $\exp(i\mathbf{k}\cdot\mathbf{R})$, permits us to define the wave-vector $\mathbf{k}$. Because the translation operator and the crystal Hamiltonian commute we can label the Bloch functions $\varphi(\mathbf{k}, \mathbf{r})$ (eigenvector of H^0) and the band structure $\mathcal{E}(\mathbf{k})$ (eigenvalue of H^0) using the wave-vector $\mathbf{k}$. The Bloch functions $\varphi(\mathbf{k}, \mathbf{r})$ and the band structure $\mathcal{E}(\mathbf{k})$ are then both periodic functions in $\mathbf{k}$ space. This means that they are invariant under a translation of a reciprocal-lattice vector $\mathbf{K}$:

$$\varphi(\mathbf{k}, \mathbf{r}) = \varphi(\mathbf{k} + \mathbf{K}, \mathbf{r}),$$

$$\mathcal{E}(\mathbf{k}) = \mathcal{E}(\mathbf{k} + \mathbf{K}).$$

It results from the eigenvalue $\exp(i\mathbf{k} \cdot \mathbf{R})$ of the translation operator $T(\mathbf{R})$ that the reciprocal-lattice vector $\mathbf{K}$ must satisfy $\mathbf{K} \cdot \mathbf{R} = n2\pi$ with n an integer. For the fcc lattice the reciprocal-lattice vector $\mathbf{K}$ that satisfies this relation is given in the orthonormal coordinates of the Bravais cell by

$$\mathbf{K} = \frac{2\pi}{a} \begin{bmatrix} -1 & 1 & 1 \\ 1 & -1 & 1 \\ 1 & 1 & -1 \end{bmatrix} \begin{bmatrix} m_1 \\ m_2 \\ m_3 \end{bmatrix}, \tag{3.7}$$

where m_1, m_2, and m_3 are integers. An inspection of $\mathbf{K}$ for the fcc lattice (in $\mathbf{r}$ space) reveals that the reciprocal lattice (in $\mathbf{k}$ space) is a body-centered cubic (bcc) lattice (see Figure 1.6(b)). Its associated three-dimensional Brillouin zone (Wigner–Seitz cell in reciprocal space) is shown in Figure 1.8.

Assume now that the uniform crystal (empty superlattice) considered is oriented along the direction $\hat{\mathbf{d}}$ (the superlattice axis). Let us call $\mathbf{a}_{SL}$ the smallest lattice vector $\mathbf{R}$ parallel to the direction $\hat{\mathbf{d}}$. The amplitude $a_{SL} = |\mathbf{a}_{SL}|$ is the superlattice parameter. Any lattice parameter parallel to $\hat{\mathbf{d}}$ is written as $\mathbf{R} = n\mathbf{a}_{SL}$, with n an integer.

It is easy to verify that for cubic lattices there always exists a reciprocal-lattice vector $\mathbf{K}$ parallel to a given direct lattice vector $\mathbf{R}$. This is equivalent to finding a constant c such that $\mathbf{K} = c\mathbf{R}$. A possible solution is $c = 8\pi/a_{SL}^2$, which leads to

$$\begin{bmatrix} m_1 \\ m_2 \\ m_3 \end{bmatrix} = \begin{bmatrix} 2 & 1 & 1 \\ 1 & 2 & 1 \\ 1 & 1 & 2 \end{bmatrix} \begin{bmatrix} n_1 \\ n_2 \\ n_3 \end{bmatrix}. \tag{3.8}$$

Let us denote by q the amplitude of the smallest reciprocal-lattice vector $\mathbf{q}$ parallel to $\hat{\mathbf{d}}$, so that any reciprocal lattice vector parallel to $\hat{\mathbf{d}}$ can be written $\mathbf{K} = m\mathbf{q}$, with m an integer. Note that $\mathbf{q}$ and $\mathbf{a}_{SL}$ satisfy $\mathbf{q} \cdot \mathbf{a}_{SL} = 2\pi p$, with p an integer, since we have

$$\mathbf{R} \cdot \mathbf{K} = 2\pi \mathbf{n} \cdot \mathbf{m} = 2\pi [n_1, n_2, n_3] \begin{bmatrix} m_1 \\ m_2 \\ m_3 \end{bmatrix}. \tag{3.9}$$

For the $\langle 100 \rangle$ direction ($\mathbf{R} = a(1, 0, 0)$), we have $[n_1, n_2, n_3] = [-1, 1, 1]$, and the smallest reciprocal vector is given by $[m_1, m_2, m_3] = [0, 1, 1]$, so that we have $a_{SL} = a$ and $p = \mathbf{n} \cdot \mathbf{m} = 2$. We can now introduce a one-dimensional wave-vector k along the superlattice axis $\hat{\mathbf{d}}$ defined by

$$\mathbf{k} = \mathbf{k}_\perp + k\hat{\mathbf{d}}, \tag{3.10}$$

with $\mathbf{k}_\perp$ the momentum transverse to the direction $\hat{\mathbf{d}}$. It follows that the three-dimensional Bloch function and the band structure are one-dimensional periodic functions of k with period q:

$$\varphi(\mathbf{k}_\perp + k\hat{\mathbf{d}}, \mathbf{r}) = \varphi(\mathbf{k}_\perp + (k + nq)\hat{\mathbf{d}}, \mathbf{r}),$$

$$\mathcal{E}(\mathbf{k}_\perp + k\hat{\mathbf{d}}) = \mathcal{E}(\mathbf{k}_\perp + (k + nq)\hat{\mathbf{d}})$$

for n, an integer. Since $\mathcal{E}(\mathbf{k})$ and $\varphi(\mathbf{k}, \mathbf{r})$ are periodic functions of k we can express them in terms of a Fourier series:

$$\varphi(k, \mathbf{k}_\perp, \mathbf{r}) = \frac{1}{\sqrt{q}} \sum_{m=-\infty}^{\infty} w(m, \mathbf{r}, \mathbf{k}_\perp) \exp\left(\frac{ikm2\pi}{q}\right),$$

$$\mathcal{E}(\mathbf{k}) = \sum_{m=-\infty}^{\infty} \mathcal{E}_m(\mathbf{k}_\perp) \exp\left(\frac{ikm2\pi}{q}\right),$$

where $w(m, \mathbf{r}, \mathbf{k}_\perp)$ and $\mathcal{E}_m(\mathbf{k}_\perp)$ are the Fourier coefficients:

$$w(m, \mathbf{r}, \mathbf{k}_\perp) = \frac{1}{\sqrt{q}} \int_{-\frac{q}{2}}^{\frac{q}{2}} \exp\left(\frac{-ikm2\pi}{q}\right) \varphi(\mathbf{k}_\perp + k\hat{\mathbf{d}}, \mathbf{r})\, dk,$$

$$\mathcal{E}_m(\mathbf{k}_\perp) = \frac{1}{q} \int_{-\frac{q}{2}}^{\frac{q}{2}} \exp\left(\frac{-ikm2\pi}{q}\right) \mathcal{E}(\mathbf{k}_\perp + k\hat{\mathbf{d}})\, dk.$$

The one-dimensional Wannier function $w(m, \mathbf{r}, \mathbf{k}_\perp)$ that we have defined as the Fourier coefficients of the Bloch waves along the superlattice axis $\hat{\mathbf{d}}$ is, in fact, a hybrid state consisting of a Wannier function $|m\rangle$ along the superlattice axis and a Bloch function $|\mathbf{k}_\perp\rangle$ in the transverse direction. This hybrid state will be symbolically denoted $|m, \mathbf{k}_\perp\rangle$ for arbitrary representation so that we have in the position representation

$$\langle \mathbf{r} \,|\, m, \mathbf{k}_\perp \rangle = w(m, \mathbf{r}, \mathbf{k}_\perp). \tag{3.11}$$

Despite the lengthy derivation, essentially the same Wannier picture is obtained for the three-dimensional crystal.

As an example, consider in Figure 3.4 the band structure of a GaAs, AlAs and $Al_{0.3}Ga_{0.7}As$ conduction band for wave-vectors k running from Γ to X along the $\langle 100 \rangle$ direction. One recognizes the Γ valley and the upper X valley. These band structures are also periodic (note, however, that the period is $4\pi/a$ along $\langle 100 \rangle$). In Figure 3.5 we see that their Fourier series expansion can be approximately truncated to ten Fourier coefficients. *This is the number of lattice layers required in an epitaxial layer to create these band structures along the* $\langle 100 \rangle$ *direction.*

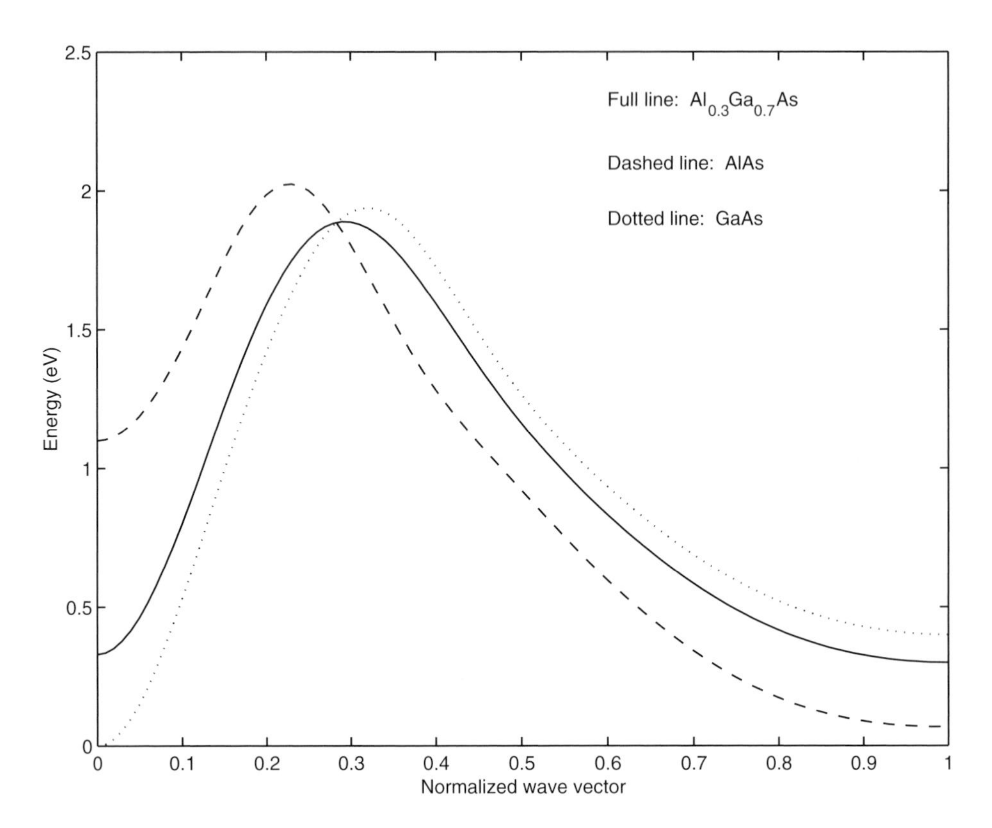

Fig. 3.4. Conduction-band structure along Δ (Γ–X) for GaAs, $Al_{0.3}Ga_{0.7}As$, and AlAs.

3.3 Spatially-varying band

The Wannier picture introduced in the previous section provided a natural picture for studying a lattice subject to a smooth, spatially-varying potential energy (for example, the electrostatic potential $-qV(x)$). The simplicity of the Wannier picture hinged on the properties of the Wannier functions, namely that the Wannier functions are tightly localized at the lattice sites and orthogonal. Also, in the Wannier picture, the Hamiltonian can be readily evaluated from the band structure.

In this section we shall use the Wannier picture to analyze spatially-varying crystals. Spatially-varying crystals are defined as consisting of different crystal layers of various thicknesses. Such a structure is sometimes referred to as a superlattice in the general sense of the term.

In spatially-varying crystals, the long-range periodicity of the lattice is destroyed and the translation operator no longer commutes with the lattice Hamiltonian. Therefore the wave-vector k cannot be defined (k is not a good quantum number) and the Bloch functions do not exist. As a consequence, it is no longer possible to define

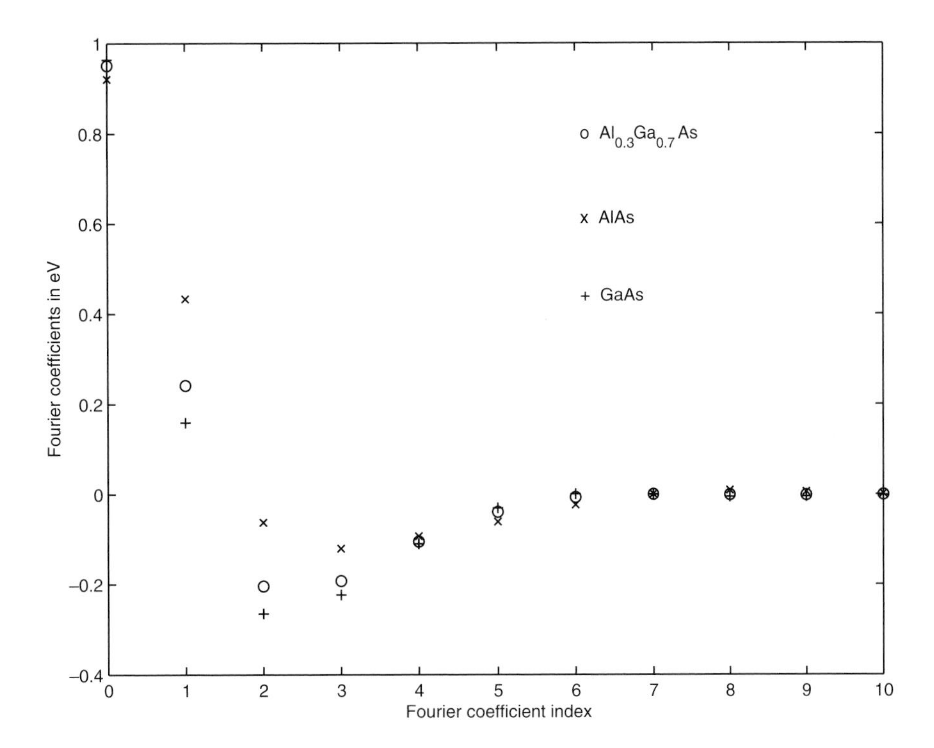

Fig. 3.5. Fourier coefficients of the Γ to X band structure shown in Figure 3.4.

the Wannier functions as the Fourier coefficients of the Bloch functions. However, if we consider a lattice site far from the interfaces we know intuitively that it should still be possible to use the Wannier functions of the crystal, since these functions are exponentially localized and do not 'see' the interface. However Wannier functions cannot be defined near the interface. Kohn and Onffroy [3] have shown that it is theoretically possible to define new orthogonal functions ($\langle n|m\rangle = \delta_{nm}$) with properties similar to the Wannier functions even near the interface. These functions are called generalized Wannier functions. Like the Wannier functions these generalized Wannier functions are tightly localized at a lattice site n and exponentially decay for large $x - na$. Consequently, the generalized Wannier functions are also labeled by their lattice site $|n\rangle$. Obviously the generalized Wannier functions are no longer translation invariant and we have $\langle x|n\rangle = w(x, n) \neq w_0(x - na)$. The generalized Wannier functions $w(x, n)$ away from the interfaces approach the Wannier function $w_0(x - na)$ of the perfect lattice in the following exponential manner:

$$\exp(2hna)\,[w(x, n) - w_0(x - na)] \to 0 \quad \text{for } n \to 0,$$

where h is a constant which measures the exponential decay. The reader is referred

Table 3.1. *Properties of Wannier and generalized Wannier functions.*

Properties	Wannier functions:	Generalized Wannier functions:
Notation:	$\|n\rangle$	$\|n\rangle$
Localization:	exponentially localized	exponentially localized
Orthonormal properties:	$\langle m', b' \| m, b\rangle = \delta_{nn'}\delta_{bb'}$	$\langle m', b' \| m, b\rangle = \delta_{nn'}\delta_{bb'}$
Translation invariance:	$\langle x \| n, b\rangle = w_b(x - na)$	NONE: $\langle x \| n, b\rangle = w_b(x, n)$
Index b:	bands	generalized bands & surface states

to [2] for an example of calculated generalized Wannier functions near an interface showing their rapid relaxation to the bulk lattice Wannier function away from the interface. Like for the Wannier functions, the generalized Wannier functions are also labeled according to bands of energies. Note that surface states can introduce novel virtual lattice sites. The generalized Wannier functions are still orthonormal and also form a complete set. Again we shall rely on the one-band approximation which assumes that the generalized Wannier functions of the band of interest form a complete set.

For comparison a summary of the properties of the multi-band Wannier functions and generalized Wannier functions is given in Table 3.1.

3.3.1 Heterojunction case (tight-binding approximation)

As an application of the generalized Wannier picture we now wish to consider the problem of the heterojunction between two semiconductors (1 and 2). For simplicity we shall assume that both have a tight-binding band structure and that no electrostatic potential is applied. On the left-hand side of the heterojunction, semiconductor 1 has the band structure:

$$\mathcal{E}_1(k) = \mathcal{E}_{1,0} - A_1 \cos(k_1 a)$$

and on the right-hand side of the heterojunction, semiconductor 2 has the band structure

$$\mathcal{E}_2(k) = \mathcal{E}_{2,0} - A_2 \cos(k_2 a).$$

We shall assume that due to the exponential localization of the Wannier functions, the Hamiltonian is only perturbed at the interface. The Hamiltonian matrix is therefore

$$H = \begin{bmatrix} \ddots & \ddots & \ddots & 0 & & & & \\ 0 & -\dfrac{A_1}{2} & \mathcal{E}_{1,0} & -\dfrac{A_1}{2} & 0 & & & \\ & 0 & -\dfrac{A_1}{2} & \mathcal{E}_{1,0} & -\dfrac{A_{12}}{2} & 0 & & \\ & & 0 & -\dfrac{A_{21}}{2} & \mathcal{E}_{2,0} & -\dfrac{A_2}{2} & 0 & \\ & & & 0 & -\dfrac{A_2}{2} & \mathcal{E}_{2,0} & -\dfrac{A_2}{2} & 0 \\ & & & & 0 & \ddots & \ddots & \ddots \end{bmatrix}$$

The only unknown terms are therefore the overlap Hamiltonian matrix elements A_{12} and A_{21} at the interface between the semiconductors, namely

$$A_{12} = -\frac{1}{2}\langle i|H|i+1\rangle,$$

$$A_{21} = -\frac{1}{2}\langle i+1|H|i\rangle,$$

where the interface is assumed to be located between the sites i and $i+1$. Quantum mechanics postulates that a physical system is described by a Hermitian Hamiltonian. Using the hermiticity of the Hamiltonian we must have

$$H_{nm} = H_{mn}^*$$

and we have therefore $A_{12} = A_{21}^*$. In this tight-binding model only one Hamiltonian matrix element is therefore unknown. Using a similar approach one can verify that for a band structure represented by two Fourier coefficients at least three matrix elements are unknown [5]. For three Fourier coefficients, at least five are unknown and so on. The unknown matrix element is therefore a characteristic of the interface as is the band structure for the lattice.

In Problem 3.4 a simple theory is developed for the tight-binding problem which permits us to derive the matrix element A_{12} for the case of an ideal (transparent) heterojunction. An ideal heterojunction is defined as a heterojunction which is as little reflective as possible for an incident electron of arbitrary energy E. It is demonstrated in Problem 3.4 that this occurs when we have

$$A_{12} = (A_1 A_2)^{1/2}.$$

The generalization to a band structure with N coefficients will be discussed in Section 3.3.3. We need first to introduce a definition of current in a general heterostructure.

3.3.2 Definition of the electron particle current (flux)

Given a potential energy usually controlled by an applied electrostatic potential, we can use the Wannier picture to solve for the Wannier envelope $f(n, E)$. The latter provides us with the electron distribution $|f(n, E)|^2$ in the device. In vertical devices which rely on the quantum transport properties of the structure we also need to calculate the electron current. For this purpose we need to introduce a definition of current to measure the flux of an electron in a spatially-varying band structure.

Conservation equation

First let us start by verifying that the Schrödinger equation enforces the conservation of particles. This property results from the hermiticity of the Hamiltonian. We first write the Schrödinger equation and its complex conjugate:

$$i\hbar \frac{d}{dt} f(n, t) = \sum_m H_{nm} f(m, t),$$

$$-i\hbar \frac{d}{dt} f^*(n, t) = \sum_m H_{nm}^* f^*(m, t).$$

We then multiply the above equations by $f^*(n, t)$ and $f(n, t)$, respectively:

$$i\hbar f^*(n, t) \frac{d}{dt} f(n, t) = \sum_m H_{nm} f^*(n, t) f(m, t),$$

$$-i\hbar f(n, t) \frac{d}{dt} f^*(n, t) = \sum_m H_{nm}^* f(n, t) f^*(m, t).$$

By subtracting these equations we obtain

$$\begin{aligned} i\hbar \frac{d}{dt} |f(n, t)|^2 &= \sum_m H_{nm} f^*(n, t) f(m, t) - H_{nm}^* f(n, t) f^*(m, t) \\ &= 2i \, \mathrm{Im} \left[\sum_m H_{nm} f^*(n, t) f(m, t) \right]. \end{aligned}$$

This equation describes the conservation of particles. Let us rewrite it as

$$\frac{d}{dt} |f(n, t)|^2 = -\frac{2}{a} \sum_p j(n, n + p, t)$$

with $j(n, m, t)$ given by

$$j(n, m, t) = -\frac{a}{\hbar} \, \mathrm{Im} \left[f^*(n, t) f(m, t) H_{nm} \right].$$

$j(n, m, t)$ is identified to be an elemental particle current from lattice site n to lattice site m. Note that these elemental particle currents satisfy the following obvious

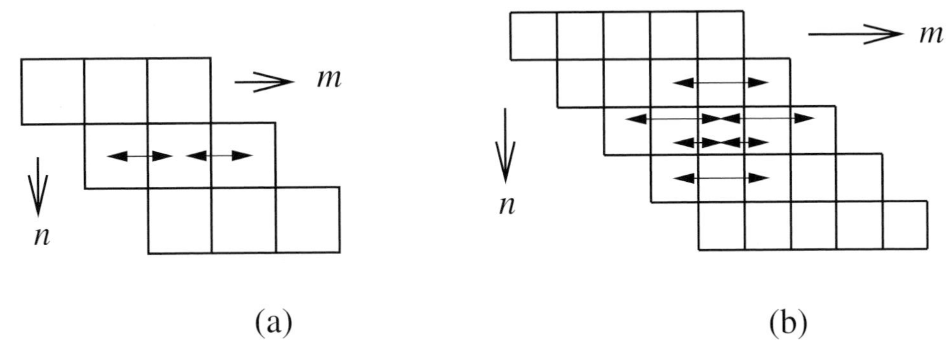

Fig. 3.6. (a) Symbolic representation of the two elemental currents contributing to the total current in a band Hamiltonian with one Fourier coefficient. (b) Symbolic representation of the six elemental currents contributing to the total current in a band Hamiltonian with two Fourier coefficients.

property:

$$\begin{aligned} j(m,n,t) &= -\frac{a}{\hbar}\,\mathrm{Im}\left[f^*(m,t)f(n,t)H_{mn}\right] \\ &= -\frac{a}{\hbar}\,\mathrm{Im}\left[f(n,t)f^*(m,t)H^*_{nm}\right] \\ &= -j(n,m,t). \end{aligned}$$

Total particle current definition

A total particle current (flux) can be defined by summing these elemental currents in the following way:

$$J(n,t) = \sum_{m=1}^{N}\sum_{p=0}^{m-1} j(n-p, n-p+m, t) + j(n+p-m, n+p, t).$$

Let us consider the tight-binding band structure for which $N = 1$. The total current is then simply

$$J(n,t) = j(n,n+1,t) + j(n-1,n,t) \quad \text{for } p = 0 \ m = 1$$

A graphical representation of this is given in Figure 3.6(a). For a band structure with two Fourier coefficients $N = 2$

$$\begin{aligned} J(n,t) = \ & j(n,n+1,t) + j(n-1,n,t) && (\text{for } p = 0 \ m = 1\,) \\ & + j(n,n+2,t) + j(n-2,n,t) && (\text{for } p = 0 \ m = 2\,) \\ & + j(n-1,n+1,t) + j(n-1,n+1,t) && (\text{for } p = 1 \ m = 2\,). \end{aligned}$$

A graphical representation of this is given in Figure 3.6(b).

As an example let us demonstrate that this total particle current is conserved for an eigenstate (stationary) solution in the case of two Fourier coefficients. To do this we need only to demonstrate that $J(n+1)$ is equal to $J(n)$: $J(n+1)$ is given by

$$\begin{aligned} J(n+1) = {} & j(n+1, n+2) + j(n, n+1) \\ & + j(n+1, n+3) + j(n-1, n+1) \\ & + j(n, n+2) + j(n, n+2). \end{aligned}$$

$J(n+1)$ and $J(n)$ are equal because both can be written as

$$J(n+1) = J(n) = 2\,j(n, n+1) + 2\,j(n-1, n+1) + 2\,j(n, n+2) \tag{3.12}$$

if we use the following identities derived from the particle conservation equation (with $(d/dt)|f(n,t)|^2 = 0$ for an eigenstate solution):

$$j(n-1, n) + j(n-2, n) = j(n, n+1) + j(n, n+2), \tag{3.13}$$

$$j(n, n+1) + j(n-1, n+1) = j(n+1, n+2) + j(n+1, n+3), \tag{3.14}$$

together with $j(m,n) = -j(n,m)$.

Flat-band case

The definition of the total particle current (flux) holds for an arbitrary spatially-varying one-band system. Let us verify that in the case of a uniform band it reduces to a well-known result.

We assume that the electron is in a propagating Bloch state

$$f_k(n) = \exp(ikna).$$

Note that we have $|f_k(n)| = 1$, so that this state is not normalized $\sum_{-\infty}^{\infty} |f(n)|^2 \neq 1$. In a uniform band case we have

$$H_{nm} = \mathcal{E}_{m-n}.$$

Let us calculate the following elemental current

$$\begin{aligned} j(n-p, n-p+m) &= -\frac{a}{\hbar}\,\mathrm{Im}\big\{\exp[-ika(n-p)] \\ &\quad \times \exp[ika(n-p+m)]H_{n-p\;n-p+m}\big\} \\ &= -\frac{a}{\hbar}\,\mathrm{Im}\big[\exp(+ikam)\mathcal{E}_m\big] \\ &= j(n, n+m). \end{aligned}$$

As we can see this elemental current is independent of the lattice site n considered and the index p can be dropped. Therefore the total particle current is given by

$$J(n) = \sum_{m=1}^{N} m\,j(n, n+m) - m\,j(n, n-m)$$

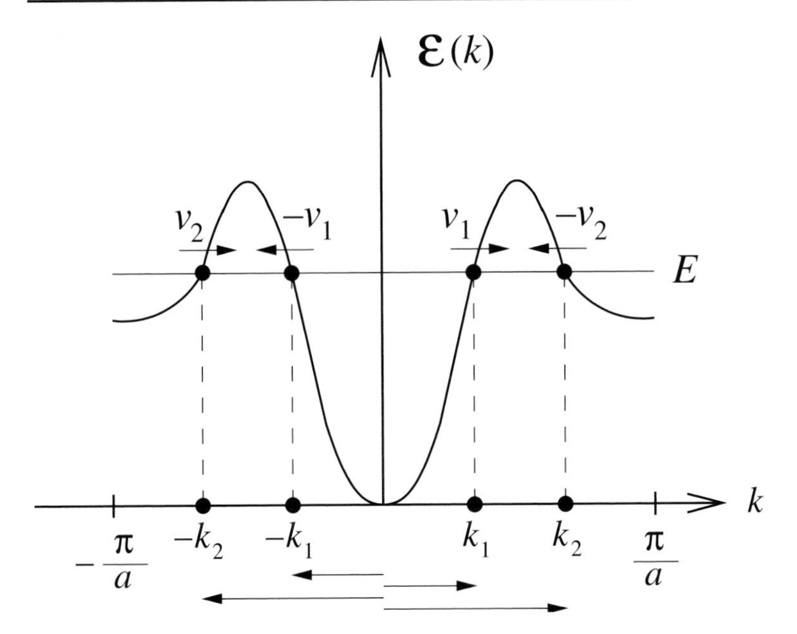

Fig. 3.7. Direction of the electron velocity $v(k)$ and the wave-vector k for the various propagating waves.

$$
\begin{aligned}
&= \sum_{m=-N}^{N} m\, j(n, n+m) \\
&= -\frac{a}{\hbar} \sum_{m=-n}^{N} m \,\mathrm{Im}\left[\exp(ikam)\mathcal{E}_m\right] \\
&= -\frac{a}{\hbar} \mathrm{Im}\left[\sum_{m=-N}^{N} m \exp(ikam)\mathcal{E}_m\right] \\
&= -\frac{a}{\hbar} \mathrm{Im}\left[\frac{1}{ia}\frac{\partial}{\partial k}\sum_{m=-N}^{N} \mathcal{E}_m \exp(ikam)\right] \\
&= \frac{1}{\hbar}\frac{\partial}{\partial k}\mathcal{E}(k).
\end{aligned}
$$

Since we selected the Wannier envelope function $f_k(n)$ to be an unnormalized Bloch state $|k\rangle$ with $|f_k(n)|^2 = 1$, the particle current $J(n) = |f_k(n)|^2 v(k)$ (flux = number of electrons × velocity) we calculated is the electron velocity $v(k)$. The velocity $v(k)$ of an electron in the Bloch state $|k\rangle$ in a uniform band, is therefore given by the gradient of the band structure:

$$
v(k) = \frac{1}{\hbar}\frac{\partial}{\partial k}\mathcal{E}(k). \tag{3.15}
$$

Note that the direction of the electron velocity is not necessarily the same as the wave-vector k as can be seen in Figure 3.7.

3.3.3 Matching theory

Let us return now to the heterojunction problem discussed in Section 3.3.1. Near the interface of a heterojunction the matrix elements of the Hamiltonian are unknown.

It is of interest to consider the case of a heterojunction which simulates a homojunction. A heterojunction is said to be transparent if like a homojunction it enforces the conservation of the elemental currents at an interface:

$$\left.\begin{aligned} j_1(n, m) &= j_2(n, m), \\ H_{1nm} f_1^*(n) f_1(m) &= H_{2nm} f_2^*(n) f_2(m), \end{aligned}\right\} \tag{3.16}$$

where $f_1(n)$ is the Wannier envelope of semiconductor 1 and $f_2(n)$ is the Wannier envelope of semiconductor 2. Both of these envelopes and Hamiltonians are extended across the interface as is done in Problem 3.4. In the case of a homojunction we simply have $f_1(n) = f_2(n)$ and $H_{1nm} = H_{2nm}$ and the conservation of the elemental current is obvious. Note that the tight-binding heterojunction is always transparent (see Problem 3.1).

A transparent heterojunction is said to be maximally transparent when the matrix elements are selected such that an electron of arbitrary energy undergoes to a minimum reflection at the heterojunction. Maximum transparency is achieved in a transparent heterojunction located between site i and $i + 1$ when we have

$$f_2(n) = \lambda f_1(n), \tag{3.17}$$

where n are the $2N$ sites around the interface (e.g., for $N = 2$ $n = i-1,\ i,\ i+1,\ i+2$). As is demonstrated in [5] for $N = 2$ and $N = 3$ maximum transparency can only be achieved if the band structures of both semiconductors satisfy

$$\mathcal{E}_2(k) - \mathcal{E}_{2,0} = \lambda^2[\mathcal{E}_1(k) - \mathcal{E}_{1,0}],$$

where $\mathcal{E}_{1,0}$ and $\mathcal{E}_{2,0}$ are the DC (zeroth order) Fourier coefficients of the band structures $\mathcal{E}_1(k)$ and $\mathcal{E}_2(k)$, respectively.

For the tight-binding band ($N = 1$), maximum transparency is achieved (see [9] and Problem 3.4) for

$$A_{12} = (A_1 \times A_2)^{1/2}.$$

Similar results can be obtained for higher order band structures. The Hamiltonian for $N = 2$ is shown in Figure 3.8. Transparent matching (see Equation (3.16)) is achieved for [5]

$$\left.\begin{aligned} B_{12} &= A_{12}\frac{B_1}{A_1}, \\ B_{21} &= A_{12}\frac{B_2}{A_2}. \end{aligned}\right\} \tag{3.18}$$

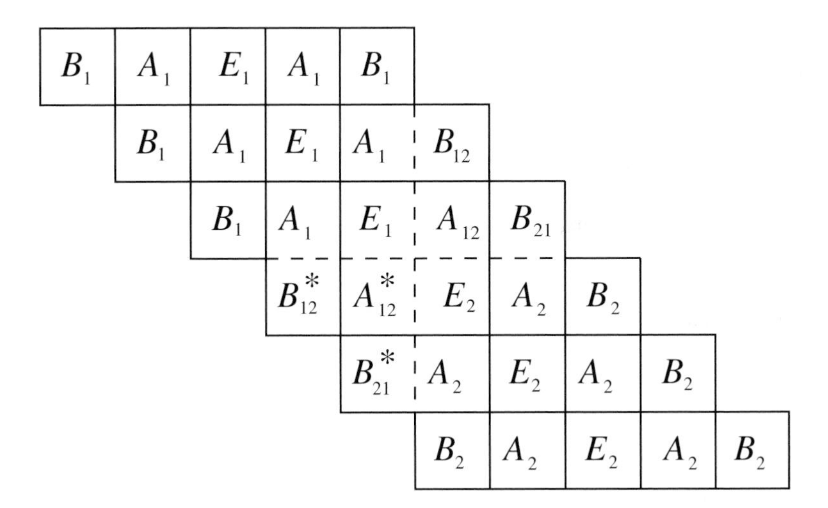

Fig. 3.8. Matching for $N = 2$.

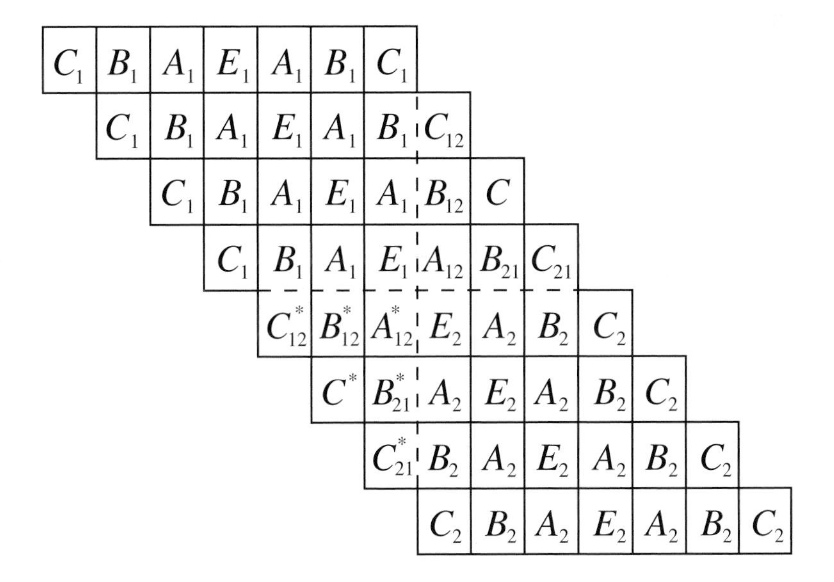

Fig. 3.9. Matching for $N = 3$.

Maximum transparency is achieved for $A_{12} = (A_1 \times A_2)^{1/2}$.

The Hamiltonian for $N = 3$ is shown in Figure 3.9. Transparent matching (see Equation (3.16)) is achieved for [5]

$$B_{12} = A_{12}\frac{B_1}{A_1} \qquad\qquad C_{12} = A_{12}\frac{C_1}{A_1},$$

$$B_{21} = A_{12}\frac{B_2}{A_2} \qquad C_{21} = A_{12}\frac{C_2}{A_2}.$$

Maximum transparency is achieved for

$$A_{12} = (A_1 \times A_2)^{1/2} \qquad C = (C_1 \times C_2)^{1/2}.$$

3.3.4 Three-dimensional effects

The existence of the generalized Wannier functions in heterostructures has only been rigorously demonstrated in the case of one-dimensional crystals [1,2]. Their extension to a three-dimensional layered system should not, however, raise any problem. The layered heterostructures considered here are spatially varying along the superlattice axis x and are uniform in the perpendicular direction (y and z). In such three-dimensional heterostructures the generalized Wannier functions can still be labeled with the perpendicular wave-vector $\mathbf{k}_\perp$ because the heterostructure Hamiltonian is translation invariant in the transverse direction and therefore commutes with the transverse translation operator (see Section 3.2.2).

In a spatially-varying crystal the state $|\mathbf{k}_\perp, n\rangle$ is therefore a generalized Wannier function at the lattice site n along the superlattice direction and a quasi-Bloch state $\mathbf{k}_\perp$ in the perpendicular direction.

Like the Wannier functions these generalized Wannier functions are orthogonal:

$$\langle \mathbf{k}'_\perp, n' | \mathbf{k}_\perp, n\rangle = \delta_{n'\,n}\,\delta(\mathbf{k}'_\perp - \mathbf{k}_\perp),$$

and are assumed here to form a complete basis (generalized one-band approximation).

In this three-dimensional picture the electron wave-function $|\Psi\rangle$ is expanded in terms of the generalized Wannier functions $|\mathbf{k}_\perp, n\rangle$ using the envelope function $f(\mathbf{k}_\perp, n, t)$

$$|\Psi\rangle = \sum_n \int f(\mathbf{k}_\perp, n, t)|\mathbf{k}_\perp, n\rangle\, d\mathbf{k}_\perp. \tag{3.19}$$

As for Wannier functions, the matrix element of the heterostructure Hamiltonian H_e in the generalized Wannier function basis is

$$\langle \mathbf{k}'_\perp,\ n' \mid H_e \mid \mathbf{k}_\perp,\ n\rangle = H^e_{n'n}(\mathbf{k}_\perp)\delta(\mathbf{k}'_\perp - \mathbf{k}_\perp)$$

with (assuming slowly varying $V(x)$)

$$H^e_{n'\,n}(\mathbf{k}_\perp) = H^0_{n'\,n}(\mathbf{k}_\perp) - eV(na)\delta_{n'\,n},$$

where $H^0_{n'\,n}(\mathbf{k}_\perp)$ is the matrix element of the unbiased heterostructure (generalized band structure) and $V(x)$ is the applied electrostatic potential sampled at the lattice site $x = na$.

In a general superlattice device the generalized band structure varies with position. For simplicity of presentation we shall now assume that the generalized band structure is well represented by an effective-mass approximation in the transverse direction so that we have

$$H^e_{n'\,n}(\mathbf{k}_\perp) = H_{n'\,n} + \left(\frac{\hbar^2 k_\perp^2}{2m^*(n)}\right)\delta_{n'\,n}.$$

In order to handle the longitudinal variation of the transversal mass it is convenient to treat the effective spatial variation of the transversal kinetic energy like an effective longitudinal potential:

$$H^e_{n'\,n}(\mathbf{k}_\perp) = \tilde{H}_{n'\,n} + \left[\frac{\hbar^2 k_\perp^2}{2m^*(0)}\right]\delta_{n'\,n},$$

where $\tilde{H}_{n'\,n}$ is defined using

$$\tilde{H}_{n'\,n} = H_{n'\,n} + \frac{\hbar^2 k_\perp^2}{2m^*(0)}\left[\frac{m^*(0)}{m^(n)} - 1\right]\delta_{n'\,n}$$

with the perpendicular energy at site 0 arbitrarily selected as a reference energy.

The three-dimensional Wannier recurrence equation is now

$$\jmath\hbar\frac{d}{dt}f(\mathbf{k}_\perp, n, t) = \frac{\hbar^2 {\mathbf{k}_\perp}^2}{2m^*(0)}f(\mathbf{k}_\perp, n, t) + \sum_{n'=-N_B}^{N_B} \tilde{H}_{n\,n'}f(\mathbf{k}_\perp, n', t). \qquad (3.20)$$

Therefore the effective reference potential $\tilde{H}_{n\,n}$ seen by an electron in a heterostructure will be dependent upon the perpendicular momentum of the electron and therefore the angle of incidence of the electron upon the layered heterostructure. In particular, electrons of total energy $E = E_x + E_\perp$ with the same longitudinal energy E_x but different perpendicular energies $E_\perp$ will experience different quantum heterostructures due to the variation of the (transverse) effective mass $m^*(n)$ from layer to layer as is shown graphically in Figure 3.10.

3.4 Multi-band tridiagonal Wannier picture

It is advantageous to model full-band structures with a simpler multi-band tridiagonal Wannier system. This is presented in detail below.

3.4.1 Multi-band tridiagonal Wannier system

The Hamiltonian of a multi-band tridiagonal system is

$$\sum_{b'=1}^{N_B}\sum_{n'=n-1}^{n+1} H(b, b', n, n', \mathbf{k}_\perp)f(b', n, E, \mathbf{k}_\perp) = E\ f(b, n, E, \mathbf{k}_\perp), \qquad (3.21)$$

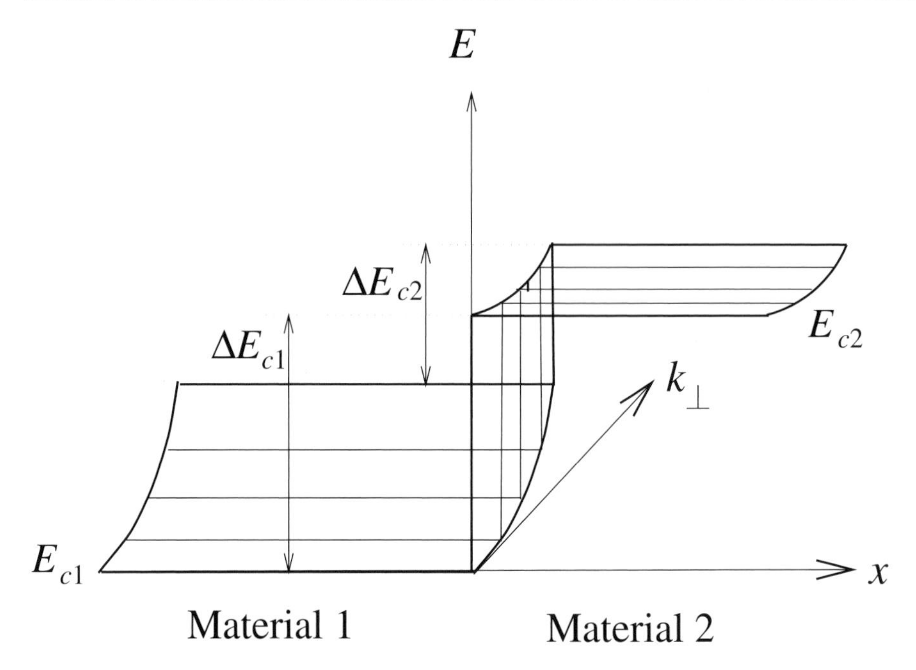

Fig. 3.10. Dependence of the conduction-band discontinuity ΔE_c upon the transverse momentum $k_\perp$ due to the variation of the transverse effective mass.

where b is the band index, n is the Wannier lattice site, and $f()$ is the Wannier envelope function. This Hamiltonian therefore represents a system of N_B tridiagonal Wannier bands and N_B envelope functions, which are coupled by the Hamiltonian at each heterojunction present in the heterostructure system. In the flat-band regions of the left- and right-hand contacts these N_B Wannier bands are decoupled. As for any physical system the Hamiltonian is Hermitian and we have

$$H(b, b', n, n', \mathbf{k}_\perp) = H^*(b', b, n', n, \mathbf{k}_\perp).$$

The multi-band electron current $J_T(n)$ through the heterostructure is

$$J_T(n) = \sum_b \sum_{b'} \left[j(b, b', n, n+1) + j(b', b, n-1, n) \right], \tag{3.22}$$

where $j(b', b, n', n)$ is the elemental current

$$j(b', b, n', n) = \mathrm{Im}[H(b', b, n', n, \mathbf{k}_\perp) f^*(b', n', E, \mathbf{k}_\perp) f(b, n, E, \mathbf{k}_\perp)]. \tag{3.23}$$

In Problem 3.5, the reader is invited to verify that the hermiticity of the Hamiltonian leads to the conservation of this current through the heterostructure. This, in turn, demonstrates the validity of the proposed current definition since this definition relaxes to the conventional current definition in contacts where the bands are decoupled (no interband current).

In the flat-band contacts as well as in the semiconductor regions away from a material interface, the Hamiltonian consists simply of N_B uncoupled tridiagonal Wannier bands. At a site n the local band structure of the band b is therefore of the form:

$$E(b, \mathbf{k}, n) = \mathcal{E}_c(b, n) - eV(n) - \frac{\hbar^2}{2a^2 m_b^*(n)}[1 \pm \cos(k_x a)] + \frac{\hbar^2 k_\perp^2}{2m_{b\perp}^*(n)}, \tag{3.24}$$

where $m_b^*(n)$ and $m_{b\perp}^*(n)$ are the longitudinal and transverse effective masses of band b at site n, $V(n)$ is the electrostatic potential, and $\mathcal{E}_c(b, n)$ is the bottom edge of the band b. Note that the $-$ sign is used when the band minimum is at Γ and the $+$ sign when the band minimum is at X.

3.4.2 Effective-mass wave-matching for a two-band Wannier system

The effective-mass matching technique presented earlier for a single Wannier band system can be applied to a system of two coupled Wannier bands (1 and 2) to establish the Hamiltonian at the junction of two different materials A and B.

Coupling between the two Wannier bands will take place at each heterointerface. Let us introduce the following notation for the matrix elements associated with the tridiagonal band of Equation (3.24):

$$\left.\begin{aligned} A_b &= H(b, b, n_A, n_A + 1) = H(b, b, n_A, n_A - 1) = \mp\frac{\hbar^2}{2a^2 m_{A,b}^*}, \\ B_b &= H(b, b, n_B, n_B + 1) = H(b, b, n_B, n_B - 1) = \mp\frac{\hbar^2}{2a^2 m_{B,b}^*}, \\ H_{A/B,b}(n) &= H(b, b, n_{A/B}, n_{A/B}, \mathbf{k}_\perp) \\ &= \frac{\hbar^2}{a^2 m_{A/B,b}^*} + \mathcal{E}_{c,A/B}(b) - eV(n) + \frac{\hbar^2 k_\perp^2}{2m_{A/B,b\perp}^*}, \end{aligned}\right\} \tag{3.25}$$

where $n_{A/B}$ is a site in material A/B, $m_{A/B,b}$ is the effective mass in the band b of material A/B, and $\mathcal{E}_{c,A/B}$ is the bottom of the band b in material A/B.

The two-band Hamiltonian system associated with the interface located at sites n and $n+1$ between materials A and B is:

$$H = \left[\begin{array}{cccccccc|cccccccc}
\ddots & \ddots & \ddots & & & & & & & & & & & & & \\
& A_1 & H_{A,1} & A_1 & & & & & & & & & & & & \\
& & A_1 & H_{A,1} & C_{12} & & & & & & C_{13} & C_{14} & & & & \\
& & & C_{21} & H_{B,1} & B_1 & & & & & C_{23} & C_{24} & & & & \\
& & & & B_1 & H_{B,1} & B_1 & & & & & & & & & \\
& & & & & \ddots & \ddots & \ddots & & & & & & & & \\
\hline
& & & & & & & & \ddots & \ddots & \ddots & & & & & \\
& & & & & & & & & A_2 & H_{A,2} & A_2 & & & & \\
& & & C_{31} & C_{32} & & & & & & A_2 & H_{A,2} & C_{34} & & & \\
& & & C_{41} & C_{42} & & & & & & & C_{43} & H_{B,2} & B_2 & & \\
& & & & & & & & & & & & B_2 & H_{B,2} & B_2 & \\
& & & & & & & & & & & & & \ddots & \ddots & \ddots
\end{array}\right].$$

The Wannier equations at the interface are then given by

$$\left.\begin{array}{l}
A_1 f_A(1, n-1) + (H_{A,1}(n) - E) f_A(1, n) + C_{12} f_B(1, n+1) \\
\quad + C_{13} f_A(2, n) + C_{14} f_B(2, n+1) = 0, \\
B_1 f_B(1, n+2) + (H_{B,1}(n+1) - E) f_B(1, n+1) + C_{21} f_A(1, n) \\
\quad + C_{23} f_A(2, n) + C_{24} f_B(2, n+1) = 0, \\
A_2 f_A(2, n-1) + (H_{A,2}(n) - E) f_A(2, n) + C_{34} f_B(2, n+1) \\
\quad + C_{32} f_B(1, n+1) + C_{31} f_A(1, n) = 0, \\
B_2 f_B(2, n+2) + (H_{B,2}(n+1) - E) f_B(2, n+1) + C_{43} f_A(2, n) \\
\quad + C_{42} f_B(1, n+1) + C_{41} f_A(1, n) = 0,
\end{array}\right\} \quad (3.26)$$

where the C_{ij} coefficients are the unknown coupling coefficients.

The application of the transparency wave-matching technique to this two-band system is outlined below. The complete derivation is left as an exercise (see Problem 3.6) [12]. Applied to this two-band system, transparency matching requires that the following linear dependence:

$$f_A(1, m) = \lambda_{12} f_B(1, m) + \lambda_{14} f_B(2, m),$$
$$f_A(2, m) = \lambda_{32} f_B(1, m) + \lambda_{34} f_B(2, m),$$

holds for both $m = n$ and $m = n + 1$ across the interface of materials A and B.

Extending the waves f_A and f_B across the interface and enforcing linear dependence one can verify that we must have $C_{13} = C_{31} = C_{24} = C_{42} = 0$ and that the λ_{ij}

coefficients are given by

$$\lambda_{12} = \frac{C_{12}}{A_1} = \frac{C_{43}B_1}{\Delta}, \tag{3.27}$$

$$\lambda_{14} = \frac{C_{14}}{A_1} = -\frac{C_{23}B_2}{\Delta}, \tag{3.28}$$

$$\lambda_{32} = \frac{C_{32}}{A_2} = -\frac{C_{41}B_1}{\Delta}, \tag{3.29}$$

$$\lambda_{34} = \frac{C_{34}}{A_2} = \frac{C_{21}B_2}{\Delta}, \tag{3.30}$$

with $\Delta = C_{21}C_{43} - C_{41}C_{23}$. Using the hermiticity property $C_{ij} = C_{ji}^*$ and the fact that the coefficients A_i are real, it is found that

$$\Delta = \pm(A_1B_1A_2B_2)^{1/2}. \tag{3.31}$$

Note that $A_1B_1A_2B_2$ *must be a positive number*. Inspection of AlAs, AsSb, InAs, InP and GaAs bands (see Figure 1.13 and [13]) reveals that their conduction bands have a minimum at Γ and an extremum at X so that this theory is applicable to these materials and their alloys.

The hermiticity of the Hamiltonian leads to the requirement

$$1 = \frac{|C_{12}|^2}{A_1B_1} + \frac{|C_{14}|^2}{A_1B_2}, \tag{3.32}$$

which establishes the following relation between $|C_{14}|$ and $|C_{12}|$:

$$|C_{14}|^2 = A_1B_2\left(1 - \frac{|C_{12}|^2}{A_1B_1}\right). \tag{3.33}$$

In practice we can select a real solution for the C_{ij} since the coefficients A_i are real and a phase shift is without consequence on the current. Let us now consider the two cases in which band 2 has a minimum at Γ or at X while band 1 keeps its minimum at Γ for both.

Case 1: Bands 1 and 2 have a minimum at Γ

In this case all the A_i are negative and the products A_iA_j are positive. Now since $|C_{14}|^2$ must be positive, we have from Equation (3.33) that

$$0 \leq |C_{12}|^2 \leq A_1B_1,$$

such that in turn we have

$$0 \leq |C_{14}|^2 \leq A_1B_2.$$

When the coefficients A_i are all negative, it is then natural to select

$$C_{14} = -|C_{14}| \qquad \text{and} \qquad C_{12} = -|C_{12}|.$$

Case 2: Band 1 has a minimum at Γ*, band 2 at* X

In this case A_1 and B_1 are negative and A_2 and B_2 are positive. Since the product $A_1 B_2$ is negative and the product $A_1 B_1$ is positive, the property that $|C_{14}|$ must be positive requires that we have

$$|C_{12}|^2 \geq A_1 B_1 \geq 0.$$

Once C_{12} has been selected, C_{14} is known and the remaining coefficients can then be calculated. Substituting Δ from Equation (3.31) in Equations (3.27) and (3.28) gives then the final expression for the C_{43} and C_{23} coefficients which depends on the sign of Δ:

$$C_{43} = \pm C_{12} \left(\frac{A_2 B_2}{A_1 B_1} \right)^{1/2}, \quad \text{and} \quad C_{23} = \mp C_{14} \left(\frac{B_1 A_2}{A_1 B_2} \right)^{1/2}. \tag{3.34}$$

It can be verified that the sign of Δ does not affect the amplitude of the transmission coefficient. It is natural, however, to select C_{43} to have the same sign as A_2 and B_2.

Note that in the absence of band coupling ($C_{13} = C_{23} = 0$), the coupling theory for C_{12} and C_{43} reduces to the one-band effective-mass matching theory.

3.4.3 Comparison with a full-band model

In this section we shall apply the effective-mass matching theory developed in the previous section for a two-band system, to the modeling of the Γ and X valleys of the conduction band. To test the accuracy of this model we will compare simulation results obtained with the simple two tridiagonal Wannier band system with those obtained with a single full Wannier band model.

The band structure along the $\langle 100 \rangle$ direction is written

$$\mathcal{E}(\mathbf{k}) = \sum_{m=1}^{N_W} \mathcal{E}_m(\mathbf{k}_\perp) \cos\left(\frac{p k_x a}{2} \right), \tag{3.35}$$

with N_W the number of Fourier coefficients $\mathcal{E}_m$ used.

The GaAs, AlAs and $Al_{0.3}Ga_{0.7}As$ band structures along the $\langle 100 \rangle$ direction used in our comparison are those shown in Figure 3.4 for $\mathbf{k}_\perp = 0$. Ten Fourier coefficients $\mathcal{E}_m(0)$ ($N_W = 10$) are used to fit the bands. These Wannier GaAs and AlAs bands are obtained from a least square fit of band structure data obtained using a pseudopotential calculation (see Figure 1.13 and [13]). The $Al_{0.3}Ga_{0.7}As$ band is obtained using a linear weighted average of the GaAs and AlAs bands. Because of the limit of this approximation we only consider here GaAs–$Al_{0.3}Ga_{0.7}As$ resonant tunneling structures (RTDs, see next chapter), test structures for which this interface Hamiltonian is a reasonable approximation due to the low mole fraction of its barrier.

To account in the two-band model for the non-effective-mass effects in both the Γ and X valleys of the bands shown in Figure 3.4 it is necessary to use an energy-

and position-dependent coefficient $A(b, n, E)$ in each of the tridiagonal bands b of dispersion $\mathcal{E}(k_x) = A(b, n, E)[1 \mp \cos(k_x a/2)]$. For energies for which the wave is propagating in the valley b (wave-vector $k_x(b, n)$ with a zero imaginary part), $A(b, n, E)$ is selected to be

$$A(b, n, E) = \frac{E - \mathcal{E}_{c,b}(n) - eV(n)}{1 \mp \cos\{\mathrm{Re}[k_x(b, n)]a/2\}}. \tag{3.36}$$

For energies for which the wave is damped in the valley b (wave-vector $k_x(b, n)$ with a non-zero imaginary part), $A(b, n, E)$ is selected to be

$$A(b, n, E) = \frac{E - \mathcal{E}_{c,b}(n) - eV(n)}{1 - \cosh\{\mathrm{Im}[k_x(b, n)]a/2\}}. \tag{3.37}$$

Note that the same expression is used for both the Γ and X damped waves. This originates from the fact that the real part of the complex wave-vector k_x is equal to π in the X valley and 0 in the Γ valley. In both Equations (3.36) and (3.37) $k_x(b, n)$ are the Γ and X wave-vectors associated with the solution of

$$E = \mathcal{E}(k_x, \mathbf{k}_\perp) - eV(n), \tag{3.38}$$

where $E(k_x, \mathbf{k}_\perp)$ is the full-band structure corresponding to either GaAs or AlAs at the lattice site n. For a Wannier model using ten cosine harmonics to represent the GaAs, AlAs and $Al_{0.3}Ga_{0.7}As$ band structures, the solution of Equation (3.38) using analytical continuation leads to 20 complex numbers for the wave-vector. The correct Γ and X (pure imaginary) roots are to be identified for use in Equations (3.36) and (3.37) for bands $b = 1$ and $b = 2$, respectively (the other roots are evanescent waves with a very fast decay). This energy-dependent effective-mass implements a non-effective-mass correction which is of critical importance if the resonances and anti-resonances in the coupled-band model are to occur at energies similar to those of the full-band model.

Figure 3.11 compares the various plots of both Γ to Γ and Γ to X left-to-right transmission coefficients versus incident energy, obtained with the full single-band model (full line) and the effective-mass matching theory (dotted line). The test structure consists of an RTD with seven-monolayer GaAs spacers, eight-monolayer $Al_{0.3}Ga_{0.7}As$ barriers and an eight-monolayer GaAs well. The two-band model result is obtained using the coupling coefficients of Case 2 (A_2 and B_2 positive) since band 2 has a minimum at X, and using a weak Γ to X coupling $(A_1 B_1)^{1/2}/C_{12} = 0.999$.

The Γ–X resonances are observed to occur at approximately the same energies. An improved model is, however, achieved by resetting the backward coupling coefficients to zero ($C_{23} = C_{32} = 0$), leaving all the other coefficients unchanged. The resulting Γ and X transmission coefficients (dashed-dotted lines) are then found to be in even better agreement at high energies.

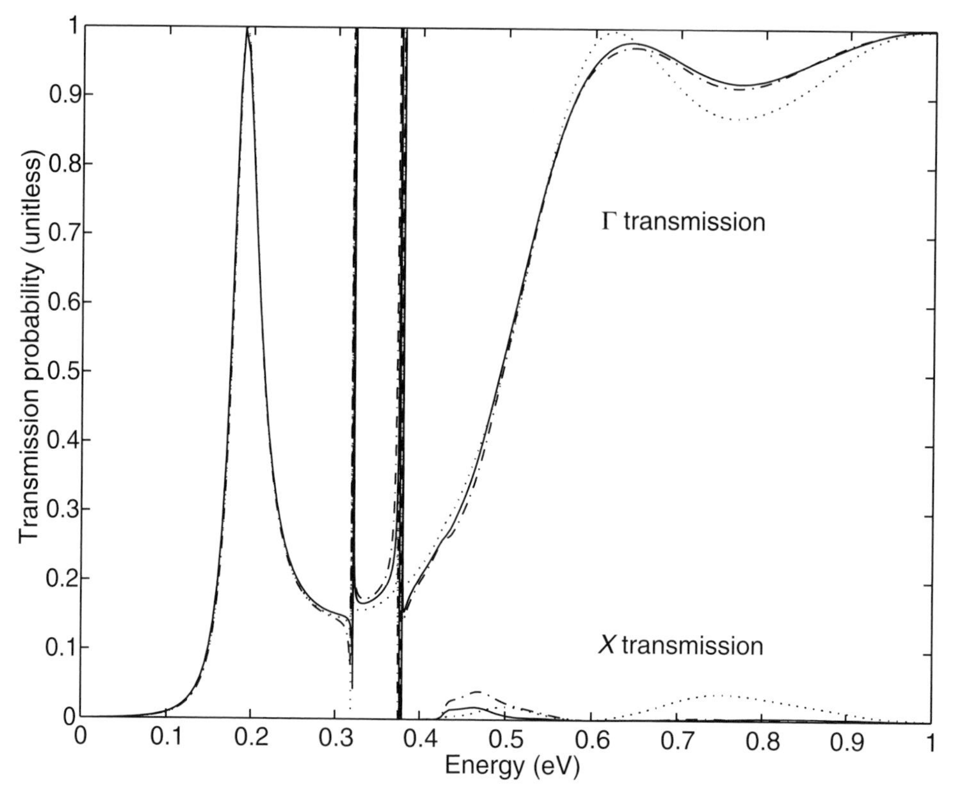

Fig. 3.11. Γ and X transmission coefficients versus incident energies obtained with the full single-band model (full line), the effective-mass matching theory (dotted line), and the modified effective-mass matching theory (dashed-dotted line). (P. Roblin, P. Sotirelis, and G. Cao, *Physics Review B*, Vol. 58, No. 19, pp. 13 103–13 114, November 15 1998. Copyright 1998 by the American Physical Society.)

The modified coupled-band model and the full-band model predict the same type of resonances ($T = 1$) and anti-resonances ($T = 0$) as are demonstrated by the more detailed comparison in Figure 3.12. This indicates that the modified coupled-band model correctly implements the physical processes of the Γ and X valley coupling. However, the improved fit obtained by setting $C_{23} = C_{32} = 0$ points toward the limit of applicability of the effective-mass matching theory. Such departures from the ideal effective-mass matching theory are, however, to be expected in real heterostructures.

3.5 Multi-band density of states

The quantum structure analyzed in the previous section exhibited some complex resonances and anti-resonances into the electron transmission coefficient. To develop more insights in the electronic characteristics of a quantum structure it is useful to

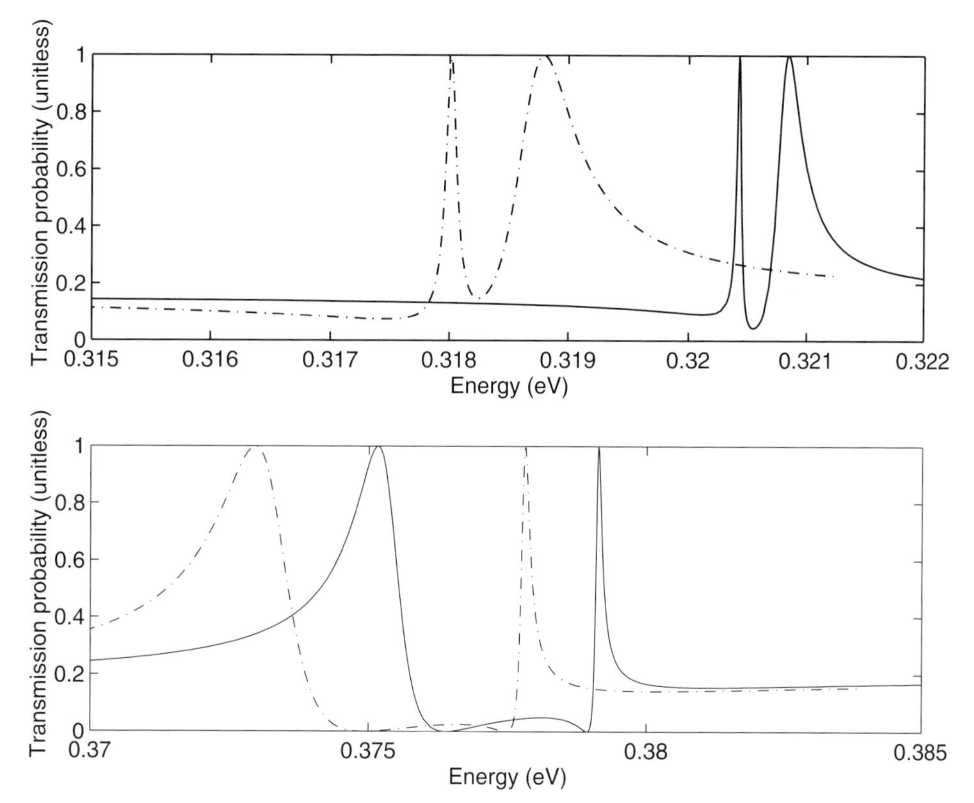

Fig. 3.12. Details of the Γ and X transmission coefficients for incident energies around the Γ–X resonance/anti-resonance obtained using the full single-band model (solid line), and the modified effective-mass matching theory (dashed-dotted line). (P. Roblin, P. Sotirelis, and G. Cao, *Physics Review B*, Vol. 58, No. 19, pp. 13 103–13 114, November 15 1998. Copyright 1998 by the American Physical Society.)

introduce a local density of states $N(E, n)$ at the lattice site n which will permit us to locate in both position and energy the resonances and anti-resonances in that quantum structure.

The density of states $N(E, n)$ at site n is defined by

$$\rho(n) = \int f_D(E) N(E, n)\, dE, \tag{3.39}$$

with $f_D(E)$ the Fermi–Dirac function. We can derive the local multi-band density function $N(E, n)$ by calculating the local total charge $\rho(n)$ *in thermal equilibrium*. In general, the charge distribution in a ballistic quantum system is given by summing over all the individual charge distributions associated with the electrons injected in all the bands b' at the left-hand L and right-hand R contacts:

$$\rho(x) = \sum_{b'} \left[\rho_L(x, b') + \rho_R(x, b')\right],$$

with $\rho_{L/R}(x, b)$ given by

$$\rho_{L/R}(x, b') = \frac{2}{(2\pi)^3} \iiint_{k_{x,L/R}} f_{D,L/R}(E)\, |\psi_{L/R}(\mathbf{k}, x, b')|^2\, d\mathbf{k},$$

where $f_{D,L/R}$ is the Fermi–Dirac distribution in the left- and right-hand contacts.

In our multi-band system the wave-function $\psi_{L/R}(\mathbf{k}, x, b')$ is expanded in terms of Wannier functions $w(b, n, \mathbf{k}_\perp)$ of the band b:

$$\psi_{L/R}(\mathbf{k}, x, b') = \sum_b f(b, n, b', \mathbf{k}) w(b, n, \mathbf{r}, \mathbf{k}_\perp),$$

where we have $\mathbf{k} = (k_x, \mathbf{k}_\perp)$ with k_x either the longitudinal wave-vector incident on the left-hand ($k_{x,L}$) or right-hand ($k_{x,R}$) flat-band contact for ψ_L and ψ_R, respectively. Indeed the contact wave-vectors $k_{x,L/R}$ are not translation invariant throughout the device unlike the transverse wave-vector $\mathbf{k}_\perp$.

A site-average probability of presence can then be obtained by integrating the wave-function ψ over x in the site interval $[a(n-\frac{1}{2}),\ a\left(n+\frac{1}{2}\right)]$. Using the orthogonality of the Wannier functions of different bands together with the reasonable approximation of locality around a single site, we obtain the following identity

$$\left|\psi_{L/R}(\mathbf{k}, n, b')\right|^2 = \frac{1}{a} \int_{a(n-1/2)}^{a(n+1/2)} |\psi(\mathbf{k}, x, b')|^2\, dx = \sum_b \left|f(b, n, b', \mathbf{k})\right|^2.$$

The site average charge distribution can now be rewritten as a summation over the coupled-bands index b and the incident band index b' of the various wave-function contributions to the charge:

$$\rho(n) = \sum_{b'} \sum_b \left[\rho_L(n, b, b') + \rho_R(n, b, b')\right],$$

where $\rho_{L/R}(n, b, b')$ is given by

$$\rho_{L/R}(n, b, b') = \frac{2}{(2\pi)^3} \iiint_{k_{x,L}} f_{D,L/R}(E)\, |f(b, n, b', \mathbf{k})|^2\, d\mathbf{k}. \tag{3.40}$$

In Problem 3.7 the reader is invited to verify that the envelope $f(b, n, b', \mathbf{k})$ can be expressed in terms of the impulse response $h(b, n, b', n', \mathbf{k})$. For waves $f(b, n, b', \mathbf{k}_L)$ injected in the left-hand contact (site N_L) or right-hand contact (site N_R) in band b', we have

$$f(b, n, b', \mathbf{k}_{L/R}) = j \frac{\hbar v_{L/R}(b', E, \mathbf{k}_\perp)}{a} h(b, n, b', N_{L/R}, E, \mathbf{k}_\perp) \tag{3.41}$$
$$\text{for} \quad n \geq N_L \quad \text{or} \quad n \leq N_R,$$

where $v_{L/R}$ is the velocity (selected positive) of the wave of total energy E injected into the device in band b', from the left-hand (L) or right-hand (R) contact.

The impulse response $h(b, n, b'', n'', E, \mathbf{k}_\perp)$ with total energy E and transverse wave-vector $\mathbf{k}_\perp$ for an excitation at site n'' in band b'' is by definition obtained from the following multi-band tridiagonal equation

$$\begin{aligned}\sum_{b'=1}^{N_B}&[H_-(b, b', n)h(b', n-1, b'', n'', E, \mathbf{k}_\perp)\\&+ H_o(b, b', n, \mathbf{k}_\perp)h(b', n, b'', n'', E, \mathbf{k}_\perp)\\&+ H_+(b, b', n)h(b', n+1, b'', n'', E, \mathbf{k}_\perp)]\\&+ \delta_{nn''}\delta_{bb''} = E\, h(b, n, b'', n'', E, \mathbf{k}_\perp).\end{aligned}$$

Using this identity in Equation (3.40), the charge distribution $\rho_{L/R}(n, b, b')$ is

$$\begin{aligned}\rho_{L/R}(n, b, b') &= \frac{2}{(2\pi)^3}\iiint_{k_{x,L/R}} f_{D,L/R}(E)\left|f(b, n, b', \mathbf{k})\right|^2\, d\mathbf{k}\\&= \frac{2}{(2\pi)^3}\iiint_{k_{x,L/R}} f_{D,L/R}(E)\frac{\hbar^2 v_{L/R}^2(b', E, \mathbf{k}_\perp)}{a^2}\left|h(b, n, b', N_{R/L}, \mathbf{k})\right|^2\, d\mathbf{k}\\&= \frac{1}{a^2}\frac{2}{(2\pi)^3}\iint d\mathbf{k}_\perp \int_{-\infty}^{\infty} dE\, f_{D,L/R}(E)\,\hbar v_{L/R}(b', E, \mathbf{k}_\perp)\\&\quad\times \left|h(b, n, b', N_{R/L}, E, \mathbf{k}_\perp)\right|^2.\end{aligned}$$

Using the hermiticity of the Hamiltonian, the reader is invited in Problem 3.7 to verify that the impulse response satisfies the following current-conservation property:

$$\begin{aligned}-\frac{a}{\hbar}2\,\mathrm{Im}[h(b, n, b, n, E, \mathbf{k}_\perp)] &= \sum_{b'} v_L(b', k_{x,L})\left|h(b, n, b', N_L, E, \mathbf{k}_\perp)\right|^2\\&\quad+ v_R(b', k_{x,R})\left|h(b, n, b', N_R, E, \mathbf{k}_\perp)\right|^2,\end{aligned} \tag{3.42}$$

with $k_{x,R/L}$ obtained from $E = E(b, k_{x,R/L}, \mathbf{k}_\perp, N_{L/R})$. The charge distribution in thermal equilibrium obtained for $f_{D,L}(E) = f_{D,R}(E) = f_D(E)$ is then given by

$$\begin{aligned}\rho(n) &= \sum_{b'}\left[\sum_b \rho_L(n, b, b') + \rho_R(n, b, b')\right]\\&= -\frac{2}{a}\sum_b \frac{2}{(2\pi)^3}\iint d\mathbf{k}_\perp \int_{-\infty}^{\infty} dE\, f_D(E)\,\mathrm{Im}[h(b, n, b, n, E, \mathbf{k}_\perp)]\\&= -\frac{1}{a\pi}\int_{-\infty}^{\infty} dE\, f_D(E)\sum_b D_b\int_0^{\infty} dE_\perp\,\mathrm{Im}[h(b, n, b, n, E, E_\perp)],\end{aligned}$$

where we used cylindrical symmetry to introduce $E_\perp = \hbar k_\perp^2/(2m_b^*)$, satisfying $D_b\, dE_\perp = 2/(2\pi)^2\, d\mathbf{k}_\perp$ with $D_b = m{*}_b/\hbar^2\pi$ the two-dimensional electron gas density of states in the band b.

Using Equation (3.39) which relates the local charge $\rho(n)$ to the local density of states, we can now identify the local density of states $N(E, n)$ as

$$N(E, n) = \int_0^\infty N(E, E_\perp, n)\, dE_\perp,$$

with $N(E, E_\perp, n)$ the partial density of states in the channel $E_\perp$ given by

$$N(E, E_\perp, n) = -\frac{1}{a\pi} \sum_b D_b \ \mathrm{Im}[h(b, n, b, n, E, E_\perp)].$$

Note that in the case of a heterostructure consisting of a single flat band, the impulse response $h(n, n', E, E_\perp)$ is easily evaluated to be $h(n, n', E, E_\perp) = ja/[\hbar v(k_x)]$, with k_x obtained from $E = E(b = 1, k_x, k_\perp, n)$. The single-band density of states is then simply

$$N(E, E_\perp) = \frac{D_1}{\pi} \frac{1}{\hbar v(k_x)}.$$

The partial multi-band density of states formula derived above is quite helpful for establishing the location and the origin of resonances ($T = 1$) and anti-resonances ($T = 0$) in a quantum structure. Consider the RTD structure (to be discussed in more detail in the next chapter), analyzed in Section 3.4.2. For the purpose of the identification of the various Γ and X valley resonances and anti-resonances we will plot separately the impulse response $\mathrm{Im}[h(b, n, b, n, E, E_\perp)]$ of each band (valleys) b.

Figure 3.13 shows the Γ valley density of states versus longitudinal energy and position for a zero transverse wave-vector. Figure 3.14 shows the X valley density of states versus longitudinal energy and position for a zero transverse wave-vector. Art and science overlap in the Γ density of states in Figure 3.13, which has the appearance of a human face. Notice the resonant ground state (no node) revealed as the bottom high-density structures (mouth) at about 0.2 eV and the first excited state (one node) as the two top structures (eyes) at about 0.75 eV. One also notices a new structure located mid-way (nose) at about 0.3 eV corresponding to the coupling between the Γ valley of the GaAs well and the X valley of the AlAs barrier as is clearly revealed by the X density of states shown in Figure 3.14. The X density of states also features an additional resonance just above the GaAs well and spacers.

The multi-band Γ and X density of states is seen to find an important application in the additional identification of both the energy band and spatial origin of the resonance and anti-resonance, which can arise from the coupling of the Γ and X valleys. We will find that the density of states introduced above is also used when calculating the occupation number as a function of position under non-equilibrium conditions for which the Fermi–Dirac distribution f_D is no longer applicable.

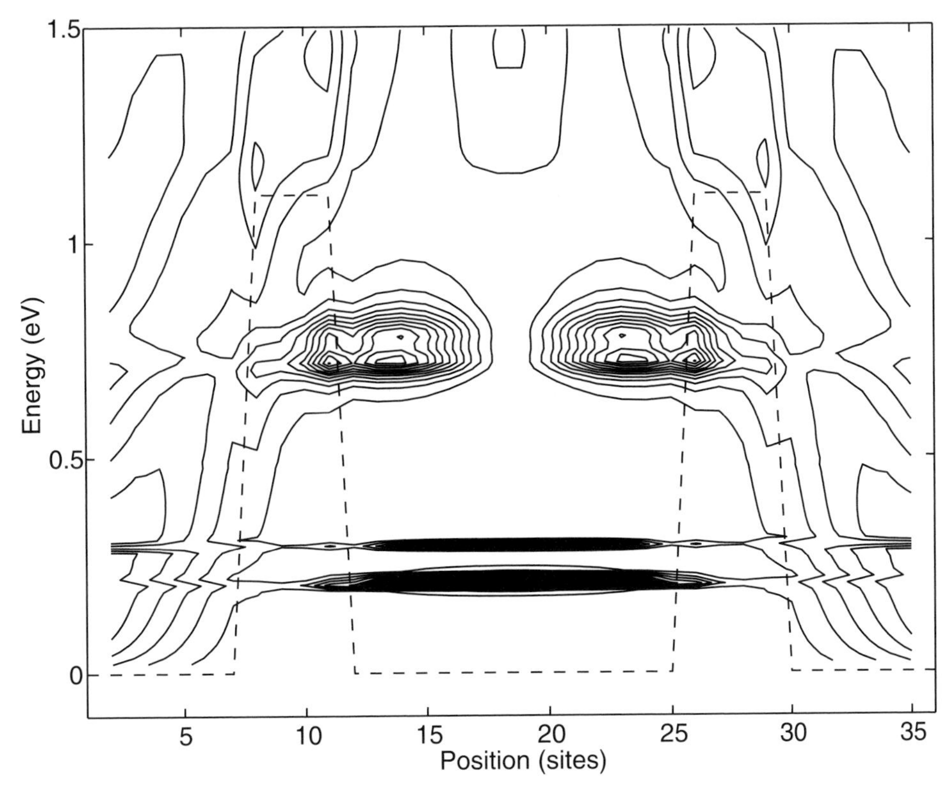

Fig. 3.13. Γ valley density of states versus longitudinal energy and position for a zero transverse wave-vector. The Γ conduction-band edge is shown using a dashed line. (P. Roblin, P. Sotirelis, and G. Cao, *Physics Review B*, Vol. 58, No. 19, pp. 13 103–13 114, November 15 1998. Copyright 1998 by the American Physical Society.)

3.6 Conclusion

In this chapter we have introduced a quantum picture of spatially–varying band structures using the generalized Wannier picture. A multi-band density of states based on the impulse response was also introduced for the spatial identification of quantum resonances. The principal advantage of using the Wannier picture lies in: (1) its inherent ability to account rigorously for both the spatial variation of the band structure and its periodicity in **k** space, and (2) its representation in terms of difference equations which are easily amenable to a numerical solution.

Armed with this quantum picture we shall present in the next chapter the principal one-dimensional quantum devices which can be realized using semiconductor heterostructures.

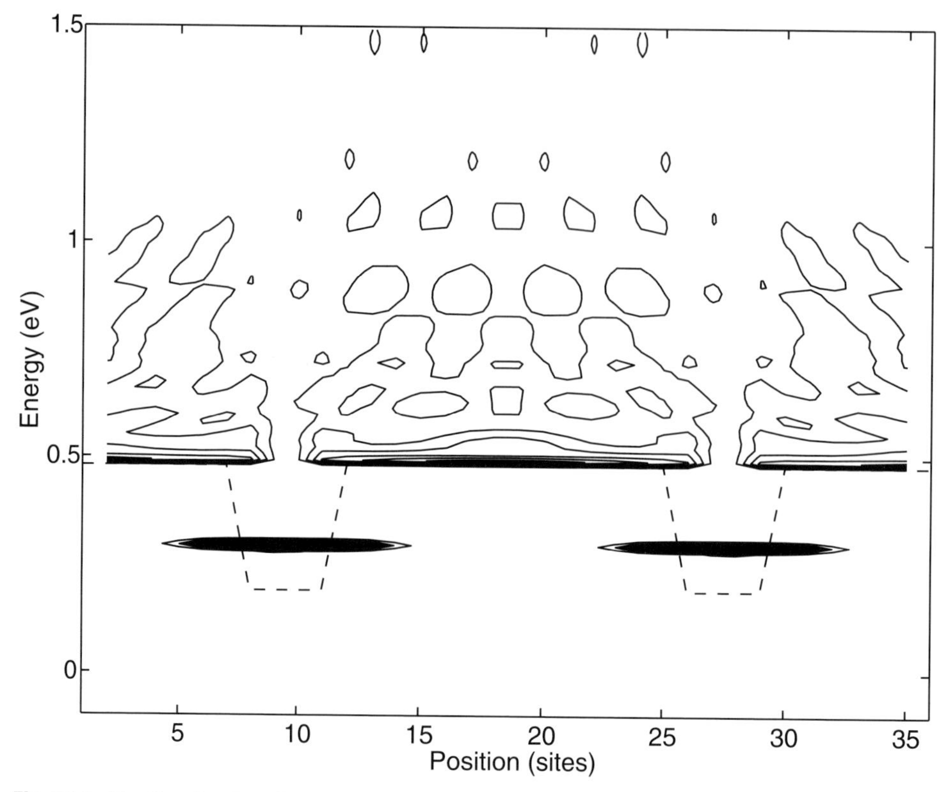

Fig. 3.14. X valley density of states versus longitudinal energy and position for a zero transverse wave-vector. The X conduction-band edge is shown using a dashed line. (P. Roblin, P. Sotirelis, and G. Cao, *Physics Review B*, Vol. 58, No. 19, pp. 13 103–13 114, November 15 1998. Copyright 1998 by the American Physical Society.)

3.7 Bibliography

3.7.1 Recommended reading

G. Bastard, *Wave Mechanics Applied to Semiconductor Heterostructures*, Les Editions de Physique, Les Ulis, France, 1990.

S. Datta, *Quantum Phenomena*, Addison Wesley, Reading, 1989.

3.7.2 References

[1] W. Kohn, 'Analytic properties of Bloch waves and Wannier functions', *Physical Review*, Vol. 115, No. 4, p. 809, 15 August 1959.

[2] J. G. Gay and J. R. Smith, 'Generalized-Wannier-function solutions to model surface potentials', *Physical Review B*, Vol. 11, No. 12, p. 4906, 15 June 1975.

[3] W. Kohn and J. R. Onffroy, 'Wannier functions in a simple nonperiodic system', *Physical Review B* Vol. 8, No. 6, p. 2485, 15 September 1973.

[4] P. Roblin and M. W. Muller, 'Spatially-varying band structures', *Physical Review B*, Vol. 32, No. 8, p. 5222, 15 October 1985.

[5] P. Roblin, 'Band structure matching at a heterojunction interface', *Superlattices and Microstructures*, Vol. 4, No. 3, pp. 363–370, 1988.

[6] P. Roblin, 'Electron states in submicron semiconductor heterostructures', D. Sc. Dissertation, Washington University, August 1984.

[7] G. H. Wannier, 'Dynamics of band electrons in electric and magnetic fields', *Reviews of Modern Physics*, Vol. 34, No. 4, October 1962.

[8] J. M. Ziman, *Principle of the Theory of Solids*, Cambridge University Press, New York, 1972.

[9] H. Kroemer and Qi-Gao Zhu, 'On the interface connection rules for effective-mass wave-functions at an abrupt heterojunction between two semiconductors with different effective mass', *Journal of Vacuum Science Technology*, Vol. 21, p. 551, 1982.

[10] D. J. Ben Daniel and C. B. Duke, 'Space-charge effects in electron tunneling', *Physical Review*, Vol. 152, p. 683, 1966.

[11] G. Bastard, *Wave mechanics applied to semiconductor heterostructures*, Les editions de Physique, Les Ulis, France, 1990.

[12] P. Roblin, P. Sotirelis, and G. Cao, 'Effective-mass wave-matching theory for two-bands Wannier systems', *Physical Review B*, Vol. 58, No. 19, pp. 13 103–13 114, Nov. 15 1998.

[13] M.L. Cohen and T.K. Bergstresser, *Physical Review* Vol. 141, No. 2, p. 789, 1963.

3.8 Problems

3.1 (a) Verify that the Wannier functions are orthonormal:

$$\langle m|n\rangle = \int_{-\infty}^{\infty} w^*(m,x)w(n,x)\,dx = \delta_{mn} = \begin{cases} 1 & n=m \\ 0 & n \neq m. \end{cases}$$

(b) Verify that the crystal Hamiltonian matrix element in the Wannier representation reduces to the Fourier coefficients of the band structure.

$$H^0_{nm} = \langle n|H^0|m\rangle = \int_{-\infty}^{\infty} w^*(n,x)H^0 w(m,x)\,dx$$

$$= \mathcal{E}_{m-n} = \frac{a}{2\pi}\int_{-\frac{\pi}{a}}^{\frac{\pi}{a}} \exp[-ika(m-n)]\mathcal{E}(k)\,dk$$

3.2 Consider a quantum well of length $L = Na$ with a flat bottom and infinite walls. Assume that the potential matrix element is diagonal

$$V_{nm} = V_n\,\delta_{nm}.$$

The well potential V_n is then given by

$$V_n = \infty \qquad \text{for } n \leq 0, \tag{3.43}$$

$$V_n = 0 \qquad \text{for } 1 \leq n \leq N-1, \tag{3.44}$$

$$V_n = \infty \qquad \text{for } n \geq N. \tag{3.45}$$

Assume that the energy band considered is a tight-binding band $\mathcal{E}(k) = A - A\cos(ka)$

(a) Using the Wannier recurrence equation obtain the Wannier envelope $f(n, E_p)$ and the discrete energies E_p for this quantum well.
(b) Verify that the tight band structure can be approximated by a parabolic band structure (effective-mass approximation) $\hbar^2k^2/2m^\star$, where $m^\star$ is the effective mass for small wave-vector k.
(c) Compare the first twelve energy levels calculated with the effective-mass approximation with those allowed by the tight-binding band for $A = 2$ eV and $N = 10$.

3.3 Consider a flat band. The band structure is the tight-binding band structure $\mathcal{E}(k) = A - A\cos(ka)$.

(a) Verify that r^n is a solution of the Wannier recurrence equation and solve for r in terms of the energy E.
(b) Verify that we have: $|r| = 1$ for $0 \leq E \leq 2A$.

3.4 Consider an heterojunction formed by two tight-binding bands (1 and 2). The junction is assumed to be located between the lattice sites n and $n+1$. There is no electrostatic potential applied (flat band). On the left-hand side of the junction the band structure is $A_1 - A_1\cos(k_1a)$ and on the right-hand side of the junction the band structure is $A_2 - \Delta E - A_2\cos(k_2a)$.
We shall study the reflection and transmission of a wave of energy E incident on the left-hand side of the heterojunction.

$$r_1(E)^m = \exp[jk_1(E)ma] \quad \text{for } m \leq n.$$

The reflected wave is

$$b \times r_1(E)^{-m} = b \times \exp[-jk_1(E)ma] \quad \text{for } m \leq n.$$

The transmitted wave is

$$c \times r_2(E)^m = c \times \exp[jk_2(E)ma] \quad \text{for } m \geq n+1.$$

b and c are unknown coefficients. Let us call f_1 the total wave on the left-hand side

$$f_1(m) = r_1^m + b\,r_1^{-m}$$

Let us call f_2 the total wave on the right-hand side

$$f_2(m) = c\,r_2^m.$$

(a) Wave-function matching at the heterojunction: Verify that the matching at the heterojunction can be expressed by the following set of equations:

$$f_1(n) = \lambda_1 f_2(n), \tag{3.46}$$
$$f_1(n+1) = \lambda_2 f_2(n+1), \tag{3.47}$$

with

$$\lambda_1 = \frac{A_2}{A_{12}^*},$$
$$\lambda_2 = \frac{A_{12}}{A_1}, \tag{3.48}$$

where A_{12} is the overlap Hamiltonian element that appears in the generalized Wannier equations at the junction:

$$\left(-\frac{A_1}{2}\right) f_1(n-1) + (A_1 - E) f_1(n) + \left(\frac{-A_{12}}{2}\right) f_2(n+1) = 0,$$

$$\left(-\frac{A_{12}^*}{2}\right) f_1(n) + (A_2 - \Delta E - E) f_2(n+1) + \left(\frac{-A_2}{2}\right) f_2(n+2) = 0.$$

Hint: First write the flat-band Wannier equations defining $f_1(n+1)$ and $f_2(n)$ across the heterojunction.

(b) Transparent heterojunctions: Note that A_{12} can be written as

$$|A_{12}|^2 = \lambda A_1 A_2,$$

where λ is defined as $\lambda = \lambda_2/\lambda_1$. We wish to establish the value of λ which results in a maximum transparency of the heterojunction (maximum transmission or equivalently minimum reflection).

(i) Using Equations (3.46) and (3.47) calculate the unknown coefficient b in terms of λ, r_1, r_2, and n.

(ii) Calculate $|b|^2$.

(iii) Verify that λ is a real number and calculate the value of λ for which we have

$$\frac{d|b|^2}{d\lambda} = 0.$$

Assume that A_1 and A_2 have the same sign.

(c) Current conservation: We know that the current is always conserved. This results from the hermiticity of the Hamiltonian. Let us verify that the conservation of current is indeed independent on the value of λ. Note: do not substitute the value for λ derived in the previous question.

(i) Calculate c using the expression obtained for b. Replace r_1 and r_2 by their exponential equivalents and set $n = 0$ for simplicity.

(ii) Calculate $|c|^2$.

(iii) Calculate the ratio

$$\frac{1 - |b|^2}{|c|^2}$$

and verify that it is equal to v_2/v_1, where v_1 and v_2 are the electron velocities in bands 1 and 2, respectively. Note that this ratio is independent of λ.

3.5 Consider the multi-band Hamiltonian of Equation (3.21).

(a) Following the procedure used for the one-band Wannier picture in Section 3.3.2 derive an equation for the conservation of particle and current for an eigenstate solution of the multi-band Hamiltonian of Equation (3.21).

(b) The multi-band electron current $J_T(n)$ through the heterostructures is defined by Equations (3.22) and (3.23). Verify that the hermiticity of the Hamiltonian and the current conservation equation derived above lead to the conservation of this current through the heterostructure.

3.6 Derive Equations (3.27)–(3.30) and verify the expression given for Δ in Equation (3.31).

3.7 In this problem we derive two important properties of the impulse response given by Equations (3.42) and (3.42). For simplicity we shall limit the derivation to a single Wannier.

(a) Consider the following equation W satisfied by the impulse $h(n, i, E)$

$$W = H_{nn-1}h(n-1, i, E) + H_{nn}h(n, i, E) + H_{nn+1}h(n+1, i, E) + \delta_{ni}$$
$$= Eh(n, i, E).$$

Using the hermiticity of the Hamiltonian demonstrate that the impulse response verifies the following current-conservation equation:

$$-\frac{a}{\hbar}\, 2\,\mathrm{Im}[h(i, i, E)] = v_L\, |h(n_L, i, E)|^2 + v_R\, |h(n_R, i, E)|^2,$$

where v_L and v_R are the electron velocities in the left- and right-hand flat-band contacts located at N_L and N_R respectively. Hint: Evaluate $Wh^*(n, i) - W^*h(n, i)$.

(b) Verify that the envelope function $f(n, E)$ solution of

$$H_{nn-1}f(n-1, E) + H_{nn}f(n, E) + H_{nn+1}f(n+1, E) = Ef(n, E)$$

can be written as (using $i = 0$)

$$f(n, E) = j\frac{\hbar v(k_x)}{a}\, h(n, 0, E) = h'(n, 0, E) \qquad \text{for} \quad n \geq N_L = 0.$$

Hint: Note that the envelope equation for $f(n, E)$ and the impulse equation for $h(n, 0, E)$ are the same everywhere except at $n = 0$. Since these difference equations are of the second order, they will admit the same solution for $n \geq 1$ if we have $f(0, E) = h'(0, 0, E)$ and $f(1, E) = h'(1, 0, E)$.

4 Quantum heterostructure devices

I cannot do it without comp[u]ters.

The Winter's Tale IV, William Shakespeare

4.1 Introduction

New devices can now be realized with thin crystalline epitaxial layers of different semiconductors. These epitaxial layers can be as thin as a few lattice parameters; where this occurs, quantum effects become dominant. In the previous chapter we developed a quantum formalism, the generalized Wannier picture, for the analysis of quantum heterostructures. In particular, this formalism was shown to account for both the periodicity in **k** space of the band structure and its spatial variation. Armed with these tools we shall now study a variety of quantum devices, literally taking the electrons through different aerobic exercises. We will start with the fundamental problem of an electron in a band which is accelerated by a uniform electric field. Both stationary and time-dependent states will be discussed. Next, we will study the confinement of electrons in quantum wells and the formation of a two-dimensional electron gas (2DEG). We will then place a quantum well between two barriers and study the resonant tunneling of electrons through this system. Finally, we will study the diffraction of electrons in periodic or aperiodic structures called superlattices.

Before starting we must mention that the observation of quantum effects in devices requires that the electron wave-function (here the Wannier envelope) interacts coherently within the device heterostructure. This is possible if the electron's mean free path is large compared with the main features of the device heterostructures. Usually this criterion is met for structures smaller than 200 Å. An in-depth study of the impact of scattering upon the electron wave-function will be given in Chapter 6.

4.2 The accelerated band electron

We shall now consider the problem of an electron in a uniform band accelerated by an electric field $-F_0$. The electrostatic potential energy is given by

$$V(x) = -(-F_0)x = F_0 x.$$

For simplicity we assume that the electron is in a band with a tridiagonal band structure given by $\mathcal{E}(k) = A - A\cos(ka)$. The resulting band diagram for this system is given in Figure 4.1(a). The Wannier recurrence equation for this system is simply

$$-\frac{A}{2} f(n-1,t) + (A - qaF_0 n) f(n,t) - \frac{A}{2} f(n+1,t) = i\hbar \frac{df(n,t)}{dt}.$$

We shall study the various solutions of this Wannier recurrence equation which describes this crystal electron accelerated by the external field F_0.

4.2.1 Stark states and the Wannier ladder

Eigenstate solutions

First we study the eigenstate solutions. The Wannier recurrence equation for this system reduces to

$$-\frac{A}{2} f(n-1,E) + (A - qaF_0 n) f(n,E) - \frac{A}{2} f(n+1,E) = E\, f(n,E).$$

We can rewrite it as

$$f(n-1,E) + f(n+1,E) = \frac{2qaF_0}{A}\left(\frac{A-E}{qF_0 a} - n\right) f(n,E)$$

$$f(n-1,E) + f(n+1,E) = \frac{2}{X}(\gamma - n) f(n,E),$$

with

$$X = \frac{A}{qaF_0},$$

$$\gamma(E) = \frac{A-E}{qaF_0} = X - \frac{E}{qaF_0}.$$

As shown in Figure 4.1(a), $2X$ is the width of the tilted band. The preceding difference equation for $f(n,E)$ is recognized as the recurrence equation of the Bessel functions. This second order difference equation admits two independent solutions given by

$J_{\gamma-n}(X)$ and $(-1)^n J_{n-\gamma}(X)$ for γ non-integer,
$J_{\gamma-n}(X)$ and $N_{\gamma-n}(X)$ for γ integer.

$J_\nu(X)$ are the Bessel functions and $N_\nu(X)$ the Newman functions. Figure 4.1(b) shows a plot of the Bessel function $J_\nu(X)$ for $X = 10$ as a continuous function of

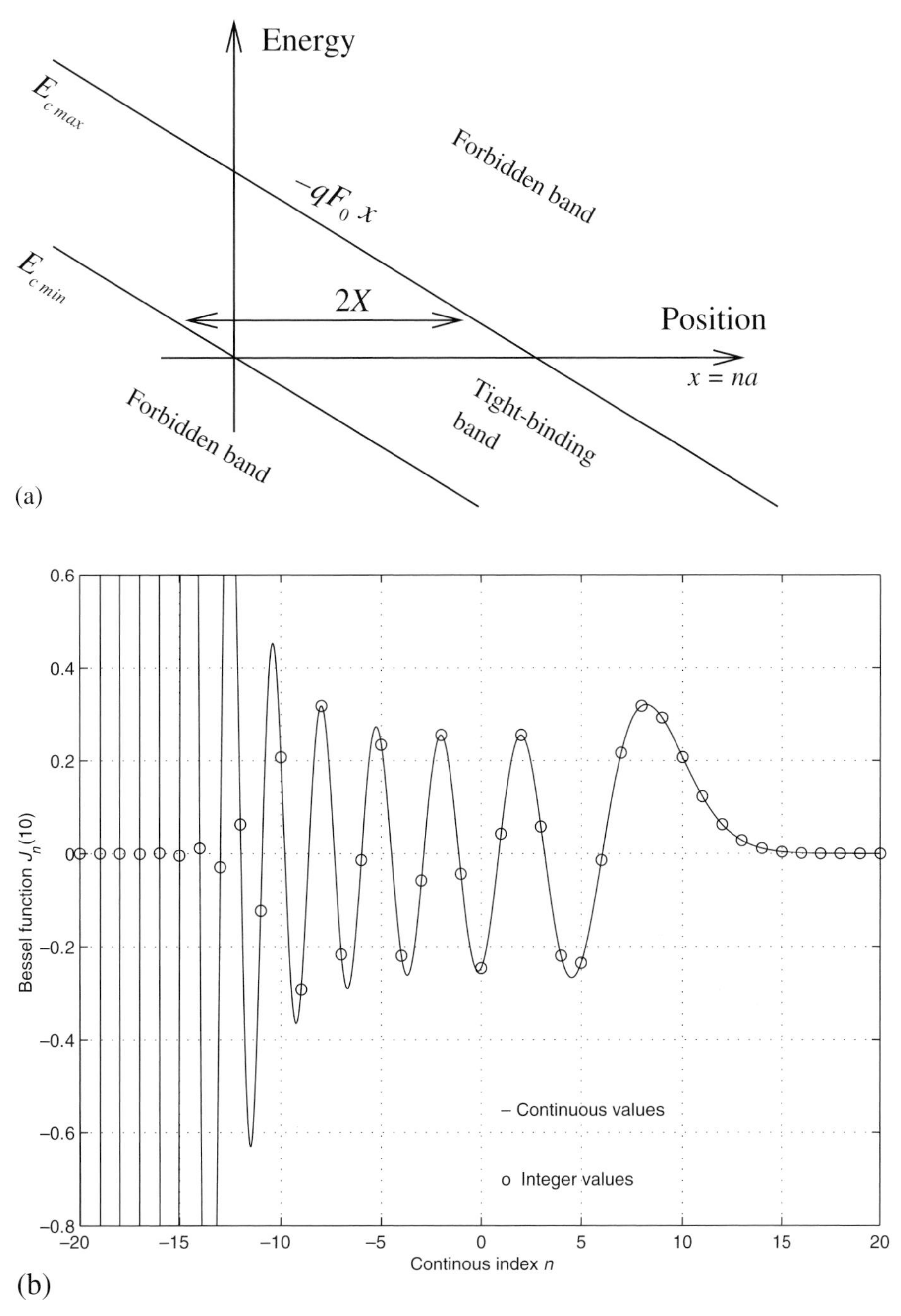

Fig. 4.1. (a) Band diagram of a tridiagonal band tilted by a uniform electric field. (b) Bessel Function $J_\nu(10)$ for continuous values of the index ν and argument X equal to 10.

its real index ν. Notice that for large negative values of ν the Bessel function only converges to zero for integer values of ν. On the other hand the Neuman function is known to diverge for both large positive and negative integer values of ν.

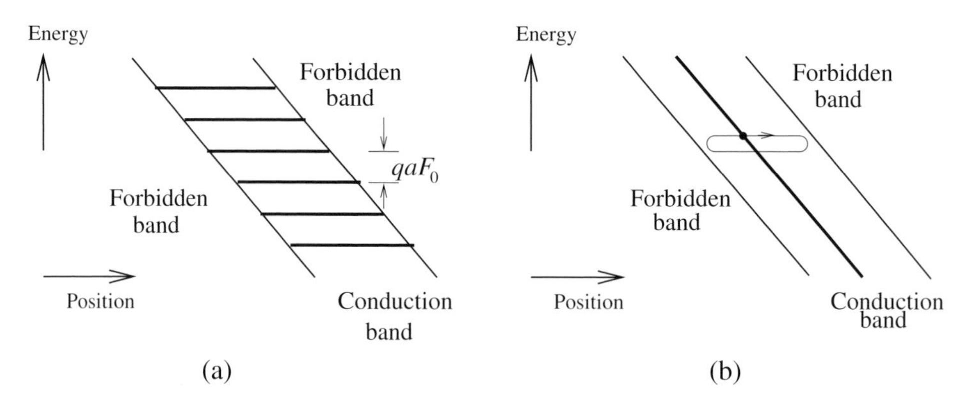

Fig. 4.2. (a) Symbolic representation of the Stark states forming the so-called Wannier ladder. (b) Symbolic representation of a Houston state. The Houston state is represented by a diagonal line moving back and forth in the band structure.

The only physical eigenstate solutions of the Wannier recurrence equation are those satisfying the boundary conditions $f(n = \pm\infty) = 0$ for which the electron wavefunction vanishes in the forbidden regions. These physical boundary conditions are only satisfied when γ is an integer p. Let us call p the integer value of γ. We can then write

$$p = \gamma = \frac{A - E_p}{qaF_0}.$$

The electron energy is then quantized

$$E_p = A - p \times qaF_0.$$

These quantized energy levels form the Wannier ladder. A symbolic representation of this is presented in Figure 4.2.

The eigenstate Wannier envelope functions are then

$$f_p(n, t) = J_{p-n}(X) \exp\left(-i\frac{E_p t}{\hbar}\right).$$

Such a state is called a Stark state. Each Stark state is centered on a lattice site. Note that the Bessel function satisfies the property

$$\sum_{n=-\infty}^{\infty} J_n^2(X) = 1, \tag{4.1}$$

so that the Stark state given above is already normalized.

The Wannier ladder has never been experimentally observed in bulk crystals, at least not without controversy [23]. Phonon scattering, which limits the Stark state lifetime, is thought to destroy the Wannier ladder by broadening (smearing) the energy levels. According to the uncertainty principle the energy width ΔE is given approximately by $\hbar/\Delta t$, where Δt is the electron lifetime. For the Wannier ladder to exist the energy width ΔE must be much larger than the energy broadening associated with scattering. Most of the experimental searches for the Wannier ladder have been performed for large electric fields $F_0 > 100$ kV/cm when the energy levels qF_0a are spaced far apart. For such large fields, competing processes such as interband tunneling can then prevent the observation of Wanner ladders. There has been a revival of interest in the Wannier ladder. Indeed as we shall see in Section 4.5, smaller bands of energy with band structures similar to the tight-binding model used in our analysis can be created in a superlattice. The existence of the Wannier ladder in a superlattice band tilted by a uniform electric field has been confirmed by optical experiments [22]. There are, however, no electron devices which make use of it as of yet. However, a device called the Zener oscillator has been proposed in [27], and its ideal DC quantum states are described below.

Zener resonant tunneling

In Figure 4.3(a) we show a heterostructure device which has been proposed for feeding a Zener resonator. We limit our discussion here to the static case in which the applied electric field is not time-varying.

The wave-function on the left-hand side is simply a plane wave:

$$f(n) = \exp[ik(E)na] + b(E)\exp[-ik(E)na],$$

where $k(E)$ is the electron wave-vector and E the electron energy. The solution of right-hand side which converges properly for all values of $\gamma(E)$ is

$$f(n) = c(E)(-1)^n J_{n-\gamma(E)}(X).$$

The matching of the left- and right-hand sides at $n = 0$ and $n = 1$ leads to

$$1 + b(E) = c(E)J_{-\gamma(E)}(X),$$

$$\exp[ik(E)a] + b(E)\exp[-ik(E)a] = -c(E)J_{1-\gamma(E)}(X),$$

giving the reflection coefficient

$$b(E) = -\frac{\dfrac{J_{1-\gamma(E)}(X)}{J_{-\gamma(E)}(X)} + \exp(ika)}{\dfrac{J_{1-\gamma(E)}(X)}{J_{-\gamma(E)}(X)} + \exp(-ika)}$$

Since no scattering is allowed for in this DC model we have $|b(E)| = 1$.

Of interest is the amplitude $H(E)$ of the standing wave inside the tilted band. From the normalization property of the Bessel function (see Equation (4.1)), the amplitude of the standing wave $J_n(X)$ is approximately $1/(2X)^{1/2}$ since the standing wave extends only over $2X$ lattice sites. It follows that the amplitude $H(E)$ of the standing wave inside the tilted band is given approximately by

$$H(E) = \frac{|c(E)|}{(2X)^{1/2}} = \frac{1}{(2X)^{1/2}} \left| \frac{1 + b(E)}{J_{-\gamma(E)}(X)} \right|.$$

The amplitude squared $[H(E)]^2$ is plotted in Figure 4.3(b) for the case of $\Delta E =$ 0.05 eV, $X = 200$, and $A = 1$ eV. This allows $\Delta E/(qaF_0) = 10$ resonant peaks. The higher the energy, the smaller the effective barrier length and the broader and lower the peaks. In addition, a shift of the resonant peak from the initial Wannier ladder is observed. This process can be referred to as Zener resonant tunneling. Such a quantum heterostructure subjected to a time-varying electric field has been proposed to induce time-varying Zener oscillations. A discussion of time-varying Zener oscillations is given next.

4.2.2 Time-dependent solutions and the Houston state

Let us come back to the time-dependent Wannier recurrence equation of an electron accelerated by an electric field F_0

$$-\frac{A}{2} f(n-1,t) + (A - qaF_0 n) f(n,t) - \frac{A}{2} f(n+1,t) = i\hbar \frac{d}{dt} f(n,t). \tag{4.2}$$

This equation admits a time-dependent solution of the following form

$$f_{k(0)}(n,t) = \left(\frac{2\pi}{a}\right)^{1/2} \exp\left[ik(t)na\right] \exp\left(-i\frac{At}{\hbar}\right) \times \exp\left\{\frac{iA}{qaF_0}\left[\sin k(t)a - \sin k(0)a\right]\right\},$$

with $k(t)$ given by

$$k(t) = k(0) + qF_0 t/\hbar.$$

We can rewrite this function in a more general form which holds for any type of band structure (tridiagonal or not)

$$f_{k(0)}(n,t) = \left(\frac{2\pi}{a}\right) \exp[ik(t)na] \exp\left[-\frac{i}{qF_0}\int_{k(0)}^{k(t)} \mathcal{E}(k)\, dk\right].$$

This time-dependent solution is called a Houston state. A Houston state is simply a phase-modulated Bloch state whose wave-vector $k(t)$ is drifting in reciprocal space. The Houston state is represented symbolically in Figure 4.2(b) by a line since like a Bloch function the Houston function extends over the entire crystal.

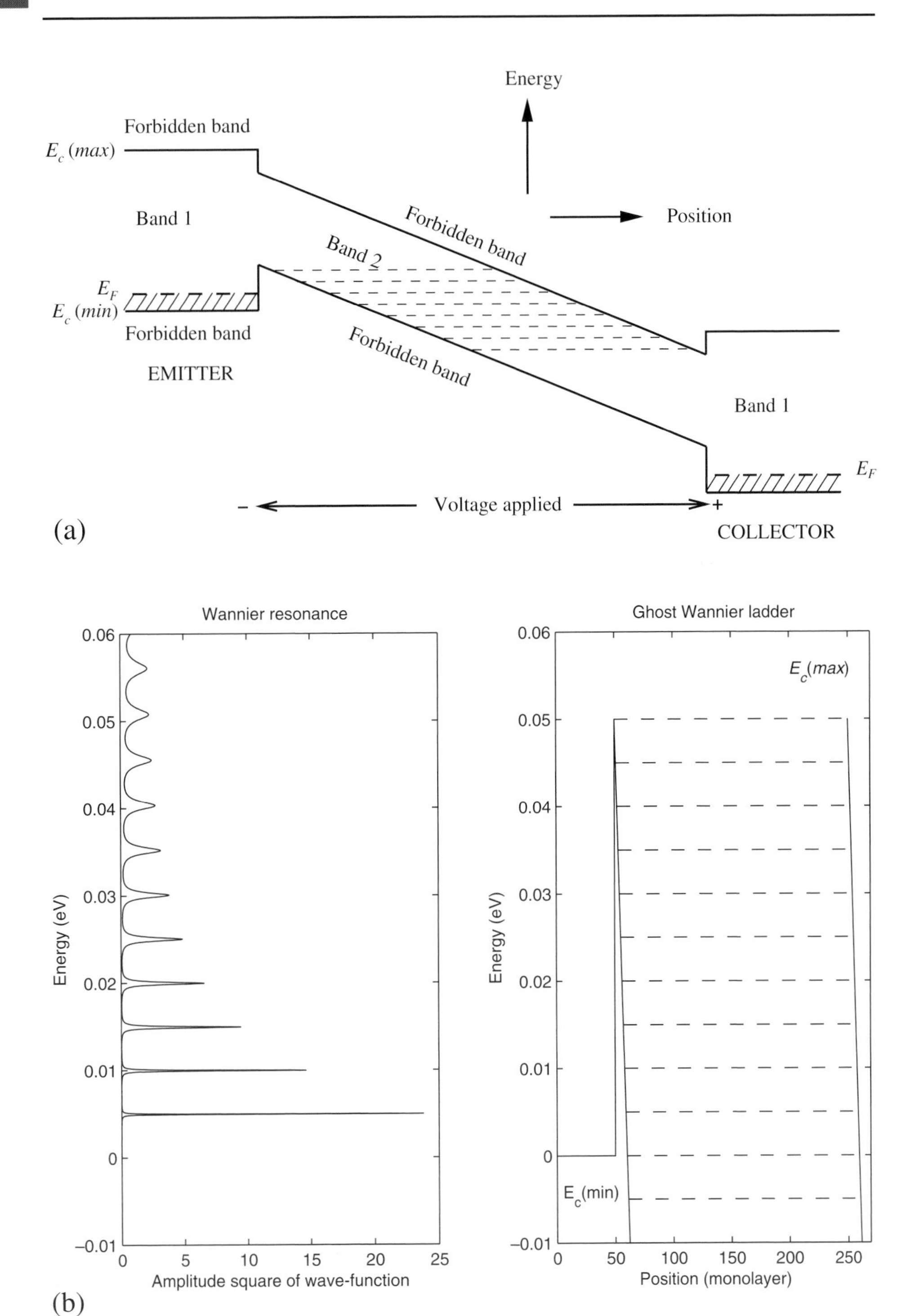

Fig. 4.3. (a) Band-diagram of a Zener resonator and its feed (b) Spectrum and calculated band diagram of the electron standing waves inside the confines of the tilted band near the emitter.

Note that the time evolution of the quasi-momentum $\hbar k(t)$ obeys Newton's law for a classical particle accelerated by the electric field $-F_0$:

$$\frac{d}{dt}\hbar k(t) = qF_0.$$

This result is referred to as the *acceleration theorem* in a crystal. Note, however, that the quasi-momentum assumes a different meaning in a crystal compared to a free electron system. Indeed, the quasi-momentum $\hbar k(t)$ is not the expected value of the momentum operator in the crystal. The quasi-momentum $\hbar k(t)$ is not usually the product of a velocity times a mass except in the vicinity of an extremum of the band structure, where we indeed have $\hbar k(t) = m^* v(t)$. Elsewhere, the concept of mass has no practical meaning, and we must resort to another approach to calculate the electron velocity.

Using the definition of the current given in Section 3.3.2 one can easily verify that the velocity of the electron in a Houston state is the same as that of an electron in the Bloch state $|k(t)\rangle$ at a time t:

$$v(t) = \frac{1}{\hbar}\frac{\partial}{\partial k}\mathcal{E}(k(t)).$$

Let us assume that the electron is located at the position $x(t)$ in the lattice. By integrating the equation

$$v(t) = \frac{dx(t)}{dt}$$

we obtain that the electron position is given by

$$x(t) = x(0) + \frac{1}{qF_0}\left[\mathcal{E}(k(t)) - \mathcal{E}(k(0))\right].$$

As a result of the periodicity of the band structure in k space, the electron oscillates back and forth in the tilted band. However, the actual position $x(0)$ of the electron is unknown. This oscillation is called a Bloch oscillation.

4.2.3 The Bloch oscillator

In bulk semiconductors the electron does not remain ballistic long enough for this oscillation to be observed. The ballistic trajectory of the electron is interrupted by scattering on the vibrations (phonons) or impurities and defects of the lattice. Esaki and Tsu have proposed a superlattice device, the Bloch oscillator, which provides a negative differential resistance if the electrons complete a large enough portion of the Bloch orbit in the miniband of a superlattice [20]. The electron velocity reduces when the electrons reach the inflection point of the miniband structure ($k = \pi/2a$) and is even inverted if they reach the top of the miniband structure. Using a classical analysis, Esaki and Tsu estimated the average electron velocity v_d for a scattering time τ to be

$$v_d = \frac{\pi\hbar}{m^*_{SL} a_{SL}} \frac{\zeta^2}{1+\pi^2\zeta^2},$$

where m^*_{SL} and a_{SL} are the superlattice mass and lattice parameters and where $\zeta = qF_0 a_{SL}\tau/\pi\hbar$ is a normalized field. We assume a tight-binding band structure for the superlattice. The velocity–field relation v_d/F_0 reaches a maximum for $\pi\zeta = 1$. For larger electric fields a negative differential mobility is therefore expected. This in turn induces a negative differential resistance in the I–V characteristics of the superlattice. The condition for negative differential resistance is easier to achieve than the requirement for ballistic Bloch oscillations. The observation of negative differential resistance in the Bloch oscillator has been reported by Sibille *et al.* [29]. More work is necessary, however, to improve the performance of this device. The reader is referred to [52] for further theoretical discussion.

4.2.4 Coherent and squeezed Zener oscillations

In a Houston state, the electron wave-function extends over the entire crystal. Due to the presence of phonon scattering and impurity scattering, it is more realistic to picture the electron in a wave-packet of Houston states. Houston states of different initial wave-vectors $f_{k(0)}(n,t)$ form a complete basis in the one-band approximation as do also the Stark states $f_p(n,t)$. Furthermore, both Houston states and Stark states are orthonormal

$$\sum_{n=-\infty}^{\infty} f_{k(0)}(n,t) f^*_{k'(0)}(n,t) = \delta(k(0) - k'(0)),$$

$$\sum_{n=-\infty}^{\infty} f_p(n,t) f^*_{p'}(n,t) = \delta_{p\,p'}.$$

Any electron state $f(n,t)$ solution of Equation (4.2) can therefore be written in terms of a wave-packet s_p of Stark states $f_p(n,t)$ or a wave-packet $h[k(0)]$ of Houston states $f_{k(0)}(n,t)$

$$\begin{aligned} f(n,t) &= \sum_{p=-\infty}^{\infty} s_p\, f_p(n,t) \\ &= \int_{-\frac{\pi}{a}}^{\frac{\pi}{a}} h(k(0))\, f_{k(0)}\, dk(0). \end{aligned}$$

Wave-packets of Houston states and Wannier states have been studied in detail in [26] (see also [2] for the case in which no electric field is applied). A wave-packet of Houston states permits us to create a classical-like electron with an effective 'size' (range of the electron distribution) calculated using

$$\Delta n(t) = \left(\langle x^2\rangle - \langle x\rangle^2\right)^{\frac{1}{2}}$$

with

$$\langle x^2\rangle = \sum_{n=-\infty}^{\infty} |f(n,t)|^2\, n^2$$

$$\langle x\rangle = \sum_{n=-\infty}^{\infty} |f(n,t)|^2\, n.$$

The effective size $\Delta n(t)$ is a periodic function of time for the ballistic electron. Therefore, unlike the plane wave of an electron in free space, the wave-function of the ballistic electron in a crystal band does not spread forever since $\Delta n(t)$ is periodic in time. It is shown in [26] that there exist special wave-packets for which the electron 'size' (a) is a fraction of the Zener orbit and (b) remains nearly constant in time. Such states for which both position and velocity are known may be called coherent states. They bear similarities to the Glauber states [1,2] (to be discussed in Chapter 5), which describe the states of the light generated by a laser for which both the amplitude and phase of the light are known. Note, however, that Glauber states are derived as the states realizing the minimum uncertainty product possible in the uncertainty principle governing the phase and amplitude uncertainties. For accelerated electrons it can be verified [26] that the equality in the uncertainty principle $\Delta k \Delta x = 1$ is only achieved by the Houston state $\Delta k = 0$.

It is also shown in [26] that there exist special wave-packets for the accelerated electron for which the electron wave-function extends over only a few lattice parameters for a short duration. This kind of state is called a squeezed state. The period over which the electron wave-functions can maintain a tight (minimum) spatial extension is short and usually compares with the lifetime of a Houston function.

So far our quantum analysis of an electron accelerated by an electric field has not included the scattering in three-dimensional of this electron by vibrations or defects. A realistic and definitive theory of quantum transport requires the inclusion of these scattering processes. We postpone such a realistic analysis to Chapter 6 and consider below other important quantum systems.

4.3 Quantum wells

A quantum well, in the general use of this term, is a potential structure which spatially confines the electron. The quantum wells grown by MBE most readily provide a one-dimensional confinement along the superlattice axis. According to quantum

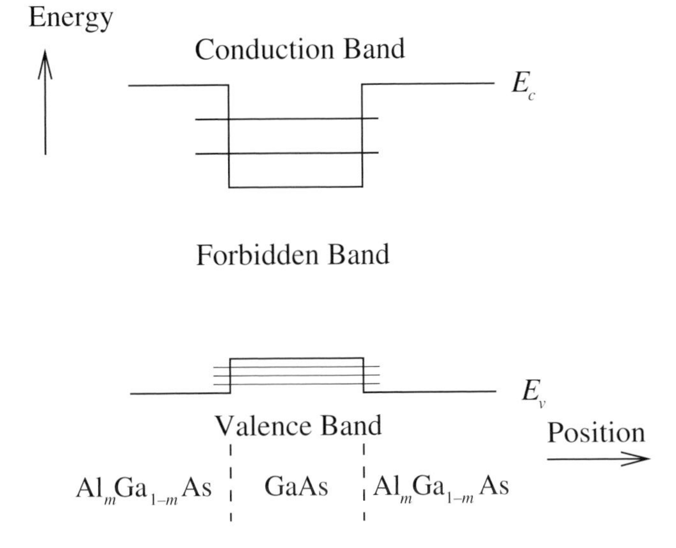

Fig. 4.4. AlGaAs/GaAs/AlGaAs quantum well.

mechanics, an electron subjected to potential confinement has its energy quantized and a discrete energy spectrum would be expected for the electron system. However, the electron remains free to move in the perpendicular direction. As we shall see this results in the creation of a two-dimensional electron gas.

4.3.1 Rectangular quantum wells

The case of a potential well with infinite walls was studied in detail in Problem 3.2 for a tight-binding band structure. In practice such a quantum well can be realized at the bottom of the conduction band and at the top of the valence band by using a narrow-bandwidth semiconductor sandwiched between two wide-band semiconductors.

An example is the AlGaAs/GaAs/AlGaAs structure shown in Figure 4.4. The bandgap of $\mathrm{Al}_m\mathrm{Ga4}_{1-m}\mathrm{As}$ for an Al mole fraction m is given by

$$E_g = 1.424 + 1.247m(\mathrm{eV}).$$

Note that only 68% of the variation of the bandgap ΔE_g which results between AlGaAs and GaAs appears as a shift of the conduction band edge ΔE_c

$$\Delta E_c = 0.68 \Delta E_g.$$

A quantum well is present in both the conduction and valence bands. In the valence band, holes instead of electrons are now trapped in the quantum well. The quantum well confinement enhances the radiative recombination of electron–hole pairs between the conduction and valence quantum wells in two ways. It prevents the diffusion of

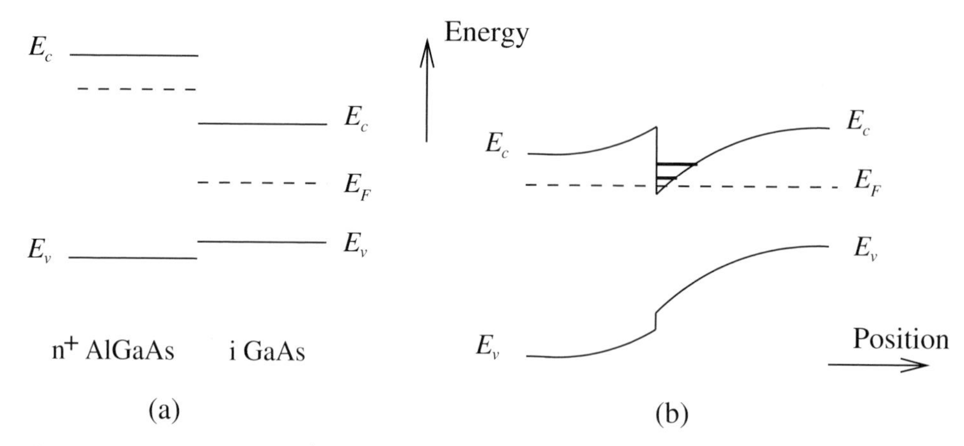

Fig. 4.5. Band diagram of an n^+-AlGaAs/i-GaAs heterojunction (a) before equilibrium and (b) at equilibrium. A triangular quantum well is created in the conduction band at the interface.

carriers and enhances the concentration of electron–hole pairs. It provides a dielectric wave-guide which confines the light emitted in the plane of the quantum well and in turn enhances stimulated emission. As a result, the threshold current for stimulated emission is reduced by several orders of magnitude and the lifetime of the device is increased. The use of a quantum well was therefore an important step in the development of usable semiconductor lasers.

4.3.2 Quantum well induced by an electric field

Quantum wells can also be realized by the confines of the built-in electrostatic barrier and the band discontinuity at the interface of a heterojunction. Consider the example shown in Figure 4.5(a) of a heterojunction formed by growing a strongly doped n^+-AlGaAs layer on top of an intrinsic i-GaAs layer. Due to the large difference in chemical potential some of the electrons transfer from the n^+-AlGaAs layer to the i-GaAs layer. This creates an electrostatic potential which balances the chemical potential until the Fermi levels (electrochemical potential) are aligned at equilibrium (Figure 4.5(b)).

The fabrication of such a structure is made possible thanks to the rapid variation of the doping concentration which can be realized in MBE growth. This technique, referred to as modulation doping, was first demonstrated for the n^+-AlGaAs/i-GaAS heterostructure by Dingle *et al.* [30].

A large built-in electric field F_0 results at the interface. The potential at the interface can be approximated by that of a triangular well:

$$E_p(x) = \begin{cases} qF_0x & \text{for} \quad x > 0 \\ \infty & \text{for} \quad x \leq 0 \, . \end{cases}$$

According to our discussion of the accelerated electron, the eigenstate inside the triangular well ($n \geq 0$) is a Bessel function $J_{\gamma(E)+n}(X)$, with E the eigenvalue. The eigenvalue E is obtained by imposing the boundary condition of an infinite wall at $n = 0$:

$$J_{\gamma(E)}(X) = 0.$$

In the effective-mass limit (vanishing lattice parameter a) X is large, and the zeros of the Bessel function for small energy E reduce to the zeros of the Airy function, which are approximately given by

$$z = -\left[\frac{3\pi}{2}\left(p + \frac{3}{4}\right)\right]^{2/3},$$

with

$$z = (\gamma(E) - X)\left(\frac{1}{X}\right)^{1/3}.$$

The energies E_p referenced to the bottom of the well are then found to be given for small indices p by

$$E_p = \left(\frac{\hbar^2}{2m^*}\right)^{\frac{1}{3}}\left[\left(\frac{3}{2}\pi q F_0\right)\left(p + \frac{3}{4}\right)\right]^{\frac{2}{3}},$$

where m^* is the electron effective mass. In Chapter 8 we will make use of the triangular quantum well to analyze the MODFET capacitor.

4.3.3 Quantum wells of arbitrary shapes

Note that calculation of the quantized energy levels in a quantum well of arbitrary shape is the solution of an eigenvalue problem. The eigenvalue problem can be greatly simplified if we can introduce infinite walls as boundary conditions. It is usually possible to do this far from the quantum well itself (after all the sample always has a finite size). In that case the eigenvalue problem reduces to that of the calculation of the eigenvalues of the Hamiltonian matrix $[H]$

$$[H][f] = E[f],$$

where the matrix elements of $[H]$ are $H_{nm} = \langle n|H|m\rangle$.

Note that for a quantum structure N_{SL} lattice parameters long, the matrix H is an $N_{SL} \times N_{SL}$ matrix. The calculation of the eigenvalues for large matrices requires special numerical techniques. In the case of a tight-binding Hamiltonian, the Hamiltonian matrix is a tridiagonal matrix, and very powerful techniques are available to calculate the eigenvalues [38] of large tridiagonal matrices.

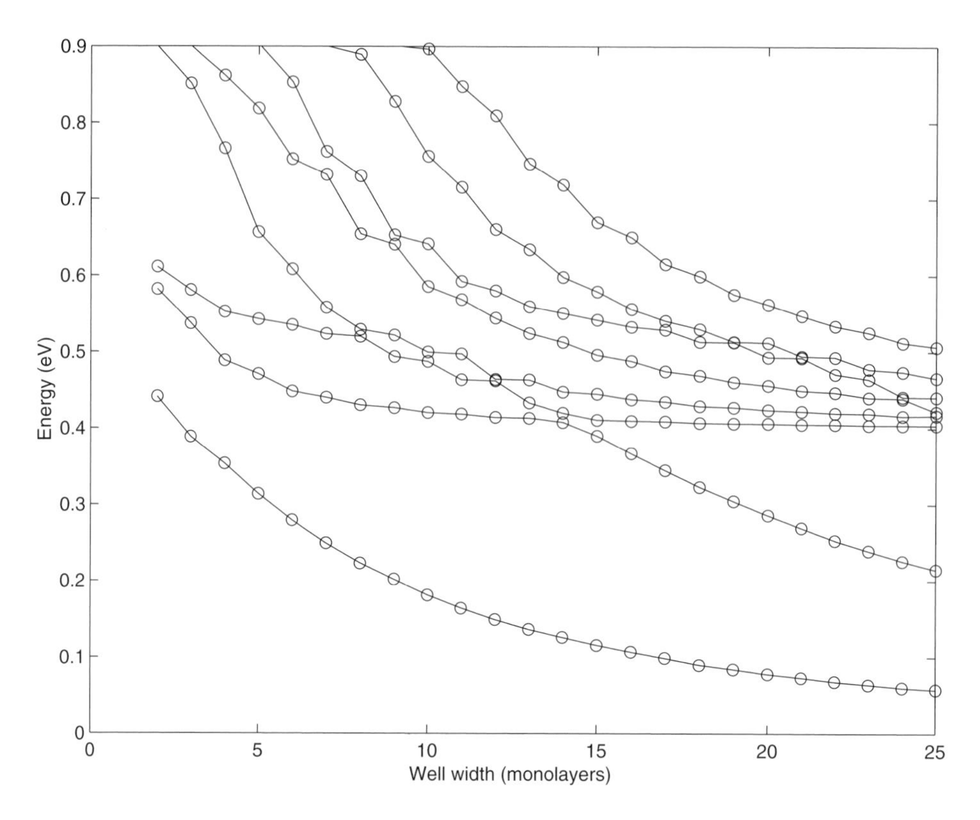

Fig. 4.6. Locus of the energies of a 0.5 eV GaAs well as a function of well width.

4.3.4 Full-band structure effects

To visualize full-band structure effects, consider now a quantum well 0.5 eV deep but with the same GaAs band structure in the well and the barriers. The locus of the energies obtained for such a GaAs well is shown in Figure 4.6 as a function of the well width. The perpendicular wave-vector of the electron is zero. For high energy, one can see that the quantum well energies associated with the Γ valley interact with the quantum well energies of the X valley. These interactions are revealed by the mutual repulsion of the energy locus at their expected crossings.

4.3.5 2DEG

So far we have only discussed the extension of the electron wave-function along the superlattice direction. We will now account for the three-dimensional nature of the electron and the lattice.

We have seen in Chapter 3 that we can assume the electron to be in a Bloch state $|\mathbf{k}_\perp\rangle$ in the direction perpendicular to the superlattice axis x. Let us consider a three-dimensional band structure model consisting of the tight-binding model $A - A\cos(k_x a)$ along the superlattice direction x and the effective-mass band structure perpendicular to the superlattice. The total energy of the electron in a flat band is therefore given by the crystal band structure $\mathcal{E}(\mathbf{k})$:

$$\mathcal{E}(\mathbf{k}) = \frac{\hbar^2}{m^* a^2}[1 - \cos(k_x a)] + \frac{\hbar^2}{2m^*}\left(k_y^2 + k_z^2\right)$$
$$= \frac{\hbar^2}{m^* a^2}[1 - \cos(k_x a)] + \frac{\hbar^2}{2m^*}\left(k_\perp^2\right),$$

using $k_\perp^2 = k_y^2 + k_z^2$. Note that the Fourier coefficient A of the band structure can be selected here so that we have for small k_x values an isotropic effective-mass band structure:

$$\mathcal{E}(\mathbf{k}) \simeq \frac{\hbar^2}{2m^*}\left(k_x^2 + k_y^2 + k_z^2\right) \quad \text{for small } k_x.$$

Let us first consider the case in which the superlattice in the x direction is a quantum well of arbitrary shape. The electrons are confined in the quantum well along the x direction but remain free to move perpendicular to the superlattice. The three-dimensional electron gas has been virtually reduced to a 2DEG. Inside the quantum well, the longitudinal one-dimensional component $\mathcal{E}(k_x)$ of the total three-dimensional electron energy E is quantized and we label it E_p. The total electron energy E is therefore

$$E = E_p + E(k_\perp) = E_p + \frac{\hbar^2}{2m^*}(k_\perp)^2.$$

We see that each energy level E_p is associated with a two-dimensional energy subband. The superlattice therefore has transformed the three-dimensional band structure into as many two-dimensional subbands as there are quantized energy levels E_p. This is represented graphically in Figure 4.7(a).

We now wish to calculate the electron concentration (population) of these subbands when the 2DEG system in the quantum well is in thermal equilibrium.

Density of states in $\mathbf{k}_\perp$ space

First, let us calculate the density of states in the $\mathbf{k}_\perp$ space. The density of states $N(k_y)$ along the y direction is defined as

$$N(k_y) = \frac{\text{number of electron states in the interval } [k_y, k_y + dk_y]}{dk_y}.$$

This is usually calculated using the periodic boundary condition. The periodic boundary condition is a mathematical device which permits us to switch from the continuous k_y wave-vector of the infinite ($-\infty \leq y \leq \infty$) crystal to a discrete set of k_y wave-vectors for a finite crystal of length L_y with $0 \leq y \leq L_y = N_y a$.

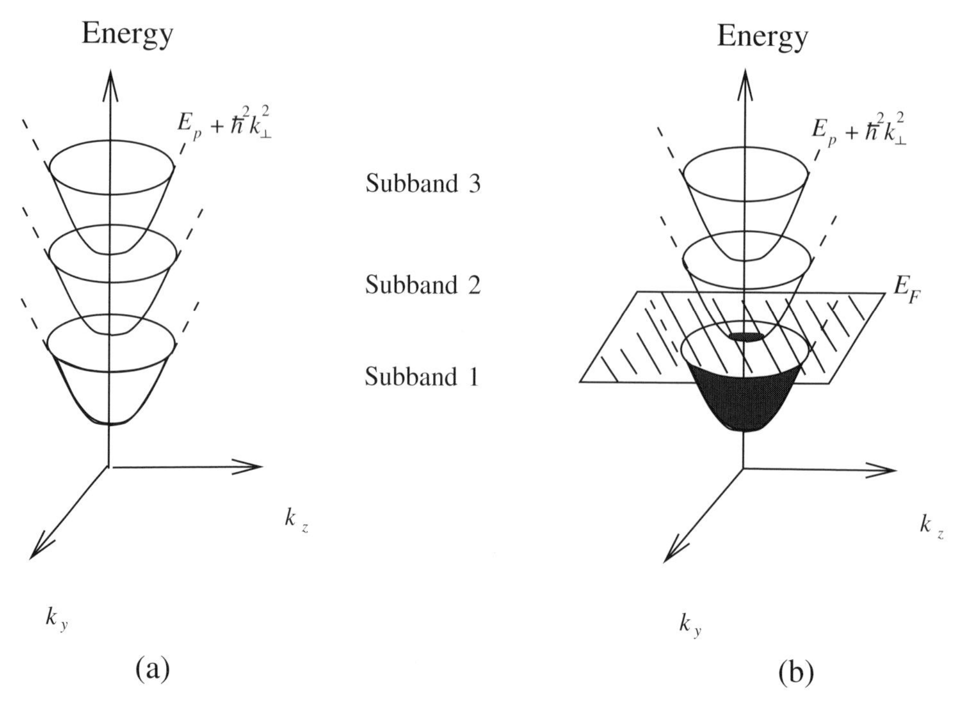

Fig. 4.7. (a) Two-dimensional subband associated to each energy level E_p. (b) Population of the subbands at $T = 0$ K for a Fermi energy E_F inside the conduction band.

A sample of length $L_y = N_y a$ is made of N_y atomic layers. Let us assume that it can be bent into a circle with a circumference L_y. The electron wave-function is now periodic. In the Wannier picture the envelope function of a Bloch wave k_y satisfies

$$f(k_y, n) = f(k_y, n + N_y),$$
$$\exp(ik_y na) = \exp[ik_y(n + N_y)a]. \tag{4.3}$$

Therefore the wave-vector k_y is quantized

$$k_{y,p}\, N_y a = 2p\pi \tag{4.4}$$
$$k_{y,p} = \frac{2p\pi}{N_y a} = \frac{2p\pi}{L_y}.$$

Since the wave-vector $k_{y,p}$ is limited to the Brillouin zone

$$-\frac{\pi}{a} \leq k_y \leq \frac{\pi}{a},$$

the integer p is limited to the range

$$\left.\begin{aligned} -\frac{L_y}{2a} &\leq p \leq \frac{L_y}{2a}, \\ -\frac{N_y}{2} &\leq p \leq \frac{N_y}{2}. \end{aligned}\right\} \tag{4.5}$$

There are consequently N_y Bloch waves $k_{y,p}$. Indeed there are as many Bloch states $|k_y\rangle$ as there are lattice layers along the y axis.

Since the wave-vectors are uniformly distributed in the Brillouin zone, the density of states N_y is simply

$$N(k_y) = \frac{N_y}{2\pi/a} = \frac{L_y}{2\pi}.$$

Note that a quantum well of length L_y can be used instead of a periodic boundary condition to calculate the density of states. It is easily verified that the same density of states $N(k_y)$ results although the quantized stationary waves of the quantum well have slightly different distributions in **k** space than the quantized Bloch waves of the periodic boundary model.

The 2DEG density of states $N(\mathbf{k}_\perp)$ is now given by

$$N(\mathbf{k}_\perp) = 2\frac{L_y}{2\pi}\frac{L_z}{2\pi} = \frac{A}{2\pi^2},$$

where a factor 2 is used to account for the electron spin degeneracy and where $A = L_y L_z$ is the sample cross-section perpendicular to the superlattice axis.

Density of states in E space

Next we shall calculate the density of states in energy space. The density of states $N(E)$ at the energy E is defined as

$$N(E) = \frac{\text{number of electron states in the interval } [E, E + dE]}{dE}$$

$$N(E)\, dE = N(\mathbf{k}_\perp)\, dS(k_\perp),$$

where we have

$$E = E_p + \frac{\hbar^2 k_\perp^2}{2m^*},$$

$$S(k_\perp) = \pi k_\perp^2.$$

Using the following identities:

$$dE = \frac{\hbar^2}{m^*} k_\perp\, dk_\perp,$$

$$dS(k_\perp) = 2\pi k_\perp\, dk_\perp,$$

we obtain

$$N(E)\frac{\hbar^2}{m^*} k_\perp\, dk_\perp = \frac{A}{2\pi^2} 2\pi k_\perp\, dk_\perp.$$

After canceling identical terms the 2DEG density of states $N(E)$ is found to be

$$N(E) = \frac{m^*}{\hbar^2} \frac{A}{\pi}.$$

The density of states D per unit area, of the 2DEG,

$$D = \frac{N(E)}{A} = \frac{m^*}{\hbar^2 \pi},$$

is seen to be a constant.

Fermi–Dirac statistics in a 2DEG

Let us now evaluate the 2DEG population of a subband p in thermal equilibrium. The number of electrons per unit area $n_{S,p}$ populating the subband E_p is calculated using Fermi–Dirac statistics:

$$\begin{aligned} n_{S,p} &= \int_{E_p}^{\infty} D \frac{1}{\exp\left(\dfrac{E - E_F}{k_B T}\right) + 1} dE \\ &= D k_B T \ln\left[\exp\left(\frac{E_F - E_p}{k_B T}\right) + 1\right] \end{aligned}$$

where E_F is the Fermi level. The total electron concentration per unit area n_S of the 2DEG is then given by summing over all the subbands p:

$$\begin{aligned} n_S &= \sum_{p=1}^{\infty} n_{S,p} \\ &= D k_B T \sum_{p=1}^{\infty} \ln\left[\exp\left(\frac{E_F - E_p}{k_B T}\right) + 1\right]. \end{aligned}$$

This result permits us to calculate the population of a 2DEG system once the discrete energy levels E_p of the quantum well under consideration have been obtained by solving the Wannier recurrence equation. Note that since we have used Fermi–Dirac statistics, the n_S–E_F relation obtained is valid for both degenerate (metallic) and non-degenerate (diluted solution) populations of these energy levels. The population of the energy subbands at 0 K is represented graphically in Figure 4.7(b) for a Fermi energy E_F in the conduction band.

Note that for each subband, the distribution along the x (superlattice) axis of the 2DEG concentration $n_{S,p}$ is given by the normalized eigenvector $|f(E_p, n)|^2$ of the subband considered. The 2DEG charge distribution in turn affects the built-in potential seen by each electron, and the Schrödinger and the Poisson equations must therefore be solved self-consistently. A more complete discussion of the analysis of such a numerical problem is postponed to Chapter 8.

The formation of a 2DEG by modulation doping in a heterostructure has been used to create a new field-effect transistor (FET) often referred to as the Modulation Doped FET (MODFET). For obvious reasons, the MODFET is also referred to as a Two-dimensional Electron Gas FET (TEGFET) or a High Electron Mobility Transistor (HEMT). Indeed, in a MODFET the 2DEG is used as the channel of a field-effect transistor. The electrons in the 2DEG channel move in an intrinsic material and experience reduced scattering. The resulting high mobility in the channel has permitted impressive low-noise and high-speed performance. In Chapter 8 we discuss the control of this 2DEG channel by a Schottky-barrier contact.

4.4 Resonant tunneling

We shall now study a vertical device which relies on the wave properties of the electron for its operation. Consider the heterostructure shown in Figure 4.8(a). The conduction band-edge of this device forms two potential barriers. The region between the two barriers defines a virtual quantum well since the electrons can escape the well confinement by tunneling. We shall see that for electrons with an energy corresponding approximately to the virtual resonant energy level of the quantum well, the transmission coefficient is 1. That is, an electron with this resonant energy can cross the double barrier without being reflected. This resonance phenomenon is similar to that taking place in the optical Fabry–Perot resonator or in a microwave capacitively-coupled transmission-line resonator.

4.4.1 Double-barrier system

The double-barrier structure is readily studied using the Wannier recurrence equation. Let us assume that a Bloch wave of energy E is incident on the left-hand side of this quantum structure. An arbitrary amplitude of 1 is selected for the incident wave. No attempt is made to normalize the wave-function since this of no consequence for the calculated transmission and reflection coefficients we calculate below. A reflected wave of the same energy E and unknown amplitude $b(E)$ is expected. The resulting envelope function on the left-hand side of the quantum structure is then

$$f_1(n, E) = \exp[ik_1(E)na] + b(E)\exp[-ik_1(E)na].$$

On the right-hand side of the quantum structure we expect a transmitted wave of unknown amplitude $c(E)$:

$$f_2(n, E) = c(E)\exp[ik_2(E)na].$$

Note that the Wannier recurrence equation provides us with the means to calculate

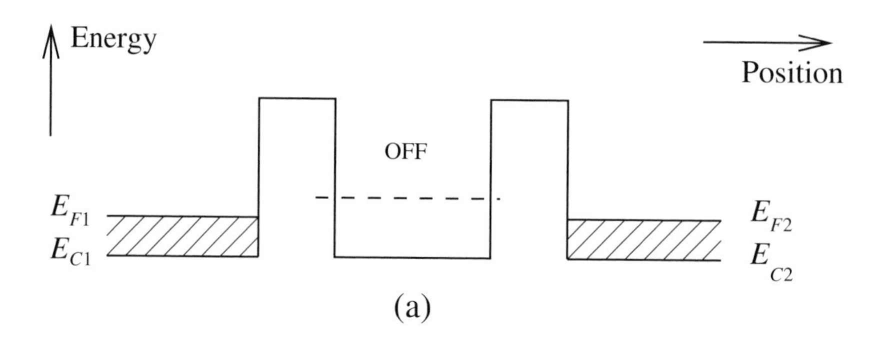

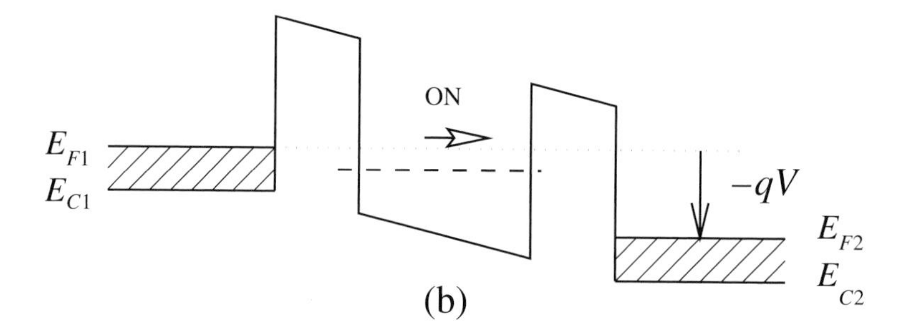

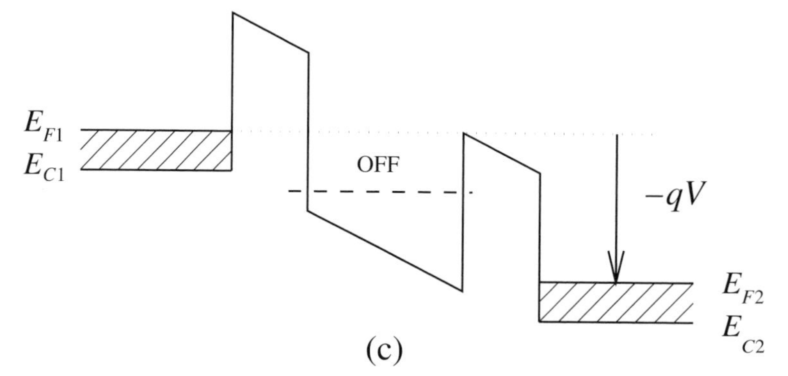

Fig. 4.8. Band diagram of a double barrier: (a) in equilibrium ($V = 0$), (b) when the current is maximum (the left-hand Fermi level is aligned with the transmission peak), and (c) when the current is quenched (the transmission peak is below the left-hand conduction band).

$f_1(n+1, E)$ from $f_1(n-1, E)$ and $f_1(n, E)$

$$f_1(n+1, E) = -f_1(n-1, E) + 2\left(\frac{A + V_n - E}{A}\right) f(n, E),$$

where V_n is the diagonal potential of the quantum structure considered. Assume now that the quantum structure is located between site 0 and N_{SL}. We can apply the Wannier recurrence equation to cross the heterostructure from $n = 0$ to $n = N_{SL} + 1$ and calculate $f_1(N_{SL} + 1, E)$ and $f_1(N_{SL} + 2, E)$ in terms of $f_1(-2, E)$ and $f_1(-1, E)$. The unknown coefficients $b(E)$ and $c(E)$ are then obtained by solving the following system of equations:

$$f_1(N_{SL} + 1, E) = f_2(N_{SL} + 1, E),$$
$$f_1(N_{SL} + 2, E) = f_2(N_{SL} + 2, E).$$

From our discussion of the conservation of electron current we know that we must have

$$\left[1 - |b(E)|^2\right]\frac{1}{\hbar}\frac{\partial E_1(k_1)}{\partial k_1} = |c(E)|^2\frac{1}{\hbar}\frac{\partial E_2(k_2)}{\partial k_2}.$$

Introducing v_1 and v_2, the electron velocities in semiconductors 1 and 2, we can simply rewrite the current conservation as

$$v_1(1 - |b(E)|^2) = v_2|c(E)|^2$$
$$1 - |b(E)|^2 = \frac{v_2}{v_1}|c(E)|^2$$
$$1 - R(E) = T(E),$$

where we have introduced the reflection coefficient R and the transmission coefficient T defined respectively as

$$R(E) = |b(E)|^2,$$
$$T(E) = \frac{v_2}{v_1}|c(E)|^2.$$

For a symmetrical double-barrier structure with a well width Ma and with no voltage applied, the following analytic expression (see Problem 4.1) can be obtained:

$$T_{2B}(E) = |t_{2B}|^2 = \frac{1}{1 + 4\dfrac{R_B(E)}{T_B^2(E)}\sin^2[k_1(E)Ma - \theta(E)]}, \tag{4.6}$$

where $k_1(E)$ is the wave-vector in the well for an incident electron of energy E; R_B and $T_B = 1 - R_B$ are the reflection and transmission coefficients respectively of the barriers; θ is the angle of the scattering parameter of the barrier $r_B = \sqrt{R_B}\exp(-i\theta)$ (see Problem 4.1).

The transmission is a function of the energy E of the incident electron and a plot of the transmission coefficient $T(E)$ versus E is shown on Figure 4.9(a) for various biases. The transmission coefficient varies between 1 and $T_{2B}(min)$. The peaks ($T_{2B} = 1$) occur for $k_1(E)Ma - \theta(E) = \pi$. For small values of θ (large barriers) these peaks occur for energies corresponding to the resonant energy levels of the

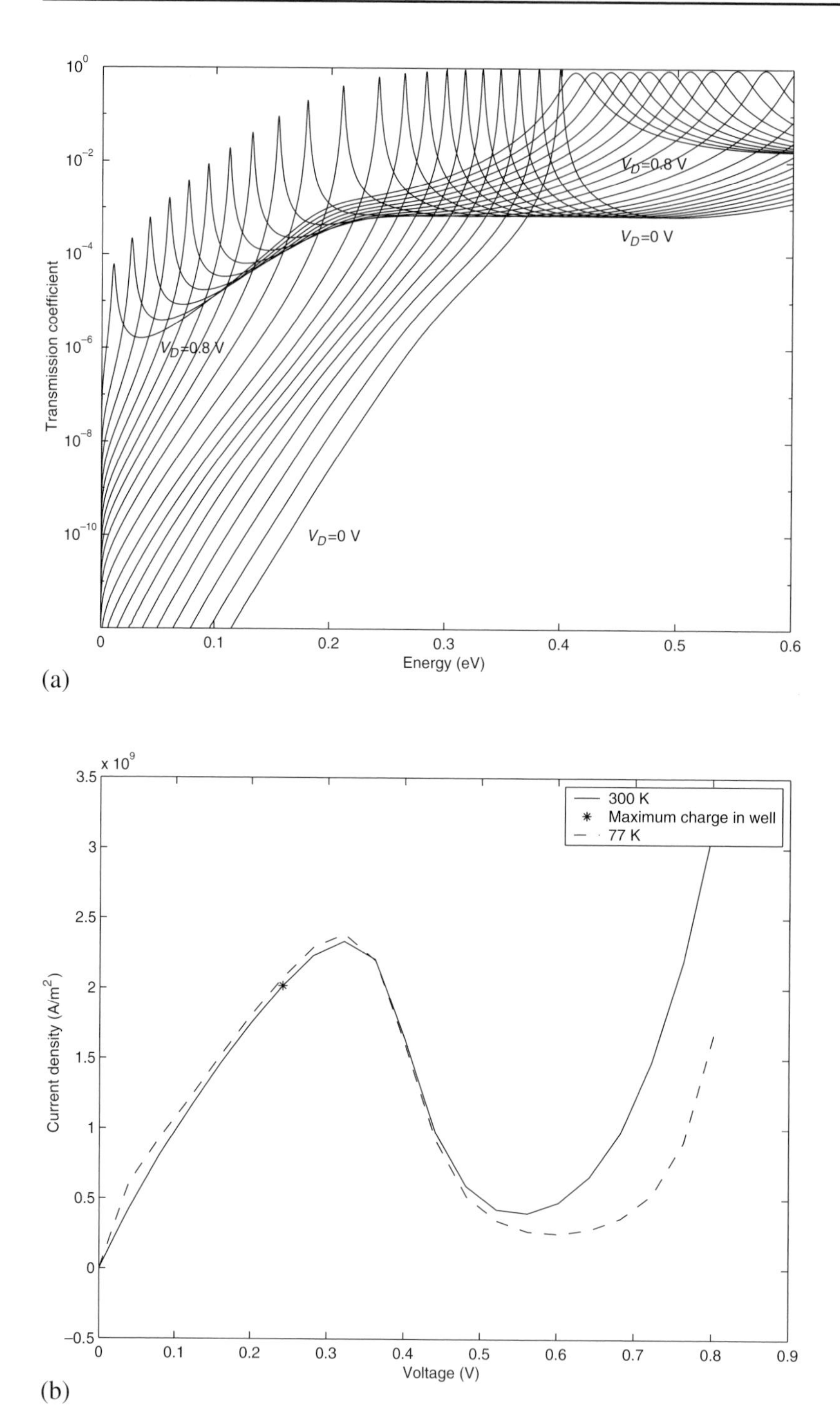

Fig. 4.9. (a) Transmission coefficient versus incident energy of a symmetric double barrier. (b) I–V characteristic of this double-barrier diode at 300 K (full line) and 77 K (dashed line). In both plots the applied voltage varies from 0 to 0.8 V in steps of 0.08 V.

inner quantum well $k_1(E)Ma = \pi$. The minimum transmission coefficient $T_{2B}(min)$ (off-resonance) can be reduced by using a large barrier width ($R_B = 1 - T_B$ large). However, if the barriers are too large, the transmission peaks become so narrow that scattering can easily suppress them by broadening. In this analysis we have assumed that the electron is ballistic, that is, the electron energy is conserved. In Chapter 6 we will discuss the impact of phonon scattering upon the transmission coefficient $T(E)$.

For more complex structures and for applied voltages, the transmission coefficient is readily calculated numerically using the Wannier recurrence equation. The transmission coefficient can only reach unity transmission at resonance when the potential barriers are symmetric. Therefore, a unity transmission is not achieved in a symmetric double barrier when a voltage is applied.

For larger wells, the first resonant energies are reduced in energies and smaller applied voltages are required to suppress the resonant tunneling such that the asymmetry is reduced. Sometimes a larger barrier is also used on the right-hand side (where the positive polarity is applied) to compensate for the asymmetry introduced by the applied electrostatic potential.

A formula which allows us to calculate the total tunneling current is derived in the next section.

4.4.2 Tunneling current and resonant tunneling

The resonant tunneling effect can be used to make a diode that exhibits a large negative differential resistance in its I–V characteristics (see Figure 4.9(b)). This was proposed by Tsu and Esaki [32] and verified experimentally by Chang *et al.* [19] and more recently by Sollner *et al.* [31]. Since then, this device has been extensively studied because of its high-speed performance and its potential use in the design of high-speed electronic circuits.

Consider the simplified band diagram shown in Figure 4.8(a). A degenerate (strongly doped) semiconductor is used on each side of the double-barrier structure. When no voltage is applied (Figure 4.8(a)) electrons are incident from the left and the right, and due to the symmetry of the device, no current results as should be the case in equilibrium. When a small voltage is applied (see Figure 4.8(b)) the symmetry is broken. Now only electrons from semiconductor 1 can tunnel through the resonant energy level of the quantum well to semiconductor 2, and a large current is observed. When a larger voltage is applied (see Figure 4.8(c)), the resonant energy level of the quantum well is lowered below the energy band of semiconductor 1. Resonant tunneling is no longer possible and the diode current rapidly drops. For even larger applied voltages, thermal emission over the barrier and Fowler–Nordheim tunneling through the barrier become important and the diode current again rises rapidly. Note that this device is symmetric as are its I–V characteristics.

We now wish to derive an expression that enables us to calculate the current through a vertical quantum device such as a double barrier or a superlattice. We start by calculating an expression relating the three-dimensional electron gas concentration to the 2DEG concentration.

The three-dimensional electron gas concentration is given by

$$N = \int_{-\infty}^{\infty} N(\mathbf{k}) \frac{1}{\exp\left(\dfrac{E - E_F}{k_B T}\right) + 1} d\mathbf{k},$$

where $N(\mathbf{k})$ is the density of states in $\mathbf{k}$ space

$$N(\mathbf{k}) = 2\frac{V}{(2\pi)^3} = \frac{V}{4\pi^3} = \frac{1}{2\pi}\frac{V}{2\pi^2}$$

and V is the crystal volume.

Using

$$n = \frac{N}{V},$$

$$E = \mathcal{E}_{1,0} + A - A\cos(k_x) + \mathcal{E}(k_\perp),$$

$$k_\perp^2 = k_y^2 + k_z^2,$$

$$\mathcal{E}(k_\perp) = \frac{\hbar^2 k_\perp^2}{2m^*}, \tag{4.7}$$

we can readily rewrite the three-dimensional electron gas concentration n per unit of volume as

$$n = \frac{1}{2\pi}\int_{-\pi/a}^{\pi/a} dk_x \left\{ \frac{1}{2\pi^2}\int_{-\infty}^{\infty}\int_{-\infty}^{\infty} dk_y\, dk_z \frac{1}{\exp\left[\dfrac{\mathcal{E}(k_x) + \mathcal{E}(k_\perp) - E_F}{k_B T}\right] + 1} \right\}$$

$$n = \frac{1}{2\pi}\int_{-\pi/a}^{\pi/a} dk_x \left\{ \frac{m^*}{\pi\hbar^2}\int_{0}^{\infty} d\mathcal{E}(k_\perp) \frac{1}{\exp\left[\dfrac{\mathcal{E}(k_x) + \mathcal{E}(k_\perp) - E_F}{k_B T}\right] + 1} \right\}$$

$$n = D k_B T \frac{1}{2\pi}\int_{-\pi/a}^{\pi/a} \left[\ln\left\{1 + \exp\left[-\frac{\mathcal{E}(k_x) + \mathcal{E}(k_\perp) - E_F}{k_B T}\right]\right\}\right]_0^{\infty} dk_x$$

$$n = D k_B T \frac{1}{2\pi}\int_{-\pi/a}^{\pi/a} \ln\left\{\exp\left[\frac{E_F - \mathcal{E}(k_x)}{k_B T}\right] + 1\right\} dk_x$$

Consider now a quantum structure starting with the band 1 $\mathcal{E}_1(k_{1x})$ on the left-hand side and ending with the band 2 $\mathcal{E}_2(k_{2x})$ on the right. Electrons incident ($k_{1,x} > 0$) on the left-hand side (1) generate the following current:

$$J_{\rightarrow} = \frac{Dk_BT}{2\pi}\int_0^{\pi/a} qT_{12}(\mathcal{E}_1(k_{1x}))v_1(\mathcal{E}_1(k_{1x}))\ln\left\{\exp\left[\frac{E_{F1}-\mathcal{E}_1(k_{1x})}{k_BT}\right]+1\right\}dk_{1x}$$

$$J_{\rightarrow} = \frac{Dk_BT}{2\pi}\frac{q}{\hbar}\int_0^{\pi/a} T_{12}(\mathcal{E}_1(k_{1x}))\ln\left\{\exp\left[\frac{E_{F1}-\mathcal{E}_1(k_{1x})}{k_BT}\right]+1\right\}\frac{d\mathcal{E}_1(k_{1x})}{dk_{1x}}dk_{1x}$$

$$J_{\rightarrow} = \frac{Dk_BT}{2\pi}\frac{q}{\hbar}\int_{\mathcal{E}_1(0)}^{\mathcal{E}_1(\pi/a)} T_{12}(\mathcal{E}_1)\ln\left[\exp\left(\frac{E_{F1}-\mathcal{E}_1}{k_BT}\right)+1\right]d\mathcal{E}_1,$$

where E_{F1} is the Fermi level of semiconductor 1.

The electrons incident on the right-hand side (2) contribute the current:

$$J_{\leftarrow} = \frac{Dk_BT}{2\pi}\frac{q}{\hbar}\int_{\mathcal{E}_2(0)}^{\mathcal{E}_2(\pi/a)} T_{21}(\mathcal{E}_2)\ln\left[\exp\left(\frac{E_{F2}-\mathcal{E}_2}{k_BT}\right)+1\right]d\mathcal{E}_2,$$

where E_{F2} is the Fermi level of semiconductor 2. The total diode current per unit area, J_D, is then given by the difference between the left- and right-hand current densities:

$$J_D = J_{\rightarrow}(E_{F1}) - J_{\leftarrow}(E_{F2}).$$

As a consequence of time-reversal symmetry, the transmission coefficient satisfies the following property (reciprocity):

$$T_{12}(E) = T_{21}(E).$$

Assuming that an electrostatic potential V is applied between semiconductors 2 and 1 we have

$$E_{F2} - E_{F1} = -qV.$$

The diode current per unit area can then be written

$$J_D = \frac{Dk_BT}{2\pi}\frac{q}{\hbar}\int_{\mathcal{E}_1(0)}^{\mathcal{E}_1(\pi/a)} T_{12}(\mathcal{E}_1, V)\ln\left[\frac{\exp\left(\frac{E_{F1}-\mathcal{E}_1}{k_BT}\right)+1}{\exp\left(\frac{E_{F1}-qV-\mathcal{E}_1}{k_BT}\right)+1}\right]d\mathcal{E}_1,$$

which is the original expression reported by Tsu and Esaki [32]. A rule of thumb which is often used estimates the voltage at which the peak current is obtained to be twice the energy in electron volts at which resonance occurs.

4.4.3 Charge distribution inside the well

As in the case of quantum wells, the electron distribution of electrons inside a double-barrier structure modifies the electrostatic potential barrier. This is particularly the case at resonance when the electron amplitude of the wave-function (and therefore the probability of presence) in the double barrier is very large. The electron distribution is then required to solve the Poisson equation. The electron distribution per unit area A can be obtained from the wave-function using

$$\left.\begin{aligned}
\rho(n) &= \rho_{\rightarrow}(n) + \rho_{\leftarrow}(n),\\
\rho_{\rightarrow}(n) &= q\frac{Dk_BT}{2\pi}\int_0^{\pi/a} |f(n, \mathcal{E}_1(k_{1x}))|^2 \ln\left\{\exp\left[\frac{E_{F1} - \mathcal{E}_1(k_{1x})}{k_BT}\right] + 1\right\} dk_{1x},\\
\rho_{\leftarrow}(n) &= q\frac{Dk_BT}{2\pi}\int_0^{\pi/a} |f(n, \mathcal{E}_2(k_{2x}))|^2 \ln\left\{\exp\left[\frac{E_{F2} - \mathcal{E}_2(k_{2x})}{k_BT}\right] + 1\right\} dk_{2x}.
\end{aligned}\right\} \quad (4.8)$$

This formula is derived assuming that a Bloch wave of amplitude 1 is incident on the device and that the flat band on the left-hand side extends for a large length, essentially equal to the length of the device $L_x = N_x a$. The amplitude of the wave-function then approximately satisfies

$$\sum_{all\ n} |f(n, E)|^2 \simeq N_x.$$

The term N_x cancels since the density of states is given in the periodic boundary condition model by

$$N(\mathbf{k}) = 2\frac{N_x}{2\pi}\frac{N_y}{2\pi}\frac{N_z}{2\pi} = 2\frac{N_x}{2\pi}\frac{A}{4\pi^2}.$$

A self-consistent solution of the Schrödinger and Poisson [33] equations is required and can be implemented by iteratively solving these equations using the charge distribution of Equation (4.8). Figure 4.10(a) shows the charge distribution at resonance and the initial and final potential barriers.

The electron charge in the double-barrier diode will effectively screen the applied potential and shift the resonance peak of the transmission coefficient to a higher energy. As a result the current peak of the I–V characteristic occurs at a higher voltage. This is illustrated in Figures 4.10 by plotting the conduction-band edge and charge distribution for an $In_{0.53}Ga_{0.47}As/AlAs$ RTD with a 30-monolayer undoped spacer on each side of the double barrier and a donor doping of 10^{25} m^3 in the emitter and collector. Notice the potential hump when no bias is applied in Figure 4.10(a).

The resonance in the well (see full line in Figure 4.10(b)) occurs for a voltage smaller than the peak voltage in this double-barrier diode (see $*$ in the I–V characteristic Figure 4.9(b)). Note also in Figure 4.10(b) the presence of an accumulation of charge (monolayer [60,70]) in the undoped region of the emitter contact and the presence of a depletion of charge (monolayer [120,140]) in the undoped region of the collector contact that occurs for higher voltages. This accumulation of charge arises from the potential hump in the emitter region (monolayer [60,70]) in Figure 4.10(a). This has motivated the bandgap engineering of this region to reduce the hump and increase the RTD current [34].

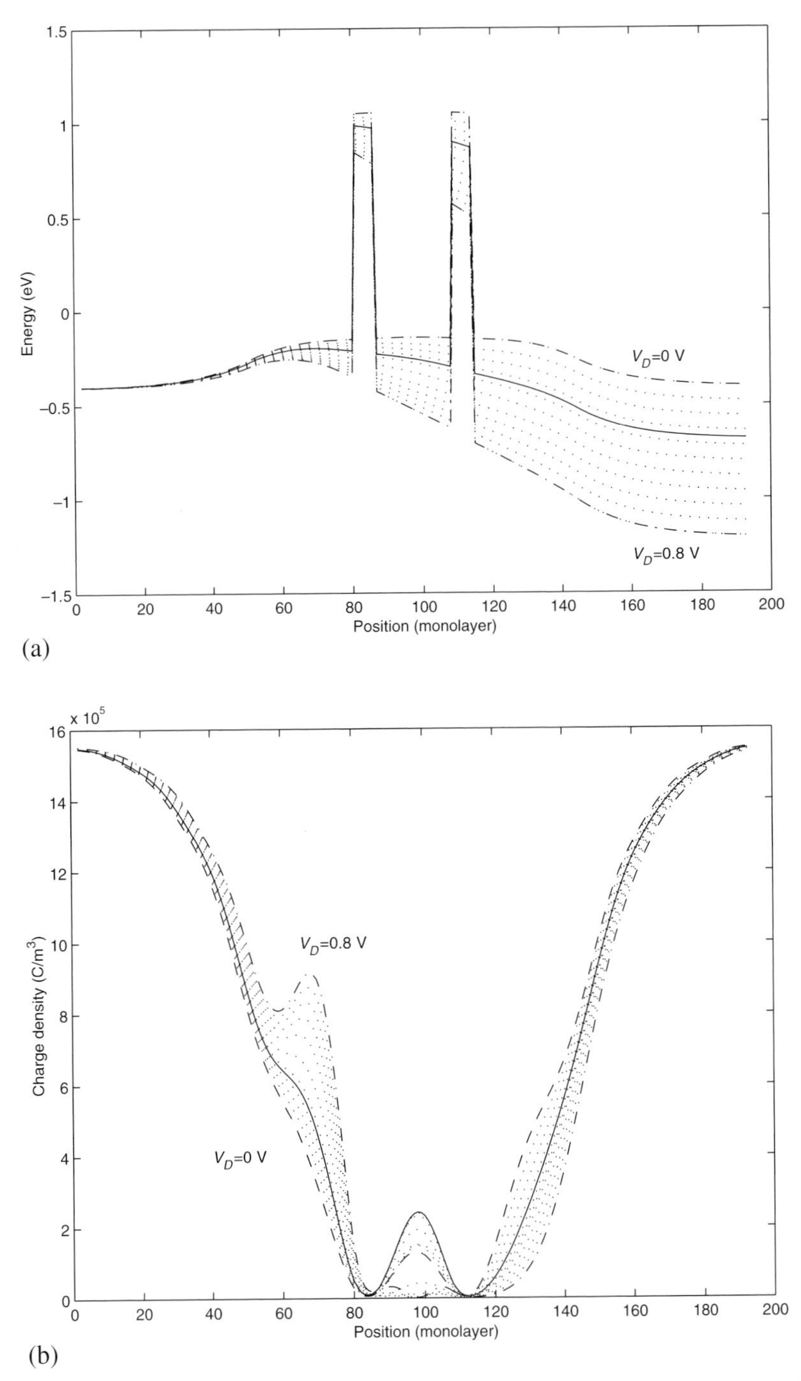

Fig. 4.10. (a) Conduction-band edge (dotted line) and (b) charge distribution (dotted line) obtained using a self-consistent solution of the Schrödinger and Poisson equations. In both plots the voltage applied varies from 0 to 0.8 V in steps of 0.08 V, and a full line is used to indicate the potential (marked with $*$ on Figure 4.9) for which the charge distribution is maximum in the well.

4.4.4 Exchange correlation

Another many-body effect to be accounted for is the exchange correlation which is responsible for the bandgap narrowing in degenerate semiconductors. Exchange correlation can be implemented using the local density functional formalism following the approach of Gawlinski *et al.* [35]. In the local function density approach the exchange correlation potential is expressed in terms of the local charge distribution using the formula given in Equations 8 and 9 in [35]. Note, however, that the charge distribution, like the current, requires only the left- and right-hand contact Fermi levels. The introduction of a spatially-varying Fermi level to define the charge distribution in the RTD is incorrect. The validity of the current and charge definition is verified by applying a time-varying potential. Indeed the total AC charge and AC current calculated must satisfy the continuity equation (see Chapter 7).

Exchange correlation is a second order many-body effect compared to the contribution of the self-consistent potential and usually only brings a small correction. Note that exchange correlation should be differentiated from electron–electron scattering.

4.4.5 Scattering induced broadening

We have assumed so far that the electron is ballistic and therefore conserves its energy and perpendicular momentum. However, in real devices the electron is scattered by the lattice vibrations and other types of defects. As a result of scattering an electron can acquire or lose energy in the scattering process and change perpendicular momentum $\mathbf{k}_\perp$. These scattering processes can permit an electron with a longitudinal energy for which transmission is not possible to scatter to a state of energy for which resonant tunneling is possible. This leads to an effective broadening of the transmission coefficient $T(\mathcal{E}_1)$. Such a broadening can be represented phenomenologically by a new transmission coefficient $T^{'}(\mathcal{E}_1)$ given by

$$T^{'}(\mathcal{E}_1) = \int_{\mathcal{E}_{1,0}}^{\infty} T(\mathcal{E}_1)\frac{\Gamma/\pi}{(\mathcal{E}_1^{'} - \mathcal{E}_1)^2 + \Gamma^2}\, d\mathcal{E}_1^{'},$$

where Γ is the energy broadening which is typically a few millielectron volts at room temperature. The main consequence of this energy broadening is to reduce the peak-to-valley ratio of the I–V characteristics of the RTD as is shown in Figure 4.11. A more rigorous treatment of scattering-assisted tunneling is postponed to Chapter 6.

4.4.6 Full-band structure effects

We have so far limited our analysis to a system with a tight-binding band structure. For the range of energies typically considered in these devices, this is equivalent to an effective-mass approximation. The spatial variation of the mass or both the longitudinal and transversal mass can easily be handled with the formalism introduced at the

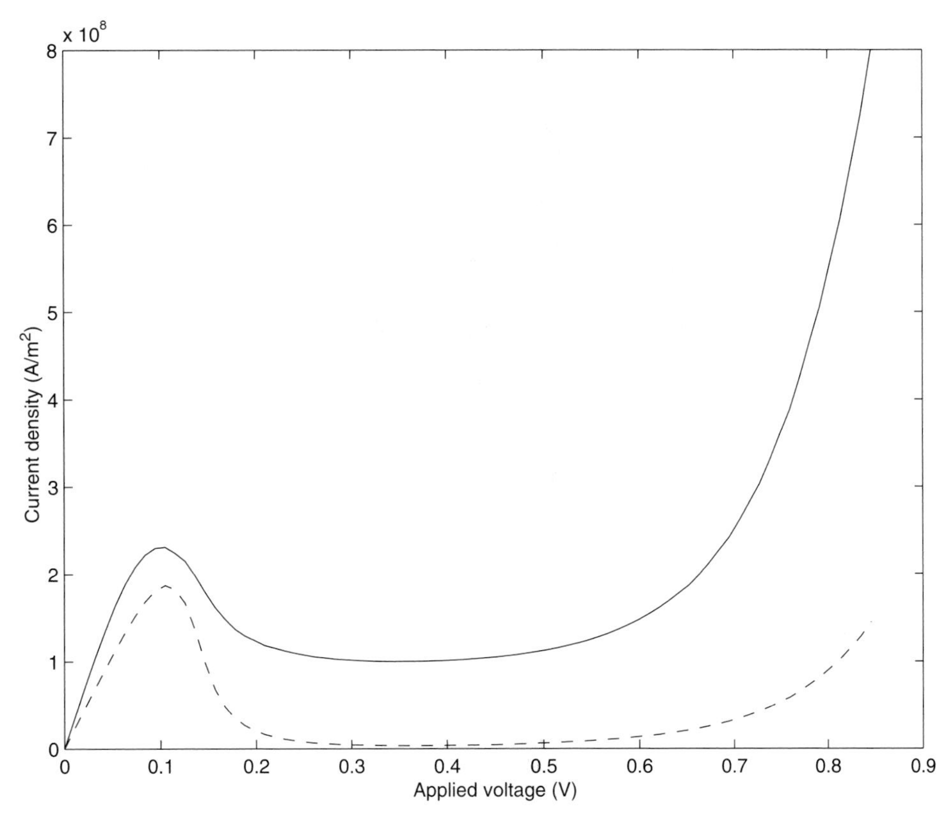

Fig. 4.11. Comparison of the I–V characteristics obtained in the absence (dashed line) and the presence (full line) of scattering.

end of Chapter 3. The transmission coefficient becomes dependent on the transverse momentum of the electron and the integration over the transverse coordinates must be performed numerically.

For large applied voltages the bottom of the X valley in the well (see Figure 3.14) becomes aligned with the bottom of the Γ valley in the emitter (see Figure 3.13) and it becomes necessary to consider the impact of the full-band structure on the transmission coefficient.

To illustrate the coupling between the X and Γ valleys consider first the case in which no voltage is applied on a double $Al_{0.3}Ga_{0.7}As$ barrier: GaAs–$Al_{0.3}Ga_{0.7}As$. An RTD device with a $Ga_{0.7}Al_{0.3}As$ barrier width and a GaAs well width of eight monolayers is used. The band structures of GaAs and AlAs in $\langle 100 \rangle$ direction for $\mathbf{k}_\perp = 0$, were computed with the empirical pseudopotential method and optimized using a least squares fit (see Chapter 3). Because $Ga_{0.7}Al_{0.3}As$ band is fairly close to GaAs, an arithmetic average can be used to calculate the Hamiltonian matrix elements at the interface.

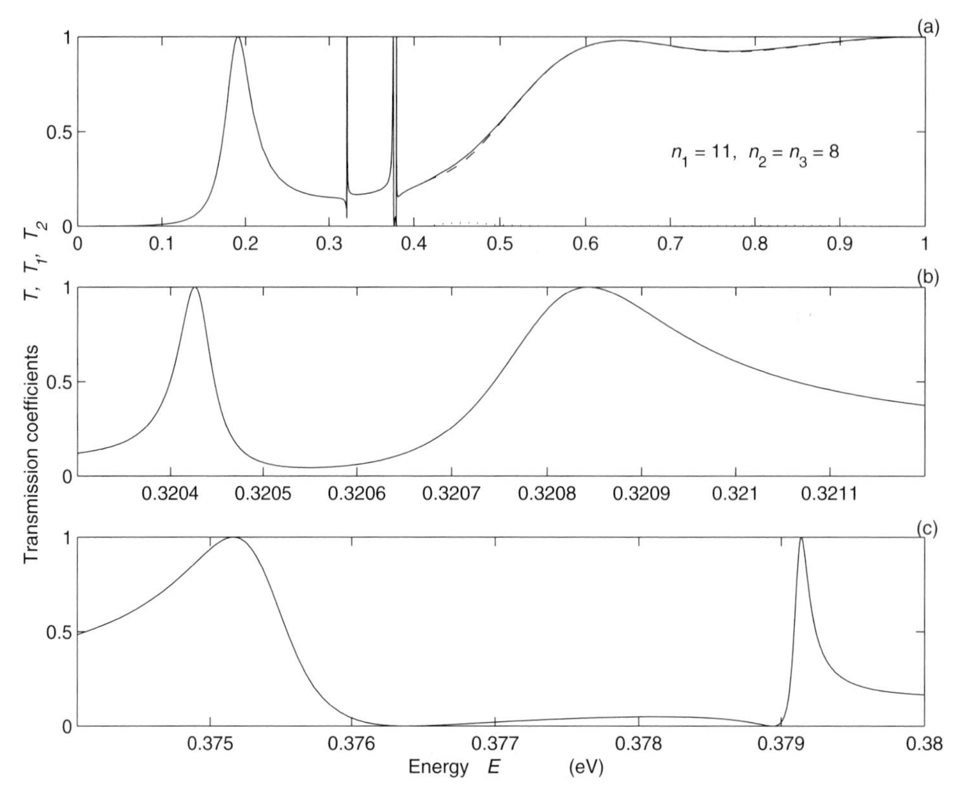

Fig. 4.12. Transmission coefficients for $\mathbf{k}_\perp = 0$ corresponding to the band structure with the $Al_{0.3}Ga_{0.7}As$ barriers, specified by $(1-x)E_{GaAs} + xE_{AlAs}$ for $x = 0.3$, and no voltage is applied ($V_D = 0.0$ eV). Full line: T, the transmission coefficient; dashed line: T_1, the transmission coefficient in the Γ band; dotted line: T_2, the transmission coefficient in the X band. Part (a) plots the energy ranging from 0 to 1 eV, part (b) zooms in the two resonant states between 0.3204 eV and 0.3209 eV, and part (c) further zooms in the two switching points between 0.375 eV.

In Figure 4.12, we show transmission coefficients for $\mathbf{k}_\perp = 0$ as a function of incident energy for two RTDs. We notice that electrons can go through the double barrier without any reflection (100% transmission) when their energy is about 0.190 91 eV, 0.320 43 eV, 0.320 84 eV, 0.375 16 eV, and 0.379 14 eV (resonant states). However, almost no electrons are transmitted when their energy level is lower than 0.1 eV and around 0.3764 eV and 0.3789 eV. The transmission probability reaches a (local) minimum of about 4% at an energy level around 0.320 55 eV, which is between the resonant energies of 0.320 43 eV and 0.320 84 eV. The transmission probability remains less than 5% between 0.3764 eV and 0.3789 eV (two off-states). For this structure the behavior is apparently very similar to conventional resonant tunneling in the effective-mass approximation (100% transmission at resonance). However, a closer inspection reveals the presence of sharp variations of the transmission coefficient from 100% to 4%, and then back to 100% when the incident energy

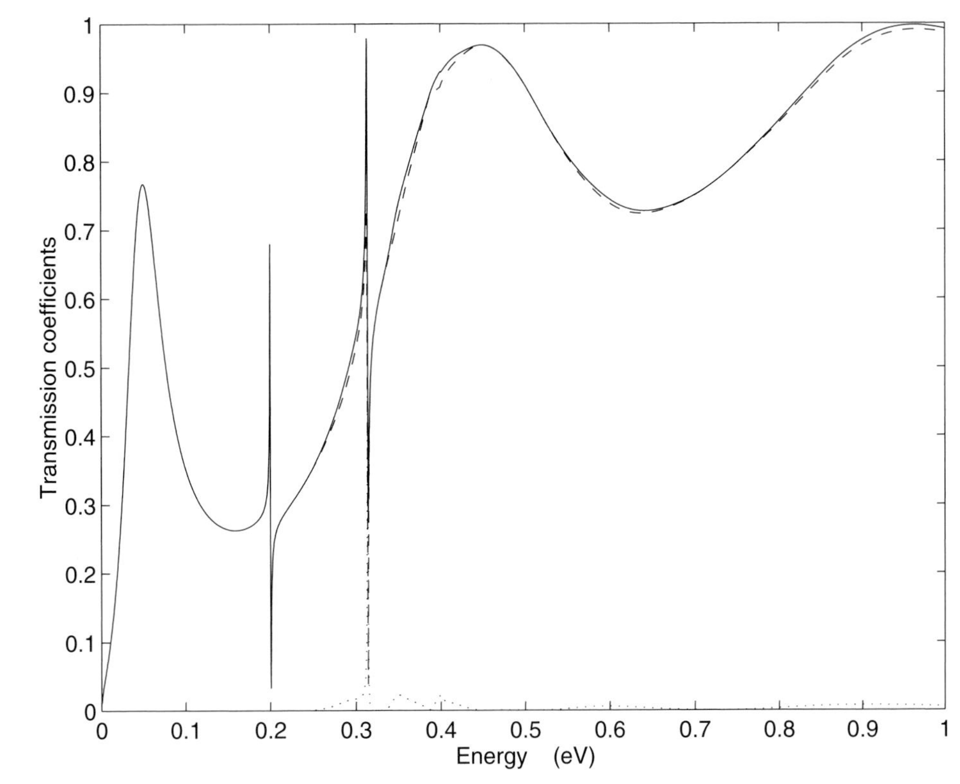

Fig. 4.13. Transmission coefficients for $\mathbf{k}_\perp = 0$ corresponding to the band structure with the $Al_{0.3}Ga_{0.7}As$ barriers, specified by $(1 - x)E_{\mathrm{GaAs}} + xE_{\mathrm{AlAs}}$ for $x = 0.3$, and with $V_D = 0.15$ V applied. Full line: T, the transmission coefficient; dashed line: T_1, the transmission coefficient in the Γ band; dotted line: T_2, the transmission coefficient in the X band.

changes from 0.320 43 eV to 0.320 84 eV, an energy variation of only 0.000 41 eV. Even more dramatically, the transmission coefficient varies from 100% to 0% when the incident energy increases from 0.375 16 eV to 0.376 40 eV or decreases from 0.379 14 eV to 0.378 94 eV. These fine structures are associated with a resonance and anti-resonance between the Γ valley in the GaAs well and the X valley in the $Al_{0.3}Ga_{0.7}As$ barriers, as was initially documented for RTDs in [40,25,41]. Indeed this resonance/anti-resonance is made possible by the fact that the bottom of the X valley of $Al_{0.3}Ga_{0.7}As$ is at 0.3003 eV in the band structure considered.

When a voltage is applied the upper valley in the collector is lowered compared to the Γ valley in the emitter. The anti-resonance coupling between the Γ and X valleys then occurs at lower energies. This is illustrated in Figure 4.13 for a bias voltage of 0.15 V.

4.4.7 High-frequency and high-speed response

The RTD is expected to operate up to terahertz (1000 GHz) [31]. Microwave oscillators operating around 712 GHz have been fabricated by Brown and coworkers [4,5]. The maximum power which can be generated from such a microwave oscillator is approximately limited by the product

$$\frac{1}{8}(V_{valley} - V_{peak})(I_{peak} - I_{valley}).$$

Diamond *et al.* [7] have analyzed the large-signal switching characteristics of an RTD. To the first order the switching time τ from peak to valley can be estimated as

$$\tau = \frac{C_{RTD}(V_{valley} - V_{peak})}{I_{peak} - I_{valley}} = \frac{C_{RTD}}{I_{peak}} \frac{V_{valley} - V_{peak}}{1 - \dfrac{I_{valley}}{I_{peak}}}, \tag{4.9}$$

where C_{RTD}, the RTD capacitance, is assumed to be constant and C_{RTD}/I_{peak} is referred to as the speed index. Switching times of a few picoseconds can be achieved with currently available devices. A more complete discussion of accurate techniques used to simulate the high-frequency response of RTDs is postponed to Chapter 7. The reader is also referred to [8] for a review of the application of RTDs in triggering, pulse forming and sampling, low-power memory cell, analog-to-digital conversion, high-speed logic and oscillator circuits. The same reference also provides a discussion of the fabrication process and a comparison of the device performances obtained using various combinations of materials and widths for the tunnel barrier, quantum well, spacers and substrate of RTDs.

4.4.8 Resonant interband tunneling diodes (RITDs)

The resonant tunneling diode bears some similarity to the tunnel or Esaki diode [9,10] The tunnel diode is a bipolar device which relies on the tunneling through the bandgap of electrons from the conduction band to the valence band. Peak-to-valley current ratios of 4 and 8.3 have been reported for the alloyed Esaki tunnel diode in Si [11] and Ge [12]. The combination of the operating principles of the RTD and tunnel diodes leads to the conception of the RITD. Several approaches for implementing RITDs are possible, as is illustrated on Figure 4.14. Figure 4.14(a) shows an RITD for which resonant interband tunneling takes place between the 2DEG of a quantum well in the conduction band and the two-dimensional hole gas of a quantum well in the valence band [13,14]. Figure 4.14(b) shows an RITD for which resonant tunneling takes place from the n^+-InAs left-hand emitter to the n^+-InAs via the 2DHG (primarily the light-hole) of the valence band [15]. The barriers are provided by the bandgap of the AlSb barriers. When a large enough biasing voltage is applied, interband resonant tunneling is quenched when the valence-band quantum well is lowered below the conduction band of the emitter region. Finally a third RITD structure can be realized using spike doping in a silicon semiconductor [16] as is sketched conceptually in Figure 4.14(c). The spike doped regions induce a triangular quantum well in the conduction and valence bands of both the emitter (left-hand side) and collector (right-hand side) regions. Resonant tunneling then takes place from the emitter 2DEG to the collector 2DHG. Sb and B are possible dopants for the spike doped layer in the emitter and collector, respectively. As usual the tunneling region (barrier/bandgap) should be kept intrinsic to avoid impurity-assisted tunneling which increases the diode leakage current. $Si_{0.5}Ge_{0.5}$ instead of Si can be beneficially used in the intrinsic tunneling region [16] to prevent the diffusion of impurities in this region during subsequent high-temperature fabrication processes. A peak-to-valley current ratio of 5.5 with a peak current density of 8 kA/cm^2 has been reported for the Si/SiGe/Si template [17]. The possibility of building RITDs in Si technology is obviously very exciting, as it offers the prospect of integrating CMOS transistors with tunnel diodes to realize high-speed and low-power circuits. For example, it has been demonstrated for memory cells that RTDs can boost the performance of a transistor technology by reducing by a factor of 3 the number of transistors required [18].

4.5 Superlattice

In the general sense of the term, a superlattice is an MBE heterostructure whose electronic characteristics result from quantum interferences. In a more restrictive sense of the term, a superlattice is an artificial lattice fabricated by a periodic epitaxial growth. A periodic superlattice can be realized by growing alternate layers of two different semiconductors on top of each other, each semiconductor being grown to the same thickness and mole fraction each time. Alternatively, a superlattice can be fabricated in a uniform semiconductor using a periodic modulation of the doping concentration.

The superlattices fabricated by MBE growth are periodic in one dimension only, although two- or three-dimensional superlattices can be conceived. Superlattices were first proposed by Esaki and Tsu [20] and fabricated by Esaki *et al.* [21].

Note that superlattices are also used in MBE growth to block the diffusion of impurities from the substrate and to improve the smoothness of the wafer. For example, an AlAs/GaAs superlattice can be used to remove the roughness of the GaAs substrate.

There are three general types of superlattices: (a) periodic superlattices, (b) random superlattices, and (c) quasi-periodic superlattices. We will start our discussion with the important case of the periodic superlattice.

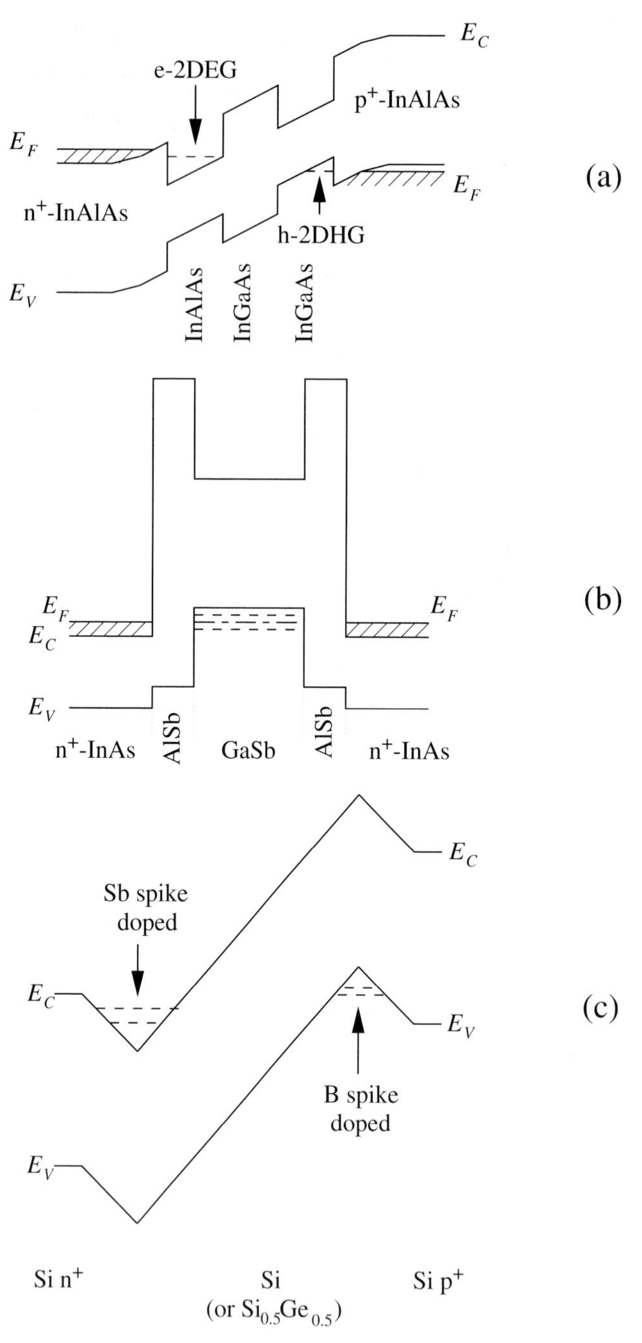

Fig. 4.14. Various approaches for implementing RITDs. In (a) and (c) resonant interband tunneling takes place between the 2DEG in the conduction band of the emitter and the 2DHG in the valence band of the collector [13, 14, 16]. In (b) resonant tunneling takes place between the conduction band of the emitter and the collector via the 2DHG in the valence band [15]. Note that in (a) the 2DEG and 2DHG are created using bandgap engineering [13, 14] whereas spike doping is used in (c) [16].

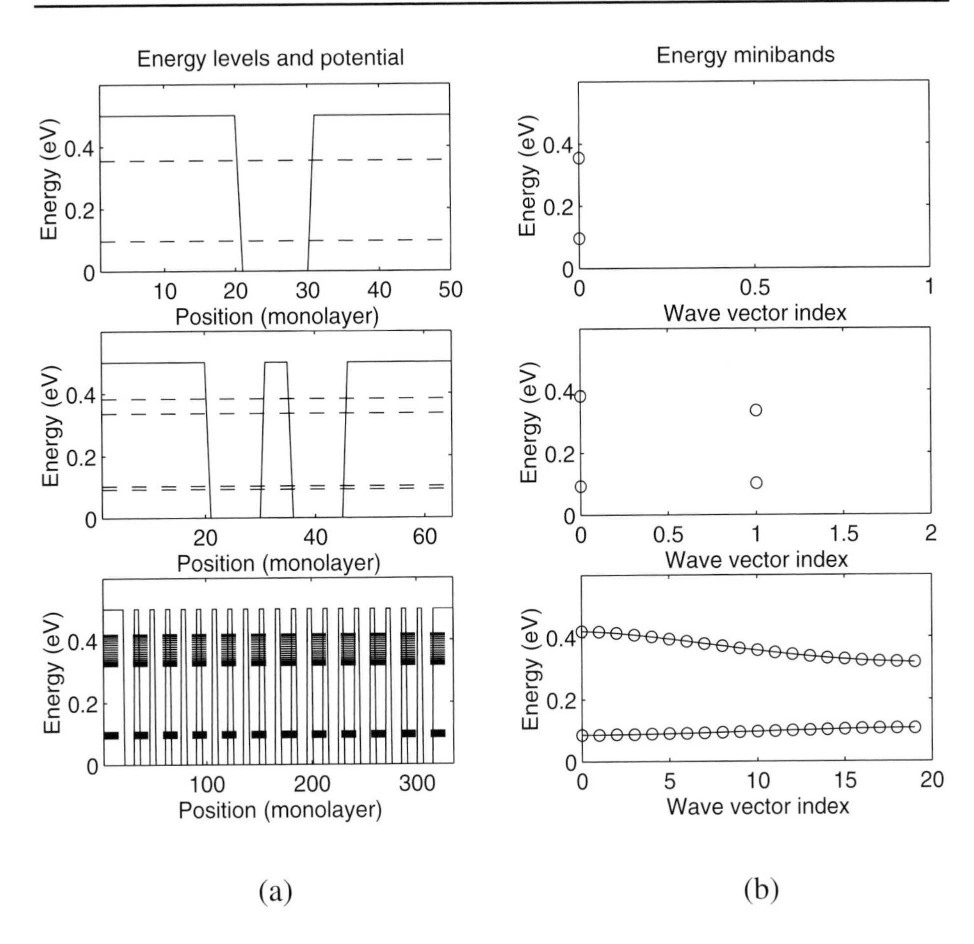

Fig. 4.15. The superlattice (a) band diagram and (b) band structures realized with 1, 2 and 20 wells.

4.5.1 Periodic superlattices

Let us consider the properties of periodic superlattices. As a result of the periodicity of the potential of a superlattice, we expect the formation of minibands of energies and miniforbidden-bands within both the conduction and valence-band structures of the semiconductors involved.

Consider the formation of a superlattice by $N_W = 1$, 2, and 20 adjacent quantum wells in the conduction band (see Figure 4.15(a)). Assume that when the wells are spaced far apart they have two discrete energies, and when the wells are adjacent, these energies are split into two bands of N_W energies. This is in agreement with the Pauli exclusion principle which states that the electrons cannot all be in the same electron states at the same time. Consider the case of quantum wells, 10 lattice parameters wide $(10a)$ separated by a potential barrier 10 lattice parameters wide $(10a)$. The resulting band structure for $N = 20$ wells and an effective mass $m = 0.12m_0$ is shown in Figure 4.15(b). Note that these band structures can be approximated by nearest-neighbor band

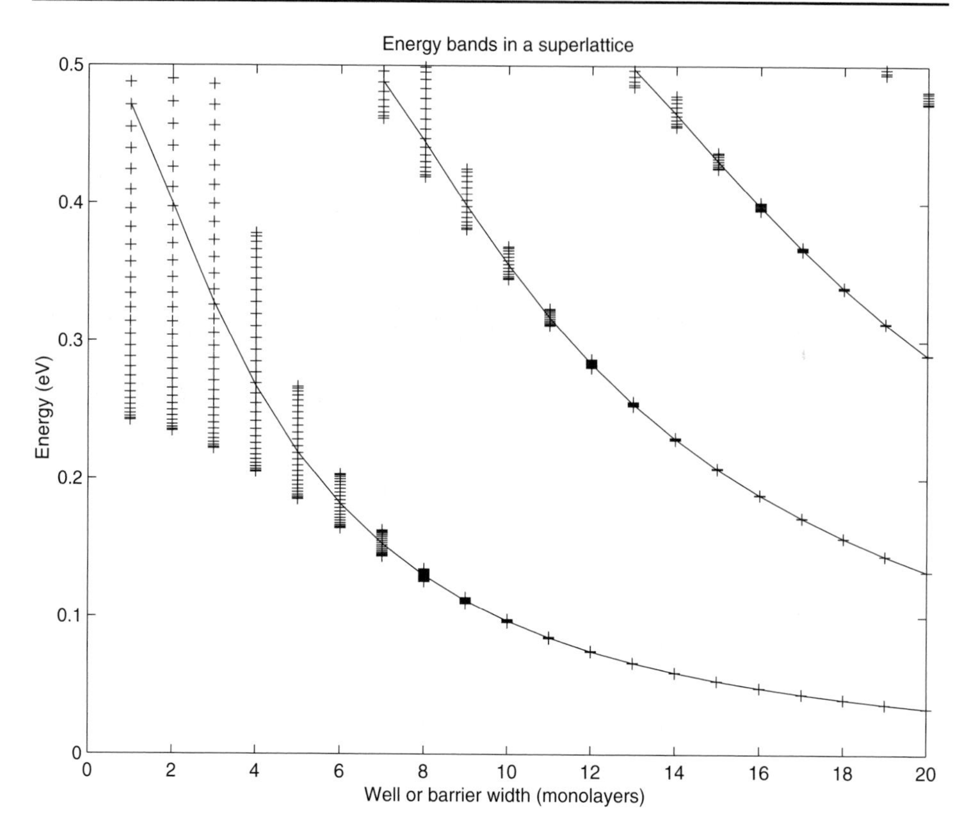

Fig. 4.16. Superlattice bands versus the barrier/well width.

structures of the form $\mathcal{E}_{1,0} - A_1 \cos(k_1 a_{SL})$ and $\mathcal{E}_{2,0} + A_2 \cos(k_2 a_{SL})$ for the first and second bands, respectively. Also note that the superlattice parameter is $a_{SL} = 20a$. As a result of the superlattice periodicity, the Brillouin zone of the lattice has been divided by the superlattice into $a_{SL}/a = 20$ regions of width $2\pi/20a$.

A plot of the energy bands versus the thickness of the barrier/well is shown in Figure 4.16 for a superlattice with a barrier height of 0.5 eV and the same well width and barrier width. Note in Figure 4.16 that when the wells are far apart they become weakly coupled and that the band structure width $2A$ reduces to the energy level of a single quantum well.

As discussed at the beginning of this chapter, periodic superlattices find an important application in the realization of the Bloch oscillator proposed by Esaki [20] and demonstrated by Sibille *et al.* [29]. The existence of the Wannier ladder in a superlattice has also been verified in optical experiments [22]. Figure 4.17 shows (see [50]) the Stark states calculated in a superlattice with an applied voltage of 0.2 V. The first ($n = 1$) and last ($n = 19$) states are surface states of this finite superlattice.

This cartoon from [51] illustrates well the diffraction of electrons/Bloch waves in a superlattice. In this cartoon, Mr. Tompkins explores a fantasy world in which the Planck's constant is much larger than normal. The caption reads, 'Sir Richard was ready to shoot, when the professor stopped him.' The professor explained, 'there is very little chance of hitting an animal when it is moving in a diffraction pattern'.

4.5.2 Random superlattice

In a superlattice the unavoidable fluctuations of the potential energies of the barriers and wells and the interface position can prevent the formation of crystal bands. Instead

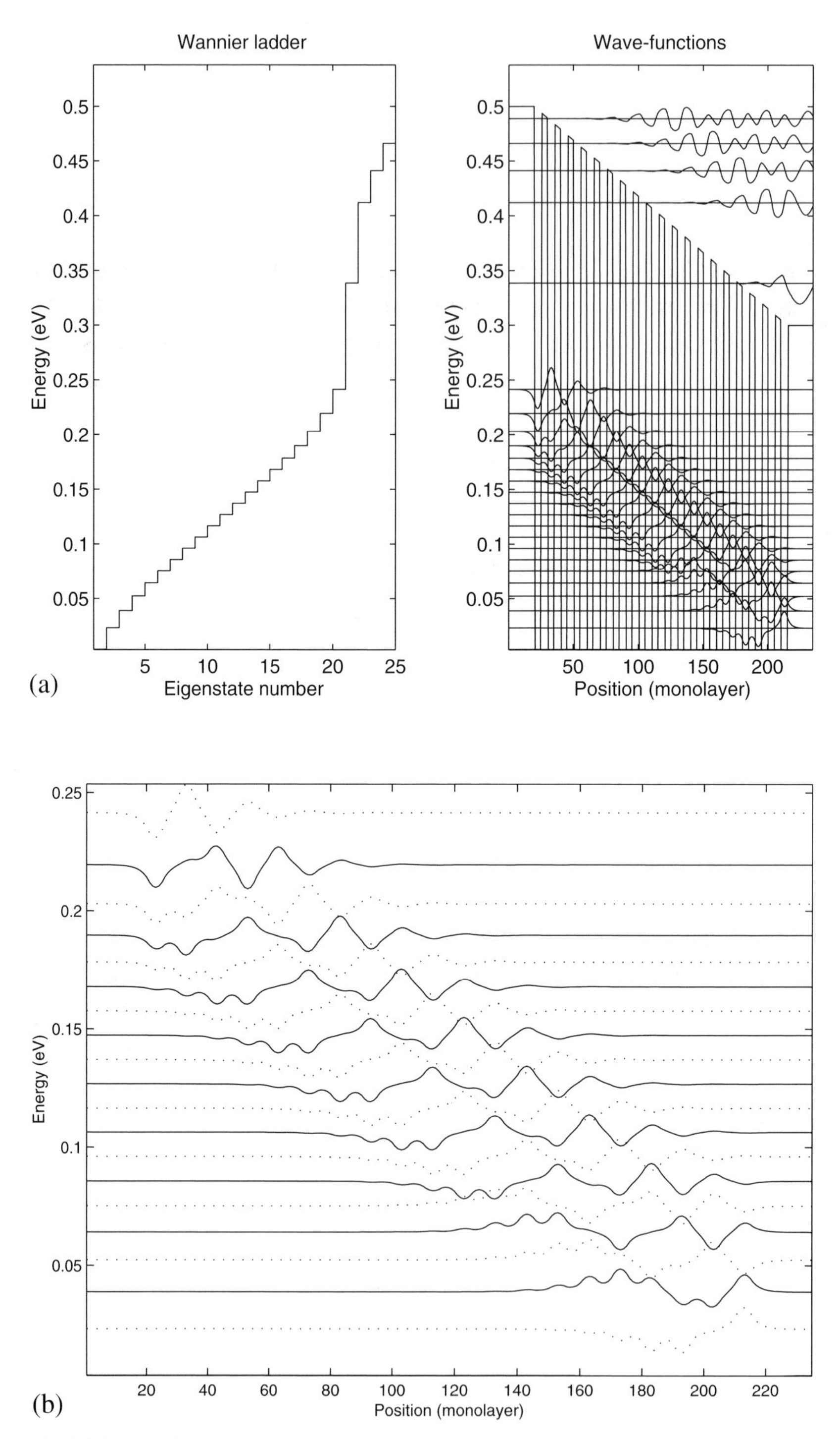

Fig. 4.17. Wannier Stark ladder in a superlattice: (a) band diagram, energy levels and wave-functions; (b) wave-functions.

of propagating Bloch states, non-propagating localized states arise. This phenomenon is called Anderson localization. It occurs when the fluctuations W of the reference energy are larger than B, the width of the superlattice miniband (see Problem 4.6):

$$\text{Bloch waves for } B > W,$$
$$\text{localized states for } B < W.$$

A study of this effect was initially reported by Schmidt [44] for one-dimensional structures. The existence of localization in three dimensions was reported by Anderson (Nobel Prize winner) in 1958 in a paper entitled 'Absence of Diffusion in Certain Random Lattices' [3]. The experimental verification of this effect in a one-dimensional superlattice was given by Chomette *et al.* [43].

Figure 4.18 shows (see [50]) the creation of localized states in a superlattice when the standard deviation of the barrier width is of three monolayers. For small disorders, the wave-functions extend over the entire superlattice (Bloch waves). For larger disorders the wave-functions become localized. Transport becomes limited by hopping (tunneling) from localized states to localized states assisted by the lattice vibrations. The electron mobility is then greatly reduced. No long-range order is observed in random superlattices.

4.5.3 Quasi-crystals and Fibonacci superlattices

Quasi-crystals were discovered in 1984 by Shetchman *et al.* [45] while studying rapidly cooled Al–Mn alloys. X-ray diffraction patterns revealed the presence of symmetry of order 5 (rotational symmetry of angle $2\pi/5$) which according to crystallography theory was previously thought to be incompatible with long-range order. Quasi-crystals are not periodic, however, but do exhibit long-range order.

Quasi-crystals can be realized in one dimension by using the Fibonacci* sequence [47]. The first experimental study of a GaAs/AlAs quasi-crystal was reported by Merlin *et al.* [46]. The Fibonacci model assumes that it takes a generation for a baby B to become an adult A and that at each new generation an adult procreates a baby B. The total number of adults A and babies B at generation l is therefore given by

$$F(l) = F(l-1) + F(l-2).$$

This recurrence equation admits a solution of the form τ^l where $\tau = (1 \pm \sqrt{5})/2$. The root $\tau = (1 + \sqrt{5})/2$ is called the Fibonacci gold number and gives the growth rate $F(l+1)/F(l)$ of the population. The relationship between the Fibonacci number τ and the rotation of order 5 is expressed by $\cos(2\pi/5) = 1/(2\tau)$.

* Fibonacci was an italian monk in the XII century. He traveled to the Orient and contributed to the transfer to Europe of the mathematical advances achieved by the Middle-East and Asian civilizations. He is known for his own work on the demography of rabbits.

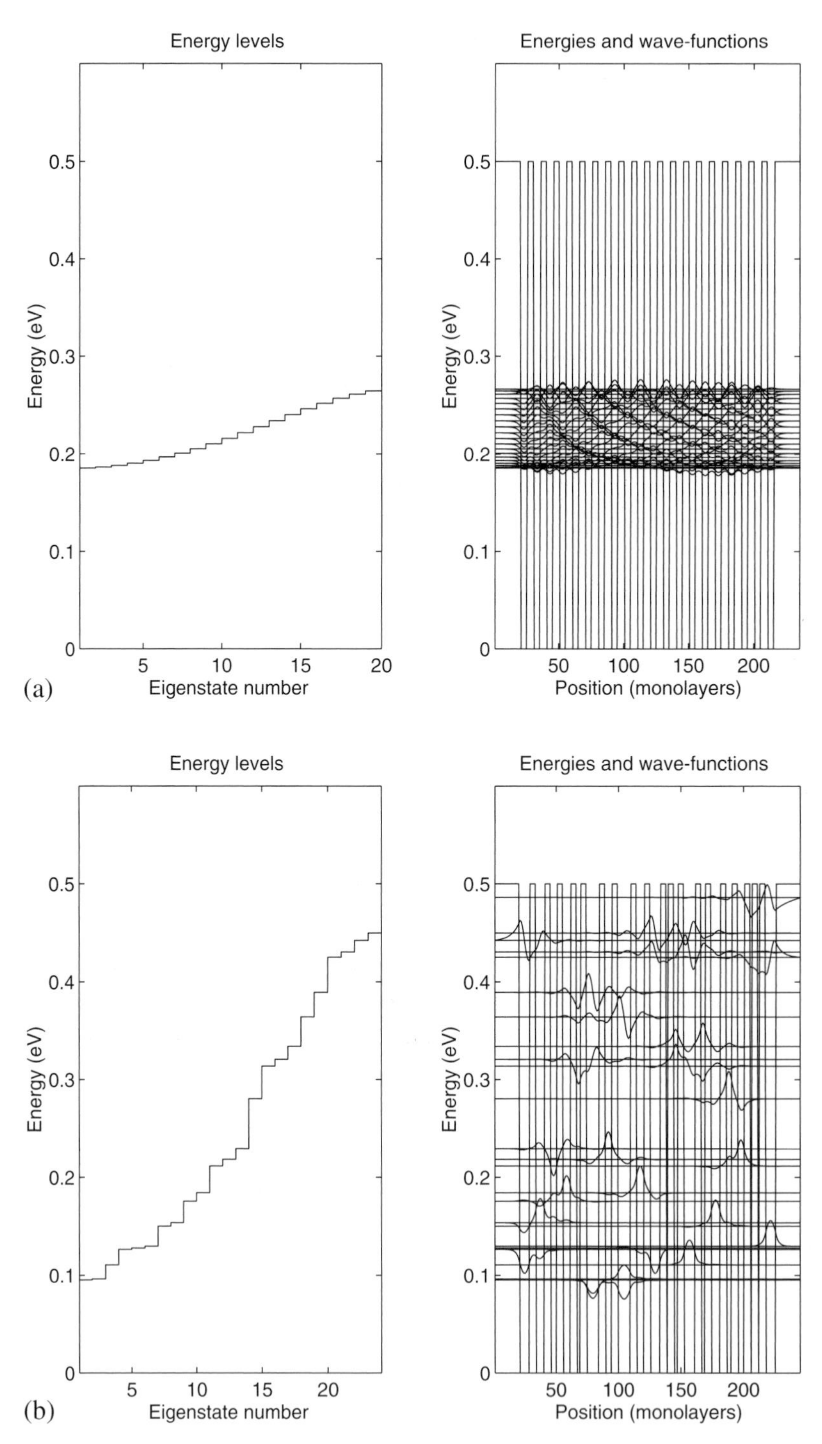

Fig. 4.18. Spectrum and wave-functions in (a) a perfect and (b) a random $GaAs/Al_{0.3}Ga_{0.7}As$ superlattice with wells and barriers of an average width of five monolayers and with a standard deviation of three monolayers in the barrier width of the random superlattice.

Assuming that we start with a baby B at generation $l = -1$ and the evolution rules are given by $B \to A$ and $A \to AB$, the adult and baby population admits the following sequence for successive generations l

$$\begin{array}{rc} -1 & B \\ 0 & A \\ 1 & AB \\ 2 & ABA \\ 3 & ABAAB \\ 4 & ABAABABA \\ 5 & ABAABABAABAAB \\ 6 & ABAABABAABAABABAABABA \end{array}$$

These are the so-called Fibonacci sequences $S(l)$. It can be verified that $S(l = \infty)$ is not invariant under a translation but is invariant under the dilatation $B \to A$ and $A \to AB$.

To build a GaAs/AlGaAs superlattice using the Fibonacci sequences, the elements A and B must each consist of a different motif of n_1 GaAs and n_2 AlGaAs layers. A periodic Fibonacci superlattice is obtained by repeating a Fibonacci sequence of order ℓ several times. The transition from a periodic superlattice to a quasi-periodic superlattice can then be studied by increasing the order ℓ of the Fibonacci sequence used. Such Fibonacci superlattices were first studied theoretically by Kohmoto and coworkers [48,49].

Figure 4.19 shows the spectrum of the energy bands obtained for the periodic approximation of a quasi-crystal. The spectrum of a Fibonacci superlattice admits a fractal character for increasing ℓ, that is, as ℓ is increased, an increase in the structure (details) of the spectrum is observed. A quasi-Fibonacci superlattice features both localized and extended states (some critical energies). The localized states can, however, be regrouped into minibands, as can be seen in Figure 4.19. Fibonacci and random superlattices are mostly of scientific interest and have not yet led to device applications.

4.6 Conclusion

In this chapter we have studied four different classes of quantum effects and associated devices: (1) Zener oscillators (accelerated electrons), (2) quantum well structures, (3) RTDs, and (4) superlattices. Special quantum resonance effects were observed when the electron wavelength was comparable to the heterostructure dimension. However, these quantum resonances can only be observed in practice in relatively small devices, when the electron mean free path is large compared with the main features of the device heterostructures. Therefore an in-depth study of quantum devices requires us

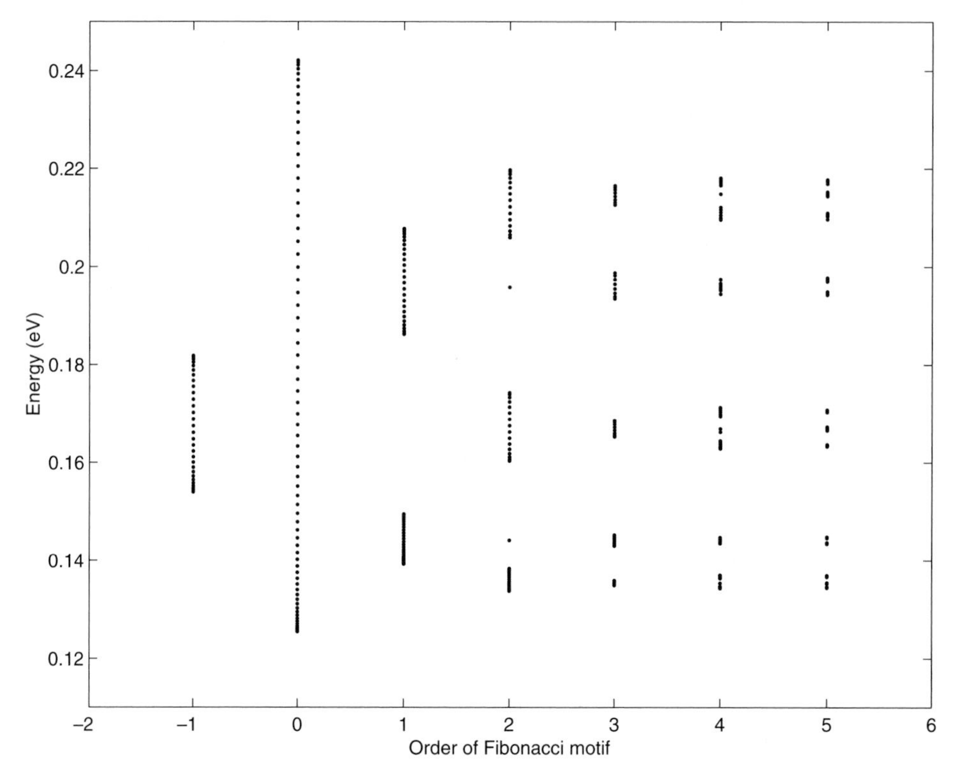

Fig. 4.19. Spectrum for Fibonacci superlattice sequences of order ℓ varying from -1 to 5 for GaAs and $Al_{0.3}Ga_{0.7}As$ layers with width (n_1, n_2) of (5,5) and (5,10) monolayers for motifs A and B, respectively.

to consider the impact of scattering on these devices. This will be pursued in Chapters 5 and 6, where we will develop a more realistic picture of quantum transport. Finally the response of these quantum devices at high frequencies, which is also of critical importance for device applications, will be investigated in Chapter 7.

4.7 Bibliography

4.7.1 Recommended reading

C. Weisbuch and B. Vinter, *Quantum Semiconductor Structures*, Academic Press, London, 1991.

4.7.2 References

[1] J. R. Glauber, 'Coherent and incoherent states in radiation fields', *Physical Review*, Vol. 131, p. 2766, 1963

[2] J. Zak, 'Coherent states in solids', *Physical Review B*, Vol. 21, No. 8, p. 3345, April 15 1980.

[3] P. W. Anderson, 'Absence of diffusion in certain random lattices', *Physical Review*, Vol. 109, No. 5, pp. 1492–1505, March 1 1958.

[4] E. R. Brown, W. D. Goodhue and T. C. L. G. Sollner, 'Fundamental oscillations up to 200 GHz in resonant tunneling diodes and new estimates of their maximum oscillation frequency from stationary-state tunneling theory', *Journal of Applied Physics*, Vol 64, pp. 1519–1529, 1988.

[5] E. R. Brown, 'Quantum transport', in N. G. Einspruch and W. R. Frensley (editors), *Heterostructures and Quantum Devices*, Academic Press, San Diego, pp. 305–350, 1994.

[6] D. Calecki, J. F. Palmier and A. Chomette, 'Hopping conduction in multiquantum well structures', *Journal de Physique*, Vol. C17, p. 5017, 1984 and Vol. 3, p. 381, 1985.

[7] S. K. Diamond, E. Ozbay, M. J. W. Rodwell, D. M. Bloom, Y. C. Pao, and J. S. Harris, 'Resonant tunneling diodes for switching applications', *Applied Physics Letters*, Vol. 54, pp. 153–155, 1989.

[8] A. Seabaugh and R. Lake, 'Resonant tunneling microscopy', in *Encyclopedia of Applied Physics*, ed. George L. Trigg, associate eds. Eduardo S. Vera, Walter Grulick, Wiley-VCH Physics, New York, Vol. 22, pp. 335–359, 1998.

[9] L. Esaki, 'New phenomenon in narrow Ge p–n junctions', *Physical Review*, Vol. 109, pp. 603–604, 1958.

[10] L. Esaki, 'Discovery of the tunnel diode', *IEEE Transactions on Electron Devices*, Vol. 23, pp. 644–647, 1976.

[11] V. M. Franks, K. F. Hulme, and J. R. Morgan, 'An alloy process for making high current density silicon tunnel diode junctions', *Solid State Electronics*, Vol. 8, p. 343, 1965.

[12] Device supplier: Germanium Power Device Corporation (1996).

[13] H. H. Tsai, Y. K. Su, H. H. Lin, R. L. Wang, and T. L. Lee, 'P–N double quantum well resonsant interband tunneling diode with peak-to-valley current ratio of 144 at room temperature', *IEEE Transactions on Electron Devices*, Vol. 15, pp. 357–359, 1994.

[14] D. J. Day, R. Q. Yang, J. Lu, J. M. Xu, 'Experimental demonstration of resonant interband tunnel diode with room temperature peak-to-valley current ratio over 100', *Journal of Applied Physics*, Vol. 73, pp. 1542–1544.

[15] J. R. Söderström, D. H. Chow, and M. C. McGill, 'New negative differential device based on resonant interband tunneling', *Applied Physics Letters*, Vol. 55, pp. 1094–1096, 1989.

[16] S. L. Rommel, T. E. Dillon, M. W. Dashiell, H. Feng, J. Kolodzey, P. R. Berger, P. E. Thompson, K. D. Hobart, R. Lake, A. C. Seabaught, G. Klimeck and D. K. Blanks, 'Room temperature operation of epitaxially grown $Si/Si_{.5}Ge_{.5}/Si$ resonant interband tunneling diodes', *Applied Physics Letters*, Vol. 73, No. 15, pp. 2191–2194, 1998.

[17] R. Duschl *et al.*, 'Epitaxially grown Si/SiGe interband tunneling diodes with high room-temperature peak-to-valley ratio', *Applied Physics Letters*, Vol. 76, p. 879, 2000.

[18] J. P. A. Vand Der Wagt, A. C. Seabaugh and E. A. Beam, 'RTD/HFET low standby power SRAM cell', *IEEE Electron Device Letters*, Vol. 19, No. 1, pp. 7–9, 1998.

[19] L. L. Chang, L. Esaki, and R. Tsu, 'Resonant tunneling in semiconductor double barriers', *Applied Physics Letters*, Vol. 24, p. 593, 1974.

[20] L. Esaki and R. Tsu, 'Superlattice and negative differential conductivity in semiconductors', *IBM Journal of Research and Development*, Vol. 14, p. 61, 1970.

[21] L. Esaki, L. L. Chang, W. E. Howard, V. L. Rideout, *Proceedings of the 11th International Conference on the Physics of Semiconductors, Warsaw, Poland*, PWN-Polish Scientific Publishers, Warsaw, p. 431, 1972.

[22] E.E. Mendez, F. Agullo-Rueda, and J. M. Hong, 'Stark localization in GaAs–GaAlAs superlattices under an electric field', *Physical Review Letters*, Vol. 60, p. 2426, (1988). P. Voisin, J. Bleuse, C. Bouche, S. Gaillard, C. Alibert and A. Regreny, 'Observation of Wannier–Stark quantization in a semiconductor superlattice', *Physical Review Letters*,Vol. 61, p. 1639, (1988).

[23] R. W. Koss and L. M. Lambert, 'Experimental observation of Wannier levels in semi-insulating gallium arsenide', *Physical Review B5*, p. 1479, 1972.

[24] P. Roblin and M. W. Muller, 'Spatially-varying band structures', *Physical Review B*, Vol. 32, No. 8, 15 October 1985.

[25] P. Roblin, 'Bandstructure matching at a heterojunction interface', *Superlattices and Microstructures*, Vol. 4 No. 3, pp. 363–370, (1988).

[26] P. Roblin and M. W. Muller, 'Coherent Zener oscillations', *Journal of Physics C: Solid State Physics*, Vol. 16, pp. 4547–4554, 1983.

[27] M. W. Muller, P. Roblin, D. L. Rode, 'Proposal for a terahertz Zener oscillator', *The physics of submicron structures: The Proceedings of a Workshop on Submicron Device Physics (1982)*, H. L. Grubin, ed., Plenum Press, New York, 1984.

[28] P. Roblin, 'Electron states in submicron semiconductor heterostructures', DSc Dissertation, Washington University, August 1984.

[29] A. Sibille, J. F. Palmier, and H. Wang, 'Observation of Esaki–Tsu negative differential velocity in GaAs/AlAs superlattices', *Physical Review Letters*, Vol. 64, Number 1, p. 52, 1990.

[30] R. Dingle, H. L. Störmer, A. C. Gossard and W. Wiegmann, 'Electron mobilities in modulation-doped semiconductor heterojunction superlattices', *Applied Physics Letters*, Vol 33, p. 665, 1978.

[31] T. C. L. G. Sollner, W. E. Goodhue, P. E. Tannenwald, C. D. Parker and D. D. Peck, 'Resonant tunneling through quantum wells at frequencies up to 2.5 THz', *Applied Physics Letters*, Vol. 43, p. 588, 1983.

[32] R. Tsu, and L. Esaki, 'Tunneling in a finite superlattice', *Journal of Applied Physics Letters*, Vol. 22, p. 562, 1973.

[33] M. Cahay, M. McLennan, S. Datta, and M. S. Lundstrom, 'Importance of space-charge effects in resonant tunneling diode', *Applied Physics Letters*, Vol. 50, p. 612, 1987.

[34] L. Yang, D. E. Mars, and M. R. T. Tan, 'Effect of electron launcher structures on AlAs/GaAs double barrier resonant tunneling diodes', *Journal of Applied Physics*, Vol. 73, No. 5, pp. 2540–2542, March 1 1993.

[35] E. Gawlinski, T Dzurak, and R. A. Tahir-Kheli, 'Direct and exchange-correlation carrier interaction effects in a resonant tunneling diode', *Journal of Applied Physics*, Vol. 72, No. 8, pp. 3562–3569, October 15 1992.

[36] R. de L. Kronig and W. G. Penney, 'Quantum mechanics of electrons in crystal lattices', *Proceedings of the Royal Society*, Vol. 130, pp. 499–513, 1930.

[37] G. H. Wannier, 'Dynamics of band electrons in electric and magnetic fields', *Reviews of Modern Physics*, Vol. 34, No. 4, October 1962.

[38] J. H. Wilkinson, *The Algebraic Eigenvalue Problem*, Clarendon, Oxford, pp. 282–307, 1965.

[39] J. M. Ziman, *Principle of the Theory of Solids*, Cambridge University Press, New York, 1972.

[40] D. Y. K. Ko and J. C. Inkson, 'Matrix method for tunneling in heterostructures: Resonant tunneling in multilayer systems', *Physical Review B*, Vol. 38, No.14, p. 9945, 1988; *Semiconductor Science and Technology*, Vol. 3, p. 791, 1988.

[41] K. V. Rousseau, K. L. Wang and J. N. Schulman, 'Gamma and X-state influences on resonant tunneling current in single and double-barrier GaAs/AlAs structures', *Applied Physics Letters*, Vol. 54., No.14, p. 1341, 1989.

[42] D. Z.-Y. Ting, E. T. Yu, and T. C. McGill, 'Multiband treatment of quantum transport in interband tunnel devices', *Physical Review B*, Vol. 45, No. 7, pp. 3583–3592, February 1992.

[43] A. Chomette, B. Deveaud, A. Regreny and G. Bastard, 'Observation of carrier localization in intentionally disordered GaAs/GaAlAs superlattices', *Physical Review Letters*, Vol. 57, p. 1464, 1986.

[44] H. Schmidt, 'Disordered one-dimensional crystals', *Physical Review*, Vol 103, p. 425, 1957.

[45] D. Shetchman, I. Blech, d. Gratias and J. W. Cahn, 'Metallic phase with long-range orientational disorder and no translational symmetry', *Physical Review Letters*, Vol. 53, p. 1951, 1984.

[46] R. Merlin, K. Bajema, R. Clarke, F. Y. Juang and P. K. Bhattacharay, 'Quasiperiodic GaAs–AlAs heterostructures', *Physical Review Letters*, Vol 55, p. 1768, 1985.

[47] D. Levine and P. J. Steihardt, 'Quasicrystals: A new class of ordered structures', *Physical Review Letters*, Vol. 53, p. 2477, 1984.

[48] M. Kohmoto, P. Kadanoff and C. Tang, 'Localization problem in one dimension: Mapping and escape', *Physical Review Letters*, Vol. 50, p. 1870, 1983.

[49] M. Kohmoto, B. Sutherland and C. Tang, 'Critical wave function and a Cantor-set spectrum of a one-dimensional quasicrystal model', *Physical Review B*, Vol. 35, p. 1020, 1987.

[50] F. Laruelle, 'Etudes des propriétés structurales et electroniques des super-réseaux apériodiques GaAs/GaAlAs', PhD dissertation, Universite de Paris-Sud, 1988.

[51] G. Gamow, *Mr. Tompkins in Paperback*, Cambridge University Press, Cambridge, 1965.

[52] H. Kroemer, 'Large-amplitude oscillation dynamics and domain suppression in a superlattice Bloch oscillator', cond-mat/0009311 (http://xxx.langl.gov/abs/cond-mat/0009311)

4.8 Problems

'Why', said the Dodo,'the best way to explain it is to do it.'

Alice in Wonderland, LEWIS CARROLL

4.1 In this problem we shall derive an analytic formula for the transmission coefficients of a single and a double barrier (when no voltage is applied). For this purpose we shall introduce scattering parameters relating the incident and reflected waves. Scattering parameters can simplify the calculation of a complex system. We can first calculate the scattering parameters of each

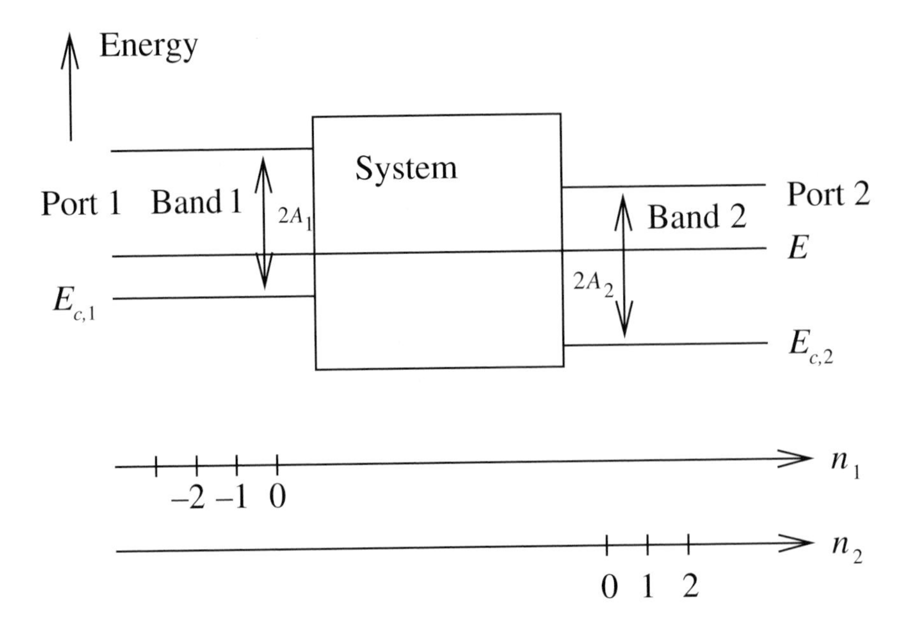

Fig. 4.20. Band diagram for the definition of the S-parameters.

individual section of a system. The total scattering parameters are then obtained by cascading each section.

Let us now define the scattering parameters of the system (black box) represented by the band diagram shown in Figure 4.20. Two different flat tridiagonal bands are used on the left- (port 1) and right-hand side (port 2). Note that the wave-functions are labeled on each side with their own n axis. The n_1 axis is directed toward the sytem. The n_2 axis is directed away from the system. The last lattice site before entering the system is $n_1 = n_2 = 0$. Consider an eigenstate of energy E:

$$E = E_{C,1} + A_1[1 - \cos(k_1 a)] = E_{C,2} + A_2[1 - \cos(k_2 a)].$$

The incident wave on port 1 is

$$a \exp[ik_1(E)n_1 a] \quad \text{for } n_1 \leq 0.$$

The incident wave on port 2 is

$$d \exp[-ik_2(E)n_2 a] \quad \text{for } n_2 \geq 0.$$

The reflected (1 to 1) or transmitted (2 to 1) wave at port 1 is

$$b \exp[-ik_1(E)n_1 a] \quad \text{for } n_1 \leq 0.$$

The reflected (2 to 2) or transmitted (1 to 2) wave at port 2 is

$$c \exp[ik_2(E)n_2 a] \quad \text{for } n_2 \geq 0.$$

The total wave f_1 at port 1 is then

$$f_1(n_1) = a \exp[ik_1(E)n_1 a] + b \exp[-ik_1(E)n_1 a].$$

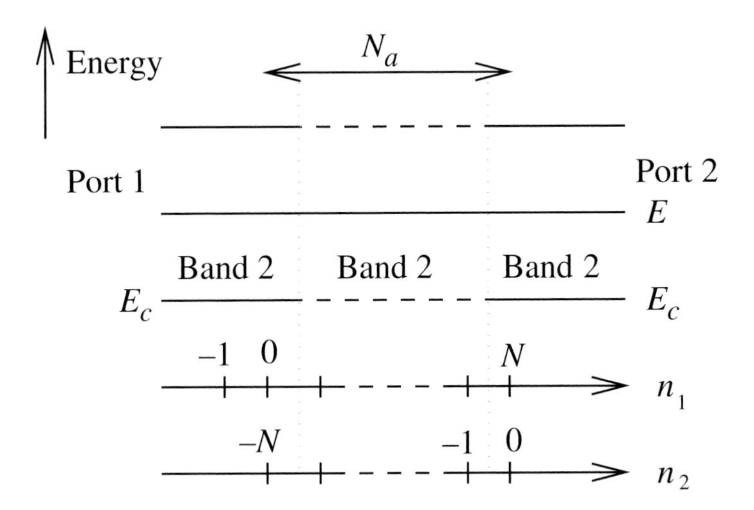

Fig. 4.21. Band diagram for a uniform potential.

The total wave f_2 at port 2 is then

$$f_2(n_2) = c \ \exp[ik_2(E)n_2 a] \ + \ d \ \exp[-ik_2(E)n_2 a].$$

We can assume that a and d are coefficients controlled by the user and that b and c are unknown coefficients to be determined. Since the Wannier equation is a linear equation for each system, we can express the reflected coefficients in terms of the transmission coefficients using a scattering matrix $\mathbf{S}$

$$\begin{bmatrix} b \\ c \end{bmatrix} = \begin{bmatrix} r_1 & t_2 \\ t_1 & r_2 \end{bmatrix} \begin{bmatrix} a \\ d \end{bmatrix} = \mathbf{S} \begin{bmatrix} a \\ d \end{bmatrix}. \tag{4.10}$$

Note that these scattering parameters allow us to calculate the transmission and reflection coefficients. The transmission coefficient T is defined as the ratio of the transmitted current over the incident current (see Chapter 3):

$$T_{12} = \frac{v_2}{v_1}|t_1|^2,$$

$$T_{21} = \frac{v_1}{v_2}|t_2|^2,$$

where v_1 and v_2 are the electron velocities in bands 1 and 2, respectively. The reflection coefficient is defined as the ratio of the reflected current over the incident current (see Chapter 3):

$$R_1 = |r_1|^2,$$

$$R_2 = |r_2|^2.$$

(a) Calculate the scattering matrix $\mathbf{S}$ of band 1 for a uniform potential of length N lattice parameters (see band diagram in Figure 4.21). Assume that the wave-vector is k_1.
(b) Calculate the scattering matrix $\mathbf{S}$ for a step potential (see band diagram in Figure 4.22). Assume that the heterojunction is maximally transparent. The wave-vector on the left-hand

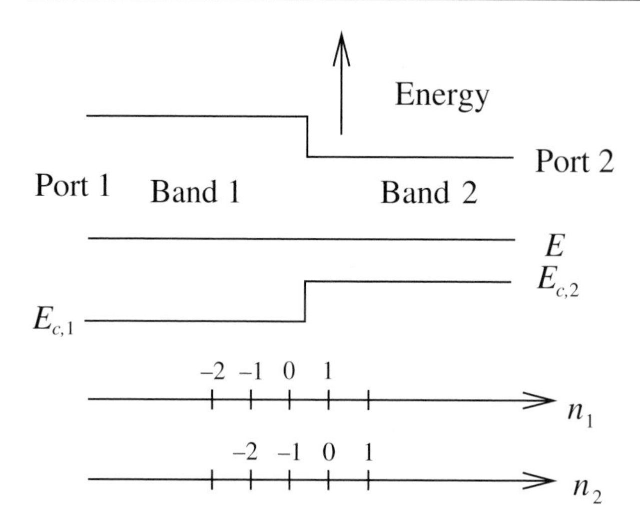

Fig. 4.22. Band diagram for the heterojunction realized with bands 1 and 2.

side is k_1 and the wave-vector (sometimes imaginary) on the right-hand side is k_2. Use the boundary conditions derived in Chapter 3 for a maximally transparent heterojunction

$$\left.\begin{aligned}\sqrt{A_2}\, f_2(n+1) &= \sqrt{A_1}\, f_1(n+1),\\ \sqrt{A_1}\, f_1(n) &= \sqrt{A_2}\, f_2(n).\end{aligned}\right\} \quad (4.11)$$

Verify that for small wave-vectors k_1 and k_2 (the effective-mass approximation) we have

$$\begin{bmatrix} r_1 & t_2 \\ t_1 & r_2 \end{bmatrix} \simeq \begin{bmatrix} \dfrac{k_1 - k_2}{k_1 + k_2} & \left(\dfrac{m_1^*}{m_2^*}\right)^{1/2} \dfrac{2k_2}{k_2 + k_1} \\ \left(\dfrac{m_2^*}{m_1^*}\right)^{1/2} \dfrac{2k_1}{k_2 + k_1} & \dfrac{k_2 - k_1}{k_1 + k_2} \end{bmatrix}. \quad (4.12)$$

Note that the junction is reciprocal and that we have

$$T_S = T_{12} = T_{21} = |t_1||t_2|,$$
$$R_S = R_1 = R_2 = |r_2|^2 = |r_1|^2 = |r_1||r_2|.$$

(c) Express the scattering matrix **S** for the reverse step potential (see band diagram in Figure 4.23) in terms of the coefficients r_1, t_1, r_2 and t_2 defined in (b).

(d) Calculate the scattering matrix $\mathbf{S}_B$ of a potential barrier of length N lattice parameters (see band diagram in Figure 4.24). For this purpose, divide the problem into three regions: a step, a flat band and a step and calculate the total scattering parameters using flow graph analysis (show the flow graph).

From the symmetry of the barrier we expect the scattering matrix to be of the form

$$\mathbf{S}_B = \begin{bmatrix} r_B & t_B \\ t_B & r_B \end{bmatrix}. \quad (4.13)$$

Verify that for $E \geq E_{C,2}$ (k_2 is real) and small wave-vectors k_1 and k_2 (r_2 is real), the

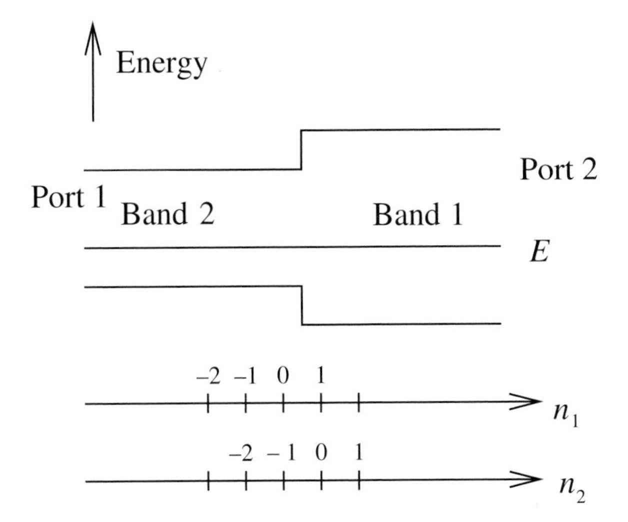

Fig. 4.23. Band diagram for the heterojunction realized with bands 2 and 1.

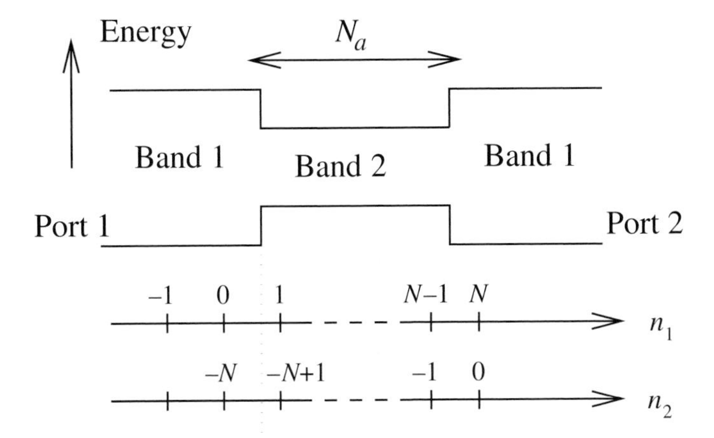

Fig. 4.24. Band diagram for the barrier structure realized with bands 1/2/1.

transmission coefficient T_B of the barrier is given by

$$T_B = |t_B|^2 = \frac{T_S^2}{1 + R_S^2 - 2R_S \cos(2k_2 Na)}.$$

Hint: Use the reciprocity property $|t_1|^2 |t_2|^2 = T_S$. The reflection coefficient is then given by $R_B = |r_B|^2 = 1 - T_B$.
For $E \geq E_{C,2}$ the transmission coefficient T_B of the barrier is seen to be a periodic function of $k_2 Na$ and oscillates between 1 and $T_B(min)$

$$T_B(min) = \frac{(1 - R_S)^2}{(1 + R_S)^2}.$$

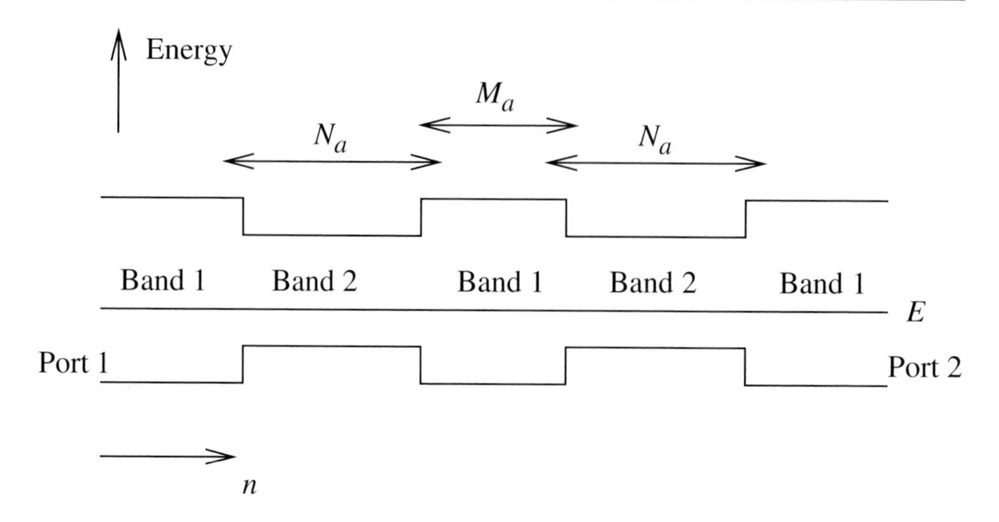

Fig. 4.25. Band diagram for the final double-barrier structure.

(e) Consider now the case $E \leq E_{C,2}$ so that k_2 is now imaginary (damped wave). Using $k_2 = i\gamma$ verify that for small wave-vectors k_1 and γ (the effective-mass approximation) t_B can be written:

$$t_B = \frac{i4k_1\gamma\exp(-\gamma Na)}{(k_1 + i\gamma)^2 - (k_1 - i\gamma)^2\exp(-2\gamma Na)} \tag{4.14}$$

and calculate r_B.

Verify that the transmission coefficient T_B of the barrier is given by

$$T_B = |t_B|^2 = 1 - |r_B|^2 = 1 - R_B = \frac{4k_1^2\gamma^2}{4k_1^2\gamma^2 + (k_1^2 + \gamma^2)^2\sinh^2(\gamma\text{Na})}.$$

(f) The scattering matrix $\mathbf{S}_{2B}$ of a double-potential barrier with a barrier width of N lattice parameters and a well width of M lattice parameters can be calculated using the results of (d) (see band diagram in Figure 4.25). From the symmetry of the double barrier we expect the scattering matrix to be of the form

$$\mathbf{S}_{2B} = \begin{bmatrix} r_{2B} & t_{2B} \\ t_{2B} & r_{2B} \end{bmatrix}. \tag{4.15}$$

Using the result derived in (d) we obtain by inspection

$$t_{2B} = \frac{t_B^2\exp(ik_1Ma)}{1 - r_B^2\exp(i2k_1Ma)}. \tag{4.16}$$

Using $r_B^2 = R_B\exp(-i2\theta)$ we can then calculate the transmission coefficient to be

$$T_{2B} = |t_{2B}|^2 = \frac{1}{1 + 4\dfrac{R_B}{T_B^2}\sin^2(k_1Ma - \theta)}. \tag{4.17}$$

The transmission is now a periodic function of k_1Ma and oscillates between 1 and $T_{2B}(min)$. Specify whether the width of the barrier N should be increased or decreased in order to minimize the value $T_{2B}(min)$.

4.2 Consider a double-barrier structure realized using a GaAs well of 51 Å width and $Al_{0.3}Ga_{0.7}As$ barriers of 34 Å width.

(a) Calculate the transmission coefficient from 0 to 1.0 eV using the result of Problem 4.1 and the effective-mass approximation.

(b) Calculate the transmission coefficient from 0 to 0.1 eV using the result of Problem 4.1 but without using the effective-mass approximation.

(c) Compare results (a) and (b)

(d) The advantage of the Wannier picture is that a numerical solution can be readily obtained for arbitrary potentials and heterostructures. Using the recurrence equations provided by the Wannier Picture calculate the transmission coefficients when a voltage of 0.1, 0.2 and 0.3 V is applied across the diode. Use the MATLAB* function *tunnel.m* available for this purpose. See the Preface for the availability of these MATLAB* tools.

4.3 Reproduce the plot of Figure 4.1(b) for X equal to 10, 20, 40, 80 and verify that $J_n(X)$ is approximately $1/(2X)^{1/2}$. Can you explain why the wave-function is larger on the edges than in the center? Use the MATLAB* function *bessel.m* available for this purpose. See the Preface for the availability of this MATLAB* tool.

4.4 The spectrum for the Zener resonator on Figure 4.3 was plotted for $A = 1$ eV and $X = A/qaF = 200$. Plot the Zener spectrum for an electric field F of twice as strong. Use the MATLAB* function *zener.m* available for this purpose. See the Preface for the availability of this MATLAB* tool. Explain the variations of the height and width of the peaks.

4.5 Generate the Wannier ladder shown in Figure 4.17 for barrier voltages of 0.1, 0.2, and 0.3 eV. Compare the simulated and theoretical energy spacing for the three Wannier ladders generated. Use the MATLAB* function *wannier.m* available for this purpose. See the Preface for the availability of this MATLAB* tool.

4.6 We now wish to study the Anderson localization effect. Use the MATLAB* function *random.m* available for this purpose. See the Preface for the availability of this MATLAB* tool. Set the standard deviation of the barrier width and height to zero. Plot the spectrum and wave-functions and establish the width B of the miniband. Then successively increase the standard deviation of the barrier height to $2B$, $3B$, and $4B$, while keeping the standard deviation of the barrier width at zero. Plot the spectrum and wave-functions for each of these three cases and determine whether the criterion for the Anderson localization of the wave-function is verified.

* MATLAB is a registered trademark of the MathWorks, Inc.

5 Scattering processes in heterostructures

Ships would be safer if they stayed in the shelter of harbors. They are however built for venturing in deep seas.

ANONYMOUS

5.1 Introduction

So far our study of quantum heterostructure devices in Chapter 4 has assumed that the devices were small compared to the mean free path of the electron. Transport in this type of device is referred to as ballistic transport. In real crystals the electron is always subjected to some type of scattering*. In Chapter 6 we will study the impact of scattering on the electron wave-function and develop a simple three-dimensional quantum transport theory. In preparation for this analysis, we must first study the scattering mechanisms to which an electron is subjected.

Various scattering mechanisms exist in semiconductors. We will consider first scattering by the lattice vibrations, and, in particular, discuss polar, acoustic, and intervalley scattering. Next, we will turn our attention to scattering processes specific to heterostructures and discuss interface roughness scattering and alloy scattering. Finally, we will conclude this chapter with a discussion of electron–electron scattering.

Note that quantum heterostructures such as resonant tunneling diodes (RTDs), superlattices, and quantum wells (e.g., modulation doped field-effect transistors (MODFETs)) are usually undoped to minimize impurity scattering. Therefore impurity scattering is usually small compared to polar scattering and interface roughness scattering.

5.2 Phonons and phonon scattering

A crystal can be represented as a network of masses connected by springs. The masses are the atoms and the springs the covalent bonds between the atoms (see Figure 5.1).

* A noted exception is the absence of scattering in superconductors under DC conditions.

```
  ||       ||       ||
= Ga  =  As  =  Ga =
  ||       ||       ||
= As  =  Ga  =  As =
  ||       ||       ||
       Bond
= Ga  =  As  =  Ga =
  ||       ||       ||
```

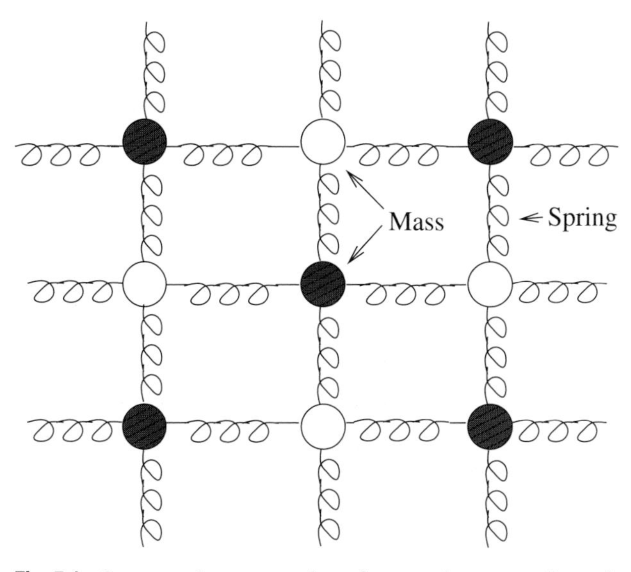

Fig. 5.1. Conceptual representation of a crystal as a two-dimensional mattress in which the springs are the covalent bonds between atoms.

Such a system is represented in classical mechanics by a Hamiltonian (kinetic energy + potential energy) of the form

$$H_{ph} = \sum_i \left[\frac{\mathbf{p}_i^2}{2M_i} + \sum_j V_{bond}(\mathbf{r}_i - \mathbf{r}_j) \right],$$

where $\mathbf{p}_i = M_i(d\mathbf{r}_i/dt)$ and $\mathbf{r}_i$ are respectively the momentum and position of the atom of mass M_i at the lattice site $\mathbf{R}_i$. The difference between the atom's position $\mathbf{r}_i$

and its equilibrium lattice site position $\mathbf{R}_i$ is called the lattice displacement $\mathbf{U}$:

$$\mathbf{U}_i = \mathbf{r}_i - \mathbf{R}_i.$$

For small lattice displacements, the bond potential V_{bond} can be expanded in a Taylor series around the equilibrium position $\mathbf{R}_i$. At this position, the bond potential reaches its minimum $V_{bond}(0)$ and the crystal Hamiltonian H_{ph} can then be rewritten as (using $V_{bond}(0) = 0$)

$$H_{ph} = \sum_i \left[\frac{\mathbf{p}_i^2}{2M_i} + \frac{1}{2} \sum_j \mathbf{U}_i \cdot \frac{\partial^2 V_{bond}(\mathbf{R}_i - \mathbf{R}_j)}{\partial \mathbf{R}_i \partial \mathbf{R}_j} \cdot \mathbf{U}_j \right]. \tag{5.1}$$

Let us now study the vibrations allowed by this crystal Hamiltonian H_{ph}. We assume that the vibrations are propagating waves of the form

$$\mathbf{U}_i = \mathbf{w_q} \cos(\omega_{\mathbf{q}} t - \mathbf{q} \cdot \mathbf{r}_i + \phi_{\mathbf{q}}),$$

with frequency ω, wave-vector $\mathbf{q}$, polarization $\mathbf{w_q}$, and phase $\phi_{\mathbf{q}}$. Substituting the proposed solution in the Hamiltonian leads to an eigenvalue problem that associates each vibration mode $\mathbf{q}$ with a specific vibration frequency $\omega_{\mathbf{q}}$ (eigenvalue). This occurs because the crystal Hamiltonian commutes with the translation operator and the vectors $\mathbf{q}$ can be used to label the eigenvalues (frequencies). This also implies that the dispersion relation $\omega_{\mathbf{q}}$ is a periodic function of $\mathbf{q}$ in reciprocal space and is left invariant by a translation of a reciprocal-lattice vector $\mathbf{K}$ (see Chapter 1).

The dispersion relation $\omega_{\mathbf{q}}$ depends on the nature of the bonds which form the crystal (see Harrison [4]) and is therefore a characteristic of the crystal. The phonon-spectrum of GaAs is given in Figure 5.2.

The polarization $\mathbf{w_q}$ is a normalized vector indicating the direction of the vibration. A vibration with arbitrary polarization can be represented as the superposition of two fundamental polarizations: transversal and longitudinal. In a transversal mode, the atoms vibrate in a direction perpendicular to the wave-vector $\mathbf{q}$. In a longitudinal mode, the atoms vibrate in a direction parallel to the wave-vector $\mathbf{q}$ and we have

$$\mathbf{w_q} = \frac{\mathbf{q}}{q} = \hat{\mathbf{q}}.$$

An arbitrary displacement $\mathbf{U}$ of the atoms from their equilibrium position can be expressed as a superposition with weight $U_{\mathbf{q}}$ of the vibrations modes $\mathbf{q}$:

$$\mathbf{U}_i = \frac{2}{\sqrt{\Omega}} \sum_{\mathbf{q}} U_{\mathbf{q}} \mathbf{w_q} \cos(\omega_{\mathbf{q}} t - \mathbf{q} \cdot \mathbf{r}_i + \phi_{\mathbf{q}}). \tag{5.2}$$

To simplify the scattering analysis we will use a continuous displacement wave approximation (see Kittel [2])

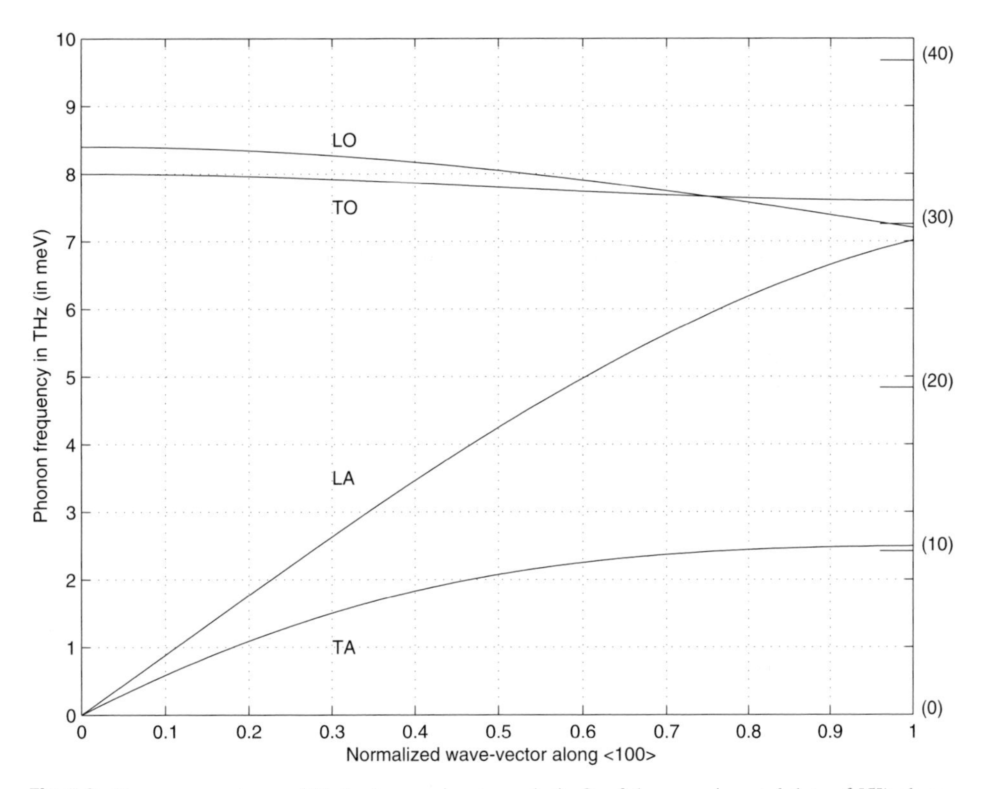

Fig. 5.2. Phonon–spectrum of GaAs (approximate analytic fit of the experimental data of [5]) along the Γ to X direction.

$$\mathbf{U}(\mathbf{r}) = \frac{2}{\sqrt{\Omega}} \sum_{\mathbf{q}} U_{\mathbf{q}} \mathbf{w}_{\mathbf{q}} \cos(\omega_{\mathbf{q}} t - \mathbf{q} \cdot \mathbf{r} + \phi_{\mathbf{q}}), \tag{5.3}$$

which is accurate for long-wavelength vibrations.

In compound semiconductors two different masses are involved, e.g., Ga and As in GaAs. This allows us to understand the formation of two major different modes of vibration in a crystal: the acoustic mode, and the optical mode. In the acoustic mode, both masses move in the same direction, but in the optical mode, they move in opposite directions. In Figure 5.2 one can distinguish the longitudinal acoustic mode (LA), transversal acoustic mode (TA), longitudinal optical mode (LO), and transversal optical mode (TO).

Consider the acoustic modes. When the wave-vector $\mathbf{q}$ is small, the wavelength (period in space) $2\pi/|\mathbf{q}|$ of the vibration is large and the frequency of vibration is small. Conversely, when the wave-vector $\mathbf{q}$ is large, the wavelength $2\pi/|\mathbf{q}|$ of the vibration is small and the frequency of vibration is large. In our analysis we shall rely on a simple linear dispersion relation

$$\hbar\omega_{\mathbf{q}} \simeq v_s |\mathbf{q}|,$$

where v_s is recognized as the sound velocity since it can be calculated using the group velocity formula

$$v_s = \left| \frac{\partial \omega_{\mathbf{q}}}{\partial \mathbf{q}} \right|.$$

Now consider the optical phonons. Note that the phonon frequency in the optical branch varies little with the wave-vector $\mathbf{q}$. The optical phonon frequency for the longitudinal mode is about $\omega_{LO} \simeq 8$ THz in GaAs. We will consider only polar scattering by longitudinal optical phonons and will use the so-called Einstein model which neglects the wave-vector dependence:

$$\omega_{\mathbf{q}} \simeq \omega_{LO}.$$

Note that in order to introduce the vibration modes, we have assumed that the displacement of the atoms from their equilibrium positions was small and we retained only the first order term in the expansion of the bond potential. The crystal then reduces to a harmonic oscillator. Inclusion of the high-order terms in the bond potential introduces mechanisms which permit the collision of vibrations with other vibrations and therefore limit their lifetime.

5.2.1 Phonons

We have introduced the vibration of the lattice using a classical Hamiltonian and this was found sufficient to predict the vibration spectrum $\omega_{\mathbf{q}}$. The analysis of the scattering of an electron by vibrations, however, requires us to treat the crystal vibrations using quantum mechanics. We shall see that a quantum treatment is necessary in order to introduce the concepts of spontaneous emission and stimulated emission and absorption which are of critical importance for a correct picture of scattering. Once these quantum effects have been established in this section, we shall develop an intermediate semiclassical picture which incorporates them approximately.

A quantum analysis of the crystal vibration is obtained by replacing the momentum $\mathbf{p}_i$ in the Hamiltonian Equation (5.1), by an operator $-i\hbar(\partial/\partial\mathbf{r}_i)$. As a result of this quantum analysis (see [2]) (called second quantization) the vibrations of a crystal must also be treated as particles called phonons.

As particles, the phonons are attributed an energy E and a momentum $\mathbf{p}$ given by

$$E = \hbar\omega_{\mathbf{q}},$$
$$\mathbf{p} = \hbar\mathbf{q}.$$

The particle nature of the phonons is revealed in the interaction (collision) of electrons and phonons. It can be verified that in the collision of an electron of energy E_1 with a phonon of energy $\hbar\omega_{\mathbf{q}}$ of wave-vector $\mathbf{q}$, the electron either absorbs or emits a phonon and the final electron energy E_2 is given by $E_1 \pm \hbar\omega_{\mathbf{q}}$.

In a uniform semiconductor (flat band) one can also derive a three-dimensional wave-vector conservation rule, $\mathbf{k}_2 = \mathbf{k}_1 \pm \mathbf{q}$, if we assume the electron to be in the Bloch states $|\mathbf{k}_1\rangle$ and $|\mathbf{k}_2\rangle$ respectively before and after the electron–phonon collision. A generalization of this formula including Umklapp processes is discussed in Ziman [1].

Note that for heterostructures varying along the x direction, we shall derive in Chapter 6 an alternative two-dimensional wave-vector conservation rule:

$$\mathbf{k}_{2,\perp} = \mathbf{k}_{1,\perp} \pm \mathbf{q}_{\perp},$$

where $\mathbf{k}_{1,\perp}$ and $\mathbf{k}_{2,\perp}$ are respectively the *perpendicular* wave-vectors of the electron before and after the electron–phonon collision.

The conservation of energy and momentum is therefore a manifestation of the particle nature of both the electrons and the phonons. Other important quantum effects are the processes of stimulated and spontaneous emission. To introduce these concepts it is necessary to discuss the quantum picture of the phonons in more detail. A summary of the quantum treatment of phonons is presented below. The reader is referred to Loudon [6], or Kittel [2] for a derivation.

Using the mode expansion, the crystal Hamiltonian H_{ph} can be written as the sum of harmonic oscillators $H_{\mathbf{q}}$ for each mode of vibration $\mathbf{q}$ (see Ferry [10], Ziman [1])

$$H_{ph} = \sum_{\mathbf{q}} H_{\mathbf{q}}. \tag{5.4}$$

The crystal state can be obtained by solving the eigenvalue problem for the harmonic oscillator Hamiltonian $H_{\mathbf{q}}$ for each mode $\mathbf{q}$:

$$H_{\mathbf{q}}|n_{\mathbf{q}}\rangle = E_{n_{\mathbf{q}}}|n_{\mathbf{q}}\rangle.$$

A discrete spectrum is obtained (see Figure 5.3) and the eigenvalues (energy) $E_{n_{\mathbf{q}}}$ are labeled with the integer $n_{\mathbf{q}}$ and the eigenstates $|n_{\mathbf{q}}\rangle$. As for any eigenvalue problem, the eigenstates are orthogonal

$$\langle n'_{\mathbf{q}}|n_{\mathbf{q}}\rangle = \delta_{n'_{\mathbf{q}}\, n_{\mathbf{q}}}. \tag{5.5}$$

It can be demonstrated that the harmonic oscillator Hamiltonian of the mode $\mathbf{q}$ can be written

$$H_{\mathbf{q}} = \hbar\omega_{\mathbf{q}}\left(a_{\mathbf{q}}^{+} a_{\mathbf{q}} + \frac{1}{2}\right), \tag{5.6}$$

where $a_{\mathbf{q}}^{+}$ and $a_{\mathbf{q}}$ are the creation and annihilation operators of phonons in the mode $\mathbf{q}$. The creation and annihilation operators for the phonon of mode $\mathbf{q}$ will be defined here by their operation upon the phonon state $|n_{\mathbf{q}}\rangle$:

$$a_{\mathbf{q}}\,|n_{\mathbf{q}}\rangle = \sqrt{n_{\mathbf{q}}}\;|n_{\mathbf{q}} - 1\rangle \tag{5.7}$$

$$a_{\mathbf{q}}^{+}\,|n_{\mathbf{q}}\rangle = \sqrt{n_{\mathbf{q}} + 1}\;|n_{\mathbf{q}} + 1\rangle. \tag{5.8}$$

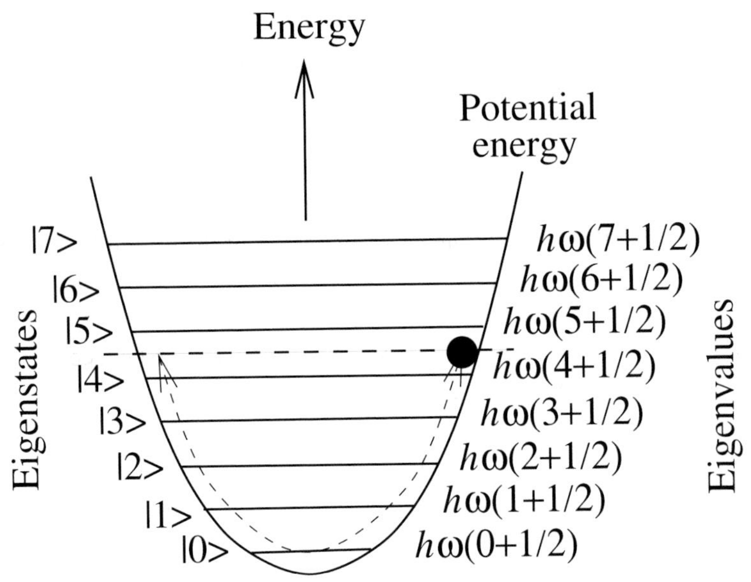

Fig. 5.3. Eigenvalues and potential energy of the harmonic oscillator for a given mode **q**. Multiple phonons can populate the same vibration mode **q**. The dashed line shows the classical trajectory of a particle which is approximated by the Glauber state.

We can easily verify that creation and annihilation operators of the same mode **q** do not commute. However, the creation and annihilation operators of different modes do commute. This is summarized by the property

$$a_{\mathbf{q}} a_{\mathbf{q}'}^{+} - a_{\mathbf{q}'}^{+} a_{\mathbf{q}} = \delta_{\mathbf{q}\mathbf{q}'}. \tag{5.9}$$

Therefore we have

$$a_{\mathbf{q}}^{+} a_{\mathbf{q}} \left| n_{\mathbf{q}} \right\rangle = n_{\mathbf{q}} \left| n_{\mathbf{q}} \right\rangle \tag{5.10}$$

and the energy of a photon state $|n_{\mathbf{q}}\rangle$ is then given by

$$E_{n_{\mathbf{q}}} = \hbar \omega_{\mathbf{q}} \left(n_{\mathbf{q}} + \frac{1}{2} \right). \tag{5.11}$$

We can now identify the state $|n_{\mathbf{q}}\rangle$ as a state of lattice vibration for the mode **q** involving $n_{\mathbf{q}}$ quanta of energy $\hbar\omega_{\mathbf{q}}$. A quantum of energy $\hbar\omega_{\mathbf{q}}$ is called a phonon. The state $|n_{\mathbf{q}}\rangle$ is therefore a vibration state involving $n_{\mathbf{q}}$ phonons.

In this quantum picture, the lattice displacement (the displacement of the atoms relative to their equilibrium crystal site) is given by the following operator [2]:

$$\mathbf{U} = \frac{1}{\Omega^{1/2}} \sum_{\mathbf{q}} \mathbf{w}_{\mathbf{q}} \left(\frac{\hbar}{2\rho\omega_{\mathbf{q}}} \right)^{1/2} \left[a_{\mathbf{q}} \exp(i\mathbf{q} \cdot \mathbf{r}) + a_{\mathbf{q}}^{+} \exp(-i\mathbf{q} \cdot \mathbf{r}) \right], \tag{5.12}$$

where Ω is the crystal volume, ρ the semiconductor density, and $\mathbf{w}_{\mathbf{q}}$ the polarization of the phonon.

Note that the lattice displacement operator of Equation (5.12) is similar to the classical lattice displacement of Equation (5.3). In both cases, we are summing over all the modes and are using a continuous position approximation. However, the amplitude $U_{\mathbf{q}}$ is now replaced by the creation and annihilation operators $a_{\mathbf{q}}^{+}$ and $a_{\mathbf{q}}$.

What is a phonon?

We have introduced the phonon state as an eigenstate of the crystal Hamiltonian for a given vibration mode **q**. Phonon states are labeled by their phonon numbers in a given mode **q**. Indeed, several phonons can occupy the same phonon state. This classifies the phonon as a Boson particle (after the scientist Bose who first conceived of them).

The phonon state $|n_{\mathbf{q}}\rangle$ is an eigenstate and therefore corresponds to a vibration whose energy is exactly known. However, the phase of this vibration is not known. This is verified by calculating the expected value of the lattice displacement operator for a phonon of mode **q**:

$$\begin{aligned}\langle n_{\mathbf{q}}|U(\mathbf{r})|n_{\mathbf{q}}\rangle &= \frac{1}{\Omega^{1/2}} \sum_{\mathbf{q}} \mathbf{w}_{\mathbf{q}} \left(\frac{\hbar}{2\rho\omega_{\mathbf{q}}}\right)^{1/2} \left[\langle n_{\mathbf{q}}|a_{\mathbf{q}}|n_{\mathbf{q}}\rangle \exp(i\mathbf{q}\cdot\mathbf{r}) \right. \\ &\quad \left. + \langle n_{\mathbf{q}}|a_{\mathbf{q}}^{+}|n_{\mathbf{q}}\rangle \exp(-i\mathbf{q}\cdot\mathbf{r})\right] \\ &= 0, \end{aligned} \tag{5.13}$$

where we have used the orthogonality property of the phonon states. The average lattice displacement is zero for a phonon state. This is rather different from the classical vibration mode and is simply due to the fact that the phase $\phi_{\mathbf{q}}$ of the vibration is unknown and can assume all possible values (see Figure 5.4).

It is, however, possible to construct a classical vibration for the mode **q** by constructing a state of vibration which is a superposition (wave-packet) $P(n_{\mathbf{q}})$ of various phonon states of the mode **q**:

$$|\psi_{\mathbf{q}}\rangle = \sum_{n_{\mathbf{q}}=0}^{\infty} P_{\psi}(n_{\mathbf{q}}) \exp(-in_{\mathbf{q}}\omega_{\mathbf{q}}t)|n_{\mathbf{q}}\rangle.$$

The phonon distribution that yields the closest approximation to a classical state is the Glauber state $|N\rangle$, which is generated by

$$P_{N_{\mathbf{q}}}(n) = \exp\left(-\frac{1}{2}|N_{\mathbf{q}}|^2\right) \frac{N_{\mathbf{q}}^n}{(n!)^{1/2}}$$

where $N_{\mathbf{q}}$ is the average number of phonons in this state. The phonon distribution $|\langle n|N_{\mathbf{q}}\rangle|^2 = P_{N_{\mathbf{q}}}^2(n)$ is recognized to be a Poisson distribution (see Figure 5.5). In a Glauber state $|N_{\mathbf{q}}\rangle$, the phase and amplitude of the vibrations are both known with the minimum possible uncertainty allowed by the uncertainty principle. In the case of light (photons), this is indeed the state which is approximately generated by a laser.

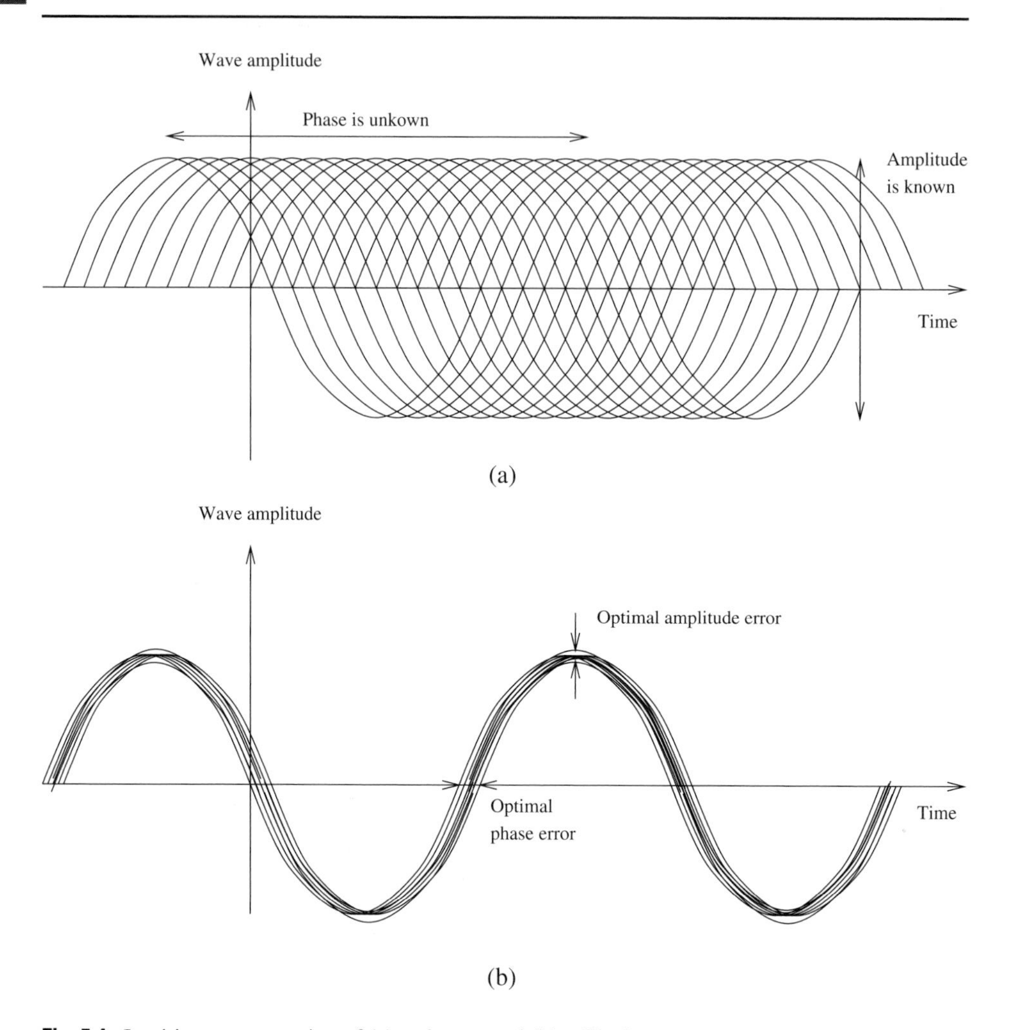

Fig. 5.4. Intuitive representation of (a) a phonon and (b) a Glauber state.

For a crystal in equilibrium at temperature T, $P(n)$ is randomly distributed and time-varying since it is updated by the collision of the phonons with other phonons or electrons. The ensemble average ($\langle \cdots \rangle_{E.A.}$) of the phonon distribution is of the form (maximum disorder)

$$\langle |P_{Th,\mathbf{q}}(n)|^2 \rangle_{E.A.} = \frac{N_\mathbf{q}^n}{(1+N_\mathbf{q})^{n+1}},$$

where the average number of phonons $N_\mathbf{q}$ in the mode $\mathbf{q}$ is given by the Bose–Einstein distribution:

$$N_\mathbf{q} = \frac{1}{\exp\left(\dfrac{\hbar\omega_\mathbf{q}}{k_B T}\right) - 1}$$

Note that the average number of phonons, $N_\mathbf{q}$, becomes infinite for phonons of low

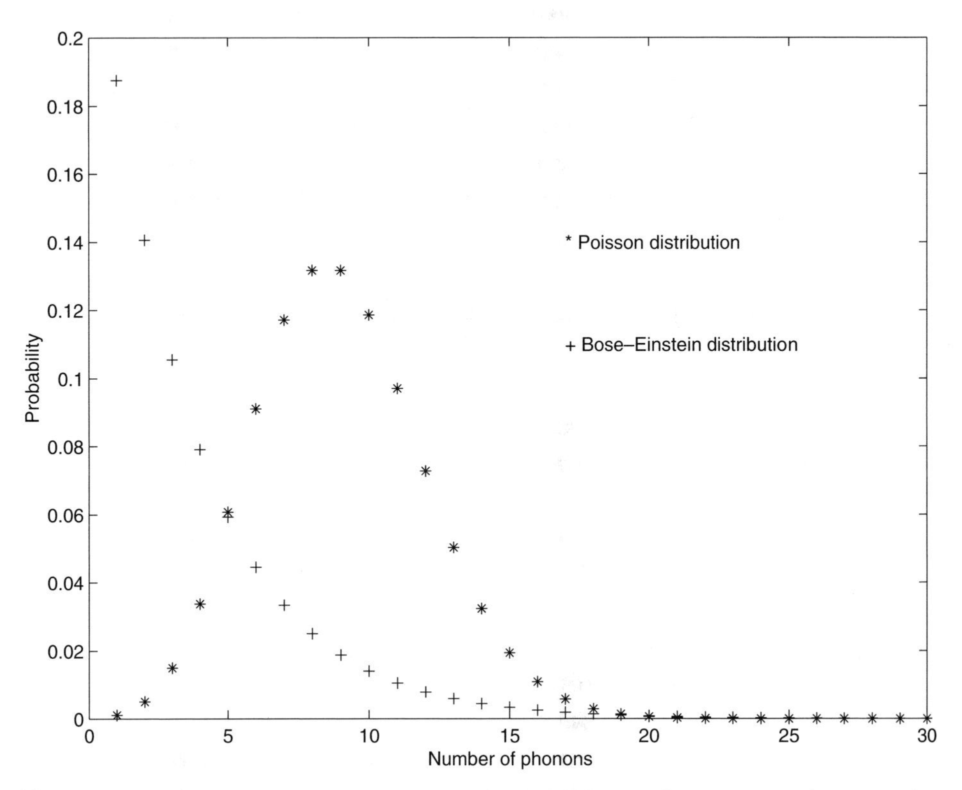

Fig. 5.5. Poisson distribution (∗) and Bose–Einstein distribution (+) for an average phonon number of 3.

frequency (acoustic mode). For optical phonons in GaAs, the number of phonons $N_{\mathbf{q}}$ in each mode $\mathbf{q}$ is about 0.34 at room temperature. It is with such a thermal wave-packet of phonons (vibrations) that the electron is colliding in a crystal.

5.2.2 Spontaneous and stimulated emissions

A rigorous quantum mechanical analysis of the interaction of electrons and phonons requires the study of the total electron–phonon system. The total Hamiltonian of this electron–phonon system includes the electron and lattice Hamiltonians and the interaction Hamiltonians H_{e-ph}

$$H_{total} = H_{e^-} + H_{ph} + H_{e-ph}.$$

In our discussion of the scattering of electrons by phonons (Sections 5.3–5.5), we shall see that the electron–phonon interaction term is of the form

$$H_{e-ph} = \frac{i}{\sqrt{\Omega}} \sum_{\mathbf{q}} \alpha(\mathbf{q}) \left[a_{\mathbf{q}} \exp(i\mathbf{q}\cdot\mathbf{r}) - a_{\mathbf{q}}^{+} \exp(-i\mathbf{q}\cdot\mathbf{r}) \right],$$

where Ω is the lattice volume and $\alpha(\mathbf{q})$ is the interaction weight for the phonon wavevector $\mathbf{q}$ (see Sections 5.3–5.5). This interaction Hamiltonian is similar to the lattice displacement, which is not surprising since it is the lattice displacement which induces the scattering.

We wish now to discuss the spontaneous and stimulated emission processes. The electron–phonon Hamiltonian couples together the combined electron–phonon states $|n, \mathbf{q}_\perp, n_\mathbf{q}\rangle$ of the non-interacting system. The interaction Hamiltonian H_{e-ph} of the phonon states written in the phonon basis involves for the mode $\mathbf{q}$ the matrix element:

$$\langle n'_\mathbf{q}|a^+_\mathbf{q}|n_\mathbf{q}\rangle = (n_\mathbf{q}+1)^{1/2}\,\delta_{n'_\mathbf{q}\,n_\mathbf{q}+1}, \tag{5.14}$$

$$\langle n'_\mathbf{q}|a_\mathbf{q}|n_\mathbf{q}\rangle = (n_\mathbf{q})^{1/2}\,\delta_{n'_\mathbf{q}\,n_\mathbf{q}-1}. \tag{5.15}$$

The matrix element of Equation (5.14) couples a phonon state $|n_\mathbf{q}\rangle$ to a phonon state $|n_\mathbf{q}+1\rangle$ and corresponds therefore to the emission of a phonon of mode $\mathbf{q}$. The matrix element of Equation (5.15) couples a phonon $|n_\mathbf{q}\rangle$ to a phonon state $|n_\mathbf{q}-1\rangle$ and corresponds therefore to the absorption of a phonon of mode $\mathbf{q}$. According to Fermi's golden rule, the rates of absorption and emission are proportional to the amplitude squared of these matrix elements. The emission rate of photons is therefore proportional to $n_\mathbf{q}+1$, whereas the absorption rate of photons is proportional to $n_\mathbf{q}$.

First, note that the emission and absorption rates of phonons of mode $\mathbf{q}$ is enhanced by the presence of $n_\mathbf{q}$ phonons. Emission and absorption are said to be stimulated. The phonon scattering rate will therefore increase when the temperature is raised, since the Bose–Einstein distribution predicts that the average number of phonons increases. Conversely, it is possible to decrease the phonon population and therefore the scattering rate by reducing the temperature. One can then actually eliminate the absorption of phonons by the electrons at 0 K, since the rate of absorption of phonons is proportional to the number of phonons $n_\mathbf{q}$. However, it is not possible to completely eliminate the emission of phonons since even in the absence of phonons ($n_\mathbf{q}=0$) the emission rate remains finite. This is referred to as spontaneous emission. An electron can spontaneously emit a phonon. This is the physical mechanism which allows an excited band electron to relax toward the bottom of the band even at 0 K.

A critical point in these absorption and emission processes is that their rates are different. Note that the coupling coefficient associated with the emission is larger by a factor of $(n_\mathbf{q}+1)/n_\mathbf{q}$ than the coupling coefficient associated with absorption. This means that the emission of a phonon by an electron has a higher probability of occurring than the absorption of a phonon by an electron. This explains why electrons in equilibrium (electron gas) preferentially populate states with lower energy rather than states with higher energy.

The processes of spontaneous emission and stimulated emission and absorption are quantum features that are of critical importance for the development of a scattering picture which includes the mechanisms which restore equilibrium. In the next section, we shall see how we can incorporate these quantum effects in a simpler semiclassical model.

5.2.3 Semiclassical phonon model

In our treatment of scattering, we shall assume that the electron is coupled to classical lattice vibrations. However, we use the coupling coefficients derived when the electron is coupled to quantized lattice vibrations or phonons. This semiclassical treatment neglects the quantum noise associated with the random thermal distribution of the lattice vibration over the various phonon states $|n_{\mathbf{q}}\rangle$ for a given mode $\mathbf{q}$.

To develop this classical model, we start from the electron–phonon interaction Hamiltonian

$$H_{e-ph} = \frac{i}{\Omega^{1/2}} \sum_{\mathbf{q}} \alpha_{\mathbf{q}} \left[a_{\mathbf{q}} \exp(i\mathbf{q} \cdot \mathbf{r}) - a_{\mathbf{q}}^{+} \exp(-i\mathbf{q} \cdot \mathbf{r}) \right]. \tag{5.16}$$

The semiclassical picture for the phonon interaction Hamiltonian is obtained by performing the following substitutions:

$$a_{\mathbf{q}} = A_{\mathbf{q}\pm} \exp[-i(\omega_{\mathbf{q}} t + \phi_{\mathbf{q}})]$$

$$a_{\mathbf{q}}^{+} = A_{\mathbf{q}\pm} \exp[i(\omega_{\mathbf{q}} t + \phi_{\mathbf{q}})]$$

where $A_{\mathbf{q}\pm}$ is the lattice vibration amplitude of mode $\mathbf{q}$ with frequency $\omega_{\mathbf{q}}$ and phase $\phi_{\mathbf{q}}$. Note that in this substitution, we consider two possible amplitudes $A_{\mathbf{q}+}$ and $A_{\mathbf{q}-}$ because we intend to use the quantum result that the phonon emission and absorption rates are weighted by the factors $1 + N_{\mathbf{q}}$ and $N_{\mathbf{q}}$, respectively. $A_{\mathbf{q}+}$ will be used for both $a_{\mathbf{q}}$ (scattering) and $a_{\mathbf{q}}^{+}$ (backscattering) when these operators contribute to the absorption of a phonon by the incident electron. $A_{\mathbf{q}-}$ will be used for both $a_{\mathbf{q}}^{+}$ (scattering) and $a_{\mathbf{q}}$ (backscattering) when these operators contribute to the emission of a phonon by the incident electron.

After substitution the following semiclassical interaction Hamiltonian is obtained:

$$H_{e-ph} = \frac{2}{\Omega^{1/2}} \sum_{\mathbf{q}} \alpha_{\mathbf{q}} A_{\mathbf{q}\pm} \sin(\omega_{\mathbf{q}} t - \mathbf{q} \cdot \mathbf{r} + \phi_{\mathbf{q}}). \tag{5.17}$$

Note that the amplitude $A_{\mathbf{q}\pm}$ and the phase $\phi_{\mathbf{q}}$ are randomly updated at times set by the phonon lifetime. For optical phonons we have $\Delta\omega = 0.012\omega_{LO}$ [7]. The corresponding phonon lifetime is 8.3 ps and the resulting energy broadening is on the order of 0.5 meV. For a given device, this introduces a noise in the DC current. Since we are only interested here in the average current, a simpler model is obtained by assuming that the phonon amplitude $A_{\mathbf{q}\pm}$ and phonon phase $\phi_{\mathbf{q}}$ are randomly distributed and mutually uncorrelated. The average current is then obtained from an ensemble average over the phonon amplitude and phase using

$$\left.\begin{aligned} \langle A_{\mathbf{q}'\pm}\exp(i\phi_{\mathbf{q}'})A_{\mathbf{q}\pm}\exp(-i\phi_{\mathbf{q}})\rangle_{E.A.} &= \langle A^2_{\mathbf{q}\pm}\rangle_{E.A.}\delta_{\mathbf{q}'\mathbf{q}} \\ &= \left(N_{\mathbf{q}} + \frac{1}{2} \mp \frac{1}{2}\right)\delta_{\mathbf{q}'\mathbf{q}}, \\ \langle A_{\mathbf{q}'\pm}\exp(i\phi_{\mathbf{q}'})A_{\mathbf{q}\pm}\exp(i\phi_{\mathbf{q}})\rangle_{E.A.} &= 0. \end{aligned}\right\} \tag{5.18}$$

The notation $\langle\cdots\rangle_{E.A.}$ indicates that an ensemble average is performed over the scattering events. $N_{\mathbf{q}}$ is the average number of phonons in the mode $\mathbf{q}$ given by the Bose–Einstein thermal distribution:

$$N_{\mathbf{q}} = \frac{1}{\exp\left(\dfrac{\hbar\omega_{\mathbf{q}}}{k_B T}\right) - 1}.$$

The phonon field extends over the entire heterostructure system (sample) of length L_x. Note, however, that we shall only study the interaction between electrons and phonons in the superlattice region (e.g., double barrier) of the heterostructure whose length L_{SL} is generally much smaller than L_x. It is convenient to assume that the lattice vibrations vanish at the edges of the sample. Using $\phi_{q_x,\mathbf{q}_\perp} = \phi_{-q_x,\mathbf{q}_\perp} + \pi$ we obtain the desired standing waves:

$$H_{e-ph} = \frac{2}{\Omega^{1/2}} \sum_{q_x>0} \sum_{\mathbf{q}_\perp} C_\pm(\mathbf{q}) \cos(\omega t - \mathbf{q}_\perp \cdot \mathbf{r}_\perp + \phi_{\mathbf{q}}) \sin(q_x x), \tag{5.19}$$

where we have introduced the constant

$$C_\pm(\mathbf{q}) = 2\alpha_{\mathbf{q}} A_{\mathbf{q}\pm}. \tag{5.20}$$

q_x is now quantized $q_x L = q_x N a = p\pi$ and given by

$$q_x = \frac{p\pi}{L} = \frac{p}{N}\frac{\pi}{a} \quad \text{for} \quad 0 \le p \le \frac{N}{2}. \tag{5.21}$$

The number of q_x modes in the sample is given by half the number of lattice sites. If the sample length L_x is large enough compared to L_{SL}, the phonon momentum quantization does not affect the phonon-assisted tunneling process. The length of the sample L_x then becomes arbitrary. Indeed, the increase in the number of modes q_x resulting from an increase of L_x is compensated in H_{e-ph} by the decrease in the polar interaction strength since we have $\Omega = L_x S$, where $S = L_y L_z$ is the superlattice area and L_x its length.

Note that only bulk phonons have been considered here. Scattering by localized and interface phonon modes is also expected to contribute [8,9]. In the next sections, we will calculate the constant $\alpha_{\mathbf{q}}$ for two important phonon scattering processes.

5.3 Polar scattering by optical phonons

The optical vibration mode induces a polarization in compound semiconductor crystals which is an efficient scatterer for the electron. The result is called polar scattering in which an electron is scattered by the longitudinal optical phonons (LO) through the interaction of its Coulomb field with the polarization waves of the lattice. We shall see that the vibration–electron coupling is the strongest for long wavelength vibrations (small $\mathbf{q}$).

We shall now derive the interaction Hamiltonian by considering the interaction of a charge distribution $\rho(\mathbf{r})$ with the polarization generated by the lattice displacement. We derive the interaction potential in SI units. A lattice displacement $\mathbf{X}$ has associated with it a lattice polarization

$$\mathbf{P}_{lattice} = F\mathbf{X},$$

where F ($m^{1/2}C$)is the Fröhlich constant [3] (see Kittel [2] for a derivation) given by

$$F = \left[\frac{\hbar\omega_{LO}}{2}\epsilon_v^2\left(\frac{1}{\epsilon_{opt}} - \frac{1}{\epsilon_{stat}}\right)\right]^{\frac{1}{2}}$$

where ϵ_{opt}, ϵ_{stat}, and ϵ_v are, respectively, the optical, static, and vacuum dielectric constants.

A charge distribution $\rho(\mathbf{r})$ generates an external electric displacement $\mathbf{D}_{ext}$ according to

$$\nabla \cdot \mathbf{D}_{ext} = \rho(\mathbf{r}). \tag{5.22}$$

The external electric displacement in turn induces an external electronic polarization $\mathbf{P}_{ext}$

$$\mathbf{P}_{ext} = \left(\frac{\epsilon_{opt} - \epsilon_v}{\epsilon_{opt}}\right)\mathbf{D}_{ext}.$$

The internal electric field $\mathbf{E}_{int}$ that is built up is related to the external electric displacement $\mathbf{D}_{ext}$ by

$$\mathbf{D}_{ext} = \epsilon_v\mathbf{E}_{int} + \mathbf{P}_{lattice} + \mathbf{P}_{ext},$$

so that

$$-\mathbf{E}_{int} = \frac{F\mathbf{X}}{\epsilon_v} - \frac{1}{\epsilon_{opt}}\mathbf{D}_{ext}.$$

The total electrostatic energy U is then

$$\begin{aligned} U &= \frac{1}{2}\epsilon_{opt}\mathbf{E}_{int}^2 \\ &= \frac{1}{2}\epsilon_{opt}\left[\frac{F}{\epsilon_v}\mathbf{X}\right]^2 + \frac{1}{2}\epsilon_{opt}\left[\frac{\mathbf{D}_{ext}}{\epsilon_{opt}}\right]^2 - \frac{F}{\epsilon_v}\mathbf{X}\cdot\mathbf{D}_{ext}, \end{aligned}$$

from which the interaction potential between the external electric displacement and the lattice polarization is obtained as

$$\Delta U = -\frac{F}{\epsilon_v}\mathbf{X}\cdot\mathbf{D}_{ext} = -\frac{1}{\epsilon_v}\mathbf{P}_{lattice}\cdot\mathbf{D}_{ext},$$

with

$$\mathbf{X} = \frac{1}{\sqrt{\Omega}}\sum_{\mathbf{q}}\hat{\mathbf{q}}\left[a_{\mathbf{q}}\exp(i\mathbf{q}\cdot\mathbf{r}) + a_{\mathbf{q}}^{+}\exp(-\mathbf{q}\cdot\mathbf{r})\right].$$

The total interaction is the summation over all the dipoles generated at each lattice site. For small lattice vectors $\mathbf{q}$ (the dominant contribution), the summation can be approximated by the integral

$$H_{e-ph}^{LO} = \int \Delta U\, d\mathbf{r} = -\frac{F}{\epsilon_v}\int \mathbf{X}\cdot\mathbf{D}_{ext}\, d\mathbf{r}. \tag{5.23}$$

The electron-generated displacement $\mathbf{D}(\mathbf{r}')$ at $\mathbf{r}'$ by an electron located at $\mathbf{r}$ is

$$\mathbf{D}(\mathbf{r}') = -\frac{e}{4\pi}\frac{\mathbf{r}'-\mathbf{r}}{|\mathbf{r}'-\mathbf{r}|^3}.$$

The total interaction potential is then

$$H_{e-ph}^{LO} = \frac{eF}{4\pi\epsilon_v}\frac{1}{\sqrt{\Omega}}\sum_{\mathbf{q}}\hat{\mathbf{q}}\left[a_{\mathbf{q}}\int \exp(i\mathbf{q}\cdot\mathbf{r})\frac{\mathbf{r}'-\mathbf{r}}{|\mathbf{r}'-\mathbf{r}|^3}d\mathbf{r}' + cc\right].$$

Consequently after integration the interaction Hamiltonian is

$$\begin{aligned} H_{e-ph}^{LO} &= \frac{-ieF}{\epsilon_v\sqrt{\Omega}}\sum_{\mathbf{q}}\frac{1}{q}\left(a_{\mathbf{q}}\exp(i\mathbf{q}\cdot\mathbf{r}) - a_{\mathbf{q}}^{+}\exp(-i\mathbf{q}\cdot\mathbf{r})\right), \\ &= \frac{-i}{\sqrt{\Omega}}\sum_{\mathbf{q},LO}\alpha(\mathbf{q})\left\{a_{\mathbf{q}}\exp(i\mathbf{q}\cdot\mathbf{r}) - a_{\mathbf{q}}^{+}\exp(-i\mathbf{q}\cdot\mathbf{r})\right\} \end{aligned}$$

where $\alpha_{\mathbf{q},\mathbf{LO}}$ (J m$^{3/2}$) in terms of α_{LO} (J m$^{1/2}$) is given by

$$\alpha_{\mathbf{q},LO} = e\left[\frac{\hbar\omega_{LO}}{2}\left(\frac{1}{\epsilon_{opt}} - \frac{1}{\epsilon_{stat}}\right)\right]^{1/2}\frac{1}{q} = \frac{\alpha_{LO}}{q}. \tag{5.24}$$

These are the coupling coefficients we shall use in Chapter 6 to study polar scattering by LO phonons.

5.4 Deformation potential scattering by acoustic phonons

Electrons in a crystal are scattered by any deviation of the crystal potential from the ideal periodic potential. Such deviations can occur due to the displacement of the atoms from their lattice sites induced by the crystal vibrations. Such a displacement of the atoms is equivalent to a local change of the lattice parameter and therefore induces a modification of the band structure.

For electrons in the conduction band, the interaction potential H_{e-ph}^{AC} of the acoustic phonon with the lattice will be given by the variation of the conduction-band edge E_C:

$$H_{e-ph}^{AC} = \delta E_C,$$

induced by the acoustic phonons. Since the phonons deform the crystal in three dimensions, we can assume for small stress and for an isotropic crystal that δE_C is given by

$$\delta E_C = \Xi \frac{\delta\Omega}{\Omega} = \Xi\Delta,$$

where Ξ is the so-called deformation potential and $\Delta = \delta\Omega/\Omega$ is the dilatation of the lattice with Ω the crystal volume and $\delta\Omega$ its variation. The deformation potential is on the order of 10 eV in compound semiconductors.

In a crystal the phonons introduce a local dilatation $\Delta(\mathbf{r})$. Let us relate this dilatation to the lattice displacement $\mathbf{U}(\mathbf{r})$ of the position r in the lattice

$$\mathbf{U}(\mathbf{r}) = \mathbf{r}' - \mathbf{r}.$$

First, we calculate the local variation of the volume resulting from the displacement $\mathbf{U}(\mathbf{r})$. For simplicity, we select $\mathbf{r} = 0$. Consider the cube generated by the orthogonal vectors $\mathbf{a}$, $\mathbf{b}$, $\mathbf{c}$:

$$\mathbf{a} = \begin{Bmatrix} dx \\ 0 \\ 0 \end{Bmatrix}, \qquad \mathbf{b} = \begin{Bmatrix} 0 \\ dy \\ 0 \end{Bmatrix}, \qquad \mathbf{c} = \begin{Bmatrix} 0 \\ 0 \\ dz \end{Bmatrix}.$$

The volume of the cube is

$$\Delta\Omega = \mathbf{a} \cdot (\mathbf{b} \times \mathbf{c}) = dx\, dy\, dz$$

Upon the displacement $\mathbf{U}(0)$, the cube is distorted and the vectors $\mathbf{a}$, $\mathbf{b}$, and $\mathbf{c}$ are transformed into $\mathbf{a}'$, $\mathbf{b}'$, and $\mathbf{c}'$

$$\mathbf{a}' = \begin{Bmatrix} dx + \dfrac{\partial u_x}{\partial x} dx \\ \dfrac{\partial u_y}{\partial x} dx \\ \dfrac{\partial u_z}{\partial x} dx \end{Bmatrix}, \qquad \mathbf{b}' = \begin{Bmatrix} \dfrac{\partial u_x}{\partial y} dy \\ dy + \dfrac{\partial u_y}{\partial y} dy \\ \dfrac{\partial u_z}{\partial y} dz \end{Bmatrix}, \qquad \mathbf{c}' = \begin{Bmatrix} \dfrac{\partial u_x}{\partial z} dz \\ \dfrac{\partial u_y}{\partial z} dz \\ dz + \dfrac{\partial u_z}{\partial z} dz \end{Bmatrix}.$$

The new volume of the distorted cube is now

$$\Delta\Omega' = \mathbf{a}' \cdot (\mathbf{b}' \times \mathbf{c}') = \begin{vmatrix} dx + \dfrac{\partial u_x}{\partial x} & \dfrac{\partial u_x}{\partial y}dy & \dfrac{\partial u_x}{\partial z}dz \\ \dfrac{\partial u_y}{\partial x} & dy + \dfrac{\partial u_y}{\partial y}dy & \dfrac{\partial u_y}{\partial z}dz \\ \dfrac{\partial u_z}{\partial x} & \dfrac{\partial u_z}{\partial y}dy & dz + \dfrac{\partial u_z}{\partial z}dz \end{vmatrix}.$$

For small lattice displacements $\mathbf{U}(\mathbf{r})$, $\Delta\Omega'$ is approximately

$$\Delta\Omega' = dx\, dy\, dz \left(1 + \frac{\partial u_x}{\partial x} + \frac{\partial u_y}{\partial y} + \frac{\partial u_z}{\partial z} + \cdots\right).$$

Therefore the local dilatation Δ is

$$\Delta(\mathbf{r}) = \frac{\delta\Delta\Omega}{\Delta\Omega} = \frac{\Delta\Omega' - \Delta\Omega}{\Delta\Omega} = \frac{\partial u_x}{\partial x} + \frac{\partial u_y}{\partial y} + \frac{\partial u_z}{\partial z} = \nabla \cdot \mathbf{U}(\mathbf{r})$$

and we have

$$H_{e-ph}^{AC} = \delta E_c = \Xi\ \Delta(\mathbf{r}) = \Xi\ \nabla \cdot \mathbf{U}(\mathbf{r}).$$

The lattice displacement $\mathbf{U}(\mathbf{r})$ for long wavelength phonons (slowly varying $\mathbf{U}(\mathbf{r})$) is given by

$$\mathbf{U}(\mathbf{r}) = \frac{1}{\Omega^{1/2}} \sum_{\mathbf{q}} \mathbf{w}_{\mathbf{q}} \left(\frac{\hbar}{2\rho\omega_{\mathbf{q}}}\right)^{1/2} \left[a_{\mathbf{q}} \exp(i\mathbf{q} \cdot \mathbf{r}) + a_{\mathbf{q}}^{+} \exp(-i\mathbf{q} \cdot \mathbf{r})\right],$$

where ρ is the semiconductor density and $\mathbf{w}_q$ is the polarization vector. For longitudinal phonons, the polarization vector is

$$\mathbf{w}_q = \frac{\mathbf{q}}{q} = \hat{\mathbf{q}}.$$

The acoustic deformation interaction potential is then

$$\begin{aligned} H_{e-ph}^{AC} &= \Xi\, \nabla \cdot \mathbf{U} \\ &= i\frac{1}{\Omega^{1/2}} \sum_{\mathbf{q}} \mathbf{w}_{\mathbf{q}} \cdot \mathbf{q} \left(\frac{\hbar\Xi^2}{2\rho\omega_{\mathbf{q}}}\right)^{1/2} \left[a_{\mathbf{q}} \exp(i\mathbf{q} \cdot \mathbf{r}) - a_{\mathbf{q}}^{+} \exp(-i\mathbf{q} \cdot \mathbf{r})\right] \\ &= i\frac{1}{\Omega^{1/2}} \sum_{\mathbf{q}} \alpha_{\mathbf{q},AC} \left[a_{\mathbf{q}} \exp(i\mathbf{q} \cdot \mathbf{r}) - a_{\mathbf{q}}^{+} \exp(-i\mathbf{q} \cdot \mathbf{r})\right], \end{aligned}$$

with

$$\alpha_{\mathbf{q},AC} = \mathbf{w}_{\mathbf{q}} \cdot \mathbf{q} \left(\frac{\hbar\Xi^2}{2\rho\omega_{\mathbf{q}}}\right)^{1/2}.$$

This is the coupling coefficient we shall use in Chapter 6 to study deformation potential scattering by acoustic phonons.

5.5 Intervalley scattering by LO phonons

We have seen in the previous chapters that electrons in the conduction band can elastically transfer from the Γ valley to the X valley. Γ to X transfer can also be induced by inelastic phonon scattering. From symmetry considerations, the only phonons contributing to the Γ to X transfer are the LOX phonons [11]. The dependence of the phonon energy $\omega_{\mathbf{q}}$ on the phonon wave-vector $\mathbf{q}$ is usually neglected (see [12] for the associated phonon energies). Intervalley scattering arises from the deformation potential induced by the LOX phonons and we can therefore rely on the calculation done for deformation potential scattering for acoustic phonons. Assuming that the deformation potential is isotropic the coupling coefficient of its interaction potential is then given by [13]

$$\alpha_{\mathbf{q},IV} = \left(\frac{\hbar \, \Xi_{LOX}^2}{2\rho\omega_{LOX}} \right)^{1/2},$$

where the reader is refered to [12] for the deformation potentials Ξ_{LOX} of various bulk materials. This is the coupling coefficient $\alpha_{\mathbf{q},IV}$ we shall use in Chapter 6 to study intervalley scattering by phonons.

5.6 Interface roughness scattering

An important scattering process in a heterostructure is the scattering of electrons by the roughness of the interface of two different semiconductors. A distribution of terraces, typically of a monolayer thickness, is indeed present at the interface [14]. The electron is scattered elastically by these terraces, i.e., the total energy of the electron is conserved $E_0 = E_1$.

Interface roughness scattering can be represented by a potential V^{IR} at the interface which extends the interface by a monolayer

$$V^{IR}(\mathbf{r}) = V_B(N_{IR}\, a) \; \mathrm{rect}\left(\frac{x - N_{IR}\, a}{a} \right) \; F(\mathbf{r}_\perp)$$

where the function rectangle (rect) is defined as

$$\mathrm{rect}(x) = \begin{cases} 1 \;\text{ for } \; |x| < 1/2 \\ 0 \;\text{ for } \; |x| > 1/2 \end{cases}.$$

The function $F(\mathbf{r}_\perp)$ gives the distribution of the terraces and is therefore either 0 or 1 depending on whether or not a terrace is present at the position $\mathbf{r}_\perp$. The barrier height V_B arises from the discontinuity of the conduction-band edge at the interface.

In the mixed Wannier and Bloch function representation, the matrix element of V^{IR} reduces to

$$\langle n', \mathbf{k}'_\perp | V^{IR}(\mathbf{r}) | n, \mathbf{k}_\perp \rangle = V_B(N_{IR}a)\delta_{n\,N_{IR}} \langle \mathbf{k}'_\perp | F(\mathbf{r}_\perp) | \mathbf{k}_\perp \rangle \delta_{n'n},$$

where the matrix element of $F(\mathbf{r}_\perp)$ remains to be calculated. As we shall see, the ensemble-average value of the amplitude squared of the matrix element of $F(\mathbf{r}_\perp)$ can be evaluated from the autocorrelation of $F(\mathbf{r}_\perp)$.

Following Leo and MacDonald [15], we can represent $F(r_\perp)$ as a superposition of square terraces (see Figure 5.6):

$$F(\mathbf{r}_\perp) = \sum_T \ \text{rect}\left(\frac{\mathbf{r}_\perp - \mathbf{r}_{\perp T}}{L_T}\right),$$

where $\mathbf{r}_\perp$ is the position of the terrace center and L_T the terrace width. The average terrace is now assumed to be of the Gaussian type

$$\frac{1}{N_T} \sum_{T in L_y L_z} \text{rect}\left(\frac{\mathbf{r}_\perp}{L_T}\right) = \exp\left(-\frac{|\mathbf{r}_\perp|^2}{2\sigma^2}\right) = N(\mathbf{r}_\perp),$$

where σ^2 is the average area of the terraces and N_T is the number of terraces in the volume $L_y L_z$.

Let us now calculate the autocorrelation function R_{yz} of $F(\mathbf{r}_\perp)$ defined as

$$R_{yz} = \lim_{\substack{L_y \to \infty \\ L_z \to \infty}} \frac{1}{L_y L_z} \int_{-L_y/2}^{L_y/2} \int_{-L_z/2}^{L_z/2} F(\mathbf{r}_\perp) F(\mathbf{r}'_\perp)\, d\mathbf{r}_\perp.$$

We assume that the cross-correlation between different terraces is zero (the terraces are uncorrelated). The autocorrelation is then

$$R_{yz} = \lim_{\substack{L_y \to \infty \\ L_z \to \infty}} \frac{N_T}{L_y L_z} \int_{-L_y/2}^{L_y/2} \int_{-L_z/2}^{L_z/2} N(\mathbf{r}_\perp) N(\mathbf{r}_\perp)\, d\mathbf{r}_\perp.$$

The autocorrelation of $N(\mathbf{r}_\perp)$ can be estimated using the integral

$$\int_{-\infty}^{\infty} \exp\left[-\frac{(x - a/2)^2}{\sigma^2}\right] dx = \sqrt{\pi}\,\sigma\, \exp\left(-\frac{a^2}{4\sigma^2}\right).$$

Therefore

$$R_{yz} = D_T\, \pi\sigma^2\, \exp\left(-\frac{|\mathbf{r}_\perp - \mathbf{r}'_\perp|^2}{4\sigma^2}\right),$$

where D_T is the terrace density

$$D_T = \frac{N_T}{L_y L_z} = \frac{N_T}{\text{Area}}.$$

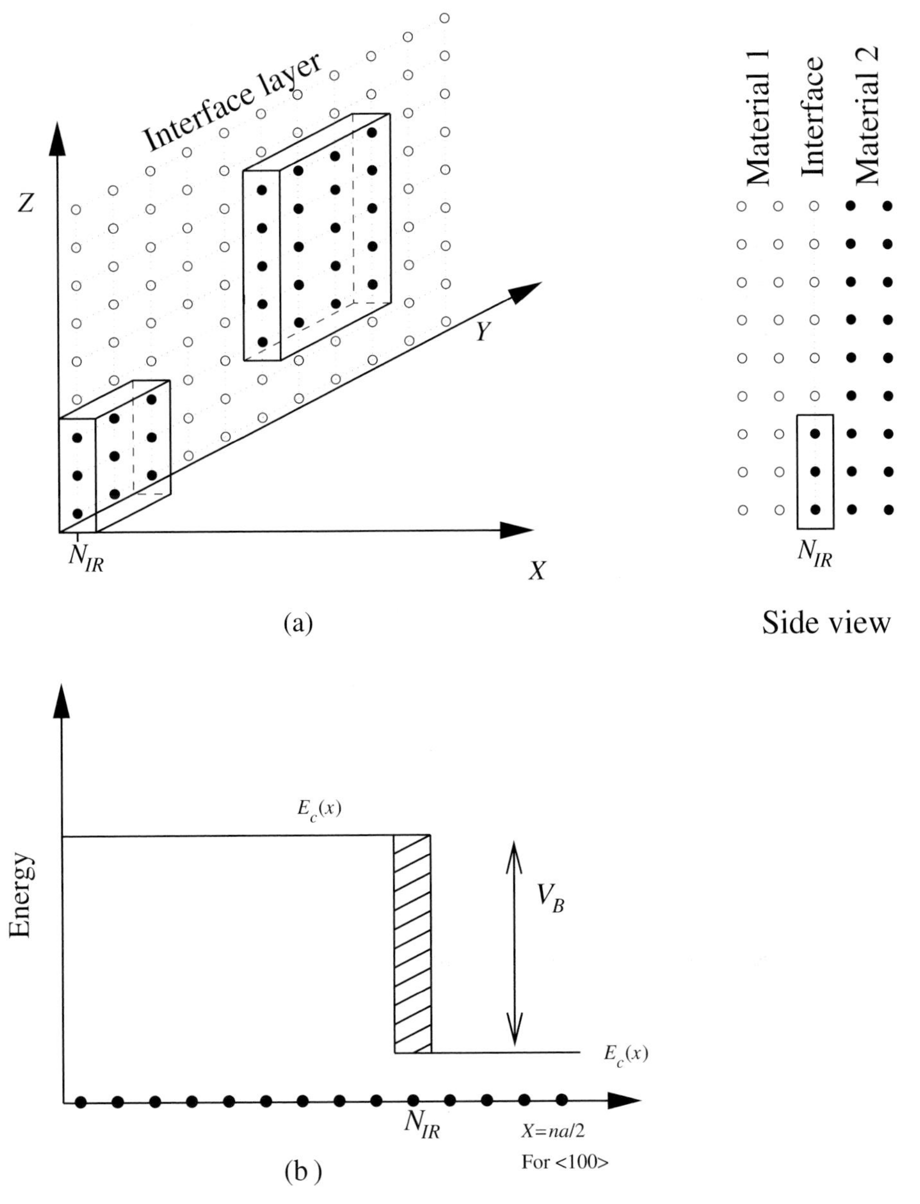

Fig. 5.6. (a) Three-dimensional and (b) one-dimensional representation of terraces at the interface of two different semiconductor materials. White and black circles are used to represent materials 1 and 2 respectively. For an interface which is a monolayer deep each circle actually represents two atoms.

This model reduces to the phenomenological model of Prange and Nee [16] who assume

$$R_{yz} = \exp\left(-\frac{|\mathbf{r}_\perp - \mathbf{r}'_\perp|^2}{\Lambda^2}\right)$$

if we use

$$\Lambda = 2\sigma,$$
$$D_T = \frac{N_T}{Area} = \frac{4}{\pi\Lambda^2} \simeq \frac{1}{\Lambda^2}.$$

This is equivalent to assuming that the terrace size σ is comparable to the distance between terraces.

Let us assume that $F(\mathbf{r}_\perp)$ is a periodic function

$$F(y + L_y,\ z + L_z) = F(y, z).$$

The function $F(\mathbf{r}_\perp)$ can therefore be expanded in a Fourier series

$$F(\mathbf{r}_\perp) = \sum_{\mathbf{q}_\perp} F_{\mathbf{q}_\perp} \exp(i\mathbf{q}_\perp \cdot \mathbf{r}_\perp).$$

Note that the Fourier coefficient $F_{\mathbf{q}_\perp}$ is equal to the matrix element of $F(\mathbf{r}_\perp)$ (approximating the Bloch waves by plane waves) if we use $\mathbf{q}_\perp = \mathbf{k}_\perp - \mathbf{k}'_\perp$:

$$\begin{aligned}\langle \mathbf{k}'_\perp | F(\mathbf{r}_\perp) | \mathbf{k}_\perp \rangle &= \frac{1}{L_y L_z} \int_{-L_y/2}^{L_y/2} \int_{-L_z/2}^{L_z/2} \exp[i(\mathbf{k}_\perp - \mathbf{k}'_\perp) \cdot \mathbf{r}_\perp] F(\mathbf{r}_\perp)\, d\mathbf{r}_\perp \\ &= F_{\mathbf{q}_\perp}.\end{aligned}$$

The amplitude squared of the matrix element $F(\mathbf{r}_\perp)$ is then

$$\begin{aligned}|F_{\mathbf{q}_\perp}|^2 &= |\langle \mathbf{k}'_\perp | F(\mathbf{r}_\perp) | \mathbf{k}_\perp \rangle|^2 \\ &= \frac{1}{L_y L_z} \int_{-L_y/2}^{L_y/2} \int_{-L_z/2}^{L_z/2} \exp[i(\mathbf{k}_\perp - \mathbf{k}'_\perp) \cdot \mathbf{r}_\perp]\ F(\mathbf{r}_\perp)\, d\mathbf{r}_\perp \\ &\quad \times \frac{1}{L_y L_z} \int_{-L_y/2}^{L_y/2} \int_{-L_z/2}^{L_z/2} \exp[-i(\mathbf{k}_\perp - \mathbf{k}'_\perp) \cdot \mathbf{r}'_\perp] F(\mathbf{r}'_\perp)\, d\mathbf{r}'_\perp \\ &= \frac{1}{L_y^2 L_z^2} \iint_{-L_y/2}^{L_y/2} \iint_{-L_z/2}^{L_z/2} \exp[i(\mathbf{k}_\perp - \mathbf{k}'_\perp) \cdot (\mathbf{r}_\perp - \mathbf{r}'_\perp)] F(\mathbf{r}_\perp) F(\mathbf{r}'_\perp)\, d\mathbf{r}_\perp\, d\mathbf{r}'_\perp.\end{aligned}$$

We can now evaluate the desired ensemble-average value of this matrix element. To do this we use the ergodic principle

$$\langle F(\mathbf{r}_\perp) F(\mathbf{r}'_\perp) \rangle_{E.A.} = R_{yz},$$

where $\langle \cdots \rangle_{E.A.}$ indicates that an ensemble average is performed. This gives

$$\begin{aligned}\langle |F_{\mathbf{q}_\perp}|^2 \rangle_{E.A.} &= \langle |\langle \mathbf{k}'_\perp | F(\mathbf{r}_\perp) | \mathbf{k}'_\perp \rangle \rangle|^2 \rangle_{E.A} \\ &= \frac{1}{L_y^2 L_z^2} \iint_{-L_y/2}^{L_y/2} \iint_{-L_z/2}^{L_z/2} \exp[i(\mathbf{k}_\perp - \mathbf{k}'_\perp) \cdot (\mathbf{r}_\perp - \mathbf{r}'_\perp)] \\ &\quad \times \exp\left(-\frac{|\mathbf{r}_\perp - \mathbf{r}'_\perp|^2}{\Lambda^2}\right) d\mathbf{r}_\perp\, d\mathbf{r}'_\perp.\end{aligned}$$

Using $\mathbf{R}_\perp = \mathbf{r}_\perp - \mathbf{r}'_\perp$ and $R_\perp = |\mathbf{R}_\perp|$ we have

$$\langle |F_{\mathbf{q}_\perp}|^2 \rangle_{E.A.} = \frac{1}{L_y L_z} \int_{-L_y/2}^{L_y/2} \int_{-L_z/2}^{L_z/2} \exp(i\mathbf{q}_\perp \cdot \mathbf{R}_\perp) \exp\left[-\frac{R_\perp^2}{\Lambda^2}\right] d\mathbf{R}_\perp.$$

We can evaluate this integral by extending the limits of integration to infinity:

$$\langle |F(\mathbf{q}_\perp)|^2 \rangle_{E.A.} = \frac{1}{L_y L_z} \iint_{-\infty}^{\infty} \exp\left[-\frac{R_\perp^2}{\Lambda^2}\right] \exp(i\mathbf{R}_\perp \cdot \mathbf{q}_\perp)\, d\mathbf{R}_\perp$$

$$= \frac{1}{L_y L_z} \int_{-\infty}^{\infty} \exp\left[-\frac{R_y^2}{\Lambda^2}\right] \exp(i\, R_y Q_y)\, dR_y$$

$$\times \int_{-\infty}^{\infty} \exp\left(-\frac{R_z^2}{\Lambda^2}\right) \exp(i\, R_z Q_z)\, dR_z.$$

Using the identity

$$\int_{-\infty}^{\infty} \exp(-a^2 x^2) \exp(i\xi x)\, dx = \sqrt{2\pi}\, \frac{1}{a\sqrt{2}} \exp(-\xi^2/4a^2),$$

with $a^2 = 1/\Lambda^2$ and $\xi = q_y$ we can evaluate the integrals and obtain

$$\langle |F(\mathbf{q}_\perp)|^2 \rangle_{E.A} = \frac{1}{L_y L_z} \sqrt{\pi}\, \Lambda \exp(-q_y^2/4\,\Lambda^2)\, \sqrt{\pi}\, \Lambda \exp(-q_z^2/4\,\Lambda^2)$$

$$= \frac{1}{L_y L_z} \pi\, \Lambda^2 \exp\left(-\frac{q_\perp^2}{4}\, \Lambda^2\right), \tag{5.25}$$

with $q_\perp^2 = |\mathbf{q}_\perp|^2$. The resulting ensemble-average value of the matrix element for interface roughness scattering is seen to favor variations $\mathbf{q}_\perp = \mathbf{k}_\perp - \mathbf{k}'_\perp$ of the perpendicular electron wave-vectors of the order of $1/\Lambda$.

The terrace size Λ is found experimentally to be on the order of 70 Å [14]. The ensemble-average value of the matrix element calculated above will be used in Chapter 6 to study interface roughness scattering.

5.7 Alloy scattering

In an alloy $A_\alpha B_{1-\alpha} C$ (e.g., $Al_\alpha Ga_{1-\alpha} As$), the atoms A and B are distributed randomly over the same sites. If atoms A and B have the same valence the crystal structure can be assumed to be preserved. The crystal potential $V(\mathbf{r})$, however, is no longer periodic. It is possible to represent the crystal potential of the alloy in terms of an average potential, $\bar{V}(\mathbf{r})$, which is periodic, plus a fluctuating potential $\delta V(\mathbf{r})$, which accounts for the local departure of the actual alloy potential $V(\mathbf{r})$ from the average potential $\bar{V}(\mathbf{r})$. This fluctuating potential $\delta V(\mathbf{r})$ introduces an effective scattering process referred to as alloy scattering.

For the alloy $A_\alpha B_{1-\alpha}C$, the crystal potential is the superposition of the potential contributed by the atoms A, B, and C (see Bastard [17])

$$V(\mathbf{r}) = \sum_{\mathbf{R}_A} V_A(\mathbf{r} - \mathbf{R}_A) + \sum_{\mathbf{R}_B} V_B(\mathbf{r} - \mathbf{R}_B) + \sum_{\mathbf{R}_C} V_C(\mathbf{r} - \mathbf{R}_C).$$

Let us introduce the mathematical identities

$$\begin{aligned} V_A(\mathbf{r} - \mathbf{R}_A) &= \alpha V_A(\mathbf{r} - \mathbf{R}_A) + (1-\alpha) V_B(\mathbf{r} - \mathbf{R}_A) \\ &\quad + (1-\alpha) V_A(\mathbf{r} - \mathbf{R}_A) - (1-\alpha) V_B(\mathbf{r} - \mathbf{R}_A), \\ V_B(\mathbf{r} - \mathbf{R}_B) &= \alpha V_A(\mathbf{r} - \mathbf{R}_B) + (1-\alpha) V_B(\mathbf{r} - \mathbf{R}_B) \\ &\quad - \alpha V_A(\mathbf{r} - \mathbf{R}_B) + \alpha V_B(\mathbf{r} - \mathbf{R}_B). \end{aligned}$$

Using these identities, the crystal potential can be written as the superposition of

$$V(\mathbf{r}) = \bar{V}(\mathbf{r}) + \delta V(\mathbf{r}),$$

where $\bar{V}(\mathbf{r})$ is the average lattice potential:

$$\bar{V}(\mathbf{r}) = \sum_{R_i = \mathbf{R}_A \& \mathbf{R}_B} \alpha V_A(\mathbf{r} - \mathbf{R}_i) + (1-\alpha) V_B(\mathbf{r} - \mathbf{R}_i) + \sum_{\mathbf{R}_C} V_C(\mathbf{r} - \mathbf{R}_C),$$

and $\delta V(\mathbf{r})$ is the fluctuating potential:

$$\begin{aligned} \delta V(\mathbf{r}) &= \sum_{\mathbf{R}_A} (1-\alpha) \left[V_A(\mathbf{r} - \mathbf{R}_A) - V_B(\mathbf{r} - \mathbf{R}_A) \right] \\ &\quad + \sum_{\mathbf{R}_B} \alpha \left[V_B(\mathbf{r} - \mathbf{R}_B) - V_A(\mathbf{r} - \mathbf{R}_B) \right]. \end{aligned}$$

Using $\delta V_{AB}(\mathbf{r}) = V_A(\mathbf{r}) - V_B(\mathbf{r})$ we can rewrite $\delta V(\mathbf{r})$ as

$$\delta V(\mathbf{r}) = \sum_{\mathbf{R}_A} (1-\alpha) \delta V_{AB}(\mathbf{r} - \mathbf{R}_A) - \sum_{\mathbf{R}_B} \alpha \delta V_{AB}(\mathbf{r} - \mathbf{R}_B).$$

Let us assume that we can approximate $\delta V_{AB}(\mathbf{r})$ by an impulse function

$$\delta V_{AB}(\mathbf{r}) = \Omega_0 \, \Delta V_{AB} \;\; \delta(\mathbf{r}),$$

with Ω_0 the volume of a unit cell and $\delta(\mathbf{r})$ the impulse function. We can now calculate the correlation function

$$R_{xyz} = \lim_{L_x = L_y = L_z \to \infty} \frac{1}{L_x L_y L_z} \int_{-\frac{L_x}{2} - \frac{L_y}{2} - \frac{L_z}{2}}^{\frac{L_x}{2} \frac{L_y}{2} \frac{L_z}{2}} \delta V(\mathbf{r}) \, \delta V(\mathbf{r}') \, d\mathbf{r}$$

$$= \lim_{L_x=L_y=L_z\to\infty} \frac{1}{L_xL_yL_z}\left\{\sum_{\mathbf{R}_A}(1-\alpha)^2\delta(\mathbf{r}-\mathbf{r}')\Omega_0^2\Delta V_{AB}^2\right.$$
$$\left. + \sum_{\mathbf{R}_B} \alpha^2\delta(\mathbf{r}-\mathbf{r}')\Omega_0^2\Delta V_{AB}^2\right\}.$$

Considering that of $L_xL_yL_z/\Omega_0$ unit cells $\mathbf{R}_i$ we have a fraction α at the $\mathbf{R}_A$ sites and $(1-\alpha)$ being at the $\mathbf{R}_B$ sites, we can rewrite the correlation function as

$$R_{xyz} = \lim_{L_x=L_y=L_z\to\infty} \frac{1}{L_xL_zL_z}\,\Omega_0^2\,\Delta V_{AB}^2\,\delta(\mathbf{r}-\mathbf{r}')$$
$$\times\left\{\alpha\cdot\frac{L_xL_yL_z}{\Omega_0}(1-\alpha)^2+(1-\alpha)\frac{L_xL_yL_z}{\Omega_0}\alpha^2\right\}$$
$$=\Omega_0\Delta V_{AB}^2\,\alpha(1-\alpha)\,\delta(\mathbf{r}-\mathbf{r}').$$

We assume now that $\delta V(\mathbf{r})$ is a periodic function:

$$\delta V(x,y,z)=\delta V(x+L_x,\ y+L_y,\ z+L_z),$$

so that the function $\delta V(\mathbf{r})$ can be expanded in a Fourier series

$$\delta V(\mathbf{r})=\sum_{\mathbf{q}} V_{\mathbf{q}}\exp(i\mathbf{q}\cdot\mathbf{r}).$$

The amplitude squared of this matrix element $\delta V(\mathbf{r})$ is then

$$|V_{\mathbf{q}}|^2=|\langle\mathbf{k}'|\delta V(\mathbf{r})|\mathbf{k}\rangle|^2$$
$$=\frac{1}{L_x^2L_y^2L_z^2}$$
$$\times\iint_{-L_x/2}^{L_x/2}\iint_{-L_y/2}^{L_y/2}\iint_{-L_z/2}^{L_z/2}\exp[i(\mathbf{k}-\mathbf{k}')\cdot(\mathbf{r}-\mathbf{r}')]\,\delta V(\mathbf{r})\,\delta V(\mathbf{r}')\,d\mathbf{r}\,d\mathbf{r}'.$$

We can now calculate the ensemble average of this matrix element. To do this we use the ergodic principle

$$\langle\delta V(\mathbf{r})\delta V(\mathbf{r}')\rangle_{E.A.}=R_{xyz},$$

where $\langle\cdots\rangle_{E.A}$ indicates that an ensemble average is performed. This gives

$$\langle|\langle\mathbf{k}'|\delta V(\mathbf{r})|\mathbf{k}\rangle|^2\rangle_{E.A.}=\langle|V_{\mathbf{q}}|^2\rangle_{E.A.}$$
$$=\frac{1}{L_x^2L_y^2L_z^2}\iint_{-L_x/2}^{L_x/2}\iint_{-L_y/2}^{L_y/2}\iint_{-L_z/2}^{L_z/2}\exp[i(\mathbf{k}-\mathbf{k}')\cdot(\mathbf{r}-\mathbf{r}')]$$
$$\times\Omega\cdot\Delta V_{AB}^2\alpha(1-\alpha)\delta(\mathbf{r}-\mathbf{r}')\,d\mathbf{r}\,d\mathbf{r}'$$
$$=\frac{1}{L_xL_yL_z}\,\Omega_0\,\Delta V_{AB}^2\alpha(1-\alpha),$$

where α is the alloy mole fraction. The ensemble-average matrix element for alloy scattering is found to be independent of the initial and final electron wave-vectors $\mathbf{k}$ and $\mathbf{k}'$. Note that Ω_0, the volume of the elementary cell, is given in terms of the elementary crystal axis by

$$\Omega_0 = |\mathbf{a} \times \mathbf{b} \cdot \mathbf{c}| = \begin{vmatrix} 0 & \frac{a}{2} & \frac{a}{2} \\ \frac{a}{2} & 0 & \frac{a}{2} \\ \frac{a}{2} & \frac{a}{2} & 0 \end{vmatrix} = \frac{a^3}{8} + \frac{a^3}{8} = \frac{a^3}{4},$$

where $\mathbf{a}$, $\mathbf{b}$, and $\mathbf{c}$ are the unit cell vectors which in Bravais cell coordinates are given by

$$\mathbf{a} = \frac{a}{2}\begin{Bmatrix} 0 \\ 1 \\ 1 \end{Bmatrix}, \quad \mathbf{b} = \frac{a}{2}\begin{Bmatrix} 1 \\ 0 \\ 1 \end{Bmatrix}, \quad \mathbf{c} = \frac{a}{2}\begin{Bmatrix} 1 \\ 1 \\ 0 \end{Bmatrix}.$$

ΔV_{AB} is typically selected to be the conduction-band discontinuity between the crystals AC and BC.

The ensemble-average value of the matrix element derived above will be used in Chapter 6 to study alloy scattering.

5.8 Electron–electron scattering

To study electron–electron scattering we consider a single electron interacting with a many-electron system through the Coulomb interaction potential. The Hamiltonian of a given electron is then

$$H(\mathbf{r}_1) = H_1(\mathbf{r}_1) + H_{int}(\mathbf{r}_1).$$

The Hamiltonian, $H_1(\mathbf{r}_1)$, is the non-interacting single-electron Hamiltonian and $H_{int}(\mathbf{r}_1)$ is the screened Coulomb interaction potential given by

$$H_{int}(\mathbf{r}_1) = \sum_j V(\mathbf{r}_j - \mathbf{r}_1) = \frac{e^2}{4\pi\epsilon} \sum_j \frac{\exp(-\alpha|\mathbf{r}_j - \mathbf{r}_1|)}{|\mathbf{r}_j - \mathbf{r}_1|}. \tag{5.26}$$

Here, α is the screening constant and $V(\mathbf{r}_j - \mathbf{r}_1)$ is the two-body screened Coulomb potential between electrons at $\mathbf{r}_1$ and $\mathbf{r}_j$.

It is convenient to rewrite the screened Coulomb interaction potential, $H_{int}(\mathbf{r}_1)$, using the second quantization formalism, which accounts for the Pauli exclusion principle. To do this, we introduce the many-body field operator defined as

$$\Psi(\mathbf{r}) = \left[\frac{2}{(2\pi)^3}\right]^{1/2} \int d^3\tilde{\mathbf{k}}\, \phi_{\tilde{\mathbf{k}}}(\mathbf{r})\, b_{\tilde{\mathbf{k}}},$$

where $b_{\tilde{\mathbf{k}}}$ is the creation operator (and $b^+_{\tilde{\mathbf{k}}}$ the destruction operator) for the state $\phi_{\tilde{\mathbf{k}}}(\mathbf{r})$, which is the eigenstate corresponding to the single-body Hamiltonian $H_1(\mathbf{r})$. The creation and destruction operators $b_{\tilde{\mathbf{k}}}$ and $b^+_{\tilde{\mathbf{k}}}$ satisfy the same properties as the phonon creation and destruction operators $a_{\mathbf{q}}$ and $a^+_{\mathbf{q}}$.

The state $\phi_{\tilde{\mathbf{k}}}(\mathbf{r})$ is normalized according to

$$\int d^3\mathbf{r}\, \phi^*_{\tilde{\mathbf{k}}'}(\mathbf{r})\phi_{\tilde{\mathbf{k}}}(\mathbf{r}) = \delta(\tilde{\mathbf{k}}' - \tilde{\mathbf{k}}). \tag{5.27}$$

The notation $\tilde{\mathbf{k}}$ is used to label the above states as energy eigenstates. This is to differentiate them from the plane wave Bloch states labeled without the tilde.

Rewriting the interaction Hamiltonian, $H_{int}(\mathbf{r}_1)$, using the second quantization formalism results in the following expression:

$$\begin{aligned} H_{int}(\mathbf{r}_1) &= \Omega \int d^3\mathbf{r}_2\ \Psi^+(\mathbf{r}_2)\ V(\mathbf{r}_2 - \mathbf{r}_1)\ \Psi(\mathbf{r}_2) \\ &= \frac{2\Omega}{(2\pi)^3} \int d^3\tilde{\mathbf{k}}'_2 \int d^3\tilde{\mathbf{k}}_2\ V(\tilde{\mathbf{k}}_2, \tilde{\mathbf{k}}'_2) b^+_{\tilde{\mathbf{k}}'_2}\, b_{\tilde{\mathbf{k}}_2}, \end{aligned} \tag{5.28}$$

where

$$V(\tilde{\mathbf{k}}_2, \tilde{\mathbf{k}}'_2) = \int d\mathbf{r}_2\, \phi^*_{\tilde{\mathbf{k}}'_2}(\mathbf{r}_2)\ V(\mathbf{r}_2 - \mathbf{r}_1)\ \phi_{\tilde{\mathbf{k}}_2}(\mathbf{r}_2).$$

The normalization volume is given by Ω. The calculation of the matrix element $V(\tilde{\mathbf{k}}_2, \tilde{\mathbf{k}}'_2)$ in the Wannier representation is left as an exercise for the reader. Note that the conservation of particles expected for a many-body electron system is enforced by the terms $b^+_{\tilde{\mathbf{k}}'_2}\, b_{\tilde{\mathbf{k}}_2}$ which occur in pairs in the interaction Hamiltonian, indicating that the destruction and creation of states takes place simultaneously when the single-body electron 1 scatters on electron 2 of the many-body system.

Let us demonstrate now how the semiclassical approximation introduced for the scattering of a single electron by the many-body phonon field can also be applied to the scattering of a single electron by the many-body electron field. To do this, we use the semiclassical equivalent of the creation and annihilation operators which is obtained by replacing the creation and annihilation operators by *complex* numbers

$$\begin{aligned} b_{\tilde{\mathbf{k}}} &\Rightarrow A_{\tilde{\mathbf{k}}}\ \exp[-i(\omega_{\tilde{\mathbf{k}}}t + \varphi_{\tilde{\mathbf{k}}})], \\ b^+_{\tilde{\mathbf{k}}'} &\Rightarrow A_{\tilde{\mathbf{k}}'}\ \exp[+i(\omega_{\tilde{\mathbf{k}}'}t + \varphi_{\tilde{\mathbf{k}}'})]. \end{aligned}$$

The energy, $\hbar\omega_{\tilde{\mathbf{k}}}$, and wave-vector $\tilde{\mathbf{k}}$ refer to the single-electron state. The argument $\varphi_{\tilde{\mathbf{k}}}$ is the classical phase of the single-electron state. We assume that the many-electron system is in thermal equilibrium which results in the phases $\varphi_{\tilde{\mathbf{k}}}$ being randomly and uniformly distributed (maximum entropy). By performing an ensemble average, a correspondence is found between the amplitudes, $A_{\tilde{\mathbf{k}}}$, and the Fermi statistical factors. Consider the operator $b^+_{\tilde{\mathbf{k}}'}b_{\tilde{\mathbf{k}}}$ operating on a state $|N_{\tilde{\mathbf{k}}'}\rangle|N_{\tilde{\mathbf{k}}}\rangle$. The result is

$$b^+_{\tilde{\mathbf{k}}'}b_{\tilde{\mathbf{k}}}|N_{\tilde{\mathbf{k}}'}\rangle|N_{\tilde{\mathbf{k}}}\rangle = (1 - N_{\tilde{\mathbf{k}}'})^{1/2}\ (N_{\tilde{\mathbf{k}}})^{1/2}\ |N_{\tilde{\mathbf{k}}'} + 1\rangle|N_{\tilde{\mathbf{k}}} - 1\rangle.$$

Only the states $|0_{\tilde{\mathbf{k}}'}\, 1_{\tilde{\mathbf{k}}}\rangle$ lead to scattering events, and the probability of such an event is then $N_{\tilde{\mathbf{k}}}(1 - N_{\tilde{\mathbf{k}}'})$.

In thermal equilibrium the average probability of scattering and backscattering is therefore given by $f_{\tilde{\mathbf{k}}}(1 - f_{\tilde{\mathbf{k}}'})$, where $f_{\tilde{\mathbf{k}}}$ is the Fermi–Dirac statistical factor given for a specific Fermi energy E_F by

$$f_{\tilde{\mathbf{k}}} = f_D(E_{\tilde{\mathbf{k}}}) = \left\{1 + \exp[(E_{\tilde{\mathbf{k}}} - E_F)/kT]\right\}^{-1}.$$

Note that backscattering is the scattering event, $b^+_{\tilde{\mathbf{k}}} b_{\tilde{\mathbf{k}}'}$, following the scattering event, $b^+_{\tilde{\mathbf{k}}'} b_{\tilde{\mathbf{k}}}$. The inclusion of backscattering enforces current conservation during the scattering process. Other correlated backscattering events are prevented if we assume that the heat bath randomizes the phases before the next scattering event occurs. The ensemble-average results are shown below. For a discrete system of finite volume, Ω, we obtain:

$$\begin{aligned}&\langle A_{\tilde{\mathbf{K}}} \exp(i\varphi_{\tilde{\mathbf{K}}})\, A_{\tilde{\mathbf{K}}'} \exp(-i\varphi_{\tilde{\mathbf{K}}'})\, A_{\tilde{\mathbf{k}}'} \exp(i\varphi_{\tilde{\mathbf{k}}'})\, A_{\tilde{\mathbf{k}}} \exp(-i\varphi_{\tilde{\mathbf{k}}})\rangle_{E.A.}\\ &\quad = \delta_{\tilde{\mathbf{K}}\,\tilde{\mathbf{k}}}\delta_{\tilde{\mathbf{K}}'\,\tilde{\mathbf{k}}'}\, f_{\tilde{\mathbf{k}}}(1 - f_{\tilde{\mathbf{k}}'}).\end{aligned}$$

The conversion to a continuous system is done by comparing the normalizations of the wave-function for a discrete and a continuous system. Using $\Omega\delta_{\tilde{\mathbf{k}}'\,\tilde{\mathbf{k}}} = (2\pi)^3\delta(\tilde{\mathbf{k}}' - \tilde{\mathbf{k}})$, the ensemble average takes the following form:

$$\begin{aligned}&\langle A_{\tilde{\mathbf{K}}} \exp(i\varphi_{\tilde{\mathbf{K}}})\, A_{\tilde{\mathbf{K}}'} \exp(-i\varphi_{\tilde{\mathbf{K}}'})\, A_{\tilde{\mathbf{k}}'} \exp(i\varphi_{\tilde{\mathbf{k}}'})\, A_{\tilde{\mathbf{k}}} \exp(-i\varphi_{\tilde{\mathbf{k}}})\rangle_{E.A.}\\ &\quad = \frac{(2\pi)^6}{\Omega^2}\delta(\tilde{\mathbf{K}} - \tilde{\mathbf{k}})\delta(\tilde{\mathbf{K}}' - \tilde{\mathbf{k}}')\, f_{\tilde{\mathbf{k}}}\,(1 - f_{\tilde{\mathbf{k}}'}).\end{aligned} \tag{5.29}$$

This is the ensemble average which is used in Chapter 6 when evaluating the self-energy corresponding to electron–electron scattering. We shall also see in Chapter 6 that an additional Pauli exclusion term is also introduced for the single-body test electron 1.

5.9 Conclusion

In this chapter we have introduced the major elastic and inelastic scattering processes which limit the performance of quantum devices: electron–phonon, electron–electron, and other scattering processes specific to heterostructure devices. The general form of the electron–phonon interaction Hamiltonian was derived and the specific matrix elements for polar, acoustic, and intervalley phonon scattering processes were evaluated. Interface roughness scattering was analyzed using a model of uncorrelated terraces with a Gaussian distribution in size, and alloy scattering was analyzed using the virtual-crystal model. Armed with these scattering models we are now in position to study in the next chapter the impact of scattering upon transport in quantum devices.

5.10 Bibliography

5.10.1 Recommended reading

J. M. Ziman, *Principle of the Theory of Solids*, 2nd Edition, Cambridge University Press, New York, 1972.

C. Kittel, *Quantum Theory of Solids*, Wiley, New York, 1963.

K. Seeger, *Semiconductor Physics, An Introduction*, 3rd Edition, Springer Verlag, Berlin, 1985.

5.10.2 References

[1] J. M. Ziman, *Principle of the Theory of Solids*, 2nd Edition, Cambridge University Press, New York, 1972.

[2] C. Kittel, *Quantum Theory of Solids*, Wiley, New York, 1963.

[3] H. Fröhlich, 'Theory of electrical breakdown in ionic crystal', *Proceedings of the Royal Society*, Vol. A160, p. 230, 1937.

[4] W. A. Harrison, *Electronic Structure and the Properties of Solids*, Freeman, San Francisco,1980.

[5] J. L. T. Waugh and G. Dolling, 'Crystal dynamics of gallium arsenide', *Physical Review*, Vol. 132, p. 2410, 1963.

[6] R. Loudon, *The Quantum Theory of Light*, Clarendon Press, Oxford,1973.

[7] P. G. Klemens, 'Anharmonic decay of optical phonons', *Physical Review*, Vol. 148, No. 2, pp. 845–848, 1966.

[8] P. J. Turley and S. W. Teitsworth, 'Electronic wave functions and electron-confined-phonon matrix elements in GaAs/AlGaAs double-barrier resonant tunneling structures', *Physical Review B*, Vol. 44, No. 7, pp. 3199–3210, August 15 1991.

[9] S. Teitsworth, P. Turley, and C. R. Wallis, W. Li, P. Bhattacharya, 'Electron-localized phonon interactions in GaAs/AlAs double barrier structures', *Proceedings of the 1991 International Semiconductor Device Research Symposium*, Omni Charlottesville Hotel, December 4–6, Engineering Academic Outreach, University of Virginia, Charlottesville, pp. 93–96, 1991.

[10] D. K. Ferry, *Semiconductors*, Macmillan, New York, 1991.

[11] S. Zollner, S. Gopalan, and M. Cardona, 'Intervalley deformation potentials and scattering rates in xinc blende', *Applied Physics Letters*, Vol. 52, p. 614, 1989.

[12] J. Q. Wang, Z. Q. Gu, M. F. Li, and W. Y. Lai, 'Intervalley gamma–x deformation potential in III–V zinc-blende semiconductors by *abinitio* pseudopotential calculation', *Physical Review B*, Vol. 46, p. 12 358, 1992.

[13] S. Zollner, S. Gopalan, and M. Cardona, 'Microscopic theory of intervalley scattering in GaAs: K dependence of deformation potentials and scattering rates', *Journal of Applied Physics*, Vol. 68, p. 1682, 1990.

[14] H. Sakaki, T. Noda, K. Hirakawa, M. Tanaka, and T. Matsusue, 'Interface roughness scattering in GaAs/AlAs quantum wells,' *Applied Physics Letters*, Vol. 51, No. 23, December 7 1987.

[15] J. Leo and A. H. MacDonald, 'Disorder-assisted tunneling through a double-barrier structure,' *Physical Review Letters*, Vol. 64, No. 8, pp. 817–820, February 19 1990.

[16] P. E. Prange and T.-W. Nee, 'Quantum spectroscopy of the low-field oscillations in the surface impedance,' *Physical Review*, Vol. 168, No.3, pp. 779–786, April 15 1968.

[17] G. Bastard, *Wave Mechanics Applied to Semiconductor Heterostructures*, Les éditions de Physique, Les Ulis, France, 1990.

5.11 Problems

5.1 Using Equations (5.5), (5.7), and (5.8) verify Equations (5.9), (5.10), (5.11), (5.13) (5.14), and (5.15).

5.2 Using Equation (5.16) derive Equations (5.17) and (5.19) for the electron–phonon interaction Hamiltonian inside a layer of width L. Assume the lattice vibrations vanish at the edges of the sample.

5.3 Consider an effective-mass electron with energy E_1 and momentum $\mathbf{k}_1$ in a uniform conduction band. Calculate the range of the electron velocity for which the electron can spontaneously emit an acoustic phonon. Use the conservation of energy and momentum equations and assume a linear dispersion model for $\omega_{\mathbf{q},AC}$.

6 Scattering-assisted tunneling

A philosopher once said 'It is necessary for the very existence of science that the same conditions always produce the same results'. Well, they do not.

The Character of Physical Law, RICHARD FEYNMAN

6.1 Introduction

Our study of quantum heterostructure devices in Chapter 4 neglected the scattering processes. Transport in the absence of scattering is referred to as ballistic transport. A true ballistic electron by definition travels without being scattered, thus conserving its total energy E and its perpendicular wave-vector $\mathbf{k}_\perp$ in layered devices. Note that the electron's longitudinal energy $E - \hbar^2 k_\perp^2/2m^*(n)$ will, however, vary from site to site if the device varies spatially along the superlattice axis.

In real crystals the electron always experiences some type of scattering.* Therefore, ballistic transport only exists as an event with a certain probability of occurence. Ballistic transport through a device can therefore be assumed to take place when the device length is short compared with the mean free path of the electron between scattering events.

The description of ballistic transport given above is semiclassical. It is indeed the nature of quantum mechanics that all electron paths (incident and scattered) must be considered simultaneously, as they affect each other. Consider an electron wave incident on a heterostructure device. As a consequence of the imperfection or vibrations of the lattice, the incident wave of total energy E_0 generates a continuum (in energy) of scattered waves of energy E_1 corresponding to new electron paths in the device (see Figure 6.1). The scattered waves in turn modify the incident wave in a process called backscattering. Backscattering is responsible for setting the mean free path of the incident wave. Backscattering also modifies the phase of the incident wave and the incident wave then interacts differently with the heterostructure potential. This is referred to as a self-energy effect, because a similar change of phase could be realized if the energy of the incident electron actually had been changed. Note that, in

* A noted exception is the absence of scattering in superconductors under DC conditions.

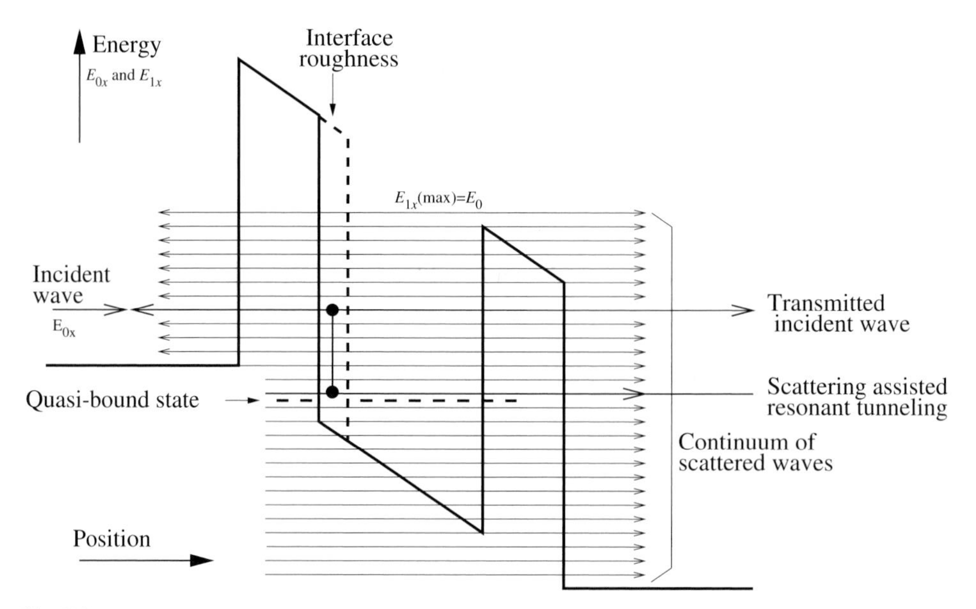

Fig. 6.1. Representation of the continuum of scattered waves generated by interface roughness (IR) scattering at one site by changing the electron direction. Electrons can then cross the double barrier by scattering-assisted resonant tunneling through a quasi-bound state even if it is located below the conduction-band edge of the emitter contact.

turn, the scattered waves spawn new scattered waves and therefore are also subjected to backscattering.

All the quantum effects discussed above indicate that in order to properly account for scattering, it is necessary to include it in the Schrödinger equation. In this chapter, we will develop a theory which will permit us to analyze the impact of scattering on the tunneling process in a heterostructure. The test device considered will be the resonant tunneling diode (RTD).

6.2 Importance of three-dimensional scattering

As we shall see, three-dimensional effects must be accounted for if one wishes to simulate the broadening of the transmission coefficient introduced by scattering processes in some modes of operation. Let us motivate it for the case of phonon scattering.

Consider a plane-wave electron with a total energy E_0 incident on a quantum structure. Let us write the total incident energy E_0 in terms of its perpendicular $E_{0\perp}$ and longitudinal E_{0x} parts

$$E_0 = \frac{\hbar^2 k_{0\perp}^2}{2m^*} + E_{0x}. \tag{6.1}$$

For a single sequential scattering event in which an electron absorbs or emits a phonon of energy $\omega_{\mathbf{q}}$, the total energy E_1 of an electron after the collision is $E_1 = E_0 \pm \hbar\omega_{\mathbf{q}}$.

Let us also write the total energy E_1 in terms of its perpendicular $E_{1\perp}$ and longitudinal E_{1x} parts:

$$E_1 = \frac{\hbar^2 k_{1\perp}^2}{2m^*} + E_{1x}. \tag{6.2}$$

We shall see that in a one-dimensional heterostructure, the perpendicular momentum satisfies the two-dimensional momentum conservation rule:

$$\mathbf{k}_{1\perp} = \mathbf{k}_{0\perp} \pm \mathbf{q}_{\perp}. \tag{6.3}$$

The longitudinal part E_{1x} of the final energy E_1 is then

$$E_{1x} = E_{0x} \pm \hbar\omega_{\mathbf{q}} + \frac{\hbar^2\left(k_{0\perp}^2 - k_{1\perp}^2\right)}{2m^*}. \tag{6.4}$$

Note that the electron can be scattered by any phonon $\mathbf{q}$. Since the magnitude $k_{1\perp}$ of the final perpendicular momentum given by Equation (6.3) must have a positive value, the longitudinal part of the energy E_1 must satisfy the condition

$$E_{1x} \leq E_{0x} \pm \hbar\omega_{\mathbf{q}} + \frac{\hbar^2 k_{0\perp}^2}{2m^*}. \tag{6.5}$$

Equations (6.4) and (6.5) state that the longitudinal electron energy E_{1x} can have continuous values within a prescribed range. In a one-dimensional treatment, the perpendicular wave-vector is assumed to remain unchanged. Therefore the longitudinal part of the energy E_1 can only assume two values $E_{1x} = E_{0x} \pm \hbar\omega_{\mathbf{q}}$ for a single phonon scattering event. However, a one-dimensional model does not account for all the possible scattered waves.

Consider the case of the double-barrier heterostructure. For an incident electron with longitudinal energy E_{0x} for which resonant tunneling is not directly possible, there exists a family of phonons which will change its perpendicular wave-vector from $\mathbf{k}_{0\perp}$ to $\mathbf{k}_{1\perp}$ such that its final longitudinal energy E_{1x} is now aligned with the virtual eigenvalue of the quantum well (provided Equation (6.5) is satisfied). This process is called phonon-assisted resonant tunneling. Phonon-assisted resonant tunneling can potentially lead to an effective broadening of the transmission coefficient $T(E_{0x})$ by increasing the transmission coefficient off resonance and decreasing it at resonance.

Note, however, that the transition probability for polar scattering is weighted by the factor $1/q^2$. This favors the scattering by small-wave-vector $\mathbf{q}$ (large-wavelength) phonons and therefore prevents too large a variation of the perpendicular momentum. As a result, phonon-assisted resonant tunneling approximately follows the simplified selection rule $E_{1x} = E_{0x} \pm \hbar\omega_{\mathbf{q}}$ for small wave-vectors. Phonon-assisted resonant tunneling introduces two additional peaks in the transmission coefficient which can be observed in the *I–V* characteristic at low temperature.

As we shall see, the three-dimensional effects discussed above for phonon scattering apply also to elastic scattering processes.

6.3 Scattering-assisted tunneling theory

As described above we need to develop a scattering formalism capable of handling both tunneling and three-dimensional scattering in layered devices. The semiclassical formalism we shall develop for this purpose is referred to as the Multiple-Sequential Scattering-Assisted Tunneling (MSSCAT) theory. We shall start, however, by developing the MSSCAT theory for single sequential scattering (1SS) processes before generalizing it to multiple-sequential scattering (MSS) processes.

6.3.1 Semiclassical scattering picture

The quantum trajectory of a single electron subjected to various elastic and inelastic scattering processes i can be described using a total semiclassical Hamiltonian H consisting of the electron Hamiltonian H_e plus the various electron-scattering interaction Hamiltonians:

$$H = H_e + \sum_i H_{elast,i} + \sum_i H_{e-ph,i}(t) + \sum_i H_{e-e,i}(t).$$

The time-independent interaction Hamiltonians $H_{elast,i}$ will induce elastic scattering processes for which the total electron energy is conserved.

The time dependence of the electron–phonon scattering process interaction term $H_{e-ph,i}(t)$ results from our assumption in Chapter 5 of a classical field of lattice vibrations which remains in thermal equilibrium despite its interaction with the electrons. Similarly electron–electron scattering is represented by a time-dependent interaction $H_{e-e,i}(t)$ as it results from the scattering of the test electron by its Coulombic interaction with all the time-varying wave-functions (see Chapter 5) associated with the electrons of the many-body system.

The time-dependent interaction Hamiltonians therefore describe the scattering of our test electron by many-body systems (phonons or electrons) which are assumed to remain in thermal equilibrium. Note that the time-dependent interaction Hamiltonians induce inelastic scattering processes, as the incident (total) energy of the test electron is no longer conserved.

Note that in the single-electron treatment of electron–electron scattering, the trajectory of the test electron studied is inelastic because it is scattered by a Fermi many-body system which is assumed not to include it.

The trajectory of the electron is then given by the solution of the time-dependent Schrödinger equation:

$$H|\Psi\rangle = i\hbar\frac{d}{dt}|\Psi\rangle, \tag{6.6}$$

in the superlattice region. Our analysis will be limited to a layered heterostructure which varies spatially along the superlattice axis x and is uniform in the perpendicular directions (y and z). To calculate the trajectory of the electron in such a one-dimensional superlattice we use the generalized Wannier picture introduced in Chapter 3. In this picture, the electron wave-function $|\Psi\rangle$ is expanded in terms of the generalized Wannier functions $|\mathbf{k}_\perp, n\rangle$ using the envelope function $f(\mathbf{k}_\perp, n, t)$:

$$|\Psi\rangle = \sum_n \int f(\mathbf{k}_\perp, n, t)|\mathbf{k}_\perp, n\rangle\, d\mathbf{k}_\perp, \tag{6.7}$$

where n is the lattice site index along the superlattice direction. The Wannier expansion is limited to a single generalized band (except for interband scattering processes).

6.3.2 Matrix elements for the heterostructure Hamiltonian

In the generalized Wannier representation, the matrix element of the heterostructure Hamiltonian H_e in the generalized Wannier function basis is

$$\langle \mathbf{k}'_\perp,\ m \mid H_e \mid \mathbf{k}_\perp,\ n\rangle = H^e_{mn}(\mathbf{k}_\perp)\delta(\mathbf{k}'_\perp - \mathbf{k}_\perp),$$

where $H^e_{mn}(\mathbf{k}_\perp)$ is the matrix element of the biased heterostructure (generalized band structure).

To simplify the presentation of the MSS theory, we use the effective-mass approximation for the band structure in the transverse direction, so that we have (see Section 3.3.4)

$$H^e_{m\,n}(\mathbf{k}_\perp) = \tilde{H}_{m\,n} + \left[\frac{\hbar^2 k_\perp^2}{2m^*(0)}\right]\delta_{mn},$$

with $\tilde{H}_{m\,n}$ given by

$$\tilde{H}_{m\,n} = H_{m\,n} + \frac{\hbar^2 k_\perp^2}{2m^*(0)}\left[\frac{m^*(0)}{m(n)} - 1\right]\delta_{mn}.$$

6.3.3 Matrix elements for the interaction Hamiltonian

The scattering processes considered here are inelastic scattering by phonons (acoustic, polar, intervalley), electron–electron scattering, and elastic scattering (interface roughness and alloy fluctuation) in the heterostructures. As we shall see below, all these scattering processes are analyzed in a similar fashion except that some of them are local (occur at one lattice site, e.g. interface roughness) while others are non-local (occur simultaneously at all the sites, e.g. phonon scattering).

We have seen in Chapter 5 that the electron–phonon interaction potential can be written in the form

$$H_{e-ph} = \frac{2}{\Omega^{1/2}} \sum_{q_x>0} \sum_{\mathbf{q}_\perp} C_{ph\pm}(\mathbf{q}) \cos(\omega t - \mathbf{q}_\perp \cdot \mathbf{r}_\perp + \phi_\mathbf{q}) \sin(q_x x),$$

where the constant $C_{ph\pm}(\mathbf{q})$ was defined in Chapter 5.

In the generalized Wannier representation, the matrix element of the electron–phonon interaction Hamiltonian is

$$\begin{aligned} \langle \mathbf{k}'_\perp, m | H_{e-ph} \mid \mathbf{k}_\perp, n\rangle = {} & \frac{1}{\Omega^{1/2}} \sum_{q_x>0} \sin(q_x m a)\, \delta_{mn} \\ & \times \sum_{\mathbf{q}_\perp} C_{ph\pm}(\mathbf{q}) \Big\{ \delta(\mathbf{k}'_\perp - \mathbf{k}_\perp + \mathbf{q}_\perp) \exp[i(\omega_\mathbf{q} t + \phi_\mathbf{q})] \\ & + \delta(\mathbf{k}'_\perp - \mathbf{k}_\perp - \mathbf{q}_\perp) \exp[-i(\omega_\mathbf{q} t + \phi_\mathbf{q})] \Big\}, \end{aligned}$$

with (see Section 5.2.3) $C_{ph\pm}(\mathbf{q}) = 2\alpha_\mathbf{q} A_{\mathbf{q}\pm}$. As discussed in Chapter 5, the amplitude $A_\mathbf{q}$ and the phase $\phi_\mathbf{q}$ of different phonons are randomly distributed and mutually uncorrelated:

$$\left.\begin{aligned} \langle A_{\mathbf{q}'\pm} \exp(i\phi_{\mathbf{q}'}) A_{\mathbf{q}\pm} \exp(-i\phi_\mathbf{q})\rangle_{E.A.} &= \langle A^2_{\mathbf{q}\pm}\rangle_{E.A.} \delta_{\mathbf{q}'\mathbf{q}} \\ &= \left(N_\mathbf{q} + \frac{1}{2} \mp \frac{1}{2}\right) \delta_{\mathbf{q}'\mathbf{q}}, \\ \langle A_{\mathbf{q}'\pm} \exp(i\phi_{\mathbf{q}'}) A_{\mathbf{q}\pm} \exp(i\phi_\mathbf{q})\rangle_{E.A.} &= 0. \end{aligned}\right\} \qquad (6.8)$$

where the notation $\langle \cdots \rangle_{E.A.}$ indicates that an ensemble average is performed over the phonon scattering events. Phonon scattering is a random process, and we will use these ensemble average relations to calculate the average transmitted and reflected current for each scattering trajectory of our test electron.

A similar model can be developed for elastic scattering. A general elastic scattering process can be represented by an interaction potential $H_{elast}(\mathbf{r})$. It is convenient to use a periodic boundary condition so that we have

$$H_{elast} = \sum_\mathbf{q} V_\mathbf{q} \exp(i\mathbf{q} \cdot \mathbf{r}) = \sum_\mathbf{q} |V_\mathbf{q}| \exp(i\phi_\mathbf{q}) \exp(i\mathbf{q} \cdot \mathbf{r}),$$

where $\mathbf{q}$ is a reciprocal space vector. The phase $\phi_\mathbf{q}$ and the amplitude of $|V_\mathbf{q}|$ vary across the device area due to the random location of the scattering centers (e.g., alloy scattering and interface roughness scattering). An ensemble average over the scatterer location will therefore be used to calculate the average transmitted and reflected currents. The analysis of elastic scattering processes is therefore similar to that of phase-breaking scattering processes.

In the generalized Wannier representation, the matrix element of the elastic interaction Hamiltonian for an elastic scattering process at the lattice N_i is

$$\langle \mathbf{k}'_\perp,\ m|H_{elast,i} \mid \mathbf{k}_\perp,\ n\rangle = \delta_{mN_i}\delta_{mn}\frac{1}{S^{1/2}}\sum_{\mathbf{q}_\perp} C(\mathbf{q}_\perp, i)\delta(\mathbf{k}'_\perp - \mathbf{k}_\perp - \mathbf{q}_\perp),$$

where hermiticity requires $C(-\mathbf{q}_\perp, i) = C^*(\mathbf{q}_\perp, i)$. In the present theory we assume that the elastic scattering events are uncorrelated.

$$\langle C(\mathbf{q}'_\perp, i')C(\mathbf{q}_\perp, i)\rangle_{E.A.} = \langle |C(\mathbf{q}_\perp, i)|^2\rangle_{E.A.}\delta_{-\mathbf{q}'_\perp\ \mathbf{q}_\perp}\delta_{i'\ i}. \tag{6.9}$$

6.3.4 Envelope equations for sequential scattering

Having calculated the Hamiltonian matrix elements we can rewrite the Schrödinger Equation (6.6) in the generalized Wannier representation using Equation (6.7). For simplicity we consider a single inelastic (optical phonon) and a single elastic scattering process. The envelope function $f(\mathbf{k}_\perp, n, t)$ is then a solution of the following general envelope equation

$$\begin{aligned} i\hbar\frac{d}{dt}f(\mathbf{k}_\perp, n, t) &= \frac{\hbar^2\mathbf{k}_\perp{}^2}{2m^*(0)}f(\mathbf{k}_\perp, n, t) + \sum_{m=-N_B}^{N_B}\tilde{H}_{n\ m}f(\mathbf{k}_\perp, m, t) \\ &+ \frac{1}{\Omega^{1/2}}\sum_{q_x>0}\sin(q_x na)\sum_{\mathbf{q}_\perp}C_{ph\pm}(\mathbf{q})\{f(\mathbf{k}_\perp + \mathbf{q}_\perp, n, t)\exp[i(\omega_\mathbf{q}t + \phi_\mathbf{q})] \\ &\qquad + f(\mathbf{k}_\perp - \mathbf{q}_\perp, n, t)\exp[-i(\omega_\mathbf{q}t + \phi_\mathbf{q})]\} \\ &+ \frac{1}{S^{1/2}}\sum_i \delta_{nN_i}\sum_{\mathbf{q}_\perp}C(\mathbf{q}_\perp, i)f(\mathbf{k}_\perp - \mathbf{q}_\perp, n, t). \end{aligned} \tag{6.10}$$

The envelope equation derived above is the Schrödinger equation in the Wannier representation that we must solve to analyze scattering-assisted tunneling in a layered heterostructure. We shall assume that the incident electron can be represented by an incident wave of total energy E_0, transverse momentum $\mathbf{k}_{\perp 0}$ and envelope $f_0(n, E_{0x})$. As the incident electron absorbs or emits optical phonons, its energy will be either increased or decreased by a multiple of $\hbar\omega_\mathbf{q}$. For elastic scattering the total energy remains unchanged. A steady-state envelope solution will therefore consist of this incident wave plus all the scattered waves of various energy and transverse momentum generated. For a 1SS event the total envelope function is therefore given by

$$\begin{aligned} f(n, \mathbf{k}_\perp, t) &= f_0(n, E_{0x})\delta(\mathbf{k}_\perp - \mathbf{k}_{\perp 0})\exp\left(\frac{E_0 t}{\hbar}\right) \\ &+ \frac{1}{\Omega^{1/2}}\exp\left(-i\frac{E_0 t}{\hbar}\right)\sum_{q_x>0}\sum_{\mathbf{q}_\perp}\exp[-i(\omega_\mathbf{q}t + \phi_\mathbf{q})]C_{ph+}(\mathbf{q}) \end{aligned}$$

$$\times f_1(n, E_{1x}(\mathbf{q}_\perp), q_x)\delta(\mathbf{k}_\perp - (\mathbf{k}_{0\perp} + \mathbf{q}_\perp))$$

$$+\frac{1}{\Omega^{1/2}} \exp\left(-i\frac{E_0 t}{\hbar}\right) \sum_{q_x>0} \sum_{\mathbf{q}_\perp} \exp[i(\omega_\mathbf{q} t + \phi_\mathbf{q})] C_{ph-}(\mathbf{q})$$

$$\times f_1(n, E_{1x}(-\mathbf{q}_\perp), q_x)\delta(\mathbf{k}_\perp - (\mathbf{k}_{0\perp} - \mathbf{q}_\perp))$$

$$+\frac{1}{S^{1/2}} \exp\left(-i\frac{E_0 t}{\hbar}\right) \sum_{i} \sum_{\mathbf{q}_\perp} C_{elast}(\mathbf{q}_\perp, i)$$

$$\times f_1(n, E_{1x}(\mathbf{q}_\perp), i)\delta(\mathbf{k}_\perp - (\mathbf{k}_{0\perp} + \mathbf{q}_\perp)), \quad (6.11)$$

with $f_0(n, E_{0x})$ the incident wave envelope, $f_1(n, E_{1x}(-\mathbf{q}_\perp))$ the scattered wave envelope for 1SS and with the longitudinal incident and scattered energies E_{0x} and E_{1x} defined to be

$$E_{0x} = E_0 - \frac{\hbar^2}{2m^*(0)} |\mathbf{k}_{0\perp}|^2,$$

$$E_{1x}(\pm\mathbf{q}_\perp) = E_0 \pm \hbar\omega_\mathbf{q} - \frac{\hbar^2}{2m^*(0)} |\mathbf{k}_{0\perp} \pm \mathbf{q}_\perp|^2 \quad \text{for phonon scattering,}$$

$$E_{1x}(\mathbf{q}_\perp) = E_0 - \frac{\hbar^2}{2m^*(0)} |\mathbf{k}_{0\perp} + \mathbf{q}_\perp|^2 \quad \text{for elastic scattering.}$$

The general envelope solution $f(\mathbf{k}_\perp, n, t)$ in Equation (6.11) for 1SS is expressed in terms of the envelope functions $f_0(n, E_{0x})$ and $f_1(n, E_{1x}(-\mathbf{q}_\perp))$ which remain to be determined.

We can verify that $f(\mathbf{k}_\perp, n, t)$ in Equation (6.11) is an ensemble-average solution of the envelope Equation (6.10) limited to a 1SS event provided that f_0 and f_1 are solutions of specific envelope equations. To obtain these envelope equations, we first substitute the proposed solution (6.11) in the envelope Equation (6.10), and equate the terms of same energy (frequency) E_0, $E_0 + \hbar\omega_\mathbf{q}$, and $E_0 - \hbar\omega_\mathbf{q}$ and weighted by the same Dirac function $\delta(\mathbf{k}_\perp - \mathbf{k}'_\perp)$ (i.e., same perpendicular Bloch wave). Next we perform an ensemble average using Equations (5.18) and (6.9) assuming that elastic and phonon scattering are uncorrelated processes

$$\langle C_{elast}(\mathbf{q}'_\perp, i) C_{ph\pm,\mp}(\mathbf{q}) \exp(\pm i\phi_{\mathbf{q}'})\rangle_{E.A.} = 0.$$

Therefore f_0 must satisfy the envelope equation

$$E_{0x} f_0(n, E_{0x}) = \sum_m \tilde{H}_{n\,m} f_0(m, E_{0x})$$

$$+ \sum_{q_x>0} \left[G_{ph+}(n, E_{0x}, q_x) + G_{ph-}(n, E_{0x}, q_x)\right] + \sum_i G_{elast}(n, E_{0x}, i), \quad (6.12)$$

where $G_{ph\pm}$ and G_{elast} are coupling terms defined below. Similarly, the envelope function f_1 is a solution of the envelope equations

$$E_{1x} f_1(n, E_{1x}, q_x) = \sum_m \tilde{H}_{n\,m} f_1(m, E_{1x}, q_x) + \sin(q_x na) f_0(n, E_{0x}) \quad (6.13)$$

for phonon scattering and

$$E_{1x} f_1(n, E_{1x}, i) = \sum_m \tilde{H}_{n\,m} f_1(m, E_{1x}, i) + \delta_{n,N_i} f_0(n, E_{0x}) \tag{6.14}$$

for elastic scattering.

The coupling terms $G_{ph\pm}$ and G_{elast} in Equation (6.12) are found to be

$$G_{ph\pm}(n, E_{0x}, q_x) = \sin(q_x na) \frac{1}{\Omega} \sum_{\mathbf{q}_\perp} \langle C^2_{ph\pm}(\mathbf{q}) \rangle_{E.A.} f_1(n, E_{0x}(\pm\mathbf{q}_\perp), q_x), \tag{6.15}$$

$$G_{elast}(n, E_{0x}, i) = \delta_{nN_i} \frac{1}{S} \sum_{\mathbf{q}_\perp} \langle |C_{elast}(\mathbf{q}_\perp, i)|^2 \rangle_{E.A.} f_1(n, E_{0x}(\mathbf{q}_\perp), i). \tag{6.16}$$

To simplify the evaluation of the G terms it is advantageous to replace the summation over the perpendicular momentum $\mathbf{q}_\perp$ by an integration. For most of the scattering processes (except acoustic scattering), the integration over the perpendicular momentum can be performed analytically and the following expressions result for the coupling terms G:

$$G_{ph\pm}(n, E_{0x}, q_x) = \sin(q_x na) \times \int_{-2A_L}^{E_0 \pm \hbar\omega_{LO}} f_1(n, E_{1x}, q_x) H_{LO\pm}(E_{1x}, q_x, E_{0x}, E_{0\perp})\, dE_{1x}$$

for optical phonon scattering and

$$G_{elast}(n, E_{0x}, i) = \delta_{nN_i} \int_{-2A_L}^{E_0} f_1(n, E_{1x}, i) H_{EL}(E_{1x}, i, E_{0x}, E_{0\perp})\, dE_{1x}$$

for elastic scattering. Note that as a result of the three-dimensional analysis the coupling function H_{scat} is now a continuous function of E_{1x}, indicating that the incident electron of energy E_{0x} couples to a continuum of scattered waves with energy E_{1x} due to the variation of its transverse momentum. In Section 6.9 we shall present the coupling functions H_{scat} for the various scatterings discussed in Chapter 5.

Note that a complete MSS solution can be obtained using the same procedure. For example the envelope function accounting for two sequential scattering events for elastic scattering only is given by

$$\begin{aligned} f(n, \mathbf{k}_\perp, t) &= f_0(n)\delta(\mathbf{k}_\perp - \mathbf{k}_{\perp 0}) \exp\left(-i\frac{E_0 t}{\hbar}\right) \\ &+ \frac{1}{S^{1/2}} \exp\left(-i\frac{E_0 t}{\hbar}\right) \sum_i \sum_{\mathbf{q}_\perp} C(\mathbf{q}_\perp, i) \Big[f_1(n, \mathbf{q}_\perp, i)\delta(\mathbf{k}_\perp - (\mathbf{k}_{0\perp} + \mathbf{q}_\perp)) \\ &+ \frac{1}{S^{1/2}} \sum_{i'} \sum_{\mathbf{q}'_\perp} C(\mathbf{q}'_\perp, i') f_2(n, \mathbf{q}'_\perp, i')\delta(\mathbf{k}_\perp - (\mathbf{k}_{0\perp} + \mathbf{q}_\perp + \mathbf{q}'_\perp)) \Big]. \end{aligned} \tag{6.17}$$

Substituting Equation (6.17) into Equation (6.10) we verify that a second sequential scattering event is implemented by introducing the coupling terms in the Wannier

Equations (6.13) and (6.14) which couple the waves f_1 to the waves f_2. The presence of the coupling terms in both the incident and the scattered Wannier equations accounts for backscattering. Backscattering, among other things, enforces the conservation of the current.

6.4 Transmission coefficient for scattering-assisted tunneling

Having established the equations for solving the wave-functions f_0 and f_1 we are now in a position to calculate the reflected or transmitted currents in the quantum device studied. We have seen that for each electron of energy E_{0x} and perpendicular momentum $\mathbf{k}_{\perp 0}$ incident on the superlattice we will have a continuous spectrum of scattered states of longitudinal energy E_{1x} and transverse energy $E_{1\perp}$. We are only interested in the average current obtained from an ensemble average over the scatterers. For mutually uncorrelated incident and scattered waves, the total current is obtained by summing of the currents carried by each of those states. As is demonstrated in Problem 6.3, the total current of the incident and scattered waves is conserved.

Let us assume for simplicity that for a wide range of energies the semiconductor can be represented by a tight-binding band. We also assume the flat-band condition in the left-hand (L) contact (sites $n < N_L$) and the right-hand (R) contact (sites $n > N_R$). Under such conditions a ballistic electron of energy E_{0x} satisfies

$$E_{0x} = A_L - A_L \cos(k_{0xL}a) = A_R - A_R \cos(k_{0xR}a),$$

where k_{0xR} and k_{0xL} (selected positive) are the electron wave-vectors on the left- and right-hand contacts. We assume below that A_L and A_R are positive numbers and select the positive values for k_{0xL} and k_{0xR}.

In the left- and right-hand contacts an incident wave $f_0(n, E_{0x})$ of energy E_{0x} can then be written:

$$f_0(n) = \begin{cases} \exp(ik_{0xL}na) + b_0 \exp(-ik_{0xL}na) & \text{for } n < N_L \\ c_0 \exp(ik_{0xR}na) & \text{for } n \geq N_R, \end{cases}$$

where 1, b_0, and c_0 are the amplitudes of the incident, reflected, and transmitted subwave components of f_0.

The incident, reflected, and transmitted currents for the envelope function f_0 are given respectively by

$$J_{I0}(E_{0x}) = e\, v_L(E_{0x}),$$
$$J_{R0}(E_{0x}) = e\, |b_0|^2\, v_L(E_{0x}),$$
$$J_{T0}(E_{0x}) = e\, |c_0|^2\, v_R(E_{0x}),$$

where $v_L(E_{0x})$ is the velocity of semiconductor L on the left-hand side of the superlattice and $v_R(E_{0x})$ is the velocity of semiconductor R on the right-hand side of the superlattice. Since we assume here a tight-binding band in the flat-band region of the L and R contacts, the electron velocity is simply given by

$$v_i(E_x) = \frac{1}{\hbar}\frac{d\mathcal{E}_i(k_x)}{dk_x} = \frac{a}{\hbar}A_i \sin(k_{0xi}a).$$

The transmitted current associated with the envelope function f_1 is given by

$$e\,|c(E_{1x}, q_x)|^2\, v_R(E_{1x}).$$

The total average transmitted current $J_{T,ph\pm}$ resulting from the emission or absorption of all the phonon-scattering events is then given by

$$J_{T,ph\pm}(E_{0x}, E_{0\perp}) = e\frac{1}{\Omega}\sum_{q_x>0}\sum_{\mathbf{q}_\perp} \langle C^2_{ph\pm}(\mathbf{q})\rangle_{E.A.}\ |f_1(N_R, E_{1x}, q_x)|^2\ v_R(E_{1x})$$

$$= e\frac{1}{L_x}\sum_{q_x>0}\int_{-2A_L}^{E_0\pm\hbar\omega_{LO}} H_{LO\pm}(E_{1x}, q_x, E_{0x}, E_{0\perp})\,|f_1(N_R, E_{1x}, q_x)|^2\, v_R(E_{1x})\, dE_{1x}.$$

The total transmitted current $J_{T,elast}$ resulting from elastic scattering is similarly given by

$$J_{T,elast}(E_{0x}, E_{0\perp}) = e\sum_i \frac{1}{S}\sum_{\mathbf{q}_\perp}\langle|C_{elast}(\mathbf{q}_\perp, i)|^2\rangle_{E.A.}\,|f_1(N_R, E_{1x}, i)|^2\, v_R(E_{1x})$$

$$= e\sum_i \int_{-2A_L}^{E_0} H_{EL}(E_{1x}, i, E_{0x}, E_{0\perp})\,|f_1(N_R, E_{1x}, i)|^2\, v_R(E_{1x})\, dE_{1x}.$$

The total forward transmission coefficient for an incident electron of longitudinal energy E_{0x} and perpendicular momentum $\mathbf{k}_{0\perp}$ is then given in the 1SS approximation by

$$T_F(E_{0x}, E_{0\perp}) = T_0 + T_{ph} + T_{elast},$$

using the elementary transmission coefficients:

$$T_0(E_{0x}, E_{0\perp}) = \frac{J_{T0}(E_{0x}, k_{0\perp})}{J_{I0}(E_{0x})},$$

$$T_{ph}(E_{0x}, E_{0\perp}) = \frac{J_{T,ph+}(E_{0x}, k_{0\perp}) + J_{T,ph-}(E_{0x}, k_{0\perp})}{J_{I0}(E_{0x})},$$

$$T_{elast}(E_{0x}, E_{0\perp}) = \frac{J_{T,elast}(E_{0x}, k_{0\perp})}{J_{I0}(E_{0x})}.$$

The various scattering processes are therefore seen to contribute to the total forward transmission coefficient T_F which is why we refer to this physical process as scattering-assisted tunneling. A total reflection coefficient R_F can be defined similarly (by replacing R by L):

$$R_F(E_{0x}, E_{0\perp}) = R_0 + R_{ph} + R_{elast}.$$

From current conservation (see Problem 6.3) we have $R_F + T_F = 1$. Note that this result was derived for a 1SS event but it will be generalized in the following sections for MSS events.

As usual, the total forward diode current per unit area from semiconductor L to semiconductor R for a given Fermi energy E_{fL} on the left-hand side (see Figure 6.1) is then obtained by summing the transmitted current over all possible incident momentums $\mathbf{k}_0$:

$$I_F(E_{fL}) = \frac{1}{4\pi^3}\int_0^{\pi/a} dk_{0x} \int_{-\pi/a}^{\pi/a}\int_{-\pi/a}^{\pi/a} d\mathbf{k}_{0\perp} \frac{e\,T_F(E_{0x}, E_{0\perp})v_L(E_{0x})}{\exp\left(\dfrac{E_0 - E_{fL}}{k_B T_0}\right) + 1}.$$

Introducing the new variables $dE_{0x} = v_L(E_{0x})dk_{0x}$ and $d\mathbf{k}_{0\perp} = 2\pi k_{0\perp}dk_{0\perp}$ the diode current density is now

$$I_F(E_{fL}) = \frac{m^*}{2\pi^2\hbar^2}\int_0^{2A_L} dE_{0x}\int_0^{2A_L} dE_{0\perp}\frac{e\,T_F(E_{0x}, E_{0\perp})}{\exp\left(\dfrac{E_0 - E_{fL}}{k_B T_0}\right) + 1}, \tag{6.18}$$

with $E_0 = E_{0x} + E_{0\perp}$.

The total backward diode current I_B per unit area contributed by the electron incident on the right-hand side of the device is obtained by a similar equation but using the Fermi energy E_{fR} on the right-hand contact and the reverse transmission coefficient T_R associated with the electron transport from right to left. As for ballistic transport the total diode current I per unit area is then given by the difference between the forward and backward currents $I = I_F(E_{fL}) - I_B(E_{fR})$.

6.5 Self-energy

We have seen in the previous sections that the ensemble-average solution of the Schrödinger equation in the presence of phase-breaking scattering for a single scattering event reduced to the solution of a set of coupled difference equations. The coupling terms G involve an integration over a continuum of energy E_{1x}. For the numerical calculation we can replace this integration over E_{1x} by a summation over a discrete set of energies $E_{1x,r}$. The solution of the Schrödinger equation in the presence of scattering then reduces to the solution of the difference Equation (6.12) coupled to a finite set of difference Equations (6.13) and (6.14). For a single scattering event this can be easily handled. However, for MSS the problem becomes formidable as the number of scattered waves grows exponentially. For example, if we need say $P = 300$ energies $E_{1x,r}$ for each scattering event at a single site then for M sequential scattering events we need to solve $1 + P + P^2 + \cdots + P^M$ coupled envelope equations. It is here that it becomes numerically advantageous to introduce the impulse (Green) function to solve this problem and reduce the computation for elastic scattering to the solution of $M \times P$ *decoupled* wave-equations. This approach is also beneficial because it introduces the physical concept of self-energy, and it will facilitate the generalization of the 1SS theory presented above to the MSS case. Finally it will also allow us to develop an extension of the MSS theory which accounts for Pauli exclusion effects.

An efficient method for solving the MSS Hamiltonian system of equations can be developed if we try to evaluate the coupling terms $G(n, \cdot)$ before the wave-function f_0 is available. Indeed, as can be expected from a linear system, the functions $f_1(n)$ can be expressed as

$$f_1(n, E_{1x}, E_{1\perp}, i) = h(n, i, E_{1x}, E_{1\perp}) f_0(i, E_{0x}, E_{0\perp})$$

for local scattering and

$$f_1(n, E_{1x}, E_{1\perp}, q_x) = \sum_{m=N_L}^{N_R} h(n, m, E_{1x}, E_{1\perp}) \sin(q_x m a)\; f_0(m, E_{0x}, E_{0\perp})$$

for non-local scattering, where the function $h(n, m, E_{1x}, E_{1\perp})$ is the impulse response solution of Equation (6.14) with $f_0(n, E_{0x}, E_{0\perp}) = 1$:

$$E_{1x}\, h(n, m, E_{1x}, E_{1\perp}) = \sum_l \tilde{H}_{n\,l}(E_{1\perp})\; h(l, m, E_{1x}, E_{1\perp}) \;+\; \delta_{nm}.$$

Substituting f_1 into Equation (6.13) or (6.14) we find that for a scattering process $scat$ (ph or $elast$), the coupling term $G_{scat}(n, E_{0x}, \cdot)$ can be written

$$G_{scat}(n, E_{0x}, \cdot) = \sum_{m=N_L}^{N_R} H^{SE,0}_{scat,n,m}(\cdot, E_{0x}, E_{0\perp}) f_0(m, E_{0x}, E_{0\perp}),$$

where $H^{SE,0}_{scat,n,m}$ is the so-called self-energy matrix for the scattering process $scat$. For inelastic ($E_1 = E_0 \pm \hbar\omega_{LO}$) non-local LO phonon scattering we have

$$\begin{aligned} H^{SE,0}_{LO\pm,n,m}(q_x, E_{0x}, E_{0\perp}) &= \sin(q_x n a)\sin(q_x m a) \\ &\times \int_{-2A_L}^{E_0 \pm \hbar\omega_{LO}} h(n, m, E_{1x}, E_{1\perp}) H_{LO\pm}(E_{1x}, q_x, E_{0x}, E_{0\perp})\, dE_{1x} \end{aligned}$$

and for elastic ($E_1 = E_0$) local scattering we have

$$\begin{aligned} H^{SE,0}_{EL,n,m}(i, E_{0x}, E_{0\perp}) &= \delta_{nm}\delta_{nN_i} \\ &\times \int_{-2A_L}^{E_0} h(n, n, E_{1x}, E_{1\perp}) H_{EL}(E_{1x}, i, E_{0x}, E_{0\perp})\, dE_{1x}. \end{aligned}$$

Once the self-energy matrix has been evaluated for all the scattering processes, the incident wave is then simply obtained from the solution of

$$E_{0x}\; f_0(n, E_{0x}, E_{0\perp}) = \sum_m \left[\tilde{H}_{n\,m}(E_\perp) + H^{SE,0}_{n\,m}(E_{0x}, E_{0\perp})\right] f_0(m, E_{0x}),$$

using the total self-energy matrix

$$H_{n,m}^{SE,0}(E_{0x}, E_{0\perp}) = \sum_i H_{EL,n,m}^{SE,0}(i, E_{0x}, E_{0\perp}) + \sum_{q_x} H_{LO+,n,m}^{SE,0}(q_x, E_{0x}, E_{0\perp})$$
$$+ \sum_{q_x} H_{LO-,n,m}^{SE,0}(q_x, E_{0x}, E_{0\perp})$$

The self-energy matrix $H^{SE,0}$ is a non-Hermitian matrix whose matrix elements are complex numbers. The real part of the self-energy matrix is responsible for shifting downward the resonant energies in a superlattice device. The imaginary part accounts for the loss of current carried by the incident wave f_0 as it spawns new scattered waves. The self-energy matrix is diagonal for local scattering processes and non-diagonal for non-local scattering processes. Note that the global current conservation ($R + T = 1$) is independent of the energies $E_{1x,r}$ used. However, a reliable evaluation of the self-energy requires an adaptive integration technique which selects the energies $E_{1x,r}$ so as to meet a targeted accuracy.

The use of the impulse (Green) function to solve the incident + scattered Hamiltonian system has decoupled the calculation of the incident and scattered waves. In the next section we will also see that this provides us with an efficient numerical method for handling the general solution for an arbitrary number of sequential scattering events.

6.6 The MSS algorithm

Our discussion, so far, has focused on a 1SS event. We are, however, now ready to generalize the scattering theory to an arbitrary number of sequential scattering events. Indeed, in practice, a single scattering event is only an acceptable assumption for small quantum devices like small RTDs and weak scattering processes. For large devices like superlattices and strong scattering processes (e.g., interface roughness scattering), the electron is most likely to be involved in MSS events.

For simplicity the MSS algorithm is presented here only for local and elastic scattering processes at various sites l. In practice for a device of finite size, a finite number M of sequential scattering events is sufficient. The incident electron wave f_0 is then subjected to the sequential uncorrelated scattering events $S = 1, 2, \ldots, M$ spawning the waves $f_1, f_2, \ldots, f_S, \ldots, f_M$ of total energies $E_1, E_2, \ldots, E_S, \ldots, E_M$. Following the methodology presented in the previous section we introduce the impulse response $h_S(n, l, E_{Sx}, E_{S\perp})$ and the self-energy $H_{EL}^{SE,S}(E_{Sx}, E_{S\perp})$ associated with the Sth elastic scattering event at site l. This impulse response $h_S(n, l, E_{Sx}, E_{S\perp})$ is a solution of

$$E_{Sx}\, h_S(n, l, E_{Sx}, E_{S\perp}) = \sum_m \left[\tilde{H}_{nm}(E_{S\perp}) + H_{EL,nm}^{SE,S}(E_{Sx}, E_{S\perp})\right]$$
$$\times\, h_S(m, l, E_{Sx}, E_{S\perp}) + \delta_{nl},$$

where the self-energy $H_{EL}^{SE,S}$ for the $(S+1)$th elastic $(E_S = E_0)$ (local) scattering event is defined as

$$H_{EL,n,m}^{SE,S}(E_{Sx}, E_{S\perp}) = \delta_{nm} \int_{-2A_L}^{E_0} h_{S+1}(n, n, E_{(S+1)x}, E_{(S+1)\perp}) \times H_{EL}(E_{(S+1)x}, n, E_{Sx}, E_{S\perp})\, dE_{(S+1)x}.$$

Clearly MSS is simply implemented by including the self-energy $H_{EL}^{SE,S}$ in the Hamiltonian of the impulse response h_S itself. This effectively allows for the Sth scattered waves to be scattered themselves and spawn the $(S+1)$ scattered waves at the $(S+1)$ sequential scattering event. Note that the self-energy $H_{EL}^{SE,S}$ is indeed calculated from the h_{S+1} impulse responses generated at the $(S+1)$th scattering event. In MSS the Mth sequential scattering event is assumed to be fully ballistic (no-scattering) so that we have $H_{EL}^{SE,M} = 0$. The calculation of the successive self-energies starts therefore from the last scattering event M. The self-energies $H_{EL}^{SE,M}$, $H_{EL}^{SE,M-1}$, $H_{EL}^{SE,M-2}, \ldots$ are therefore successively calculated until the final self-energy $H_{EL}^{SE,1}$ is reached. When a sufficiently large number M (depending on the device size and the strength of the scattering processes involved) of sequential scattering events are accounted for, the final $(S = 0)$ self-energy converges to a unique value $H^{SE}(E_{0x}, E_{0\perp})$ and the impulse response to the unique value $h(n, m, E_{Sx}, E_{S\perp})$. This indicates that events beyond the $(M+1)$th event are unlikely and therefore do not have a significant impact on the final self-energy.

This convergence property of MSS is demonstrated in Figure 6.2 which compares the transmission coefficients obtained for different sequential scattering processes. Clearly for the triple barrier structure considered, MSS is found to converge in five sequential scattering processes. Also included is the perturbation treatment in which the self-energy is neglected (set to zero). The perturbation results are seen to be unstable and diverge even though renormalization $R_{norm} = R/(R+T)$ was used to combat the absence of current conservation $(R + T > 1)$ in that method.

The algorithm described above allows us to calculate the self-energy and the impulse response for MSS processes. However, a methodology is also required for calculating the reflected and transmitted currents associated with the MSS processes. Consider, for example, a tight-binding band. The impulse response is a solution of

$$\tilde{H}_{nn-1}h_S(n-1, l, E_{Sx}, E_{S\perp}) + \left[\tilde{H}_{nn} + H_{EL,nn}^{SE,S}\right] h_S(n, l, E_{Sx}, E_{S\perp}) + \tilde{H}_{nn+1}h_S(n+1, l, E_{Sx}, E_{S\perp}) + \delta_{nl} = E_{Sx}h_S(n, l, E_{Sx}, E_{S\perp}) = F.$$

From the evaluation of $Fh_S^*(n, l) - F^*h_S(n, l)$ we obtain

$$0 = -j(n-1, n) + j(n, n+1) + \frac{a}{\hbar}\delta_{nl}\,\mathrm{Im}\,[h_S(n, l, E_{Sx}, E_{S\perp})] - \frac{a}{\hbar}\,|h_S(n, l, E_{Sx}, E_{S\perp})|^2\,\mathrm{Im}\left[H_{EL,n,n}^{SE,S}(E_{Sx}, E_{S\perp})\right],$$

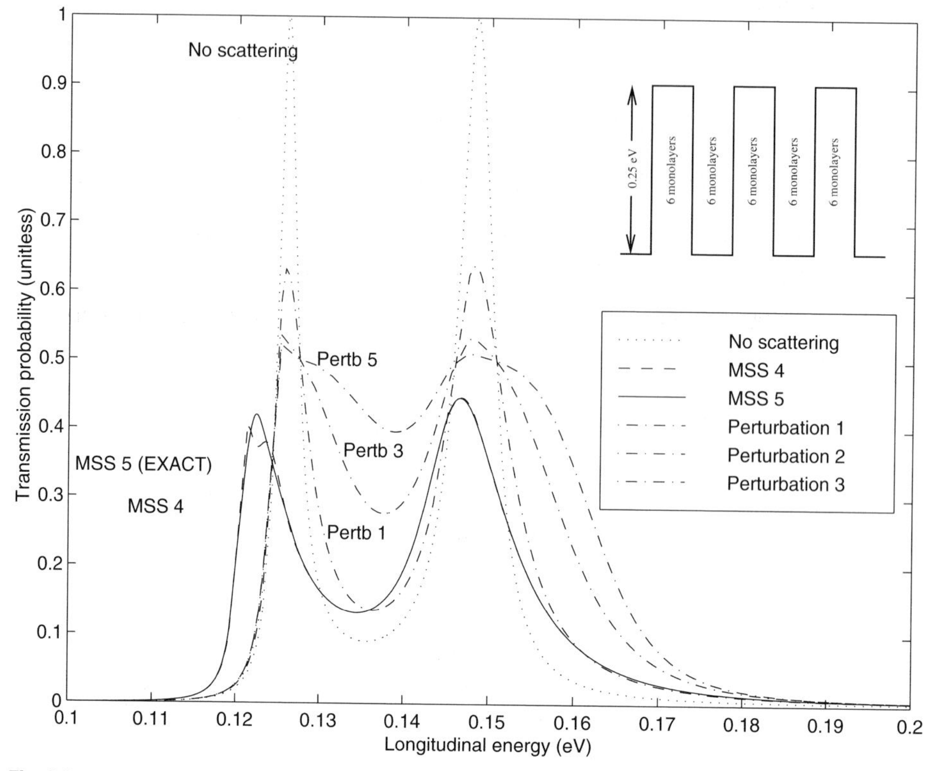

Fig. 6.2. Comparison of the transmission coefficients calculated in the presence of interface roughness scattering for 0, 4, and 5 (quasi-exact) MSS-assisted tunneling and using the renormalized perturbative calculation (Pertb 1, 3, and 5).

using the elemental (flux) current $j(n, m) = -(a/\hbar)\,\mathrm{Im}[h^*(n, \ldots)h(m, \ldots)H_{n\,m}]$. Integrating over the entire device from the left-hand contact N_L to the right-hand contact N_R, we obtain the following current-conservation equation

$$
\begin{aligned}
-\frac{2a}{\hbar}\,\mathrm{Im}\,[h_S(l, l, E_{Sx}, E_{S\perp})]&\\
&= v_L\,|h_S(N_L, l, E_{Sx}, E_{S\perp})|^2 + v_R\,|h_S(N_R, l, E_{Sx}, E_{S\perp})|^2\\
&\quad -\frac{2a}{\hbar}\sum_n |h_S(n, l, E_{Sx}, E_{S\perp})|^2\,\mathrm{Im}\left[H^{SE,S}_{EL,n,n}(E_{Sx}, E_{S\perp})\right],
\end{aligned}
\tag{6.19}
$$

where v_L and v_R are the left-hand and right-hand electron velocities. Since the last equation does not differentiate between the reflected and transmitted current components carried by the scattered wave it is advantageous to rewrite it as

$$
-\frac{2a}{\hbar}\,\mathrm{Im}\,[h_S(l, l, E_{Sx}, E_{S\perp})] = j_{R,S}(l, E_{Sx}, E_{S\perp}) + j_{T,S}(l, E_{Sx}, E_{S\perp}),
\tag{6.20}
$$

where the total reflected $j_{R,S}$ and transmitted $j_{T,S}$ currents are obtained from the summations:

$$j_{R,S}(l, E_{Sx}, E_{S\perp}) = j^0_{R,S}(l, E_{Sx}, E_{S\perp}) + \sum_n |h_S(n, l, E_{Sx}, E_{S\perp})|^2 \, j^1_{R,S}(n, E_{Sx}, E_{S\perp}),$$

$$j_{T,S}(l, E_{Sx}, E_{S\perp}) = j^0_{T,S}(n, E_{Sx}, E_{S\perp}) + \sum_n |h_S(n, l, E_{Sx}, E_{S\perp})|^2 \, j^1_{T,S}(n, E_{Sx}, E_{S\perp}),$$

involving the reflected and transmitted currents associated with the impulse function itself $h_S(n, l)$ of the scattering event S:

$$j^0_{R,S}(l, E_{Sx}, E_{S\perp}) = v_R(E_{Sx})|h_S(N_R, l, E_{Sx}, E_{S\perp})|^2,$$
$$j^0_{T,S}(l, E_{Sx}, E_{S\perp}) = v_L(E_{Sx})|h_S(N_L, l, E_{Sx}, E_{S\perp})|^2,$$

and the reflected and transmitted currents associated with the scattering states of the scattering event $(S + 1)$:

$$j^1_{R,S}(n, E_{Sx}, E_{S\perp}) = \int_{-2A_L}^{E_0} j_{R,S+1}(n, E_{(S+1)x}, E_{(S+1)\perp}) \times H_{EL}(E_{(S+1)x}, n, E_{Sx}, E_{S\perp})\, dE_{(S+1)x},$$

$$j^1_{T,S}(n, E_{Sx}, E_{S\perp}) = \int_{-2A_L}^{E_0} j_{T,S+1}(n, E_{(S+1)x}, E_{(S+1)\perp}) \times H_{EL}(E_{(S+1)x}, n, E_{Sx}, E_{S\perp})\, dE_{(S+1)x}.$$

To summarize, the MSS algorithm starts with the last scattering event $S = M$ for which a fully ballistic trajectory is assumed $H^{SE,M}_{EL} = j^1_{R,M} = j^1_{T,M} = 0$. All the preceding scattering events are then successively calculated till the incident wave is reached $(S = 0)$. At each event S, the self-energy $H^{SE,S}_{EL,nm}(E_{Sx}, E_{S\perp})$, the impulse $h_S(n, l, E_{Sx}, E_{S\perp})$, and the reflected and transmitted currents $j_{R,S}(l, E_{Sx}, E_{S\perp})$ and $j_{T,S}(l, E_{Sx}, E_{S\perp})$ for all scattering sites l are calculated is that order. When the incident state is reached ($S = 0$ event, ballistic motion) the currents contributed by all the scattered waves in the MSS processes are then calculated using the incident wave $f_0(n, E_{0x}, E_{0\perp})$ instead of the impulse $h_0(n, l, E_{0x}, E_{0\perp})$. Note that one could alternatively calculate the incident wave using the identity:

$$f_0(n, E_{0x}, E_{0\perp}) = j\frac{\hbar v_L(E_{0x})}{a} h_0(n, N_L, E_{0x}, E_{0\perp}),$$

which can be verified to hold for an electron incident on the left-hand contact (N_L). Note that the MSS algorithm has been presented here for local scattering processes but it can be extended to non-local scattering using a similar but somewhat more tedious matrix analysis.

6.7 Scattering-parameter representation

Having introduced MSS theory and its algorithm for calculating the impulse response, we will now introduce a discrete scattering parameter representation which will allow us to recast the MSS calculation of the transmission coefficient into a summation of scattering parameters over all possible scattering paths. This new symbolic representation will allow us, in turn, to address in the next section important many-body issues such as the enforcement of detailed balance and of Pauli exclusion in MSS theory.

To proceed we shall assume for simplicity that LO phonon scattering and elastic scattering can be represented by a local scattering process at a single site l. The resulting Hamiltonian for a 1SS event is then

$$E_{0x} f_0(n, E_{0x}, E_{0\perp}) = \sum_m \tilde{H}_{nm}(E_{0\perp}) f_0(m, E_{0x}, E_{0\perp}) + \sum_p G_p(n, E_{0x}, E_{0\perp}), \tag{6.21}$$

with the scattering terms G_p given by

$$G_p(n, E_{0x}, E_{0\perp}) = \delta_{nl} \int_{-2A_L}^{E_0+p\hbar\omega} f_1(n, E_{1x}, E_{0x}, E_{0\perp}) \times H_p(E_{1x}, E_{0x}, E_{0\perp})\, dE_{1x}, \tag{6.22}$$

where $f_1(n)$ is obtained from

$$E_{1x} f_1(n, E_{1x}, E_{0x}, E_{0\perp}) = \sum_m \tilde{H}_{nm}(E_{1\perp}) f_1(m, E_{1x}, E_{0x}, E_{0\perp}) + \delta_{nl}\, f_0(l, E_{0x}, E_{0\perp}), \tag{6.23}$$

with

$$E_{1p} = E_{1x} + E_{1\perp} = E_0 + p\hbar\omega_{LO} = E_{0x} + E_{0\perp} + p\hbar\omega_{LO} \quad \text{with } p = -1, 0, +1.$$

Note that p is -1 for phonon emission, 0 for elastic scattering and 1 for phonon absorption.

For such a 1SS process the resulting forward current is

$$I_F = \frac{m^*}{2\pi^2\hbar^2} \int_0^{2A_L} dE_{0x} \int_0^{2A_L} dE_{0\perp}\, f_D(E_0 - E_{fL})\, T_F(E_{0x}, E_{0\perp}),$$

with

$$T_F(E_{0x}, E_{0\perp}) = T_0 + T_{01,-1} + T_{01,0} + T_{01,+1},$$

where we have

$$T_0 = |f_0(N_R, E_{0x}, E_{0\perp})|^2 \frac{v_R(E_{0x})}{v_L(E_{0x})},$$

$$T_{01,p} = \int_{-2A_L}^{E_0+p\hbar\omega_{LO}} H_p(E_{1x}, E_{0x}, E_{0\perp})|f_1(N_R, E_{1x}, E_{0x}, E_{0\perp})|^2 \frac{v_R(E_{1x})}{v_L(E_{0x})} \, dE_{1x}.$$

Let us now replace the integration by a summation using the integration weight $S(E_{1x})$. We have

$$G_p(n) = \delta_{nl} \sum_{E_{1x}} S(E_{1x})\, H_p(E_{1x}, E_{0x}, E_{0\perp})\, f_1(l, E_{1x}, E_{0x}, E_{0\perp}),$$

$$T_{01,p} = \sum_{E_{1x}} S(E_{1x})\, H_p(E_{1x}, E_{0x}, E_{0\perp})\, |f_1(N_R, E_{1x}, E_{0x}, E_{0\perp})|^2 \frac{v_R(E_{1x})}{v_L(E_{0x})}.$$

Let us introduce the generalized scattering coupling coefficient

$$C_p(E_{0x}, E_{0\perp}, E_{1x}) = \left[S(E_{1x})\, H_p(E_{1x}, E_{0x}, E_{0\perp})\right]^{1/2}$$

and the renormalized wave-function

$$f_{1,p}(n, E_{1x}, E_{0x}, E_{0\perp}) = C_p\,(E_{0x}, E_{0\perp}, E_{1x}) f_1(n, E_{1x}, E_{0x}, E_{0\perp}). \tag{6.24}$$

The system of Equations (6.21), (6.22), and (6.23) is now

$$E_{0x} f_0(n, E_{0x}, E_{0\perp}) = \sum_m \tilde{H}_{nm}(E_{0\perp}) f_0(m, E_{0x}, E_{0\perp}) + \delta_{nl} \sum_{p=-1}^{1} \sum_{E_{1x,p}} C_p(E_{0x}, E_{0\perp}, E_{1x}) f_{1,p}(l, E_{1x}, E_{0x}, E_{0\perp}), \tag{6.25}$$

$$E_{1x} f_{1,p}\,(n, E_{1x}, E_{0x}, E_{0\perp}) = \sum_m \tilde{H}_{nm}(E_{1\perp}) f_{1,p}(m, E_{1x}, E_{0x}, E_{0\perp}) + \delta_{nl}\, C_p(E_{0x}, E_{0\perp}, E_{1x}) f_0(l, E_{0x}, E_{0\perp}). \tag{6.26}$$

Therefore the 1SS contribution to the transmission coefficient by the scattered waves is

$$T_{01,p} = \sum_{E_{1x,p}} |f_{1,p}(N_R, E_{1x}, E_{0x}, E_{0\perp})|^2 \frac{v_R(E_{1x})}{v_L(E_{0x})}.$$

The advantage of this new discrete representation is that it allows us to introduce an S-matrix representation. Indeed we can attribute a port index i to the channel of the incident ($S_i = 0$) wave or to each of the 1SS scattered ($S_i = 1$) waves $f_i = f_{S_i,p_i}(n_i, E_{(S_i)x,r})$ associated with the scattering process p_i and the energy $E_{(S_i)x,r}$ emerging on the left- ($n_i = N_L$) or right-hand ($n_i = N_R$) side of the device. The port index i therefore maps in a unique fashion all the various index combinations (S_i, p_i, r_i, n_i) possible. We will reserve the port $i = 1$ for the incident left-hand wave $f_0(N_L, E_{0x}, E_{0\perp})$ and the port $i = 2$ for the incident right-hand wave $f_0(N_R, E_{0x}, E_{0\perp})$.

Each of the waves f_i can be represented in terms of an incident wave of amplitude A_i and a reflected/transmitted wave of amplitude B_i:

$$f_i = A_i \exp(jk_{x,i}\, n_i a) + B_i \exp(-jk_{x,i}\, n_i a).$$

For the incident wave ($i = 1$) we usually select $A_1 = 1$ and $A_2 = 0$. Note that we are using the contacts (sites N_L or N_R) as the origin $n_i = 0$ of the axes n_i which are taken as exiting the quantum region under study. We can now introduce normalized incident $a_i = \sqrt{v_i}\, A_i$ and reflected $b_i = \sqrt{v_i}\, B_i$ amplitudes using the velocity v_i of the state $k_{x,i}$ at the port i. The scattering parameter S_{ij} is then defined as

$$S_{ji} = \left.\frac{v_i^{1/2} B_i}{v_j^{1/2} A_j}\right|_{A_{k\neq j}=0} = \left.\frac{b_i}{a_j}\right|_{a_{k\neq j}=0},$$

such that using matrix notation we have $\mathbf{b} = \mathbf{S}\,\mathbf{a}$.

Since the Hamiltonian system (6.25) and (6.26) is Hermitian, the incident and reflected waves satisfy the current-conservation property

$$\sum_i |b_i|^2 = \sum_i |a_i|^2,$$

which in turn implies that $\mathbf{S}$ must be unitary $\mathbf{S}\,\mathbf{S}_t^* = \mathbf{I}$ (with $\mathbf{I}$ the identity matrix) or equivalently $\mathbf{S}_t^* = \mathbf{S}^{-1}$.

Note that the Hamiltonian system (6.25) and (6.26) is real and therefore satisfies the time-reversal property which states that the complex conjugate of any solution is also a solution. Thus if the incident waves $\mathbf{a}$ generate the reflected waves $\mathbf{b}$, then the incident waves $\mathbf{b}^*$ will, in turn, generate the reflected waves $\mathbf{a}^*$ (see Problem 6.6 for a proof), such that we have

$$\mathbf{a}^* = \mathbf{S}\,\mathbf{b}^* = \mathbf{S}(\mathbf{S}\,\mathbf{a})^* = \mathbf{S}\,\mathbf{S}^*\,\mathbf{a}^*,$$

which implies that $\mathbf{S}^* = \mathbf{S}^{-1}$ and combined with the unitarity gives the reciprocity property $\mathbf{S} = \mathbf{S}_t$. Using the scattering-parameter representation we can now rewrite the 1SS transmission coefficient for the scattered waves as

$$T_{01,p}(E_{0x}, E_{0\perp}) = \sum_{i \text{ for } n_i = N_R, S_i = 1, p_i = p} |S_{L-R}(\mathcal{E}_1(i))|^2,$$

where $S_{L-R}(\mathcal{E}_1)$ is the left-to-right scattering parameter

$$S_{L-R}(\mathcal{E}_1(i)) = \left[\frac{v_R(E_{1x,r_i})}{v_L(E_{0x})}\right]^{1/2} f_{1,p_i}(N_R, E_{1x,r_i}, E_{0x}, E_{0\perp}) \tag{6.27}$$

associated with the scattering event $\mathcal{E}_1(i)$ and process $p_{01}(i)$:

$$\mathcal{E}_1 = (E_{0x}, E_{0\perp}) \xrightarrow{p_{01}} (E_{1x}, E_{1\perp}).$$

So far we have only considered a 1SS event ($S_i = 1$). The S parameters can also be used for an arbitrary number of sequential scattering events. The same scattering-parameter definition can be used, but one must use the final scattered wave-functions $f_{S_i,\mathcal{E}(i)}$ generated after the S_i sequential scattering events. For example, to the event sequence $\mathcal{E}_2$:

$$\mathcal{E}_2 = (E_{0x}, E_{0\perp}) \xrightarrow{p_{01}} (E_{1x}, E_{1\perp}) \xrightarrow{p_{12}} (E_{2x}, E_{2\perp})$$

correspond the scattering parameters for the various energy paths i

$$S_{L-R}(\mathcal{E}_2(i)) = \left[\frac{v_R(E_{2x,r_{2,i}})}{v_L(E_{0x})}\right]^{1/2} f_{2,\mathcal{E}_2(i)}(N_R, E_{2x,r_{2,i}}, E_{1x,r_{1,i}}, E_{0x}, E_{0\perp}), \tag{6.28}$$

where $p_{01}(i)$ and $p_{12}(i)$ are the specific scattering processes involved in the scattering path $\mathcal{E}(i)$. The transmission coefficient for all the S sequential scattering events is then

$$T_{0S}(E_{0x}, E_{0\perp}) = \sum_{i \text{ for } n_i=N_R, S_i=S} \left|S_{L-R}(\mathcal{E}_{S_i}(i))\right|^2 .$$

We can symbolically write the total transmission coefficient as the sum of the forward (L–R) scattering parameters over all the scattering paths $\mathcal{E}$ possible:

$$T(E_{0x}, E_{0\perp}) = \sum_{\mathcal{E}} |S_{L-R}(\mathcal{E})|^2 .$$

Similarly the total reflection coefficient is the sum of the reflected (L–L) scattering parameters over all the scattering paths $\mathcal{E}$ possible:

$$R(E_{0x}, E_{0\perp}) = \sum_{\mathcal{E}} |S_{L-L}(\mathcal{E})|^2 .$$

We will now find it particularly useful to introduce the impulse function in the calculation of the scattering parameters for arbitrary MSS events. For the local scattering example selected with a single scatterer located at site l, the impulse response h_S at the Sth scattering event is obtained from

$$E_{Sx} h_S(n, l, E_{Sx}, E_{S\perp}) = \sum_m \tilde{H}_{nm}(E_{S\perp}) h_S(m, l, E_{Sx}, E_{S\perp}) + \delta_{nl}$$
$$+ H_{ll}^{SE,S}(E_{Sx}, E_{S\perp})\, h_S(l, l, E_{Sx}, E_{S\perp}),$$

where $H_{ll}^{SE,S}$ is the self-energy at the scatterer site l:

$$H_{ll}^{SE,S}(E_{Sx}, E_{S\perp}) = \sum_{p=-1}^{1} \sum_{E_{(S+1)x,p}} |C_p(E_{Sx}, E_{S\perp}, E_{(S+1)x})|^2$$
$$\times\, h_{S+1}(l, l, E_{(S+1)x}, E_{(S+1)\perp}).$$

The magnitude of the S parameter of Equation (6.27) associated with the scattering event $\mathcal{E}_1$ is now:

$$|S_{L-R}(\mathcal{E}_1)|^2 = \frac{\hbar^2}{a^2}\, v_L(E_{0x})\, |h_0(l, N_L, E_{0x}, E_{0\perp})|^2$$
$$\times\, |C_{p01}(E_{0x}, E_{0\perp}, E_{1x})|^2\, |h_1(N_R, l, E_{1x}, E_{1\perp})|^2\, v_R(E_{1x}),$$

where we have used for $f_{1,p}$ the expression given in Equation (6.24):

$$\begin{aligned} f_{1,p}(n, E_{1x}, E_{0x}, E_{0\perp}) &= C_p(E_{0x}, E_{0\perp}, E_{1x}) f_1(n, E_{1x}, E_{0x}, E_{0\perp}) \\ &= C_p(E_{0x}, E_{0\perp}, E_{1x}) f_0(l, E_{0x}, E_{0\perp}) h_1(n, l, E_{1x}, E_{1\perp}) \end{aligned}$$

as well as (see Chapter 3)

$$f_0(n, E_{0x}, E_{0\perp}) = j\frac{\hbar v_L(E_{0x})}{a}\, h_0(n, N_L, E_{0x}, E_{0\perp}).$$

The S parameter of Equation (6.28) associated with the event $\mathcal{E}_2$ is similarly derived to be:

$$S_{L-R}(\mathcal{E}_2) = \frac{\hbar^2}{a^2}\, v_L(E_{0x})\, |h_0(l, N_L, E_{0x}, E_{0\perp})|^2\, |C_{p01}(E_{0x}, E_{0\perp}, E_{1x})|^2$$
$$\times\, |h_1(l, l, E_{1x}, E_{1\perp})|^2\, |C_{p12}(E_{1x}, E_{1\perp}, E_{2x})|^2\, |h_2(N_R, l, E_{2x}, E_{2\perp})|^2 v_R(E_{2x}).$$

It is now obvious that the scattering parameters of any sequence of scattering events can be simply written by inspection in terms of the successive impulse responses and scattering coupling coefficients involved at each scattering site. We are now armed with an efficient evaluation technique, which will be used in the next section to address important many-body issues.

6.8 Detailed balance and Pauli exclusion in MSS

In equilibrium (i.e. when there is no applied voltage, temperature gradient or light) a zero average total current is expected from a device and must therefore be predicted by the MSS theory even for non-symmetric structures. We shall see in Chapter 9 while studying the Boltzmann equation that a zero total current in equilibrium results from detailed balance which is usually established in momentum space (see Chapter 9). To establish detailed balance in MSS we will need instead to inspect the paths from one contact to another. In MSS many paths (scattering events) connecting an incident state $(E_{ix}, E_{i\perp})$ to a final state $(E_{fx}, E_{f\perp})$ are possible. However, detailed balance will also be achieved if we demonstrate that for each path connecting $(E_{ix}, E_{i\perp})$ to $(E_{fx}, E_{f\perp})$ there exists a reciprocal path, connecting $(E_{fx}, E_{f\perp})$ to $(E_{ix}, E_{i\perp})$, which will carry the same current.

In the previous section we saw that time reversal was responsible for the reciprocity property. However, time reversal is not at the origin of the detailed balance in scattering-assisted tunneling. Indeed, the time-reversal argument can only be applied when the same Hamiltonian is used to describe both the forward path from i to f and the backward path from f to i. Consider a system with three energy levels. The 1SS Hamiltonian system connecting $f_0(E_1)$ to $f_1(E_1)$, $f_1(E_2)$, and $f_1(E_3)$ is different from the 1SS Hamiltonian system connecting $f_1(E_1)$ to $f_0(E_1)$, $f_0(E_2)$, and $f_0(E_3)$. Now recall that these Hamiltonians are created using the ensemble average. It is therefore the ensemble average which has destroyed the time-reversal property for the many-body system. However, it is still present for a single ballistic electron, and we have

$$T_{0,F} = |f_{0,F}(N_R, E_{0x}, E_{0\perp})|^2 \frac{v_R(E_{0x})}{v_L(E_{0x})}$$

$$= \frac{\hbar^2 v_L^2(E_{0x})}{a^2} |h_0(N_R, N_L, E_{0x}, E_{0\perp})|^2 \frac{v_R(E_{0x})}{v_L(E_{0x})}$$

$$= |f_{0,R}(N_L, E_{0x}, E_{0\perp})|^2 \frac{v_L(E_{0x})}{v_R(E_{0x})} = T_{0,R}$$

after we make use of the property:

$$|h_0(N_L, N_R, E_{0x}, E_{0\perp})|^2 = |h_0(N_R, N_L, E_{0x}, E_{0\perp})|^2.$$

Even though time reversibility cannot be applied to the many-body system as a whole, it is applicable to each individual one-electron scattering event. This guarantees that for each sequence of scattering events there exists a reverse sequence of scattering events returning the final scattered state to its initial state. In MSS such a process also contributes to current conservation in a process referred to as backscattering.

Let us now study under which conditions detailed balance is established. Indeed in equilibrium the total current must be zero:

$$0 = I_F - I_B$$

$$= \frac{m^*}{2\pi^2\hbar^2} \int_0^{2A_L} dE_{0x} \int_0^{2A_L} dE_{0\perp} f_D(E_0 - E_F)[T_F(E_{0x}, E_{0\perp}) - T_B(E_{0x}, E_{0\perp})].$$

We have seen that the transmission coefficient is obtained from a summation of the scattering parameters over all the scattering paths ($\mathcal{E}$) possible. For an electron incident on the left-hand contact with energy $(E_{0x}, E_{0\perp})$ the forward transmission coefficient is

$$T_F(E_{0x}, E_{0\perp}) = \sum_{\mathcal{E}} |S_{L-R}(\mathcal{E})|^2.$$

"AS I MENTIONED NEXT WEEK IN MY TALK ON REVERSIBLE TIME..."

'As I mentioned next week in my talk on reversible time . . . ' by Sidney Harris.

Similarly for an electron incident on the right-hand contact with energy $(E_{0x}, E_{0\perp})$ the backward transmission coefficient is

$$T_B(E_{0x}, E_{0\perp}) = \sum_{\mathcal{E}'} |S_{R-L}(\mathcal{E}')|^2.$$

Now since to any event $\mathcal{E}$ corresponds an event $\mathcal{E}'$ corresponding to the reverse event, symbolically written $-\mathcal{E}$, we can write the zero total equilibrium current as

$$0 = \frac{m^*}{2\pi^2\hbar^2} \int_0^{2A_L} dE_{0xi} \int_0^{2A_L} dE_{0\perp i} \tag{6.29}$$

$$\times \sum_{\mathcal{E}(E_{0xi}, E_{0\perp i})} \left[f_D(E_{0i} - E_F)\,|S_{L-R}(\mathcal{E})|^2 - f_D(E_{0f} - E_F)\,|S_{R-L}(-\mathcal{E})|^2 \right],$$

where $E_{0i}(\mathcal{E}) = (E_{0xi}, E_{0\perp i})$ and $E_{0f}(\mathcal{E}) = (E_{0xf}, E_{0\perp f})$ are, respectively, the initial and final total energies for the $|S_{L-R}(\mathcal{E})|^2$ path. Note that E_{0f} is the initial total energy and E_{0i} is the final total energy of the event $-\mathcal{E}$.

Thus detailed balance will be enforced for MSS if we have

$$f_D(E_{0i} - E_F)\,|S_{L-R}(\mathcal{E})|^2 = |S_{R-L}(-\mathcal{E})|^2\, f_D(E_{0f} - E_F). \tag{6.30}$$

As was stated earlier, time reversal is not applicable here to demonstrate this reciprocity requirement because the same Hamiltonian is not used for electrons injected in the left- and right-hand sides. Instead, we shall demonstrate for elastic scattering that detailed balance results from the convergence of the impulse response when an infinite number of 1SS events are used for MSS. When we generalize this proof to the inelastic case we will establish the impact of the Fermi–Dirac boundary condition upon the matrix element to be used in MSS.

In theory, we need to demonstrate the reciprocity for all events. In practice, it is sufficient to demonstrate the procedure for a few sequential events. We consider here the two events $\mathcal{E}_1$ and $\mathcal{E}_2$ discussed earlier. The transmission coefficients for the reciprocal events $-\mathcal{E}_1$, $-\mathcal{E}_2$ are:

$$|S_{R-L}(-\mathcal{E}_1)|^2 = \frac{\hbar^2}{a^2} v_L(E_{0x}) v_R(E_{1x})\, |C_{p10}(E_{1x}, E_{1\perp}, E_{0x})|^2$$
$$\times\, |h_1(N_L, l, E_{0x}, E_{0\perp})|^2\, |h_0(l, N_R, E_{1x}, E_{1\perp})|^2,$$

$$|S_{R-L}(-\mathcal{E}_2)|^2 = \frac{\hbar^2}{a^2} v_L(E_{0x})\, v_R(E_{2x})\, |C_{p21}(E_{2x}, E_{2\perp} E_{1x})|^2$$
$$\times\, |C_{p10}(E_{1x}, E_{1\perp}, E_{0x})|^2\, |h_0(l, N_R, E_{2x}, E_{2\perp})|^2$$
$$\times\, |h_1(l, l, E_{1x}, E_{1\perp})|^2\, |h_2(N_L, l, E_{0x}, E_{0\perp})|^2.$$

First let us demonstrate that the impulse response $h_S(j, E_{Sx}, E_{S\perp}, i)$ is independent of the sequential event S for an infinite number of 1SS events. Indeed we have seen that when a sufficiently large number of sequential scattering events is used, the impulse response and the self-energy converge. Under such conditions we have

$$h_0(j, i, E_{Sx}, E_{S\perp}) = h_1(j, i, E_{Sx}, E_{S\perp}) = h_S(j, i, E_{Sx}, E_{S\perp}) = h(j, i, E_{Sx}, E_{S\perp})$$

so that the Sth index can be dropped from the impulse response notation. Thus for event $\mathcal{E}_1$and $-\mathcal{E}_1$ we have

$$f_D(E_1 - E_F)\, |S_{R-L}(-\mathcal{E}_1)|^2 = |S_{L-R}(\mathcal{E}_1)|^2\, f_D(E_0 - E_F),$$

if we have

$$|C_{p10}(E_{1x}, E_{1\perp}, E_{0x})|^2\, f_D(E_1 - E_F) = |C_{p01}(E_{0x}, E_{0\perp}, E_{1x})|^2\, f_D(E_0 - E_F).$$

For elastic scattering ($p_{10} = p_{01} = 0$), the total energy is constant $E_1 = E_0$ and reciprocity results from the time-reversal property of elastic scattering:

$$|C_{p10}(E_{1x}, E_{1\perp}, E_{0x})|^2 = |C_{p01}(E_{0x}, E_{0\perp}, E_{1x})|^2.$$

For inelastic scattering ($p = \pm 1$) the original matrix element used in MSS for phonon scattering satisfied

$$\frac{1}{n_{LO}+1}\,|C_{-1}(E_{Sx}, E_{S'x})|^2 = \frac{1}{n_{LO}}\,|C_1(E_{S'x}, E_{Sx})|^2,$$

where $p = -1$ is used for phonon emission and $p = 1$ is used for phonon absorption, and where n_{LO} is the average number of phonons of energy $\hbar\omega_{LO}$:

$$n_{LO} = \frac{1}{\exp(\hbar\omega_{LO}/k_B T) - 1}.$$

Hence the detailed balance condition of Equation (6.30) is achieved for $p_{01} = -1$ if we have

$$(n_{LO}+1)\ f_D(E_0 - E_F) = n_{LO}\ f_D(E_0 - \hbar\omega_{LO} - E_F),$$

which is satisfied if a Boltzmann distribution is used as a boundary condition:

$$f_D(E_0 - E_F) = \exp\left(\frac{E_F - E_0}{k_B T}\right).$$

To allow for having a Fermi–Dirac distribution as boundary conditions it is necessary to add the Pauli exclusion factor in the definition of the $C_p(E_{Sx}, E_{S\perp}, E_{(S+1)x})$ matrix element such that it satisfies

$$\begin{aligned}\frac{1}{n_{LO}+1}\,\frac{1}{1 - f_D(E_{S+1} - E_F)}\,|C_{-1}(E_{Sx}, E_{(S+1)x})|^2 \\ = \frac{1}{n_{LO}}\,\frac{1}{1 - f_D(E_S - E_F)}\,|C_1(E_{(S+1)x}, E_{Sx})|^2.\end{aligned}$$

The detailed balance condition of Equation (6.30) is achieved for event $\mathcal{E}_1$ with $p_{01} = -1$ if we have

$$\begin{aligned}(n_{LO}+1) f_D(E_0 - E_F)[1 - f_D(E_0 - \hbar\omega_{LO} - E_F)] \\ = n_{LO} f_D(E_0 - \hbar\omega_{LO} - E_F)[1 - f_D(E_0 - E_F)],\end{aligned}$$

which admits as a solution the Fermi–Dirac distribution as desired

$$f_D(E_0 - E_F) = \frac{1}{\exp\left(\frac{E_0 - E_F}{k_B T}\right) + 1}.$$

For event $\mathcal{E}_2$ with $p_{01} = p_{12} = -1$, we have

$$f_D(E_0 - E_F)(n_{LO}+1)^2[1 - f_D(E_0 - \hbar\omega_{LO} - E_F)][1 - f_D(E_0 - 2\hbar\omega_{LO} - E_F)]$$
$$= f_D(E_0 - 2\hbar\omega_{LO} - E_F)n_{LO}^2[1 - f_D(E_0 - \hbar\omega_{LO} - E_F)][1 - f_D(E_0 - E_F)]$$

and detailed balance is also enforced.

Having established the detailed balance condition for inelastic and elastic scattering, let us now extend the formalism to non-equilibrium. The total current is given by using the left- and right-hand Fermi levels E_{FL} and E_{FR}

$$I = \frac{m^*}{2\pi^2\hbar^2}\int_0^{2A_L} dE_{0xi}\int_0^{2A_L} dE_{0\perp i}$$
$$\times \sum_{\mathcal{E}(E_{0xi},E_{0\perp i})} f_D(E_{0i} - E_{FL})\,|S_{L-R}(\mathcal{E})|^2 - f_D(E_{0f}(\mathcal{E}) - E_{FR})\,|S_{R-L}(-\mathcal{E})|^2$$
$$= \frac{m^*}{2\pi^2\hbar^2}\int_0^{2A_L} dE_{0xi}\int_0^{2A_L} dE_{0\perp i}$$
$$\times \sum_{\mathcal{E}(E_{0xi},E_{0\perp i})} |S_{L-R}(\mathcal{E})|^2\,[f_D(E_{0i} - E_{FL}) - f_D(E_{0i} - E_{FR})], \tag{6.31}$$

where the equilibrium detailed balance identity has been used to rewrite the non-equilibrium backward current in terms of the non-equilibrium forward current. Note that the transmission coefficient $|S_{L-R}(\mathcal{E})|^2$ must be calculated using the occupation number f_{ON} at site i so that the phonon matrix elements are related by the relation:

$$\frac{1}{n_{LO}+1}\,\frac{1}{1 - f_{ON}(i, E_{S'x}, E_{S'\perp})}\,|C_{-1}(E_{Sx}, E_{S'x})|^2$$
$$= \frac{1}{n_{LO}}\,\frac{1}{1 - f_{ON}(i, E_{Sx}, E_{S\perp})}\,|C_1(E_{S'x}, E_{Sx})|^2.$$

The calculation of the occupation number f_{ON} at position i can be done using the non-equilibrium occupation number f_{ON} in the absence of scattering obtained from the charge [6]

$$\rho(i) = \frac{m^*}{\pi\hbar^2}\int_0^{2A_L} dE_{0\perp}\int_0^{2A_L} dE_{0x}\, f_{ON}(i, E_{0x}, E_{0\perp})\, D(i, E_{0x}, E_{0\perp})$$

where D is the density of states

$$D(i, E_{0x}, E_{0\perp}) = -\frac{1}{a\pi}\,\mathrm{Im}[h(i, i, E_{0x}, E_{0\perp})]$$

and where the spectral charge density (see Chapter 3) is

$$f_{ON}(i, E_{0x}, E_{0\perp})\, D(i, E_{0x}, E_{0\perp}) = \frac{\hbar}{2\pi a^2}\left[v_F(E_{0x})|h(N_L, i, E_{0x}, E_{0\perp})|^2\right.$$
$$\left.\times\, f_D(E_0 - E_{FL}) - v_R(E_{0x})|h(N_R, i, E_{0x}, E_{0\perp})|^2 f_D(E_0 - E_{FR})\right],$$

so that the occupation number is

$$f_{ON}(i, E_{0x}, E_{0\perp}) = -\frac{\hbar}{2a}\left\{\frac{f_D(E_0 - E_{FL})v_L(E_{0x})\,|h(N_L, i, E_{0x}, E_{0\perp})|^2}{\mathrm{Im}\,[h(i, i, E_{0x}, E_{0\perp})]} - \frac{f_D(E_0 - E_{FR})v_R(E_{0x}\,|h(N_R, i, E_{0x}, E_{0\perp})|^2}{\mathrm{Im}\,[h(i, i, E_{0x}, E_{0\perp})]}\right\}.$$

The analysis of detailed balance has allowed us to extend MSS theory so that it now includes the Pauli exclusion effect. With the introduction of the local Pauli exclusion weight factors $(1 - f_{ON}(n))$ MSS-assisted tunneling theory now enforces detailed balance for the general case of Fermi–Dirac statistic as boundary conditions.

6.9 Coupling functions for various scattering processes

In this section we calculate the self-energy for the various scattering processes discussed in Chapter 5.

The first scattering process considered is polar scattering, in which an electron is scattered by the LO phonons through the interaction of its Coulomb field with the polarization waves of the lattice. The coupling constant was derived in Chapter 5 to be [1]

$$\alpha_{\mathbf{q},LO} = e\left[\frac{\hbar\omega_{LO}}{2}\left(\frac{1}{\epsilon_{opt}} - \frac{1}{\epsilon_{stat}}\right)\right]^{1/2}\frac{1}{q} = \frac{\alpha_{LO}}{q}, \tag{6.32}$$

where q is the amplitude of the phonon wave-vector $\mathbf{q}$. For GaAs the optical phonon frequency $\omega_{LO}/(2\pi)$, which is assumed to be independent of $\mathbf{q}$ (the Einstein model), is 8.55 THz. The coupling function H obtained after integration of $\langle C_{LO}^2\rangle_{E.A.}$ (see Equations (5.19) and (6.15)) over the perpendicular momentum is

$$H_{LO,\pm}(E_{1x}, q_x) = \frac{m^*}{2\pi\hbar^2 L_x}\frac{4\alpha_{LO}^2\left(N_{\mathbf{q}} + \frac{1}{2} \mp \frac{1}{2}\right)}{\left\{\left[q_x^2 + \frac{2m}{\hbar^2}(E_{0x} \pm \hbar\omega_{LO} - E_{1x})\right]^2 + 4q_x^2k_{0\perp}^2\right\}^{1/2}}. \tag{6.33}$$

We have seen that electrons in a crystal are also scattered by the displacement of the atoms from their lattice sites. The displacement of the atoms induces a local change of the bandgap, which acts as a potential, scattering the electrons. For longitudinal acoustic phonons the coupling constant was found in Chapter 5 to be [2]

$$\alpha_{\mathbf{q},AC} = \left(\frac{\hbar\Xi^2}{2\rho\omega_{\mathbf{q}}}\right)^{1/2},$$

where Ξ is the so-called deformation potential and ρ is the semiconductor density. For GaAs Ξ is 7 eV and ρ is 5.37 g/cm^3.

The number of phonons $N_{\mathbf{q}}$ given by the Bose–Einstein distribution is approximated using the so-called equipartition approximation by

$$N_{\mathbf{q}} \simeq \frac{k_B T_0}{\hbar\omega_{\mathbf{q}}} \quad \text{for } N_{\mathbf{q}} \gg 1.$$

For long wavelengths (q small) we have $\omega_{\mathbf{q}} = q v_s$ where v_s is the sound velocity. Despite the simplicity of the model used, the wave-vector-dependent frequency $\omega_{\mathbf{q}}$ prevents the integration over the perpendicular momentum being carried out analytically (see Conwell [3]). We can then elect to treat acoustic scattering as an elastic scattering process. This underestimates the emission process and overestimates the absorption process. The coupling constant H obtained after integration of $\langle C_{AC}^2 \rangle_{E.A.}$ (see Equations (5.19) and (6.15)) over the perpendicular momentum is simply

$$H_{AC,\pm}(E_{1x}) \simeq \frac{1}{2\pi} \frac{m^*}{\hbar^2} \left(\frac{2\Xi^2 \, k_B T_0}{\rho v_s^2} \right)^{1/2}. \tag{6.34}$$

Another phonon-scattering process is deformation potential LO phonon scattering which can induce Γ to X intervalley transfer. The coupling constant for intervalley scattering by LOX phonons was found in Chapter 5 to be

$$\alpha_{\mathbf{q},IV} = \left(\frac{\hbar \Xi_{LOX}^2}{2\rho\omega_{LOX}} \right)^{1/2},$$

where Ξ_{LOX} is the deformation potential for LOX phonons. The coupling function H obtained after integration over the perpendicular momentum [8] is

$$H_{IV,\pm}(E_{1x}) = \frac{1}{2a} \left(N_{\omega LOX} + \frac{1}{2} \mp \frac{1}{2} \right) (2\alpha_{\mathbf{q},IV})^2 \frac{1}{(2\pi)^2} \frac{2\pi m_1^*}{\hbar^2} g(A(E_{1x})), \tag{6.35}$$

with $N_{\omega LOX}$ the number of LOX phonons and with the energy selection rule

$$A(E_{1x}) = 1 - \frac{m_X^* k_{0\perp}^2}{m_\Gamma^* K_{0\perp}^2} + \frac{2m_X^*}{\hbar^2 K_{0\perp}^2} (E_{1x} - E_{0x} + \Delta_{X\Gamma} \mp \hbar\omega_{LOX}),$$

where we have introduced $\mathbf{K}_\perp = \mathbf{k}_\perp - \mathbf{X}_\perp$, with $\mathbf{X}_\perp$ the displacement of the X valley in the transverse direction. The function $g(X)$, which ranges in value from 0 to 1, is defined by

$$2\pi g(X) = 2 \begin{cases} \pi & X < 0 \\ 2\cos^{-1}(X^{1/2}) & 0 < X < 1 \\ 0 & X > 1. \end{cases}$$

To derive the coupling term $H_{IV,\pm}(E_{1x})$ a summation over q_x is performed using the identity

$$\sum_{q_x>0} \sin(q_x na)\sin(q_x ma) = \frac{L_x}{2a}\delta_{nm}. \tag{6.36}$$

The δ_{nm} result obtained demonstrates that an isotropic phonon-scattering process with constant phonon energy (no dependence on $\mathbf{q}$) reduces to a local scattering process.

An important scattering process in a heterostructure is the scattering of electrons by the roughness of the interface between two different semiconductors. A distribution of terraces typically of a monolayer thickness is present at the interface [4]. The electron is scattered elastically by these terraces, i.e., the total energy of the electron is conserved. However, the longitudinal and perpendicular energies of the electron change in the process.

In Chapter 5 we showed that the ensemble average of the coupling coefficient was of the form proposed Prange and Nee [5]

$$\langle |C_{IR}(\mathbf{q}_\perp, i)|^2\rangle_{E.A.} = V_B^2(N_i\, a)\pi\Lambda^2 \exp\left(\frac{q_\perp^2\Lambda^2}{4}\right),$$

where V_{Bi} is the conduction-band discontinuity at the interface located at the lattice site N_i.

The coupling constant H_{IR} at the lattice site N_i obtained after integration of $\langle |C_{IR}|^2\rangle$ (see Equation (6.16)) over the perpendicular momentum is

$$H_{IR}(E_{1x}, i) = \frac{V_{Bi}^2(N_i\, a)}{2}\Lambda^2\frac{m^*}{\hbar^2}\exp\left[-\frac{m^*\Lambda^2}{2\hbar^2}(E_0 - E_{1x} + E_{0\perp})\right]$$
$$\times I_0\left\{\frac{m^*}{\hbar^2}\Lambda^2\left[E_{0\perp}^{1/2}\,(E_0 - E_{1x})^{1/2}\right]\right\}$$

where $I_0[x]$ is the modified Bessel function of order 0. Note that the terraces are usually one monolayer wide, and a should be selected to be half a lattice parameter (the normal choice for the [100] direction (see Chapter 3)). Alternatively, if a is selected to be the lattice parameter, V_{Bi} should be divided by 4.

In an alloy $A_\alpha B_{1-\alpha}C$ the crystal potential is not periodic. We saw in Chapter 5 that the crystal potential of the alloy can be represented in terms of a non-periodic fluctuating potential superposed on an average potential which is periodic. This fluctuating potential introduces an effective scattering process referred to as alloy scattering (AL). The ensemble average of the coupling constant was derived in Chapter 5 to be:

$$\langle |C_{AL}(\mathbf{q}_\perp, i)|^2\rangle_{E.A.} = \Delta V_{AB}^2\,\alpha_i(1-\alpha_i)\,\frac{\Omega_0}{a},$$

where α_i is the mole fraction at the lattice site i, Ω_0 is the volume of the elementary cell which is given in terms of the elementary crystal axis by $\Omega_0 = a^3/4$, and where ΔV_{AB} is essentially the variation of the conduction band at Γ between alloys AC and BC. After integration over the perpendicular momentum of $\langle |C_{AL}|^2 \rangle_{E.A.}$ the following coupling constant H_{AL} is obtained:

$$H_{AL}(E_{1x}, i) = \frac{1}{2\pi} \Delta V_{AB}^2 \alpha_i (1 - \alpha_i) \frac{\Omega_0}{a} \frac{m^*}{\hbar^2}.$$

Finally we conclude with electron–electron scattering. It is not possible to calculate a coupling constant for electron–electron coupling analytically without some serious approximations. We therefore provide instead only the self-energy term:

$$\begin{aligned} G(n_1) = {} & \frac{1}{(2\pi)^2} \int d^2\mathbf{q}_\perp \frac{2}{(2\pi)^3} \int d^3\tilde{\mathbf{k}}_2 \int d\tilde{k}'_{2x} \\ & \times f_1(n_1, \mathbf{q}_\perp, \tilde{\mathbf{k}}_2, \tilde{k}'_{2x})\, f_{\tilde{\mathbf{k}}_2} (1 - f_{\tilde{k}'_{2x}, \mathbf{k}_{2\perp} + \mathbf{q}_\perp}) \\ & \times 4\pi\, |R(\mathbf{q}_\perp, \tilde{\mathbf{k}}_2, \tilde{k}'_{2x}, n_1)|^2, \end{aligned} \tag{6.37}$$

where R is given by

$$\begin{aligned} R(\mathbf{q}_\perp, \tilde{\mathbf{k}}_2, \tilde{k}'_{2x}, n_1) = {} & \frac{e^2}{4\pi\epsilon} \frac{2\pi}{(q_\perp^2 + \alpha^2)^{1/2}} \\ & \times \sum_{n_2} f_U^*(n_2, \tilde{k}'_{2x}, \mathbf{k}_{2\perp} + \mathbf{q}_\perp) f_U(n_2, \tilde{\mathbf{k}}_2) \\ & \times \exp[-(q_\perp^2 + \alpha^2)^{1/2} |n_1 - n_2| a], \end{aligned} \tag{6.38}$$

in terms of the unscattered open states f_U of the device and with α the screening factor.

6.10 Results for resonant tunneling structures

We shall now demonstrate the application of the scattering-assisted tunneling theory developed in this chapter by showing some results obtained for an RTD. The device used to test the scattering-assisted tunneling theory, developed in this chapter, is the resonant tunneling structure shown in Figure 6.3. This resonant tunneling structure consists of a conventional undoped AlGaAs/GaAs/AlGaAs double-barrier structure sandwiched between two strongly doped GaAs n^+ contacts.

We shall consider two double-barrier structures with barrier and well widths of 6 and 9 lattice parameters ($\simeq$ 34 and 50 Å) respectively. A barrier height of 0.25 eV corresponding to an Al mole fraction of 0.3 is used. A donor concentration $N_D = 10^{18}$ cm^{-3} is used in the contacts of this test device.

We first analyze the individual impacts of polar, acoustic, alloy, interface roughness, intervalley, and electron–electron scattering mechanisms upon resonant tunneling before considering their combined effect.

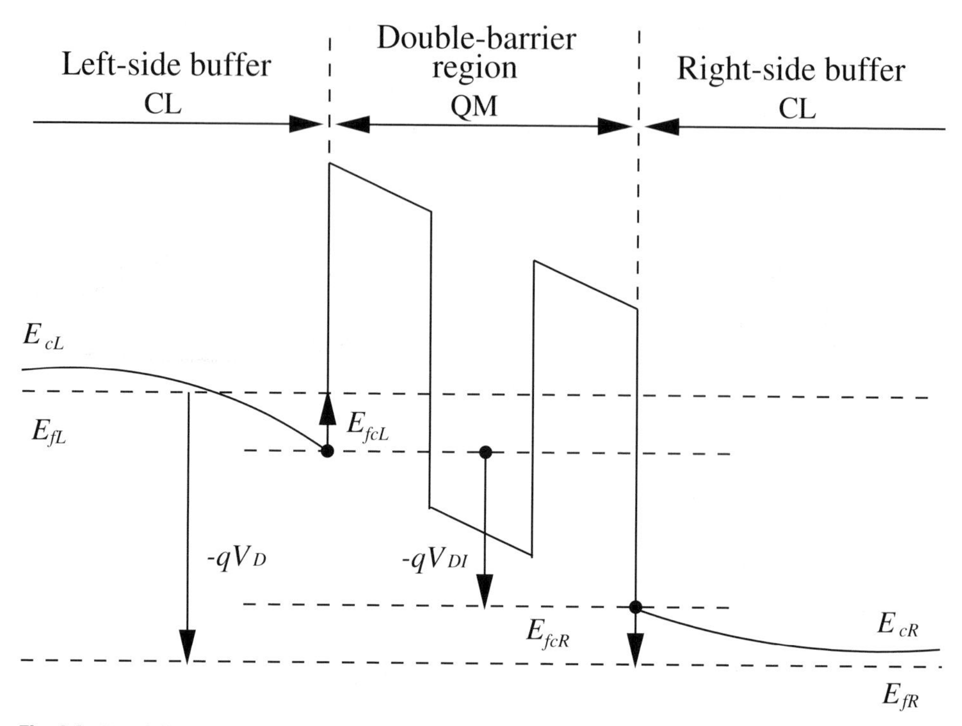

Fig. 6.3. Band diagram of the test resonant tunneling structure. The double-barrier structure is sandwiched between two degenerately doped n^+ structures.

Let us consider polar scattering. We show in Figure 6.4 the equilibrium (no bias applied) transmission coefficient $T_F(E_{0x}, E_{0\perp})$ plotted versus the incident energy E_{0x} for $E_{0\perp} = 0$ for a 50/50/50 Å diode at the lattice temperature of 4.2 K (dashed line), 100 K(dashed-dotted line), and 300 K (dotted line). Also shown is the transmission coefficient in the absence of scattering (full line).

Three different transmission peaks are observed. Indeed, the total transmission coefficient plotted results from the superposition of direct tunneling and phonon-assisted tunneling by emission and absorption of optical phonons. The main peak centered at the energy $E_{res} \simeq 86$ meV corresponds to direct resonant tunneling T_0 by unscattered electrons. Note the 3.5 meV self-energy shift of the central resonant transmission peak relative to the peak (full line) obtained in the absence of scattering $E_{res,0} \simeq 90$ meV. The right-hand peak centered upon $E_{res,0} + \hbar\omega_{LO}$ corresponds to resonant tunneling assisted by *emission* of phonons. The left-hand peak centered upon $E_{res,0} - \hbar\omega_{LO}$ corresponds to resonant tunneling assisted by *absorption* of phonons. The absorption peak is not noticeable at 4.2 K due to the negligible number of LO phonons at low temperature.

We next examine the impact of polar scattering upon the *I–V* characteristic of a 50 Å barrier diode at 4.2 K and a 34 Å barrier diode at 100 K. The *I–V* characteristic calculated in the presence (dashed line) and absence (full line) of polar scattering

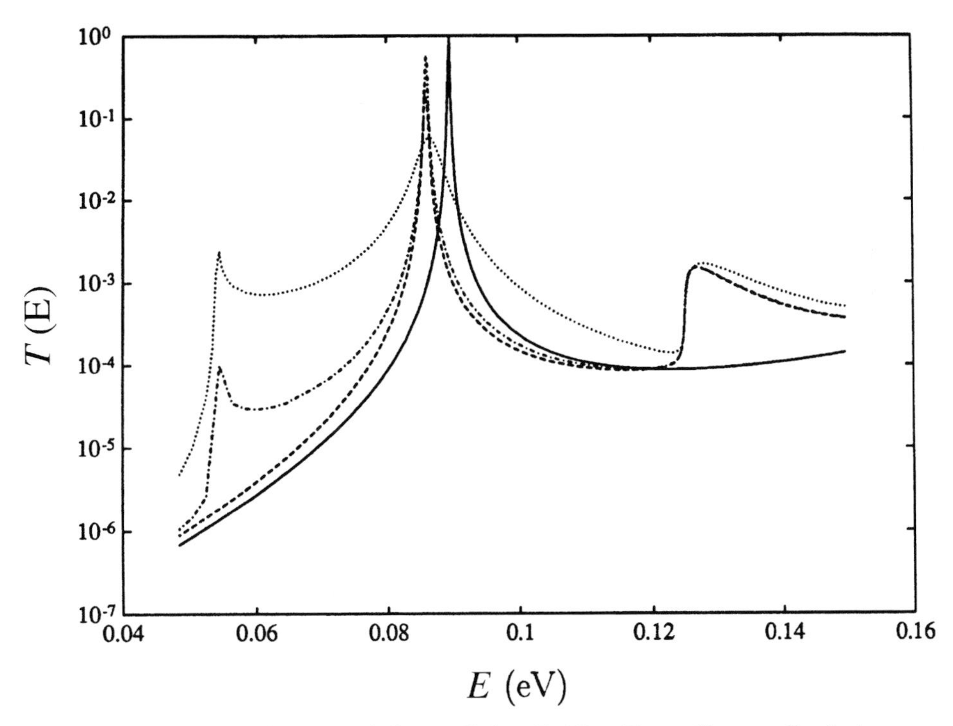

Fig. 6.4. Equilibrium (no bias) transmission coefficient $T_F(E_{0x}, E_{0\perp} = 0)$ versus E_{0x} in the presence of polar scattering for 50/50/50 Å diode at the lattice temperature of 4.2 K (dashed line), 100 K (dashed-dotted line), and 300 K (dotted line). Also shown is the transmission coefficient in the absence of polar scattering (full line). (P. Roblin and W. R. Liou, *Physics Review B*, Vol. 47, No. 4, Pt II, pp. 2146–2161, January 15 1993. Copyright 1993 by the American Physical Society.)

shown in Figures 6.5(a) and (b) is seen to induce both a decrease in the peak current and an increase in the valley current. Polar scattering therefore contributes to the reduction of the peak-to-valley current ratio. At 4.2 K the phonon emission peak of the transmission coefficient has introduced a secondary peak in the I–V characteristic at around $V_D = 0.6$ V. At higher temperature (here 100 K) the variation of the Fermi–Dirac occupation is more gradual around the Fermi energy, and a phonon peak is usually not resolved because its small contribution is smoothed out in the current integration. Note that the detection of a phonon peak in the I–V characteristic is facilitated when plotting higher order derivatives of the I–V characteristic.

The reader is referred to [7] for results on acoustic phonon scattering which usually induces a weak scattering-assisted tunneling component to the diode current.

The next phonon-scattering process considered is Γ–X intervalley scattering induced by LOX phonons. The transmission coefficient shown in Figure 6.6(a) is seen to be subjected to a self-energy shift. 5% of the current remains carried by the Γ valley. Intervalley scattering is seen to leave the valley current unchanged but does effectively increase the classical diode leakage current at high voltages.

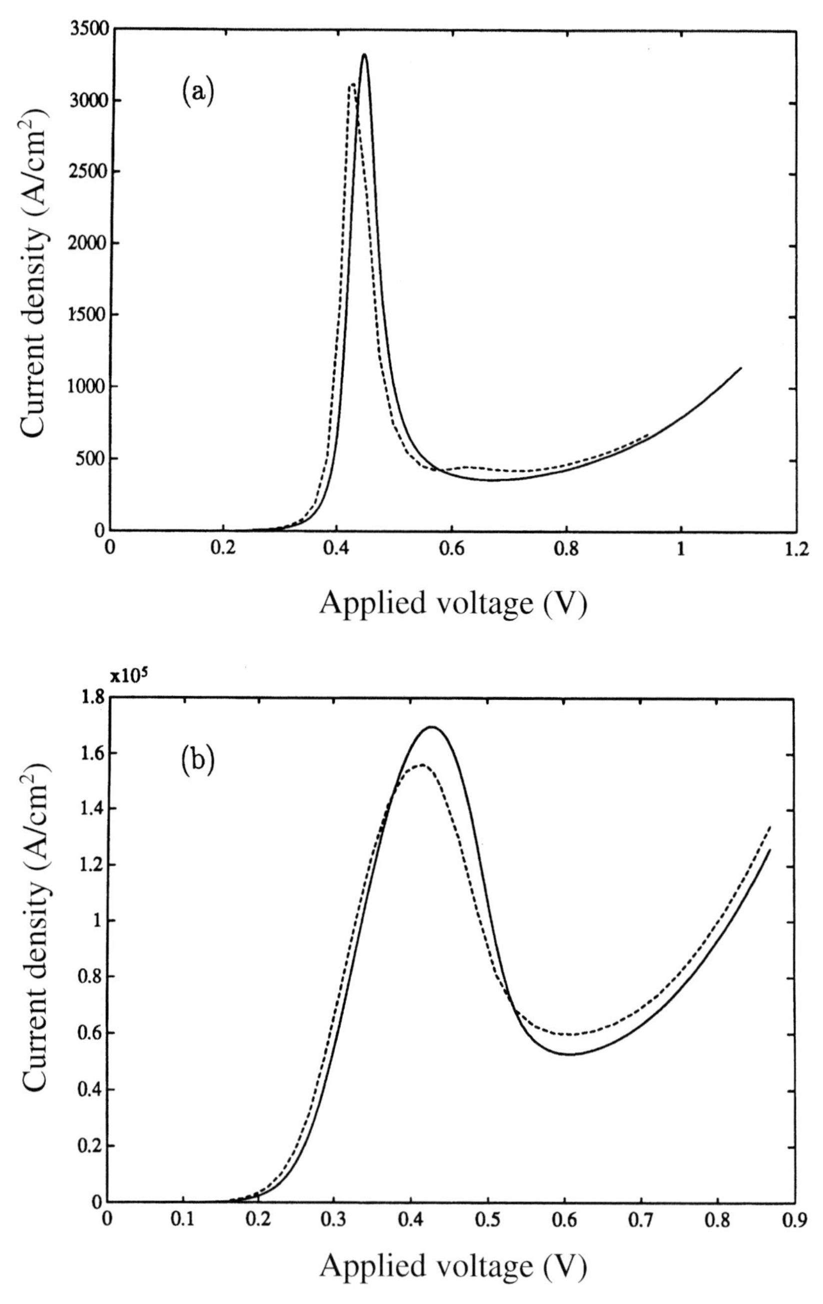

Fig. 6.5. I–V characteristic in the presence (dashed line) and absence (full line) of polar scattering calculated for (a) a 50/50/50 Å diode at 4.2 K and (b) a 34/34/34 Å diode at 100 K. (P. Roblin and W. R. Liou, *Physics Review B*, Vol. 47, No. 4, Pt II, pp. 2146–2161, January 15 1993. Copyright 1993 by the American Physical Society.)

Next we consider interface roughness scattering. We show in Figure 6.7 the equilibrium (no bias applied) transmission coefficient $T_F(E_{0x}, E_{0\perp})$ plotted versus

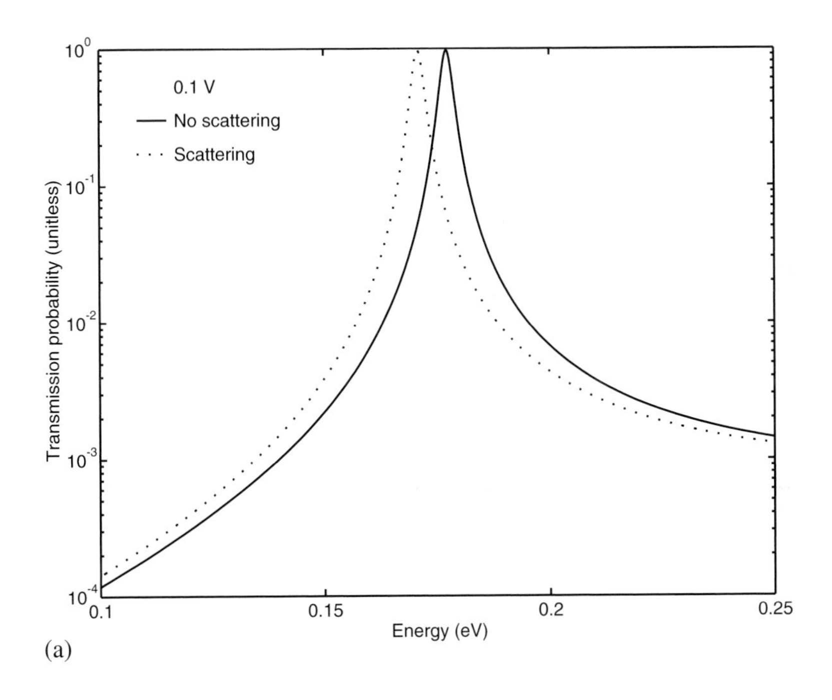

(a)

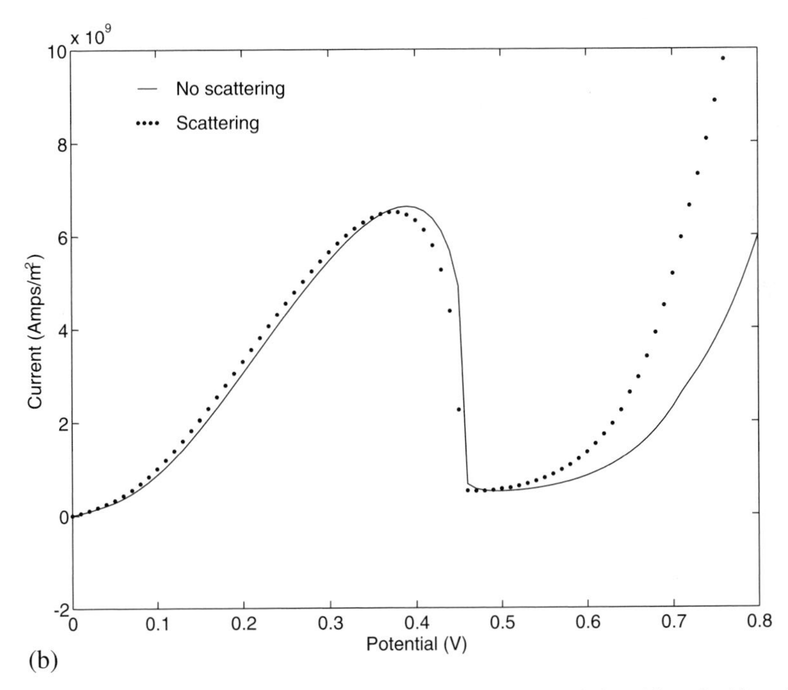

(b)

Fig. 6.6. (a) Transmission probability and (b) I–V characteristic with and without intervalley scattering from Γ to X. (P. Sotirelis and P. Roblin, *Physics Review B*, Vol. 51, No. 19, pp. 13 381–13 388, May 15 1995. Copyright 1995 by the American Physical Society.)

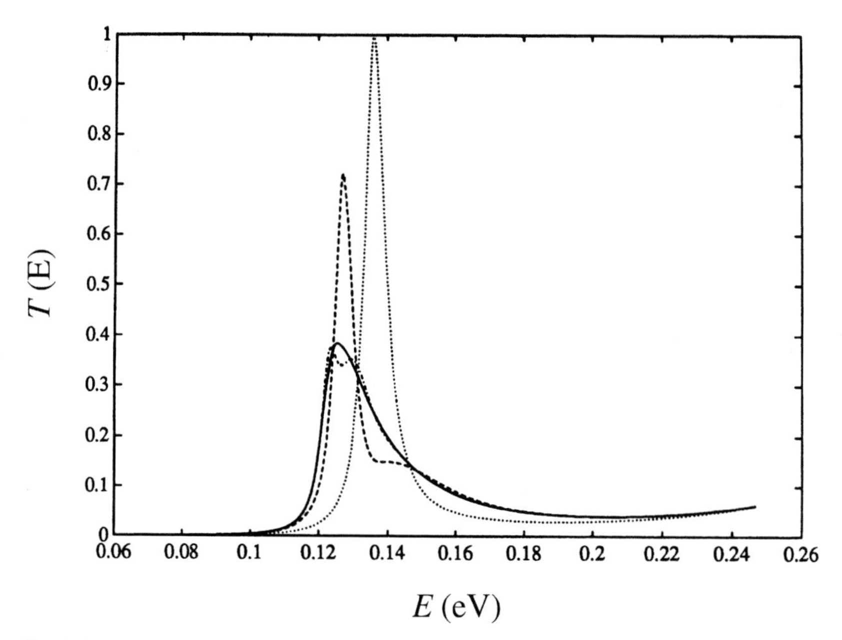

Fig. 6.7. Equilibrium (no bias) transmission coefficient $T_F(E_{0x}, E_{0\perp} = 0)$ versus E_{0x} in the presence of interface roughness scattering for a 34/34/34 Å diode for 0 (dotted line), 1 (dashed line), 3 (dashed-dotted line), and 6 (full line) sequential scattering events. (P. Roblin and W. R. Liou, *Physics Review B*, Vol. 47, No. 4, Pt II, pp. 2146–2161, January 15 1993. Copyright 1993 by the American Physical Society.)

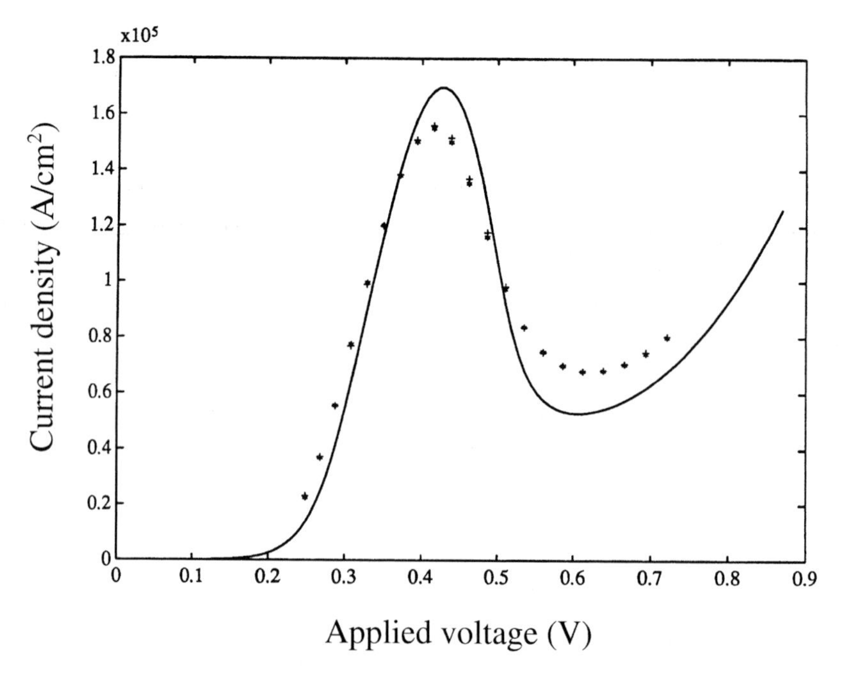

Fig. 6.8. Current–voltage characteristic of a 34/34/34 Å diode at 100 K after 0 (full line), 1(∗), and 2(+) sequential IR scattering events. (P. Roblin and W. R. Liou, *Physics Review B*, Vol. 47, No. 4, Pt II, pp. 2146–2161, January 15 1993. Copyright 1993 by the American Physical Society.)

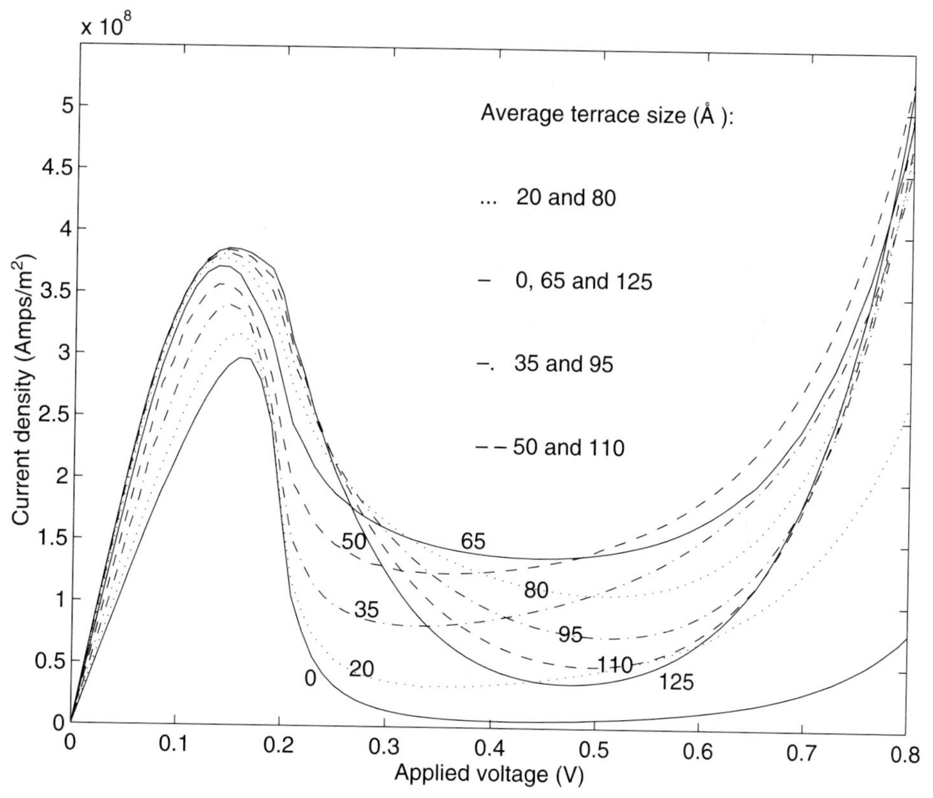

Fig. 6.9. *I–V* characteristics calculated in the presence of IR scattering with terrace widths varying from 0 to 125 Å. (P. Roblin, R. C. Potter and A. Fathimulla, *Journal of Applied Physics*, Vol. 79, No. 5, pp. 2502–2508, March 1 1996.)

E_{0x} for $E_{0\perp} = 0$ for a 34/34/34 diode for 0, 1, 3 and 6 sequential scattering events and an average terrace size of $\Lambda = 70$ Å. No temperature dependence is expected for interface roughness scattering. The transmission coefficient is seen to clearly converge after a few sequential scattering events (6 for the 34 Å diode). The importance of interface roughness scattering is measured by the large self-energy shift (about 11 meV for the 34 Å diode) it induces. The final transmission coefficient has a characteristic *shark-fin* shape. The slow decrease of the transmission coefficient at large energies can be expected to strongly increase the diode (leakage) valley current. The total area under the transmission coefficient does not, however, vary appreciably and the peak current will remain approximately constant. The resulting *I–V* characteristic supporting these predictions is shown in Figure 6.8. The *I–V* characteristic was calculated for 0, 1, 2 sequential IR scattering events at 100 K. Convergence toward the final *I–V* characteristic is found to take place for a smaller number of sequential scattering events than for the transmission coefficient. This is due to the fact that the diode current involves an integration of the transmission coefficient. Clearly

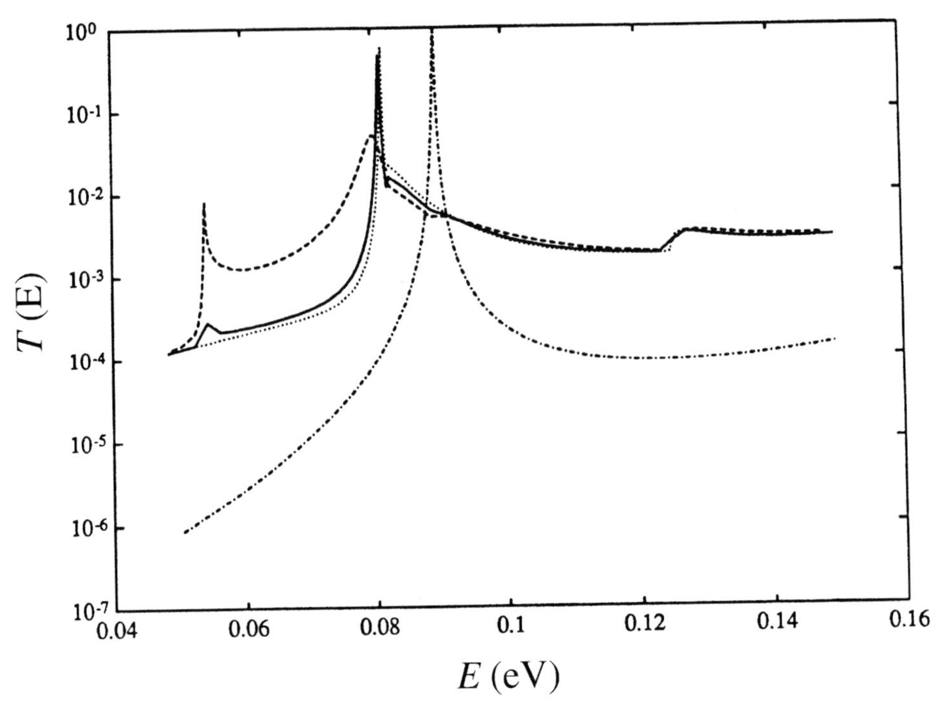

Fig. 6.10. Equilibrium (no bias) transmission coefficient $T_F(E_{0x}, E_{0\perp} = 0)$ versus E_{0x} for $E_{0\perp} = 0$ in the presence of LO, AC, IR and AL scattering for a 50/50/50 Å diode at the lattice temperature of 4.2 K (dotted line), 100 K (full line), and 300 K (dashed line). Also shown is the transmission coefficient in the absence of scattering (dotted-dashed line). (P. Roblin and W. R. Liou, *Physics Review B*, Vol. 47, No. 4, Pt II, pp. 2146–2161, January 15 1993. Copyright 1993 by the American Physical Society.)

interface roughness scattering is seen to be a very strong scattering process which greatly impacts the valley current. Figure 6.9 shows the *I–V* characteristic calculated for an AlAs/InGaAS RTD with an InAs subwell [9], using an average terrace width varying from 0 to 125 Å. The valley current is seen to reach a maximum value for an average terrace width of 65 Å. It is natural to expect the existence of such a critical value. Indeed, when the terrace size is much larger than the electron wavelength the electrons experience a smooth surface and IR scattering is suppressed. Similarly when the terrace size is much smaller than the electron wavelength the electrons cannot resolve the fast spatial variation of the interface potential and only experience an average and therefore smooth interface potential and IR scattering is suppressed. The reader is referred to [7] for results on alloy scattering, which usually induces a weak scattering-assisted tunneling component to the diode current.

We now consider the combined impact upon resonant tunneling of these four scattering mechanisms. We show in Figure 6.10 the equilibrium (no bias) transmission coefficient $T_F(E_{0x}, E_{0\perp})$ plotted versus E_{0x} for $E_{0\perp} = 0$ for 50/50/50 Å diode

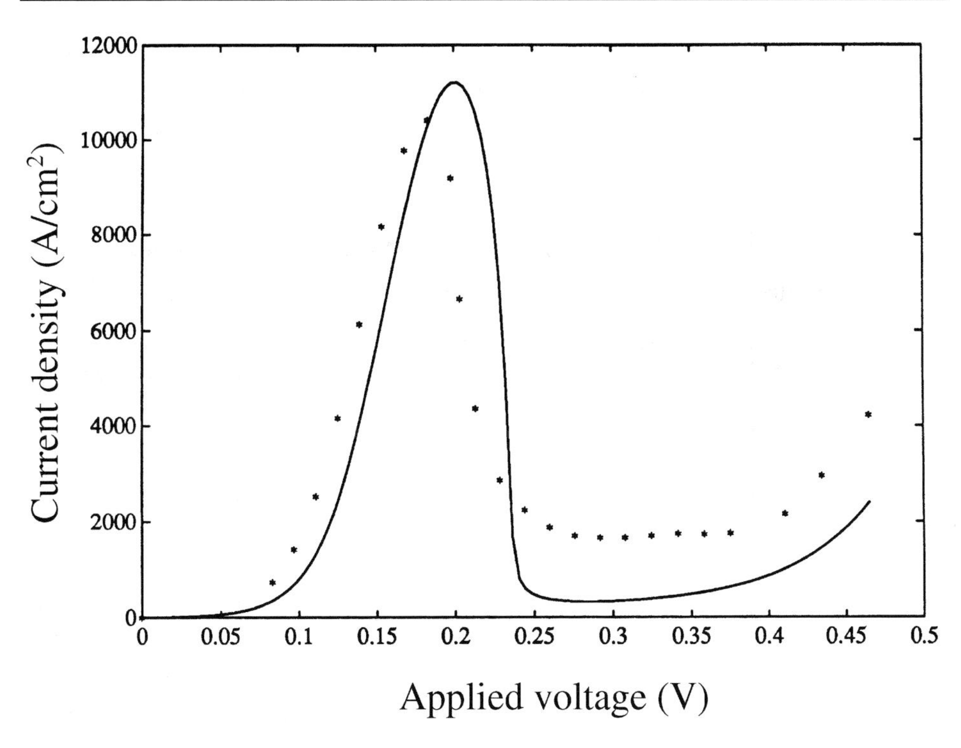

Fig. 6.11. *I–V* characteristic obtained for the 50/50/50 Å diode at 100 K in the presence of LO, AC, IR and AL scattering (stars) and in the absence of scattering (full line). (P. Roblin and W. R. Liou, *Physics Review B*, Vol. 47, No. 4, Pt II, pp. 2146–2161, January 15 1993. Copyright 1993 by the American Physical Society.)

at lattice temperatures of 4.2, 100, and 300 K. Also shown is the transmission coefficient in the absence of scattering. Clearly the transmission coefficient exhibits complex structures resulting from the superposition of each scattering mechanism. However, one can recognize the dominant contribution of both polar (LO) and interface roughness (IR) scattering.

The *I–V* characteristic obtained for the 50/50/50 Å diodes at 100 K is shown in Figure 6.11. The various scattering processes (dominated here by IR) cooperate to reduce the peak-to-valley current ratio of the diode.

6.11 Conclusion

In this chapter we have introduced the MSS formalism in order to account for scattering in quantum devices. Note that other formalisms, such as the Wigner distribution and the non-equilibrium Green function (NEGF) formalisms, have also been developed (see [10] for a review). The Wigner distribution [10], like the Boltzmann distribution, is a function of both momentum and position (see Chapter

9). Unfortunately, the Wigner distribution takes negative values and cannot therefore be equated to a probability density and its interpretation (like its calculation) is difficult. The NEGF formalism starts from a more rigorous many-body picture and its application relies on complex but controlled approximations. However, the NEGF formalism can be verified to give results equivalent to MSS [11].

The principal advantage of the MSS approach presented here is its simplicity and intuitive nature as the scattering of electrons by the many-body electron or photon systems is reduced to a single-electron problem. This simple single-electron system admits a simple wave-packet physical solution for 1SS. The extension to the MSS case then relies on the introduction of the self-energy and the impulse response (Green function). MSS exactly enforces current conservation and when Pauli exclusion weights are introduced it also enforces detailed balance. The MSS method is not, however, without limitations of its own. We have assumed for example that the phonon field remains in equilibrium. This neglects any phonon-drag effects in which the phonon field is strongly modified by the electrons. Also by treating interface and alloy scattering as phase randomizing processes we neglect any possible correlation of scattering events in interface and alloy scattering. Strong correlation effects could localize the electron as in Anderson localization in random superlattices (see Section 4.5.2). Such effects are believed to be important in wide bandgap material systems [12].

This chapter has focused on the calculation of the DC characteristics of devices which are subjected to incoherent (stochastic) excitations. In the next chapter we will focus on the calculation of the AC characteristic of quantum devices which are subjected to coherent harmonic electromagnetic excitations.

6.12 Bibliography

6.12.1 Recommended reading

D. K. Ferry, *Semiconductor*, Macmillan, New York, 1991.

Supriyo Datta, *Electronic Transport in Mesoscopic Systems*, Cambridge University Press, Cambridge, 1995.

6.12.2 References

[1] H. Fröhlich, 'Theory of electrical breakdown in ionic crystals,' *Proceedings of the Royal Society,* London, Vol. A160, p. 230 (1937).

[2] C. Kittel, *Quantum Theory of Solids*, Wiley: New York, 1963.

[3] E. M. Conwell, 'High field transport in semiconductors,' in *Solid State Physics*, eds. F. Seitz, D. Turnbull, and H. Ehrenreich, Supplement 9, Academic Press, New York, 1967.

[4] H. Sakaki, T. Noda, K. Hirakawa, M. Tanaka, and T. Matsusue, 'Interface roughness scattering in GaAs/AlAs quantum wells,' *Applied Physics Letters,* Vol. 51, No. 23, December 7 1987.

[5] P. E. Prange and T.-W. Nee, 'Quantum spectroscopy of the low-field oscillations in the surface impedance,' *Physical Review,* Vol. 168, No.3, pp. 779–786, April 15 1968.

[6] Roger Lake, T.I., personal communication.

[7] P. Roblin, and W. R. Liou, '3D scattering-assisted tunneling in resonant tunneling diodes,' *Physical Review B,* Vol. 47, No. 4, Part II, pp. 2146–2161, January 15 1993.

[8] P. Sotirelis and P. Roblin, 'Intervalley scattering in GaAs/AlAs resonant tunneling diodes,' *Physical Review B,* Vol 51, No. 19, pp. 13 381–13 388, May 15 1995.

[9] P. Roblin, R. C. Potter and A. Fathimulla, 'Interface roughness scattering in AlAs/InGaAs resonant tunneling diodes with an InAs subwell,' *Journal of Applied Physics,* Vol. 79, No. 5, pp. 2502–2508, March 1 1996.

[10] D. K. Ferry, *Semiconductor*, Macmillan, New York, 1991.

[11] R. Lake, G. Klimeck, R. C. Bowen and D. Jovanovic 'Single and multiband modeling of quantum electron transport through layered semiconductor devices,' *Journal of Applied Physics*, Vol. 81, No. 12, June 15 1997.

[12] Y. Zhang and J. Singh, 'Charge control and mobility studies for an AlGaN/GaN high electron mobility transistor,' *Journal of Applied Physics,* Vol. 85, No. 1, pp. 587–594, 1999.

6.13 Problems

6.1 The semiclassical electron–phonon interaction potential can be written in the form

$$H_{e-ph} = \frac{2}{\Omega^{1/2}} \sum_{q_x>0} \sum_{\mathbf{q}_\perp} C_\pm(\mathbf{q}) \cos(\omega_\mathbf{q} t - \mathbf{q}_\perp \cdot \mathbf{r}_\perp + \phi_\mathbf{q}) \, \sin(q_x x).$$

Verify that in the generalized Wannier function basis the matrix element of the electron–phonon interaction is

$$\begin{aligned}\langle \mathbf{k}'_\perp, m | H_{e-ph} \mid \mathbf{k}_\perp, n\rangle = {} & \frac{1}{\Omega^{1/2}} \sum_{q_x>0} \sin(q_x m a)\, \delta_{mn} \\ & \times \sum_{\mathbf{q}_\perp} C_\pm(\mathbf{q}) \Big\{ \delta(\mathbf{k}'_\perp - \mathbf{k}_\perp + \mathbf{q}_\perp) \exp[i(\omega_\mathbf{q} t + \phi_\mathbf{q})] \\ & + \delta(\mathbf{k}'_\perp - \mathbf{k}_\perp - \mathbf{q}_\perp) \exp[-i(\omega_\mathbf{q} t + \phi_\mathbf{q})] \Big\}\end{aligned}$$

Approximate the Bloch function by a plane wave

$$\varphi(\mathbf{k}_\perp, \mathbf{r}_\perp) = \frac{1}{2\pi} \exp(i\mathbf{k}_\perp \cdot \mathbf{r}_\perp).$$

Note that

$$\int_{-\infty}^{\infty} \exp[i(k_y - k'_y)y]\, dy = 2\pi\delta(k_y - k'_y).$$

6.2 Interface roughness scattering at site N_i can be represented by the potential

$$H_i^{IR}(\mathbf{r}) = V_B(N_i a)\ \mathrm{rect}\left(\frac{x - N_i\, a}{a}\right) F(\mathbf{r}_\perp),$$

where V_B is the barrier height and the function rectangle (rect) is defined as

$$\mathrm{rect}(x) = \begin{cases} 1 & \text{for } |x| < 1/2 \\ 0 & \text{for } |x| > 1/2. \end{cases}$$

The function $F(\mathbf{r}_\perp)$ gives the distribution of the terrace. Let us assume that $F(\mathbf{r}_\perp)$ is a periodic function

$$F(y + L_y,\ z + L_z) = F(y, z).$$

The function $F(\mathbf{r}_\perp)$ can therefore be expanded in a Fourier series

$$F(\mathbf{r}_\perp) = \sum_{\mathbf{q}_\perp} F_{\mathbf{q}_\perp} \exp(i\mathbf{q}_\perp \cdot \mathbf{r}_\perp).$$

Verify that in the generalized Wannier function basis the matrix element of interface roughness scattering process i at lattice N_i is

$$\langle \mathbf{k}'_\perp,\ m|H_i^{IR} \mid \mathbf{k}_\perp,\ n\rangle = \delta_{mN_i}\delta_{mn}\frac{1}{S^{1/2}}\sum_{\mathbf{q}_\perp} C(\mathbf{q}_\perp, i)\delta(\mathbf{k}'_\perp - \mathbf{k}_\perp - \mathbf{q}_\perp).$$

Calculate $C(\mathbf{q}_\perp, i)$.

6.3 In this problem we wish to demonstrate that the current is conserved in 1SS-assisted tunneling. To do this, let us consider the equivalent coupled Hamiltonian system of an incident electron f_0 coupled to multiple scattered waves f_{1r}:

$$E_{0x} f_0(n) = \sum_m H_{n\ m}\ f_0(m)\ +\ \sum_r C_r(n) f_{1r}(n), \tag{6.39}$$

$$E_{1x,r} f_{1r}(n) = \sum_m H_{n\ m}\ f_{1r}(m)\ +\ C_r^*(n) f_0(n), \tag{6.40}$$

where $C_r(n)$ is the coupling constant.

(a) Multiply Equations (6.39) and (6.40) by $f_0^*(n)$ and $f_{1r}^*(n)$ respectively, and take the imaginary parts to derive the following elemental-current conservation equations:

$$0 = \sum_m j_0(n, m) + \sum_r j_{0,1r}(n),$$
$$0 = \sum_m j_{1r}(n, m) + j_{1r,0}(n), \tag{6.41}$$

with

$$j_i(n, m) = q(-a/\hbar)\ \mathrm{Im}[H_{n\ m}\ f_i^*(n) f_i(m)],$$
$$j_{0,1r}(n) = q(-a/\hbar)\ \mathrm{Im}[C_r(n) f_0^*(n) f_{1r}(n)],$$
$$j_{1r,0}(n) = q(-a/\hbar)\ \mathrm{Im}[C_r^*(n) f_{1r}^*(n) f_0(n)].$$

$j_i(n, m)$ is the elemental electron current from lattice site n to lattice site m for the state i (0 or $1r$) and $j_{0,1r}(n)$ and $j_{1r,0}(n)$ are, respectively, the electron currents from the state 0 to the state $1r$ and from the state $1r$ to the state 0 (both at the lattice site n).

(b) For mutually uncorrelated incident and scattered states, the total ensemble-average current $\langle J_{TOT} \rangle_{E.A.}$ in the left- and right-hand sides of the quantum system is given by $\langle J_{TOT} \rangle_{E.A.} = J_0 + \sum_r J_{1r}$. Demonstrate that the total ensemble-average current for the states 0 and $1r$ is conserved. For simplicity, assume a tight-binding Hamiltonian, so that the total current for a state i is simply $J_i(n) = j_i(n, n+1) + j_i(n-1, n)$. For your demonstration rely on the continuity Equation (6.41) and the elemental-current properties $j_i(n, m) = -j_i(m, n)$ and $j_{0,1r}(n) = -j_{1r,0}(n)$.

6.4 Derive the coupling coefficients H_{scat} for: (a) phonon scattering, (b) acoustic phonon scattering, (c) interface roughness scattering, (d) alloy scattering, (e) intervalley scattering, and (f) electron–electron scattering.

6.5 Consider the impulse response (Green function) $h(n, i, E_{1x})$ solution of

$$E_{1x} h(n, i, E_{1x}) = -\frac{A}{2} h(n-1, i, E_{1x}) + \left\{ [A - qV(n)] + H_{nn}^{SE} \right\} h(n, i, E_{1x}) - \frac{A}{2} h(n+1, i, E_{1x}) + \delta_{n,i}.$$

(a) Using the hermiticity of the Hamiltonian and the associated current conservation property demonstrate that for $H_{nn}^{SE} = 0$ the impulse function satisfies the identity:

$$-\frac{a}{\hbar} 2\,\mathrm{Im}[h(i, i, E_{1x})] = v_L(b', k_{x,L})\, |h(N_L, i, E_{1x})|^2 + v_R(b', k_{x,R})\, |h(N_R, i, E_{1x})|^2, \tag{6.42}$$

where N_L and N_R are the positions of the emitter and collector on the left and right respectively of the RTD structure and where v_L and v_R are the left- and right-hand electron velocities.

(b) Demonstrate that in the presence of scattering (non-zero self-energy H_{nn}^{SE}) the current-conservation property is now given by Equation (6.19).

(c) Verify that in the presence of scattering the current conservation property can then be rewritten in the form of Equation (6.20).

6.6 We wish to demonstrate the reciprocity property of the scattering parameters introduced in Section 6.7. Assume that the following plane waves measured at port 1 (left-hand, flat-band contact) and port 2 (right-hand, flat-band contact) of the quantum system:

$$\left. \begin{aligned} f_1(n, E) &= A_1 \exp(jk_{x,1} n_1 a) + B_1 \exp(-jk_{x,1} n_1 a), \\ f_2(n, E) &= A_2 \exp(jk_{x,2} n_2 a) + B_2 \exp(-jk_{x,2} n_2 a), \end{aligned} \right\} \tag{6.43}$$

are a solution of the Hamiltonian H.

(a) Verify in the Wannier picture that if $f(n, E)$ is a solution of $Hf = Ef$ then $f(n, E)^*$ is also a solution of $Hf = Ef$. Hint: From the time-reversal invariance of the crystal Hamiltonian we know that the band structure is symmetric $\mathcal{E}(\mathbf{k}) = \mathcal{E}(-\mathbf{k})$.

(b) Demonstrate the reciprocity property $S_{21} = S_{12}$ where the scattering parameters S_{ij} are defined as

$$S_{ij} = \left. \frac{v_i^{1/2} B_i}{v_j^{1/2} A_j} \right|_{A_{k \neq j} = 0}$$

Hint: Start with the solution for $f(n)$ satisfying the boundary conditions $A_1 = 1$, $A_2 = 0$ and form a new solution $f(n) + \lambda f(n)^*$ satisfying the new boundary conditions $A_1 = 0$,

$A_2 = 1$. Calculate S_{12} and verify that it is equal to S_{21} if the current-conservation property $|S_{12}|^2 + |S_{11}|^2 = 1$ is used.

7 Frequency response of quantum devices from DC to infrared

Dauer im wechsel (Duration in change).

JOHANN WOLFGANG VON GOETHE

7.1 Introduction

In this chapter we will analyze the response of quantum devices to time-dependent excitations. Devices are only useful if they can process time-varying information. We will mostly focus our discussion on steady-state AC excitations. We will first consider the canonic case of a uniform (not spatially-varying) time-varying potential applied to an arbitrary device. Next we will consider a time-varying potential which is linearly varying in position (constant AC electric field). Finally we will introduce the formalism used to obtain a steady-state solution of a general time-varying quantum system and apply it to the case of the resonant tunneling diode (RTD). We will then discuss the frequency response of the RTDs and the importance of time-dependent space-charge-limited transport. We then conclude by studying the interaction of quantum devices with electromagnetic waves and the application to the quantum cascade lasers.

7.2 Analytic solution for a uniform time-dependent potential

Let us consider a closed quantum system whose electron envelope function $f(n, t)$ is a solution of

$$i\hbar \frac{df(n,t)}{dt} = \sum_{n=-\infty}^{\infty} H_{mn} f(n,t).$$

We shall look for eigenstate solutions of the form $\exp[-iE_p t/\hbar] f_p(n)$, where $f_p(n)$ is a solution of the following Wannier recurrence equation:

$$E_p f_p(m) = \sum_{n=-\infty}^{\infty} H_{mn} f_p(n).$$

Consider now the new time-varying quantum system obtained from the original quantum system by adding a time-varying potential $V(t)$. We assume $V(t)$ to be uniform in space. We must now solve the following Wannier recurrence equation:

$$i\hbar \frac{d}{dt} f(m,t) = \sum_{n=-\infty}^{\infty} H_{mn} f(n,t) + V(t) f(m,t). \tag{7.1}$$

One easily verifies that Equation (7.1) admits solutions of the form

$$f(m,t) = \exp\left\{-\frac{i}{\hbar}\left[E_p t + \int_0^t V(t')dt'\right]\right\} f_p(m). \tag{7.2}$$

A Fourier transform of the electron wave-function $f(n,t)$ reveals that the electron energy is now broadened by the potential $V(t)$. If a periodic potential $V(t)$ of period ω is used this time-varying solution consists of a superposition of the states of energies $E_p + r\hbar\omega$, where r is an integer.

This time-varying quantum system is not by itself a very useful system to study as the same time-varying potential is applied all over the world (uniform in space) and we would not know how to prepare such a system. Furthermore for the time variation of the potential $V(t)$ to be of interest, we need to define a reference (ground) for the potential somewhere at a specific position in the device. This therefore requires the AC potential to be spatially varying as well as time varying. The spatially-uniform time-dependent solution we have obtained will, however, be useful as a boundary condition for modeling the anode contact of a quantum system when an AC voltage is applied between the anode and cathode and we assume the cathode is grounded.

7.3 Radiation coupling with an external modulated electric field

In this section we shall study the problem of a time-dependent field of the form

$$V(x,t) = -q\,[F_{DC} + F_{AC}\cos\omega t]\,x,$$

applied over a uniform crystal. Note that we are particularly interested in the situation in which the frequency of the AC field signal corresponds to that of the Wannier ladder energy spacing $\hbar\omega = qF_{DC}a$ in order to study any resonant coupling possible [1]. It is not actually possible to treat as a perturbation a term of the form $qaF_{AC}n$, even for small values of F_{AC}, because of the divergence for large n. Indeed, for large n, the wave-function converges as $1/n!$ which does not admit any power series expansion such as would be generated by perturbation methods. An exact solution will therefore be pursued.

The Bloch functions are very suitable as a choice for the representation, and we shall expand the wave-function as a wave-packet of Houston functions. Let us write the Houston function

$$|k_x(0)\rangle = \exp\left[-\frac{i}{qF_{DC}}\int_{k_x(0)}^{k_x(t)} E(\mathbf{k})dk_x\right]|\mathbf{k}(t)\rangle,$$

with $k_x(t) = k_x(0) + qF_{DC}t/\hbar$ and $|\mathbf{k}(t)\rangle$ a Bloch function $|k_x(t),\ \mathbf{k}_\perp\rangle$. The Houston states verify the following orthogonality:

$$\langle k'_x(0)|k_x(0)\rangle = \delta[k'_x(0) - k_x(0)],$$

if the electron is assumed to be in a Gaussian wave-packet centered around $|\mathbf{k}_\perp\rangle$ which normalizes to 1. We assume a one-band representation so that

$$H_{DC}|k_x(0)\rangle = i\hbar\frac{\partial}{\partial t}|k_x(0)\rangle,$$

where H_{DC} is the Hamiltonian of the crystal plus the DC potential supported by the static field

$$H_{DC} = H_{crystal} - qF_{DC}x.$$

The total Hamiltonian H is

$$H = H_{DC} - qF_{AC}\cos(\omega t)\,x \qquad \text{with} \qquad \hbar\omega = qF_{DC}a.$$

The Schrödinger equation is

$$i\hbar\frac{\partial}{\partial t}|\psi\rangle = H\,|\psi\rangle$$

and we shall seek $|\psi\rangle$ as a wave-packet $f(k_x(0))$ of Houston functions

$$|\psi\rangle = \int_{-\pi/a}^{\pi/a} f(k_x(0), t)\,|k_x(0)\rangle\,dk_x(0).$$

Substituting this wave-packet into the total Hamiltonian we have

$$\int_{-\pi/a}^{\pi/a} i\hbar\frac{\partial}{\partial t}\left[f(k_x(0), t)\,|k_x(0)\rangle\right]dk_x(0)$$
$$= \int_{-\pi/a}^{\pi/a}\left[H_{DC} - qF_{AC}\cos(\omega t)x\right]f(k_x(0))\,|k_x(0)\rangle\,dk_x(0).$$

Multiplying by $\langle k_x'(0)|$ and integrating over the x axis we obtain

$$i\hbar\frac{\partial}{\partial t}f(k_x'(0),t) = -qF_{AC}\cos(\omega t)\int_{-\pi/a}^{\pi/a}\langle k_x'(0)|x|k_x(0)\rangle f(k_x(0),t)\,dk_x(0). \tag{7.3}$$

The matrix element in the last expression is

$$\begin{aligned}\langle k_x'(0)|x|k_x(0)\rangle &= \exp\left[-\frac{i}{qF_{DC}}\int_{k_x(0)}^{k_x(t)}E(\mathbf{k})\,dk_x + \frac{i}{qF_{DC}}\int_{k_x'(0)}^{k_x'(t)}E(\mathbf{k})dk_x\right]\\ &\quad\times\langle \mathbf{k}(t')|x|\mathbf{k}(t)\rangle,\end{aligned} \tag{7.4}$$

where approximating the Bloch states by plane waves we have

$$\langle \mathbf{k}(t')|x|\mathbf{k}(t)\rangle \simeq i\,\delta\left[k_x'(0) - k_x(0)\right]\frac{\partial}{\partial k_x(0)}.$$

We need the derivative

$$\begin{aligned}&\frac{\partial}{\partial k_x(0)}\exp\left[-\frac{i}{qF_{DC}}\int_{k_x(0)}^{k_x(0)+\frac{qF_{DC}}{\hbar}t}E(\mathbf{k})\,dk_x\right]\\ &\quad=\left[-\frac{i}{qF_{DC}}E\left(\mathbf{k}(t)\right)+\frac{i}{qF_{DC}}E\left(\mathbf{k}(0)\right)\right]\exp\left[-\frac{i}{qF_{DC}}\int_{k_x(0)}^{k_x(t)}E(\mathbf{k})\,dk_x\right].\end{aligned}$$

Equation (7.3) now reads

$$\begin{aligned}i\hbar\frac{\partial}{\partial t}f(k_x(0),t) &= -iqF_{AC}\,\cos(\omega t)\\ &\quad\times\left\{\frac{\partial}{\partial k_x(0)}f(k_x(0),t) + f(k_x(0),t)\,\frac{i}{qF_{DC}}[E(\mathbf{k}(0)) - E(\mathbf{k}(t))]\right\}\end{aligned} \tag{7.5}$$

or

$$\begin{aligned}\frac{\partial}{\partial t}f(k_x(0),t) &= -\frac{qF_{AC}}{\hbar}\cos(\omega t)\frac{\partial}{\partial k_x(0)}f(k_x(0),t)\\ &\quad-\frac{i}{\hbar}\frac{F_{AC}}{F_{DC}}\,[E(k_x(0)) - E(k_x(t))]\cos(\omega t)\,f(k_x(0),t).\end{aligned}$$

Let us introduce the constants C and B and band structure $E(k_x)$

$$C = \frac{qF_{AC}}{\hbar},\qquad B = -\frac{i}{\hbar}\frac{F_{AC}}{F_{DC}}A\qquad\text{and}\qquad E(k) = -A\,\cos(ka),$$

so that we can rewrite the master equation for the Houston wave-packet as

$$\begin{aligned}&\frac{\partial}{\partial t}f(k,t) + C\,\cos(\omega t)\,\frac{\partial}{\partial k}f(k,t)\\ &\quad+B\,\cos(\omega t)[\cos(ka) - \cos(ka+\omega t)]\,f(k,t) = 0.\end{aligned} \tag{7.6}$$

We shall now solve Equation (7.6) to obtain the evolution in time of the Houston wave-packet $f(k, t)$ assuming an initial wave-packet $f(k, 0) = f_0(k)$. For this purpose we use the method of characteristics and seek the characteristic curves $\{k(s),\ t(s)$ for $s > 0\}$ given by

$$\frac{dt}{ds} = 1 \qquad \frac{dk}{ds} = C\ \cos(\omega t),$$

so as to reduce Equation (7.6) to

$$\frac{df}{ds} + B\ \cos(\omega t)[\cos(ka) - \cos(ka + \omega t)]f = 0. \tag{7.7}$$

Let $t(s) = s$ for $t(s = 0) = 0$, then

$$k(s) = \frac{C}{\omega}\ \sin(\omega s) + \tau,$$

with τ a constant. The inversion gives

$$s(k, t) = t \qquad \text{and} \qquad \tau(k, t) = k - \frac{C}{\omega}\ \sin(\omega t). \tag{7.8}$$

The initial condition in the s space is then

$$f(0) = f(s = 0) = f(k(0), t(0)) = f(\tau,\ 0) = f_0(\tau).$$

Let us write Equation (7.7) as

$$\frac{df(s)}{ds} + f(s)\ B\ \cos(\omega s) \times \left\{\cos\left[\tau a + \frac{aC}{\omega}\ \sin\left(\omega s\right)\right] - \cos\left[\tau a + \frac{aC}{\omega}\ \sin\left(\omega s\right) + \omega s\right]\right\} = 0,$$

which takes the form

$$\frac{df(s)}{ds} + f(s)\ G(s,\ \tau) = 0$$

and integrates to

$$f(s) = f(0)\ \exp\left[-\int_0^s G(s,\ \tau)ds\right].$$

Let us rewrite $G(s,\ \tau)$ using $aC/\omega = F_{AC}/F_{DC}$

$$G(s, \tau) = -\frac{i}{\hbar}\frac{F_{AC}}{F_{DC}} A\cos(\omega s)\left\{\ [\cos(\tau a) - \cos(\tau a + \omega s)] \times \cos\left[\frac{F_{AC}}{F_{DC}}\ \sin(\omega s)\right] + [\sin(\tau a + \omega s) - \sin(\tau a)]\sin\left(\frac{F_{AC}}{F_{DC}}\ \sin(\omega s)\right)\right\}. \tag{7.9}$$

For large s (time) the integral of $G(s, \tau)$ is s times the DC component of the periodic function $G(s, \tau)$. For a small field, we evaluate $G(s, \tau)$ to the first order in F_{AC}/F_{DC}:

$$\begin{aligned}
G(s,\tau) &\simeq -\frac{i}{\hbar}\frac{F_{AC}}{F_{DC}} A\cos(\omega s)[\cos(\tau a) - \cos(\tau a + \omega s)] \\
&= -\frac{i}{\hbar}\frac{F_{AC}}{F_{DC}} A\big[\cos(\tau a)\cos(\omega s) - \cos(\tau a)\cos^2(\omega s) \\
&\quad + \sin(\tau a)\sin(\omega s)\cos(\omega s)\big] \\
&= -\frac{i}{\hbar}\frac{F_{AC}}{F_{DC}} A\left[-\frac{1}{2}\cos(\tau a) + \cos(\tau a)\cos(\omega s) - \frac{1}{2}\cos(\tau a + 2\omega s)\right].
\end{aligned}$$

The DC-component of $G(s, \tau)$ is $i\alpha\cos(\tau a)$ with

$$\alpha = \frac{1}{2}\frac{F_{AC}}{F_{DC}}\frac{A}{\hbar},$$

so that for large values of s the integral of $G(s, \tau)$ is

$$\int_0^s G(s,\tau)\, ds = i\alpha\cos(\tau a)s.$$

Then for long times the Houston wave-packet is

$$f(s) = f(0)\exp[-i\alpha\cos(\tau a)s]$$

and back in the original time and momentum space using Equations (7.8), the Houston wave-packet for long times is

$$f(k,t) = f_0\left[k - \frac{F_{AC}}{aF_{DC}}\sin(\omega t)\right]\exp\left\{-i\alpha t\cos\left[ka - \frac{F_{AC}}{F_{DC}}\sin(\omega t)\right]\right\}. \tag{7.10}$$

Having obtained a solution of the evolution of the Houston wave-packet for long times and a small F_{AC}/F_{DC} ratio, we shall now study its behavior. For this purpose we calculate the expectation value of the position $\langle x\rangle$ using

$$\begin{aligned}
\langle x\rangle &= \iint dk'\, dk\; f^*(k')f(k)\langle k'_x(0)\,|x|k_x(0)\rangle = i\int_{-\pi/a}^{\pi/a} f^*(k)\frac{d}{dk}f(k)\, dk \\
&\quad + \frac{1}{qF_{DC}}\int_{-\pi/a}^{\pi/a}|f(k)|^2\left[E\left(k + \frac{qF_{DC}}{\hbar}t\right) - E(k)\right] dk,
\end{aligned}$$

using the matrix element of Equation (7.4) and defining $k = k_x(0)$. Let $f(k) = |f|\exp(i\phi)$. Let

$$\langle(\cdots)\rangle = \int_{-\pi/a}^{\pi/a}(\cdots)\, dk$$

It follows using the periodicity of f on the limits of integration

$$\begin{aligned}
i\left\langle f^*\frac{d}{dk}f\right\rangle &= i\left\langle |f|\frac{d}{dk}|f|\right\rangle - \left\langle\frac{d\phi}{dk}|f|^2\right\rangle \\
&= -\left\langle\frac{d\phi}{dk}|f|^2\right\rangle
\end{aligned} \tag{7.11}$$

that the position expectation is

$$\langle x \rangle = -\left\langle \frac{d\phi}{dk} |f|^2 \right\rangle + \frac{1}{qF_{DC}} \left\langle |f|^2 \left[E\left(k + \frac{qF_{DC}}{\hbar} t\right) - E(k) \right] \right\rangle \tag{7.12}$$

For a small F_{AC}/F_{DC} ratio we can neglect the additional oscillation in the large time solution of Equation (7.10) and write

$$f(k,t) = f_0(k) \exp[-i\alpha t \cos(ka)] = |f| \exp(i\phi)$$

Let $f_0(k)$ be a Gaussian wave-packet as our initial condition

$$f_0(k) = (2\pi\sigma^2)^{-1/4} \exp\left[-(k-k_0)^2/4\sigma^2\right] = N^{1/2}(k_0, \sigma),$$

where $\sigma = \Delta k$ is the variance and k_0 is the center of the wave-packet. Then we have

$$|f| = f_0(k) \qquad \text{and} \qquad \phi = -\alpha t \cos(ka).$$

Let us evaluate the two terms on the right-hand side of Equation (7.12). The second term on the right-hand side of Equation (7.12) includes expressions of the form:

$$\begin{aligned} \left\langle |f|^2 E\left(k + \frac{qF_{DC}}{\hbar} t\right) \right\rangle &= -A \int N(k_0, \sigma) \cos(ka + \omega t)\, dk \\ &= -A \int N(0, \sigma) \cos(ka + k_0 a + \omega t)\, dk \\ &= -A\beta \cos(k_0 a + \omega t), \end{aligned}$$

where $\beta \simeq \exp(-\sigma^2 a^2/2)$ is the relative amplitude of the Zener oscillation. The first term on the right-hand-side of Equation (7.12) is:

$$\begin{aligned} -\langle \frac{d\phi}{dk} |f|^2 \rangle &= -\alpha\, ta \int \sin(ka)\, N(k_0, \sigma)\, dk \\ &= -\alpha\, ta \int N(0, \sigma) \sin(ka + k_0 a)\, dk \\ &= -\alpha\, ta\, \beta \sin(k_0 a). \end{aligned}$$

Therefore the expectation value of the electron position is

$$\langle x \rangle = -\alpha\, ta\, \beta \sin(k_0 a) + \frac{\beta A}{qF_{DC}} [\cos(k_0 a) - \cos(k_0 a + \omega t)]. \tag{7.13}$$

The first term on the right-hand side is the drift of the electron and the second is the Zener oscillation. The drift of the electron is weighted by $\alpha \sin(k_0 a)$, where $k_0 a$ is the phase of the Zener oscillation relative to the AC applied. Depending on this product, the Zener oscillation is moving toward either the right or the left of the device. When the Zener oscillation is moving toward the left, the electron is simultaneously gaining energy by climbing the tilted conduction band. One can say that the electron

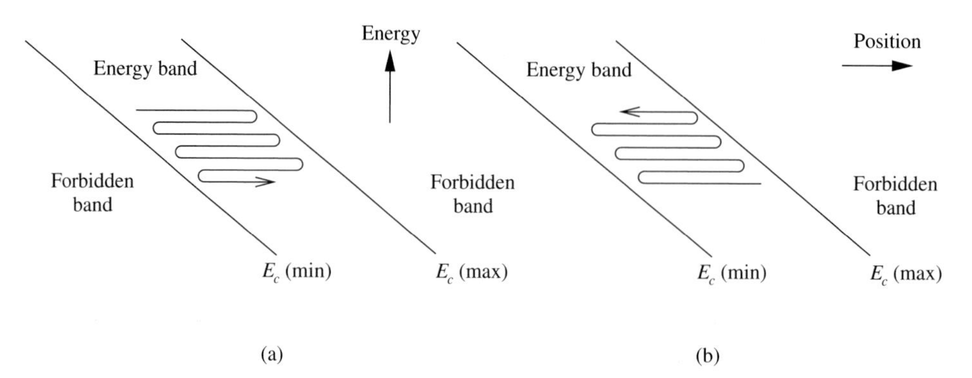

Fig. 7.1. Motion a ballistic electron confined in a band and accelerated by both a uniform DC and an AC field: (a) conducted radiation emission; (b) conducted radiation absorption.

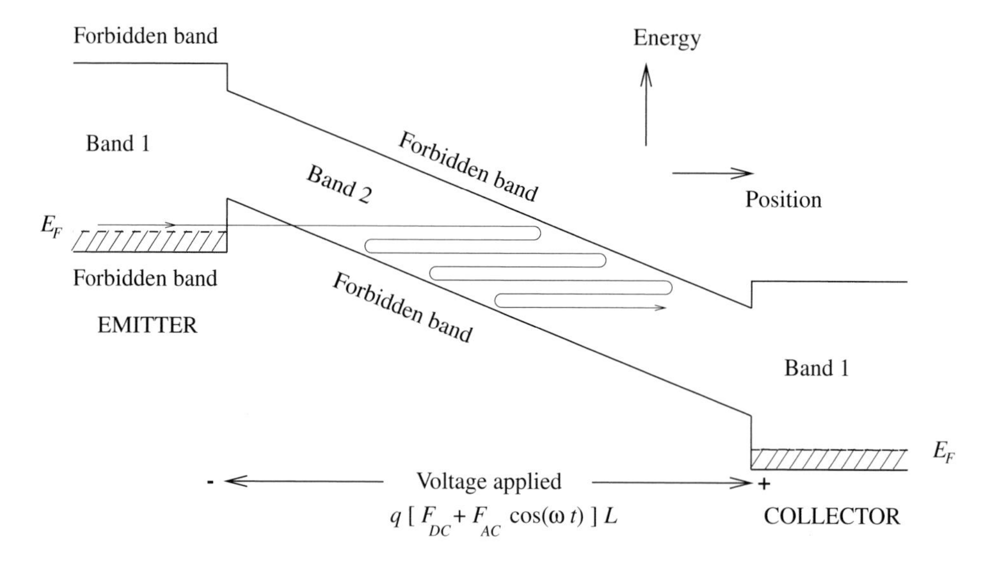

Fig. 7.2. Device for launching Zener oscillations.

is taking energy from the applied modulation field to climb the potential barrier. When the Zener oscillation is moving toward the right, the electron is converting its DC potential energy into RF energy. This is the regime under which the Zener oscillations are radiative. This is represented in Figure 7.1. A possible conceptual device implementation [2] for both launching these Zener oscillations and inducing the radiation coupling with the AC field is shown in Figure 7.2.

7.4 Time-dependent tunneling theory

We shall now consider time-varying potentials with an arbitrary position dependence and present a general numerical methodology for analyzing them. An example of such

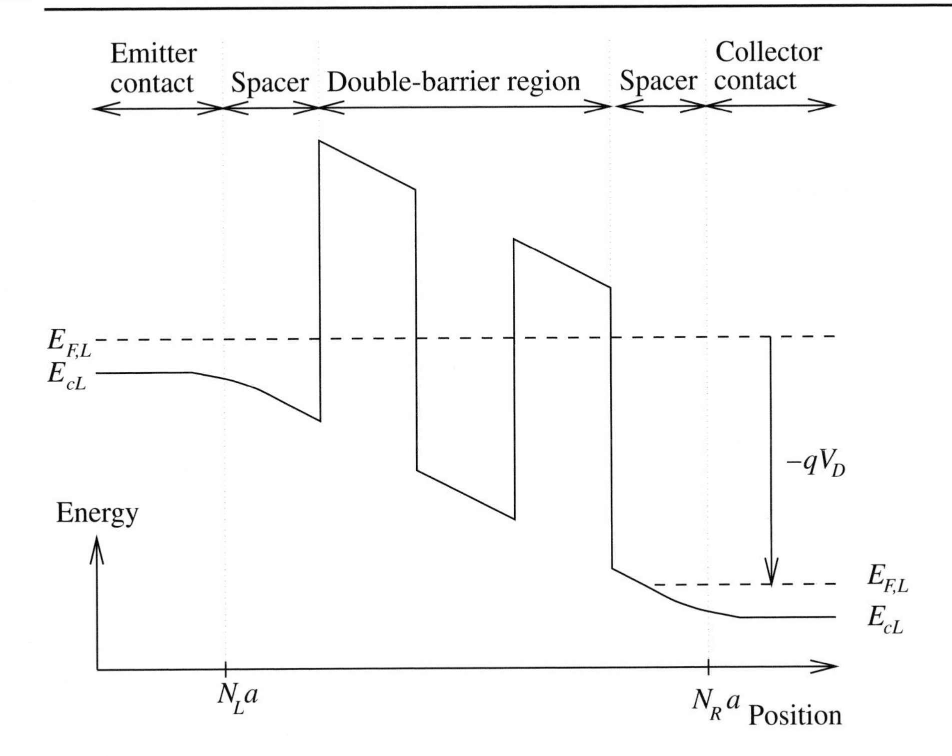

Fig. 7.3. Band diagram of a test resonant tunneling structure.

a structure is the resonant tunneling structure shown in Figure 7.3.

The analysis of time-dependent tunneling in a general layered quantum device can be obtained by solving the one-electron Schrödinger equation:

$$i\hbar \frac{\partial \Psi(x,t)}{\partial t} = [H_0(x) - qV(x,t)]\,\Psi(x,t), \tag{7.14}$$

where $H_0(x)$ is the crystal Hamiltonian and $V(x,t)$ is the DC + AC electrostatic potential.

We shall limit our analysis of time-dependent tunneling to harmonic electrostatic potentials of the form

$$V(x,t) = V_{DC}(x) + \sum_p V_p(x)\exp(ip\omega t), \tag{7.15}$$

with $V_p(x) = V^*_{-p}(x)$.

Our approach for the electron problem is based on the generalized Wannier picture [3] for which the electron wave-function $|\Psi\rangle$ is expanded in terms of the generalized Wannier functions

$$|\Psi\rangle = \sum_{n=-\infty}^{\infty} f(n,t)\,|n\mathbf{k}_\perp\rangle,$$

where $f(m,t)$ is the Wannier envelope function and m the lattice site index. Scattering is not considered and the transverse wave-vector $\mathbf{k}_\perp$ is conserved across the device.

The analysis is presented for a tight-binding band structure of the form:

$$E(\mathbf{k}, n) = E_c(n) + \frac{\hbar^2 k_\perp^2}{2m^*} + \frac{\hbar^2}{m^* a^2}[(1 - \cos(k_x a)], \tag{7.16}$$

where k_x is the electron wave-vector along the x direction from the cathode to the anode and $E_c(n)$ is the conduction-band reference energy at the position $x = na$.

The electron envelope function $f(m,t)$ is then a solution of the Wannier equation

$$\begin{aligned} i\hbar \frac{\partial}{\partial t} f(n,t) &= -\frac{A}{2} f(n-1,t) - \frac{A}{2} f(n+1,t) \\ &+ \left\{ E_c(n) + A - qV_0(na) - q\left[\sum_p V_p(na) \exp(ip\omega t)\right]\right\} f(n,t), \end{aligned} \tag{7.17}$$

with $A = \hbar^2/(m^* a^2)$. For simplicity of presentation we assume here that mass does not vary along the superlattice direction.

We are interested in the steady-state response of the quantum device for the harmonic electrostatic potential applied. In the steady state the envelope function is also periodic and can be expanded in terms of a Fourier series

$$f(n,t) = \sum_p f_p(n) \exp\left(-i\frac{E_{0x}}{\hbar}t - ip\omega t\right) \tag{7.18}$$

where to each harmonic component $f_p(n)$ is associated an energy $E_{0x} + p\hbar\omega$. Clearly the electron transport is no longer ballistic as the electron energy can increase or decrease by quantum of energy $\hbar\omega$. Substituting Equation (7.18) into Equation (7.17) we verify that the harmonic components $f_p(n)$ are themselves solution of

$$\begin{aligned} (E_{0x} + p\hbar\omega) f_p(n) &= -\frac{A}{2}[f_p(n+1) + f_p(n-1)] + [E_c(n) + A - qV_0(na)] f_p(n) \\ &- q\sum_m \left[V_{-m}(na) f_{p-m}(n) + V_m(na) f_{p+m}(n)\right]. \end{aligned} \tag{7.19}$$

The set of Equations (7.19) for all the harmonics $f_p(n)$ forms a linear system of coupled difference equations which can be readily solved numerically.

To calculate the forward (backward) current we can use an incident wave with unit amplitude injected in the left (right) flat-band contact of the device. For the forward current calculation the reference potential (ground) is selected to be on the emitter (left) side such that the conduction-band edge on the collector (right) side varies with the applied AC voltage. The analytic solution obtained in Section 7.2 can be used as a boundary condition for the transmitted waves on the time-varying collector side. However, in a numerical solution in which a finite number of harmonics is targeted inside the device, the same truncation in the Fourier series expansion should be used on the collector side.

The time-dependent electron current between sites n and $n+1$ is then given by the usual summation over incident energy E_{0x} [4]:

$$I_{F/B}(n,t) = \frac{Dk_BT}{2\pi}\frac{q}{\hbar}\int_0^{2A} T(n,t)\ln\left[\exp\left(\frac{E_{F,L/R} - E_{c,L/R} - E_x}{k_BT}\right)+1\right]dE_x, \tag{7.20}$$

where $E_{F,L/R}(t)$ is the left/right Fermi level associated with the forward/backward current $I_{F/B}$ respectively and $T(n,t)$ is a time- and position-dependent transmission coefficient defined as the ratio of the instantaneous transmitted to incident currents

$$T(n,t) = \frac{j(n,n+1,t) + j(n,n-1,t)}{\frac{aA}{\hbar}\sin(k_xa)}, \tag{7.21}$$

with $j(n,m,t) = -(a/\hbar)\,\mathrm{Im}[f^*(n,t)f(m,t)H_{n,m}]$ the elemental current. For a uniform band (constant mass) the transmission coefficient reduces to

$$T(n,t) = \frac{\mathrm{Im}\left[f^*(n,t)f(n+1,t)\right]}{\sin(k_xa)}. \tag{7.22}$$

Substitution of the Fourier expansion of the envelope function given by Equation (7.18) into Equation (7.20) permits us to expand the current in a Fourier series:

$$I_{F/B}(n,t) = I_{0,F/B}(n) + \sum_p I_{p,F/B}(n)\exp(ip\omega t) \tag{7.23}$$

where $I_{0,F/B}(n)$ is the DC forward/backward current, and $I^*_{p,F/B}(n) = I_{-p,F/B}(n)$ is the pth harmonic current component.

The electron distribution $\rho(n)$ at each lattice site n is similarly calculated using the following expression:

$$\rho_{F/B}(n,t) = \frac{Dk_BT}{2\pi}q\int_0^{\frac{\pi}{a}} |f(n,t)|^2 \times \ln\left\{\exp\left[\frac{E_{F,L/R}(t) - E_{c,L/R} - E_x(k_x)}{k_BT}\right]+1\right\}dk_x, \tag{7.24}$$

with $E_x(k_x) = A[(1-\cos(k_xa)]$. Again the substitution of the Fourier expansion of the envelope function given by Equation (7.18) into Equation (7.24) permits us to expand the electron distribution in a Fourier series:

$$\rho_{F/B}(n) = \rho_{o,F/B}(n) + \sum_p \rho_{p,F/B}(n)\exp(ip\omega t), \tag{7.25}$$

where $\rho_0(n)$ is the DC charge density, and $\rho_p(n) = \rho^*_{-p}(n)$ is the pth harmonic charge density. The current and charge derived above for each harmonic p satisfy the continuity equation

$$I_{p,F/B}(N+1) - I_{p,F/B}(1) = p\omega\sum_{n=2}^{N}\rho_{p,F/B}(n). \tag{7.26}$$

As the electrons are injected on both the emitter and collector contacts the total electron current $I_{elec,p}$ for the harmonic p is the difference between the forward (F) and backward (B) currents

$$I_{elec,p} = I_{p,F} - I_{p,B}, \tag{7.27}$$

whereas the total charge is the sum of the forward and backward charge

$$\rho_p(n) = \rho_{p,F} + \rho_{p,B}. \tag{7.28}$$

7.5 Small-signal response without self-consistent potential

To test the time-dependent tunneling theory we shall apply it to the RTD of Figure 7.3. This RTD consists of a conventional undoped AlGaAs/GaAs/AlGaAs double-barrier (DB) structure with two lightly doped spacers, one on each side, sandwiched between two strongly doped GaAs n^+ buffers. The barrier and well widths are six lattice parameters wide (34 Å). A barrier height of 0.25 eV corresponding to an Al mole fraction of 0.3 is used. A donor concentration $N_D = 10^{18}$ cm^{-3} is used in the buffers, and the area of the test device is 4.5×10^{-8} cm^2. The diode I–V characteristic is shown in Figure 7.4.

To test the quantum AC modeling technique presented above let us first apply it to the calculation of the small-signal frequency response of an RTD device model, where the electrostatic potential across the RTD is approximated by a linearly-varying electrostatic potential

$$V(n,t) = \left[V_{D,DC} + V_{D,AC}\cos(\omega t)\right]\frac{n - N_L}{N_R - N_L} \quad \text{for} \quad N_L < n < N_R. \tag{7.29}$$

In this approximate model we are neglecting the impact of the RTD DC and AC internal charges on the electrostatic potential. The small-signal admittance of the RTD in the negative differential conductivity (NDC) region is plotted versus frequency in Figure 7.5. It is seen that the device admittance (to be defined in the next section) of the intrinsic RTD remains constant up to terahertz frequencies. The small-signal frequency response obtained is consistent with the results reported by [7,8,9,10,11]. This potential for operating at very high-frequencies explains the interest in RTDs. The decrease of the negative resistance at high frequency is attributed to the inertia of the electron. Note that the small-signal admittance shown in Figure 7.5 does not include the displacement current which is required to calculate the diode maximum frequency of operation. Also the approximate DC and AC electrostatic potentials were not solved self-consistently from the DC and AC charges present in the RTD device. A more rigorous analysis of the RTD response at high frequencies is developed in the next section.

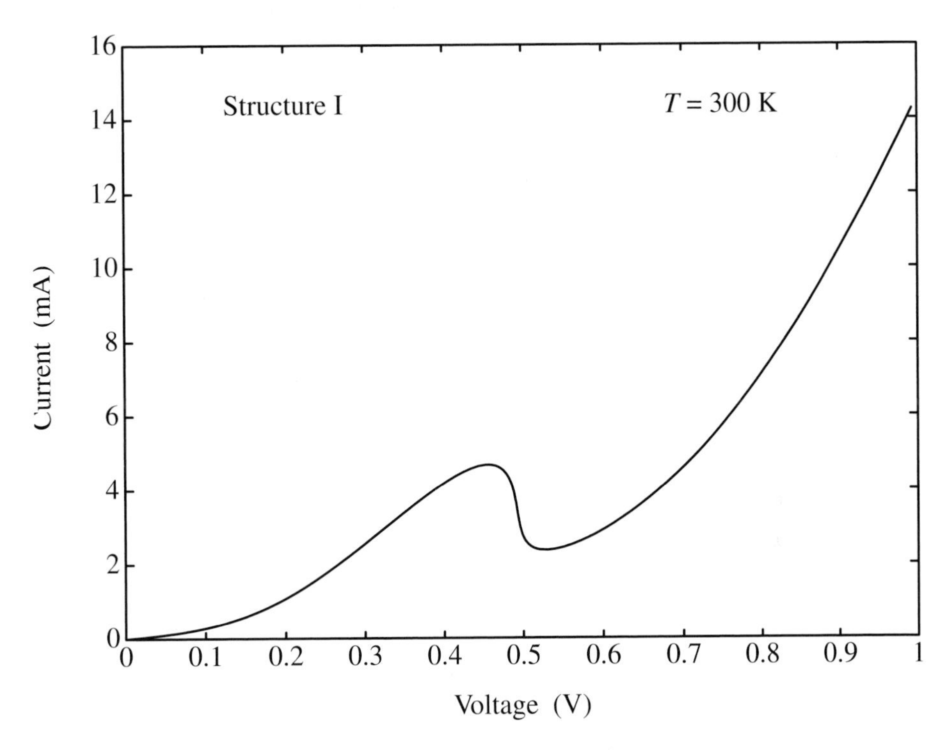

Fig. 7.4. Current versus applied voltage V_D for the 34/34/34 Å test RTD considered. (W.-R. Liou and P. Roblin, *IEEE Transactions on Electron Devices*, Vol. 41, No. 7, pp. 1098–1112, July 1994. ©1994 IEEE.)

7.6 Self-consistent solution

Transport in RTDs like in any ballistic device is space-charge-limited. That is the electrostatic potential in the device must be calculated self-consistently from the charge distribution inside the device. Indeed we have seen in Chapter 4 that a self-consistent solution of the Poisson and Schrödinger equations is required to obtain a realistic calculation of the DC characteristic of the RTD [12]. Similarly we can expect that a self-consistent solution of the Poisson and Schrödinger equations is required in order to obtain more a realistic AC response of RTDs.

As for DC self-consistency, AC self-consistency is achieved when the AC electrostatic potential for each harmonic is obtained using the Poisson equation from the calculated AC charge distribution across the device:

$$\frac{d}{dx}\left[\epsilon(x)\frac{dV_0(x)}{dx}\right] = \rho_0(x) - qN_D^+(x), \tag{7.30}$$

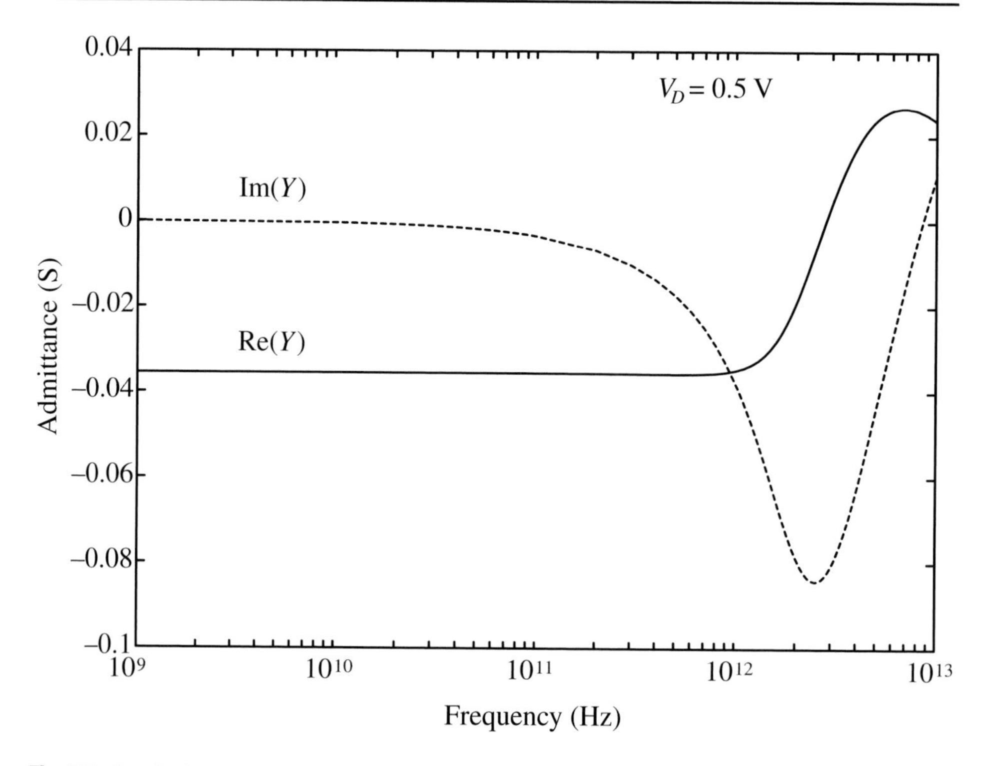

Fig. 7.5. Small-signal admittance of the RTD (structure I) in the NDC region.

$$\frac{d}{dx}\left[\epsilon(x)\frac{dV_p(x,t)}{dx}\right] = \rho_p(x,t). \tag{7.31}$$

The external boundary conditions applied across the RTD consist of a DC voltage V_D and an AC voltage V_{AC} for the first harmonic:

$$V_D(t) = V_D + V_{AC}\cos(\omega t). \tag{7.32}$$

All higher harmonic voltages are shorted outside the device but not inside the device. These boundary conditions are used to calculate all the harmonic components of the RTD current for both the small- and large-signal responses.

For small AC signals, only the first harmonic of the current and internal potential is required. For a large AC signal, higher harmonics of both the internal electrostatic potential and the current must be included due to the non-linearity of the RTD. The minimum number of harmonics required for an accurate numerical solution is given approximately by $X = qV_{AC}/(\hbar\omega)$. This can be inferred from the analytic envelope function of Equation (7.2) which admits an expansion in terms of Bessel functions $J_p(X)$ and therefore vanishes as $1/p!$ for harmonics $p > X$.

The total diode current at a position n for each harmonic p is given by

$$I_{total,p}(n) = I_{elec,p}(n) + I_{displ,p}(n), \tag{7.33}$$

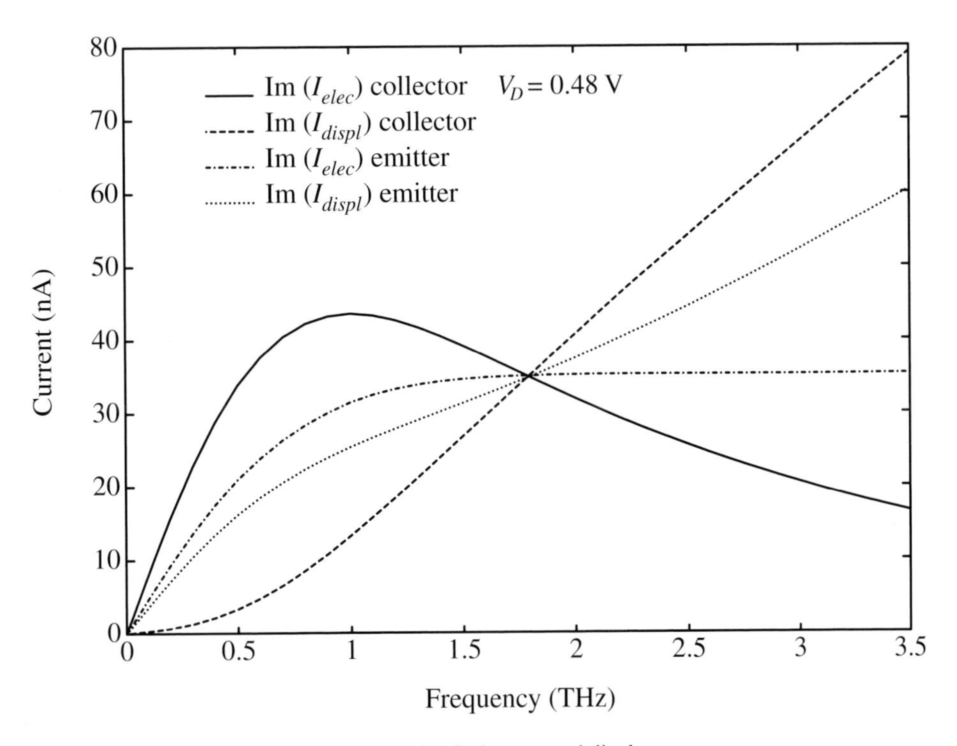

Fig. 7.6. Imaginary part of the small-signal AC electron and displacement currents versus frequency. The full and dashed lines represent the electron and displacement currents at the right-hand buffer (collector) while the dashed-dotted and dotted lines represent the electron and displacement currents at the left-hand buffer (emitter). (W.-R. Liou and P. Roblin, *IEEE Transactions on Electron Devices*, Vol. 41, No. 7, pp. 1098–1112, July 1994. ©1994 IEEE.)

i.e. the sum of the electron current $I_{elect,p}$ of Equation (7.27) and the displacement current $I_{displ,p}$, which is calculated from the AC electric field using

$$I_{displ,p}(n) = -\epsilon(n) jp\omega \frac{dV_p(na)}{dx}. \tag{7.34}$$

The small- and large-signal RTD admittance $Y(\omega, V_D, V_{AC})$ for the first harmonic for a given DC bias V_D and AC voltage V_{AC} is then calculated using Kurokawa's definition [13]:

$$Y(\omega, V_D, V_{AC}) = \frac{I_{total,1}(\omega, V_D, V_{AC})}{V_{AC}} = G(\omega, V_D, V_{AC}) + jB(\omega, V_D, V_{AC}), \tag{7.35}$$

where $G(\omega, V_D, V_{AC})$ is the device conductance and $B(\omega, V_D, V_{AC})$ the device susceptance. For small-signal AC voltages ($qV_{AC} \ll \hbar\omega$), the quantities Y, G and B are effectively independent of V_{AC} and only depend upon the DC bias V_D and the frequency ω.

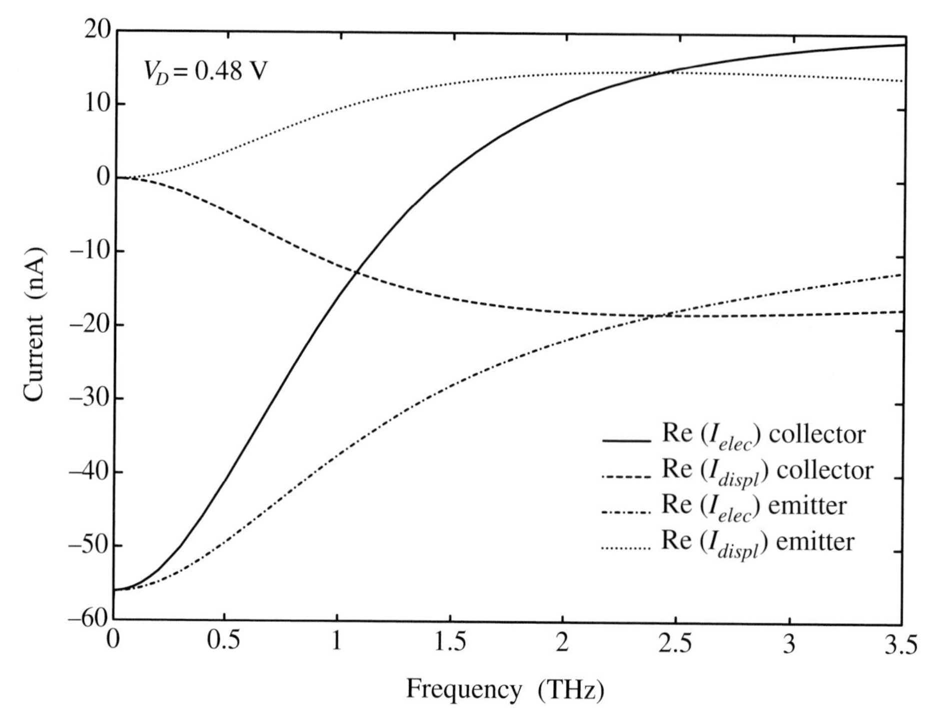

Fig. 7.7. Real part of the AC electron and displacement currents versus frequency. The full and dashed lines represent the electron and displacement currents at the right-hand buffer (collector), while the dashed-dotted and dotted lines represent the electron and displacement currents at the left-hand buffer (emitter). (W.-R. Liou and P. Roblin, *IEEE Transactions on Electron Devices*, Vol. 41, No. 7, pp. 1098–1112, July 1994. ©1994 IEEE.)

Let us now demonstrate the impact of the DC and AC self-consistent analysis on the AC current and charge distributions. Consider the case in which the RTD is biased in the negative differential resistivity (NDR) region ($V_D = 0.48$ V) with a small-signal AC excitation. Figure 7.6 shows a plot versus frequency of the imaginary part of the displacement current $\text{Im}[I_{displ,1}(n)]$ and electron current $\text{Im}[I_{elec,1}(n)]$ at the emitter side ($n = N_L$) and the collector side ($n = N_R$) of the RTD for the first harmonic. We define here the emitter and collector sides as the positions at which the buffer ends and the spacer starts.

Similarly Figure 7.7 shows a plot versus frequency of the real part of the displacement current $\text{Re}[I_{displ,1}(n)]$ and electron current $\text{Re}[I_{elec,1}(n)]$ at the emitter side ($n = N_L$) and collector side ($n = N_R$) of the RTD for the first harmonic.

Finally Figure 7.8 shows a plot versus frequency of both the real part $\text{Re}[I_{total,1}(n)]$ and imaginary part $\text{Im}[I_{total,1}(n)]$ of the total current at the emitter side ($n = N_L$) and collector side ($n = N_R$) of the RTD for the first harmonic. Clearly the left- and right-hand total currents overlap and cannot be distinguished in Figure 7.8. This

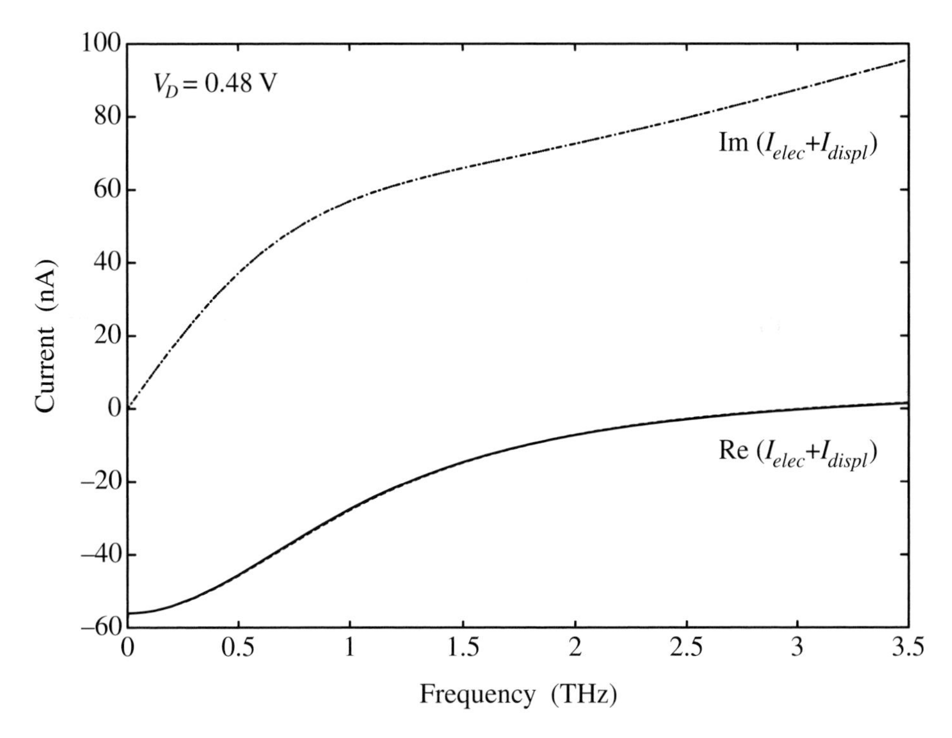

Fig. 7.8. Real and imaginary parts of the small-signal AC current versus frequency. The full and dashed lines represent the summation of real parts of the AC electron and displacement currents at the right- and left-hand buffers, respectively, while the dashed-dotted and dotted lines represent the the summation of imaginary parts of the AC electron and displacement currents at the right- and left-hand buffers, respectively. (W.-R. Liou and P. Roblin, *IEEE Transactions on Electron Devices*, Vol. 41, No. 7, pp. 1098–1112, July 1994. ©1994 IEEE.)

demonstrates the continuity of the total current. However, the displacement and electron currents on the left- and right-hand sides are clearly different, justifying the need for the self-consistent treatment used for the AC analysis. Indeed, the calculation of the displacement and electron currents should not be decoupled in a rigorous analysis of space-charge-limited devices.

7.7 RTD conductances and capacitances

Examination of Figure 7.8 indicates that for a small AC voltage V_{AC} both the imaginary and real parts of the total current $I_{total,1} = Y(\omega, V_D)V_{AC}$ and therefore the conductance $G(\omega, V_D)$ and susceptance $B(\omega, V_D)$ have a complex frequency dependence.

However, for small frequencies ($\omega \ll 100$ GHz) the conductance can be approximated by a frequency-independent conductance which is simply obtained by

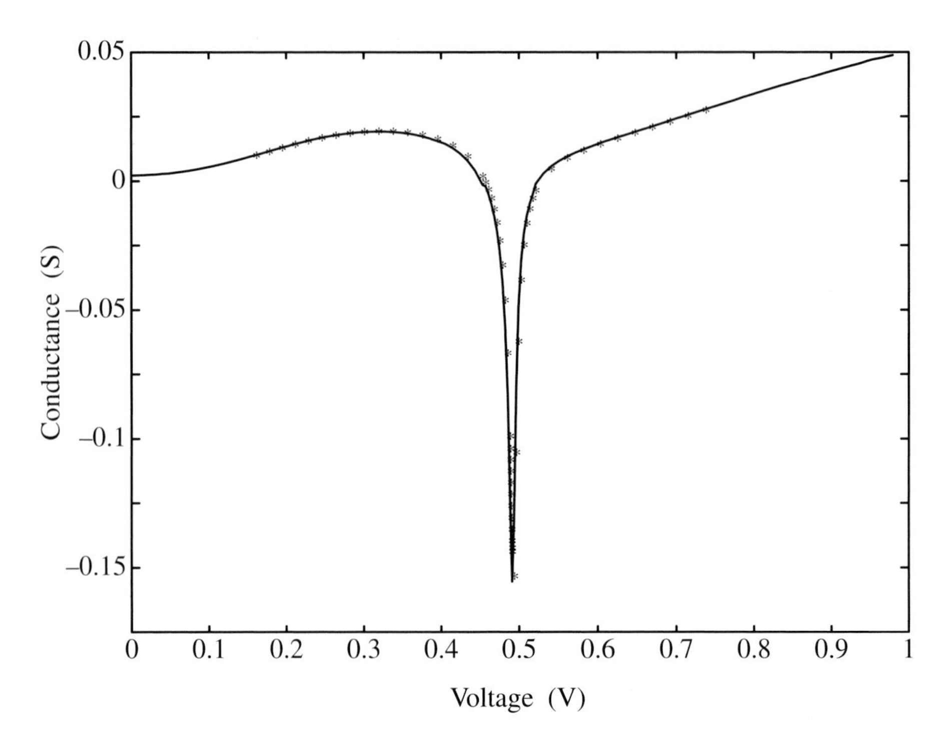

Fig. 7.9. Comparison of the conductance versus bias voltage calculated from the DC I–V characteristics (full line) with the the small-signal conductance B ($*$) at low frequencies. (W.-R. Liou and P. Roblin, *IEEE Transactions on Electron Devices*, Vol. 41, No. 7, pp. 1098–1112, July 1994. ©1994 IEEE.)

differentiating the DC I–V characteristic of the RTD

$$G(\omega, V_D) \simeq \frac{dI_D}{dV_D} \qquad \text{for small } \omega, \tag{7.36}$$

as is verified in Figure 7.9.

For small frequencies ($\omega \ll 500$ GHz) we can verify in Figure 7.8 that the RTD susceptance increases linearly with frequency and is therefore well approximated by a frequency independent capacitance C

$$B(\omega) \simeq \omega C(V_D), \tag{7.37}$$

where the capacitance C is the RTD capacitance.

Figure 7.10 shows the capacitance versus voltage for different spacer lengths calculated directly from the AC current using Equation (7.37). For each of these various RTDs the capacitance is observed to exhibit a large peak in the NDC region. The voltage dependence of the capacitance can be explained by analyzing Figure 7.11, where the imaginary parts of the electron and displacement currents are plotted on the left- ($n = N_L$) and right- ($n = N_R$) hand sides at a low frequency (1 GHz)

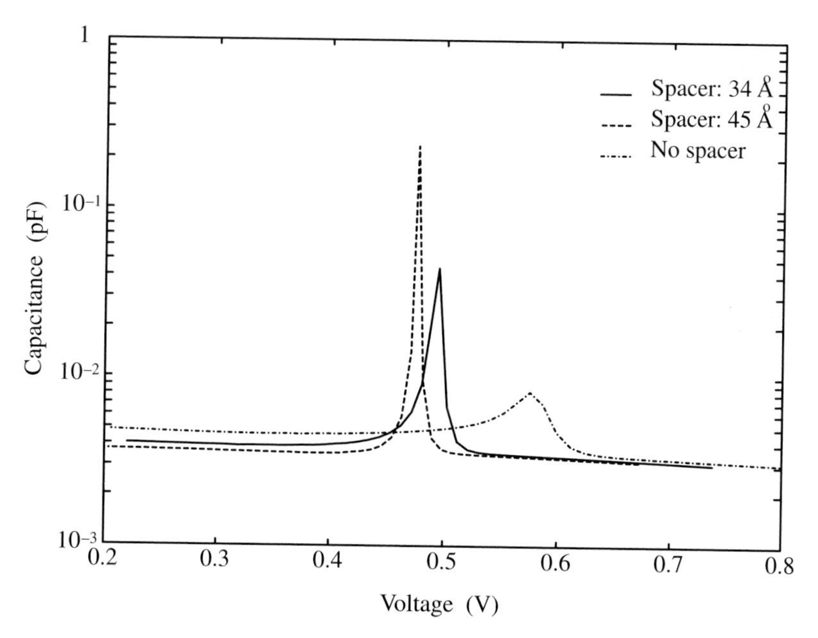

Fig. 7.10. RTD capacitance versus bias voltage for different spacer widths calculated from the AC current. The full line represents a spacer width of 34 Å, the dashed line no spacer, and the dashed-dotted line a spacer width of 45 Å. (W.-R. Liou and P. Roblin, *IEEE Transactions on Electron Devices*, Vol. 41, No. 7, pp. 1098–1112, July 1994. ©1994 IEEE.)

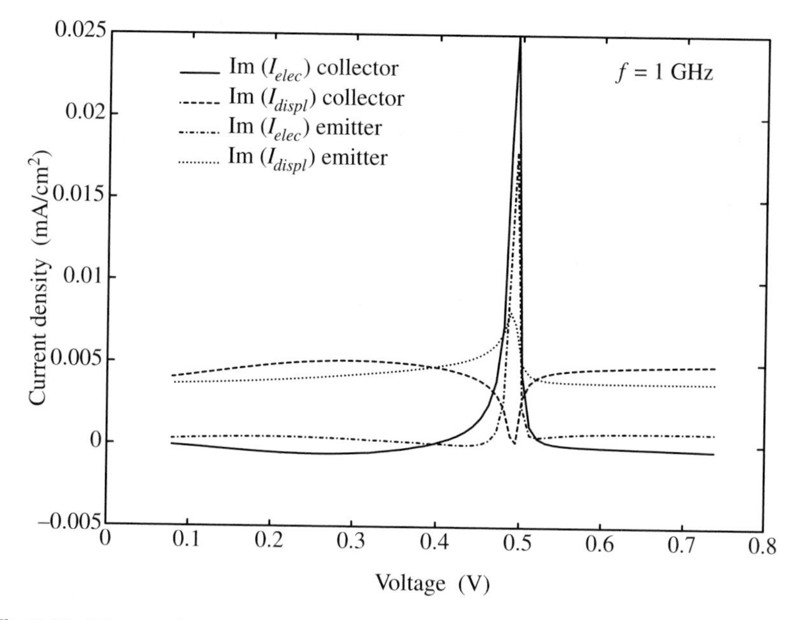

Fig. 7.11. The imaginary parts of the AC electron and displacement currents versus the bias voltage V_D. The full and dashed lines represent the electron and displacement currents at the right-hand buffer (collector) while the dashed-dotted and dotted lines represent the electron and displacement currents at the left-hand buffer (emitter). (W.-R. Liou and P. Roblin, *IEEE Transactions on Electron Devices*, Vol. 41, No. 7, pp. 1098–1112, July 1994. ©1994 IEEE.)

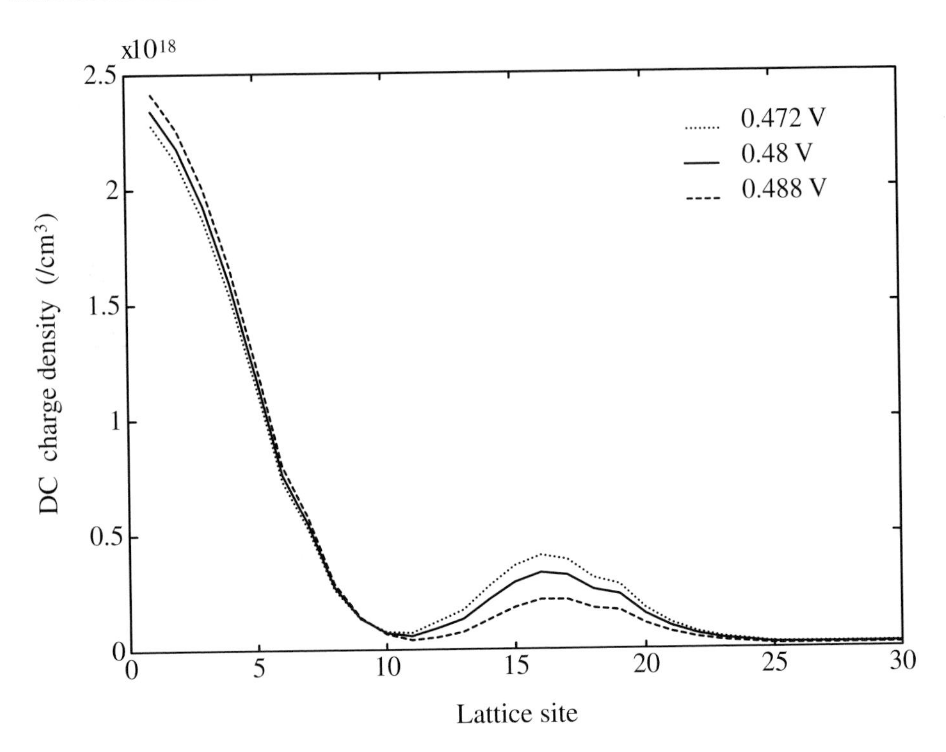

Fig. 7.12. Distribution of the DC charge across the RTD for three different biasing voltages in the NDC region. (W.-R. Liou and P. Roblin, *IEEE Transactions on Electron Devices*, Vol. 41, No. 7, pp. 1098–1112, July 1994. ©1994 IEEE.)

as a function of bias voltage. The RTD capacitance measured is contributed by the imaginary parts of both the electron and displacement currents. The contribution of the displacement current to the capacitor is dominant outside the NDC region (below 0.4 V and above 0.51 V). This indicates that the effective RTD capacitor extends approximately from the emitter buffer to the collector buffer for bias voltages outside the NDC region. However, for bias voltages inside the NDC region (0.4–0.51 V) the contribution of the electron current to the capacitance becomes dominant. Notice also that the displacement current is null on the collector side where all the current is carried by the electron component. This indicates that the charge distribution in the collector side is frozen and that the width of the RTD capacitor has shrunk. This reduction of the RTD capacitor width explains the calculated increase of the RTD capacitance. The RTD capacitor is being supported by the accumulation charge in the emitter spacer on the left-hand side and the accumulation charge inside the RTD well on the right-hand side. This AC result is supported by the analysis of the DC distribution of the charge in the RTD shown in Figure 7.12. Indeed, the charge distribution in the RTD plotted in Figure 7.12 for different DC voltages around the bias point of maximum NDC indicates that the DC charge in the RTD

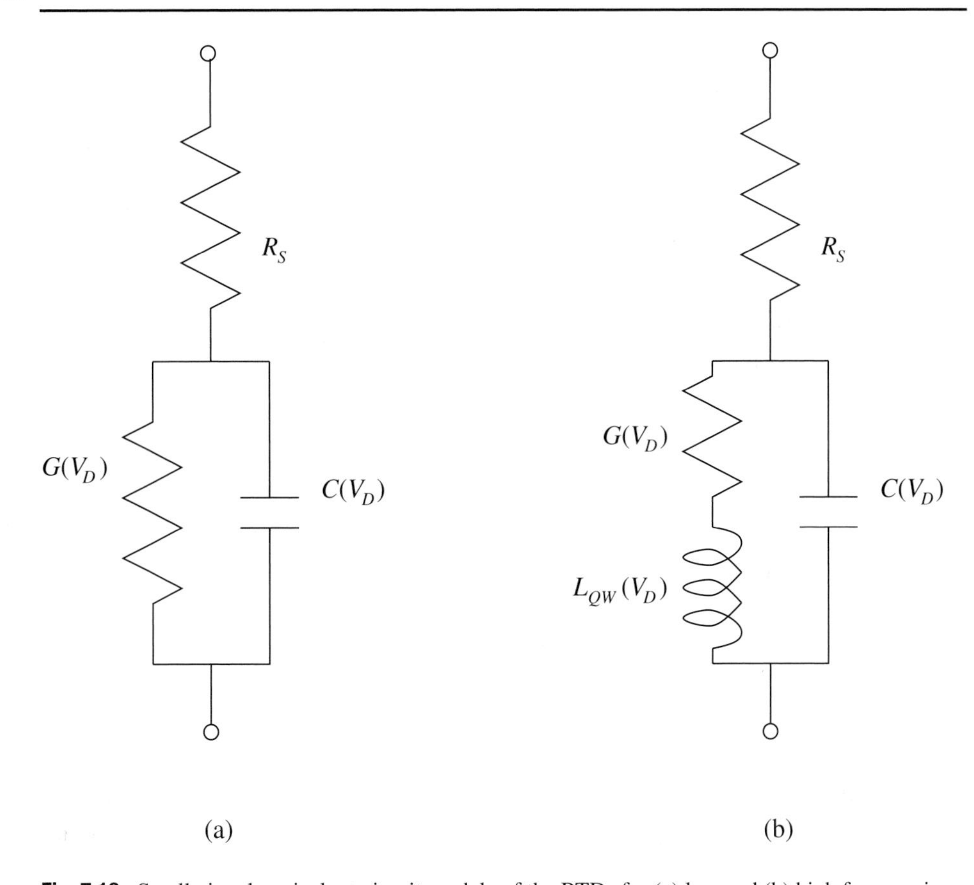

Fig. 7.13. Small-signal equivalent-circuit models of the RTDs for (a) low and (b) high frequencies.

quantum well varies in the opposite direction to that of the DC charge in the left-hand (emitter) spacer region. For example, to a differential decrease of the charge in the emitter spacer corresponds a differential increase of the charge in the quantum well.

The rapid increase of the capacitance in the region of maximum NDC predicted by this AC theory is in agreement with the experimental C–V characteristic reported by Sammut and Cronin [5] and with the experimental results presented in Section 7.9.

7.8 High-frequency response of the RTD

We have seen that at low frequencies the susceptance of the RTD in the NDC region could be modeled by a capacitor and a negative resistor in parallel. At higher frequencies the finite velocity of the electrons in the device will limit the rate at which the RTD charge can redistribute itself inside the device given the instantaneous variation of the external AC voltage. This results in the device admittance exhibiting

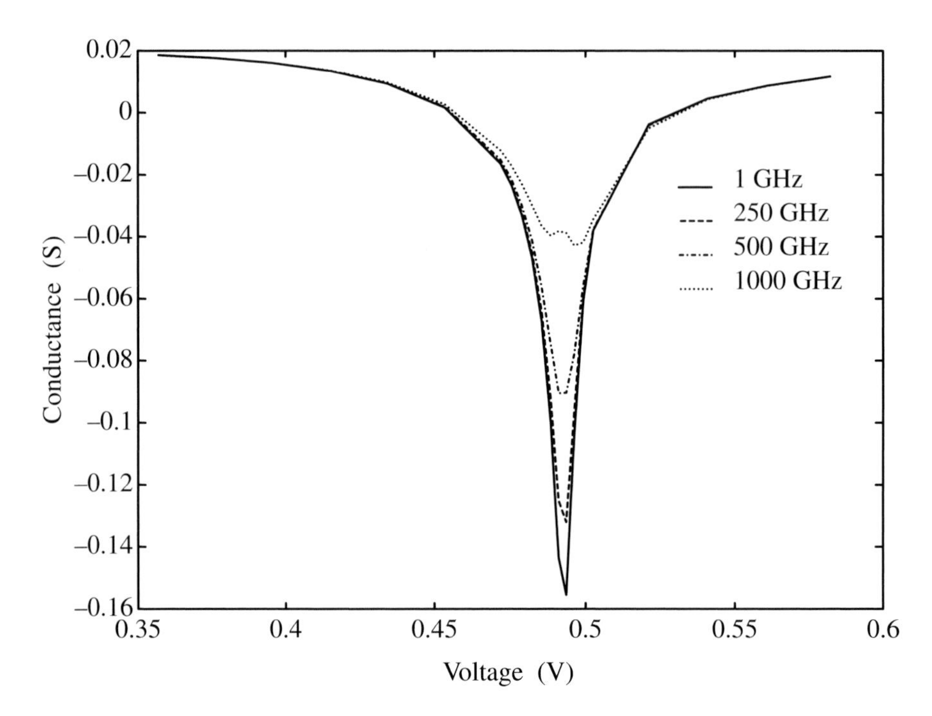

Fig. 7.14. RTD conductance versus bias voltage for 1 GHz (full line), 250 GHz (dashed line), 500 GHz (dashed-dotted line) and 1000 GHz (dotted line). (W.-R. Liou and P. Roblin, *IEEE Transactions on Electron Devices*, Vol. 41, No. 7, pp. 1098–1112, July 1994. ©1994 IEEE.)

a more complex frequency dependence at high frequencies.

At low frequencies the RTD is well represented by the circuit shown in Figure 7.13(a), where $G(V_D)$ is the RTD conductance, $C(V_D)$ the RTD capacitance and R_S the series resistance of the ohmic contact. We can rely on a frequency-dependent capacitor $C(f, V_D) = B(f, V_D)/(2\pi f)$ and $G(f, V_D)$ to study the range of validity of this model. At higher frequencies and in the bias range where these elements are frequency-dependent a more complex equivalent circuit is needed to represent the RTD admittance $Y(f, V_D) = G(f, V_D) + jB(f, V_D)$.

Figures 7.14, 7.15 and 7.16 show the real part (conductance) and imaginary part (susceptance) and capacitance of the RTD admittance versus the bias voltage for different frequencies. From these figures, we can see that the admittance is only strongly frequency-dependent in the region of maximum NDC and almost frequency-independent outside the NDC region.

We note in Figures 7.14 and 7.16 the decrease of both the RTD conductance and capacitance with increasing frequency. The reduction of the capacitance and the conductance with high frequency can be well modeled by the equivalent circuit of Figure 7.13(b), which uses a quantum inductance in series with the negative conductance of the RTD [6]. Such an equivalent circuit with a frequency-independent

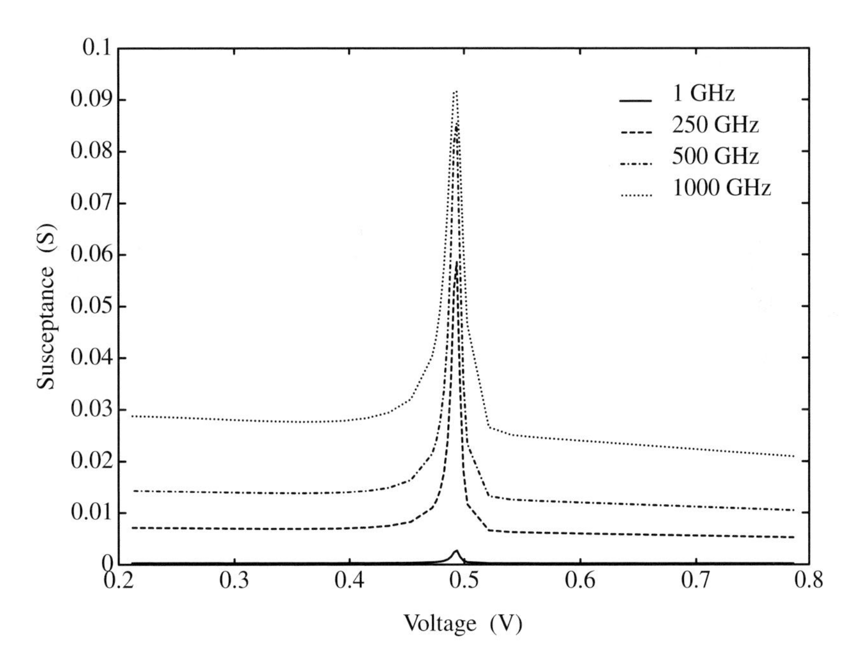

Fig. 7.15. RTD susceptance versus bias voltage for 1 GHz (full line), 250 GHz (dashed line), 500 GHz (dashed-dotted line) and 1000 GHz (dotted line). (W.-R. Liou and P. Roblin, *IEEE Transactions on Electron Devices*, Vol. 41, No. 7, pp. 1098–1112, July 1994. ©1994 IEEE.)

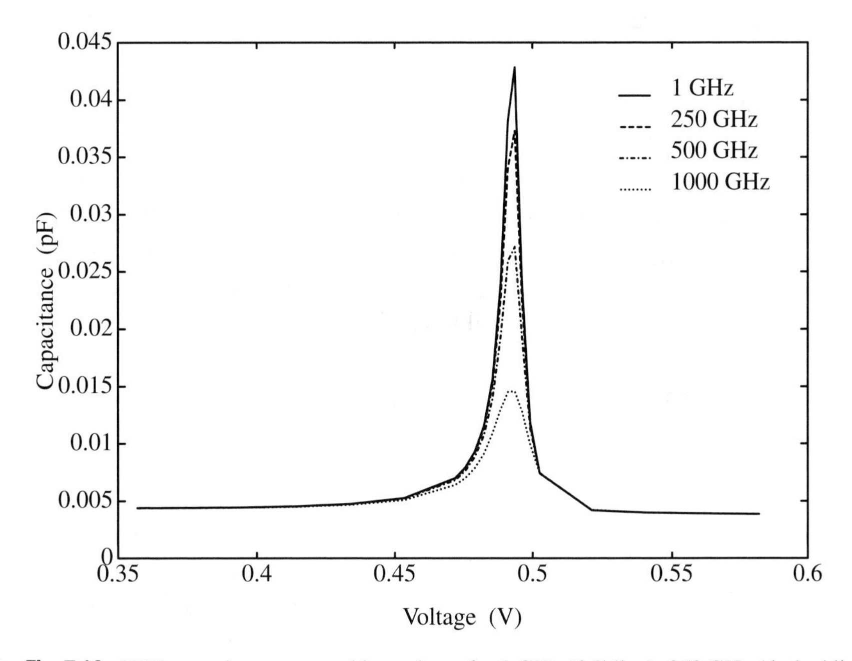

Fig. 7.16. RTD capacitance versus bias voltage for 1 GHz (full line), 250 GHz (dashed line), 500 GHz (dashed-dotted line) and 1000 GHz (dotted line). (W.-R. Liou and P. Roblin, *IEEE Transactions on Electron Devices*, Vol. 41, No. 7, pp. 1098–1112, July 1994. ©1994 IEEE.)

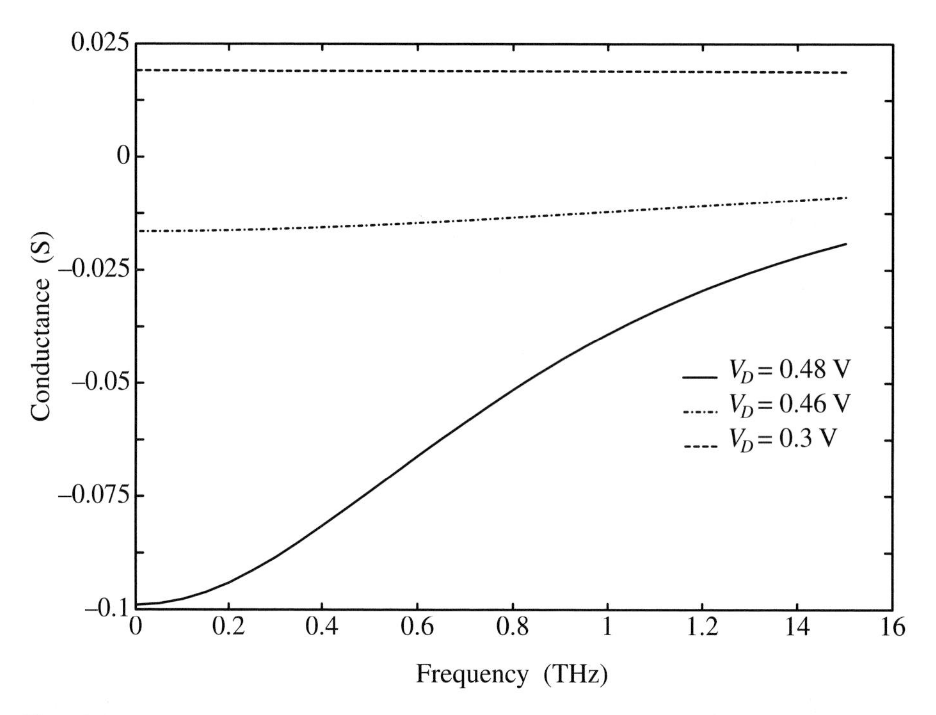

Fig. 7.17. Conductance versus frequency curve in the PDC region $V_D = 0.3$ V (dashed line) and in the NDC region 0.46 V (dashed-dotted line) and 0.48 V (full line). (W.-R. Liou and P. Roblin, *IEEE Transactions on Electron Devices*, Vol. 41, No. 7, pp. 1098–1112, July 1994. ©1994 IEEE.)

element is more physical and is applicable to large-signal simulations.

To investigate this effect further we show in Figures 7.17 and 7.18 the frequency dependence of the RTD conductance and susceptance when the RTD is biased in the NDC region at 0.46 and 0.48 V (full line) and in the positive differential region (PDC) at 0.3 V (dashed line).

Clearly the conductance for bias voltage in the NDC region is strongly frequency-dependent compared to the PDC region. Furthermore, departure of the conductance from its DC value arises at the relatively low frequency of 150 GHz and the 3-dB break frequency is 800 GHz. Note that here we define the break frequency as the frequency where the negative conductance is halved. We obtained a 3-dB break frequency of nearly 1 THz at 0.48 V. and 2 THz at 0.5 V in Figures 7.5 and 7.8, when we calculated the conductance without self-consistency. The self-consistent solution of the Schrödinger and Poisson equations for both DC and AC has reduced the 3-dB break frequency of the RTD conductance to a more realistic value. The RTD nevertheless remains a very high-frequency device.

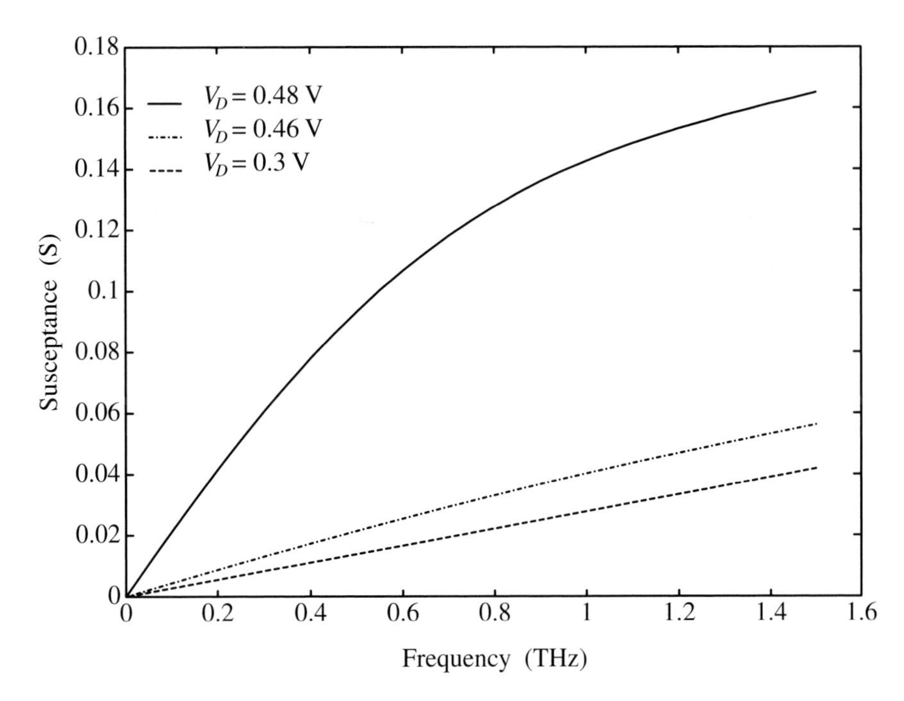

Fig. 7.18. Susceptance versus frequency curve, in the PDC region $V_D = 0.3$ V (dashed line) and in the NDC region 0.46 V (dashed-dotted line) and 0.48 V (full line). (W.-R. Liou and P. Roblin, *IEEE Transactions on Electron Devices*, Vol. 41, No. 7, pp. 1098–1112, July 1994. ©1994 IEEE.)

The maximum frequency of an RTD is given by the conventional formula [14]

$$f_{max}(V_D) = \frac{1}{2\pi C(f, V_D)} \left[\frac{-G(f, V_D)}{R_s} - G(f, V_D)^2 \right]^{1/2}. \tag{7.38}$$

At high frequencies when the conductance and capacitance are both frequency-dependent, the expression for f_{max} becomes a transcendental equation. One can verify that f_{max} is relatively weakly modified by the frequency dependence of the conductance and the susceptance. This is due to the fact that both the capacitance and conductance decrease with frequency.

As mentioned earlier the time-dependent tunneling theory presented can also handle large AC signals. To handle large signal excitations ($qV_{AC} \gg \hbar\omega$), more harmonics are required to solve the Schrödinger and Poisson equations self-consistently. Figures 7.19 and 7.20 show the variation of the large AC signal conductance $G(f, V_{AC}, V_D = 0.48 \text{ V})$ and susceptance $B(f, V_{AC}, V_D = 0.48 \text{ V})$ (see Equation (7.35) for a definition) with the frequency f for different AC voltages V_{AC}.

The negative conductance is observed to decrease in amplitude when the amplitude of the AC voltage V_{AC} increases. Also like for the small-signal response, the negative conductance decreases in amplitude when the frequency increases.

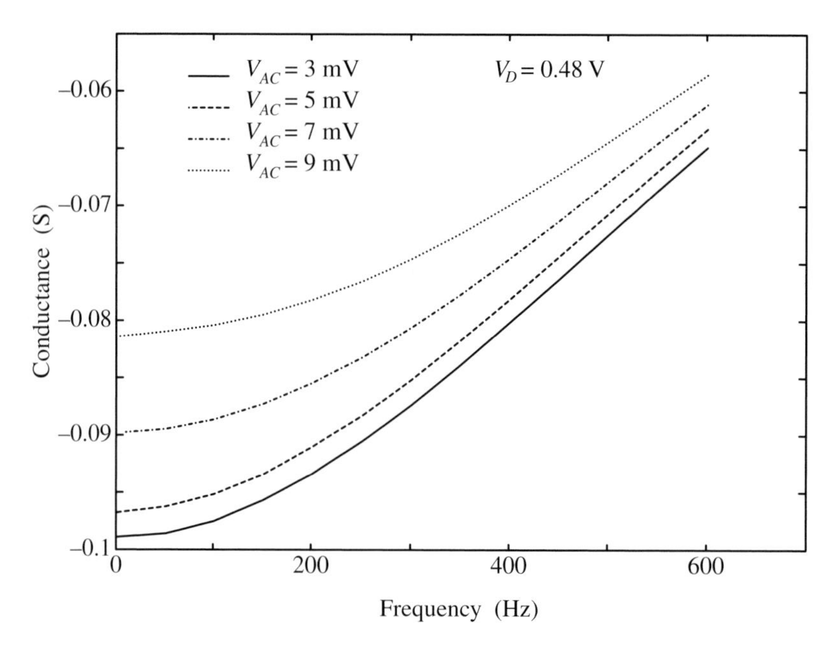

Fig. 7.19. Conductance $G(f, V_{AC})$ at $V_D = 0.48$ V as a function of frequency for different amplitudes of the AC signal V_{AC}. (W.-R. Liou and P. Roblin, *IEEE Transactions on Electron Devices*, Vol. 41, No. 7, pp. 1098–1112, July 1994. ©1994 IEEE.)

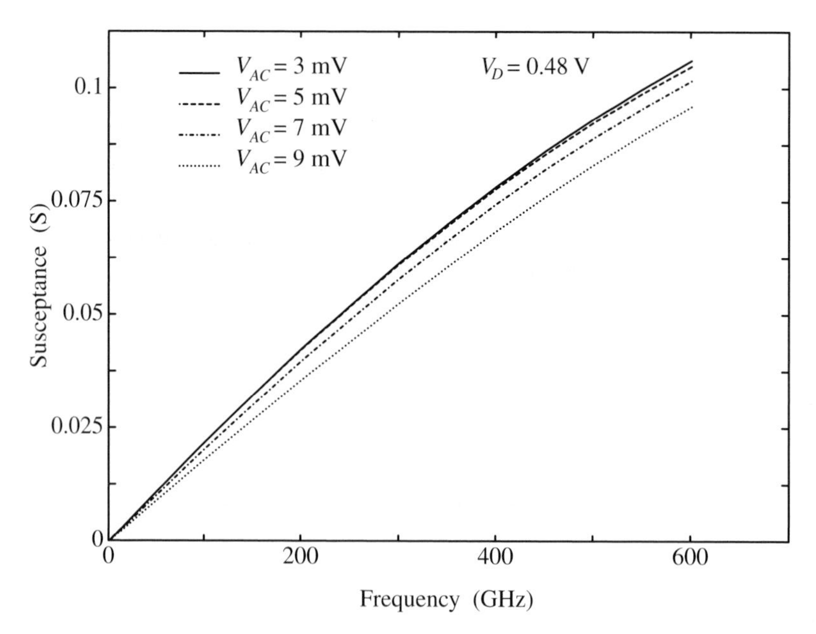

Fig. 7.20. Susceptance $B(f, V_{AC})$ at $V_D = 0.48$ V as a function of frequency for different amplitudes of the AC signal V_{AC}. (W.-R. Liou and P. Roblin, *IEEE Transactions on Electron Devices*, Vol. 41, No. 7, pp. 1098–1112, July 1994. ©1994 IEEE.)

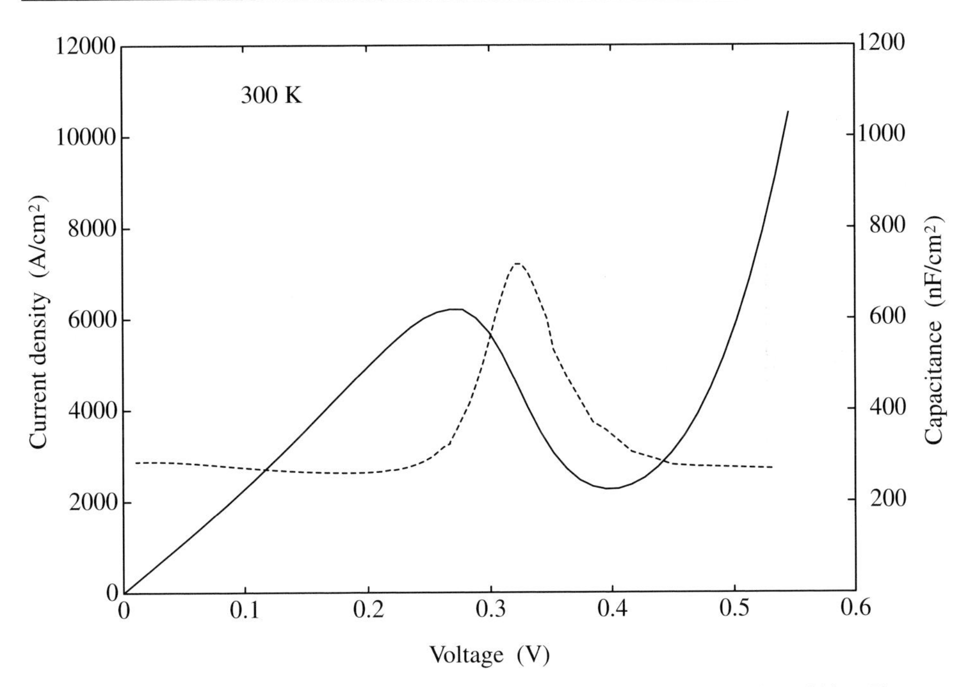

Fig. 7.21. Calculated current (full line) and capacitance (dashed line) as a function of bias. The capacitance has a maximum at the inflection-point voltage in the I–V characteristic corresponding to the maximum NDR region.

7.9 Microwave measurement of the C–V characteristics

We have presented in the previous sections a quantum simulation technique which allows one to study the high-frequency response of RTDs for both small- and large-signal AC voltages. This simulation technique calls for the solution of the Poisson and Schrödinger equations self-consistently using a harmonic balance technique. This quantum simulation technique guarantees that the total displacement plus electron current is continuous and allows us to predict effects such as the anomalous RTD capacitance effect.

To experimentally verify the peak in the C–V characteristic predicted by the theory described above (see Figure 7.21), we now discuss results obtained on an InP-based RTD. The RTD heterostructure considered was grown using molecular beam epitaxy (MBE) on an InP substrate. The undoped tunneling structure had two 45 Å InAlAs ([Al] = 0.48) barriers and a well consisting of 20 Å InGaAs, 20 Å of InAs, and 20 Å of InGaAs ([Ga] = 0.47). There is a 150-Å undoped InGaAs spacer on either side of the tunneling structure. The heavily doped InGaAs contact layers (Si at 10^{19} cm^3) are about 2000 Å thick. The I–V characteristic of a 6×60 μm^2 device is shown in Figure 7.22. To stabilize the device and keep it from oscillating [15] the DC I–V characteristic needs to be measured using a cascade probe station. The cascade probe

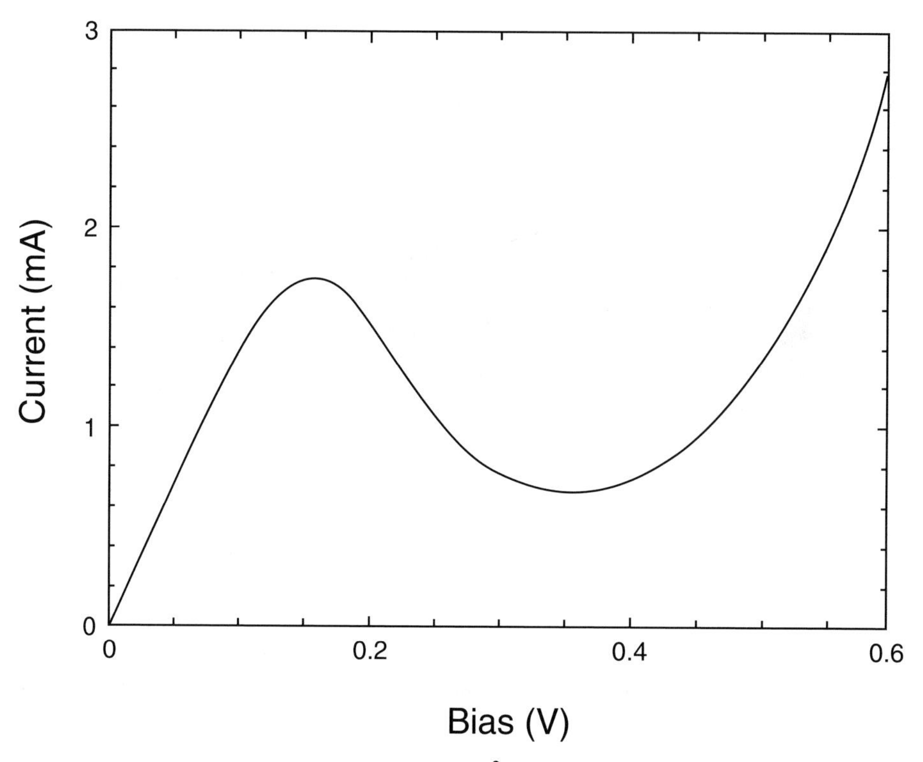

Fig. 7.22. The I–V characteristics of the $6 \times 60\ \mu m^2$ RTD measured with on-wafer probing. Note that the I–V characteristic is stable even in the NDR region.

is connected to a network analyzer via a bias tee. The network analyzer is used to characterize the diode at microwave frequencies and also provides a 50 Ω impedance termination which reduces the chances of RF and microwave instabilities in the DC I–V characteristic.

On-wafer one-port S-parameter data were taken using an HP8510 network analyzer [17] and are shown in Figure 7.23 for the $6 \times 60\ \mu m^2$ device. Each set of data represents a frequency scan from 0.1 to 18 GHz for each bias point. When the reflection coefficient is greater than 1, the device has a net NDR. Obtaining data in the NDC region can be difficult because the device may oscillate, even when it is in a 50-Ω system. The S-parameter data (at each bias point) can be fitted to the model shown in Figure 7.13(a). From this fitting, the device series resistance, capacitance and conductance are obtained for each bias point. A bias-independent series resistance of $R_S = 4\ \Omega$ is found to give a good fit. The dependence of the capacitance and conductance upon the diode bias voltage is shown in Figure 7.24. The conductance calculated from the I–V characteristic is also shown in Figure 7.24 (dashed line) and is in good agreement with the conductance (full line) extracted from the microwave data.

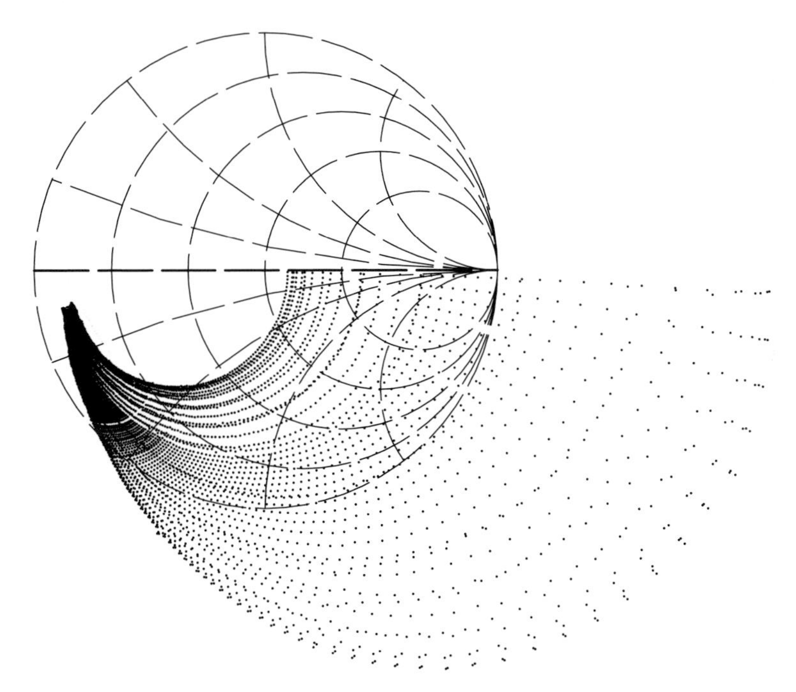

Fig. 7.23. The on-wafer, one-port S-parameter data are shown for different DC biases. When the reflection coefficient is outside of the Smith chart ($\Gamma \geq 1$), the device has a net NDR.

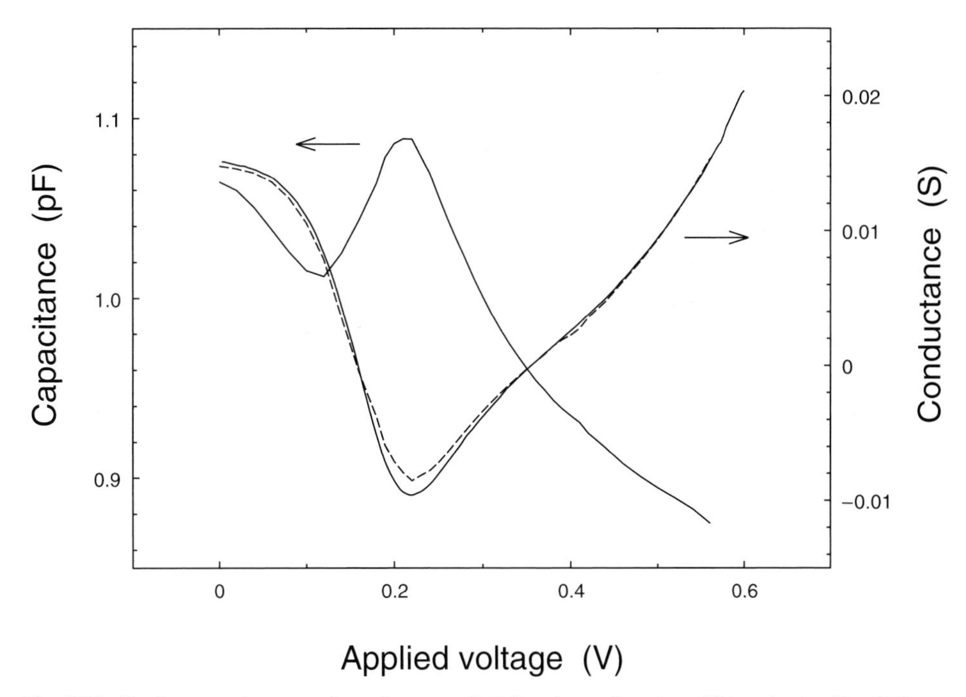

Fig. 7.24. Device capacitance and conductance (full lines) as a function of bias obtained by fitting the S-parameter data with the model shown in Figure 7.13(a). The conductance calculated from the $I-V$ characteristic is also shown (dashed line).

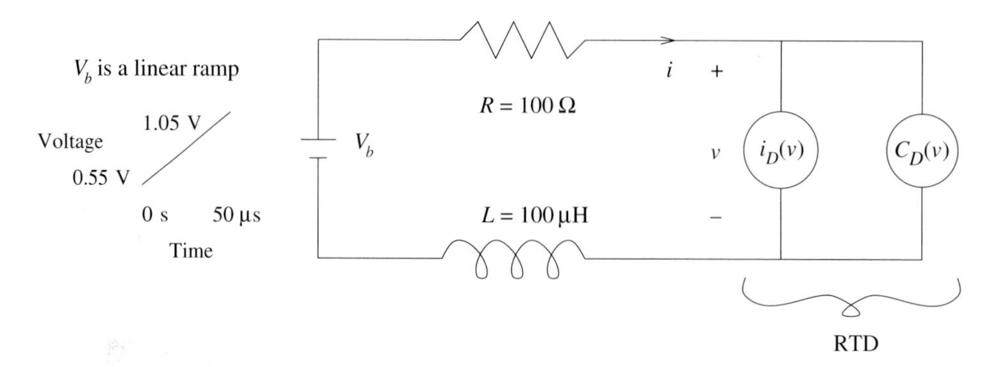

Fig. 7.25. Biasing circuit for an RTD.

The capacitance (dotted line) has a peak at the voltage that corresponds to the maximum in the device negative conductance. A similar peak in the capacitance has also been reported [5] for a double-barrier RTD made from the GaAs-based system.

7.10 DC bias instabilities

As discussed above the measurement of the I–V characteristics of an RTD can be hampered by instabilities in the NDC region.

Consider the biasing circuit shown in Figure 7.25. The inductor is a stray inductor associated with the bonding or contact. It is critically important for understanding the bias instabilities even if it is extremely small. The response of the circuit for constant C_D can be obtained by solving the non-linear differential equation [15]

$$\frac{d^2v}{dt^2} + \left(\frac{R}{L} + \frac{1}{C_D}\frac{di_D}{dv}\right)\frac{dv}{dt} + \frac{1}{LC_D}[v + Ri_D(v)] = \frac{V_b(t)}{LC_D}.$$

Analysis of this differential equation reveals that the circuit is DC stable when the circuit parameters satisfy the following conditions [16]:

$$R < \frac{V_v - V_p}{I_p - I_v},$$

$$\frac{L}{R} < C_D\left|\frac{di}{dv}\right|^{-1}.$$

If these stability conditions are not enforced, oscillations will start in the NDC region. The time-average DC measured will then be modified in the presence of these oscillations. To demonstrate this effect and to study the impact of the capacitance peak in the NDC, the transient simulation [17] of an RTD with its DC biasing circuit are presented in Figure 7.26. In these simulations the DC voltage source is swept from 0.55 to 1.05 V.

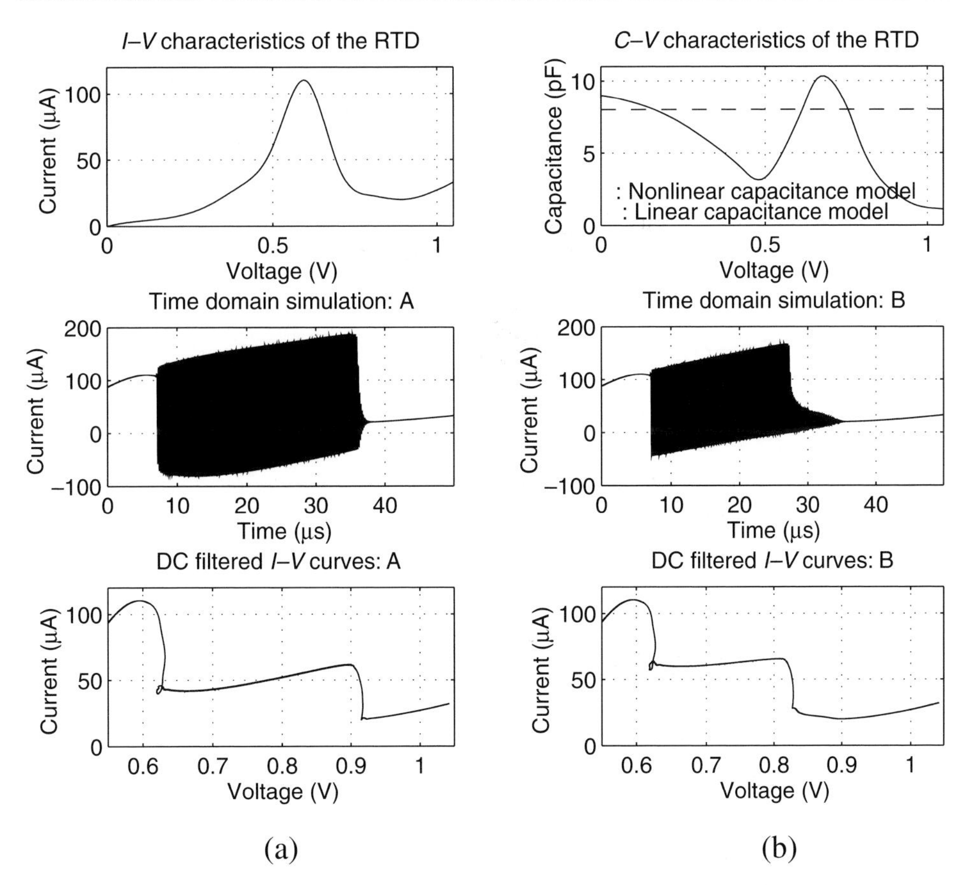

Fig. 7.26. Impact of the capacitance peak in the NDC in transient simulations of the RTD circuit shown in Figure 7.25. The top figures show the DC I–V characteristic (top left) and the C–V characteristic (top right) used for the RTD model of Figure 7.25. The middle and bottom figures compare the instantaneous (middle) and time-averaged (bottom) I–V characteristics obtained in a circuit simulator for an RTD without (left) and with (right) a capacitance peak in the NDR.

These simulation results demonstrate the development of oscillations in the NDC region when a sufficiently large parasitic inductor is placed in series with the device. The impact of the peak of the capacitance in the NDC on the time-average I–V characteristic and the microwave oscillations in the NDC region is also demonstrated.

7.11 Infrared response of quantum devices

Our discussion so far has focused on the high-frequency response of quantum devices for time-varying electrostatic potentials. In this section we generalize our discussion to that of the coupling of quantum devices to infrared electromagnetic waves. We will first briefly describe the physical wave-guide system under which the interactions can

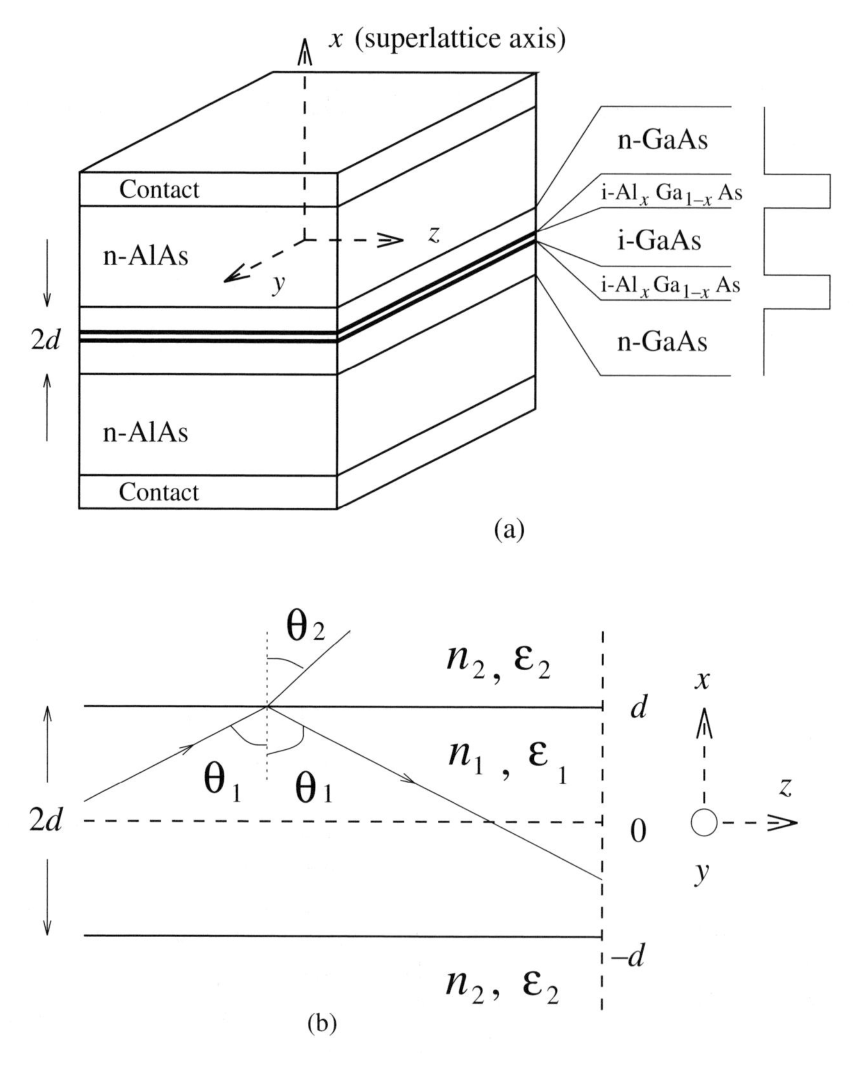

Fig. 7.27. (a) Layered structure/superlattice. (b) Wave propagation inside the active layer.

take place. We then study the coupling between the infrared electromagnetic field and the electrons, and present some simulation results. Finally we briefly describe infrared quantum cascade lasers.

7.11.1 Modeling the infrared wave-guide

The geometry of the test heterostructure used in this chapter is shown in Figure 7.27(a). The device consists of an active layer of thickness $2d$ sandwiched between two layers of n-type AlAs. The active layer itself consists of different layers as shown on the right of Figure 7.27(a). The conduction-band bottom edge is also plotted, and we clearly see that a quantum well is formed due to the bandgap discontinuity between GaAs and $Al_xGa_{1-x}As$. Such a simple RTD structure is used here instead of the more complex superlattice and quantum cascade structures to demonstrate our approach.

A cross-section of the active layer is outlined in Figure 7.27(b). This layer, which we shall refer to as 'the core', is pictured as a layer of dielectric material (actually a semiconductor) of effective refractive index n_1 and thickness $2d$, sandwiched between two layers of material of effective refractive index n_2, where $n_1 > n_2$. Those two layers, which we shall refer to as 'the cladding', are treated as being semi-infinite, and the effective refractive index is considered to be constant everywhere except at the boundaries between different layers where it changes abruptly from one value to another. In such a structure a wave can be made to propagate down the active layer through total reflections. We only consider TM waves ($H_z = 0$ and $E_z \neq 0$) which will couple with the active device. Solving Maxwell equations, assuming no electron current and charges in the slab wave-guide, we can calculate the first TM mode potential vector in the core to be:

$$\mathbf{A} = -\frac{A_0}{\omega^2 \epsilon_1}[k_{1z}\cos(k_{1x}x)\sin(\omega t - k_{1z}z)\mathbf{x} + k_{1x}\sin(k_{1x}x)\cos(\omega t - k_{1z}z)\mathbf{z}], \quad (7.39)$$

with $k_{1z} = k_1 \sin\theta_1 = n_1 k_0 \sin\theta_1$ obtained from the transcendental equation:

$$\tan\left[n_1 k_0 \cos\theta_1 d - \frac{n\pi}{2}\right] = \left[\frac{(n_1^2 - n_2^2) - n_1^2\cos^2\theta_1}{n_1^2\cos^2\theta_1}\right]^{1/2}. \quad (7.40)$$

The amplitude A_0 of the TM mode can be related to the total power flow P obtained from integration of the Poynting vector over x (core and cladding):

$$P = \int_{-\infty}^{\infty} \frac{1}{2}\mathbf{E} \times \mathbf{H}^*\, dx$$
$$= \frac{A_0^2 k_{1z}}{2\omega}\left\{\frac{1}{\epsilon_1}\left[d + \frac{\sin(2k_{1x}d)}{2k_{1x}}\right] + \frac{1}{\epsilon_2}\frac{\cos^2(k_{1x}d)}{\gamma_2}\right\}\mathbf{z} \quad (\mathrm{W/m}), \quad (7.41)$$

with P, the optical power per unit width (in the y direction), flowing only in the z direction. In a simulation P can be used to calculate the value of A_0 from Equation (7.41).

Note that a self-consistent solution of the potential vector might be required, as the assumption of zero charges and potential in the core is not satisfied by the semiconductor quantum devices. An effective dielectric constant should be sufficient to account for the presence of a superlattice in the core region. The analysis presented here stresses the need for varying the active region across the core of the dielectric wave-guide so as to account for the varying TM field.

7.11.2 Coupling of quantum transport with infrared radiation

To analyze the impact of the infrared radiation emitted or absorbed on the quantum transport, we shall limit our analysis to the 'first quantization picture' in which the electron is quantized, but the electromagnetic fields are not. This is quite justifiable for the relatively large infrared fields present in lasers, where the large stimulated emission dominates the spontaneous emission. The electron Hamiltonian is therefore given by

$$\left[H_{el}|\psi\rangle - \frac{q}{m}\mathbf{A}(x, z, t)\cdot\mathbf{p}\right]|\psi\rangle = \imath\hbar\frac{\partial}{\partial t}|\psi\rangle,$$

with $\mathbf{A}(x, z, t)$ the potential vector of the TM wave introduced in the previous section. Note that the Hamiltonian used here assumes that the electromagnetic field strength is not excessively large so that we can neglect the term $(q^2/2m)A^2$ of the exact electron–photon Hamiltonian.

To solve this Schrödinger equation we expand the electron wave-function in terms of generalized Wannier functions

$$|\psi\rangle = \sum_n \int f(\mathbf{k}_\perp, n, t)|\mathbf{k}_\perp, n\rangle\, d\mathbf{k}_\perp. \tag{7.42}$$

In the above expression $|\mathbf{k}_\perp, n\rangle$ is the generalized Wannier function (electron state) and $f(\mathbf{k}_\perp, n, t)$ is the Wannier envelope function. The generalized Wannier state $|\mathbf{k}_\perp, n\rangle$ is a hybrid state, being a localized one-dimensional state centered around lattice site n along the superlattice axis and a quasi-Bloch state with wave-vector $\mathbf{k}_\perp$ in the perpendicular direction.

An exact solution can be obtained in terms of a Fourier series expansion. When the quantum structure is located at the center of the slab wave-guide we have for the case of an expansion up to the second harmonics:

$$\begin{aligned}
f(\mathbf{k}_\perp, n, t) = {} & f_0(n)\delta(\mathbf{k}_\perp - \mathbf{k}_{\perp 0})\exp\left(-j\frac{E_0}{\hbar}t\right) \\
& + f_{-1}(n)\delta(\mathbf{k}_\perp - \mathbf{k}_{\perp 0} + \mathbf{k}_{1z})\exp(j\omega t)\exp\left(-j\frac{E_0}{\hbar}t\right) \\
& + f_{+1}(n)\delta(\mathbf{k}_\perp - \mathbf{k}_{\perp 0} - \mathbf{k}_{1z})\exp(-j\omega t)\exp\left(-j\frac{E_0}{\hbar}t\right) \\
& + f_{-2}(n)\delta(\mathbf{k}_\perp - \mathbf{k}_{\perp 0} + 2\mathbf{k}_{1z})\exp(2j\omega t)\exp\left(-j\frac{E_0}{\hbar}t\right) \\
& + f_{+2}(n)\delta(\mathbf{k}_\perp - \mathbf{k}_{\perp 0} + 2\mathbf{k}_{1z})\exp(-2j\omega t)\exp\left(-j\frac{E_0}{\hbar}t\right)
\end{aligned}$$

where the envelope functions $f_p(n)$ are obtained from the Hamiltonian equation [18]:

$$\begin{bmatrix} -\frac{A}{2} & \lambda & 0 & 0 & 0 \\ -\lambda & -\frac{A}{2} & \lambda & 0 & 0 \\ 0 & -\lambda & -\frac{A}{2} & \lambda & 0 \\ 0 & 0 & -\lambda & -\frac{A}{2} & \lambda \\ 0 & 0 & 0 & -\lambda & \frac{A}{2} \end{bmatrix} \begin{bmatrix} f_{-2}(n-1) \\ f_{-1}(n-1) \\ f_0(n-1) \\ f_{+1}(n-1) \\ f_{+2}(n-1) \end{bmatrix} = \begin{bmatrix} (E_0 - 2\hbar\omega) f_{-2}(n) \\ (E_0 - \hbar\omega) f_{-1}(n) \\ E_0 f_0(n) \\ (E_0 + \hbar\omega) f_{+1}(n) \\ (E_0 + 2\hbar\omega) f_{+2}(n) \end{bmatrix}$$

$$- \begin{bmatrix} \left(\mathcal{E}_c(na) + \frac{\hbar^2 |\mathbf{k}_{\perp 0} - 2\mathbf{k}_{1z}|^2}{2m^*} + A - qV(na) \right) f_{-2}(n) \\ \left(\mathcal{E}_c(na) + \frac{\hbar^2 |\mathbf{k}_{\perp 0} - \mathbf{k}_{1z}|^2}{2m^*} + A - qV(na) \right) f_{-1}(n) \\ \left(\mathcal{E}_c(na) + \frac{\hbar^2 k_{\perp 0}^2}{2m^*} + A - qV(na) \right) f_0(n) \\ \left(\mathcal{E}_c(na) + \frac{\hbar^2 |\mathbf{k}_{\perp} + \mathbf{k}_{1z}|^2}{2m^*} + A - qV(na) \right) f_{+1}(n) \\ \left(\mathcal{E}_c(na) + \frac{\hbar^2 |\mathbf{k}_{\perp} + 2\mathbf{k}_{1z}|^2}{2m^*} + A - qV(na) \right) f_{+2}(n) \end{bmatrix}$$

$$- \begin{bmatrix} -\frac{A}{2} & -\lambda & 0 & 0 & 0 \\ \lambda & -\frac{A}{2} & -\lambda & 0 & 0 \\ 0 & \lambda & -\frac{A}{2} & -\lambda & 0 \\ 0 & 0 & \lambda & -\frac{A}{2} & -\lambda \\ 0 & 0 & 0 & \lambda & -\frac{A}{2} \end{bmatrix} \begin{bmatrix} f_{-2}(n+1) \\ f_{-1}(n+1) \\ f_0(n+1) \\ f_{+1}(n+1) \\ f_{+2}(n+1) \end{bmatrix},$$

with the electron–photon coupling constant $\lambda = |q| A_0 k_{1z} \hbar / (4\omega^2 \epsilon_1 m^* a)$ [18]. The results presented here hold for the case of a single tight-binding Wannier band of the form

$$\mathcal{E}(\mathbf{k}'_{\perp}, n) = \mathcal{E}_c(na) + \frac{\hbar^2 k_{\perp}^2}{2m^*} + A[1 - \cos(k_x a)], \tag{7.43}$$

where $\mathcal{E}_c(na)$ is the energy of the conduction-band edge at the lattice site n and $A = \hbar^2/(m^* a^2)$ with m^* the effective mass of the electron.

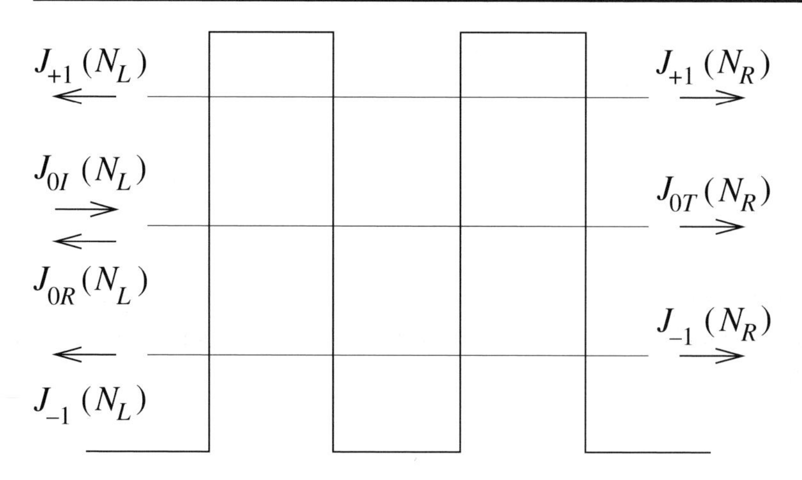

Fig. 7.28. The different currents for an electron incident on the left side.

7.11.3 Optical absorption/emission coefficient

The Schrödinger equation presented above allows us to calculate the quantum ballistic trajectory of an electron in the presence of an induced or applied infrared radiation field. One can verify both theoretically and numerically that the electron current is conserved:

$$\sum_b J_b(N_L) = \sum_b J_b(N_R),$$

with J_b the current associated with the harmonics b. Also conserved is the power flow between the electron and photon via $H_{int} = -q\mathbf{p} \cdot \mathbf{A}/m$:

$$0 = -\sum_b \left[\frac{J_b(N_L)}{2} - \frac{J_b(N_R)}{2} \right] \delta(\mathbf{0}_\perp)(b\hbar\omega) + \langle\psi|\frac{\partial H_{int}}{\partial t}|\psi\rangle \tag{7.44}$$

In Figure 7.28 the different current density components are outlined for an electron incident on the left-hand side of the quantum well.

Writing $J_b(N_L) = -J_{bR}$ and $J_b(N_R) = J_{bT}$ we define three attenuation factors:

$$\alpha(E_{0x}, E_{0\perp}, P) = -\frac{1}{\delta(\mathbf{0}_\perp)}\langle\psi|\frac{\partial H_{int}}{\partial t}|\psi\rangle = \alpha_a(E_{0x}, E_{0\perp}, P) - \alpha_e(E_{0x}, E_{0\perp}, P)$$

where α_a and α_e are coefficients for the absorption and the emission processes, respectively, and are defined as

$$\alpha_a(E_{0x}, E_{0\perp}, P) = \frac{1}{P}\sum_{b>0}(J_{bR} + J_{bT})\frac{b\hbar\omega}{2},$$

$$\alpha_e(E_{0x}, E_{0\perp}, P) = \frac{1}{P}\sum_{b<0}(J_{bR} + J_{bT})\frac{b\hbar\omega}{2}.$$

We note that both α_a and α_e take positive values. The total absorption coefficient for injection on the left side is obtained by integration over k_x and $\mathbf{k}_\perp$:

$$\alpha_{total,L}(P) = \frac{1}{4\pi^3}\int_0^{\pi/a} dk_x \int_{-\pi/a}^{\pi/a}\int_{-\pi/a}^{\pi/a} d\mathbf{k}_\perp \frac{1}{1+\exp\left(\frac{E_{F,L}-E_0}{k_BT}\right)}\alpha(E_{0x}, E_{0\perp}, P). \tag{7.45}$$

The total absorption coefficient is finally given as the sum of the absorption coefficients for electrons incident on the left-hand side and for electrons incident on the right-hand side, that is,

$$\alpha_{total}(P) = \alpha_{total,L}(P) + \alpha_{total,R}(P). \tag{7.46}$$

This total absorption coefficient becomes negative when the device is amplifying the optical field. The laser condition is therefore given by the usually lasing condition

$$0 = \alpha_{total}(P) + \alpha_{loss} + \frac{1}{L}\ln\frac{1}{R} \tag{7.47}$$

with α_{loss} the sum of the unaccounted losses per unit length (contacts, cladding), L the resonator length, and R the reflectivity of the mirrors. The solution of Equation (7.47) yields therefore the power flow for both the forward and backward waves since in our approach $\alpha_{total}(P)$ is dependent on the optical power P. Note that our approach also treats the photon semiclassically but the electron transport is treated fully quantum mechanically.

Simulation verification

Consider a RTD (two periods of a superlattice) with a barrier of six monolayers and a well of twelve monolayers. The I–V characteristic for this structure is shown in Figure 7.29(a). In Figure 7.29(b) the transmission coefficient is plotted for each harmonic for a potential of 0 V across the quantum region. We observe in Figure 7.29(b) that the transmission probability T_0 presents two peaks, each of which corresponds to a resonant energy level of the quantum well. For the other harmonics the transmission coefficients T_{-1} and T_{+1} present two pairs of peaks around each peak of T_0. One of the peaks in each pair occurs at the same longitudinal energy as the peak of T_0. The other peak occurs at energy values $\hbar\omega$ above (for the case of T_{-1}) or below (for the case of T_{+1}) the energy for which the peak of T_0 occurs. These new peaks arise from the stimulated emission (T_{-1}, full line) and stimulated absorption (T_{+1}, dotted line) of infrared photons.

The attenuation factors are shown in Figure 7.30. As Figure 7.30 clearly outlines, *there is a voltage range for which emission dominates and the total attenuation is negative*. Maximum emission is observed around 0.35 V. However, the value of the negative attenuation coefficient (gain) is very small ($-\alpha_{total} = 10^{-3}$ m^{-1}) since only a narrow active region of about 100 Å out of a 10 μm core region was modeled. The

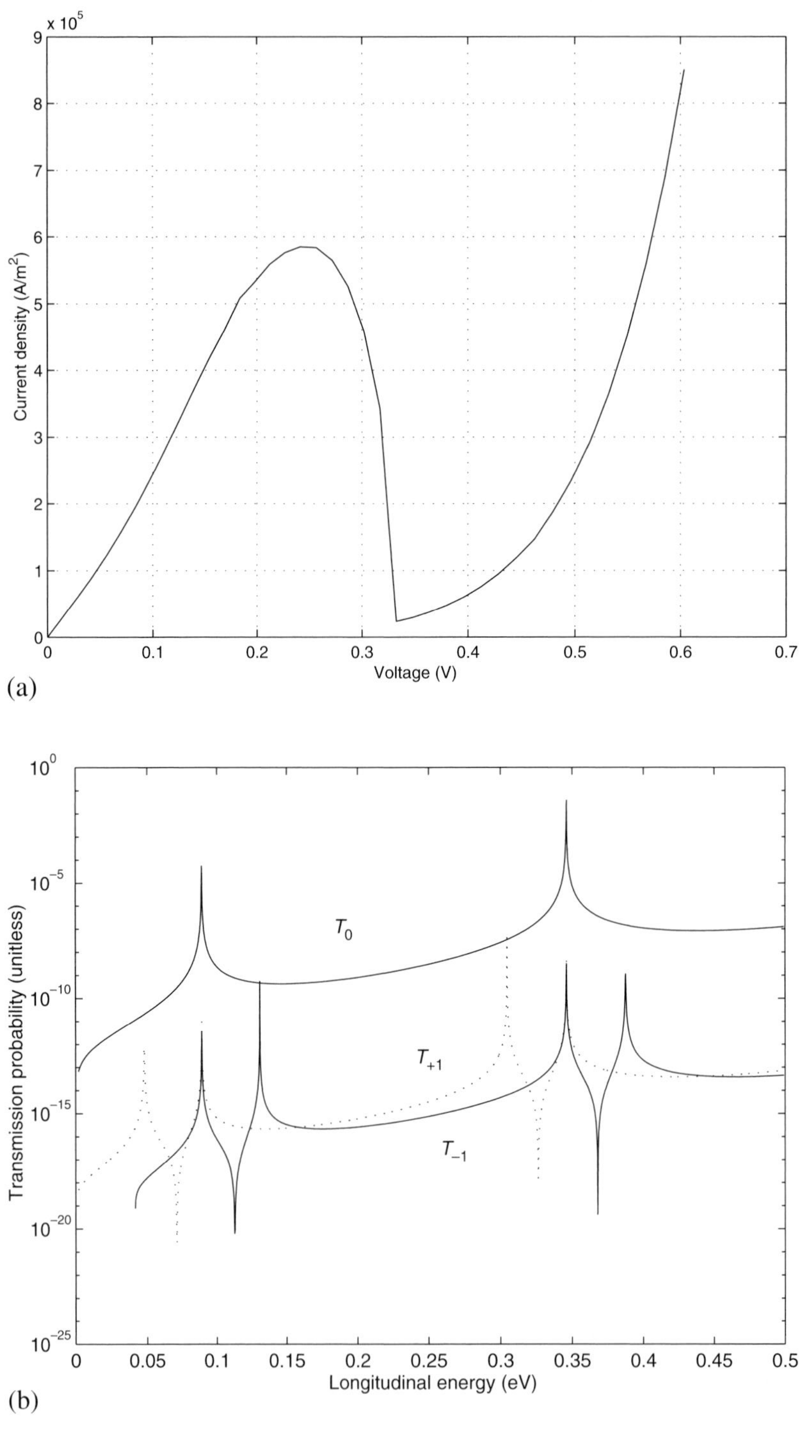

Fig. 7.29. (a) I–V characteristic. (b) Transmission coefficients. The potential across the quantum region is 0 V.

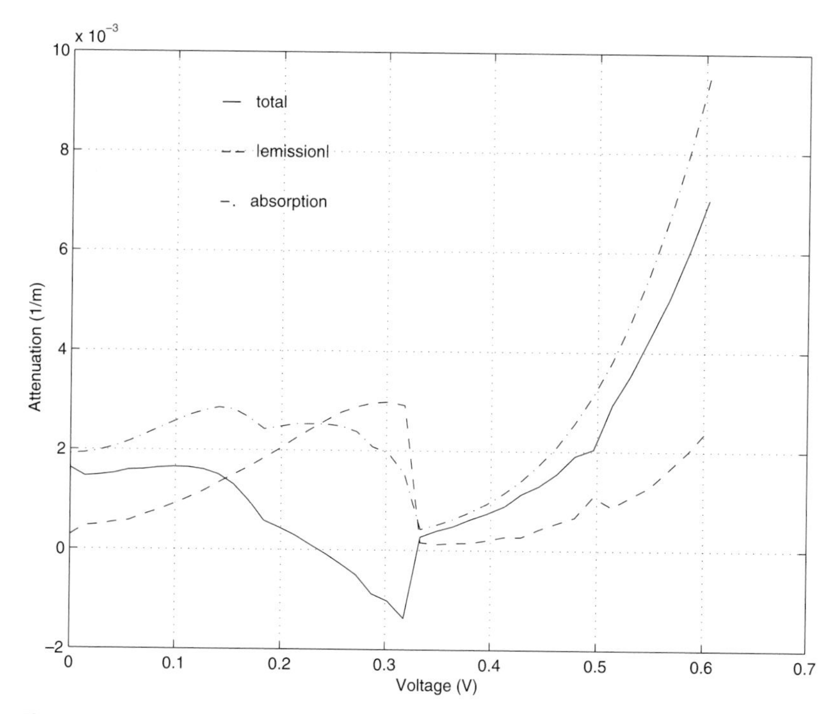

Fig. 7.30. Total (full line), emission (dashed line) and absorption (dashed-dotted line) power attenuation factors.

filling factor is therefore about 1000. If the active region is filled with such RTD structures, the absorption coefficient scaled by the filling factor gives about a total gain of $1\ \text{m}^{-1} = 10^{-2}\ \text{cm}^{-1}$. The emission and absorption coefficients remain small. Only one resonant tunneling energy level was involved in the stimulated emission and absorption processes in Figure 7.30. We consider now a structure where, as in a laser, the infrared photon energy is comparable to the difference between the two resonant energy levels. As is shown in the middle plot of Figure 7.31, when the photon energy exactly satisfies $\hbar\omega = E_2 - E_1$, the emission coefficient T_{-1} (full line) then reaches a resonant emission peak, because its emission peak is aligned with the second resonance (E_2) in T_0 (dashed-dotted line). A smaller transmission peak is observed when the photon energy is slightly smaller (left-hand plot) or larger (right-hand plot) than $E_2 - E_1$. Similar results hold for the resonant absorption (dotted lines).

7.11.4 Quantum cascade laser

In this section we now describe the quantum cascade laser which has been developed for the generation of laser light in the infrared from 4 to 13 μm [19,20]. This laser relies on the quantum engineering of the device to create the laser levels inside the conduction bands. The principle of the quantum cascade laser is described

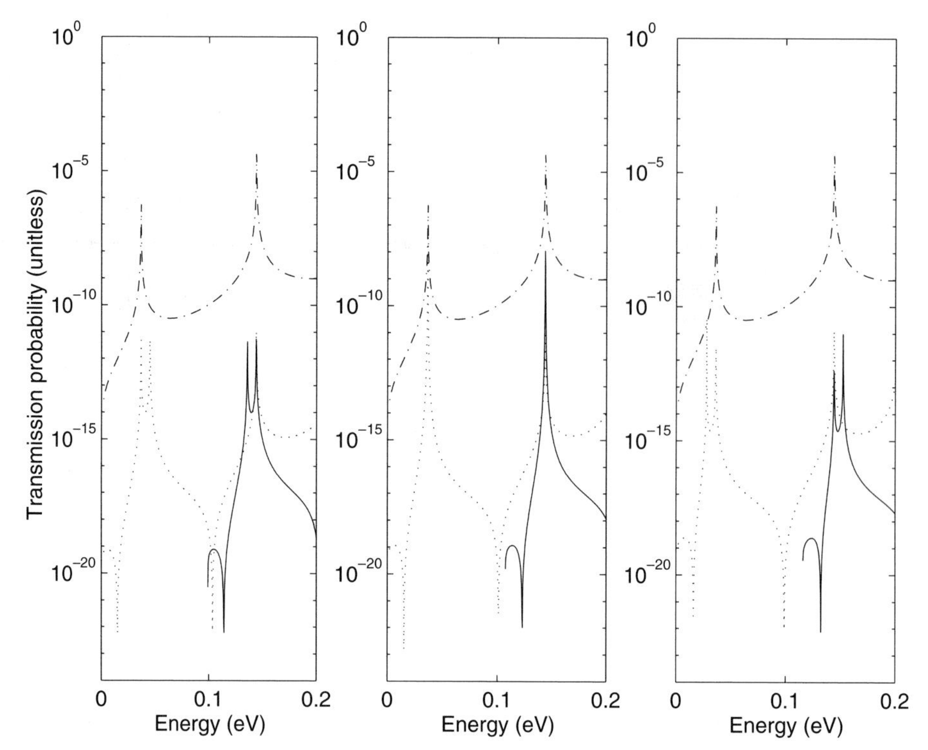

Fig. 7.31. Transmission coefficients for a structure where the infrared photon energy is comparable to the difference between the first two resonant energy levels (E_1 and E_2). The emission coefficient T_{-1} (full line), the absorption coefficient T_{+1} (dotted line) and the direct transmission coefficient T_0 (dashed-dotted line) are plotted for cases where the photon energy $\hbar\omega$ is smaller than (left), equal to (center) and larger than (right) the energy difference $E_1 - E_2$.

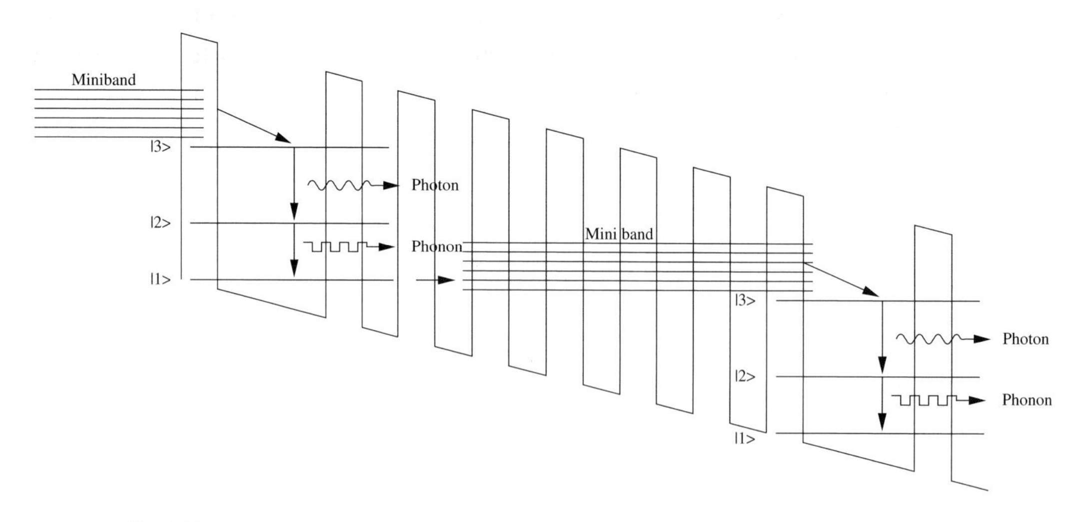

Fig. 7.32. Schematic description of the operation of the quantum cascade laser.

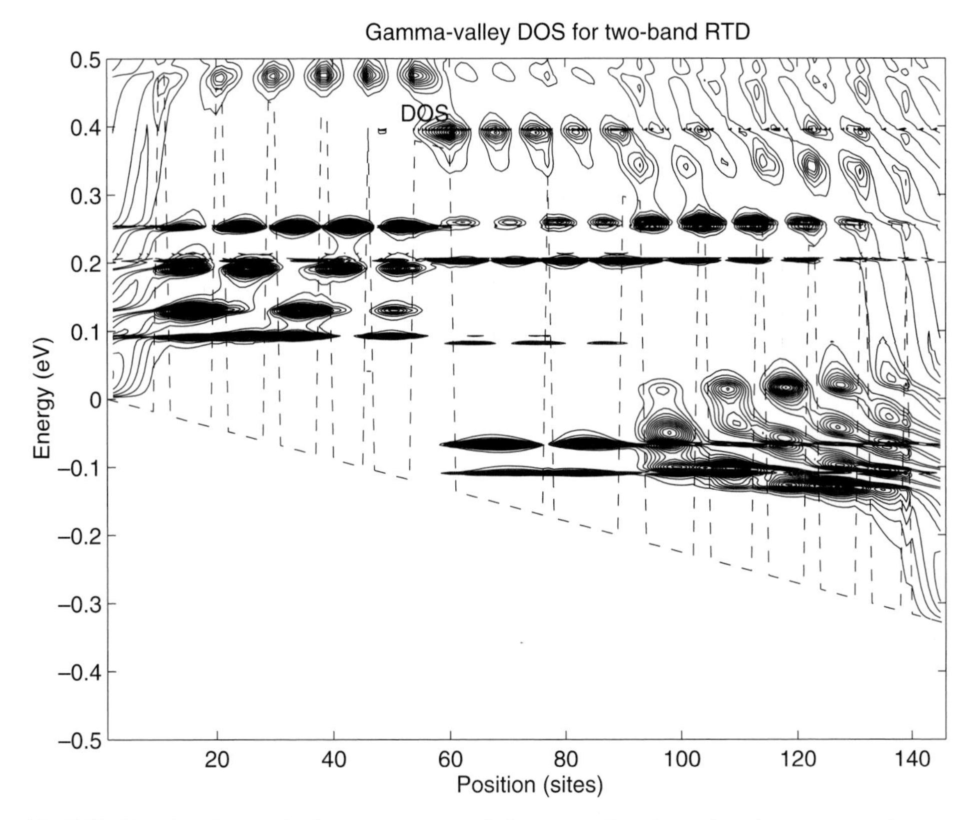

Fig. 7.33. Density of states in the quantum cascade laser revealing the various laser states and minibands.

schematically in Figure 7.32. The quantum laser involves multiple wells where the photons are generated. These wells are separated by a superlattice to create a miniband between the wells with regions of forbidden energies around the state $|3\rangle$. The electrons originating from the previous well are injected via the miniband into level $|3\rangle$. The electrons relax from the state $|3\rangle$ to the energy state $|2\rangle$ by radiating infrared photons of energy $\hbar\omega = E_3 - E_2$. The electrons in state $|2\rangle$ must then be rapidly transfered to state $|1\rangle$ and the miniband so as to establish a population inversion between state $|3\rangle$ and $|2\rangle$. This is achieved by selecting the energy spacing between state $|2\rangle$ and $|1\rangle$ to correspond to the optical phonon energy $\hbar\omega_{LO} = E_2 - E_1$.

We show in Figure 7.33 the density of states calculated for a quantum cascade structure. The density of states reveals the various laser states and minibands in the well and superlattice of the quantum cascade laser, respectively.

The advantage of the quantum cascade laser is that a large band of infrared frequencies can be realized with the same material by engineering the superlattice. Wavelengths from 4 to 13 μm have been realized. The power obtained with such a laser is also very high because for a laser with N wells, an electron will emit N

photons. The power generated can be expressed [21] by

$$P = \eta N \hbar \omega \frac{I - I_{threshold}}{q},$$

where η is the collection efficiency, I is the laser current and $I_{threshold}$ is the current required to create a sufficiently large population for lasing to take place. Note that the threshold current is very high (on the order of kA/cm^2) in a quantum cascade laser due to the relatively short lifetime (on the order of tenth of picoseconds) of the state $|3\rangle$. The power generated by these lasers is typically on the order of 100 mW. The development of sources working at even longer wavelengths (far infrared) remains a technical challenge.

7.12 Conclusion

In this chapter we have studied several canonical problems of quantum devices subjected to external time-varying excitations.

Using an exact wave-packet solution we established the condition for conducted radiation emission or absorption of the electron undergoing Bloch oscillations. We also derived a methodology for obtaining the steady-state response of quantum devices (RTD, quantum cascade) subjected to harmonic electrostatic or electromagnetic excitations.

A direct measure of the speed of devices is established by the decrease of their gain or negative resistance at high frequencies. For the RTD, the resonant tunneling process is intrinsically limited to terahertz frequencies as a result of the inertia associated with the electron mass. This effect is measured by the so-called quantum inductance.

However, we verified that in practice the limitation of quantum devices arises from their capacitances. Indeed, transport in quantum devices is fundamentally space-charge-limited. In fact we saw that the self-consistency between the field and transport equations must be enforced for both DC and AC, as this led to the prediction of the experimentally observed anomalous-capacitance effect in RTDs.

The final topic of this chapter was the generalization of steady-state analysis to quantum devices subjected to electromagnetic excitations. It is important to note that in this chapter our modeling of the electromagnetic excitations relied on a classical electromagnetic field. This classical field approximation is justified by the fact that our analysis was indeed targeted to the practical case of coherent (classical-like) electromagnetic fields (see the Glauber state in Section 5.2.1) which are achieved in lasers. In such classical fields the number of photons is very large, and stimulated emission or absorption is therefore largely dominant over spontaneous emission.

Spontaneous emission of photons or phonons remains, however, of critical importance as it is the key force permitting the relaxation of quantum or semiclassical systems toward equilibrium, as is discussed respectively in both Chapters 6 and 9.

The reader might then wonder how incoherent phonon scattering and coherent photon scattering could be simultaneously considered? As can be inferred by the inquisitive reader, the impact of phase-breaking scattering on the *steady-state* response of time-dependent systems can be accounted for with the use of a self-energy (see Chapter 6) calculated using the *steady-state* impulse response of the time-dependent quantum system.

7.13 Bibliography

[1] P. Roblin and M. W. Muller, 'Time-dependent tunneling and the injection of coherent zener oscillations', *Semiconductor Science and Technology*, Vol. 1 No. 3, pp. 218–225, 1986.

[2] M. W. Muller, P. Roblin and D. L. Rode, 'Proposal for a terahertz Zener oscillator', in *The Physics of Submicron Structures*, ed. H. L. Grubin, Plenum, New York, 1982.

[3] W.-R. Liou and P. Roblin, 'High frequency simulation of resonant tunneling diodes', *IEEE Transactions on Electron Devices*, Vol. 41, No. 7, pp. 1098–1112, July 1994.

[4] L. L. Chang, L. Esaki and R. Tsu, 'Resonant tunneling in semiconductor double barriers', *Applied Physics Letters*, p. 593, 1974.

[5] C. V. Sammut and N. J. Cronin, 'Comparison of measured and computed conversion loss from a resonant tunneling device multiplier', *IEEE Microwave and Guided Wave Letters*, Vol. 2, No. 12, pp. 486–488, December 1992.

[6] E. R. Brown, C. D. Parker and T. C. L. G. Sollner, 'Effect of quasibound-state lifetime on the oscillation power of resonant tunneling diodes', *Applied Physics Letters*, Vol. 54, No. 10, pp. 934–936, March 1989.

[7] W. R. Frensley, 'Quantum transport calculation of the small-signal response of a resonant tunneling diode', *Applied Physics Letters*, Vol. 51, No. 6, pp. 448–450, 1987.

[8] R. K. Mains and G. I. Haddad, 'Wigner function modeling resonant tunneling diodes with high peak-to-valley ratios', *Journal Applied Physics*, Vol. 64, No. 10, pp. 5041–5044, 1988.

[9] L. Y. Chen and C. S. Ting, 'AC conductance of a double-barrier resonant tunneling system under a dc-bias voltage', *Physical Review Letters*, Vol. 64, No. 26, pp. 3159–3162, 1990.

[10] V. Kislov and A. Kamenev, 'High-frequency properties of resonant tunneling devices', *Applied Physics Letters*, Vol. 59, No. 12, pp. 1500–1502, 1991.

[11] Yaotian Fu and S. C. Dudley, 'Quantum inductance within linear response theory', *Physical Review Letters*, Vol. 70, No. 1, pp. 65–68, January 4 1993.

[12] M. Cahay, M. McLennan, S. Datta and M. S. Lundstrom, 'Importance of space-charge effects in resonant tunneling diode', *Applied Physics Letters*, Vol. 50, p. 612, 1987.

[13] K. Kurokawa, 'Some basic characteristics of broadband negative resistance oscillator circuits', *Bell System Technical Journal*, Vol. 48, p. 1937, 1969.

[14] T. C. L. G. Sollner, E. R. Brown, W. D. Goodhue and H. Q. Le, 'Microwave and millimeter-wave resonant tunneling devices', *Physics of Quantum Electron Devices*, ed. F. Capasso, Springer, New York, 1990.

[15] C. Y. Belhadj, K. P. Martin, S. Ben Amor, J. J. L. Rascol, R. C. Potter, H. Hier and E. Hempfling, 'Bias circuit effects on the current-voltage characteristic of double-barrier tunneling structures: Experimental and theoretical results', *Applied Physics Letters*, Vol. 57, No. 1, pp. 58–60, July 2 1990.

[16] M. E. Hines, 'High frequency negative-resistance circuit principles for Esaki Diode applications', *Bell System Technical Journal*, Vol. 39, p. 477, 1960.

[17] P. Roblin, S. Akhtar, R. Potter and A. Fathimulla, *Analysis and Measurement of the Capacitance–Voltage Characteristic of a Resonant Tunneling Diode*, Bulletin of the American Physical Society, March Meeting, p. 361, 1995.

[18] J. Zohios, Infrared Radiation Response of Layered Quantum Structures, MSc thesis, The Ohio State University, August 1996.

[19] J. Faist, F. Capasso, D. L. Sivco, C. Sirtori, A. L. Hutchinson and A. Y. Cho, 'Quantum cascade laser', *Science*, Vol. 263, p. 553, 1994.

[20] C. Sirtori, J. Faist, F. Capasso, D. L. Sivco, A. L. Hutchinson and A. Y. Cho, 'Pulsed and continuous-wave operation of long wavelength ($\lambda = 9.3$ μm) quantum cascade lasers', *IEEE Journal of Quantum Electronics*, Vol. QE33, p. 89, 1997.

[21] E. Rosencher and B. Vinter, *Optoelectronique*, Thomson-CSF, Masson, Paris, 1998.

8 Charge control of the two-dimensional electron gas

Asked by a politician about the practical worth of electricity, Michael Faraday answered: 'One day, Sir, you may tax it.'

Discovery, R. A. GREGORY

8.1 Introduction

As was discussed in Chapter 4, the confinement of electrons in a quantum well along the superlattice direction leads to the formation of a two-dimensional electron gas (2DEG). Such a quantum well can directly arise at the interface of a heterojunction due to the combined confinement provided by the heterojunction band structure discontinuities on one side and the potential barrier created by the large built-in electric field on the other side. In this chapter we shall calculate the equilibrium population in such a heterojunction and see how the 2DEG population can be controlled by a Schottky junction. As we shall see in Chapter 10, such a controllable 2DEG is the basis for a high-speed FET, the HEMT (also called MODFET or TEGFET), (see chapter 10 for the definitions of the abbreviations).

8.2 2DEG population as a function of the Fermi energy

Let us consider the 2DEG system shown in Figure 8.1. The region on the left-hand side of the heterostructure is labeled 2 and the region on the right is labeled 1. For simplicity we assume that the semiconductor materials of regions 1 and 2 are both lattice-matched to the substrate and are therefore not strained.

For a general heterostructure, the eigenstates in a quantum well are obtained from the Wannier eigenstate equation

$$\sum_{n=-N_B}^{N_B} H_{mn} f_p(n) = E_p f_p(m).$$

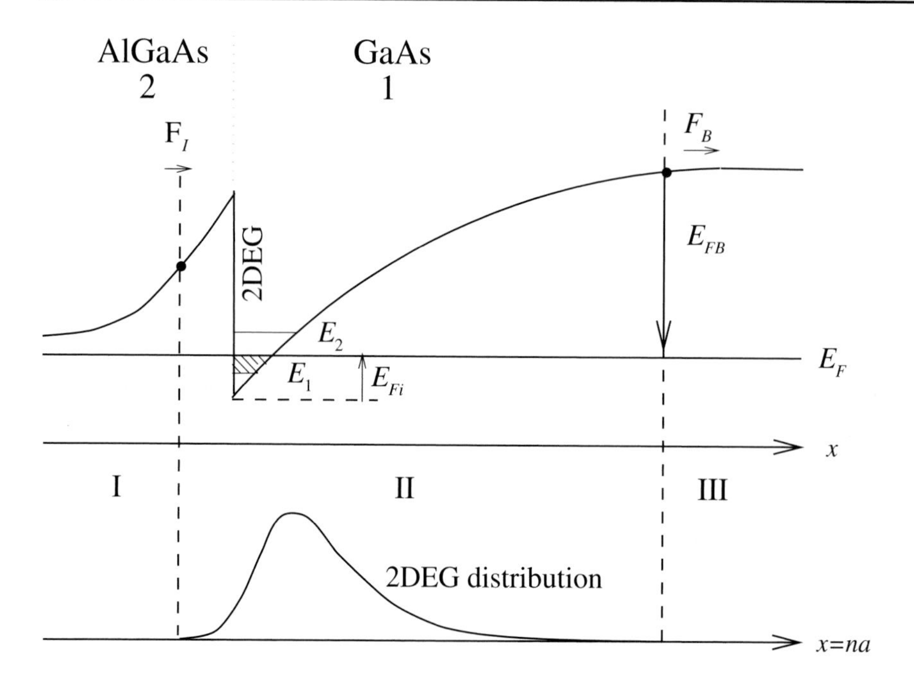

Fig. 8.1. The 2DEG system at the interface of a heterojunction.

In the tight-binding approximation, the Wannier equation reduces to

$$-\frac{A}{2} f_p(m+1) + [A + \mathcal{E}_c(m) - qV(ma)]\, f_p(m) - \frac{A}{2} f_p(m-1) = E_p\, f_p(m)$$

where $A = \hbar^2/(m^* a^2)$ and $\mathcal{E}_c(m)$ is the conduction band edge which includes the conduction-band discontinuities. $V(na)$ is the electrostatic potential which is obtained by solving the Poisson equation (see Chapter 2)

$$\epsilon(x)\frac{d^2V[x]}{dx^2} + \frac{dV[x]}{dx}\frac{d\epsilon(x)}{dx} = -\rho(x).$$

The charge distribution $\rho(x)$ is given at each lattice site by the unintentional concentration of donor or acceptors and the electron distribution in the well:

$$\rho(na) = qN_D^+(na) - qN_A^-(na) - q\sum_{p=1}^{\infty} n_{S,p}|f_p(n)|^2,$$

with $n_{S,p}$ the electron concentration in the subband p. In Chapter 4 we found $n_{S,p}$ to be given by

$$n_{S,p} = Dk_BT \ln\left[\exp\left(\frac{E_F - E_p}{k_BT}\right) + 1\right],$$

where $D = m^*/(\hbar^2\pi)$ is the 2DEG density. The use of the wave-function to calculate the electron distribution ρ is not justifiable for a single electron since the wave-function only gives the probability of the presence of the electron at a lattice site but does not guarantee its presence (an electron is indivisible and can only be at one lattice site at a time). However, in a 2DEG, the presence of many electrons with similar wave-functions guarantees that several electrons will always be present at each lattice site with the wave-function, giving the relative population of each lattice site. Obviously the wave-function of each eigenstate must be normalized

$$\sum_{n=-\infty}^{\infty} |f_p(n)|^2 = 1.$$

Note that the charge distribution can be assumed to be uniform in each lattice site n in the region of width a_n (the lattice parameter). The integration of the Poisson equation for such types of charge distribution can be expressed in a simple analytic form which is derived in Problem 2.2.

Clearly this eigenvalue problem requires the self-consistent solution of the Wannier equation and the Poisson equation. Indeed, the wave-functions $|f_p(n)|^2$ are required to solve the Poisson equation, and the potential $V(na)$ is required to obtain the wave-function from the Wannier equation.

The solution to this problem can be somewhat simplified by dividing the heterostructure shown in Figure 8.1 into three regions: Regions I, II and III. A full quantum treatment is only applied to Region II. Regions I and III are described semiclassically, i.e., no quantum wells are present in these regions. Region I will be studied in more detail in the next two sections. Wave-functions on the edge of Regions I and III can be assumed to be zero. This is a reasonable approximation, as the wave-functions die quickly inside the potential barrier reducing the eigenvalue problem to the calculation of the eigenvalues and eigenvector of a finite size square matrix ($[H_{nm}]$). Integration of the Poisson equation across the quantum Region II of width W gives

$$\epsilon_2 F_I = \epsilon_1 F_B + q\sum_p n_{S,p} + q(N_A^- - N_D^+)W,$$

where F_I is the electric field at the interface of Regions I and II, and F_B is the electric field at the interface of Regions II and III. For a uniformly doped substrate or bulk, one can easily integrate the Poisson equation in Region III and verify that the field F_B is related to the Fermi level E_{FB} (measured relative to the conduction-band edge at the II/III interface) by the relation

$$F_B = \frac{k_B T}{q}\frac{2}{L_D}\left[(U_0 - U)\sinh(U_0) + \cosh(U) - \cosh(U_0)\right]^{1/2}, \tag{8.1}$$

where L_D is the intrinsic Debye length

$$L_D = \left(\frac{\epsilon_1 k_B T}{q^2 N_i}\right)^{1/2}$$

with U_0 and U given by

$$U_0 = \sinh^{-1}\left(-\frac{N_A}{2N_i}\right), \qquad U = \frac{E_{FB}}{k_B T} - \ln\left(\frac{N_i}{N_c}\right). \tag{8.2}$$

Assume that we wish to calculate the 2DEG concentration given a Fermi level E_{Fi} specified relative to the bottom of the triangular well. The solution of such eigenvalue problems can proceed iteratively. Starting from a guess potential $V_1(na)$ the eigenenergies and wave-functions can be calculated. The Poisson equation can then be solved for the calculated charge distribution $\rho(na)$ and the boundary condition $F_B = F_B(E_{FB})$ such that the Fermi level remains constant across Regions II and III. The resulting potential $V_2(na)$ can then be used as a guess for again solving the eigenvalue problem. In practice, for this simple procedure to converge, it is necessary to damp the updated potential by averaging it with its previous guess: $0.5 \times [V_2(na) + V_1(na)]$. An improved updating technique would rely on the Jacobian associated with the Poisson equation to calculate the potential update. Repeating this iterative procedure one obtains a self-consistent numerical solution of this eigenvalue problem giving the 2DEG concentration $n_S = \sum_p n_{S,p}$, and the population of each subband, the eigenenergies E_p and the fields F_B and F_I for a given Fermi level E_{Fi}.

Note that we have assumed that the semiconductor materials of Regions 1 and 2, were both lattice-matched to the substrate and were therefore not strained. In the case of pseudomorphic materials, the tensile strain can induce a large piezoelectric field (polarization) which must be accounted for in the 2DEG calculation (see Section 2.4). The reader is referred to [4] for an AlGaN/GaN HEMT example in which the piezoelectric field induced by the strained AlGaN layer pseudomorphically grown on GaN is demonstrated to be the primary source of the 2DEG charge.

Approximate treatment

For the sake of simplicity, it is interesting to present an approximate analytic solution.

The charge stored in the buffer (Region III) is usually very small compared to the charge n_S stored in the quantum well (Region II). When this is the case we can neglect the bulk field F_B, and we have

$$\epsilon_2 F_I = q n_S.$$

For a wide region of operation the Fermi level lies between the first eigenenergies E_1 and E_2. Because these energy levels are sufficiently separated, the first subband is usually mostly populated, and the inclusion of two subbands is sufficient to describe the 2DEG system. The total electron concentration then reduces to

$$n_S = Dk_B T \ln\left[\exp\left(\frac{E_F - E_1}{k_B T}\right) + 1\right] + Dk_B T \ln\left[\exp\left(\frac{E_F - E_2}{k_B T}\right) + 1\right].$$

This relation can be easily inverted (see Kim *et al.* [5])

$$E_{Fi}(n_S) = k_B T \ln\left\{-\frac{1}{2}\left(\exp\frac{E_1}{k_B T} + \exp\frac{E_2}{k_B T}\right) + \left[\frac{1}{4}\left(\exp\frac{E_1}{k_B T} + \exp\frac{E_2}{k_B T}\right)^2 + \left(\exp\frac{n_S}{k_B T} - 1\right)\exp\frac{E_1 + E_2}{kT}\right]^{\frac{1}{2}}\right\}$$

Finally the eigenvalues E_1 and E_2 can be obtained using the triangular well approximation (see Chapter 4)

$$E_p = \left(\frac{\hbar^2}{2m^*}\right)^{\frac{1}{3}}\left[\left(\frac{3}{2}\pi q F_0\right)\left(p + \frac{3}{4}\right)\right]^{\frac{2}{3}}.$$

Given that the triangular potential is not a very accurate approximation, an improved model can be obtained [1] if only the field dependence is retained

$$E_n = \alpha_n F_I^{2/3}$$

and a fitting constant α_n is introduced. Since the field is proportional to the charge concentration we have

$$E_1 = \gamma_1 n_S^{2/3},$$

$$E_2 = \gamma_2 n_S^{2/3},$$

where γ_1 and γ_2 are two constants which are obtained experimentally. This approximate model provides us with an analytic expression allowing us to calculate the Fermi energy E_{Fi}, the field $F_I = qn_S$, the population of each subband, and the eigenenergies E_1 and E_2 given a specific 2DEG concentration n_S.

8.3 Equilibrium population of the 2DEG

So far our analysis has been limited to Regions II and III, and Region I was left unspecified. In this section we consider the case in which Region I consists of a strongly doped wide-bandgap semiconductor (labeled 2) (e.g., n-AlGaAs) as shown in Figure 8.2. Our goal is to calculate the equilibrium concentration in the 2DEG.

The built-in potential in Region I can be obtained by solving the Poisson equation in this region (see Chapter 2)

$$\epsilon_2 \frac{d^2V}{dx^2} = -\rho = -\frac{qN_D(x)}{1 + 2\exp\left(\dfrac{E_F - E_d + qV}{k_B T}\right)} + qN_c \mathrm{F}_{1/2}\left[\frac{E_F - \mathcal{E}_c + qV}{k_B T}\right] \tag{8.3}$$

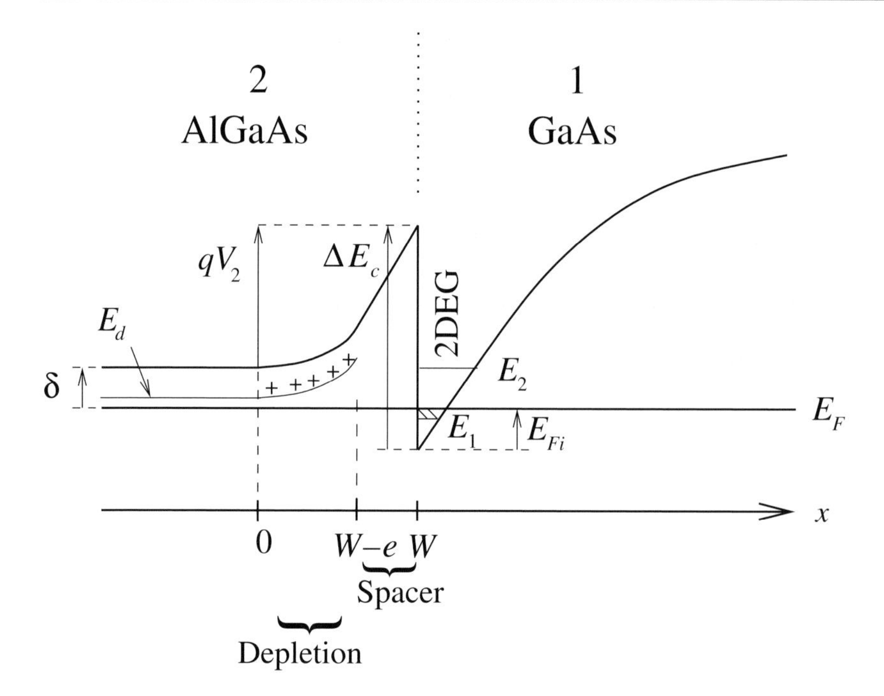

Fig. 8.2. Band diagram of an n-AlGaAs–i-GaAs heterojunction.

for non-degenerate doping or simply

$$\epsilon_2 \frac{d^2V}{dx^2} = -\rho = -qN_D(x) + qN_c \mathrm{F}_{1/2}\left[\frac{E_F - \mathcal{E}_c + qV}{k_B T}\right] \tag{8.4}$$

for degenerate doping where the donors are assumed to be fully ionized (see Section 2.2.2). In general the donor distribution N_D can be spatially varying. The interesting case of pulse doping is studied in Problem 8.2. We shall consider here the case of a uniformly doped heterostructure. A small undoped region of width e, called the spacer region, is used to separate the intrinsic GaAs from the n-doped AlGaAs material. Indeed, the wave-function of the electron penetrates a few lattice parameters inside the heterojunction barrier near the interface, and the electron in the state $|n, \mathbf{k}_\perp\rangle$ has a non-negligible probability to be scattered by the donor impurities in AlGaAs. The spacer region permits us to reduce the possibility of scattering and therefore improve the mobility of the electrons in the 2DEG.

Let us now calculate the built-in potential. Note that in the case of a uniform donor distribution N_D, the Poisson Equation (8.3) or (8.4) can be integrated analytically if we use the following approximate identity for the Fermi–Dirac integral:

$$\mathrm{F}_{1/2}[x] \simeq \frac{\exp x}{1 + \frac{1}{4}\exp x}.$$

For the sake of simplicity, we shall rely instead on the depletion approximation used in the pioneering paper of Delagebeaudeuf and Linh [1]. We assume that the depletion extends over the width $W - e$. The origin of the axis is set at the beginning of the depletion (see Figure 8.2):

$$\frac{dF}{dx} = \frac{qN_D}{\epsilon_2} \quad \text{for} \quad 0 \leq x \leq W - e,$$
$$\frac{d^2V}{dx^2} = -q\frac{N_D}{\epsilon_2} \quad \text{for} \quad 0 \leq x \leq W - e.$$

Using the boundary condition $V(0) = 0$ and $F[0] = 0$ we obtain

$$F_2[x] = \frac{qN_D}{\epsilon_2}x \quad \text{for } 0 \leq x \leq W - e.$$
$$V_2[x] = -q\frac{N_D}{2\epsilon_2}x^2 \quad \text{for } 0 \leq x \leq W - e. \tag{8.5}$$

The field $F[W^-]$ at the heterojunction interface is

$$F[W^-] = F[W - e] = \frac{qN_D}{\epsilon_2}(W - e).$$

The potential $V[W]$ at the heterojunction interface is

$$V[W - e] = -\frac{qN_D}{2\epsilon_2}(W - e)^2$$

$$\begin{aligned} V[W] &= V[W - e] - F[W - e]\, e \\ &= -\frac{qN_D}{2\epsilon_2}(W - e)^2 - \frac{qN_D}{\epsilon_2}(W - e)e \\ &= -\frac{q}{2\epsilon_2}N_D(W^2 - e^2). \end{aligned}$$

Let us define the potential V_2:

$$V_2 = -V[W] = \frac{q}{2\epsilon_2}N_D(W^2 - e^2).$$

The width W is then

$$W = \left(\frac{2\epsilon_2 V_2}{qN_D} + e^2\right)^{\frac{1}{2}}$$

and the interface field $F[W^-]$ is

$$F[W^-] = \frac{qN_D}{\epsilon_2}\left[\left(\frac{2\epsilon_2 V_2}{qN_D} + e^2\right)^{\frac{1}{2}} - e\right]$$
$$\epsilon_2 F[W^-] = \left(2\epsilon_2 q N_D V_2 + q^2 N_D^2 e^2\right)^{\frac{1}{2}} - qN_D e.$$

The interface field can be calculated once we can evaluate V_2. From the band diagram in Figure 8.2 we have

$$\Delta E_c = qV_2 + \delta + E_{Fi}[n_S],$$

with $\delta = E_c[-\infty] - E_F$ the distance between the Fermi level and the conduction band in n-AlGaAs far from the interface. The equilibrium charge concentration can then be obtained by solving the following transcendental equation

$$qn_S = \epsilon_2 F[W^-] = \left\{2\epsilon_2 N_D[\Delta E_c - \delta - E_{Fi}(n_S)] + q^2 N_D^2 e^2\right\}^{\frac{1}{2}} - qN_D e. \qquad (8.6)$$

The equilibrium charge concentration $n_{S0}[e, N_D]$ is seen to be a function of the doping level N_D and the spacer width e. Typically n_{S0} is on the order of 10^{12} cm^{-2}

Delagebeaudeuf and Linh [1], who reported this expression, verified that n_{S0} is maximum for $e = 0$. Typically $e \simeq 100$ Å is used so as to reduce Coulombic scattering. This permits us to achieve very large mobilities approaching the intrinsic mobility of GaAs.

8.4 Charge control of the 2DEG with a Schottky junction

We shall now see how the population of the 2DEG can be controlled using a Schottky junction. Consider the Au–n-AlGaAs–i-GaAs structure. The band diagram of such a structure is shown in Figure 8.3 for a thick and a thin layer of AlGaAs region. In the case of a thick AlGaAs layer (Figure 8.3(b)) the depletion regions of the Schottky junction and the 2DEG do not overlap, and a parasitic n channel is present in the AlGaAs region. For a narrower AlGaAs region, the depletion regions overlap, and the entire AlGaAs region is depleted. The electrons provided by the donors are then shared between the metal and the 2DEG. By applying a voltage between the metal and the 2DEG Fermi level one can modify this charge partitioning. Such a biasing scheme can be realized with the so-called MODFET (a discussion of the MODFET is given in Chapter 10). A negative potential on the metal increases the number of uncompensated donors imaging the charge on the metal and therefore decreases the fraction of donor contributing electrons to the 2DEG. A positive potential on the metal decreases the number of uncompensated donors imaging the charge on the metal and therefore increases the fraction of donors contributing electrons to the 2DEG. Once the depletion region of the 2DEG and the Schottky junction separate, the 2DEG population saturates to its maximum value n_{S0}, and the applied voltage ceases to control the 2DEG. The metal will be referred to as the gate and the applied voltage as the gate voltage.

We shall develop a simple model to predict the variation of the 2DEG concentration n_S with the gate voltage V_G. Again, we shall rely on a depletion approximation

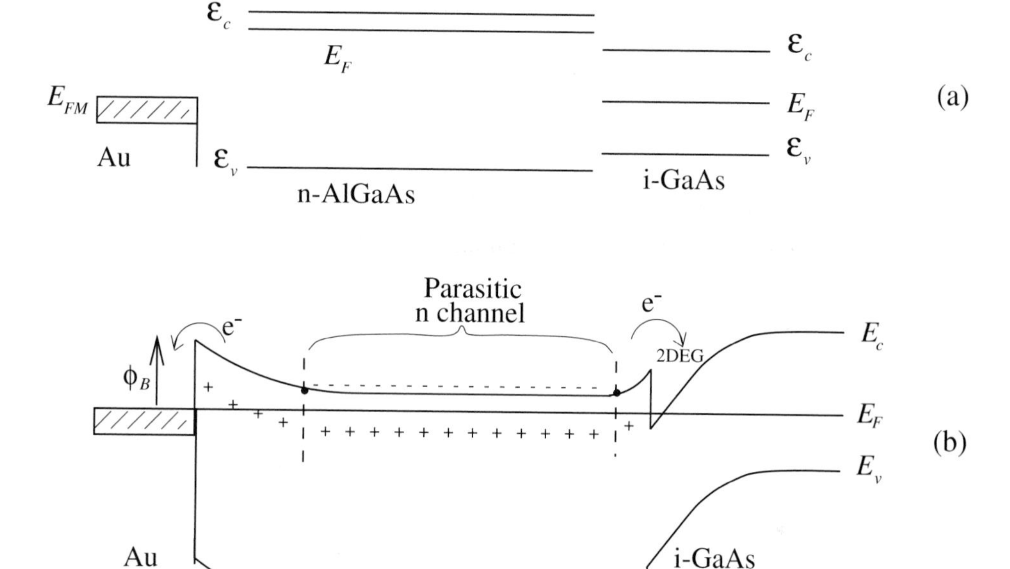

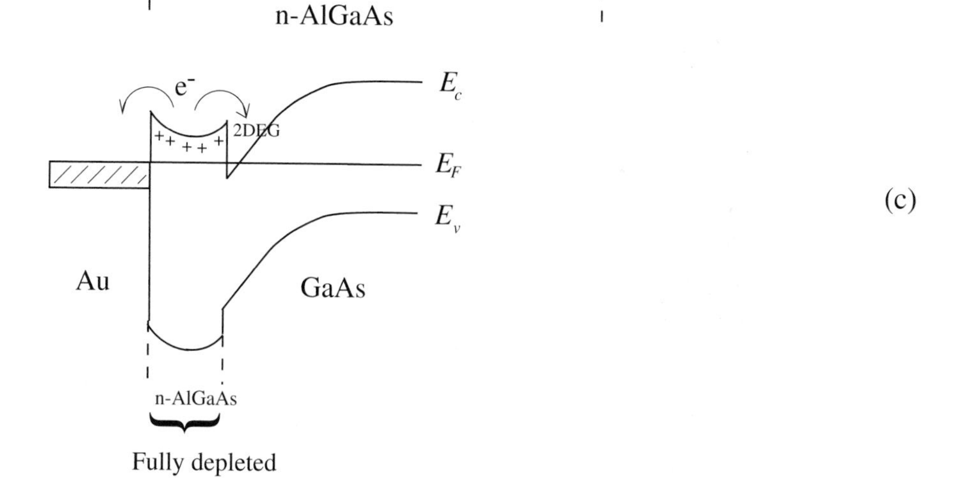

Fig. 8.3. Equilibrium band diagram of the Au–n-AlGaAs–i-GaAs structure: (a) before connection, (b) for a thick AlGaAs region, and (c) for a thin AlGaAs region.

which assumes that the entire n-AlGaAs region of width d_2 is depleted. The depletion approximation therefore holds when the depletion regions of the Schottky junction and the 2DEG overlap. We also assume that the Schottky diode is reverse-biased, so that no current is flowing through the heterostructure. The Gauss and Poisson equations in the AlGaAs region are then respectively:

$$\frac{dF}{dx} = q\frac{N_D}{\epsilon_2} \quad \text{for} \quad 0 \le x \le d_2 - e,$$

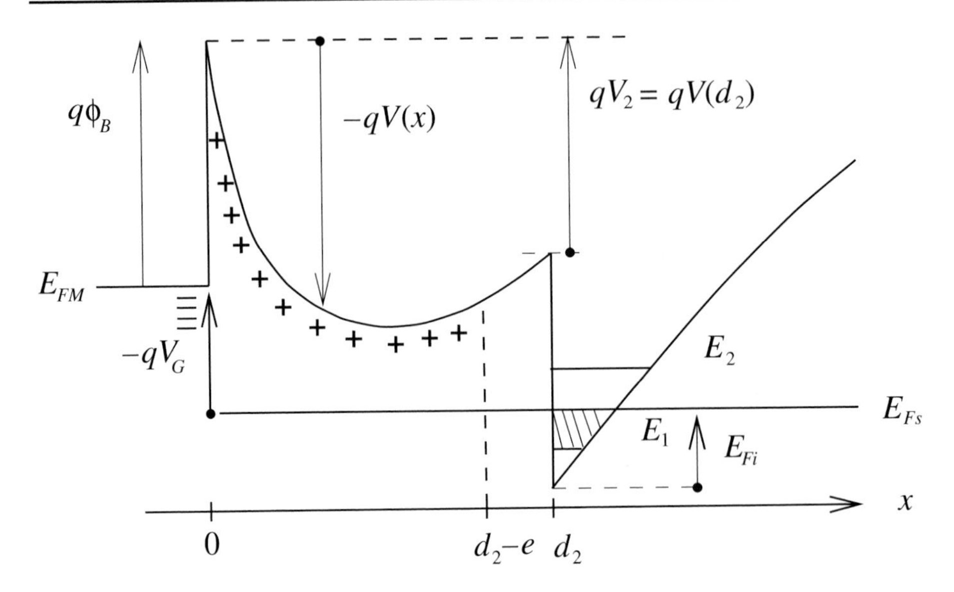

Fig. 8.4. Band diagram of the Au–n-AlGaAs–i-GaAs with a gate voltage V_G.

$$\frac{d^2V}{dx^2} = -q\frac{N_D}{\epsilon_2} \quad \text{for} \quad 0 \leq x \leq d_2 - e.$$

The origin of the x axis and the potential is selected to be the metal–semiconductor junction (see Figure 8.4). The field $F[x]$ and potential $V[x]$ are then:

$$F[x] = q\frac{N_D}{\epsilon_2}x + F[0^+],$$

$$V[x] = -\frac{1}{2}q\frac{N_D}{\epsilon_2}x^2 - F[0^+]x.$$

The field at the heterojunction interface is

$$F[d_2] = F[d_2 - e] = q\frac{N_D}{\epsilon_2}(d_2 - e) + F[0^+] \tag{8.7}$$

and the potential $V[x]$ can be rewritten

$$V[x] = -\frac{1}{2}q\frac{N_D}{\epsilon_2}x^2 + \left[q\frac{N_D}{\epsilon_2}(d_2 - e) - F[d_2]\right]x.$$

Next we evaluate the potential at $x = d_2 - e$:

$$\begin{aligned} V[d_2 - e] &= -\frac{1}{2}q\frac{N_D}{\epsilon_2}(d_2 - e)^2 + q\frac{N_D}{\epsilon_2}(d_2 - e)^2 - F[d_2](d_2 - e) \\ &= \frac{1}{2}q\frac{N_D}{\epsilon_2}(d_2 - e)^2 - F[d_2](d_2 - e). \end{aligned}$$

The potential V_2 at the heterojunction interface is then

$$V_2 = V[d_2] = -F[d_2]e + V(d_2 - e)$$

$$= \frac{1}{2} q \frac{N_D}{\epsilon_2} (d_2 - e)^2 - F[d_2] d_2$$
$$= V_P - F[d_2] d_2,$$

where V_P is the so-called pinch-off voltage:

$$V_P = \frac{1}{2} q \frac{N_D}{\epsilon_2} (d_2 - e)^2.$$

The interface field F_2 is therefore given by the relation

$$\epsilon_2 F[d_2] = \frac{\epsilon_2}{d_2} (V_P - V_2).$$

From the band diagram shown in Figure 8.4 we can obtain the following identity:

$$q V_2 + \Delta E_c = q \phi_B - q V_G + E_{Fi}(n_S).$$

The 2DEG concentration n_S is then given by

$$q n_S = \epsilon_2 F[d_2] = \frac{\epsilon_2}{d_2} \left(V_P - \phi_B + V_G + \frac{\Delta E_c}{q} - \frac{E_{Fi}[n_S]}{q} \right).$$

This is a transcendental equation, since E_{Fi} is a function of n_S. If we neglect the variation of E_F with n_S, the following linear relation is obtained:

$$q n_S = \frac{\epsilon_2}{d_2} (V_G - V_T) \quad \text{for} \quad V_T \leq V_G \leq V_{G,MAX}, \tag{8.8}$$

where V_T is the so-called threshold voltage:

$$V_T = \phi_B - \frac{\Delta E_c}{q} + \frac{E_F[n_{S0}]}{q} - V_P, \tag{8.9}$$

and where ϵ_2 / d_2 is the so-called gate capacitance. An improved calculation can be obtained by linearizing the variation of E_{Fi} with n_S [2] (see Problem 8.1). A more accurate analytic expression for the threshold voltage and gate capacitance can then be derived.

Figure 8.5 shows the variation of the 2DEG charge $q n_S$ with the applied gate voltage. As indicated in Equation (8.8), the linear charge control only applies to a limited gate voltage range. For gate voltages smaller than the threshold voltage, the exact $E_{Fi}(n_S)$ relation should be used, with the result that the 2DEG is never pinched off completely. This region is called the subthreshold region. For gate voltages larger than $V_{G,MAX} \simeq q n_{S0} d_2 / \epsilon_2$, the depletion regions of the Schottky junction and the 2DEG cease to overlap, and the depletion approximation is no longer valid and the 2DEG charge $q n_S$ saturates toward its equilibrium value n_{S0} (see previous section). Typically, the useful gate voltage range is of about 1 V. For larger gate voltages a parasitic channel is formed in the AlGaAs region.

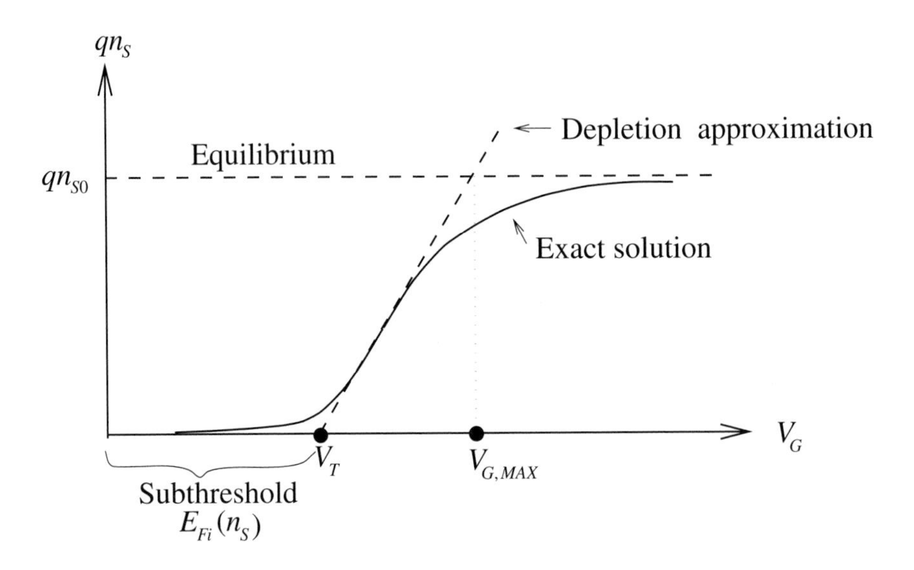

Fig. 8.5. Variation of the 2DEG charge with the gate voltage.

8.5 C–V characteristics of the MODFET capacitor

The metal–semiconductor system we have studied permits us to control the 2DEG charge with a gate voltage. Such a system is in fact a non-linear capacitor as long as the Schottky junction is reverse-biased, since the only current we would expect through the heterostructure is the displacement current. Let us now calculate the capacitance C_{GM} measured between the metal (gate) and the 2DEG Fermi level. This capacitance is given by the variation of the charge density Q_M on the metal for a variation of the gate voltage:

$$C_{GM} = \frac{dQ_M}{dV_G} = \frac{\epsilon_2 dF[0^+]}{dV_G}.$$

A plot of the capacitance C_{GM} measured experimentally is shown in Figure 8.6 (full line). The C–V characteristic can be divided into three different regions [3]: the subthreshold Region A, the 2DEG Region B and the Shottky diode Region C.

In the subthreshold Region A, the AlGaAs is fully depleted, the 2DEG is essentially unpopulated ($n_S \simeq 0$), and the field $F[0^+]$ at the Schottky diode is essentially only supported by the charge in the GaAs bulk (Region III). The capacitance C_{GM} measured at the gate is then the capacitance of the AlGaAs region ϵ_2/d_2 in series with the capacitance of the GaAs buffer (Region III) C_B:

$$C_{GM} = \frac{1}{d_2/\epsilon_2} + \frac{1}{C_B},$$

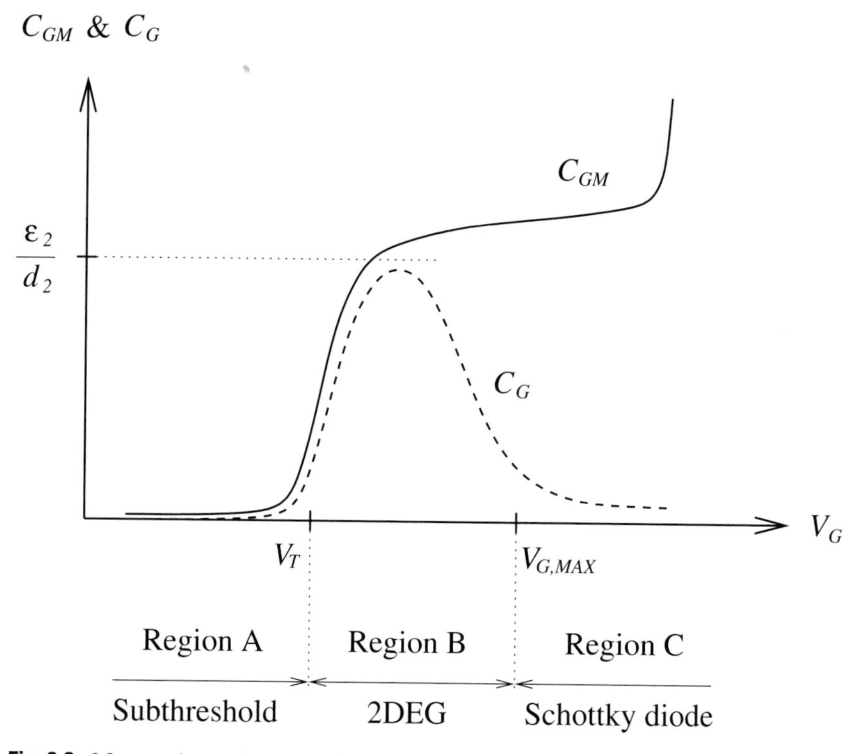

Fig. 8.6. Measured capacitance at the gate versus applied voltage (full line) and 2DEG capacitance (dashed line).

where C_B is defined from the bulk field F_B (see Equation (8.1)) by

$$C_B = \epsilon_1 q \frac{dF_B[U]}{dU}.$$

Since the buffer is undoped, the bulk capacitance C_B is quite small. One can verify [6] that the minimum value of the bulk capacitance C_B is obtained in the subthreshold region for U equal to

$$U_{min} \simeq -U_0 - 2\ln 2 - \ln|U_0|.$$

The capacitance C_{GM} measured at the gate therefore reaches a minimum value in the subthreshold region. The measurement of this minimum capacitance permits us to estimate the effective unintentional doping in the buffer [6].

In Region B of the C–V characteristic (see Figure 8.6), the population of the 2DEG is rapidly varying with the applied gate voltage. Indeed, if we assume that the AlGaAs Region II is fully depleted, Equation (8.7) holds and the capacitance C_{GM} measured at the gate is the 2DEG capacitance C_G:

$$C_{GM} = \frac{\epsilon_2 dF[0^+]}{dV_G} = \frac{\epsilon_1 dF[d_2]}{dV_G} = \frac{q dn_S}{dV_G} = C_G.$$

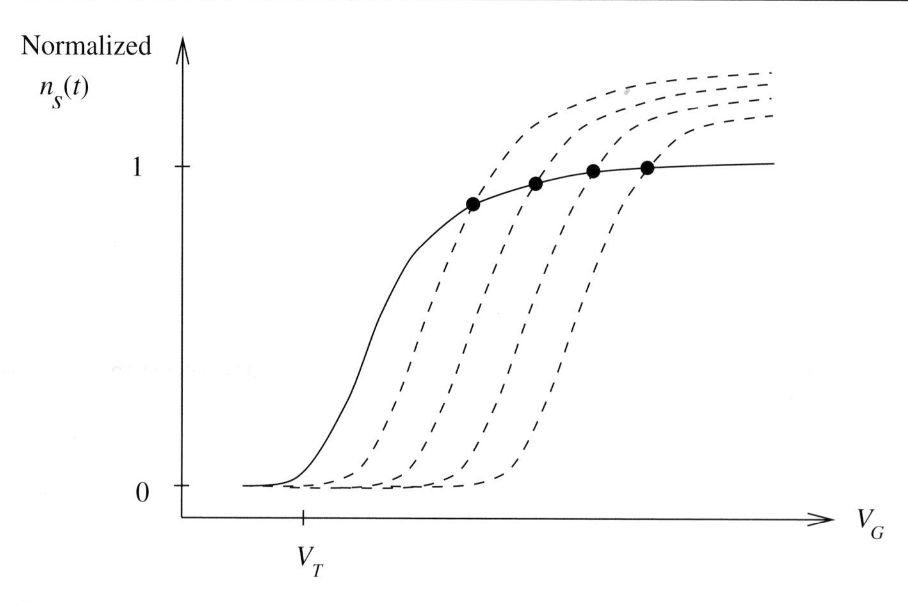

Fig. 8.7. Sketch of the DC (full line) and pulsed (dashed line) n_s(2DEG)–V_G characteristics at four different biases.

This is demonstrated in Figure 8.6, where the 2DEG C–V characteristic (C_G–V_G) is plotted (dashed line) in addition to the measured C_{GM}–V_G characteristic (full line).

Note that the 2DEG capacitance curve C_G–V_G is slightly shifted from the C_{GM}–V_G curve. This results from the modulation of donor charges in the AlGaAs region. This modulation was neglected by the depletion approximation which assumes that all the donors are fully ionized.

For larger gate voltages (Region C), the 2DEG capacitance C_G decreases when the 2DEG approaches its equilibrium concentration n_{S0}, but, the measured capacitance C_{GM} continues to increase. This originates from the variation of the ionized donor population in AlGaAs, which occurs when the depletion region of the 2DEG and the Schottky junction cease to overlap.

As the 2DEG capacitance vanishes for large gate voltages, the capacitance measured at the gate is uniquely that of the Schottky diode. For large positive gate voltages, the capacitance is actually seen to diverge when the Schottky diode turns on.

So far, our discussion has been limited to quasi-static (low-frequency) conditions. Let us now discuss the frequency dependence of the C–V characteristic of the MODFET capacitor. At high frequency, the population of the deep donors present in AlGaAs (see Section 2.2.2) does not have time to respond. Indeed, the capture and emission times of deep donors is much larger than that of shallow donors. Furthermore, the AlGaAs region, when depleted, provides a highly resistive path so that relaxation toward equilibrium is slow. As a result of the difficulty in modulating the charge of the donors in AlGaAs the measured capacitance is decreased. However, since the deep donor population is frozen, the donors do not shield the AC potential

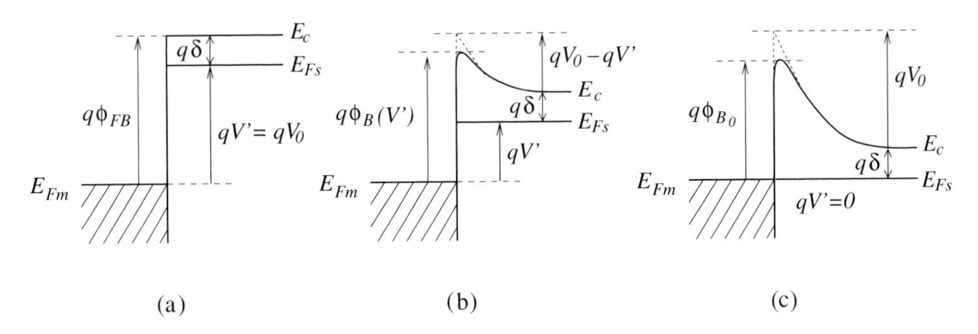

Fig. 8.8. Band diagram of a Schottky diode: (a) under the flat-band condition; (b) with a voltage applied, and (c) with zero voltage applied.

applied at the gate, and the resulting AC field will modulate the 2DEG charge. Note that the 2DEG charge can easily respond since the 2DEG channel provides a high-speed path (quasi-metallic) for the AC voltage applied. As a result at microwave frequencies we must use the maximum 2DEG capacitance $C_G = \epsilon_2/d_2$ even for large gate voltages in order to account for the transient response of the 2DEG. This is illustrated in Figure 8.7 with a sketch of the transient response of the 2DEG population for various initial biasing conditions (see Chandra and Foisy [8] for the calculated results). As is discussed in Chapter 13, this effect is also of consequence for parameter extraction, as the low-frequency dispersion introduced by deep-donors or traps in FETs can contribute to the discrepancy in values between the DC and microwave g_m's and g_d's extracted.

8.6 $I-V$ modeling of the Schottky junction

There exist two types of MODFET capacitors: the normally on and the normally off. For a normally on MODFET capacitor, the AlGaAs region is large enough so that the 2DEG population is already maximum (n_{S0}) for zero applied gate voltage. The quenching of the 2DEG population in a normally on MODFET capacitor is achieved with a negative gate voltage. The Schottky diode therefore remains reverse-biased. For a normally off MODFET capacitor, a small AlGaAs region is used for the 2DEG population to be a negligible fraction of n_{S0} for zero applied gate voltage. The population of the 2DEG can then be increased by applying a positive voltage at the gate. However, this will also forward bias the Schottky junction.

To conclude this chapter on the MODFET capacitor it is appropriate to discuss the modeling of the I–V characteristic of the Schottky junction. The band diagram of a Schottky diode is shown in Figure 8.8.

The gate current through the Schottky junction can be calculated starting from the general thermionic diode current derived in Chapter 4 (with current inverted)

$$J_G = \frac{Dk_BT}{2\pi}\frac{q}{\hbar}\int_{\mathcal{E}(min)}^{\mathcal{E}(max)} T[\mathcal{E}, V] \ln\left[\frac{\exp\left(\frac{E_{Fm} - qV_G - \mathcal{E}}{k_BT}\right) + 1}{\exp\left(\frac{E_{Fm} - \mathcal{E}}{k_BT}\right) + 1}\right] d\mathcal{E},$$

where E_{Fm} is the Fermi level on the metal side. A classical approximation provides a reasonable initial model. In the classical approximation, the transmission coefficient $T(\mathcal{E}, V)$ is 0 for energies smaller than the electron affinity χ_S of the semiconductor and is 1 for larger energies (see Figure 8.8). As the Fermi level is usually several k_BT below the transmitted energies, we can use the approximation $\ln(1 + x) \simeq x$ (the Boltzmann approximation). The diode current then reduces to

$$\begin{aligned} J_G &= \frac{Dk_BT}{2\pi}\frac{q}{\hbar}\int_{\chi_S}^{\infty}\left[\exp\left(\frac{E_{Fm} - qV_G - \mathcal{E}}{k_BT}\right) - \exp\left(\frac{E_{Fm} - \mathcal{E}}{k_BT}\right)\right] d\mathcal{E} \\ &= A^*T^2 \exp\left(-\frac{q\phi_B}{k_BT}\right)\left[\exp\left(\frac{qV_G}{kT}\right) - 1\right], \end{aligned} \tag{8.10}$$

where $q\phi_B = \chi_S - E_{Fm}$ is the barrier height and A^* is the so-called Richardson constant:

$$A^* = \frac{Dk_B^2}{2\pi}\frac{q}{\hbar} = \frac{4\pi m^* q k_B^2}{h^3}.$$

The ideal diode characteristic we have derived neglects the contribution of tunneling and the image force (the barrier lowering due to the repulsive Coulombic potential of the electron). Non-ideal effects cannot be neglected in practice. Typically the non-ideal I–V characteristic of a Schottky diode of area S_G can be modeled using the modified expression

$$I_G = S_G A^* T^2 \exp\left(-\frac{q\phi_{B_0}}{k_BT}\right)\left\{\exp\left[\frac{q(V_G - RI_G)}{nk_BT}\right] - 1\right\}, \tag{8.11}$$

where ϕ_{B_0} is the zero voltage barrier height, n is the ideality factor and R the series resistance.

ϕ_{B_0}, n and R can then be obtained using a least-square fit of the measured I_G–V_G characteristic. It is possible to interpret the ideality factor as a lowering of the effective barrier ϕ_B introduced by the applied diode voltage V_G [9]. The gate (Schottky junction) current can be written

$$I_G = S_G A^* T^2 \exp\left(-\frac{q\phi_B[V']}{k_BT}\right)\left[\exp\left(\frac{qV'}{k_BT}\right) - 1\right],$$

with $V' = V_G - I_G R$ and with the barrier height given by

$$\phi_B[V'] = \phi_{B_0} + \left(\frac{n-1}{n}\right) V'.$$

According to Figure 8.8, the flat-band barrier $q\phi_{BF} = \chi_S - E_{Fm}$ is obtained when a voltage $V' = V_0$ is applied

$$\phi_B[V_0] = \phi_{B_0} + \left(\frac{n-1}{n}\right) V_0 = \phi_{BF}.$$

From the band diagram of Figure 8.6 we also have

$$V_0 + \delta = \phi_{BF},$$

where $q\delta = E_c - E_{Fs}$ is the separation between the conduction-band edge and the Fermi level in the semiconductor side. Consequently we have

$$\phi_{B_0} + \left(\frac{n-1}{n}\right)(\phi_{BF} - \delta) = \phi_{BF}$$

$$\phi_{B_0} = \frac{\phi_{BF}}{n} + \left(\frac{n-1}{n}\right)\delta.$$

This last formula states that the effective barrier ϕ_{B_0} obtained at zero voltage is essentially the flat-band barrier $q\phi_{BF} = \chi_S - E_{Fm}$ divided by the ideality factor n, assuming that we can neglect δ (the strong doping limit). This simple theory, proposed by [9], permits us to qualitatively justify the use of an ideality factor n in Equation (8.11) on the basis of the possible voltage dependence of the effective barrier height ϕ_B. The latter could result from quantum effects (e.g., tunneling) or image force barrier lowering. However, the use of an ideality factor n that strongly departs from 1 (e.g., 2) is usually a sign of a poor metal–semiconductor junction, possibly due to structural defects or the presence of an anomalous oxide layer at the interface. These defects could change the diode electrical characteristics by introducing surface states or built-in dipoles at the junction.

The band diagram in Figure 8.4 really indicates the presence of two barriers: one at the Au–AlGaAs junction and the other at the AlGaAs–GaAs heterojunction. The current from the gate to the 2DEG channel therefore is crossing two back-to-back Schottky diodes. The barrier height of the 2DEG, however, is much smaller than that of the Au–AlGaAs junction and the AlGaAs–GaAs junction can be represented by a small resistance. Therefore there is a negligible Fermi level bending in the AlGaAs–GaAs heterojunction and most of the applied voltage is dropped across the Au–AlGaAs junction. However, under strong forward bias Ponse *et al.* [10] have demonstrated for a normally off 2DEG (MODFET) that the Fermi level bends in the spacer region near the AlGaAs–GaAs heterojunction (see Figure 8.9). For gate voltages larger than $q\phi_{B_0} - \Delta_C - E_{Fi}$, the 2DEG can be increased above its equilibrium value.

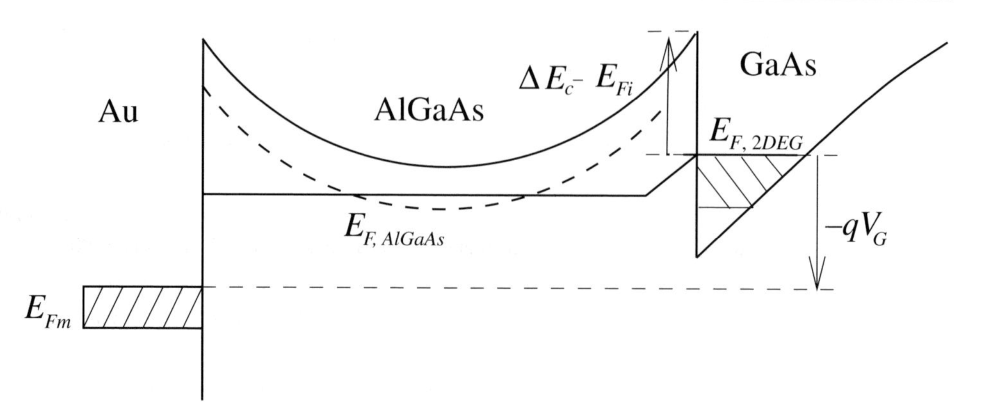

Fig. 8.9. Calculated band diagram of a MODFET under extreme forward bias. Notice the Fermi level bending near the AlGaAs–GaAs heterojunction.

8.7 Conclusion

In this chapter we have studied the 2DEG and its control with a gate contact. Our approach was based on the original models reported in the literature in order to obtain simple analytical results. A simple piece-wise linear charge-control model will be used in Chapters 10–12 for our initial study of the DC and AC responses of MODFETs. A more accurate charge-control model will then be presented in Chapter 14, in order to obtain a more realistic fit of the MODFET DC characteristics. The impact on FETs of low-frequency dispersions like those generated by deep donors and traps will be addressed in Chapter 13.

A simple model for the Schottky gate diode in MODFETs was also introduced in this chapter. In Chapter 16, we will return to this topic with an in-depth treatment of Schottky contacts and the gate resistance, as parasitics play a critical role in limiting the performance of high-speed devices.

8.8 Bibliography

[1] D. Delagebeaudeuf and N. Linh, 'Metal-(n) AlGaAs-GaAs two-dimensional electron gas FET,' *IEEE Transactions on Electron Devices,* Vol. ED-29, No. 6, pp. 955–960, June 1982.

[2] K. Lee, M. S. Shur, T. Drummond and H. Morkoç, 'Current–voltage and capacitance–voltage characteristics of modulation-doped field effect transistors,' *IEEE Transactions on Electron Devices,* Vol. ED-30, No. 3, pp. 207–212, March 1983.

[3] M. Moloney, F. Ponse, and H. Morkoç, 'Gate capacitance–voltage characteristic of MODFETs: Its effect on transconductance,' *IEEE Transactions on Electron Devices,* Vol. ED-32, No. 9, pp. 1675–1684, September 1985.

[4] Y. Zhang and J. Singh, 'Charge control and mobility studies for an AlGaN/GaN high electron mobility transistor,' *Journal of Applied Physics,* Vol. 85, No. 1, pp. 587–594, 1999.

[5] Y. M. Kim and P. Roblin, 'Two-dimensional charge control model for the MODFETs,' *IEEE Transactions on Electron Devices*, Vol. ED-33, No. 11, pp. 1644–1651, 1986.

[6] P. Roblin, H. Rohdin, C.J. Hung and S. W. Chiu, 'Capacitance–voltage analysis and current modeling of pulse-doped MODFETs,' *IEEE Transactions on Electron Devices*, Vol. ED-36, No. 11, pp. 2394–2404, November 1989.

[7] M. C. Foisy, 'A physical model for the bias dependence of the modulated-doped field-effect transistor's high-frequency performance', PhD Thesis, Cornell University 1990.

[8] A. Chandra and M. Foisy, 'Modeling of short pulse threshold voltage shifts due to DX centers in Al_xGa_{1-x}/GaAs and $Al_xGa_{1-x}/In_yGa_{1-y}As$ MODFETs,' *IEEE Transactions on Electron Devices,* Vol. 38, pp. 1238–1245, 1991.

[9] L. F. Wagner, R. W. Young and A. Sugerman 'A note on the correlation between the Schottky-diode barrier height and the ideality factor as determined from I–V measurements,' *IEEE Electron Device Letters,* Vol. EDL-4, No. 9, September 1983.

[10] F. Ponse, W. T. Masselink and H. Morkoç, 'Quasi-Fermi level bending in MODFETs and its effect on FET transfer characteristics,' *IEEE Transactions on Electron Devices,* Vol. ED-32, No. 6, pp. 1017–1023, June 1985.

8.9 Problems

8.1 Consider the MODFET structure analyzed by Delagebeaudeuf and Linh [1]. The variation of the chemical potential (Fermi level) E_{Fi} with the 2DEG concentration n_S was neglected in the calculation of the equilibrium 2DEG concentration n_{S0} and the gate capacitance C_G,

(a) Calculate a more accurate gate capacitance $C_G = d(qn_S)/dV_G$ which accounts for the dependence of the chemical potential (Fermi level) E_{Fi} upon the 2DEG concentration n_S. Verify that the gate capacitance can be written

$$C_G = \frac{\epsilon_2}{d_2 + \Delta d},$$

where d_2 is the AlGaAs width and with Δd a term to be derived. Note that the function $E_{Fi}(n_S)$ has not been specified at this point.

Elaborate on the physical origin and contribution of the term Δd.

(b) Let us consider the impact of the variation of the chemical potential (Fermi level) E_{Fi} with the 2DEG concentration n_S upon the threshold voltage V_T for the range of n_S considered. For this purpose we linearize $E_{Fi} - n_S$ for values of n_S in the vicinity of a chosen 2DEG concentration n_{S1}:

$$E_{Fi}[n_S] = E_{Fi}[n_{S1}] + \frac{dE_{fi}[n_{S1}]}{dn_S}(n_S - n_{S1}) = b + an_S.$$

Derive the new threshold voltage V_T. Note that the threshold voltage must be a constant, and therefore is independent of n_S or V_G!

(c) Assume that the self-consistent solutions of the Schrödinger and Poisson equations yields an $E_{Fi} - n_S$ relationship which can be approximated by

$$E_{Fi}(\text{eV}) = 0.234\left(\frac{n_S}{10^{16}}\right)^{\frac{1}{3}} - 0.153,$$

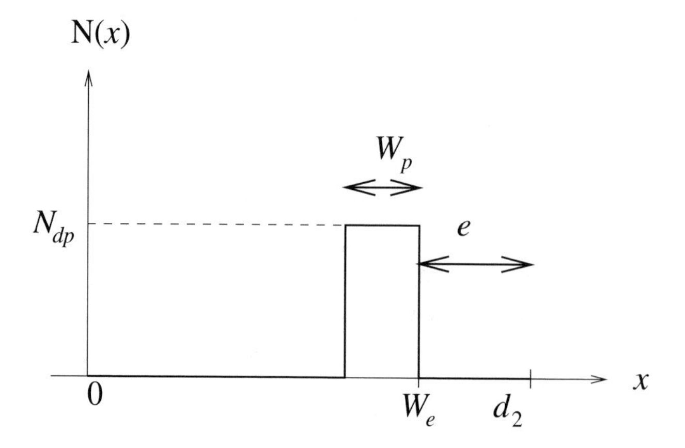

Fig. 8.10. Silicon doping profile.

where n_S is the number of electrons per m^2.

Calculate the 2DEG concentration n_{S0} using the transcendental equation derived by Delagebeaudeuf and Linh [1]:

$$qn_S = \left[2\epsilon_2\, N_D(\Delta E_c - \delta - E_{Fi}[n_S]) + q^2\, N_D^2 e^2\right]^{\frac{1}{2}} - qN_D e.$$

Use the following parameters:

Silicon doping: $N_D = 10^{24}$ m^{-3}

Dielectric constant: $\epsilon_2 = 13.1\epsilon$

Conduction-band-edge shift: $\Delta E_c = 0.25$ eV

Fermi level position: $\delta = 0.005$ eV

Spacer width: $e = 100$ Å

(d) Using the expression for E_{Fi} given in part c, calculate a, b and Δd when n_{S1} is the equilibrium 2DEG concentration n_{S0}, 5×10^{15} m^{-2} and 1×10^{15} m^{-2}. Calculate also the fractional capacitance reduction $d_2/(d_2 + \Delta d)$ using an AlGaAs width $d_2 = 300$ Å.

8.2 Consider an Au–AlGaAs–GaAs MODFET capacitor in which only a narrow region of width W_p is doped with silicon impurities in the AlGaAs region (see Figure 8.10). Such a structure is called a pulse-doped MODFET.

(a) Calculate the potential drop between the gate and the interface $V_2 = V[d_2] - V[0]$.

(b) Demonstrate that the threshold voltage can be written

$$V_T = \phi_B - \frac{\Delta E_c}{q} + \frac{E_{F0}}{q} - V_P,$$

where the pinch-off voltage is now

$$V_P \propto W_p\left(W_e - \frac{W_p}{2}\right),$$

with $W_e = d_2 - e$. Calculate V_P.

(c) Elaborate on the reason why an improved control of the threshold voltage results with the pulse-doped MODFET. For this purpose compare the derivative

$$\frac{\partial V_P}{\partial(N_{Dp}W_p)}$$

for the pulse-doped MODFET and the derivative

$$\frac{\partial V_P}{\partial(N_D W_e)}$$

for the uniformly doped MODFET assuming that the same areal doping is used ($N_{Dp}W_p = N_D W_e$).

9 High electric field transport

$S = k \log \Omega.$

Carved on the tombstone of Ludwig Boltzmann

9.1 Introduction

In Chapter 8 we studied the two-dimensional electron gas (2DEG) and its control with a gate electrode. As we shall see in Chapter 10, the 2DEG is used as the channel of a high-speed FET, the MODFET. We therefore need to develop a picture of horizontal transport in the 2DEG before studying the MODFET. The transport equations developed in this chapter will also be used for the analysis of the heterojunction bipolar transistor (HBT) in Chapter 18

Our analysis of transport in Chapters 4–7 assumed that the electron transport was mostly ballistic, i.e., the mean free path was longer than or comparable to the quantum device length. In this chapter we shall assume instead that the scale upon which the device variation takes place is large compared to the electron mean free path so that no appreciable quantum effects are expected. Indeed, multiple scattering events randomize the phase of the electron so that neglecting quantum interferences is a reasonable approximation in devices of length larger than 1000 Å. As a result a semiclassical analysis that describes the electrons as a gas of classical (known position and momentum) particles in a band (e.g., the conduction band) should be sufficient to study horizontal transport in submicron gate-length FETs (0.1–1 μm).

In this chapter we shall review the existing picture of transport developed for the three-dimensional electron gas (3DEG) based on the Boltzmann equation formalism. We will see, for example, how we can derive the semiclassical transport equations introduced in Chapter 2 for heterostructures. But the focus of this chapter is on the development of a transport model for high and non-uniform electric fields. Indeed the electrostatic potential varies rapidly from source to drain in submicron MODFETs leading to very high electric fields in the drain region. To handle transport in such a non-uniform high electric field we shall see that modified drift and diffusion equations are required.

Our analysis of high-field transport will be limited to electrons in the conduction band since we are mostly interested in n-channel FETs. Although the derivations are carried out for a 3DEG, to first order the application to a 2DEG of the simple transport models derived here will simply involve replacing the 3/2 equipartition energy factor by the 2/2 factor.

9.2 The Boltzmann equation

In thermal equilibrium the electron distribution in the conduction band of a uniform semiconductor is given by the Fermi–Dirac distribution:

$$f_{FD} = \frac{1}{\exp\left(\frac{E - E_F}{k_B T_0}\right) + 1},$$

where E is the total electron energy, T_0 is the lattice temperature and E_F is the Fermi level. In the conduction band, the total electron energy is $E = E_c + \frac{1}{2}m^* v^2 = \mathcal{E}_c - qV + \frac{1}{2}m^* v^2$, where E_c is the bottom of the conduction band, $v = |\mathbf{v}|$ is the electron velocity, and m^* is the electron effective mass. We assume that the conduction band is parabolic so that we have $\hbar\mathbf{k} = m^*\mathbf{v}$.

For a non-degenerate system ($E_c - E_F \gg kT_0$), f_{FD} reduces to the Maxwell–Boltzmann distribution

$$f_{MB} = \exp\left(-\frac{m^* v^2}{2k_B T_0}\right) \exp\left(\frac{E_F - E_c}{k_B T_0}\right).$$

The total number of electrons n in the conduction band is then given by

$$n = \int_{-\infty}^{\infty} D(\mathbf{k})\, f_{MB}(v)\, d\mathbf{k} = \frac{m^{*3}}{4\pi^3\hbar^3} \int_{-\infty}^{\infty} f_{MB}(v)\, d\mathbf{v} = \int_{-\infty}^{\infty} f_0(v)\, d\mathbf{v},$$

with f_0 given by

$$\begin{aligned} f_0(v) &= \frac{2m^{*3}}{h^3} f_{MB} = \frac{2m^{*3}}{h^3} \exp\left(-\frac{m^* v^2}{2k_B T_0}\right) \exp\left(\frac{E_F - E_c}{k_B T_0}\right) \\ &= n\left(\frac{m^*}{2\pi k_B T_0}\right)^{\frac{3}{2}} \exp\left(-\frac{m^* v^2}{2k_B T_0}\right). \end{aligned} \tag{9.1}$$

$f_0(v)$, which gives the electron distribution in velocity space, is called the Boltzmann function.

The average electron kinetic energy for such a distribution is

$$\left\langle \frac{1}{2}m^* v^2 \right\rangle = \frac{\int_{-\infty}^{\infty} \frac{1}{2}m^* v^2 f_0\, d\mathbf{v}}{\int_{-\infty}^{\infty} f_0\, d\mathbf{v}} = \frac{3}{2}k_B T_0.$$

The average thermal velocity v_{th} of the 3DEG is defined from the average electron energy

$$\frac{1}{2}m^* v_{\text{th}}^2 = \frac{3}{2}k_B T_0.$$

Note, however, that the average electron velocity along any direction is zero. For example the average electron velocity $\langle v_x \rangle$ is

$$\langle v_x \rangle = \frac{\int_{-\infty}^{\infty} v_x f_0 \, d\mathbf{v}}{\int_{-\infty}^{\infty} f_0 \, d\mathbf{v}} = 0.$$

The Maxwell–Boltzmann electron distribution we have introduced holds only for the uniform semiconductor under equilibrium conditions. In a non-uniform semiconductor under transient conditions, we can also describe the semiclassical electron gas with its distribution f in velocity space $\mathbf{v} = (v_x, v_y, v_z)$. However, this distribution is now a function of position $\mathbf{r} = (x, y, z)$ and time t:

$$f = f(x, y, z, v_x, v_y, v_z, t) = f(\mathbf{r}, \mathbf{v}, t).$$

The total number of electrons at position $\mathbf{r} = (x, y, z)$ and time t is still given by

$$n(\mathbf{r}, t) = \int_{-\infty}^{\infty} f(\mathbf{r}, \mathbf{v}, t) \, d\mathbf{v}.$$

The net current density J_x flowing along the x axis is obtained from the average electron velocity v_x by

$$J_x = -qn\langle v_x \rangle = -q \int_{-\infty}^{\infty} v_x f(\mathbf{r}, \mathbf{v}, t) \, d\mathbf{v}.$$

The conservation of particles is expressed in the absence of scattering by the total differential:

$$df = \sum_{i=x,y,z} \frac{\partial f}{\partial v_i} dv_i + \sum_{i=x,y,z} \frac{\partial f}{\partial x_i} dx_i + \frac{\partial f}{\partial t} dt = 0. \tag{9.2}$$

In the presence of scattering, the conservation of particles is governed by

$$\begin{aligned} \frac{df(\mathbf{v}, \mathbf{r}, t)}{dt} &= \int_{-\infty}^{\infty} S(\mathbf{v}', \mathbf{v}) \; [1 - f(\mathbf{v}, \mathbf{r}, t)] \, f(\mathbf{v}', \mathbf{r}, t) \, d\mathbf{v}' \\ &\quad - \int_{-\infty}^{\infty} S(\mathbf{v}, \mathbf{v}') \; \left[1 - f(\mathbf{v}', \mathbf{r}, t)\right] f(\mathbf{v}, \mathbf{r}, t) \, d\mathbf{v}', \end{aligned}$$

where $S(\mathbf{v}', \mathbf{v})$ is the probability of scattering per unit time from the initial state $\mathbf{v}'$ to the final state $\mathbf{v}$. Note that the density of states factor $m^{*3}/(4\pi^3\hbar^3)$ is included in the scattering probability S for notational simplicity. This equation states that the number of electrons in the state $(\mathbf{v}, \mathbf{r})$ is increased by the electrons scattered in this cell and decreased by the electrons scattered out of this cell. Note that the integrand of the scattering-in term from $\mathbf{v}'$ to $\mathbf{v}$ is proportional to the number of electrons $f(\mathbf{v}', \mathbf{r}, t)$ available in the state $\mathbf{v}'$. Similarly, the scattering-out term from $\mathbf{v}$ to $\mathbf{v}'$ is proportional to the number of electrons $f(\mathbf{v}, \mathbf{r}, t)$ available in the state $\mathbf{v}$. Both the scattering-in and scattering-out terms include a factor $(1 - f)$ associated with the final state so as to satisfy the Pauli exclusion principle which does not allow for more than one electron per state. We shall assume in the rest of this chapter that the electron gas is not degenerate (dilute solution) so that the electron state occupation is much smaller than 1 and the Pauli exclusion terms can be neglected. Under such dilute solution conditions we can then simply write

$$\frac{df(\mathbf{v}, \mathbf{r}, t)}{dt} = \int_{-\infty}^{\infty} S(\mathbf{v}', \mathbf{v}) f(\mathbf{v}', \mathbf{r}, t)\, d\mathbf{v}' - \int_{-\infty}^{\infty} S(\mathbf{v}, \mathbf{v}') f(\mathbf{v}, \mathbf{r}, t)\, d\mathbf{v}'. \tag{9.3}$$

Combining Equations (9.2) and (9.3), we obtain the so-called Boltzmann equation:

$$\sum_{i=x,y,z} \frac{\partial f}{\partial v_i} \frac{dv_i}{dt} + \sum_{i=x,y,z} \frac{\partial f}{\partial x_i} \frac{dx_i}{dt} + \frac{\partial f}{\partial t}$$
$$= \int_{-\infty}^{\infty} \left[S(\mathbf{v}', \mathbf{v}) f(\mathbf{v}', \mathbf{r}, t) - S(\mathbf{v}, \mathbf{v}') f(\mathbf{v}, \mathbf{r}, t) \right] d\mathbf{v}'.$$

Let us now consider a semiconductor with an electric field $F(x)$ applied along the x axis. The electron distribution $f(x, \mathbf{v}, t)$ then varies in space along the x axis only. Furthermore, from the acceleration theorem we have

$$\frac{dv_x}{dt} = -\frac{qF(x)}{m^*} \quad \text{and} \quad \frac{dv_{y,z}}{dt} = 0.$$

The Boltzmann equation for this system reduces to

$$\frac{\partial f}{\partial t} = \frac{qF(x)}{m^*} \frac{\partial f}{\partial v_x} - v_x \frac{\partial f}{\partial x} + \int_{-\infty}^{\infty} \left[S(\mathbf{v}', \mathbf{v}) f(\mathbf{v}') - S(\mathbf{v}, \mathbf{v}') f(\mathbf{v}) \right] d\mathbf{v}'. \tag{9.4}$$

The solution of this Boltzmann equation gives the electron distribution $f(x, \mathbf{v}, t)$ in a semiconductor subjected to the spatially-varying electric field $F(x)$. Let us note that if we integrate this Boltzmann equation in velocity space we obtain the so-called continuity equation

$$\frac{\partial n}{\partial t} = -\frac{\partial}{\partial x}(n \langle v_x \rangle) = \frac{\partial}{\partial x}\left(\frac{J}{q}\right),$$

which enforces the macroscopic conservation of particles.

Note that for a uniform semiconductor in the absence of an electric field, the steady-state Boltzmann equation reduces to

$$\int_{-\infty}^{\infty} \left[S(\mathbf{v}', \mathbf{v}) f_0(\mathbf{v}') - S(\mathbf{v}, \mathbf{v}') f_0(\mathbf{v}) \right] d\mathbf{v}' = 0,$$

where $f_0(\mathbf{v}')$ is the Boltzmann distribution. By virtue of the principle of detailed balance this identity is satisfied provided that we have

$$\frac{S(\mathbf{v},\mathbf{v}')}{S(\mathbf{v}',\mathbf{v})} = \frac{f_0(\mathbf{v}')}{f_0(\mathbf{v})} = \exp\left[\frac{m^*(v^2 - v'^2)}{2k_B T_0}\right]. \tag{9.5}$$

For elastic scattering, the kinetic energy of the electron does not change in the scattering process and the identity results from the Hermiticity of the matrix element. For inelastic scattering the kinetic energy of the electron changes and the different scattering rate results from the fact that stimulated emission is favored over stimulated absorption. Assume, for example, that the scattering process $S(\mathbf{v},\mathbf{v}')$ corresponds to the emission of a phonon $\omega_{\mathbf{q}}$. The scattering process $S(\mathbf{v}',\mathbf{v})$ then corresponds to the absorption of a phonon and we have (see Chapter 5)

$$\frac{S(\mathbf{v},\mathbf{v}')}{S(\mathbf{v}',\mathbf{v})} = \frac{N_{\mathbf{q}}+1}{N_{\mathbf{q}}} = \exp\left(\frac{\hbar\omega_{\mathbf{q}}}{k_B T_0}\right) \tag{9.6}$$

with $N_{\mathbf{q}}$, the average number of phonons, given by the Bose–Einstein distribution. The identification of Equation (9.5) with Equation (9.6) is completed when we use the conservation of energy identity $\frac{1}{2}m^*v^2 = \frac{1}{2}m^*v'^2 + \hbar\omega_{\mathbf{q}}$.

9.3 Electron transport in small electric fields

9.3.1 Uniform semiconductor case

Let us now consider the case of a uniform semiconductor subjected to a constant electric field F applied along the x axis. We shall now search for a steady-state solution of the Boltzmann equation. The electron distribution $f(\mathbf{v})$ is uniquely a function of velocity $\mathbf{v}$:

$$\frac{\partial f}{\partial t} = 0 \quad \text{and} \quad \frac{\partial f}{\partial x_i} = 0.$$

The Boltzmann equation then reduces to

$$0 = \frac{qF}{m^*}\frac{\partial f(\mathbf{v})}{\partial v_x} + \int_{-\infty}^{\infty} \left[S(\mathbf{v}',\mathbf{v})f(\mathbf{v}') - S(\mathbf{v},\mathbf{v}')f(\mathbf{v})\right] d\mathbf{v}'.$$

In the limit of a small electric field F, a solution can be obtained using a Taylor series expansion

$$f(\mathbf{v}) = f_0(v) + f_1(\mathbf{v}) + \cdots = f_0 + Fh + F^2 g + \cdots,$$

where f_0 is the equilibrium distribution and $f_1 = Fh$ a small perturbation (first order in F). The first order equation (i.e., $\propto F$) obtained is

$$0 = \frac{qF}{m^*}\frac{df_0(v)}{dv_x} + \int_{-\infty}^{\infty} \left[S(\mathbf{v}',\mathbf{v})f_1(\mathbf{v}') - S(\mathbf{v},\mathbf{v}')f_1(\mathbf{v})\right] d\mathbf{v}'.$$

For isotropic scattering, $S(\mathbf{v}', \mathbf{v})$ is even in $\mathbf{v}'$ and $\mathbf{v}$. Since f_0 is even in $\mathbf{v}$, f_1 must be odd in $\mathbf{v}$, and we have

$$\int_{-\infty}^{\infty} S(\mathbf{v}', \mathbf{v}) f_1(\mathbf{v}')\, d\mathbf{v}' = 0.$$

The Boltzmann equation then reduces to the following equation:

$$\frac{qF}{m^*}\frac{df_0(v)}{dv_x} = f_1(\mathbf{v}) \int_{-\infty}^{\infty} S(\mathbf{v}', \mathbf{v})\, d\mathbf{v}' = \frac{f_1(\mathbf{v})}{\tau_m(v)},$$

where we define $\tau_m = \tau_{TOT}$ as the integral

$$\frac{1}{\tau_{TOT}(v)} = \int_{-\infty}^{\infty} S(\mathbf{v}, \mathbf{v}')\, d\mathbf{v}'.$$

τ_{TOT} is the lifetime of an electron in the state $\mathbf{v}$ since $1/\tau_{TOT}$ is the total scattering rate. For non-randomizing elastic scattering, the momentum scattering rate τ_m can be demonstrated to be [3]

$$\frac{1}{\tau_m(\mathbf{v})} = \int_{-\infty}^{\infty} S(\mathbf{v}, \mathbf{v}')(1 - \cos\theta)\, d\mathbf{v}'$$

where θ is the angle between $\mathbf{v}$ and $\mathbf{v}'$.

We can now calculate f_1, the first order correction to the electron distribution,

$$f_1(\mathbf{v}) = \tau_m(\mathbf{v})\frac{qF}{m^*}\frac{df_0(v)}{dv_x} = -\frac{\tau_m(\mathbf{v})qF}{k_B T_0} f_0(v) v_x. \tag{9.7}$$

Once the electron distribution is known, the drift current J is calculated using

$$J = -qn\langle v_x \rangle = -q\int_{-\infty}^{\infty} v_x f\, d\mathbf{v} = -q\int_{-\infty}^{\infty} v_x f_1\, d\mathbf{v}.$$

It is customary to rewrite J as $J = qnv_d = q\mu_n nF$, where the drift velocity is defined by $v_d = -\langle v_x \rangle$ and the electron mobility μ_n is defined by

$$\mu_n = -\frac{\langle v_x \rangle}{F} = \frac{v_d}{F} = -\frac{1}{F}\frac{\int_{-\infty}^{\infty} v_x f(\mathbf{v})\, d\mathbf{v}}{\int_{-\infty}^{\infty} f(\mathbf{v})\, d\mathbf{v}}. \tag{9.8}$$

Substituting Equation (9.7) into the mobility definition of Equation (9.8) we obtain the following low-field mobility:

$$\mu_n = \frac{q}{k_B T_0}\frac{\int_{-\infty}^{\infty} \tau_m(\mathbf{v}) v_x^2 f_0\, d\mathbf{v}}{\int_{-\infty}^{\infty} f_0\, d\mathbf{v}}$$

For isotropic $\tau_m(v)$ the low-field mobility can be written after integration in the form

$$\mu_n = \frac{q}{m^*}\langle \tau_m \rangle,$$

where the average relaxation time $\langle \tau_m \rangle$ is defined to be

$$\langle \tau_m \rangle = \frac{1}{\Gamma(\frac{5}{2})} \int_0^\infty \tau_m(v) X^{3/2} \exp(-X)\, dX,$$

with $X = m^* v^2/(2k_B T_0)$. When the momentum relaxation time τ_m follows the power law

$$\tau_m = \alpha v^r = \tau_0(T_0) X^{\frac{r}{2}} \qquad \text{with} \qquad \tau_0(T_e) = \alpha \left(\frac{2k_B T_e}{m^*} \right)^{\frac{r}{2}}$$

($r = -1$ for acoustic deformation potential phonon scattering and $r = 3$ for ionized impurity scattering) one can verify that we have

$$\langle \tau_m \rangle = \tau_0 \frac{\Gamma(\frac{r}{2} + \frac{5}{2})}{\Gamma(\frac{5}{2})}, \tag{9.9}$$

where the Gamma function is defined by

$$\Gamma(p) = \int_0^\infty X^{p-1} \exp(-X)\, dX = (p-1)\Gamma(p-1).$$

For a small electric field, the total electron distribution f is well represented by the equilibrium distribution f_0 shifted by the drift velocity $v_d = \mu_n F$ along the x axis. Indeed, one can easily verify by a Taylor series expansion that we have

$$f = f_0 + f_1 \simeq n \left(\frac{m^*}{2\pi k_B T_0} \right)^{\frac{3}{2}} \exp\left(-\frac{m^*}{2k_B T_0} |\mathbf{v} + \mathbf{v}_d|^2 \right),$$

with $\mathbf{v}_d = -\langle \mathbf{v} \rangle = v_d \hat{x}$.

9.3.2 Non-uniform semiconductor case

Let us now solve the Boltzmann equation for a non-uniform semiconductor in a steady state for small gradients of the forces applied. We start from the Boltzmann equation derived in Section 9.2:

$$0 = \frac{qF(x)}{m^*} \frac{\partial f(x, \mathbf{v})}{\partial v_x} - v_x \frac{\partial f(x, \mathbf{v})}{\partial x} - \frac{f(x, \mathbf{v}) - f_0(x, v)}{\tau_m},$$

where we have replaced the collision integral by the relaxation expression $-(f - f_0)/\tau_m = -f_1/\tau_m$ which holds for small departures of f from f_0 (see previous section).

For small gradients of the potentials, we can use the following approximation:

$$v_x \frac{\partial f(x, \mathbf{v})}{\partial x} \simeq v_x \frac{\partial f_0(x, v)}{\partial x}.$$

According to Equation (9.1) the equilibrium distribution f_0 in a spatially-varying semiconductor is given by

$$f_0(x, v) = \frac{2m^{*3}}{h^3} \exp\left[\frac{E_{Fi}(x) - \frac{1}{2}m^* v^2}{k_B T_0}\right],$$

where $E_{Fi}(x)$, the chemical potential, is given by $E_{Fi}(x) = E_F(x) - E_c(x)$ with $E_c(x) = \mathcal{E}_c(x) - qV(x)$ the absolute conduction band edge at position x. This leads to

$$v_x \frac{\partial f(x, \mathbf{v})}{\partial x} \simeq v_x f_0(x, v) \frac{1}{k_B T_0} \frac{dE_{Fi}(x)}{dx}.$$

We can then use the method of perturbation ($f = f_0 + f_1$) and evaluate f_1 to be

$$f_1(x, \mathbf{v}) = -\tau_m \frac{q}{k_B T_0}\left[F(x) + \frac{1}{q}\frac{dE_{Fi}(x)}{dx}\right] v_x f_0(x, v).$$

We see that f_1 has the same form as in the previous section if we introduce the generalized electric field $\mathcal{F}$

$$\mathcal{F}(x) = F(x) + \frac{1}{q}\frac{dE_{Fi}(x)}{dx}.$$

Since the chemical potential E_{Fi} is related to the Fermi level (electrochemical potential) by $E_{Fi} = E_F - E_c = E_F - \mathcal{E}_c + qV$, we have $F = (1/q)(dE_c/dx)$ and we can express the generalized electric field $\mathcal{F}$ in terms of the Fermi energy:

$$\mathcal{F}(x) = \frac{1}{q}\frac{dE_F(x)}{dx}.$$

The current density $J = qn\mu_n\mathcal{F}$ is then given by

$$J = \mu_n(x)\, n(x) \frac{dE_F(x)}{dx},$$

where the Fermi energy $E_F(x)$ is obtained from

$$n(x) = N_c(x) \exp\left[\frac{E_{Fi}(x)}{k_B T_0}\right] = N_c(x) \exp\left[\frac{E_F(x) - \mathcal{E}_c(x) + qV(x)}{k_B T_0}\right]. \tag{9.10}$$

Our derivation was general and this equation applies also to semiconductor heterostructures (see Marshak *et al.* [1] for a deeper discussion). Extracting the Fermi level from Equation (9.10) we obtain

$$E_F(x) = \mathcal{E}_c(x) - qV(x) + kT_0 \ln n(x) - kT_0 \ln N_c(x).$$

The current density obtained is then

$$J = -q\mu_n n \frac{dV}{dx} + qD_n \frac{dn}{dx} + \mu n \frac{d\mathcal{E}_c}{dx} - kT_0 \mu_n \frac{n}{N_c}\frac{dN_c}{dx},$$

where the first term is the drift current, the second is the diffusion current with the diffusion constant $qD_n = kT_0\mu_n$, the third is a quasi-drift current arising from the variation of electron affinity $\mathcal{E}_c$, and the fourth is a quasi-diffusion current arising from the variation of the effective carrier density N_c. This current equation was used in Chapter 2 to study the p–n heterojunction.

9.4 Electron transport in a large electric field

We shall now consider the solution of the Boltzmann equation when a large electric field is applied. We shall first study the case of a uniform semiconductor before considering the non-uniform semiconductor.

9.4.1 Uniform semiconductor case

For a uniform material subjected to a constant electric field F we found that the Boltzmann equation under steady-state conditions was

$$-\frac{qF}{m^*}\frac{df(\mathbf{v})}{dv_x} = \left.\frac{\partial f}{\partial t}\right|_{coll} = \int_{-\infty}^{\infty}\left[S(\mathbf{v}',\mathbf{v})f(\mathbf{v}') - S(\mathbf{v},\mathbf{v}')f(\mathbf{v})\right]d\mathbf{v}'.$$

Let us now write f in terms of its even part f_{even} and its odd part f_{odd}:

$$f = f_{even} + f_{odd}.$$

Consequently, we can symbolically split the collision term $(\partial f/\partial t)|_{coll}$

$$\left.\frac{\partial f}{\partial t}\right|_{coll} = \left.\frac{\partial f_{odd}}{\partial t}\right|_{coll} + \left.\frac{\partial f_{even}}{\partial t}\right|_{coll}.$$

In equilibrium when $f_{even} = f_0$, electrons should be scattered in and out of state $\mathbf{v}$ at the same rate so that

$$\left.\frac{\partial f}{\partial t}\right|_{coll} = \int_{-\infty}^{\infty}\left[S(\mathbf{v}',\mathbf{v})f_0(\mathbf{v}') - S(\mathbf{v},\mathbf{v}')f_0(\mathbf{v})\right]d\mathbf{v}' = 0.$$

This is the so-called principle of detailed balance. Since the departure of f_{even} from f_0 introduces a new scattering contribution in the even collision integral $(\partial f_{even}/\partial t)|_{coll}$ we shall approximate it by

$$\left.\frac{\partial f_{even}}{\partial t}\right|_{coll} = -\frac{f_{even} - f_0}{\tau_E},$$

where τ_E will be called the energy relaxation time. τ_E is typically on the order of picoseconds.

Following the low-field analysis, the odd collision integral can be written

$$\left.\frac{\partial f_{odd}}{\partial t}\right|_{coll} = -\frac{f_{odd}}{\tau_m},$$

where the momentum relaxation time τ_m is on the order of 0.1–0.01 ps. Substituting $f = f_{even} + f_{odd}$ into the Boltzmann equation and equating the even and odd terms, we obtain

$$\left.\begin{aligned} -\frac{qF}{m^*}\frac{df_{even}}{dv_x} &= -\frac{f_{odd}}{\tau_m}, \\ -\frac{qF}{m^*}\frac{df_{odd}}{dv_x} &= -\frac{f_{even} - f_0}{\tau_E}. \end{aligned}\right\} \tag{9.11}$$

Substituting the first equation into the second, we obtain the following equation for f_{even}

$$\frac{q^2F^2}{m^{*2}}\tau_m\frac{d^2 f_{even}}{dv_x^2} = \frac{f_{even} - f_0}{\tau_E}. \tag{9.12}$$

Let us try the following approximate solution for f_{even}

$$f_{even} = n\left(\frac{m^*}{2\pi k_B T_e}\right)^{\frac{3}{2}} \exp\left(-\frac{m^* v^2}{2k_B T_e}\right).$$

Using this approximate solution and the equation for f_{even} one can demonstrate (see Problem 9.1) that the electron temperature T_e must satisfy the following energy balance equation:

$$\frac{\frac{3}{2}k_B(T_e - T_0)}{\langle\tau_E\rangle} = qF^2\frac{q\langle\tau_m\rangle}{m^*} = qFv_d.$$

This equation states that the electron gas is heated by the electric field by the Joule effect. The electron energy gained is then dissipated by the lattice at a rate set by the energy relaxation time $\langle\tau_E\rangle$. When the Joule effect is suppressed ($F = 0$) the electron temperature T_e relaxes to the lattice temperature T_0.

The odd part of the distribution can then be calculated from the even part of the distribution. The total electron distribution can still be approximated by a displaced Maxwell–Boltzmann distribution (see Problem 9.2) but now with temperature T_e

$$f = f_{even} + f_{odd} \simeq n\left(\frac{m^*}{2\pi k_B T_e}\right)^{\frac{3}{2}} \exp\left(-\frac{m^*|\mathbf{v} + \mathbf{v}_d|^2}{2k_B T_e}\right).$$

The drift velocity v_d is calculated from the energy balance equation. In silicon the energy relaxation time τ_E can be approximated by [2]

$$\langle\tau_E\rangle \simeq 4\times 10^{-12}\left(\frac{T_e}{T_0}\right)^{1/2} = \alpha\left(\frac{T_e}{T_0}\right)^{1/2}.$$

Assuming that the dominant scattering process is the acoustic deformation potential, τ_m varies as $1/v$ [3] and one can easily derive (see Problem 9.3) that the dependence of μ_n upon the electron temperature T_e is given by

$$\mu_n(T_e) = \mu_n(T_0)\left(\frac{T_0}{T_e}\right)^{1/2}.$$

This equation states that the electron mobility will decrease with the electron temperature if the electron gas has a Maxwellian distribution. We can rewrite the energy conservation as follows:

$$\frac{3}{2}k_B\frac{(T_e - T_0)}{\alpha\,(T_e/T_0)^{1/2}} = q\mu_n(T_0)\left(\frac{T_e}{T_0}\right)^{1/2}F^2,$$

from which we obtain the electron temperature T_e

$$T_e = T_0 + \frac{2}{3}\frac{\alpha}{k_B}\mu_n(T_0)F^2.$$

The electron drift velocity is then

$$v_d = \mu_n(T_e)F = \frac{\mu_n(T_0)F}{\left[1 + \frac{2}{3}\frac{\alpha}{k_B T_0}\mu_n(T_0)F^2\right]^{1/2}}$$

$$v_d = \mu_n(F)F = \frac{\mu_n(T_0)F}{\left(1 + F^2/F_c^2\right)^{1/2}}.$$

The resulting velocity-field relation is plotted in Figure 9.1. One observes that the drift velocity saturates for large electric fields. Note that *the concept of a field-dependent drift velocity* $v_d(F)$ and a field-dependent electron mobility $\mu_n(F) = v_d(F)/F$, *is only valid for a uniform semiconductor* when there is no diffusion current. This will be demonstrated in the next section.

9.4.2 Non-uniform semiconductor case

Under high-field conditions the even and odd Boltzmann equations are easily verified to be

$$v_x\frac{\partial f_{even}}{\partial x} - \frac{qF}{m^*}\frac{df_{even}}{dv_x} = -\frac{f_{odd}}{\tau_m},$$

$$v_x\frac{\partial f_{odd}}{\partial x} - \frac{qF}{m^*}\frac{df_{odd}}{dv_x} = -\frac{f_{even} - f_0}{\tau_E}.$$

Let us assume (the temperature model) that f_{even} is of the form:

$$f_{even} = \frac{2m^{*3}}{h^3}\exp\left[-\frac{m^* v^2}{2k_B T_e(x)}\right]\exp\left[\frac{E_{Fi}(x)}{k_B T_e(x)}\right]$$

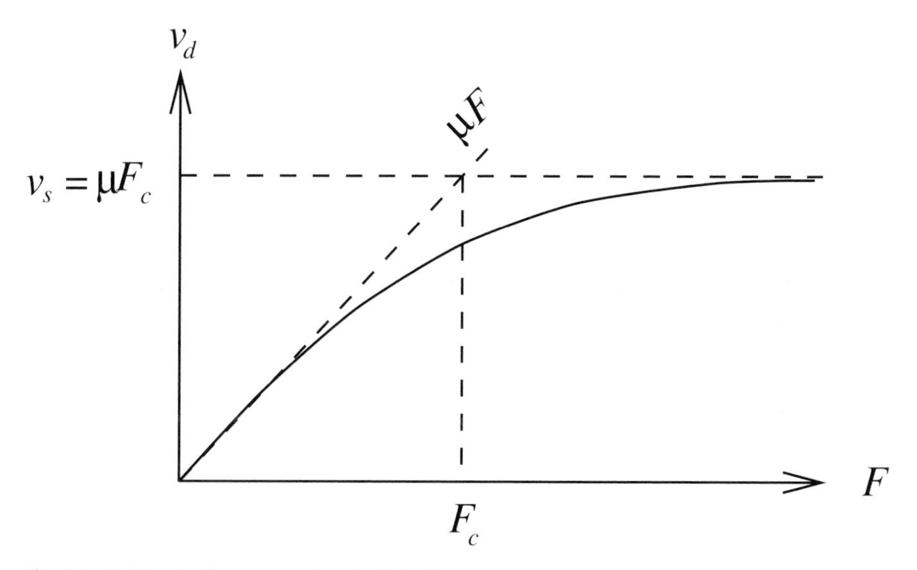

Fig. 9.1. Drift velocity versus electric field for a uniform (bulk) semiconductor in the single-valley approximation.

and let us evaluate the following term:

$$v_x \frac{\partial f_{even}}{\partial x} = v_x f_{even} \left[\frac{1}{k_B T_e} \frac{dE_{Fi}}{dx} - \left(\frac{E_{Fi} - \frac{1}{2} m^* v^2}{k_B} \right) \frac{1}{T_e^2} \frac{dT_e}{dx} \right].$$

We recognize the diffusion term involving dE_{Fi}/dx and a new term involving the temperature gradient dT_e/dx. This new term introduces a new component into f_{odd}:

$$f_{odd} = f_{odd}|_{drift+diffusion} + \tau_m \left(\frac{E_{Fi} - \frac{1}{2} m^* v^2}{k_B} \right) \frac{1}{T_e^2} \frac{dT_e}{dx} f_{even}\, v_x.$$

This in turn introduces a new component into the average velocity $\langle v_x \rangle$:

$$n\langle v_x \rangle = n\langle v_x \rangle|_{drift+diffusion} + \frac{1}{T_e^2} \frac{dT_e}{dx} \int_{-\infty}^{\infty} \tau_m \left(\frac{E_{Fi} - \frac{1}{2} m^* v^2}{k_B} \right) f_{even} v_x^2 \, d\mathbf{v}. \tag{9.13}$$

Let us now assume that the momentum relaxation time τ_m follows the power law $\tau_m = \alpha v^r = \tau_0(T_e) X^{\frac{r}{2}}$ with $X = m^* v^2/(2k_B T_e)$. For acoustic deformation potential phonon-scattering $r = -1$ and for ionized impurity scattering $r = 3$. Using Equation (9.9) and the identity

$$\Gamma\left(\frac{r}{2} + \frac{7}{2}\right) = \left(\frac{r}{2} + \frac{5}{2}\right) \Gamma\left(\frac{r}{2} + \frac{5}{2}\right)$$

we obtain after integrating Equation (9.13)

$$n\langle v_x \rangle = n\langle v_x \rangle|_{drift+diffusion} + n\mu_n(T_e) \frac{1}{q} \left[\frac{E_{Fi}}{T_e} - k_B \left(\frac{r}{2} + \frac{5}{2} \right) \right] \frac{dT_e}{dx}.$$

The conduction current J is then defined as

$$J = -qn\langle v_x\rangle$$
$$= \mu_n(T_e)n(x)\left\{\frac{dE_F(x)}{dx} + \left[\left(\frac{r}{2}+\frac{5}{2}\right) - \frac{E_{Fi}}{k_B T_e}\right]k_B\frac{dT_e(x)}{dx}\right\}, \quad (9.14)$$

where $E_F(x) = E_{Fi}(x) + \mathcal{E}_c(x) - qV(x)$ is the Fermi energy (the electrochemical potential) and $E_{Fi}(x)$ the chemical potential obtained from

$$n = N_c(T_e)\exp\left[\frac{E_{Fi}(x)}{k_B T_e(x)}\right],$$

with $N_c(T_e)$ the effective density of states:

$$N_c(T_e) = 2\left(\frac{2\pi m_n^* k_B T_e}{h^2}\right)^{3/2}.$$

Substituting E_F in Equation (9.14) one can easily verify (see Problem 9.4) that for a uniform material ($\mathcal{E}_c$ not spatially varying) the current density J can be rewritten as

$$J = -q\mu_n(T_e)n(x)\frac{dV(x)}{dx} + q\frac{d}{dx}[D_n(T_e)n(x)], \quad (9.15)$$

where the diffusion constant $D_n(T_e)$ is defined by the Einstein relation

$$\frac{D_n(T_e)}{\mu_n(T_e)} = \frac{k_B T_e}{q}.$$

This equation is similar to the regular drift-diffusion equation except that now the diffusion constant is *inside* the spatial derivative and we must know the electron temperature T_e to calculate the mobility and the diffusion constants.

Note that Equation (9.15) can be rewritten in a form showing explicitly the contribution of the temperature or energy gradient

$$J = -q\mu_n(T_e)n(x)\frac{dV(x)}{dx} + qD_n(T_e)\frac{d}{dx}[n(x)] + (r+2)\mu_n(T_e)n(x)\frac{du}{dx}, \quad (9.16)$$

where $u = \frac{1}{2}k_B T_e$.

In the non-uniform semiconductor, the electron temperature T_e is given by the following approximate energy balance equation

$$\frac{\partial}{\partial t}\left(n\frac{3}{2}k_B T_e\right) = \frac{\partial}{\partial x}\left[-\frac{J}{(-q)}\frac{3}{2}k_B T_e\right] + J\cdot F - n\frac{\frac{3}{2}k_B(T_e - T_0)}{\tau_E(T_e)}. \quad (9.17)$$

This equation is obtained using a procedure similar to the one employed in the uniform case (see Problem 9.1). This energy balance equation gives the rate of variation of the electron gas energy (left-hand side). The terms on the right-hand side are recognized to be respectively the gradient of energy flux, the Joule effect term and the energy relaxation term.

As we can see this transport picture proposed for non-uniform semiconductors relies on a temperature-dependent mobility $\mu_n(T_e)$ and diffusion constant $D_n(T_e)$ instead of a field-dependent mobility $\mu_n(F)$ and diffusion constant $D_n(F)$. In fact, the use of field-dependent mobility $\mu_n(F)$ and diffusion constant $D_n(F)$ which we introduced for a bulk semiconductor is incorrect for non-uniform semiconductors. To demonstrate this, consider a semiconductor in equilibrium. By definition the lattice and the electron gas are in equilibrium when the electron temperature T_e relaxes to the lattice temperature T_0. However, a large electric field can be present inside non-uniform semiconductors. For example, consider the case of a p–n junction in equilibrium. A large built-in electric field is usually present at the p–n junction. The electron gas is not heated however by the large built-in electric field via the Joule effect $J \cdot F$ since the total current J is zero. The drift and diffusion currents therefore are evaluated using the low-field mobility $\mu_n(T_0)$ and diffusion constant $D_n(T_0)$, respectively. The use of the field-dependent mobility and diffusion constants to calculate the equilibrium drift and diffusion currents is therefore incorrect.

Note, however, that as soon as the current J departs from zero, a large Joule effect $J \cdot F$ results in the presence of a large built-in electric field F_0. Gunn [4] has indeed demonstrated rigorously that in the presence of an equilibrium built-in field F_0 the effective mobility of the electron for a small departure from equilibrium is the chordal electron mobility which would result if the equilibrium built-in field F_0 was applied by some external means. Therefore if the total electric field is $F = F_0 + F_1$ with F_0 the equilibrium built-in electric field and F_1 a small DC applied perturbation electric field, the resulting current is given by

$$J = qn\mu_n(F_0)F_1$$

with $\mu_n(F_0)$ the chordal mobility given by

$$\mu_n(F_0) = \frac{v_d(F_0)}{F_0}.$$

9.5 High-field transport: two-valley model

In compound semiconductors such as GaAs, low-field conduction by electrons takes place in the Γ central valley. However, under high-field conditions some of the electrons can acquire enough energy from the electric field to transfer to the upper X and L valleys (whichever is the lowest) (see Figure 9.2). The electrons transfer from the Γ valley to the upper valleys by collision with optical phonons through the deformation potential.

Consider now such a two-valley system. We shall assume that the semiconductor is uniform. Let us call n_1 and n_2 the number of electrons in bands 1 and 2, respectively.

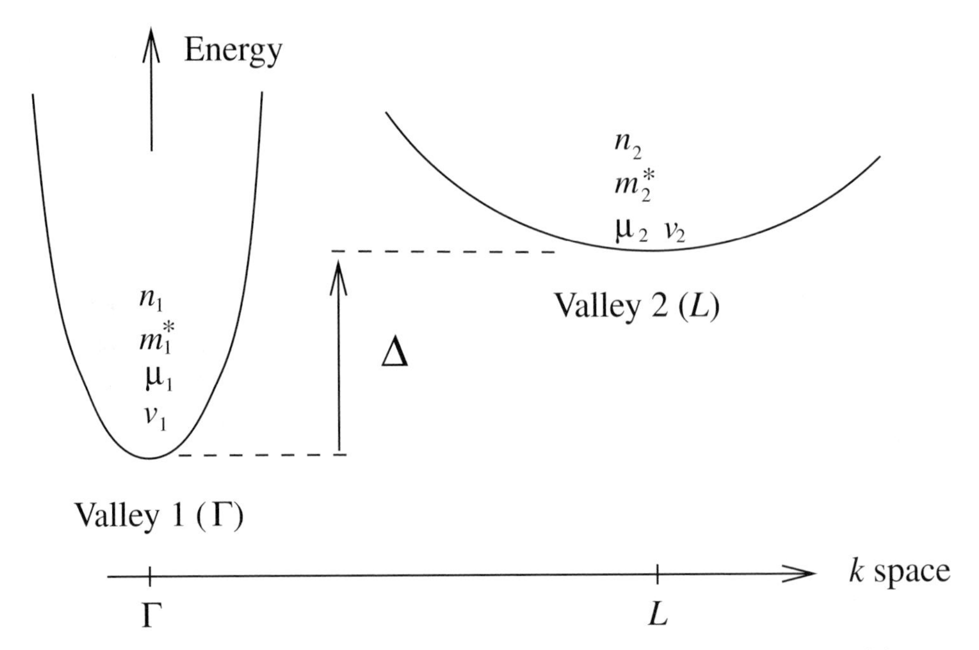

Fig. 9.2. Two-valley model of band structure. In GaAs the lower valley is the Γ valley, and the upper valley is the L valley.

Conservation of particles in this uniform two-valley system is expressed by the following continuity equations written by inspection:

$$\frac{\partial n_1}{\partial t} = -\frac{n_1}{\tau_{12}} + \frac{n_2}{\tau_{21}},$$
$$\frac{\partial n_2}{\partial t} = -\frac{n_2}{\tau_{21}} + \frac{n_1}{\tau_{12}},$$

where τ_{ij} is the scattering time from valley i to valley j. Let us assume that the electrons in valleys 1 and 2 are in equilibrium and therefore have the same electron temperature. This assumption is not quite correct but greatly simplifies the analysis. The populations of the valleys 1 and 2 are given by the Boltzmann distribution:

$$n_1 = N_{c1} M_1 \exp\left(\frac{E_F - \mathcal{E}_c}{k_B T_e}\right),$$
$$n_2 = N_{c2} M_2 \exp\left(\frac{E_F - \mathcal{E}_c - \Delta}{k_B T_e}\right),$$

with M_1 and M_2 the number of equivalent valleys 1 and 2. For example, in GaAs, we have $M_1 = 1$ since there is only one Γ valley and $M_2 = \frac{1}{2}8 = 4$ since there are 8 L valleys shared by each adjacent Brillouin zone. The ratio of the population of valley 2 to that of valley 1 is then

$$\frac{n_2}{n_1} = \left(\frac{N_{c2} M_2}{N_{c1} M_1}\right) \exp\left(-\frac{\Delta}{k_B T_e}\right) = \frac{\tau_{21}}{\tau_{12}} = R \exp\left(-\frac{\Delta}{k_B T_e}\right).$$

Momentum conservation in this two-valley system is enforced by the following equations:

$$\frac{dp_1}{dt} = qF - \frac{p_1}{\tau_{m1}(T_e)} - \frac{p_1}{\tau_{12}(T_e)},$$
$$\frac{dp_2}{dt} = qF - \frac{p_2}{\tau_{m2}(T_e)} - \frac{p_2}{\tau_{21}(T_e)},$$

where $p_i = m_i^* v_i$ is the momentum and τ_{mi} the momentum scattering rate in the valley i. These momentum conservation equations assume that the intervalley scattering process is randomizing so that in the transfer from valley i to valley j there is a loss of momentum p_i/τ_{ij} in valley i but no gain of momentum on average in valley j.

In steady state, the electron mobilities in valleys 1 and 2 are then

$$\mu_1(T_e) = \frac{p_1}{m_1^* F} = \frac{q}{m_1^*}\left[\frac{1}{\tau_{m1}(T_e)} + \frac{1}{\tau_{12}(T_e)}\right]^{-1},$$
$$\mu_2(T_e) = \frac{p_2}{m_2^* F} = \frac{q}{m_2^*}\left[\frac{1}{\tau_{m2}(T_e)} + \frac{1}{\tau_{21}(T_e)}\right]^{-1}.$$

The electron temperature T_e is obtained from the energy balance equation. Since we are assuming that valleys 1 and 2 are in equilibrium we shall rely for simplicity on the simple one-valley energy conservation:

$$\frac{\partial}{\partial t}\left(n\frac{3}{2}k_B T_e\right) = nqv_d F - n\,\frac{3}{2}k_B\frac{T_e - T_0}{\tau_E},$$

with $n = n_1 + n_2$ and $J = qnv_d = q(n_1 v_1 + n_2 v_2)$.

In steady state, the electron temperature T_e of the two-valley system subjected to the uniform electric field F is

$$T_e(v_d, F) = T_0 + \tau_E \frac{2}{3k_B} qFv_d.$$

The drift velocity is then the average electron velocity in the two-valley system:

$$v_d = \frac{n_1}{n}v_1 + \frac{n_2}{n}v_2.$$

Assuming as is the case in GaAs that v_2 is much smaller than v_1, the drift velocity is

$$v_d(F) \simeq \frac{n_1}{n}v_1 = \frac{\mu_1 F}{1 + \dfrac{n_2}{n_1}} = \frac{\mu_1 F}{1 + R\exp\left[-\dfrac{\Delta}{k_B T_e(v_d, F)}\right]}. \tag{9.18}$$

This is a transcendental equation which is easily solved by iteration. The resulting velocity-field relations calculated for GaAs are given in Figure 9.3 for the temperatures T_0 of 350 K, 300 K , 250 K, and 200 K. The parameters used are $\mu_1 = 0.8$ m^2/(V s), $\Delta = 0.3$ eV, $\tau_E = 10^{-12}$ s and $R = 100$. The velocity–field relation exhibits a region

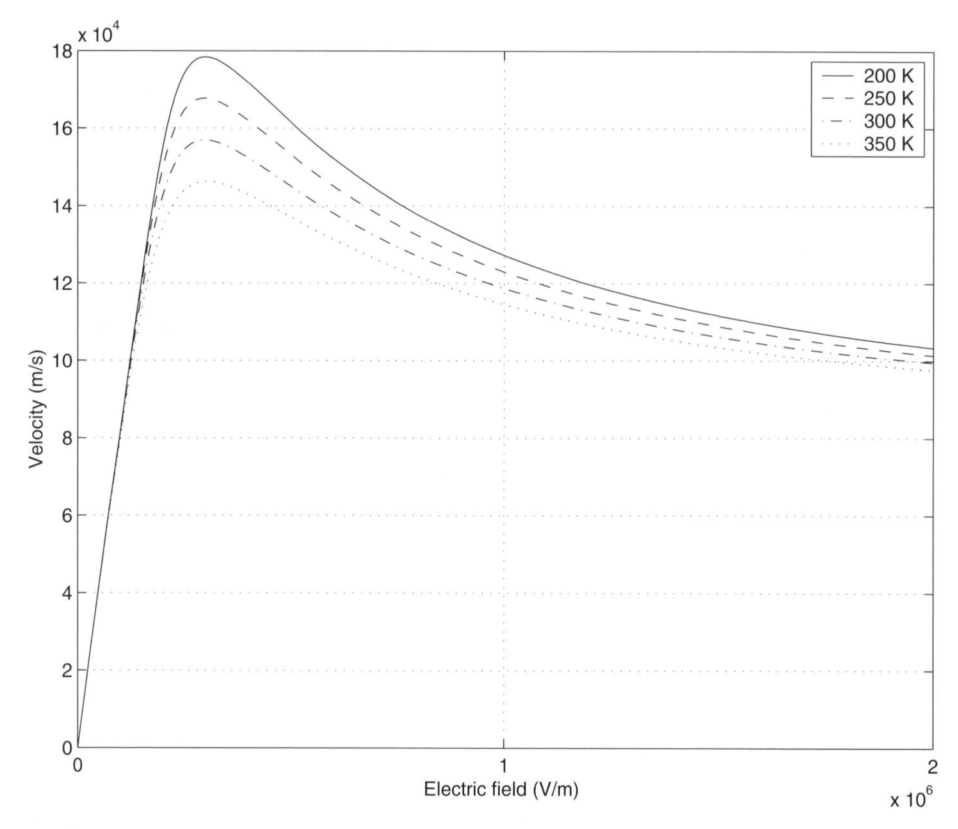

Fig. 9.3. Drift velocity versus electric field predicted by the two-valley one-temperature model for bulk GaAs.

of negative differential mobility. Such a velocity–field relation can be the source of instabilities as is discussed in the next section.

Obviously, the simple model developed here only provides a rough estimate of the effective velocity relation. The following parameters have been measured at 300 K in pure GaAs:

$$\mu_0 = 0.8 \text{ m}^2/\text{V s},$$
$$v_p = 2.2 \times 10^7 \text{ cm/s},$$
$$v_S = 1.14 \times 10^7 \text{ cm/s},$$
$$F_T = 3.2 \text{ kV/cm}.$$

Note that the peak electron velocity v_p is not much smaller that the equilibrium thermal velocity $v_{th}(T_e)$:

$$v_{th}(T_e) = \left(\frac{3kT_e}{m^*}\right)^{1/2} \simeq 4.4 \times 10^7 \left(\frac{T_e}{T_0}\right)^{1/2} \text{ cm/s} \quad \text{for } T_0 = 300 \text{ K}.$$

Therefore the use of the simple drifted Maxwellian distribution is not fully justified.

Nonetheless, the temperature model permits us to develop insight into the physical process. This transport model was developed using the effective-mass model for the Γ valley which does not set a limit on the electron velocity. As was established in Chapter 3, the maximum velocity of an electron in a band is limited by the band structure. The maximum velocity of an electron in the conduction band of GaAs is

$$v_{max}(\text{GaAs}) = \frac{1}{\hbar}\frac{\partial E(\mathbf{k})}{\partial \mathbf{k}}\bigg|_{max} = 9.5 \times 10^7 \text{ cm/s},$$

which is much larger than v_p. The effective-mass approximation is therefore acceptable in this particular example.

Once again let us emphasize that the concept of a velocity–field relation does not hold for spatially-varying systems in which large built-in electric fields are present. In fact, as we shall see in Section 9.8, non-local effects in an FET channel lead the electron velocity to overshoot or undershoot the stationary velocity predicted from $v_d(F)$ using the local value of the field F.

9.6 Negative differential mobility and the Gunn effect

Consider a sample of a material such as GaAs which features a two-valley system conduction band. This sample of length L is terminated by an ohmic contact on each side. The sample is uniformly doped with a donor concentration N_D and the equilibrium electron concentration n_0 in the sample is given by $n_0 = N_D$ assuming all donors are ionized. A voltage V is applied across the sample. The uniform electric field F_0 raised in the sample by the applied voltage V is

$$F_0 = -\frac{V}{L}.$$

One expects the current J_0 flowing through this uniform sample then to be

$$J_0 = qn_0v_d(F_0),$$

using the velocity–field relation $v_d(F)$. Note that $v_d(F_0)$ is an odd function $(v_d(-F_0) = -v_d(F_0))$ and $\mu_n(F_0)$ is an even function $(\mu_n(-F_0) = \mu_n(F_0))$.

Assume now that the electron distribution $n(x, t)$ accidentally departs from its equilibrium concentration n_0 in the interval $[x_1, x_2]$. The electric field F across the sample is given by Gauss's law:

$$\epsilon\frac{\partial F(x,t)}{\partial x} = q[n_0 - n(x,t)].$$

The electric field $F(x, t)$ integrated over the sample must satisfy the potential boundary condition

$$V = -\int_0^L F(x,t)\,dx = -\int_0^L [F(x,t) - F_0(t)]\,dx - \int_0^L F_0(t)\,dx$$
$$= -\int_{x_1}^{x_2} (F(x,t) - F_0(t))\,dx - F_0(t)L.$$

Thus the uniform field $F_0(t)$ outside the domain departs from its equilibrium value F_0. Let ϕ be the excess potential drop arising across the interval $[x_1, x_2]$:

$$\phi(t) = -\int_{x_1}^{x_2} [F(x,t) - F_0(t)]\,dx.$$

We will now study the variation in time of the excess potential $\phi(t)$ when it is subjected to the DC boundary condition

$$V = \phi(t) - F_0(t)L.$$

We must solve the continuity equation

$$q\frac{dn}{dt} = -\frac{dJ}{dx},$$

where J is the drift-diffusion current. From Gauss's law we have

$$n = n_0 - \frac{\epsilon}{q}\frac{\partial F}{\partial x}. \tag{9.19}$$

We can simultaneously solve Gauss's law and the continuity equation by introducing the total current J_T:

$$J_T = J + \epsilon\frac{\partial F_0(t)}{\partial t},$$

which satisfies the generalized continuity equation

$$\frac{dJ_T}{dx} = 0.$$

The total current through the unperturbed regions ($[0, x_1]$ and $[x_2, L]$) is

$$J_T(t) = qn_0 v_d(F_0(t)) + \epsilon\frac{\partial F_0(t)}{\partial t}.$$

The total current through the perturbed region ($[x_1, x_2]$) is

$$J_T(t) = qn(x,t)v_d(F(x,t)) + \frac{\partial}{\partial x}[qD_n(x,t)n(x,t)] + \epsilon\frac{\partial F(x,t)}{\partial t}.$$

Enforcing the continuity of the total current we have

$$\epsilon\frac{\partial(F - F_0)}{\partial t} = qn_0 v_d(F_0) - qnv_d(F) - \frac{\partial}{\partial x}(qD_n n).$$

Using Gauss's law (Equation (9.19)) the following identity is obtained:

$$\epsilon \frac{\partial (F - F_0)}{\partial t} = q n_0 [v_d(F_0) - v_d(F)] + v_d(F) \epsilon \frac{\partial F}{\partial x} - \frac{\partial}{\partial x}(q D_n n).$$

Integrating this equation from x_1 to x_2 we obtain

$$\epsilon \frac{d}{dt} \int_{x_1}^{x_2} (F - F_0)\, dx = q n_0 \int_{x_1}^{x_2} [v_d(F_0) - v_d(F)]\, dx,$$

since we have

$$\int_{x_1}^{x_2} \left[-\frac{\partial}{\partial x}(q D_n n) + v_d(F) \epsilon \frac{\partial F}{\partial x} \right] dx = 0.$$

For a small perturbation, a small-signal analysis can be performed by expanding the velocity in a Taylor series

$$v_d(F) = v_d(F_0) + \frac{d v_d(F_0)}{dF}(F - F_0) = v_d(F_0) + \mu_{diff}(F_0)(F - F_0).$$

The identity now reduces to

$$\frac{d}{dt} \int_{x_1}^{x_2} (F - F_0)\, dx = -\frac{q n_0}{t} \frac{d v_d(F_0)}{dF} \int_{x_1}^{x_2} (F - F_0)\, dx.$$

Identifying the excess potential ϕ, we obtain an equation describing its time evolution:

$$\frac{d\phi}{dt} = -\frac{q n_0}{\epsilon} \frac{d v_d(F_0)}{dF} \phi = -\frac{\phi}{\tau_d},$$

where $\tau_d = \epsilon / [q n_0 \mu_{diff}(F_0)]$ is the so-called dielectric relaxation time. A solution is then

$$\phi(t) = \phi(0) \exp\left(-\frac{t}{\tau_d} \right).$$

When the differential mobility $\mu_{diff} = d v_d(F_0)/dF$ is positive, the perturbation of the electron distribution rapidly decays and space-charge neutrality is recovered. In electromagnetic theory, the inverse of the dielectric relaxation time is called the plasma frequency. For microwave frequencies smaller than the plasma frequency, the semiconductor is transparent. For frequencies larger than the plasma frequency the semiconductor is reflective. When the differential mobility is negative as is the case in semiconductors such as GaAs for fields $F > F_p$, the perturbation will grow. The sample therefore is unstable for electric fields $F > F_p$. Such an instability will arise when the applied voltage V is larger than $F_p L$, where F_p is the peak velocity of the stationary velocity–field relation $v_d(F_0)$.

A large-signal analysis (see, for example, Carroll [5]) reveals that the perturbation grows into a domain of width d which drifts through the sample starting from the

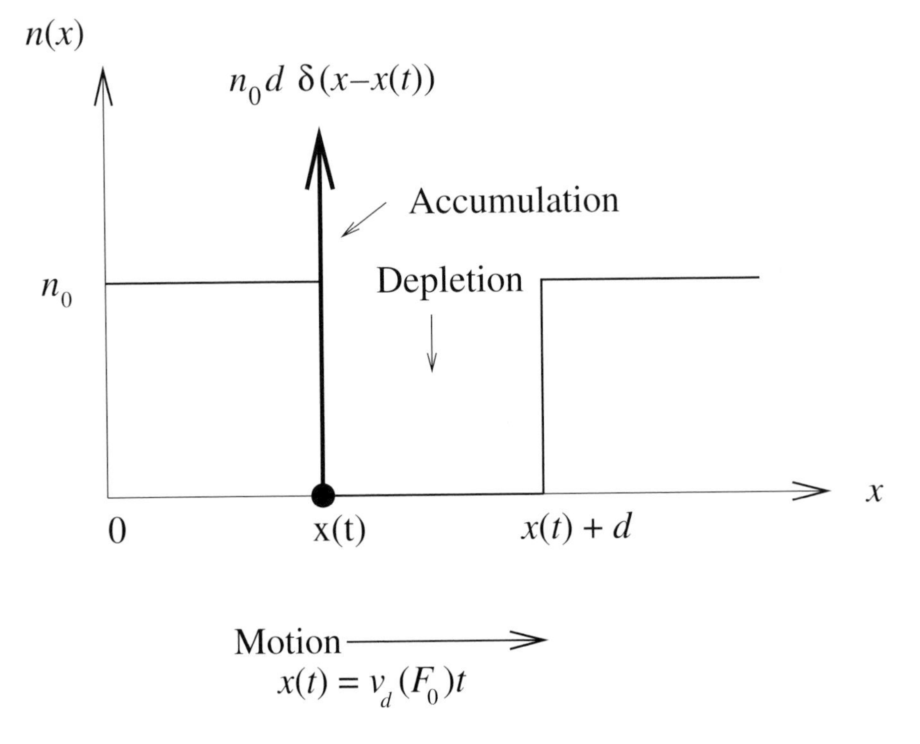

Fig. 9.4. Charge distribution in a dipole in the ideal limit of zero diffusion.

cathode and moving toward the anode. The velocity of the domain is $v_d(F_0)$, where F_0 is the field outside the domain. The domain consists of a dipole which in the limit of zero diffusion ($D_n = 0$) is represented by the following electron distribution:

$$n(x) = n_0 + n_0 d\, \delta[x - x(t)] - n_0 \{u[x - x(t)] - u[x - x(t) - d]\},$$

where $x(t) = x(0) + v(F_0)t$ is the position of the domain, $\delta[x]$ is the impulse function and $u[x]$ is the step function. The electron distribution in a domain therefore consists of an accumulation and a depletion region (see Figure 9.4) forming a dipole. The field F_0 outside the domain is given by the solution of the transcendental equation set by the boundary condition

$$V = \phi(F_0) + F_0 L.$$

The potential drop $\phi(F_0)$ across the domain is easily evaluated to be (see Problem 9.6)

$$\phi(F_0) = \frac{\epsilon}{2qn_0}[F_0 - F_{peak}(F_0)]^2, \tag{9.20}$$

where F_{peak} the peak field inside the domain is related to the field F_0 by the equal-area rule (see Carroll [5]):

$$\int_{F_0}^{F_{peak}(F_0)} [v_d(F) - v_d(F_0)]\, dF = 0.$$

The field F_0 obtained is smaller than the peak field F_p even though we have ($F_p < V/L$) so that no other instabilities can arise elsewhere once one domain has been created.

This domain drifts toward the anode and is discharged upon reception. A new domain is then launched at the cathode. The frequency of domain launching f is therefore

$$f = \frac{v_d(F_0)}{L}.$$

The generation and destruction of these domains generates a high-frequency fluctuation of the current which can be observed when a voltage is applied. The resulting 'apparent' noise was discovered by Gunn at IBM in 1963 [6] while studying high-field transport in GaAs. This effect had been predicted by Ridley and Watkins [7] and Hilsum [8] in 1962.

The differential negative resistance presented by such a device is used to make a microwave source which is called a Gunn diode. Since only one domain is formed in a Gunn diode, the frequency of operation of the Gunn diode is given by $f = v_d(F_0)/L$. The Gunn diode belongs therefore to the family of transit-time devices. A MESFET (metal–semiconductor field-effect transistor) with a resistive gate of 10–50 μm [9] has been developed for use as a microwave source operating at a frequency of 40 GHz. This device permits the generation of multiple domains and the frequency of operation is limited by the number of domains times the device transit frequency. In a regular field-effect transistor (FET), however, the generation and discharge of domains would introduce undesirable noise and could potentially burn the FET. In an FET, the channel width Δd is usually too small for the generation of traveling domains and static domains are formed instead toward the drain region. Applying the work of Kroemer [13] and Kino and Robson [14] on Gunn diodes to FETs, one can infer following [10] that the following criteria must be verified for traveling Gunn domains to be formed and sustained in a GaAs MESFET of channel doping n_0, length L_g and channel width Δd:

$$n_0 \cdot L_g > 10^{12}\ \text{cm}^{-2},$$
$$n_0 \cdot \Delta d > 2 \times 10^{11}\ \text{cm}^{-2}.$$

However, stable FET operation has been observed even when Gunn oscillations would be expected according to the above criteria. As is discussed by Yamaguchi *et al.* [10], a more accurate stability criterion for GaAs MESFETs should also be dependent on the FET gate length and the applied gate voltage. Two types of stable FET operation with and without negative output conductance are also possible [10] depending on the formation or not of a static domain in the FET. MESFETs with stable negative output conductance have been reported for a 3 μm gate by Tucker and Young [11]. Such an effect is not, however, usually observed in smaller gate-length MESFETs and MODFETs (modulation doped FETs) due to the effect of velocity overshoot. Indeed, our analysis was developed for a uniform sample and relied on the concept of a field-dependent velocity. As we have stated earlier such an analysis does not apply to devices such as the short-channel FET in which large built-in electric fields are present.

9.7 Transient velocity overshoot in a time-varying field

The velocity-field relation in Sections 9.4 and 9.6 for the one- and two-valley systems applies to uniform samples under stationary conditions. When the electric field in a uniform sample is switched abruptly at time $t = 0$, from 0 to a value F, non-stationary effects lead the electron drift velocity v_d to overshoot its stationary value $v_d(F)$ for a short period τ_E. This originates from the fact that the energy relaxation time τ_E is much larger than the momentum relaxation time τ_m.

Consider a uniform two-valley system such as GaAs. Let us assume that the upper and lower valleys are in equilibrium such that we have

$$\frac{n_2}{n_1} = R \exp\left[-\frac{\Delta}{k_B T_e(t)}\right].$$

The drift velocity is still

$$v_d = \frac{v_1(t)}{1 + R \exp\left[-\Delta/k_B T_e(t)\right]},$$

but now the electron velocity and the temperature vary in time. The electron temperature is obtained from the energy conservation equation

$$\frac{\partial T_e}{\partial t} = \frac{\partial}{\partial t}(T_e - T_0) = \frac{2 q v_d F}{3 k_B} - \frac{T_e - T_0}{\tau_E}$$

which admits the solution

$$T_e(t) = T_0 + \tau_E \frac{2}{3 k_B} q v_d F \left[1 - \exp\left(-\frac{t}{\tau_E}\right)\right].$$

The time-varying electron velocity $v_1(t)$ is obtained from the momentum conservation equation

$$\frac{d}{dt}(m_1^* v_1) = -qF + \frac{m_1^* v_1}{\tau_{M1}},$$

with $\tau_{M1}^{-1} = \tau_{m1}^{-1} + \tau_{12}^{-1}$. The electron velocity $v_1(t)$ obtained is then

$$v_1(t) = \frac{q F \tau_{M1}}{m_1^*} \left[1 - \exp\left(-\frac{t}{\tau_{M1}}\right)\right].$$

The drift velocity v_d is a solution of the following transcendental equation:

$$v_d(t) = \frac{\mu_1 F\left[1 - \exp\left(-\frac{t}{\tau_{M1}}\right)\right]}{1 + R\exp\left(\frac{k_B}{\Delta}\left\{T_0 + \tau_E\frac{2}{3k_B}qv_d(t)F\left[1 - \exp\left(-\frac{t}{\tau_E}\right)\right]\right\}\right)}.$$

The drift velocity first quickly rises to the velocity $\mu_1 F$ in a time on the order of τ_{M1} and then relaxes to the stationary velocity $v_d(F) = \mu_n(F)F$ for time on the order of τ_E. We have assumed that the electron gases in valleys 1 and 2 were in equilibrium, an assumption which should be relaxed in a more accurate model so as to account for the finite intervalley relaxation times τ_{12} and τ_{21}.

Velocity overshoot is also possible in the single-valley system since the electron mobility is also reduced as the electron temperature is increased. The effect in silicon is, however, small and can be usually neglected.

9.8 Stationary velocity overshoot in short devices

Velocity overshoot can also arise under stationary conditions in the presence of spatially-varying electric fields. Consider the energy balance equation

$$n\frac{3}{2}\frac{k_B(T_e - T_0)}{\tau_E} = \frac{d}{dx}\left(\frac{J}{q}\frac{3}{2}k_BT_e\right) + JF(x).$$

Assuming current continuity $dJ/dx = 0$ as is the case in a field-effect transistor we obtain

$$T_e = T_0 + \tau_E\frac{2}{3k_B}q\frac{JF}{n} + \Delta T_e,$$

with

$$\Delta T_e = \tau_E\frac{J}{n}\frac{dT_e}{dx}.$$

ΔT_e is the departure of the electron temperature T_e(*non-local*) from the electron temperature T_e(*local*) which would arise in a uniform field. When the electrons are accelerated by an increasing (negative) electric field as is the case in an FET channel from source to drain, the current J is negative ($J < 0$), the temperature gradient dT_e/dx is positive, and the influx of colder electrons reduces the electron temperature T_e from its local stationary value since we have $\Delta T_e < 0$. Therefore, the electrons remain longer in the lower valley (valley 1) than in the case of a uniform field. As a consequence the drift velocity overshoots its local stationary value.

In simple device modeling of short-channel devices, the following simple modified drift velocity–field relation has been used with some success:

$$v_d = \mu_n F \quad \text{for} \quad F \le F_c,$$

$$v_d = v_S \quad \text{for} \quad F \ge F_c,$$

where v_S is an effective saturation velocity–field relation and $F_c = v_S/\mu_n$. The effective saturation velocity v_S represents the saturation velocity which effectively sets the transconductance of a short-channel FET. It has been proposed by Rohdin [12] from the analysis of the high-frequency characteristics measured on MODFETs with gate lengths varying from 0.25 to 0.7 μm that the effective saturation velocity v_S is actually the stationary peak velocity $v_p = v_d(F_p)$ for a wide range of gate lengths.

9.9 Conclusion

In this chapter we have derived drift-diffusion current and energy-balance equations using a drifted and heated Maxwell–Boltzmann distribution as an approximate solution of the Boltzmann equation. These simple transport equations can be useful for the modeling of the MODFET and HBT. However, as was discussed in Chapter 1, simple transport models may no longer be applicable in devices in which ballistic transport is important.

In Chapters 3–7, ballistic transport was studied in depth in the quantum transport regime (electron wavelength comparable to the device size). For devices operating in the semiclassical transport regime, we have seen (see Figure 2.6) that ballistic effects can also strongly impact the electron energy distribution. A phenomenological model of ballistic-electron launching was introduced in Chapter 2, and its impact on the HBT was studied in Problem 2.3. A detailed discussion of ballistic transport in HBTs using a direct solution of the Boltzmann equation will be presented in Chapter 18.

9.10 Bibliography

9.10.1 Recommended reading

K. Hess, *Advanced Theory of Semiconductor Devices*, Prentice Hall, New Jersey, 1988.

M. Lundstrom, *Fundamentals of Carrier Transport*, Volume X, Modular Series on Solid State Devices, eds. G. W. Neudeck and R. F. Pierret, Addison-Wesley, Reading, MA, 1990.

9.10.2 References

[1] A. H. Marshak and K. M Van Vliet, 'Electrical current in solids with position-dependent band structure', *Solid State Electronics,* Vol. 21, pp. 417–427, 1978.

[2] K. Hess, 'Phenomenologic physics of hot carriers in semiconductors', in *Physics of Nonlinear Transport in Semiconductors*, eds. D. K. Ferry, J. R. Barker and C. Jacoboni, Plenum, New York, 1980.

[3] M. Lundstrom, *Fundamentals of Carrier Transport,* Volume X, Modular Series on Solid State Devices, eds. G. W. Neudeck and R. F. Pierret, Addison-Wesley, Reading, MA, 1990.

'Department of Entropy' by Sidney Harris.

[4] J. B. Gunn, 'Transport of electrons in a strong built-in electric field', *Journal of Applied Physics*, Vol. 39, No. 10, pp. 4602–4604, 1968.

[5] J. E. Carroll, *Hot Electron Microwave Generators*, American Elsevier Publishing Company, Inc., New York, 1970.

[6] J. B. Gunn, 'Microwave oscillations of current in III–V semiconductors', *IBM Journal of Research and Development,* Vol 8, pp. 141–159, 1964.

[7] B. K. Ridley and T. B. Watkins, 'The possibility of negative resistance in semiconductors',

Proceedings of the Physical Society, Vol. 78, pp. 293–304, 1961.

[8] C. Hilsum, 'Transferred electron amplifiers and oscillators', *Proceedings of the IRE*, Vol. 50, pp. 185–189, 1962.

[9] J. A. Cooper, Y. Yin, Balzan and A. E. Gessberger, 'Experimental verification of the contiguous domain oscillator concept', *Proceeding of the 1989 IEEE Cornell Conference on Advanced Concepts in High-Speed Semiconductor Devices and Circuits*, IEEE, Ithaca, NY.

[10] K. Yamaguchi, S. Asai and H. Kodera, 'Two-dimensional numerical analysis of stability criteria of GaAs FETs', *IEEE Transactions on Electron Devices*, Vol. ED 23, pp. 1283–1290, December 1976.

[11] T. W. Tucker and L. Young, 'GaAs negative conductance junction field effect transistor', *Solid State Electronics*, Vol.17, p. 31, 1974.

[12] H. Rohdin, 'Reverse modeling of E/D logic sub-micron MODFETs and prediction of maximum extrinsic MODFET current cutoff frequency', *IEEE Transactions on Electron Devices*, Vol. 37, No. 4, pp. 920–934, April 1990.

[13] H. Kroemer, 'Negative conductance in semiconductors', *IEEE Spectrum*, Vol. 5, pp. 47–56, 1968.

[14] G. S. Kino and P. N. Robson, 'The effect of small transverse dimension on the operation of Gunn devices', *Proceedings of the IEEE*, Vol. 56, pp. 2056–2057, 1968.

9.11 Problems

9.1 The following approximate solution f_{even} of Equation (9.12) is proposed:

$$f_{even} = n\left(\frac{m^*}{2\pi k_B T_e}\right)^{\frac{3}{2}} \exp\left(-\frac{m^* v^2}{2k_B T_e}\right).$$

Using this approximate solution verify that we obtain the following energy conservation equation:

$$\frac{3}{2}k_B(T_e - T_0) = \tau_E q F^2 \frac{q\tau_m}{m^*} = \tau_E q F v_d.$$

To do so start from Equation (9.12), rewrite f_{even} as

$$f_{even} = f_0 + \frac{q^2 F^2}{m^{*2}}\tau_E \tau_m \frac{d^2 f_{even}}{dv_x^2}$$

and calculate the average kinetic energy

$$\left\langle \frac{1}{2}m^* v^2 \right\rangle_{f_{even}} = \frac{\int_{-\infty}^{\infty} \frac{1}{2}m^* v^2 f_{even}\, d\mathbf{v}}{\int_{-\infty}^{\infty} f_{even}\, d\mathbf{v}}.$$

For simplicity, assume that the relaxation times τ_E and τ_m are constant (independent of the electron velocity v).

9.2 Calculate f_{odd} from Equation (9.11) and verify using a Taylor expansion that the displaced Maxwell–Boltzmann distribution is an approximate solution for v_d small:

$$f = f_{even} + f_{odd} \simeq n\left(\frac{m^*}{2\pi k_B T_e}\right)^{\frac{3}{2}} \exp\left(-\frac{m^*|\mathbf{v} + \mathbf{v}_d|^2}{2k_B T_e}\right).$$

9.3 Verify that the dependence of μ_n upon the electron temperature T_e is

$$\mu_n(T_e) = \mu_n(T_0)\left(\frac{T_e}{T_0}\right)^{\frac{r}{2}}$$

if $\tau_m(v)$ varies as v^r.

Hint: Start from the definition of the mobility

$$\mu_n(T_0) = -\frac{q}{k_B T_0}\frac{\int_{-\infty}^{\infty}\tau_m(v)v_x^2 f_0\, d\mathbf{v}}{\int_{-\infty}^{\infty} f_0\, d\mathbf{v}} = \frac{q}{m^*}\langle\tau_m\rangle,$$

with f_0 given by

$$f_0(v) = n\left(\frac{m^*}{2\pi k_B T_0}\right)^{\frac{3}{2}}\exp\left(-\frac{m^* v^2}{2k_B T_0}\right).$$

You do not need to evaluate this integral but can use a simple change of variables to calculate the ratio $\mu_n(T_e)/\mu_n(T_0)$.

9.4 Verify that the expressions given for the current density J in Equations (9.14), (9.15) and (9.16) are all equivalent for a uniform material ($\mathcal{E}_c$ not spatially varying). To do so verify that each of these three equations can be written

$$J = -q\mu_n(T_e)n(x)\frac{dV(x)}{dx} + qD_n(T_e)\frac{d}{dx}[n(x)] + \left(\frac{r}{2}+1\right)\mu_n(T_e)n(x)\frac{dT_e}{dx}, \tag{9.21}$$

where we assume that $\tau_m(v)$ varies as v^r. Note that the temperature dependence of the mobility $\mu(T_e)$ (see Problem 9.3) and the effective density of states $N_c(T_e)$ are used together with the Einstein relation to establish these equivalences.

9.5 Solve the transcendental Equation (9.18) at 300 and 200 K and plot the electron drift velocity v_d versus the electric field from 0 to 20 kV/cm. Use the following parameters $\mu_1 = 0.8$ m^2/(V s), $\Delta = 0.3$ eV, $\tau_E = 10^{-12}$ s and $R = 100$.

9.6 Calculate the electric field and derive the potential drop given by Equation (9.20) across the domain shown in Figure 9.4.

10 $I-V$ model of the MODFET

The best material model of a cat is another, or preferably the same, cat.

Philosophy of Science, Vol. 12, 1945. A. Rosenblueth and N. Wiener

10.1 Introduction

In Chapter 8 we studied the charge control of the 2DEG (two-dimensional electron gas). In Chapter 9 we studied high-field transport models applicable to horizontal transport in the 2DEG. The motivation for these studies was the application of the 2DEG as the channel of a field-effect transistor (FET). The resulting FET is referred under the various names of MODFET (modulation doped FET (University of Illinois, USA)), HEMT (high electron mobility FET (Japan)), TEGFET (two-dimensional electron gas FET (France)), and SDHT (segregation doping heterojunction transistor (Bell Lab., USA)) depending on the different laboratories which simultaneously developed it. The AlAs–InGaAs–InP lattice-matched MODFET [2] which provides power gain at millimeter frequencies ($f_{max} = 405$ GHz) is presently with the HBT one of the fastest semiconductor transistors. The microwave characteristic of the MODFET will be discussed in Chapters 11, 12, 13, 15 and 16.

We compare in Figure 10.1 the layout of an AlGaAs–GaAs MODFET (c) with that of that of a silicon MOSFET (a) and a GaAs MESFET (b). Although the layout of the MODFET is similar to that of the MESFET, its normal principle of operation (control of a 2DEG with a gate voltage) is similar to that of the MOSFET. Other semiconductors can be used to fabricate MODFETs (see [1] for a review). A semiconductor with a bandgap much wider than that of AlGaAs can also be used to simulate an insulator. These transistors are called MISFETs (metal–insulator semiconductor FETs). Due to the large variety of MODFET layouts and material possibilities, the general name of HFET (heterojunction FET) has been proposed. In the general sense of the term heterojunction a MOSFET is an HFET, even though the oxide is not a semiconductor.

In this chapter we shall study the ideal three-terminal MODFET or MOSFET. By ideal we mean that we will use piece-wise linear charge control and velocity-field expressions. Although no real device exists with such idealized characteristics, this

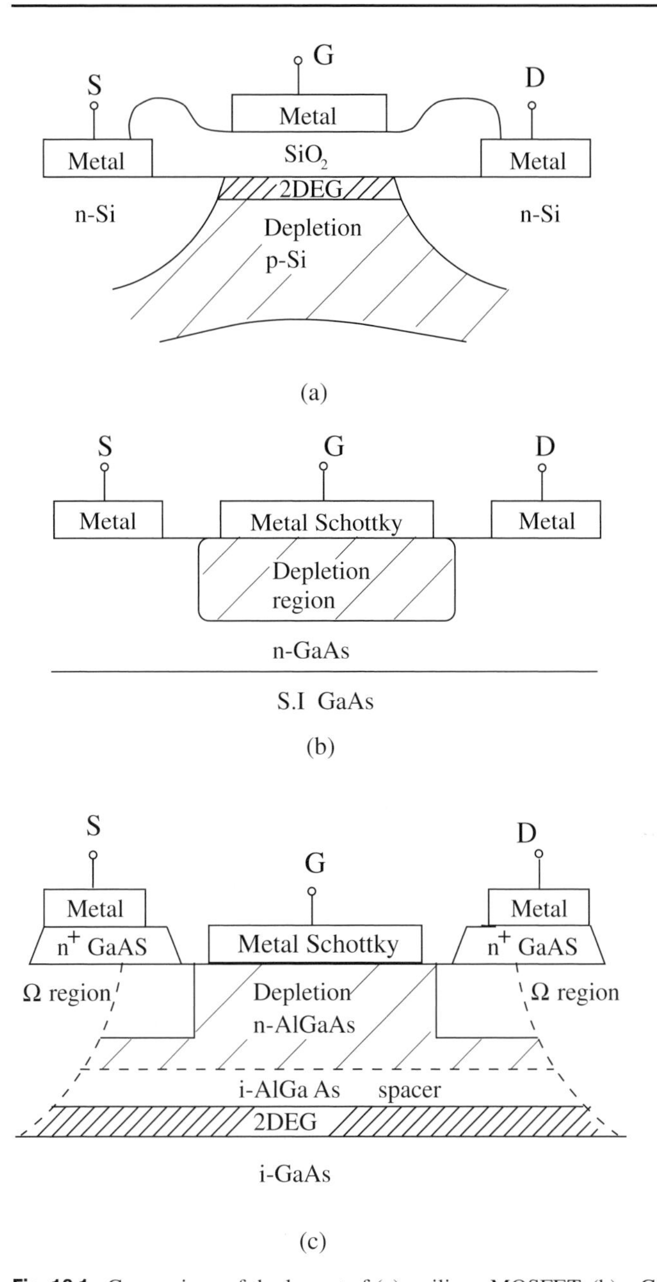

Fig. 10.1. Comparison of the layout of (a) a silicon MOSFET, (b) a GaAs MESFET and (c) an AlGaAs–GaAs MODFET.

ideal device model will permit us in this chapter and the next to obtain exact analytic solutions for its DC and AC characteristics and to study the principle of operation of FETs. In this chapter we start by studying the DC characteristics of the long- and short-channel MODFET.

10.2 Long- and short-channel MODFETs

Let us consider a long-channel MODFET. We wish to calculate the drain current I_D for a given gate-to-source voltage V_{GS} and drain-to-source voltage V_{DS}. For simplicity, we shall assume initially that the current is well described by drift alone. As we shall see later, this assumption is acceptable for the region of the channel below the gate (see Figure 10.2). Park and Kwack [9] have demonstrated that the inclusion of diffusion in the gated region of the channel introduces only a small correction in the effective threshold voltage (see Problem 10.1) of the MODFET.

In the gated region of the channel, the drain current I_D measured from drain to source (see Figure 10.2) is given by

$$I_D = qW_g N_S v_d, \tag{10.1}$$

where N_S is the DC 2DEG charge concentration, W_g the gate width and v_d the electron drift velocity. We know from our discussion of transport in Chapter 9 that the velocity versus field characteristic which approximates the velocity overshoot in FETs is

$$v_d = \mu(F)F = \begin{cases} \mu F & \text{for } F \leq F_c \\ v_S & \text{for } F \geq F_c \end{cases}, \tag{10.2}$$

where $\mu(F)$ is a field-dependent mobility, μ is a constant electron mobility (not necessarily the low-field mobility), and $F_c = v_S/\mu$ is the critical field at which the electron gas reaches its effective saturation velocity. An improved description of the short-channel MODFET characteristics can be obtained if a smoother velocity–field relation is used [4] (see Problem 10.2).

In Chapter 8 we saw that the DC 2DEG charge N_S could be approximately described by the following piece-wise linear relation:

$$qN_S = \begin{cases} 0 & \text{for } V_{GC} \leq V_T \\ C_G(V_{GC} - V_T) & \text{for } V_T \leq V_{GC} \leq V_{Gmax} \\ qN_{S0} & \text{for } V_{GC} \geq V_{Gmax} \end{cases}, \tag{10.3}$$

where C_G is the 2DEG capacitance per unit area and where $V_{GC} = V_G - V_C = V_{GS} - V_{CS}$ is the gate voltage between the Fermi level of the gate metal and the Fermi level of the 2DEG $E_{F,2DEG}$. Note that we shall use the source as a potential reference $V_S = V_C(x = 0)$ (see Figure 10.2).

This is the piece-wise linear charge-control model which was used in the first I–V model reported for the MODFET [3]. An improved description of the I–V characteristic of MODFETs for small (subthreshold regime) and large gate voltages can be achieved by using the smoother charge-control model proposed by Rohdin and Roblin [4] (see Problem 10.4).

The drain current in Equation (10.1) neglects the diffusion effect. A more general

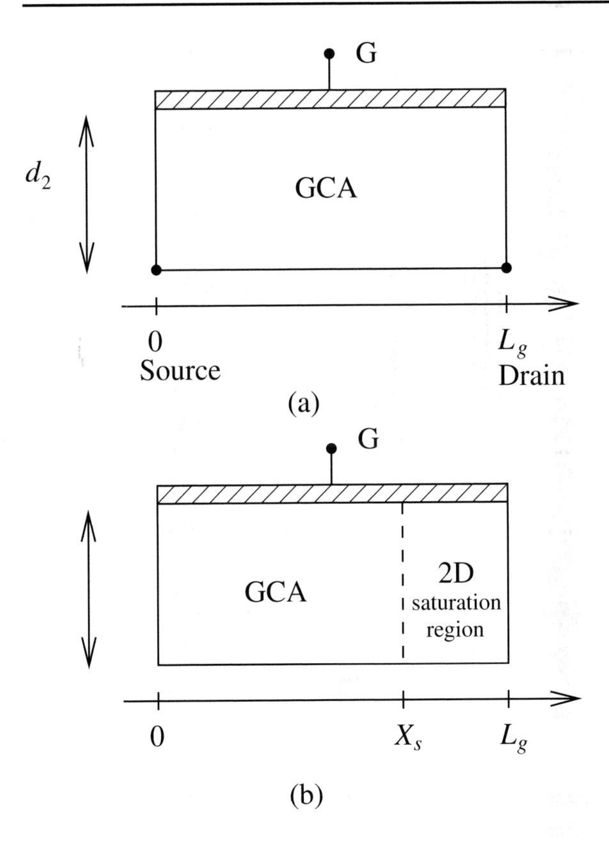

Fig. 10.2. Layout of the ideal MODFET: (a) before saturation and (b) in saturation.

expression for the drain current can be obtained from the 2DEG Fermi level using (see Chapter 8)

$$I_D = -W_g \mu(F) N_S \frac{dE_{F,2DEG}}{dx}.$$

When we neglect diffusion, the variation of the chemical potential is assumed to be small, and the spatial derivative of the Fermi level (electrochemical potential) $E_{F,2DEG} = E_{Fi} - qV_C$ of the 2DEG reduces simply to the spatial derivative of the channel potential $-qV_C$. The total drain current equation in the absence of diffusion is then the differential equation:

$$\begin{aligned} I_D &\simeq qW_g \mu(F) N_S \frac{dV_C}{dx} = qW_g N_S v_d \\ &= W_g \mu(F) C_G (V_{GS} - V_{CS} - V_T) \frac{dV_{CS}}{dx}. \end{aligned} \tag{10.4}$$

Note that we have used for N_S the charge-control expression of Equation (10.3) which was obtained by solving the Poisson equation along the y axis alone (see Chapter 8). In an FET with an applied drain-to-source voltage, the channel potential also varies along the x axis, and the Poisson equation should be solved in two dimensions (x and y axes). We therefore are assuming in Equation (10.4) that the one-dimensional solution is approximately correct. This is the so-called gradual-channel approximation (GCA) which holds if the longitudinal variation of the channel potential V_{CS} is small and can be neglected.

Let us first consider the long-channel mode where the electric field F is smaller than the critical field F_c. The mobility then reduces to $\mu(F) = \mu$. Integrating Equation (10.4) from the intrinsic source position ($x = 0$) to the channel position x (see Figure 10.2) we obtain

$$\int_0^x I_D \, dx = \beta \int_0^{V_{CS}(x)} (V_{OUT} - V_{CS}) dV_{CS}$$

$$x I_D = \beta (V_{OUT}) V_{CS}(x) - \frac{1}{2} \beta V_{CS}^2(x), \tag{10.5}$$

where we have introduced $\beta = W_g \mu C_G$ and $V_{OUT} = V_{GS} - V_T$. The channel potential V_{CS} measured from the source S is given by the second order polynomial

$$V_{CS}^2 - 2V_{OUT} V_{CS} + \frac{2x I_D}{\beta} = 0.$$

The physical GCA solution which satisfies $V_{CS}(I_D = 0) = 0$ is therefore

$$V_{CS}(x) = V_{OUT} - \left(V_{OUT}^2 - \frac{2x I_D}{\beta} \right)^{1/2} \tag{10.6}$$

and the 2DEG charge $N_S(x)$ in the channel is given by

$$q N_S(x) = C_G (V_{OUT} - V_{CS}) = C_G \left(V_{OUT}^2 - \frac{2x I_D}{\beta} \right)^{1/2}. \tag{10.7}$$

The longitudinal electric field in the channel is then

$$F(x) = -\frac{dV_{CS}}{dx} = -\frac{I_D}{\beta \left(V_{OUT}^2 - 2x I_D/\beta \right)^{1/2}}. \tag{10.8}$$

The drain current I_D is obtained from Equation (10.6) using the boundary condition $V_{CS}(L_g) = V_{DS} = V_D - V_S$ at the intrinsic drain position $x = L_g$:

$$I_D(V_{GS}, V_{DS}) = \frac{\beta}{L_g} \left(V_{OUT} V_{DS} - \frac{1}{2} V_{DS}^2 \right). \tag{10.9}$$

In Chapter 11 we will find it convenient to introduce a normalized biasing parameter k defined as:

$$k = \frac{V_{DS}}{V_{GS} - V_T} = \frac{V_{DS}}{V_{OUT}}.$$

The drain current can then be rewritten

$$I_D = \frac{\beta}{L_g} V_{OUT}^2 \frac{1}{2} (2k - k^2) \tag{10.10}$$

and the drain conductance g_d is

$$g_d = \frac{\partial I_D(V_{DS}, V_{GS})}{\partial V_{DS}} = \frac{\beta}{L_g}(V_{OUT} - V_{DS}) = \frac{\beta}{L_g} V_{OUT}\,(1-k) \tag{10.11}$$

and the transconductance g_m is

$$g_m = \frac{\partial I_D(V_{DS}, V_{GS})}{\partial V_{GS}} = \frac{\beta}{L_g} V_{DS} = \frac{\beta}{L_g} V_{OUT}\,k. \tag{10.12}$$

The channel potential can be rewritten as

$$V_{CS}(x) = V_{OUT}\left\{1 - \left[1 + (k^2 - 2k)\frac{x}{L_g}\right]^{1/2}\right\} \tag{10.13}$$

and the 2DEG charge as

$$qN_S(x) = C_G V_{OUT}\left[1 + (k^2 - 2k)\frac{x}{L_g}\right]^{1/2}. \tag{10.14}$$

Finally the channel electric field can be rewritten as

$$F(x) = -\frac{V_{OUT}}{L_g}\,\frac{2k - k^2}{2\left[1 + (k^2 - 2k)\dfrac{x}{L_g}\right]^{1/2}}. \tag{10.15}$$

The solution obtained holds for k in the range $[0, 1]$ or equivalently $0 \le V_{DS} \le V_{GS} - V_T$. Indeed for $k = 1$ or $V_{DS} = V_{GS} - V_T$, the channel concentration N_S is zero at the drain side ($x = L_g$) (see Equation (10.14)) and the channel is said to be pinched off. Simultaneously for $k = 1$, the electric field $F(x)$ at $x = L_g$ is infinite (see Equation (10.15)). Clearly the gradual-channel approximation can no longer be used in pinch-off ($k \ge 1$), and Equation (10.4) and its solution (10.9) do not apply for $k > 1$. Since the drain conductance is zero at pinch-off ($g_d(PINCH) = g_d(k = 1) = 0$), the drain current is usually assumed to remain constant for large drain voltages $V_{DS} \ge V_{GS} - V_T$. This is justified as follows. Once the MOSFET has reached pinch-off, the effective drain voltage V_{CS} at $x = L_g$ remains $V_{CS}(L_g) = V_{GS} - V_T$ even if the drain voltage is increased. The excess drain voltage $V_{DS} - V_{CS}(L_g)$ is indeed dropped across a narrow region on the drain side of the gated channel ($x \ge L_g$). The drain current in pinch-off is therefore independent of the drain voltage:

$$I_{DS}(PINCH) = \frac{\beta}{L_g}\frac{1}{2}V_{OUT}^2 = \frac{\beta}{L_g}\frac{1}{2}(V_{GS} - V_T)^2. \tag{10.16}$$

The transconductance in saturation is then given by

$$g_m(PINCH) = \frac{\partial I_{DS}(PINCH)}{\partial V_{GS}} = \frac{\beta}{L_g}V_{OUT} = \frac{\beta}{L_g}(V_{GS} - V_T). \tag{10.17}$$

The transconductance in pinch-off is seen to be inversely proportional to the gate length. This long-channel model predicts therefore that the transconductance can be

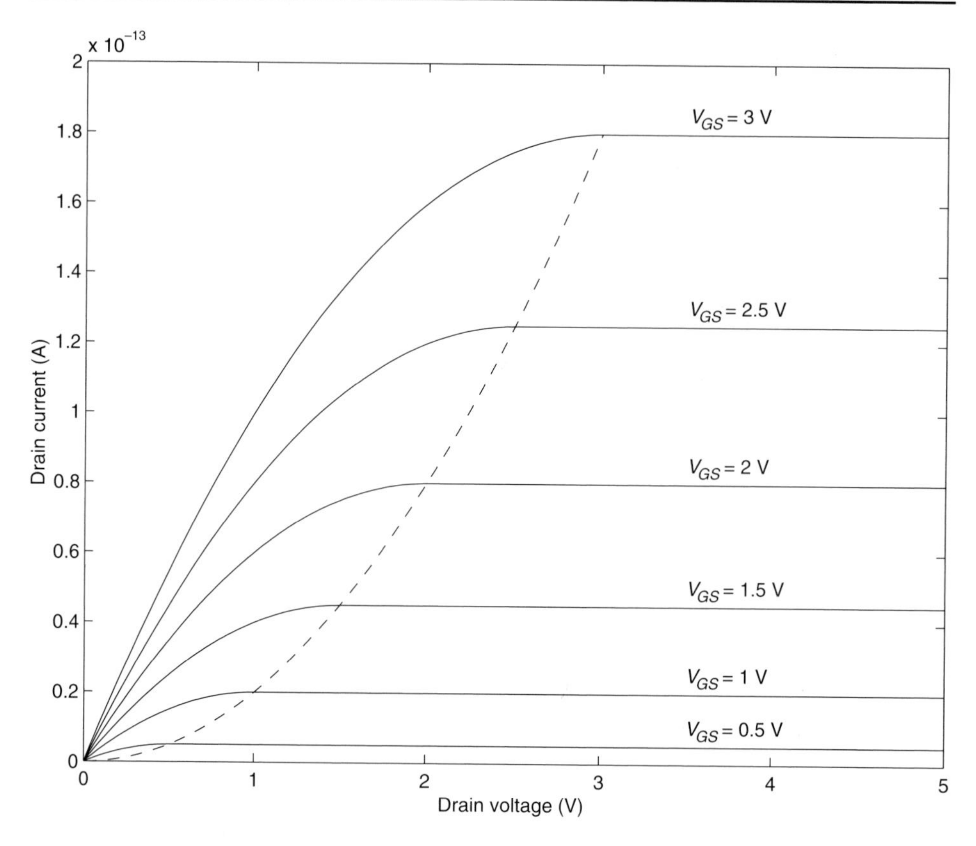

Fig. 10.3. $I-V$ characteristics of an ideal long-channel MOSFET/MODFET.

increased by decreasing the gate length of the MOSFET/MODFET. However, as the gate length is decreased below a certain value, velocity saturation takes place in the channel and the long-channel MOSFET model we are using must be replaced by a short-channel model. The $I-V$ current voltage characteristic of the ideal long-channel MOSFET/MODFET is shown in Figure 10.3.

The quadratic dependence of the drain voltage $I_{DS}(PINCH)$ in pinch-off upon V_{OUT} (see Equation (10.16)) is experimentally observed in the $I-V$ characteristics of MOSFET/MODFET with a gate length larger than 2 μm. The simple model derived above therefore provides a correct qualitative description of the long-channel MOSFET and MODFET. In the short-channel MODFET, the electric field in the drain side can become very large and induce the saturation of the electron drift velocity. As a result, the saturation of the drain current takes place before pinch-off is reached.

F_c is the electric field required for velocity saturation to take place. From Equation (10.8) we can verify that the electric field is largest at the drain side ($x = L_g$). In short-channel devices current saturation will take place at a drain current I_{DS} when the channel field at the drain side $x = L_g$ reaches the critical field F_c:

$$F(L_g) = -\frac{I_{DS}}{\beta\left(V_{OUT}^2 - \dfrac{2L_g I_{DS}}{\beta}\right)^{1/2}} = -F_c.$$

We can rewrite this equation

$$I_{DS} = \beta F_c \left(V_{OUT}^2 - \frac{2L_g I_{DS}}{\beta}\right)^{1/2}$$

$$I_{DS}^2 + (\beta F_c^2 2L_g)I_{DS} - \beta^2 F_c^2 V_{OUT}^2 = 0.$$

The positive saturation drain current I_{DS} solution is then

$$I_{DS} = -\beta F_c^2 L_g + \left[(\beta F_c^2 L_g)^2 + \beta^2 F_c^2 V_{OUT}^2\right]^{1/2} \tag{10.18}$$

and the transconductance in saturation is given by

$$g_m(SAT) = \frac{\partial I_{DS}}{\partial V_{GS}} = \frac{\beta V_{OUT}}{L_g\left(1 + \dfrac{V_{OUT}^2}{F_c^2 L_g^2}\right)^{1/2}}. \tag{10.19}$$

For large gate voltages V_{GS}, the saturation drain current becomes

$$I_{DS} \simeq \beta F_c V_{OUT} = W_g \mu F_c C_G (V_{GS} - V_T), \tag{10.20}$$

where $C_G = \epsilon_2/d_2$ is the gate capacitance per unit area (see Chapter 8). The maximum transconductance in saturation is then given by

$$g_{m,max}(SAT) = \frac{\partial I_{DS}}{\partial V_{GS}} = \beta F_c = W_g C_G v_S. \tag{10.21}$$

The maximum transconductance $g_{m,max}(SAT)$ of a short-channel MOSFET/MODFET in saturation is seen to be a constant. This is in contrast with the transconductance $g_m(PINCH)$ (see Equation (10.17)) of the long-channel MOSFET/MODFET which is proportional to V_{OUT} and inversely proportional to the gate length.

Note that as the critical field F_c is increased, the short-channel model should reduce to the long-channel model. We shall now see that in the ideal MOSFET/MODFET the transition from the long- to the short-channel mode takes place when the ratio $\alpha^{-1} = (V_{GS} - V_T)/(F_c L_g)$ becomes larger than 1. To demonstrate this let us calculate the ratio of the drain current at the onset of velocity saturation $I_{DS}(SAT)$ to the drain current in pinch-off $I_{DS}(PINCH)$ for the same gate voltage V_{GS}:

$$\frac{I_{DS}(SAT)}{I_{DS}(PINCH)} = \frac{-\beta F_c^2 L_g + \left[(\beta F_c^2 L_g)^2 + \beta^2 F_c^2 V_{OUT}^2\right]^{1/2}}{\dfrac{\beta}{L_g}\dfrac{1}{2}V_{OUT}^2}$$

$$= 2\alpha^2\left[\left(1 + \frac{1}{\alpha^2}\right)^{1/2} - 1\right] = w_1(\alpha).$$

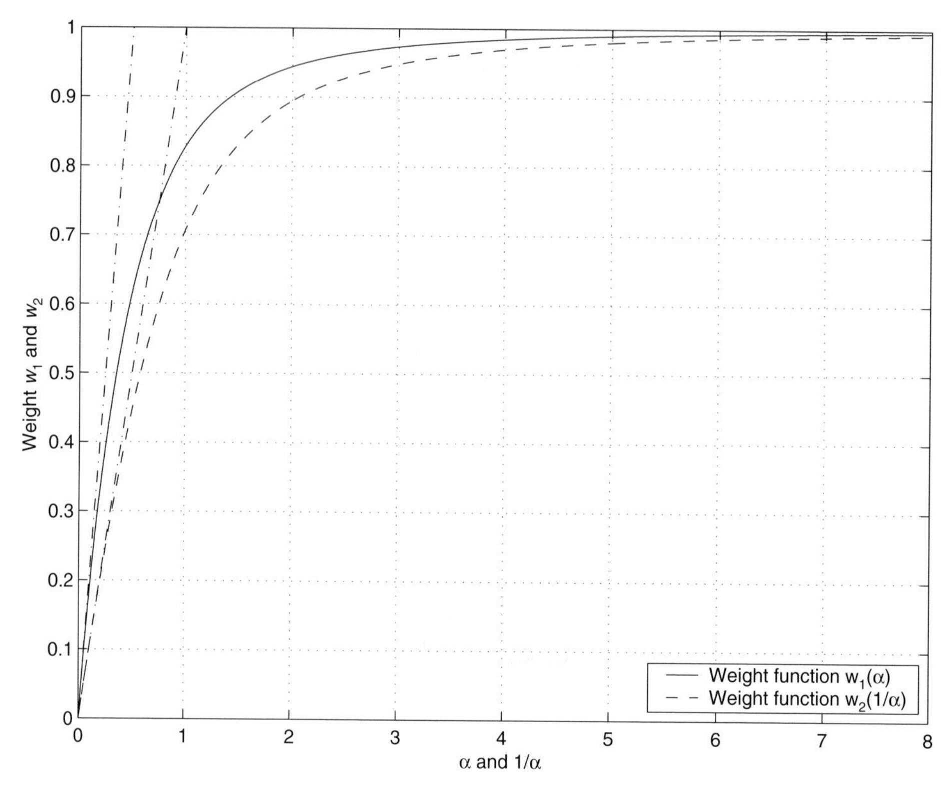

Fig. 10.4. Plot of the weight functions w_1 (full line) and w_2 (dashed line) versus α and α^{-1}, respectively.

Similarly, we can calculate the ratio of the transconductance $g_m(SAT)$ at the onset of saturation to the maximum transconductance $g_{m,max}(SAT)$

$$\frac{g_m(SAT)}{g_{m,max}(SAT)} = \frac{\beta V_{OUT}}{L_g\left(1+\dfrac{V_{OUT}^2}{F_c^2 L_g^2}\right)^{1/2}} \frac{1}{\beta F_c}$$

$$= \frac{\alpha^{-1}}{\left[1+(\alpha^{-1})^2\right]^{1/2}} = w_2(\alpha^{-1}).$$

The weight functions $w_1(\alpha)$ (full line) and $w_2(\alpha^{-1})$ (dashed line) are plotted in Figure 10.4 versus α and α^{-1}, respectively. From the tangential dashed-dotted lines shown in Figure 10.4, we see that the transition from the long- to the short-channel mode occurs for $\alpha = 1/2$ for the saturation current ratio w_1 and $\alpha^{-1} = 1$ for the saturation transconductance ratio w_2. Therefore $\alpha = 1$ gives the location of the long- to short-channel transition for the FET small-signal characteristics. Note that this criterion is only applicable in the range of validity of the GCA approximation. A GCA region is always expected in the case of high-aspect ratio (L_g/d_2) FETs, where d_2 is the gate to channel spacing. Small-aspect ratio FETs, where two-dimensional field effects are important over the entire gated channel, are not considered here. According to Equation (10.3), the capacitive charge-control model used here will fail when the gate voltage V_{GS} reaches V_{Gmax} and the channel charge N_S saturates to N_{S0}. In a MODFET this occurs when the parasitic MESFET turns on (AlGaAs (Region 2) is no longer depleted) [5]. The ratio α therefore is more correctly defined by

$$\alpha = \frac{F_c L_g}{\min[V_{Gmax}, V_{GS}] - V_T}. \tag{10.22}$$

As a consequence, it is not possible in practice to turn on the short-channel mode in a long gate-length FET (e.g., 10 μm). To summarize, we have established that the ideal MOSFET/MODFET switches from the long-channel mode to the short-channel mode when the ratio $\alpha^{-1} = (\min[V_{Gmax}, V_{GS}] - V_T)/(F_c L_g)$ becomes larger than 1. This will also be confirmed in Chapter 12 by the dependence of the current-gain cut-off frequency f_T upon α.

10.3 Saturation and two-dimensional effects in FETs

The analysis of the ideal MOSFET developed in the previous section was based on the GCA approximation. The GCA relies on a one-dimensional solution of the Poisson equation which holds as long as the longitudinal electric field F (parallel to the x axis) is small compared to the transverse electric field (parallel to the y axis) $F_\perp = qN_S/\epsilon_1$ supporting the 2DEG charge. Once the device has reached pinch-off, the perpendicular field is zero, and the transverse field is infinite. Therefore the GCA cannot be used in the drain region, and a two-dimensional solution of the Poisson equation is required to describe the FET in saturation.

10.3.1 The Grebene–Ghandhi model

Grebene and Ghandhi have developed for the MESFET an approximate two-dimensional solution of the Poisson equation which is useful in understanding the saturation regime of the MODFET.

The gated region of the channel (see Figure 10.2) is divided into two parts: the GCA region and the two-dimensional region. In the GCA region, a one-dimensional solution of the Poisson equation is sufficient, and the charge-control models developed in Chapter 8 can be used. In the two-dimensional region, an approximate two-dimensional solution of the Poisson equation will be obtained. The drain voltage applied V_{DS} is then

$$V_{DS} = V_{CS}(X_S) + V_{SAT},$$

where V_{SAT} is the potential drop across the saturation region. A main hypothesis is to assume that the two-dimensional region starts at the position X_S in the channel when velocity saturation occurs. This is justified only on the basis that both effects take place for large electric fields. From the electric field given by Equation (10.8), the position X_S is given by

$$F(X_S) = -\frac{dV_{CS}}{dx} = -\frac{I_D}{\beta\left(V_{OUT}^2 - \frac{2X_S I_D}{\beta}\right)^{1/2}} = -F_c$$

$$F_c^2\left(V_{OUT}^2 - \frac{2X_S I_D}{\beta}\right) = \frac{I_D^2}{\beta^2}$$

$$\frac{2X_S I_D}{\beta} = V_{OUT}^2 - \frac{I_D^2}{\beta^2 F_c^2}, \tag{10.23}$$

$$X_S = \frac{\beta V_{OUT}^2}{2I_D} - \frac{I_D}{2\beta F_c^2} \quad \text{for} \quad I_D \geq I_{DS} \tag{10.24}$$

The position X_S at which the GCA region stops can therefore be calculated from the drain current I_D once the device is in saturation ($I_D \geq I_{DS}$) with I_{DS} given by Equation (10.18)). The channel potential $V_{CS}(X_S)$ is then obtained by substituting Equation (10.23) into Equation (10.6)

$$V_{CS}(x) = V_{OUT} - \left(V_{OUT}^2 - \frac{2X_S I_D}{\beta}\right)^{1/2}$$
$$= V_{OUT} - \frac{I_D}{\beta F_c} \quad \text{for} \quad I_D \geq I_{DS}.$$

The current in the two-dimensional region is given by

$$I_D = qW_g N_S(X_S) v_S.$$

Since under stationary conditions the current in the channel is uniform (constant), the 2DEG concentration $N_S(X_S)$ in the channel is also constant:

$$N_S(x) = N_S(X_S) \quad \text{for} \quad X_S \leq x \leq L_g.$$

We now need to calculate the potential drop V_{SAT} across the two-dimensional region. The Poisson equation in this region is

$$\frac{\partial^2 V(x', y)}{\partial x'^2} + \frac{\partial^2 V(x', y)}{\partial y^2} = -\frac{qN_D(y)}{\epsilon_2}, \tag{10.25}$$

where V is the electrostatic potential and where the charge distribution in the depleted AlGaAs region (Region 2) consists of the ionized donor distribution. Note that the axis x' is defined by $x' = x - X_S$. The boundary conditions used for integrating the

Poisson equation in the two-dimensional region are shown in Figure 10.5(a). These boundary conditions are (setting the reference of the electrostatic potential at $y = 0$)

$$V(x', 0) = 0 \text{ (reference)}$$
$$V(0, y) = V_{GCA}(y)$$
$$V(L_G - X_S, y) = \text{ unknown}$$
$$\epsilon_2 \frac{dV(x, d_2)}{dy} = qN_S(X_S), \tag{10.26}$$

where $V_{GCA}(y)$ is the electrostatic potential (see Chapter 8) solution of the one-dimensional Poisson problem

$$\frac{d^2 V_{GCA}}{dy^2} = \begin{cases} -\dfrac{qN_D}{\epsilon_2} & \text{for } 0 \le y \le d_2 - e \\ 0 & \text{for } d_2 - e < y \le d_2 \end{cases}$$
$$V_{GCA}(0) = 0$$
$$-\frac{dV_{GCA}(d_2)}{dy} = qN_S(X_S).$$

Using the superposition principle (Equation (10.25) is linear), we can write $V(x', y)$ as the superposition of two potentials $V_1(x', y)$ and $V_2(x', y)$ with the boundary conditions shown in Figure 10.5(b).

$$V(x', y) = V_1(x', y) + V_2(x', y)$$
$$\frac{\partial^2 V_1(x', y)}{\partial x'^2} + \frac{\partial^2 V_1(x', y)}{\partial y^2} = -\frac{qN_D(y)}{\epsilon_2}$$
$$\frac{\partial^2 V_2(x', y)}{\partial x'^2} + \frac{\partial^2 V_2(x', y)}{\partial y^2} = 0. \tag{10.27}$$

The boundary conditions for $V_1(x', y)$ are

$$V_1(x', 0) = 0$$
$$V_1(0, y) = V_{GCA}(y)$$
$$V_1(L_G - X_S, y) = V_{GCA}(y)$$
$$-\epsilon_2 \frac{dV_1(x, d_2)}{dy} = qN_S(X_S). \tag{10.28}$$

The solution is simply $V_1(x', y) = V_{GCA}(y)$. The boundary conditions for $V_2(x', y)$ are

$$V_2(x', 0) = 0$$
$$V_2(0, y) = 0$$
$$V_2(L_G - X_S, y) = \text{unknown}$$
$$-\epsilon_2 \frac{dV_2(x, d_2)}{dy} = 0. \tag{10.29}$$

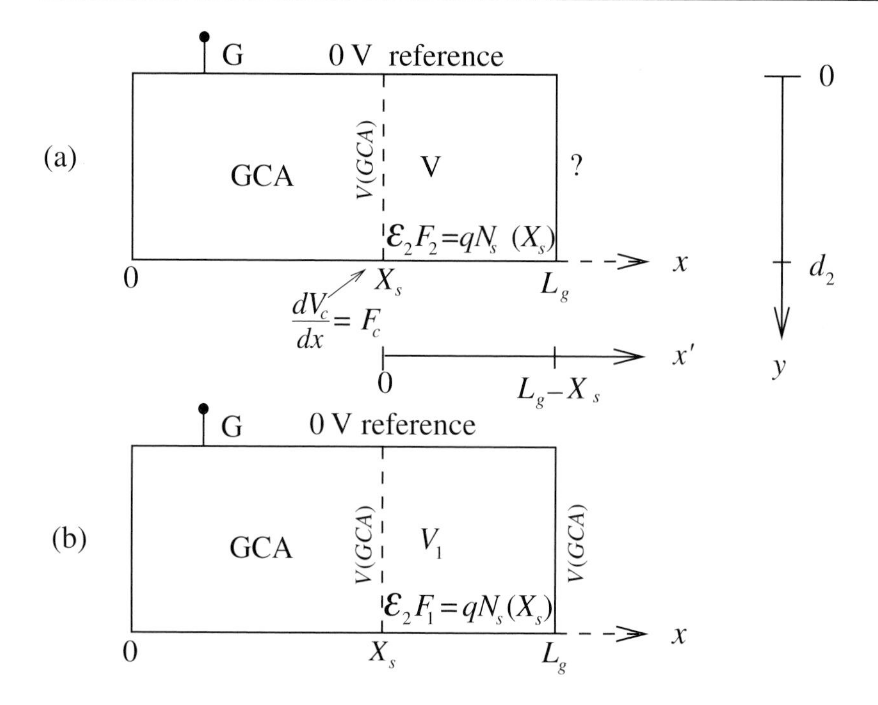

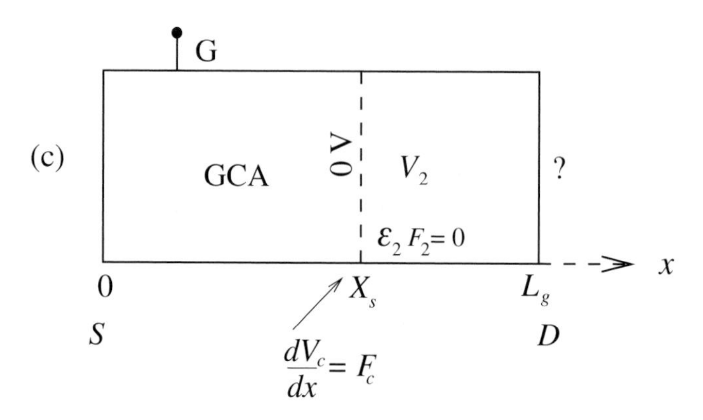

Fig. 10.5. Boundary conditions for: (a) the total potential $V(x', y)$, (b) the potential $V_1(x', y)$, and (c) the potential $V_2(x', y)$.

Note that the addition (superposition) of the boundary conditions of $V_1(x', y)$ and $V_2(x', y)$ gives the boundary conditions of $V(x', y)$.

The electrostatic potential V_2 can be obtained using the method of separation of variables:

$$V_2(x', y) = V_2(x')V_2(y)$$

$$\frac{\dfrac{d^2V_2(x')}{dx'^2}}{V_2(x')} = -\frac{\dfrac{d^2V_2(y)}{dy^2}}{V_2(y)} = K.$$

The first order solution satisfying the boundary conditions is

$$V_2(x', y) = \frac{2d_2F_c}{\pi} \sin\left(\frac{\pi y}{2d_2}\right) \sinh\left(\frac{\pi x'}{2d_2}\right).$$

We have used the boundary condition (F_c is positive)

$$-\frac{dV_2(0, d_2)}{dy} = -F_c$$

to replace the unknown boundary condition at $x = L_g$. The potential drop in the two-dimensional region is then

$$V_{SAT} = V(L_g - X_S, d_2) - V(0, d_2) = \frac{2d_2F_c}{\pi} \sinh\left[\frac{\pi(L_g - X_S)}{2d_2}\right].$$

The drain voltage V_{DS} in saturation can then be obtained from the drain current I_D using

$$V_{DS}[I_D] = V_{OUT} - \frac{I_D}{\beta F_c} + \frac{2d_2F_c}{\pi} \sinh\left[\frac{\pi(L_g - X_S)}{2d_2}\right].$$

The simple Grebene–Ghandhi model [6] predicts that as the drain current increases slightly above I_{DS}, the voltage across the two-dimensional region will increase at an exponential rate (because of the hyperbolic sine). Note that the channel potential $V_{CS}(X_S)$ across the GCA region remains approximately constant for a small increase in drain current. Conversely, as the drain voltage is increased, the drain current and the potential $V_{CS}(X_S)$ remain approximately constant and the excess potential $V_{DS} - V_{CS}(X_S)$ is dropped across the two-dimensional region. The Grebene–Ghandhi model therefore predicts a drain conductance g_d in saturation which is essentially zero.

It is valuable to verify whether a more accurate analysis including diffusion as well as drift and a full two-dimensional solution of the Poisson equation predicts the same results as the Grebene–Ghandhi model. Such a two-dimensional charge-control model was reported by Kim and Roblin [7]. In this model the channel current is obtained using the transport model

$$I_D = -W_g\mu(F)N_S\frac{dE_{F,2DEG}}{dx}, \tag{10.30}$$

where the 2DEG Fermi level $E_{F,2DEG} = E_{Fi}(N_S) - qV(x, d_2)$ is the chemical potential $E_{Fi}(N_S)$ calculated using the analytic expression derived in Chapter 8 plus the electrostatic potential energy $-qV(x, d_2)$. The electrostatic potential V is calculated using the charge distribution $\rho(x, y)$ shown in Figure 10.6(a).

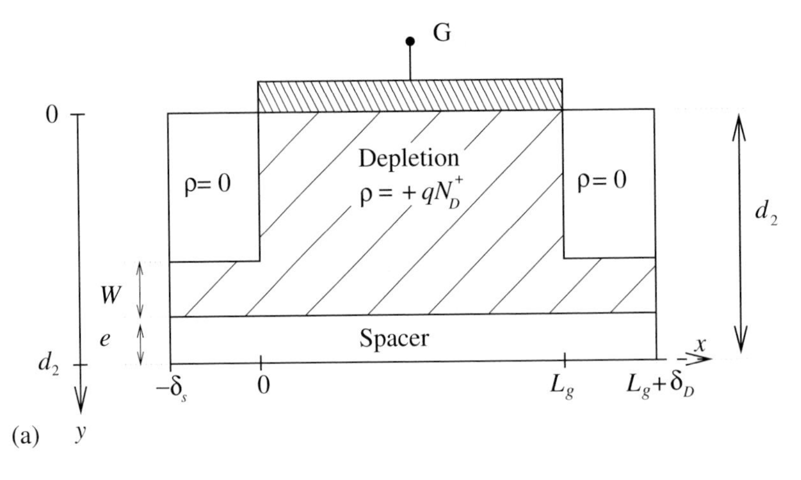

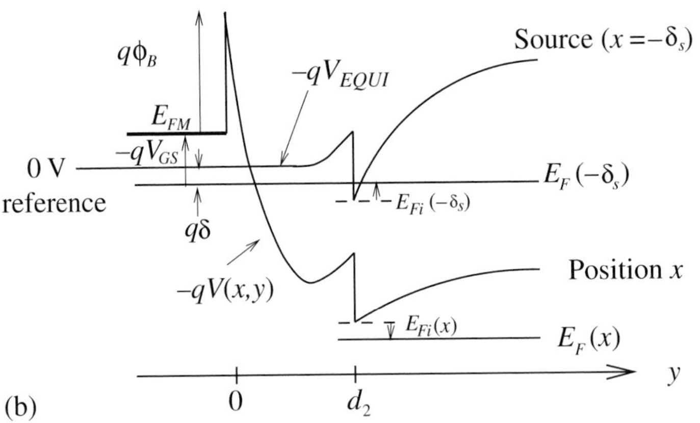

Fig. 10.6. (a) Charge distribution used for the two-dimensional charge control model and (b) band diagram defining the electrostatic potential boundary conditions.

As can be verified, the region studied now extends beyond the gated region of the channel on the source and drain sides. The Poisson equation in this region is

$$\frac{\partial^2 V(x,y)}{\partial x^2} + \frac{\partial^2 V(x,y)}{\partial y^2} = -\frac{\rho(x,y)}{\epsilon_2}. \tag{10.31}$$

These boundary conditions (see band diagram in Figure 10.6(b)), setting the potential reference at the origin $V(0,0) = 0$, are

$$V(x,0) = \begin{cases} 0 & \text{for } \delta_S \le x < 0 \\ V_{GS} - \phi_B - \delta & \text{for } 0 \le x \le L_g \\ V_{DS} & \text{for } L_g < x \le L_g + \delta_D \end{cases}$$

$$V(-\delta_S, y) = V_{EQUI}(y)$$

$$V(L_g + \delta_D, y) = V_{EQUI}(y) + V_{DS}$$

$$-\frac{dV(x,d_2)}{dy} = \frac{qN_S(x)}{\epsilon_2},$$

where $V_{EQUI}(y)$ is the electrostatic potential in the ungated 2DEG structure (see Figure 10.6(b)) obtained from the solution of the following one-dimensional Poisson problem (see Section 8.3)

$$\frac{d^2 V_{EQUI}}{dy^2} = \begin{cases} 0 & \text{for} \quad 0 \le y < d_2 - W \\ -\dfrac{qN_D}{\epsilon_2} & \text{for} \quad d_2 - W \le y \le d_2 - e \\ 0 & \text{for} \quad d_2 - e < y \le d_2 \end{cases}$$

$$V_{EQUI}(0) = 0$$

$$-\frac{dV_{EQUI}(d_2)}{dy} = qN_{S0}.$$

For a large aspect ratio d_2/L_g, the channel potential $V(x, y)$ is given approximately by

$$\begin{aligned} V_C(x) = V(x, d_2) &\simeq V_{2D}(x, d_2) - d_2 F_I + \frac{qd_2 N_{S0}}{\epsilon_2} \\ &= V_{2D}(x, d_2) - \frac{qd_2(N_S(x) - N_{S0})}{\epsilon_2}, \end{aligned} \tag{10.32}$$

$$qN_S = \epsilon_2 F_I, \tag{10.33}$$

where the two-dimensional potential $V_{2D}(x, y)$ satisfies the same boundary condition as $V(x, y)$ except for the electric field at $y = d_2$:

$$V_{2D}(x, 0) = \begin{cases} V_S & \text{for} \quad \delta_S \le x < 0 \\ V_G - \phi_B - \delta & \text{for} \quad 0 \le x \le L_g \\ V_D & \text{for} \quad L_g < x \le L_g + \delta_D \end{cases}$$

$$V_{2D}(0, y) = V_{EQUI}(y)$$

$$V_{2D}(L_G + \delta_D, y) = V_{EQUI}(y) + V_{DS}$$

$$-\frac{dV_{2D}(x, d_2)}{dy} = \frac{qN_{S0}}{\epsilon_2}.$$

Note that it is assumed that the channel concentration satisfies

$$N_S(-\delta_S) = N_S(L_g + \delta_D) \simeq N_{S0}.$$

Plots of the electrostatic potential $V_{2D}(x, y)$ and of its value in the channel (d_2) for various drain and gate voltages are given in Figure 10.7 for a 1 μm MODFET.

Once $V_{2D}(x, d_2)$ is known, the integration of the current equation reduces to

$$I_D = -W_g \mu(F) \left[\frac{dE_{Fi}(N_S(x)}{dx} - q\frac{dV_{2D}(x, d_2)}{dx} + q\frac{qd_2}{\epsilon_2}\frac{dN_S(x)}{dx} \right].$$

A numerical solution can be facilitated if this non-linear differential equation is first transformed into an integral equation [7]. The advantage of the partitioning used for

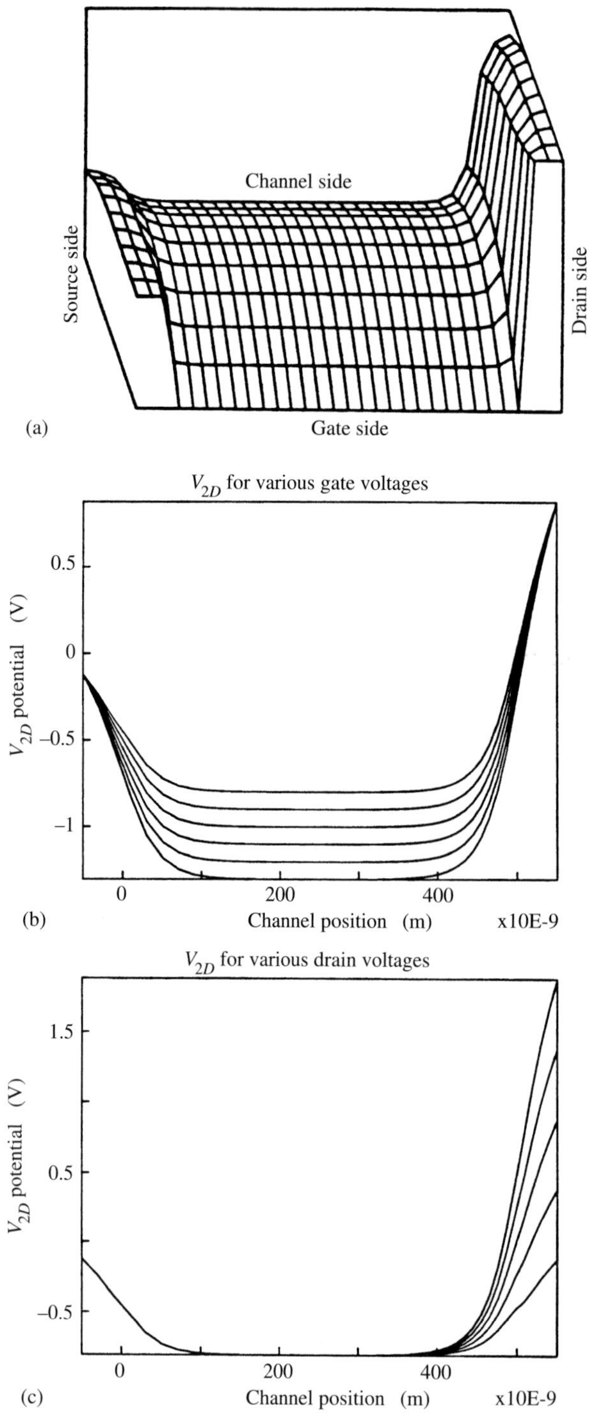

Fig. 10.7. Plot of: (a) $V_{2D}(x, y)$, (b) $V_{2D}(x, d_2)$ for various drain voltages and (c) $V_{2D}(x, d_2)$ for various gate voltages. (Y. M. Kim and P. Roblin, *IEEE Transactions on Electron Devices*, Vol. ED-33 No. 11, pp. 1644–1651, 1986. ©1986 IEEE.)

the electrostatic potential V is that the two-dimensional potential V_{2D} needs only to be calculated once for each gate-to-source voltage and drain-to-source voltage. This greatly reduces the calculation time, since a time-consuming numerical technique such as the finite difference method must be used to calculate $V_{2D}(x, y)$.

The channel potential, the channel 2DEG charge and the channel electric field calculated with this model for a 1 μm MODFET and for various drain-to-source voltages are shown in Figure 10.8. The inclusion of the diffusion current now permits us to take into account part of the ungated channel region and the built-in potential appearing on both sides of the gated channel. Indeed, the MODFET channel is an effective n^+–n–n^+ structure when the channel charge N_S controlled by the gate is smaller than the equilibrium concentration N_{S0} expected on both sides of the gate. As in the p–n junction, a built-in potential is then raised to balance the diffusion from the n^+ regions into the n region. The resulting potential barrier in the source region is seen in Figure 10.8(a) to remain constant for all the various drain-to-source voltages applied. On the drain side the built-in potentials remain constant for small drain-to-source voltages. However, once the MODFET enters in saturation for say $V_{DS} > V_{DS,SAT}$, the excess drain voltage $V_{DS} - V_{DS,SAT}$ is mostly dropped across the built-in potential barrier region on the drain side such that the channel potential under the gate (the GCA region) remains essentially unchanged. The 2DEG concentration $N_S(x)$ in the channel shown in Figure 10.8(b) shows the pinch-down of the channel for increasing drain-to-source voltages. Current saturation, however, is reached before pinch-off. In saturation the 2DEG distribution in the channel varies little. Finally, the channel electric field shown in Figure 10.8(c) demonstrates the rapid increase with increasing drain-to-source voltages of the electric field in the drain region compared to the gated part of the channel and the source region. A negligible increase of the drain current at saturation is then observed in the simulated I–V characteristics shown in Figure 10.9 (full line)

10.3.2 Channel opening: MOSFET saturation model

As we shall see in the next chapter a small drain conductance improves the power gain of FETs at microwave frequencies. The drain conductance is therefore an important figure of merit after the transconductance. The drain conductance predicted by the Grebene–Ghandhi model is essentially null. However, the DC drain conductance measured for FETs is larger than predicted by the Grebene–Ghandhi model. A larger DC drain conductance can be obtained if we account for the channel opening, that is the width of the channel.

The boundary condition we have used for the electric field at d_2 is

$$F_I = -\frac{dV(x, d_2^-)}{dy} = \frac{qN_S(x)}{\epsilon_2}.$$

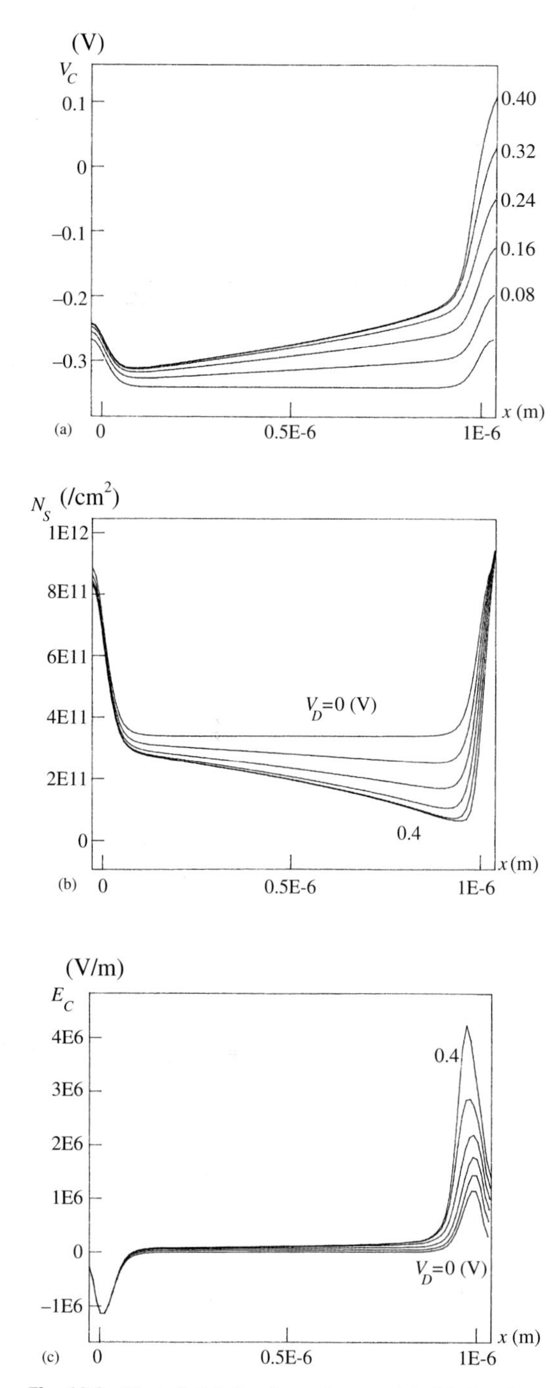

Fig. 10.8. Plot of: (a) the channel potential, (b) the channel 2DEG charge and (c) the channel electric field versus position for various drain-to-source voltages. (Y. M. Kim and P. Roblin, *IEEE Transactions on Electron Devices*, Vol. ED-33 No. 11, pp. 1644–1651, 1986. ©1986 IEEE.)

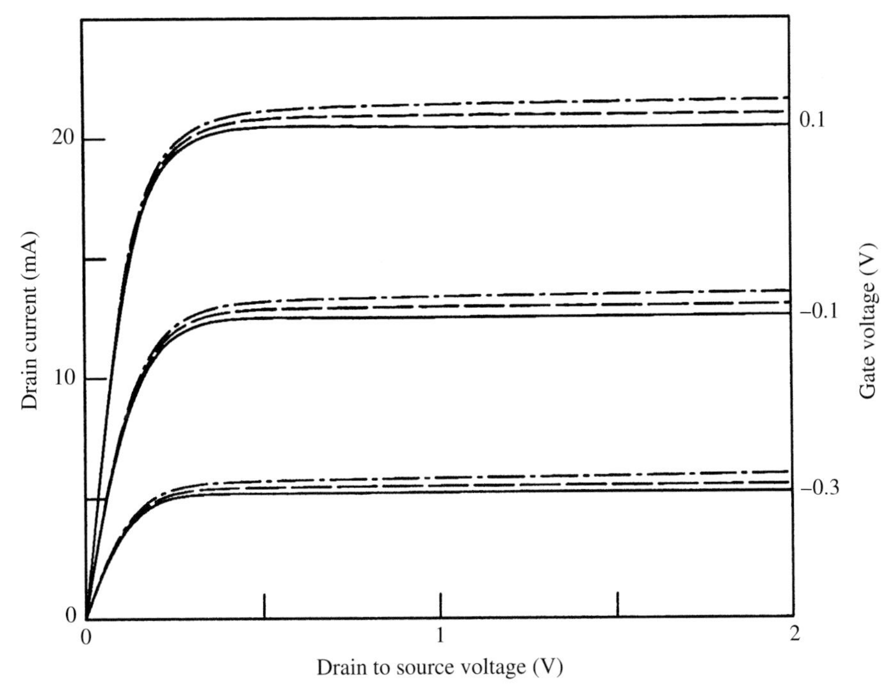

Fig. 10.9. I–V characteristics calculated using the two-dimensional charge-control model (full line, $\Delta y = 0$) and for channel widths Δy of 168 (dashed line) and 400 Å (dashed-dotted line). (Y. M. Kim and P. Roblin, *IEEE Transactions on Electron Devices*, Vol. ED-33 No. 11, pp. 1644–1651, 1986. ©1986 IEEE.)

In Chapter 8 we saw that a uniform 2DEG in equilibrium satisfied (Gauss's law)

$$\epsilon_2 F_I = \epsilon_1 F_B + q N_S, \tag{10.34}$$

where F_B is the bulk field which is negligible when the 2DEG is on.

In a MODFET, the 2DEG concentration $N_S(x)$ varies with position, and this boundary condition is only valid if we treat the 2DEG as a surface charge and therefore neglect its thickness. In a MODFET the width Δy of the 2DEG depends upon the 2DEG concentration N_S. Here we arbitrarily define the width Δy of the 2DEG as the width of the AlGaAs–GaAs triangular well (Region 1) including 95% of the 2DEG charge. For high N_S concentrations the electric field is high, and the 2DEG width is small (see Figure 10.10(a)). For low N_S concentrations the electric field is small, and the 2DEG width is large (see Figure 10.10(b)). A plot of the variation of Δy versus N_S is given in Figure 10.11 for different GaAs bulk doping N_A.

As can be seen, 95% of the 2DEG charge extends up to 400 Å in GaAs. It therefore becomes necessary to solve the Poisson equation in the 2DEG region. An approximate treatment is proposed below. We start by writing the Poisson equation in the 2DEG region (Region 1) assuming that the 2DEG concentration $N_S(x)$ is

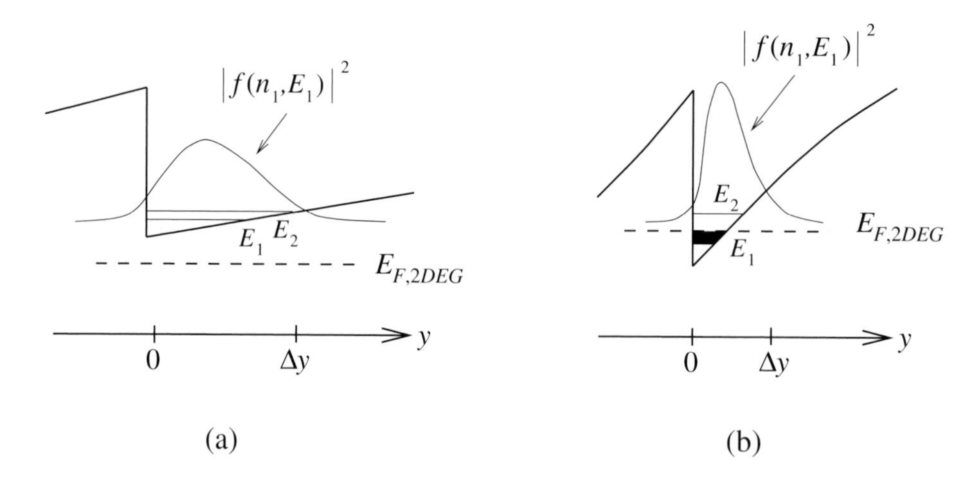

Fig. 10.10. Distribution of the 2DEG for: (a) a low, and (b) a high 2DEG concentration.

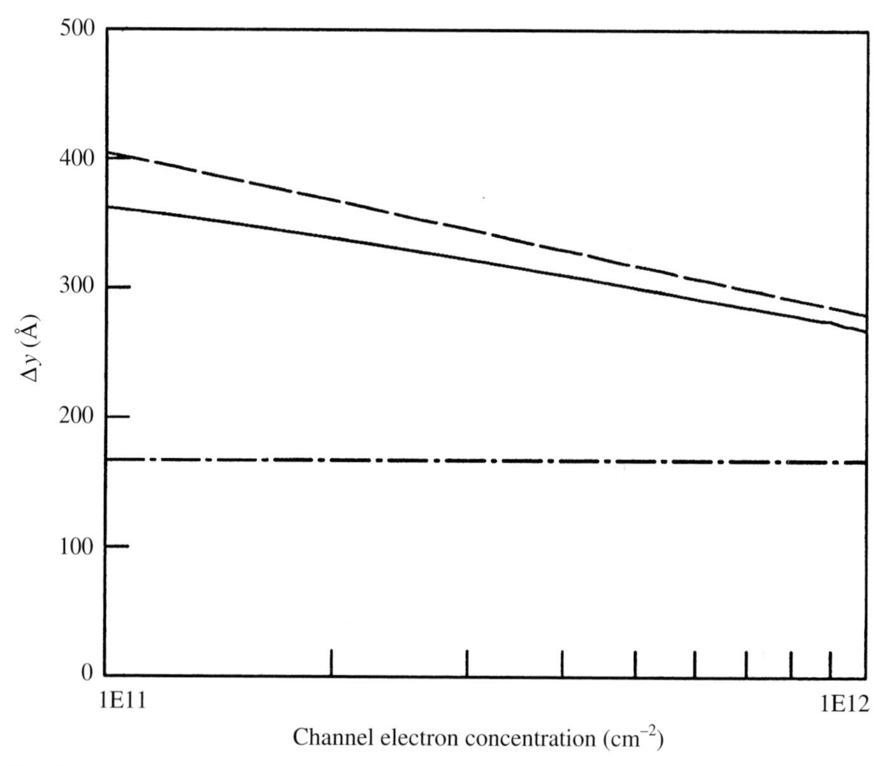

Fig. 10.11. Plot of the 2DEG width Δy as a function of the channel concentration N_S for different GaAs bulk doping of $N_A = 3 \times 10^{14}$ cm^{-3} (full line) and $N_A = 3.14 \times 10^{13}$ cm^{-3} (dashed line). When a quantum well is used to confine the 2DEG (see [8]), the 2DEG width Δy is set by the quantum well width (dashed line) and is consequently insensitive to the bulk doping. (Y. M. Kim and P. Roblin, *IEEE Transactions on Electron Devices*, Vol. ED-33 No. 11, pp. 1644–1651, 1986. ©1986 IEEE.)

uniformly distributed over a region of width $\Delta y(x)$

$$\frac{\partial^2 V(x,y)}{\partial x^2} + \frac{\partial^2 V(x,y)}{\partial y^2} = \frac{qN_S(x)}{\Delta y(x)\epsilon_1}. \tag{10.35}$$

Next we integrate this Poisson equation across the 2DEG width:

$$\int_{d_2}^{d_2+\Delta y} \frac{\partial^2 V(x,y)}{\partial x^2}\,dy + \int_{d_2}^{d_2+\Delta y} \frac{\partial^2 V(x,y)}{\partial y^2}\,dy = \int_{d_2}^{d_2+\Delta y} \frac{qN_S(x)}{\Delta y(x)\epsilon_1}\,dy$$

$$\Delta y \frac{d^2 V_C(x)}{dx^2} + \frac{dV(x, d_2+\Delta y)}{dy} - \frac{dV(x, d_2^+)}{dy} = \frac{qN_S(x)}{\epsilon_1}$$

$$\epsilon_1 \Delta y \frac{d^2 V_C(x)}{dx^2} + \epsilon_2 F_I - \epsilon_1 F_B = qN_S(x), \tag{10.36}$$

where we define the channel potential $V_C(x)$ as the average of the potential $V(x, y)$ over the 2DEG channel width

$$V_C(x) = \frac{\int_{d_2}^{d_2+\Delta y} V(x,y)\,dy}{\Delta y},$$

assuming that $\Delta y(x)$ is slowly varying with x. The boundary equation $F_I = qN_S/\epsilon_2$ is now replaced by the differential Equation (10.36). Note that for $\Delta y = 0$ the boundary condition (10.36) reduces to the boundary condition (10.34).

Using this new boundary condition (10.36) instead of (10.34), we can numerically integrate the current Equation (10.30) and obtain the I–V characteristics shown in Figure 10.9 [8].

One observes that the drain conductance is essentially null for a channel width of 0 Å (full line) and increases as the channel width is increased (dashed and dashed-dotted lines). Note that the fitting of the drain conductance of a real MODFET might require even larger channel widths than predicted from the equilibrium theory. Further increases of the channel width could result from the heating of the 2DEG. Indeed for a three-dimensional gas in a triangular potential well

$$\begin{aligned} N_S &= N_c \exp\left[\frac{E_F - E_c(y)}{k_B T_e}\right] \\ &= N_c \exp\left[\frac{E_{Fi} - qF_I\,y}{k_B T_e}\right] \\ &\propto \exp\left[\frac{-y}{\Delta y}\right], \end{aligned} \tag{10.37}$$

with $\Delta y = k_B T_e/(qF_I)$. We can see from this simple model that the channel width will increase with the electron temperature T_e.

An analytic model can be developed if we neglect the bulk field F_B and assume that for a small aspect ratio d_2/W_g the interface field F_I is obtained using a one-dimensional solution of the Poisson equation in the AlGaAs region (Region 2):

$$\epsilon_2 F_I = C_G(V_{GC} - V_T).$$

Equation (10.36) can now be rewritten

$$\epsilon_1 \Delta y(x) \frac{d^2 V_C(x)}{dx^2} + C_G(V_{GC} - V_T) = q N_S(X_S). \tag{10.38}$$

The reader is referred to Problem 10.4 for an analytic solution of this equation.

Here we shall limit our analysis to the saturation region and assume that the interface field F_I is negligible compared to the longitudinal field. Equation (10.36) can then be rewritten

$$\epsilon_1 d_S \frac{d^2 V_C(x)}{dx^2} = q N_S(X_S). \tag{10.39}$$

We have also further assumed that in the saturation region the channel width is constant $\Delta y = d_S$. The carrier $N_S(X_S)$ in saturation is obtained from the drain current I_D using

$$q N_S(X_S) = \frac{I_D}{W_g v_S}.$$

The saturation voltage drop across the two-dimensional region is therefore

$$V_{SAT} = V_C(L_g) - V_C(X_S) = \frac{I_D}{2W_g \epsilon_1 d_S v_S}(L_g - X_S)^2 + F_c(L_g - X_S),$$

using the boundary condition $dV_C(X_S)/dx = F_c$. The total drain-to-source potential is then

$$V_{DS} = V_{OUT} + \frac{I_D}{\beta F_c} + \frac{I_D}{2W_g \epsilon_1 d_S v_S}(L_g - X_S)^2 + F_c(L_g - X_S).$$

Because a parabola increases less rapidly than a hyperbolic sine, the potential drop across the saturation region is smaller for a given current I_D than in the Grebene–Ghandhi model. Conversely, an increase of the drain current does not require a large applied drain-to-source voltage. The drain conductance predicted by the channel opening model is therefore larger. The channel opening model thus provides a better fit to the experimental drain conductance than the Grebene–Ghandhi model although both two-dimensional effects are present. Such a model was first reported by Park and Kwack [9] for the MODFET. Note that an increase of the drain current requires a decrease of the GCA region width X_S. The saturation region can therefore occupy a large fraction of the gate length in saturation when the current increases above $I_{DS}(SAT)$. This effect is called gate-length modulation.

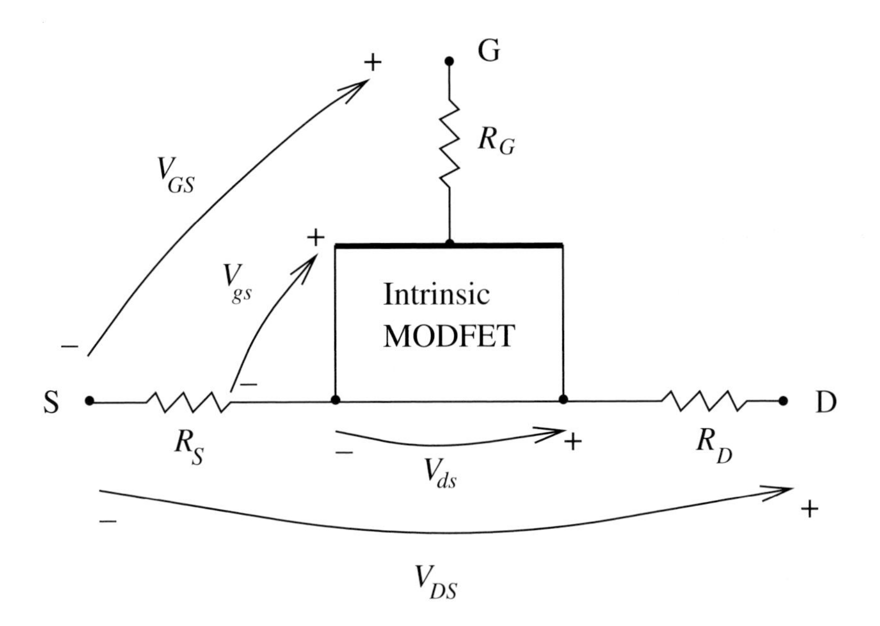

Fig. 10.12. Extrinsic and intrinsic MODFET.

10.4 The extrinsic MODFET

Our analysis of the ideal MODFET has so far been concerned with the intrinsic MODFET, that is no parasitics have been included. Figure 10.12 shows an extrinsic FET consisting of the intrinsic FET with a source resistance R_S, a drain resistance R_D, and a gate resistance R_G in series with the intrinsic source, drain and gate terminals.

The extrinsic drain and source voltages V_{DS} and V_{GS} can then be expressed in terms of the intrinsic drain and source voltages using

$$V_{GS} = V_{gs} + I_D R_S,$$

$$V_{DS} = V_{ds} + I_D(R_S + R_D).$$

Note that because the gate is insulated, the DC characteristics are not affected by the gate resistance R_G. To demonstrate the impact of the source and drain resistances on the FET characteristics, we can express the extrinsic drain conductance and the transconductance in terms of the intrinsic drain conductance and transconductance [10]. One can easily obtain the following relations (see Problem 10.5):

$$g_D = \frac{g_d}{1 + g_d(R_S + R_D) + g_m R_S},$$

$$g_M = \frac{g_m}{1 + g_d(R_S + R_D) + g_m R_S},$$

where we have used the definitions:

$$g_D = \frac{\partial I_D(V_{DS}, V_{GS})}{\partial V_{DS}},$$

$$g_M = \frac{\partial I_D(V_{DS}, V_{GS})}{\partial V_{GS}},$$

$$g_d = \frac{\partial I_D(V_{ds}, V_{gs})}{\partial V_{ds}},$$

$$g_m = \frac{\partial I_D(V_{ds}, V_{gs})}{\partial V_{gs}}.$$

Clearly both the drain conductance and the transconductance are reduced by the parasitics. It is therefore critical to reduce the source and drain resistances for the extrinsic MODFET to maintain the high performance of the intrinsic MODFET. Note that the source and drain resistance model used here assumes that the $I-V$ characteristics of the parasitics are linear. However, experimental and theoretical evidence points to the fact that the source and drain resistances increase with the drain current [11]. Non-linear source and drain resistances seem to account for the decrease of the transconductance measured at large gate voltages in MODFETs.

10.5 Conclusion

This chapter was concerned with the DC $I-V$ modeling of the MODFET. The current voltage characteristic MOSFET/MODFET was studied using simple piece-wise charge-control and transport models. A more accurate and realistic modeling of the MODFET will be developed in Chapter 14. The motivation for our approach in this chapter was to introduce the short-channel field and velocity-saturation effects, and the thresholds for their occurence. Armed with these tools we are now ready to study in the next two chapters the AC responses of the long- and short-channel MODFETs.

10.6 Bibliography

10.6.1 Recommended reading

H. Morkoç, H. Unlu, G. Ji, *Principles and Technology of MODFETs*, Vols. I and II, John Wiley & Sons, Chichester, 1991.

10.6.2 References

[1] U. K. Mishra, A. S. Brown, M. J. Delaney, P. T. Greiling and C. F. Krumm, 'The AlInAs–GaInAs HEMT for microwave and millimeter-wave applications', *IEEE Transactions*

on Microwave Theory and Techniques, Vol. 37, No. 9, pp. 1279–1285, 1989.

[2] A. J. Tessmer, P. C. Chao, K. H. G. Duh, P. Ho, M. Y. Kao, S. M. J. Liu, P. M Smith, J. M. Ballingall, A. A. Jabra and T. H. Yu, 'Very high performance 0.15 μm gate-length InAlAs/InGaAs/InP lattice-matched HEMTs', *Proceedings of the 12th Cornell Conference on Advanced Concepts in High-Speed Semiconductor Devices and Circuits*, Cornell University, Ithaca, NY, 1989.

[3] D. Delagebeaudeuf and N. Linh, 'Metal-(n) AlGaAs–GaAs two-dimensional electron gas FET', *IEEE Transactions on Electron Devices*, Vol. ED-29, No. 6, pp. 955–960, June 1982.

[4] H. Rohdin and P. Roblin, 'A MODFET DC model with improved pinchoff and saturation characteristics', *IEEE Transactions on Electron Devices*, Vol. ED-33, pp. 664–672, 1986.

[5] K. Lee, M. S. Shur, T. J. Drummond and H. Morkoç, 'Parasitic MESFET in (Al,Ga)As/GaAs modulation doped FETs and MODFET characterization.' *IEEE Transactions on Electron Devices*, Vol. ED-31, pp. 29–35, January 1984.

[6] A. B. Grebene and S. K. Ghandhi, 'General theory for pinched operation of the junction-gate FET', *Solid-State Electronics*, Vol. 12, p. 573, 1969.

[7] Y. M. Kim and P. Roblin, 'Two-dimensional charge control model for the MODFETs', *IEEE Transactions on Electron Devices*, Vol. ED-33 No. 11, pp. 1644–1651, 1986.

[8] P. Roblin, H. Rohdin, C.J. Hung and S. W. Chiu, 'Capacitance–voltage analysis and current modeling of pulse-doped MODFETs', *IEEE Transactions on Electron Devices*, Vol. ED-36, No. 11, pp. 2394–2404, November 1989.

[9] K. Park and K. D. Kwack, 'A model for the current–voltage characteristics of MODFETs', *IEEE Transactions on Electron Devices*, Vol. ED-33, pp.673–676, 1986.

[10] S. Y. Chou and D. A. Antoniadis 'Relationship between measured and intrinsic transconductances of FET's', *IEEE Transactions on Electron Devices*, Vol. ED-34, pp. 448–450, February 1987.

[11] P. Roblin, L. Rice and S. Bibyk, 'Non-linear parasitics in MODFETs and the MODFET IV characteristics' *IEEE Transactions on Electron Devices*, Vol. ED-35, No. 8, pp. 1207–1214, (1988)

10.7 Problems

10.1 In this problem we shall calculate the I–V characteristics in the GCA approximation of a MODFET using the current equation

$$I_D = -W_g \mu N_S \frac{dE_{F,2DEG}}{dx}, \tag{10.40}$$

where $E_{F,2DEG}$ is the 2DEG Fermi level. The 2DEG Fermi level is given by

$$E_{F,2DEG} = E_{Fi}(N_S) - qV_C(x),$$

where $E_{Fi}(N_S)$ is the chemical potential (see Chapter 8) and V_C is the channel potential.

(a) Verify that the ratio D/μ can be expressed in terms of the chemical potential E_{Fi}. We shall assume for the remainder of the problem that this ratio can be linearized:

$$\frac{D(N_S)}{\mu} = \frac{D(N_{S1})}{\mu} + \frac{1}{\mu}\frac{dD(N_{S1})}{dN_S}(N_S - N_{S1}) = \beta + \alpha N_S.$$

Calculate α and β at $N_{S1} = 10^{12}$ cm^{-2} for the chemical potential given by

$$E_{Fi}(\text{eV}) = 0.234\left[\frac{N_S}{10^{12}}\right]^{\frac{1}{3}} - 0.153$$

with N_S in cm^{-2}.

(b) We now wish to calculate the I–V characteristics in the GCA approximation. We use the piece-wise linear charge-control model given by Equation (10.3) and use a constant mobility model $\mu(F) = \mu$. For this purpose it is sufficient to verify that the drain current can be written in the form

$$I_D = W_g\mu' C_G(V_{GS} - V_{CS} - V_T')\frac{dV_{CS}}{dx},$$

where μ' and V_T' are, respectively, the new mobility and threshold voltage to be expressed in terms of μ, V_T, C_G, α, β and q. Calculate the ratio μ'/μ and the difference $V_T' - V_T$ for a MODFET with a gate capacitance $\epsilon_2/(d_2 + \Delta d)$ using $d_2 + \Delta d = 400\times$ Å.

10.2 Calculate the I–V characteristics in the GCA approximation of a MODFET using the silicon velocity-field relation:

$$v_d = \mu(F)F = \frac{\mu F}{1 + F/F_c}.$$

Assume that the drain current is given by the drift component only:

$$I_D = qW_g N_S v_d.$$

Use the piece-wise linear charge-control model given by Equation (10.3).

10.3 Calculate the I–V characteristics of a MODFET in the GCA approximation using the following charge model:

$$N_S = N_{S0}\left[\alpha + (1-\alpha)\tanh\left(\frac{V_{GC} - V_{GM}}{V_1}\right)\right].$$

Assume that N_{S0}, α, V_1 and V_{GM} are known constants used to fit the N_S–V_{GS} relation obtained using the numerical techniques described in Chapter 8. Assume that the drain current is given by the drift component only:

$$I_D = qW_g N_S v_d$$

and use a constant mobility model $v_d = \mu F$.
Hint: $\int \tanh x\, dx = \ln[\cosh(x)]$.

10.4 Consider the equation for the channel potential V_C in the saturation region:

$$\epsilon_1 d_S \frac{d^2 V_{CS}(x)}{dx^2} + C_G(V_{GS} - V_{CS} - V_T) = qN_S(X_S), \tag{10.41}$$

where d_S is the channel width in the saturation region.
Assume that the GCA approximation holds at the edge of the saturation region:

$$qN_S(X_S) = C_G(V_{GC}(X_S) - V_T).$$

Integrate this equation across the saturation (two-dimensional) region of length $L_g - X_S$ and calculate the saturation potential V_{SAT} in terms of the drain current. Verify that the saturation voltage is of the form

$$V_{SAT} = \alpha^{1/2} F_c \sinh\left(\frac{L_g - X_S}{\alpha^{1/2}}\right).$$

10.5 Consider an extrinsic FET consisting of an intrinsic FET with a source resistance R_S, a drain resistance R_D and a gate resistance R_G in series with the intrinsic source, drain and gate terminals (see Figure 10.12).

(a) Express the extrinsic drain conductance g_D and extrinsic transconductance g_M defined by

$$g_D = \frac{\partial I_D(V_{DS}, V_{GS})}{\partial V_{DS}},$$

$$g_M = \frac{\partial I_D(V_{DS}, V_{GS})}{\partial V_{GS}},$$

in terms of the intrinsic drain conductance g_d and transconductance g_m defined by

$$g_d = \frac{\partial I_D(V_{ds}, V_{gs})}{\partial V_{ds}},$$

$$g_m = \frac{\partial I_D(V_{ds}, V_{gs})}{\partial V_{gs}}.$$

(b) Express the ratio g_D/g_M in terms of the ratio g_d/g_m.

11 Small- and large-signal AC models for the long-channel MODFET

Music is a hidden arithmetic exercise of the soul ....

From The World is Sound: Nada Brahma, Music and the Landscaper of Consciousness, Joachim-Ernst Berendt.

GOTTFRIED WILHELM LEIBNIZ

11.1 Introduction

In this chapter we shall study the high-frequency modeling of the MODFET (modulation doped field-effect transistor). Like any other FET, the MODFET is a transit-time device and its high-frequency performance is controlled by its gate length. For a given gate length, the MODFET has achieved higher frequencies of operation than other FETs [1]. This originates from the high mobility and high effective velocity saturation of the 2DEG (two-dimensional electron gas), which leads to a high transconductance and low source and drain resistances. However, the high-frequency principle of operation of the MODFET/MOSFET does not differ from that of other FETs. It should therefore be possible to generalize the results of the analysis given in this chapter to other FETs such as the MESFET (metal semiconductor FET).

In this chapter we shall focus on the long-channel MODFET and study both its small- and large-signal modeling.

11.1.1 f_T and f_{max} figures of merit

Traditionally, transistors are characterized using figures of merit such as the unity current-gain cut-off frequency f_T and the unilateral power gain cut-off frequency f_{max}. It is therefore appropriate to first discuss these figures of merit.

Consider a transistor characterized by the following small-signal y parameters

$$i_1 = y_{11}(\omega)\, v_1 + y_{12}(\omega)\, v_2,$$
$$i_2 = y_{21}(\omega)\, v_1 + y_{22}(\omega)\, v_2,$$

or z-parameters

$$v_1 = z_{11}(\omega)\, i_1 + z_{12}(\omega)\, i_2,$$
$$v_2 = z_{21}(\omega)\, i_1 + z_{22}(\omega)\, i_2.$$

The currents and voltages are defined in Figure 11.1(a). For example, we shall use the y parameters of an FET in the common source configuration (see Figure 11.1(b)):

$$i_g = y_{gg}(\omega)\, v_{gs} + y_{gd}(\omega)\, v_{ds},$$
$$i_d = y_{dg}(\omega)\, v_{gs} + y_{dd}(\omega)\, v_{ds}.$$

The unity short-circuit current-gain cut-off frequency f_T is defined as the frequency at which the short-circuit current gain is unity:

$$h_{21}(\omega_T) = \frac{|y_{21}(\omega_T)|}{|y_{11}(\omega_T)|} = \frac{|z_{21}(\omega_T)|}{|z_{22}(\omega_T)|} = 1. \tag{11.1}$$

For the MODFET in the common source configuration, the maximum short-circuit current gain can be approximated by

$$\frac{|y_{21}(\omega)|}{|y_{11}(\omega)|} = \frac{|y_{dg}(\omega)|}{|y_{gg}(\omega)|} \simeq \frac{g_m}{\omega C_G W_g L_g}.$$

Notice the $1/\omega$ decrease with frequency (20 dB per decade) using $20\log(|y_{21}|/|y_{11}|)$ of the short-circuit current gain. In the long-channel mode ($L_g \geq (V_{G,max} - V_T)/F_c$), the maximum transconductance is obtained for $k = 1$ and $V_{GS} = V_{G,max}$ (see Chapter 10)

$$g_{m,max}(PINCH) = g_0 = \frac{\mu C_G W_g}{L_g}(V_{G,max} - V_T).$$

The maximum unity current-gain cut-off frequency of the long-channel MODFET is therefore

$$\omega_T = 2\pi f_T = \frac{\mu}{L_g^2}(V_{G,max} - V_T),$$

which is varying as $1/L_g^2$. In the short-channel mode ($L_g < (V_{G,max} - V_T)/F_c$), the maximum transconductance saturates to

$$g_{m,max}(SAT) = \mu C_G W_g F_c = v_S C_G W_g.$$

The maximum transconductance is set by the effective saturation velocity in the FET channel. The maximum unity current-gain cut-off frequency of the short-channel MODFET is therefore

$$\omega_T = 2\pi f_T = \frac{v_S}{L_g},$$

which is varying as $1/L_g$. As stated above, the current-gain cut-off frequency is seen to be controlled by the gate length. However, in the short-channel MODFET, the saturation velocity and parasitics also play an important role.

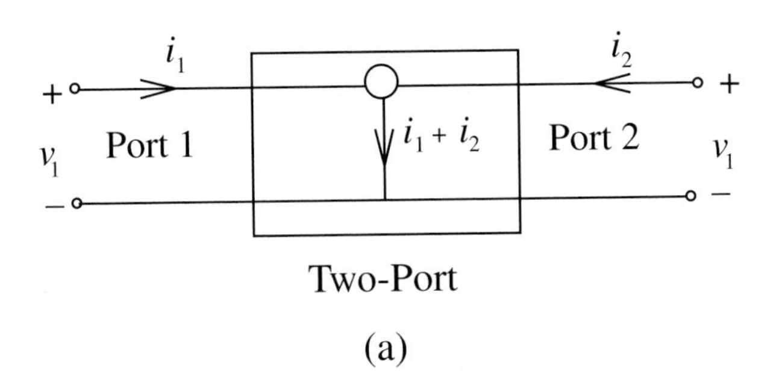

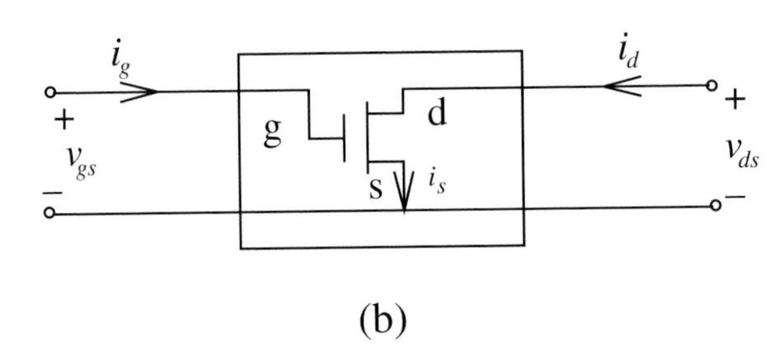

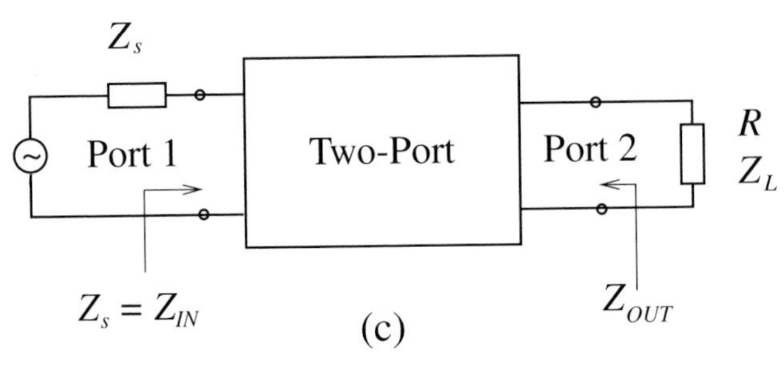

Fig. 11.1. Small-signal representation with a two-port network (a) of a three-terminal transistor and (b) of a common source FET. (c) Two-port network with source and load impedances.

11.1.2 MAG and MSG

The microwave performance of a transistor is usually characterized by its maximum transducer power gain as a function of frequency (the same as both the maximum available power gain, MAG, and the maximum power gain). The transducer power gain G_T is defined as the ratio of the power delivered to a load P_L (see Figure 11.1(c)) given the power available from the source P_S:

$$G_T = \frac{P_L}{P_S}.$$

The maximum power gain is obtained (when it exists) by simultaneously matching the input and output to obtain a conjugate match. Conjugate match means that the source impedance Z_S and the load impedance Z_L satisfy simultaneously:

$$\left.\begin{aligned} Z_S &= Z_{IN}^*, \\ Z_L &= Z_{OUT}^*, \end{aligned}\right\} \tag{11.2}$$

where Z_{IN} is the input impedance of the two-port network measured at port 1 with the load impedance Z_L connected at port 2 and where Z_{OUT} is the output impedance of the two-port network measured at port 2 with the source impedance Z_S connected at port 1.

The maximum transducer power gain ($G_{T,max}$ or MAG) is found to be [2]

$$\begin{aligned} MAG &= \\ &\frac{|y_{21}|^2}{2\,\mathrm{Re}(y_{11})\,\mathrm{Re}(y_{22}) - \mathrm{Re}(y_{12}y_{21}) + \{[2\,\mathrm{Re}(y_{11})\,\mathrm{Re}(y_{22}) - \mathrm{Re}(y_{12}y_{21})]^2 - |y_{12}y_{21}|^2\}^{1/2}} \\ &= \frac{|z_{21}|^2}{2\,\mathrm{Re}(z_{11})\,\mathrm{Re}(z_{22}) - \mathrm{Re}(z_{12}z_{21}) + \{[2\,\mathrm{Re}(z_{11})\,\mathrm{Re}(z_{22}) - \mathrm{Re}(z_{12}z_{21})]^2 - |z_{12}y_{21}|^2\}^{1/2}} \end{aligned}$$

The maximum transducer power gain (or MAG) can be rewritten

$$\begin{aligned} MAG = G_{T,max} &= \frac{|y_{21}|}{|y_{12}|}\left[K - \left(K^2 - 1\right)^{1/2}\right] \\ &= \frac{|z_{21}|}{|z_{12}|}\left[K - \left(K^2 - 1\right)^{1/2}\right], \end{aligned}$$

where K is the Rollett stability factor [2]

$$\begin{aligned} K &= \frac{2\,\mathrm{Re}(y_{11})\,\mathrm{Re}(y_{22}) - \mathrm{Re}(y_{12}y_{21})}{|y_{21}y_{12}|} \\ &= \frac{2\,\mathrm{Re}(z_{11})\,\mathrm{Re}(z_{22}) - \mathrm{Re}(z_{12}z_{21})}{|z_{21}z_{12}|}. \end{aligned}$$

This maximum transducer gain (or MAG) only exists when the stability factor K is larger than 1. When the stability factor is smaller than 1, the transistor is unstable for the conjugate matched loads and must be stabilized ($K \leq 1$) using, for example, resistive loading (a series or shunt resistor at the input or the output port) or feedback. The maximum stable gain (MSG), defined as the MAG for $K = 1$, is then used to characterize the transistor:

$$MSG = \frac{|y_{21}|}{|y_{12}|} = \frac{|z_{21}|}{|z_{12}|}.$$

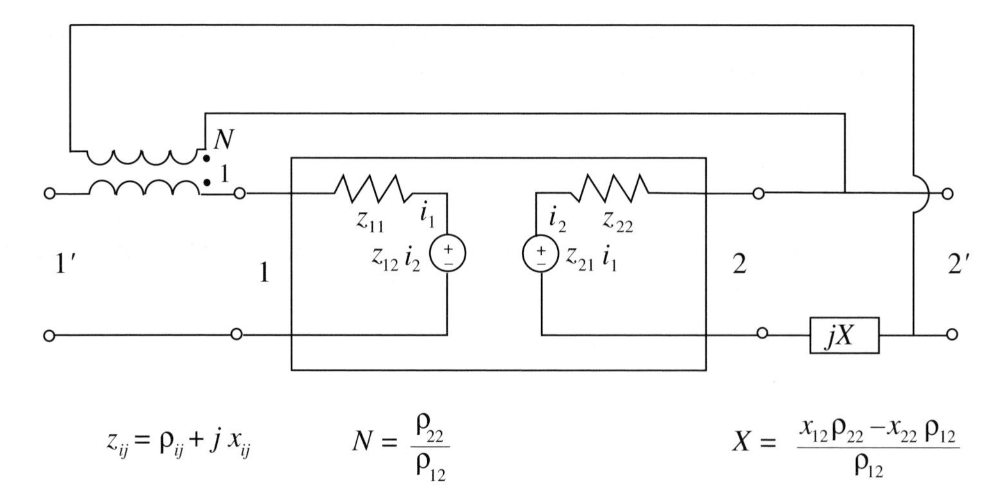

Fig. 11.2. Example of the unilateralization of a two-port device by lossless feedback (from Mason).

11.1.3 Unilateral power gain of the wave-equation model

The comparison of the high-frequency performance of two-port devices is usually done using the unilateral power gain U derived by Mason [3]:

$$\begin{aligned} U &= \frac{|y_{21} - y_{12}|^2}{4[\mathrm{Re}(y_{11})\,\mathrm{Re}(y_{22}) - \mathrm{Re}(y_{12})\,\mathrm{Re}(y_{21})]} \\ &= \frac{|z_{21} - z_{12}|^2}{4[\mathrm{Re}(z_{11})\,\mathrm{Re}(z_{22}) - \mathrm{Re}(z_{12})\,\mathrm{Re}(z_{21})]}. \end{aligned} \tag{11.3}$$

U is the maximum available power gain (MAG introduced in the previous section) of a device once it has been been unilateralized ($y_{12} = z_{12} = 0$) using lossless feedback techniques. Figure 11.2 shows a possible feedback circuit (proposed by Mason himself) to unilateralize a three-terminal device.

Let us estimate the frequency dependence of the unilateral power gain. In our study of the y parameters of the MODFET we shall verify the well-known fact that y_{gg} and y_{gd} are essentially capacitive ($j\omega C_{gg}$ and $j\omega C_{gd}$) and that their real components originate from the channel resistance in series with these capacitors:

$$\begin{aligned} y_{gg} &= \frac{j\omega C_{gg}}{1 + j\omega R_{gg} C_{gg}}, \\ y_{gd} &= -\frac{j\omega C_{gd}}{1 + j\omega R_{gd} C_{gd}}. \end{aligned}$$

Thus for small frequencies ω the real parts of y_{gg} and y_{gd}, are the second order terms $-(j\omega)^2 R_{gg} C_{gg}^2$ or $(j\omega)^2 R_{gd} C_{gd}^2$, respectively:

$$\begin{aligned} \mathrm{Re}(y_{11}) &= \mathrm{Re}(y_{gg}) \propto \omega^2, \\ \mathrm{Re}(y_{12}) &= \mathrm{Re}(y_{gd}) \propto -\omega^2. \end{aligned}$$

Using the approximate identities $y_{22}(\omega) \simeq g_d$, and $y_{21} - y_{12} \simeq y_{21} \simeq g_m$, the unilateral power gain is seen to satisfy

$$U(\omega) \propto \frac{1}{\omega^2}.$$

The maximum frequency of oscillation f_{max} is then defined as the frequency at which U is unity. f_{max} is often referred to as the frequency at which a three-port device switches from active to passive. U can then be rewritten

$$U(\omega) = \left(\frac{\omega_{max}}{\omega}\right)^2.$$

The unilateral power gain will then decrease at the rate of 20 dB per decade (using 10 log U) like the short-circuit current gain.

The importance of U and f_{max} for characterizing a device hinges on their invariance upon lossless coupling (feedback and loading) (see Mason's derivation [3]). However, because the feedback network required to unilateralize a device can only be achieved at a single frequency with lossless passive components, f_{max} is a narrow-band figure of merit. A narrow-band figure of merit is useful in classifying transistors for the design of tuned amplifiers and oscillators. f_{max} is therefore used as an RF or microwave figure of merit. This is in contrast with f_T which is a broad-band figure of merit and is therefore more relevant for classifying transistors for the design of broad-band and large-signal circuits.

11.1.4 On the ordering of f_T and f_{max}

Let us now address the issue of the ordering of f_T and f_{max}. We show in Figure 11.3 that by the use of a sufficiently large series gate resistance, it is possible to reduce the extrinsic unilateral power gain U while maintaining a constant extrinsic $f_T(ext)$ until $f_{max}(ext)$ is smaller than $f_T(ext)$. Large gate resistances are indeed a problem in submicron gate FETs and mushroom T- and L-shaped gates are used to circumvent it.

As mentioned above, f_T is a broad-band figure of merit and f_{max} a narrow-band figure of merit, and they are therefore essentially decoupled figures of merit. It is possible, however, to introduce a narrow-band f_T which can be meaningfully compared with f_{max}. For this purpose we shall introduce a narrow-band short-circuit current gain and open-circuit voltage gain. By definition, a narrow-band figure of merit is a quantity tuned to give the maximum figure of merit possible. Using lossless loading (a series or shunt lossless load at ports 1 and 2) the tuned short-circuit current gain $h_{21}(tuned)$ is

$$h_{21}(tuned) = \frac{|y_{21}|}{\mathrm{Re}[y_{11}]} = \frac{|z_{21}|}{\mathrm{Re}[z_{22}]} \tag{11.4}$$

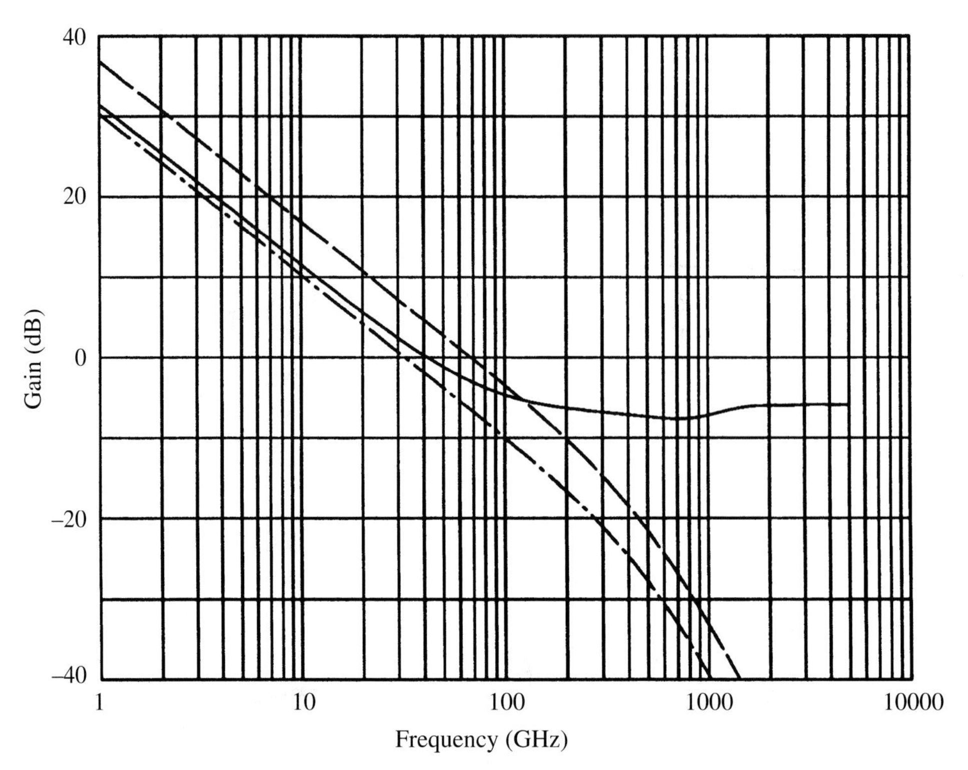

Fig. 11.3. Unilateral power gain (dashed and dotted-dashed lines) and short-circuit current gain (full line) versus frequency for a 0.3 μm extrinsic MODFET ($V_{GS} = 0$ V and $V_{DS} = 1$ V) with parasitics $R_S = R_D = 2\,\Omega$, and $C_{GS} = C_{DS} = 50$ fF using two different gate resistances $R_G = 5$ Ω (dashed line) and 25 Ω (dotted-dashed line). f_{max} is smaller than f_T for the largest gate resistance. (P. Roblin, S. Kang and H. Morkoç, *Proceedings of the 1990 International Symposium on Circuits and Systems*, Vol. 2, pp. 1501–1504, IEEE, Picataway, 1990. ©1990 IEEE.)

and the tuned open-circuit voltage gain $A_{21}(tuned)$ is

$$A_{21}(tuned) = \frac{|y_{21}|}{\mathrm{Re}[y_{22}]} = \frac{|z_{21}|}{\mathrm{Re}[z_{11}]}. \tag{11.5}$$

These narrow-band quantities are invariant under lossless loading but not under lossless feedback. Improved figures of merit can be obtained if the device is first unilateralized using the feedback circuit of Figure 11.2. Note that this unilateralization is not unique. For example, we could invert ports 1 and 2 and obtain a different unilateralized device. However, the feedback circuit selected by Mason has the advantage of transforming a bilateral transconductance amplifier into a similar unilateral transconductance amplifier. The resulting unilateralized z'-parameters are given in terms of the z-parameters ($z_{ij} = \rho_{ij} + jx_{ij}$) by

$$\left.\begin{aligned} z'_{11} &= \frac{\rho_{22}z_{11} - \rho_{12}z_{21}}{\rho_{22}}, \\ z'_{12} &= 0, \\ z'_{21} &= z_{21} - z_{12}, \\ z'_{22} &= \frac{\rho_{22}}{\rho_{12}} z_{12}. \end{aligned}\right\} \quad (11.6)$$

The unilateralized FET now has the following tuned short-circuit current gain Uh_{21}:

$$Uh_{21} = \frac{|z'_{21}|}{\mathrm{Re}[z'_{22}]} = \frac{|z_{21} - z_{12}|}{\mathrm{Re}[z_{22}]} = \frac{|y_{21} - y_{12}|}{\mathrm{Re}[y_{11}]} \quad (11.7)$$

and the following tuned open-circuit voltage gain UA_{21}:

$$\begin{aligned} UA_{21} &= \frac{|z'_{21}|}{\mathrm{Re}[z'_{11}]} = \frac{\mathrm{Re}[z_{22}]\,|z_{21} - z_{12}|}{\mathrm{Re}[z_{11}]\,\mathrm{Re}[z_{22}] - \mathrm{Re}[z_{12}]\,\mathrm{Re}[z_{21}]} \\ &= \frac{\mathrm{Re}[y_{11}]\,|y_{21} - y_{12}|}{\mathrm{Re}[y_{11}]\,\mathrm{Re}[y_{22}] - \mathrm{Re}[y_{12}]\,\mathrm{Re}[y_{21}]}. \end{aligned} \quad (11.8)$$

Note that these gains can only be meaningfully defined if we have U larger than 1 and $\mathrm{Re}[z_{22}]$ larger than 0.

One notices that the product of UA_{21} and Uh_{21} is invariant under lossless feedback/loading since it is related to the unilateral gain. Indeed, we have

$$U = \left(\frac{1}{2}\, Uh_{21}\right) \times \left(\frac{1}{2}\, UA_{21}\right). \quad (11.9)$$

This is demonstrated graphically in Figure 11.4 for an extrinsic MODFET.

Unilateral cut-off frequencies f_{UT} and f_{UA} can now be defined from the halved unilateral current and voltage gains:

$$\left.\begin{aligned} \frac{1}{2}\, Uh_{21}(\omega_{UT}) &= 1, \\ \frac{1}{2}\, UA_{21}(\omega_{UA}) &= 1. \end{aligned}\right\} \quad (11.10)$$

The unilateral power gain is usually monotonously decreasing with increasing frequency and these unilateral cut-off frequencies must now satisfy either one of the following ordering relations:

$$\left.\begin{aligned} &f_{UT} \le f_{max} \le f_{UA}, \\ &f_{UA} \le f_{max} \le f_{UT}. \end{aligned}\right\} \quad (11.11)$$

An FET is usually expected to follow the first relation (see Figure 11.4) if the gate resistance and drain conductance are not too large.

In the remainder of this chapter, we shall make use of f_T and f_{max} to characterize the high-frequency performance of the ideal MODFET.

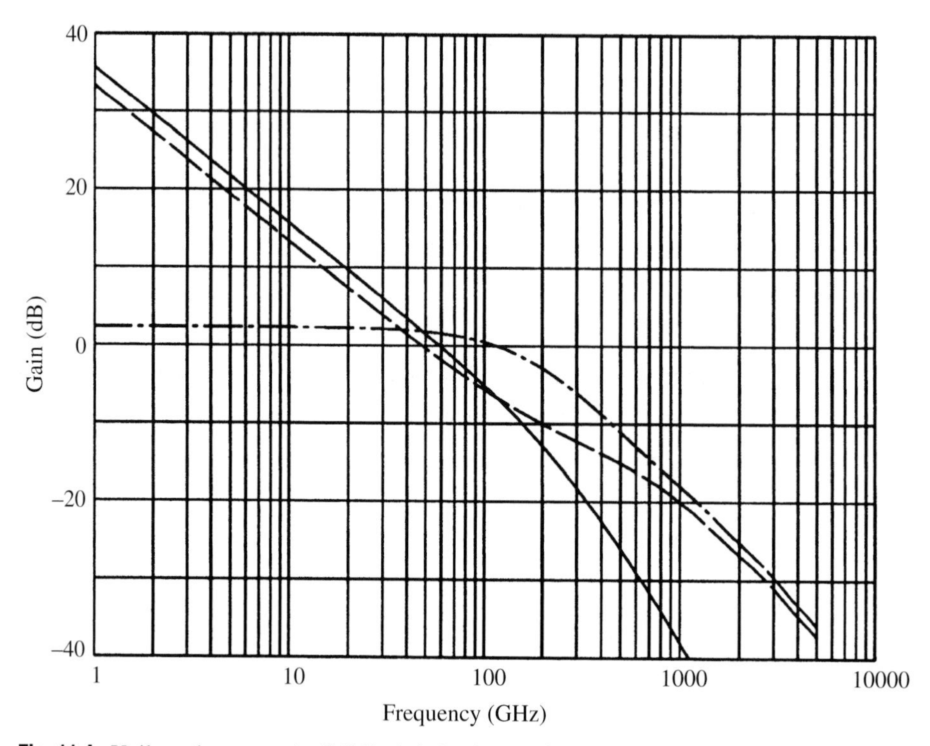

Fig. 11.4. Unilateral power gain (full line), halved open-circuit voltage gain $UA_{21}/2$ (dotted-dashed line) and halved short-circuit current gain $Uh_{21}/2$ (dashed line) versus frequency for a 0.3 μm extrinsic MODFET ($V_{GS} = 0$ V and $V_{DS} = 1$ V) with parasitics $R_S = R_G = R_D = 5\,\Omega$, and $C_{GS} = C_{DS} = 50$ fF. (P. Roblin, S. Kang and H. Morkoç, *Proceedings of the 1990 International Symposium on Circuits and Systems*, Vol. 2, pp. 1501–1504, IEEE, Picataway, 1990. ©1990 IEEE.)

11.2 The MOSFET wave-equation (long-channel case)

11.2.1 The large-signal MOSFET wave-equation

Our study of the high-frequency response of the MODFET will be based on the development of a wave-equation which accounts for the transit-time delay in the channel of the MODFET. The wave-equation for the ideal long-channel MODFET/MOSFET is referred to as the MOSFET wave-equation. The derivation and the solution of the MOSFET wave-equation given below is based on the original calculation of Burns [14] for the MOSFET in pinch-off ($k = 1$). Here his calculation is generalized to hold for both the linear ($0 \leq k < 1$) as well as the pinch-off regime ($k = 1$).

The MOSFET wave-equation is derived from the continuity equation and the current equation. The channel current $I(x, t)$ is given by the drift equation as in Chapter 10

$$I(x, t) = -\mu C_G W_g v_{GC}(x, t)\frac{\partial v_{GC}(x, t)}{\partial x}, \tag{11.12}$$

with $v_{GC}(x, t)$ the gate-to-channel voltage.

The continuity equation is

$$\frac{\partial I(x,t)}{\partial x} = -qW_g \frac{\partial n_S(x,t)}{\partial t} = -C_G W_g \frac{\partial v_{GC}(x,t)}{\partial t}. \tag{11.13}$$

Differentiating Equation (11.12) on both sides with respect to x and substituting in Equation (11.13) yields the following equation for $v_{GC}(x, t)$:

$$\frac{\partial^2}{\partial x^2}[v_{GC}^2(x,t)] = \frac{2}{\mu}\frac{\partial v_{GC}(x,t)}{\partial t}. \tag{11.14}$$

This is the large-signal MOSFET wave-equation. This equation will permit us to characterize both the small-signal response and large-signal response of a long-channel MOSFET.

11.2.2 Exact small-signal solution of the MOSFET wave-equation

We shall now study the small-signal AC response of the long-channel MOSFET. Our goal is to calculate the y parameters of the MOSFET in the common-source configuration:

$$i_g = y_{gg}(\omega)\, v_{gs} + y_{gd}(\omega)\, v_{ds},$$
$$i_d = y_{dg}(\omega)\, v_{gs} + y_{dd}(\omega)\, v_{ds}.$$

Note that once the y parameters in a given configuration are available, the y parameters, z-parameters, and S-parameters in any configuration (common drain, source, or gate) can be readily be obtained [4].

For small-signal analysis, we need to decompose the large-signal MOSFET wave-equation (11.14) into its DC part and small-signal AC part. The total AC gate-to-channel voltage is

$$v_{GC}(x,t) = V_{GC}(x) + v_{gc}(x,t), \tag{11.15}$$

with V_{GC} the DC gate-to-channel voltage and v_{gc} the AC small signal gate-to-channel voltage. In Chapter 10 the DC gate-to-channel voltage $V_{GC}(x)$ was calculated to be given by

$$V_{GC}(x) = V_{GS} - V_T - V_{CS}(x) \tag{11.16}$$

$$= (V_{GS} - V_T)\left[1 - (2k - k^2)\frac{x}{L_g}\right]^{1/2} \tag{11.17}$$

where $k = V_{DS}/(V_{GS} - V_T)$.

In small-signal analysis the second order terms $v_{gc}^2(x, t)$ can be dropped and $v_{GC}^2(x, t)$ can be approximated by

$$v_{GC}^2(x,t) \approx V_{GC}^2(x) + 2V_{GC}(x)v_{gc}(x,t).$$

Substituting Equation (11.17) into the above equation and simplifying yields

$$\begin{aligned} v_{GC}^2(x,t) &\approx (V_{GS}-V_T)^2\left[1-(2k-k^2)\frac{x}{L_g}\right] \\ &\quad + 2(V_{GS}-V_T)\left[1-(2k-k^2)\frac{x}{L_g}\right]^{1/2} v_{gc}(x,t) \\ &= (V_{GS}-V_T)^2 Z + 2(V_G-V_T)Z^{1/2}\, v_{gc}(z,t), \end{aligned} \tag{11.18}$$

with

$$Z = 1-(2k-k^2)\frac{x}{L_g}.$$

The new variable Z is introduced to simplify the notation. We will need the derivative

$$\frac{dx}{dZ} = -\frac{L_g}{2k-k^2}.$$

The procedure to differentiate Equation (11.18) is as follows

$$\begin{aligned} \frac{\partial^2}{\partial x^2} v_{GC}^2(x,t) &= \left[-\frac{(2k-k^2)}{L_g}\right]^2 \frac{\partial^2}{\partial Z^2}[v_{GC}^2(x,t)] \\ &= 2\frac{(2k-k^2)^2(V_G-V_T)}{L_g^2}\frac{\partial^2}{\partial Z^2}[Z^{1/2}v_{gc}(Z,t)] \\ &= 2\frac{(2k-k^2)^2(V_G-V_T)}{L_g^2}\frac{\partial}{\partial Z}\left(\frac{1}{2Z^{1/2}v + Z^{1/2}}\frac{\partial v_{gc}}{\partial Z}\right) \\ &= 2\frac{(2k-k^2)^2(V_{GS}-V_T)}{L_g^2} \\ &\quad \times \left(-\frac{1}{4Z^{3/2}}v + \frac{1}{2Z^{1/2}}\frac{dv_{gc}}{dZ} + \frac{1}{2Z^{1/2}}\frac{dv_{gc}}{dZ} + Z^{1/2}\frac{d^2v_{gc}}{dZ^2}\right) \\ &= 2\frac{(2k-k^2)^2(V_{GS}-V_T)}{L_g^2}\left(Z^{1/2}\frac{d^2v_{gc}}{dZ^2} + \frac{1}{Z^{1/2}}\frac{dv_{gc}}{dZ} - \frac{1}{4Z^{3/2}}v_{gc}\right). \end{aligned} \tag{11.19}$$

Replacing Equation (11.19) in Equation (11.14) yields

$$\begin{aligned} \frac{\partial v_{gc}(x,t)}{\partial t} &= \frac{\mu(V_{GS}-V_T)(2k-k^2)^2}{L_g^2} \\ &\quad \times \left[Z^{1/2}\frac{d^2v_{gc}(x,t)}{dZ^2} + \frac{1}{Z^{1/2}}\frac{dv_{gc}(x,t)}{dZ} - \frac{1}{4Z^{3/2}}v_{gc}\right] \\ &= \omega_{0k}\left[Z^{1/2}\frac{d^2v_{gc}(x,t)}{dZ^2} + \frac{1}{Z^{1/2}}\frac{dv_{gc}(x,t)}{dZ} - \frac{1}{4Z^{3/2}}v_{gc}(x,t)\right], \end{aligned} \tag{11.20}$$

where

$$\omega_{0k} = \frac{\mu(V_{GS} - V_T)(2k - k^2)^2}{L_g^2}.$$

The Laplace transform of Equation (11.20) is

$$Z^2 \frac{\partial^2 v_{gc}(s', Z)}{dZ^2} + Z \frac{dv_{gc}(s', Z)}{dZ} - \left(\frac{1}{4} + s' Z^{3/2}\right) v_{gc}(s', Z) = 0, \tag{11.21}$$

where $s' = s/\omega_{0k}$.

Equation (11.21) is the s space representation of the MODFET wave-equation. This is a modified Bessel's differential equation so that one can find an analytic solution. The complete solution is written as

$$v_{gc}(Z, s') = C_1 I_{2/3}\left[\frac{4}{3} s^{1/2} Z^{3/4}\right] + C_2 I_{-2/3}\left[\frac{4}{3} s'^{1/2} Z^{3/4}\right], \tag{11.22}$$

where C_1 and C_2 are arbitrary constants to be determined from the boundary conditions.

Once the wave-equation is solved we need to calculate the AC small-signal drain and gate currents. To do this, we need to calculate the AC current in the channel of the MOSFET. The AC channel current is obtained by decomposing Equation (11.12) into its DC and AC components while neglecting second order terms:

$$\begin{aligned} i(Z, s') &= -\mu C_G W_g \frac{d}{dx}[v_{gc}(Z, s') V_{GC}(Z)] \\ &= -\mu C_G W_g \left(-\frac{2k - k^2}{L_g}\right)(V_{GS} - V_T) \frac{d}{dZ}\left(v_{gc} Z^{1/2}\right) = G_{d0} \frac{d}{dZ}\left(v Z^{1/2}\right) \\ &= G_{d0}\left(\frac{1}{2Z^{1/2}} v_{gc} + Z^{1/2} \frac{dv_{gc}}{dZ}\right), \end{aligned} \tag{11.23}$$

where

$$G_{d0} = \frac{\mu C_G W_g (V_{GS} - V_T)(2k - k^2)}{L_g}.$$

Equation (11.23) has to be expanded in terms of C_1 and C_2 in order to apply the boundary conditions. To simplify the calculation, we introduce the new variable y

$$y = \frac{4}{3} s'^{1/2} Z^{3/4},$$

$$\frac{dy}{dZ} = s'^{1/2} Z^{-1/4}.$$

Note that the modified Bessel function has the following properties:

$$\frac{dI_n(x)}{dx} = \frac{1}{2}[I_{n+1}(x) + I_{n-1}(x)],$$

$$I_n(y) = \frac{y}{2n}[I_{n-1}(y) - I_{n+1}(y)].$$

First dv_{gc}/dZ will be expanded in terms of C_1 and C_2

$$\frac{dv_{gc}}{dZ} = \frac{dy}{dZ}\frac{dv_{gc}}{dy} = s'^{1/2}Z^{-1/4}\frac{d}{dy}[C_1 I_{2/3}(y) + C_2 I_{2/3}(y)]$$
$$= s'^{1/2}Z^{-1/4}\left\{\frac{C_1}{2}[I_{5/3}(y) + I_{-1/3}(y)] + \frac{C_2}{2}[I_{1/3}(y) + I_{-5/3}(y)]\right\}. \quad (11.24)$$

Substituting Equations (11.24) and (11.22) into Equation (11.23) gives

$$i(Z,s') = G_{d0}\left(\frac{1}{2}Z^{-1/2}[C_1 I_{2/3}(y) + C_2 I_{-2/3}(y)]\right.$$
$$+ s'^{1/2}Z^{1/4}\left\{\frac{C_1}{2}[I_{5/3}(y) + I_{-1/3}(y)] + \frac{C_2}{2}[I_{1/3}(y) + I_{-5/3}(y)]\right\}$$
$$= G_{d0}\left(\frac{C_1}{2}Z^{-1/2}\frac{\frac{4}{3}s'^{1/2}Z^{3/4}}{2\frac{2}{3}}[I_{-1/3}(y) - I_{5/3}(y)]\right.$$
$$+ \frac{C_2}{2}Z^{-1/2}\frac{\frac{4}{3}s'^{1/2}Z^{3/4}}{2(-\frac{2}{3})}[I_{-5/3}(y) - I_{1/3}(y)]$$
$$\left. + S^{1/2}Z^{1/4}\left\{\frac{C_1}{2}[I_{5/3}(y) + I_{-1/3}(y)] + \frac{C_2}{2}[I_{1/3}(y) + I_{-5/3}(y)]\right\}\right)$$
$$= \frac{G_{d0}}{2}s'^{1/2}Z^{1/4}[C_1 I_{-1/3}(y) - C_1 I_{5/3}(y) - C_2 I_{-5/3}(y) + C_2 I_{1/3}(y)$$
$$+ C_1 I_{5/3}(y) + C_1 I_{-1/3}(y) + C_2 I_{1/3}(y) + C_2 I_{-5/3}(y)]$$
$$= G_{d0}s'^{1/2}Z^{1/4}[C_1 I_{-1/3}(y) + C_2 I_{1/3}(y)].$$

Now both the AC voltage and the AC current in the GCA channel can be obtained in terms of C_1 and C_2:

$$v_{gc}(Z,s) = C_1 I_{2/3}(y) + C_2 I_{-2/3}(y), \quad (11.25)$$
$$i(Z,s) = G_{d0}s'^{1/2}Z^{1/4}[C_1 I_{-1/3}(y) + C_2 I_{1/3}(y)]. \quad (11.26)$$

The solution of the wave-equation across the entire channel requires a set of boundary conditions to be enforced at $x = 0$ and $x = L_g$ and at the boundary between the GCA and saturation region. The boundary conditions to be used at $x = 0$ and $x = L_g$ for the common source configuration are

$$v_{gc}(0) = v_{gs}, \quad (11.27)$$
$$v_{gc}(L_g) = v_{gs} - v_{ds}. \quad (11.28)$$

For simplicity, Equations (11.27) and (11.28) may be rewritten as

$$\left.\begin{aligned} v_{gc}(L_g) &= A_{11}C_1 + A_{12}C_2 = v_{gs} - v_{ds}, \\ v_{gc}(0) &= A_{21}C_1 + A_{22}C_2 = v_{gs}, \end{aligned}\right\} \quad (11.29)$$

where we have introduced the constants

$$A_{11} = I_{2/3}(y_s)$$
$$A_{12} = I_{-2/3}(y_s)$$
$$A_{21} = I_{2/3}\left(\frac{4}{3}s'^{1/2}\right)$$
$$A_{22} = I_{-2/3}\left(\frac{4}{3}s'^{1/2}\right)$$

y_s is defined from Z_s as

$$y_s = \frac{4}{3}s'^{1/2}Z_s^{3/4},$$
$$Z_s = Z(x = L_g) = (1 - k_s)^2.$$

From Equation (11.29) it is obvious that

$$\begin{aligned} C_1 &= \frac{A_{22}(v_{gs} - v_{ds}) - A_{12}v_{gs}}{\Delta} \\ &= \frac{A_{22} - A_{12}}{\Delta}v_{gs} - \frac{A_{22}}{\Delta}v_{ds} \\ &= C_{1gs}v_{gs} + C_{1ds}v_{ds}, \\ C_2 &= \frac{A_{11}v_{gs} - A_{21}(v_{gs} - v_{ds})}{\Delta} \\ &= \frac{A_{11} - A_{21}}{\Delta}v_{gs} + \frac{A_{21}}{\Delta}v_{ds} \\ &= C_{1gs}v_{gs} + C_{1ds}v_{ds}, \end{aligned}$$

with $\Delta = A_{11}A_{22} - A_{12}A_{21}$ and with

$$\begin{aligned} C_{1gs} &= \frac{(A_{22} - A_{12})}{\Delta}, \\ C_{2gs} &= \frac{(A_{11} - A_{21})}{\Delta}, \\ C_{1ds} &= -\frac{A_{22}}{\Delta}, \\ C_{2ds} &= -\frac{A_{21}}{\Delta}. \end{aligned}$$

Using Equation (11.25) the gate and drain currents are expressed in terms of C_1 and C_2 as

$$\begin{aligned} i_d &= i(x = L_g) = i(Z_s, s') \\ &= G_{d0}s'^{1/2}Z_s^{1/4}[C_1I_{-1/3}(y_s) + C_2I_{1/3}(y_s)] \\ i_g &= i(x = 0) - i(x = L_g) \end{aligned}$$

$$
\begin{aligned}
&= G_{d0}s'^{1/2}\left[C_1 I_{-1/3}\left(\frac{4}{3}s'^{1/2}\right) + C_2 I_{1/3}\left(\frac{4}{3}s'^{1/2}\right)\right] \\
&\quad - G_{d0}s'^{1/2}Z_s^{1/4}[C_1 I_{-1/3}(y_s) + C_2 I_{1/3}(y_s)].
\end{aligned}
$$

The y-parameters of the saturated MODFET are then

$$
\begin{aligned}
y_{dg} &= \frac{i_d}{v_{gs}} \\
&= G_{d0}s'^{1/2}Z_s^{1/4}[C_{1gs}I_{-1/3}(y_s) + C_{2gs}I_{1/3}(y_s)], \\
y_{gg} &= \frac{i_g}{v_{gs}} \\
&= G_{d0}s'^{1/2}\left\{C_{1gs}I_{-1/3}\left(\frac{4}{3}s'^{1/2}\right) + C_{2gs}I_{1/3}\left(\frac{4}{3}s'^{1/2}\right)\right. \\
&\quad \left. - Z_s^{1/4}[C_{1gs}I_{-1/3}(y_s) + C_{2gs}I_{1/3}(y_s)]\right\}, \\
y_{dd} &= \frac{i_d}{v_{ds}} \\
&= G_{d0}s'^{1/2}Z_s^{1/4}[C_{1ds}I_{-1/3}(y_s) + C_{2ds}I_{1/3}(y_s)], \\
y_{gd} &= \frac{i_g}{v_{ds}} \\
&= G_{d0}s'^{1/2}\left\{C_{1ds}I_{-1/3}\left(\frac{4}{3}s'^{1/2}\right) + C_{2ds}I_{1/3}\left(\frac{4}{3}s'^{1/2}\right)\right. \\
&\quad \left. - Z_s^{1/4}[C_{1ds}I_{-1/3}(y_s) + C_{2ds}I_{1/3}(y_s)]\right\}.
\end{aligned}
$$

The modified Bessel function can be calculated numerically using its power series expansion:

$$
I_n(y) = \left(\frac{y}{2}\right)^n \sum_{p=0}^{\infty} \frac{(y/2)^{2p}}{p!\,\Gamma[n+p+1]}. \tag{11.30}
$$

11.2.3 Frequency power series expansions of the y parameters

The exact solution of the small-signal MOSFET wave-equation developed in the previous section gives little insight into the frequency response of the MOSFET and a simpler approximate solution is desirable.

Small-signal y parameters holding up to high frequencies can be obtained by expanding the exact y parameters in the frequency power series initially proposed by Van Der Ziel and Eno (see also [11]):

$$
y_{ij} = g_{ij} + j\omega\alpha_{ij} + (j\omega)^2\beta_{ij}, \tag{11.31}
$$

where g_{ij} are the DC y parameters

An expansion of the y parameters in frequency power series (11.30) yields the following results [12,7,8]:

$$y_{gg} = \frac{\omega^2 C_0^2}{g_0}\mathcal{R}_{gg}(k) + j\omega C_0 \mathcal{I}_{gg}(k)$$

$$y_{gd} = -\frac{\omega^2 C_0^2}{g_0}\mathcal{R}_{gd}(k) - j\omega C_0 \mathcal{I}_{gd}(k)$$

$$y_{dg} = k g_0 - \frac{\omega^2 C_0^2}{g_0}\mathcal{R}_{dg}(k) - j\omega C_0 \mathcal{I}_{dg}(k)$$

$$y_{dd} = (1-k)g_0 + \frac{\omega^2 C_0^2}{g_0}\mathcal{R}_{dd}(k) + j\omega C_0 \mathcal{I}_{dd}(k),$$

using

$$g_0 = \frac{\mu C_G W_g (V_{GS} - V_T)}{L_g} \quad \text{and} \quad C_0 = W_g L_g C_G.$$

The coefficients $\mathcal{I}_{ij}(k)$ and $\mathcal{R}_{ij}(k)$ are given by [12,7] (see [21] for a derivation):

$$\mathcal{R}_{gg}(k) = \frac{\frac{1}{12} - \frac{1}{6}k + \frac{9}{80}k^2 - \frac{7}{240}k^3 + \frac{1}{360}k^4}{(1-\frac{1}{2}k)^5},$$

$$\mathcal{I}_{gg}(k) = \frac{1 - k + \frac{1}{6}k^2}{(1-\frac{1}{2}k)^2},$$

$$\mathcal{R}_{gd}(k) = \frac{(1-k)(\frac{1}{24} - \frac{41}{720}k + \frac{1}{45}k^2 - \frac{1}{360}k^3)}{(1-\frac{1}{2}k)^5},$$

$$\mathcal{I}_{gd}(k) = \frac{(1-k)(1-\frac{1}{3}k)}{2(1-\frac{1}{2}k)^2},$$

$$\mathcal{R}_{dg}(k) = \frac{\frac{1}{24} - \frac{1}{10}k + \frac{43}{480}k^2 - \frac{3}{80}k^3 + \frac{11}{1440}k^4 - \frac{1}{1600}k^5}{(1-\frac{1}{2}k)^6},$$

$$\mathcal{I}_{dg}(k) = \frac{\frac{1}{2} - \frac{3}{4}k + \frac{1}{3}k^2 - \frac{1}{20}k^3}{(1-\frac{1}{2}k)^3},$$

$$\mathcal{R}_{dd}(k) = \frac{(1-k)(\frac{1}{45} - \frac{7}{180}k + \frac{17}{720}k^2 - \frac{1}{160}k^3 + \frac{9}{14\,400}k^4)}{(1-\frac{1}{2}k)^6},$$

$$\mathcal{I}_{dd}(k) = \frac{(1-k)(\frac{1}{3} - \frac{1}{4}k + \frac{1}{20}k^2)}{(1-\frac{1}{2}k)^3}.$$

A plot of the functions $\mathcal{I}_{ij}(k)$ and $\mathcal{R}_{ij}(k)$ versus the bias parameter k is shown in Figure 11.5. The frequency power expansions given above hold for a large frequency range but do not degrade gracefully at high frequencies.

An alternative expansion of the y parameters has been developed by Van Nielen [13]. It can be derived using an iterative procedure or can be obtained directly by

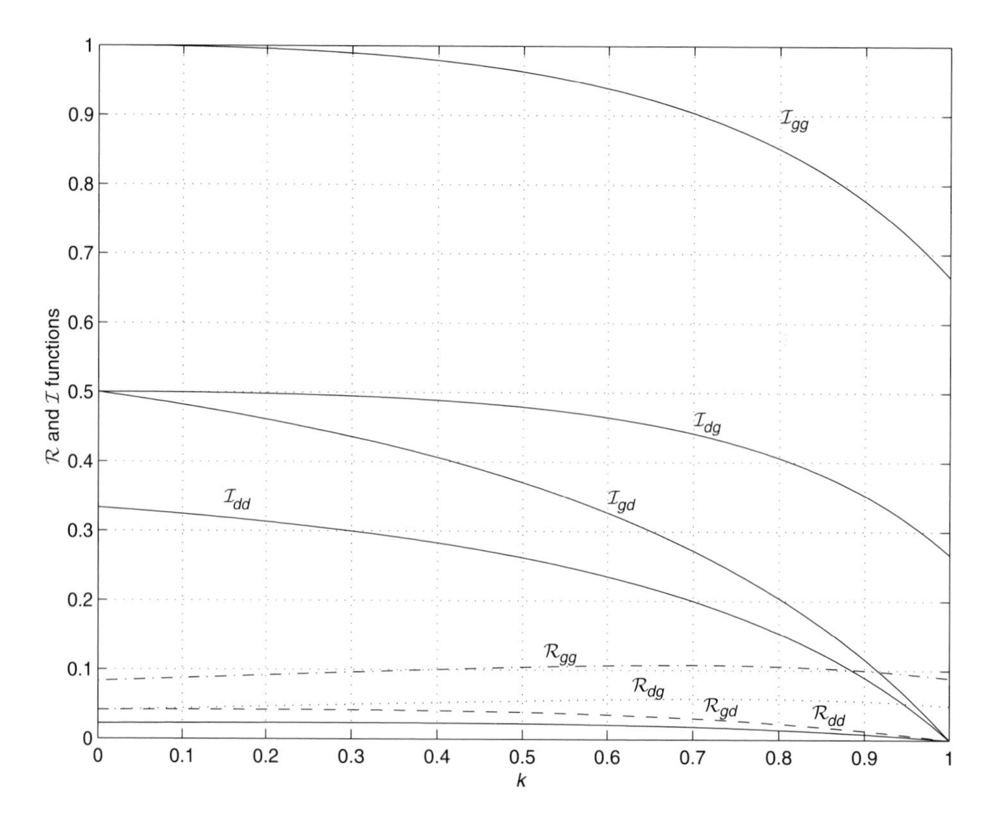

Fig. 11.5. Plot of the functions $\mathcal{I}_{ij}(k)$ and $\mathcal{R}_{ij}(k)$ versus the bias parameter k.

expanding the Bessel functions in a frequency power series in the exact solution of the small-signal MOSFET wave-equation. The first order iterative small-signal y parameters obtained are of the following form:

$$y_{ij} = \frac{g_{ij} + j\omega C_{ij}}{1 + j\omega\tau},$$

with τ a common time constant shared by all y parameters. The second order iterative small-signal y parameters obtained are of the following form:

$$y_{ij} = \frac{g_{ij} + j\omega a_{ij} + (j\omega)^2 b_{ij}}{1 + j\omega c + (j\omega)^2 d}. \tag{11.32}$$

The second order iterative small-signal y parameters admit a frequency power series expansion valid up to power 2. Compared to the y parameters obtained by the frequency power series [6], the iterative y parameters hold for higher frequencies and have the advantage of providing a more graceful degradation outside their frequency range of validity. However, we shall see in the next sections that an optimal equivalent circuit can be developed from the frequency power series y parameters expansion, using the RC topology of the first order iterative y parameters.

11.2.4 Dimensionless representation of the y parameters

One can easily verify directly from the MOSFET wave-equation or its solution that the exact small-signal y parameters obtained from the MOSFET wave-equation can be written in terms of dimensionless parameters:

$$\frac{y_{ij}}{g_0} = f_{ij}\left(k, \frac{\omega}{\omega_0}\right),$$

where $k = V_{DS}/(V_{GS} - V_T)$, g_0 is the channel conductance:

$$g_0 = \frac{\mu C_G W_g (V_{GS} - V_T)}{L_g}$$

and ω_0 is a normalization frequency given by

$$\omega_0 = 2\pi f_0 = \mu \frac{(V_{GS} - V_T)}{L_g^2}. \tag{11.33}$$

This normalization can also be applied to the frequency power series solution given in the previous section. These approximate y parameters can then be rewritten in the following normalized fashion:

$$\frac{y_{gg}}{g_0} = j\frac{\omega}{\omega_0}\mathcal{I}_{gg}(k) + \left(\frac{\omega}{\omega_0}\right)^2 \mathcal{R}_{gg}(k),$$

$$\frac{y_{gd}}{g_0} = -j\frac{\omega}{\omega_0}\mathcal{I}_{gd}(k) - \left(\frac{\omega}{\omega_0}\right)^2 \mathcal{R}_{gd}(k),$$

$$\frac{y_{dg}}{g_0} = k - j\frac{\omega}{\omega_0}\mathcal{I}_{dg}(k) - \left(\frac{\omega}{\omega_0}\right)^2 \mathcal{R}_{dg}(k),$$

$$\frac{y_{dd}}{g_0} = (1 - k) + j\frac{\omega}{\omega_0}\mathcal{I}_{dd}(k) + \left(\frac{\omega}{\omega_0}\right)^2 \mathcal{R}_{dd}(k).$$

The existence of a normalized representation is useful as it permits one to establish results which are device-independent and hold for all bias conditions. This property is used in the next section to study the range of validity of the equivalent circuit model proposed.

11.2.5 First order equivalent circuit I

To improve the degradation of the y parameters obtained from the frequency power series for frequencies ω larger than ω_0, we shall introduce a simple RC equivalent circuit model. The RC model selected consists of the DC ($\omega = 0$) small-signal parameters g_{ij} shunted by a capacitor C_{ij} in series with a charging resistor R_{ij}. The resulting intrinsic y parameters are

$$y_{gg} = \frac{j\omega C_{gg}}{1 + j\omega R_{gg} C_{gg}},$$

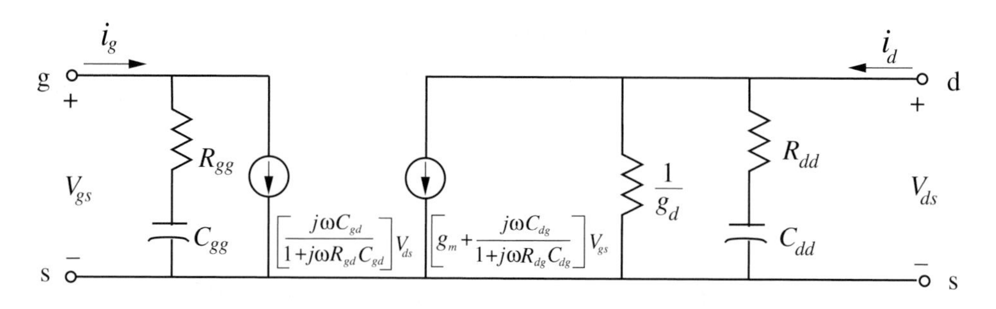

Fig. 11.6. Approximate small-signal equivalent circuit for the intrinsic long-channel MOSFET.

$$y_{gd} = \frac{j\omega C_{gd}}{1 + j\omega R_{gd}C_{gd}},$$

$$y_{dg} = g_m + \frac{j\omega C_{dg}}{1 + j\omega R_{dg}C_{dg}},$$

$$y_{dd} = g_d + \frac{j\omega C_{dd}}{1 + j\omega R_{dd}C_{dd}}.$$

The associated equivalent circuit I for the intrinsic MOSFET is shown in Figure 11.6.

For frequencies $\omega \ll 1/(R_{ij}C_{ij})$ these y parameters admit the frequency power series (11.31):

$$y_{ij} = g_{ij} + j\omega C_{ij} + \omega^2 R_{ij} C_{ij}^2. \tag{11.34}$$

We can now readily identify the resistors and capacitors to be

$$C_{gg} = \frac{g_0(V_{GS})\mathcal{I}_{gg}(k)}{\omega_0}, \qquad R_{gg} = \frac{\mathcal{R}_{gg}}{g_0(V_{GS})\mathcal{I}_{gg}^2(k)},$$

$$C_{gd} = -\frac{g_0(V_{GS})\mathcal{I}_{gd}(k)}{\omega_0}, \qquad R_{gd} = -\frac{\mathcal{R}_{gd}}{g_0(V_{GS})\mathcal{I}_{gd}^2(k)},$$

$$C_{dg} = -\frac{g_0(V_{GS})\mathcal{I}_{dg}(k)}{\omega_0}, \qquad R_{dg} = -\frac{\mathcal{R}_{dg}}{g_0(V_{GS})\mathcal{I}_{dg}^2(k)},$$

$$C_{dd} = \frac{g_0(V_{GS})\mathcal{I}_{dd}(k)}{\omega_0}, \qquad R_{dd} = \frac{\mathcal{R}_{dd}}{g_0(V_{GS})\mathcal{I}_{dd}^2(k)}.$$

The time constants $\tau_{ij} = R_{ij}C_{ij}$ in the small-signal y parameters are then given by

$$\tau_{gg} = R_{gg}C_{gg} = \frac{1}{\omega_0}\frac{60 - 120k + 81k^2 - 21k^3 + 2k^4}{15(2-k)^3(6 - 6k + k^2)},$$

$$\tau_{gd} = R_{gd}C_{gd} = \frac{1}{\omega_0}\frac{30 - 41k + 16k^2 - 2k^3}{15(2-k)^3(3-k)},$$

$$\tau_{dg} = R_{dg}C_{dg} = \frac{1}{\omega_0}\frac{600 - 1440k + 1290k^2 - 540k^3 + 110k^4 - 9k^5}{30(2-k)^3(30 - 45k + 20k^2 - 3k^3)},$$

$$\tau_{dd} = R_{dd}C_{dd} = \frac{1}{\omega_0}\frac{320 - 560k + 340k^2 - 90k^3 + 9k^4}{30(2-k)^3(20 - 15k + 3k^2)}.$$

To demonstrate the graceful degradation provided by this equivalent-circuit representation I, Figure 11.7(a) shows the magnitude and Figure 11.7(b) shows the phase of y_{gg}/g_0 for $k = 0.65$, obtained with the *RC* equivalent circuit (dashed-dotted line, EQ), the exact solution (full line, EXACT), the frequency power series (dashed line, POWER), and the second order iterative y parameters derived in [6] (dashed line, B).

11.2.6 Range of validity of the *RC* small-signal equivalent circuit I

We wish now to establish the range of validity of the *RC* circuit representation I introduced above for all bias conditions. For this purpose, we have calculated for each parameter y_{ij} the frequency $f_{5\%}(y_{ij})$ for which an error $\text{Err}(y_{ij})$ of 5% is obtained between the exact Bessel solution (see for example [9]) and the approximate results. The error $\text{Err}(y_{ij})$ is

$$\text{Err}(y_{ij}) = \frac{|y_{ij}(exact) - y_{ij}(approximate)|}{|y_{ij}(exact)|}$$

For the sake of comparison, we have plotted in Figure 11.8, $f_{5\%}(y_{ij})/f_0$ for each y_{ij} parameter as a function of the biasing parameter k for the frequency power series model (dashed line, POWER), the second order iterative results [6] (dashed line, B2), the first order iterative results [6] (dashed line, B1) and the simple *RC* circuit representation of the frequency power series model (dashed-dotted line, EQ). One observes that the simple *RC* representation I of the frequency power series holds for all bias conditions up to a higher frequency than both the frequency power series and the iterative results. On the same graph we have also plotted the unity current-gain cut-off frequency f_T/f_0 (dashed line, FT) and the maximum frequency of oscillation f_{max}/f_0 (full line, FMAX) (the frequency at which the unilateral gain is 1 [3]). Both f_T and f_{max} are calculated using the exact Bessel solution.

All approximate small-signal models except the first order iterative model hold for frequencies larger than the cut-off frequency f_T for all bias conditions. The *RC* circuit representation holds for frequencies larger than the maximum frequency of oscillation f_{max} for k smaller than $\sim$0.9. However, for k larger than $\sim$0.9, $f_{5\%}$ is smaller than f_{max}. Note that both the exact and the approximate models predict an infinite maximum frequency of oscillation at $k = 1$. Obviously in the extrinsic device the unavoidable source, drain, and gate resistances and drain output conductance will limit f_{max} to a finite value.

To conclude, note that the normalization frequency f_0 is bias-dependent. For gate voltages approaching the threshold voltage, the normalization frequency f_0 is small and none of these so-called high-frequency approximate models can account for the distributed effects arising even at low frequencies.

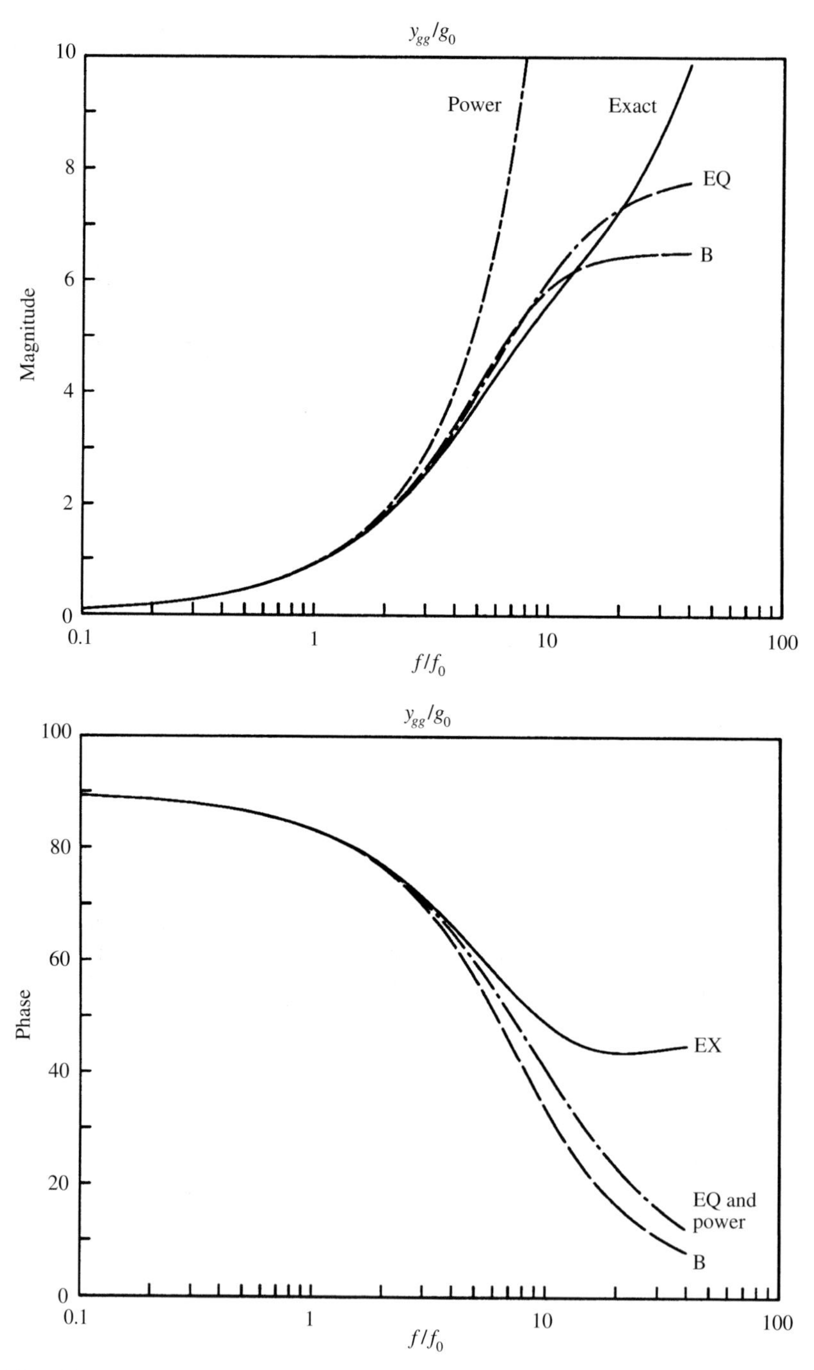

Fig. 11.7. Comparison of (a) the amplitude and (b) the phase of y_{gg}/g_0 for $k = 0.65$, obtained with the *RC* equivalent circuit I (dashed-dotted line, EQ), the exact solution (full line, EXACT), the frequency power series (dashed line, POWER), and the second-order iterative y parameters reported (dashed line, B). (P. Roblin, S. C. Kang and W. R. Liou, *IEEE Transactions of Electron Devices*, Vol. 38, No. 8, pp. 1706–1718, August 1991. ©1991 IEEE.)

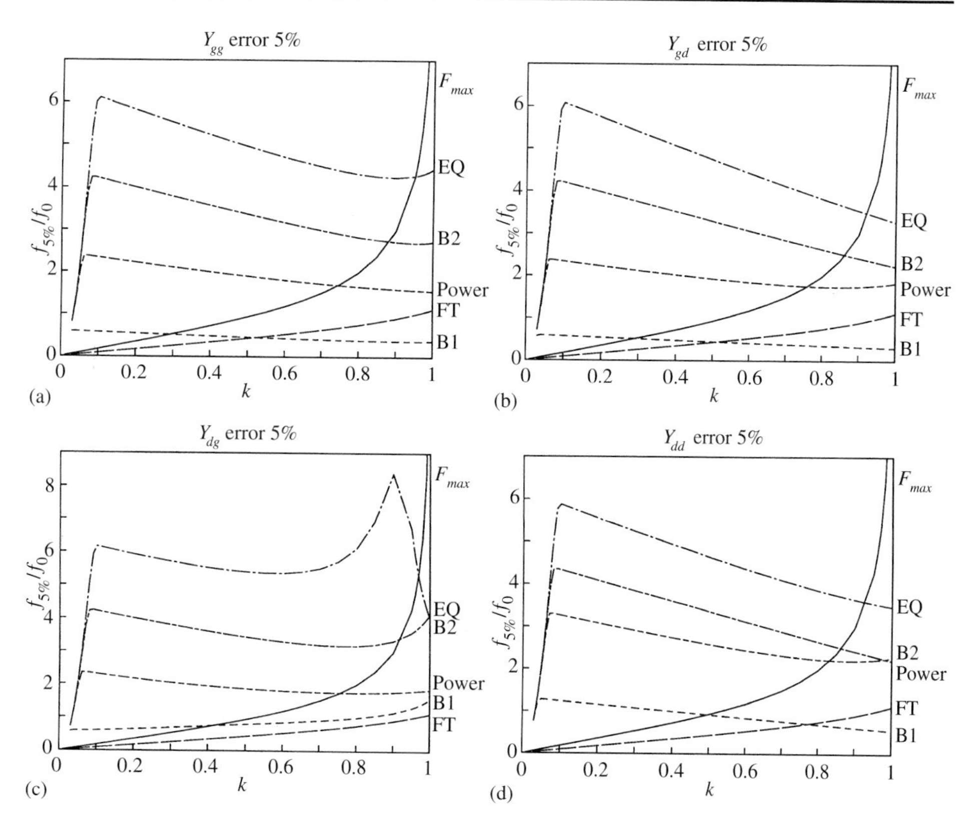

Fig. 11.8. Plot of $f_{5\%}(y_{ij})/f_0$ for a) y_{gg}, b) y_{gd}, c) y_{dg}, and (d) y_{dd} as a function of the biasing parameter k for the frequency power series model (dashed line, POWER), the first order iterative results (dashed line, B1), the second order iterative results (dashed line, B2), the *RC* equivalent circuit (dashed-dotted line, EQ). Also shown are the unity current-gain cut-off frequency f_T/f_0 (dashed line, FT), and the maximum frequency of oscillation f_{max}/f_0 (full line, FMAX). (P. Roblin, S. C. Kang and W. R. Liou, *IEEE Transactions of Electron Devices*, Vol. 38, No. 8, pp. 1706–1718, August 1991. ©1991 IEEE.)

11.2.7 Alternative equivalent circuits for the intrinsic MODFET/MOSFET

Given the frequency power series (11.31) or even the expansion (11.32), it is not possible to extract a unique small-signal equivalent circuit model. A generic equivalent circuit commonly used for FETs is given in Figure 11.9. The y parameters are given by

$$y_{gg} = y_1 + y_2,$$

$$y_{gd} = -y_2,$$

$$y_{dg} = y_3 - y_2,$$

$$y_{dd} = y_4 + y_2.$$

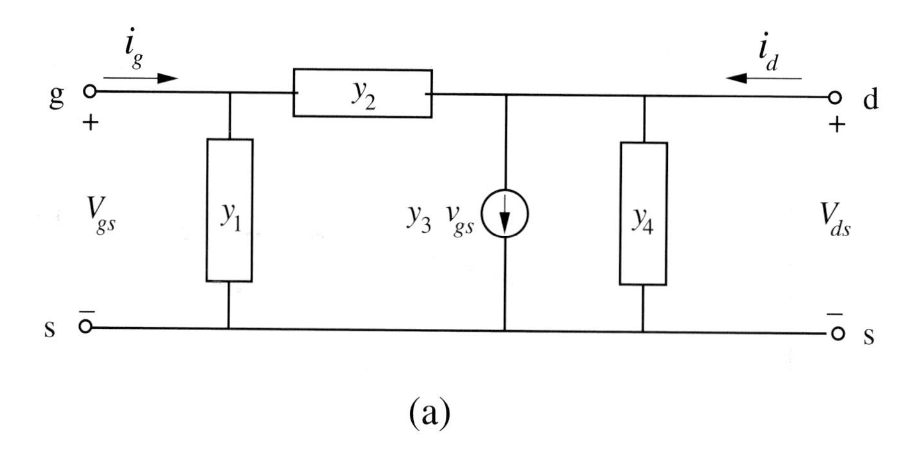

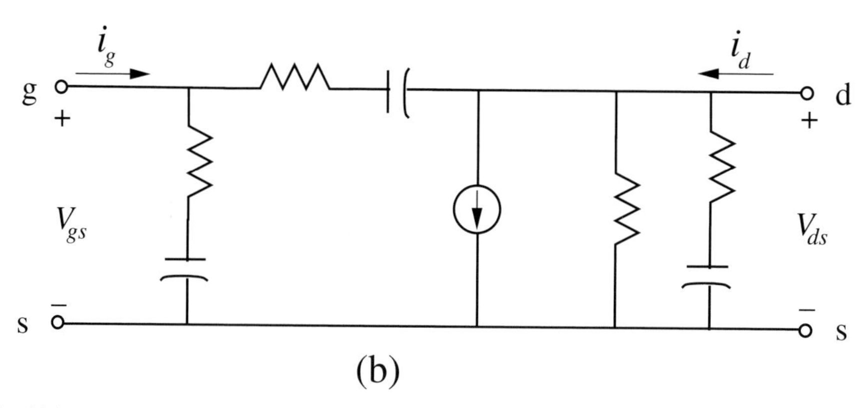

Fig. 11.9. Conventional small-signal equivalent circuit for the unsaturated intrinsic MODFET: (a) with generic impedances y_i and (b) with an RC topology for y_1, y_2, and y_3.

In the long-channel Model B, y_1, y_2, y_3, and y_4 are given by

$$y_i = g_i + \frac{j\omega C_i}{1 + j\omega R_i C_i}. \tag{11.35}$$

Alternatively in the short-channel Model A, y_3 is selected to be

$$y_3 = \frac{g_m \exp(j\omega\tau_s)}{1 + j\omega\tau_3}. \tag{11.36}$$

In Figure 11.10, we compare the performance of these various equivalent circuits against the exact solution obtained for the ideal FET wave-equation. To do so, the model elements (e.g., the capacitor and resistance or time delays τ_s and τ_3) of each y parameter are selected such as to admit the exact second order frequency power series derived for the y parameter of the MOSFET wave-equation. As can be seen, each equivalent circuit perfectly fits the exact MODFET wave-equation solution at low frequencies as intended and exhibits a graceful degradation at high frequencies. However, we shall see in the next section that the *non-quasi-static equivalent circuit*

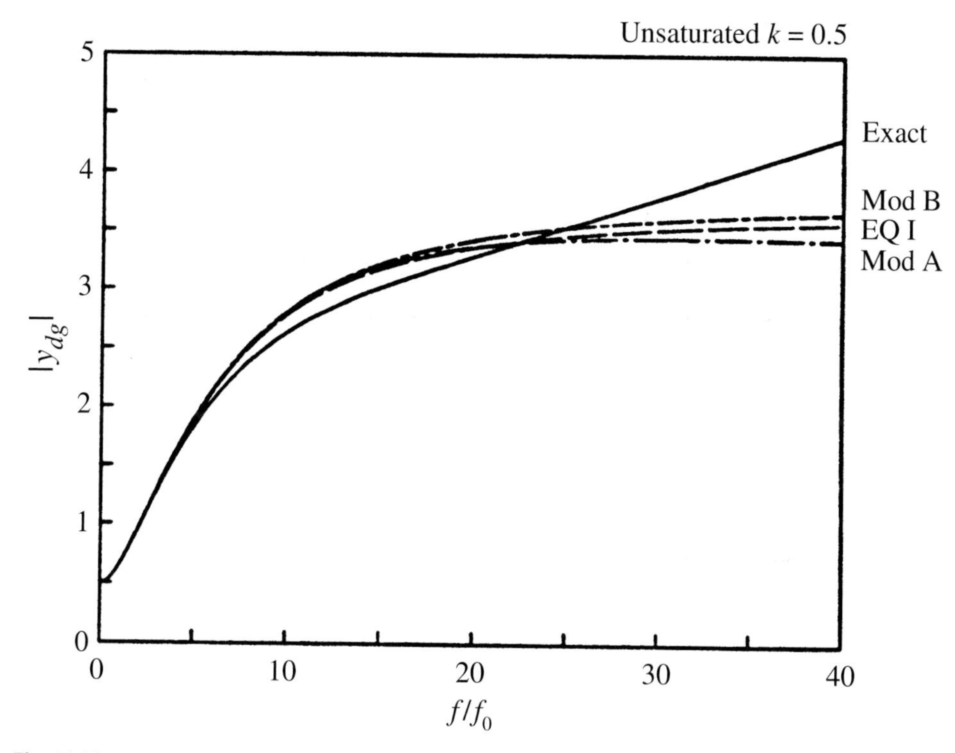

Fig. 11.10. Comparison of $|y_{dg}|$ obtained using the exact FET wave-equation solution (full line), the conventional equivalent circuit Model A (dotted-dashed line) and Model B (dashed-dashed line) and the equivalent circuit I of Figure 11.6 (dashed line).

I introduced in Figure 11.6 for the unsaturated MODFET can be readily implemented in a high-performance non-quasi-static *charge-conserving* large-signal model.

11.3 Large-signal model of the long-channel MODFET/MOSFET

The large-signal response of the ideal long-channel MOSFET is regulated by the large-signal wave-equation (11.14). This wave-equation is a non-linear partial differential equation which must be solved numerically in the general case. Burns [15], Oh *et al.* [16], and Mancini *et al.* [17] have reported a numerical solution of the transient response of the large-signal MOSFET wave-equation.

These simulations established that when the MOSFET is turned on abruptly by a step voltage applied at the gate the drain current remains zero for a time τ_d corresponding to the time for the channel charge to move from the source to the drain. This delay time τ_d was found to be approximately

$$\tau_d \simeq \frac{0.38\mu(V_{GS} - V_T)}{L_g^2}.$$

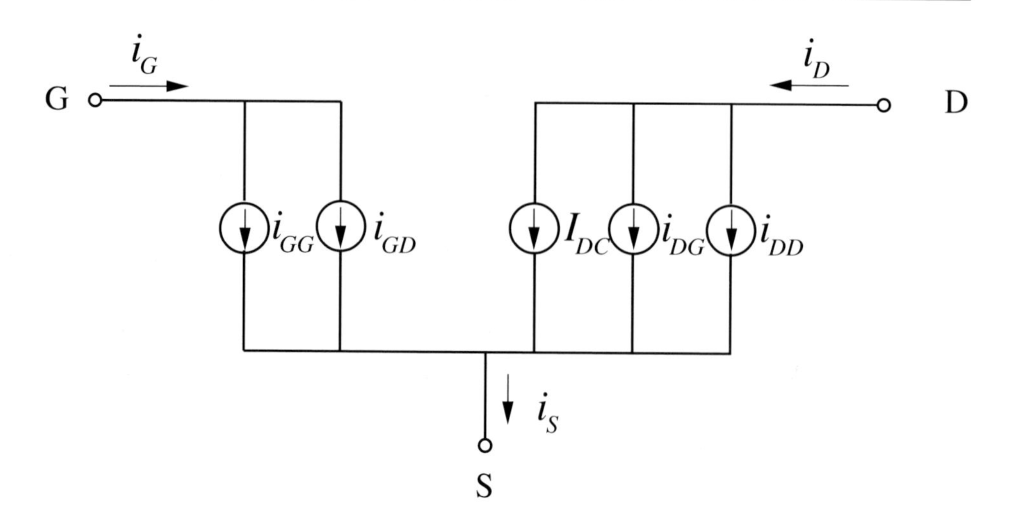

Fig. 11.11. Large-signal equivalent-circuit for the long-channel MODFET.

The numerical solution of the large-signal MOSFET wave-equation is time-consuming and therefore cannot be used in circuit simulators. Alternatively, the *RC* equivalent circuit developed for the small-signal MOSFET wave-equation can be readily transformed into a simplified large-signal model which is more suitable for circuit simulation.

This large-signal model can be obtained by simply replacing g_m and g_d with the drain current source I_D of the MOSFET *I–V* characteristics and by substituting instantaneous values of the DC gate and drain voltage V_{GS} and V_{DS} in the resistors and capacitors and in I_D. The gate, drain and source currents are then given by (see Figure 11.11)

$$i_G = i_{GG} + i_{GD},$$

$$i_D = I_D + i_{DD} + i_{DG},$$

$$i_S = i_G + i_D,$$

with

$$I_D = \frac{C_G W_g \mu}{2L_g}\left[(v_{GS} - V_T)^2 - (v_{GS} - V_T - v_{DS})^2\right],$$

$$i_{GG} = C_{gg}(v_{GS}, v_{DS})\frac{d}{dt}\left[v_{GS} - R_{gg}(v_{GS}, v_{DS})i_{GG}\right],$$

$$i_{GD} = C_{gd}(v_{GS}, v_{DS})\frac{d}{dt}\left[v_{DS} - R_{gd}(v_{GS}, v_{DS})i_{GD}\right],$$

$$i_{DG} = C_{dg}(v_{GS}, v_{DS})\frac{d}{dt}\left[v_{GS} - R_{dg}(v_{GS}, v_{DS})i_{DG}\right],$$

$$i_{DD} = C_{dd}(v_{GS}, v_{DS})\frac{d}{dt}[v_{DS} - R_{dd}(v_{GS}, v_{DS})i_{DD}].$$

To test this *RC* large-signal model, we can submit it to the four large-signal computer experiments used by Mancini *et al.* [17] and Chai and Paulos [10] for testing their four-terminal MOSFET large-signal models. The mobility ($\mu = 609\ \text{cm}^2/\text{V}$) and gate length ($L_g = 10\ \mu\text{m}$) given in [17] and [10] are used for our three-terminal MOSFET together with $V_T = 0$.

In the first test, the drain voltage is $v_{DS} = 1$ V and the gate voltage v_{GS} varies from 2 to 10 V in 1 ns. The currents calculated using the *RC* model (full lines) are shown in Figure 11.12(a) and (b). For comparison we have also plotted in Figure 11.12(a) and (b) the currents obtained using the transcapacitor model (dashed lines) which relies on the same capacitors C_{ij} but neglects the charging resistors $R_{ij} = 0$. In the second test, the drain voltage is $v_{DS} = 1$ V and the gate voltage v_{GS} varies from 10 to 2 V in 1 ns. The currents calculated using the *RC* model (full lines) and using the transcapacitor model (dashed lines) are shown in Figure 11.13(a) and (b). In the third test, the gate voltage is $v_{GS} = 10$ V and the drain voltage v_{DS} varies from 1 to 10 V in 1 ns. The currents calculated using the *RC* model (full lines) and using the transcapacitor model (dashed line) are shown in Figure 11.14(a) and (b).

The currents calculated with the three-terminal *RC* large-signal model exhibit the same type of transient obtained with the numerical results reported by Mancini *et al.* [17] for the four-terminal MOSFET. The success of the *RC* model is attributed to the fact that for these biases the MOSFET is operating in the triode region and the Cij and Rij vary slowly with the instantaneous bias.

In the fourth test, the drain voltage is $v_{DS} = 4$ V and the gate voltage v_{GS} varies from 0.0001 to 10 V in 1 ns. The currents calculated using the *RC* model (full lines) and using the transcapacitor model (dashed line) are shown in Figure 11.15(a) and (b).

When compared with the results reported by Mancini *et al.* [17] (for the four-terminal MOSFET), the (three-terminal) *RC* model suffers from the following two problems. The *RC* large-signal model predicts a negative drain current between $t = 0$ and $t \simeq 0.35$ ns, which is not present in the exact numerical solution [17]. In addition, it introduces a rapid increase of the drain and gate currents at $t = 0.4$ ns, when the MOSFET switches from the pinch-off to the triode mode. This rapid variation of the current, not observed in the exact solution [17], originates from the rapid variation of C_{ij} and R_{ij} near pinch-off. As we shall see, this problem can be great reduced if charge conservation is enforced.

Chai and Paulos [10] have reported a unified large- and small-signal model derived using an iterative technique which permitted them to reproduce the numerical results of Mancini *et al.* quite well [17]. For small-signal analysis, their first order iterative technique reduces to the first order iterative solution of the MOSFET wave-equation (see [13] and [5]). Their work suggests the use of the following alternate set of

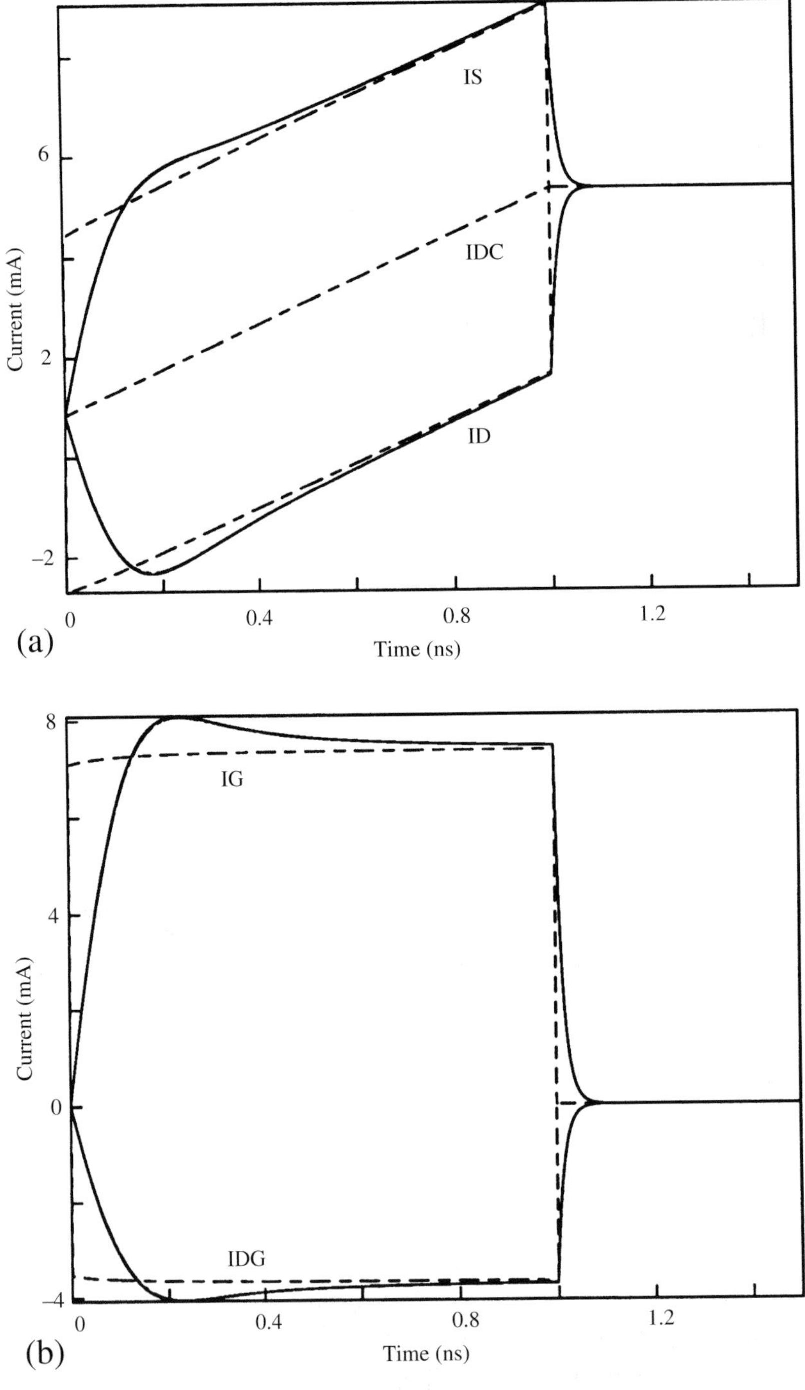

Fig. 11.12. Plot of (a) i_D, i_S and i_{DC} and (b) i_G and i_{DG} calculated for $v_{DS} = 1$ V and v_{GS} varying from 2 to 10 V in 1 ns using the *RC* model (full lines), the transcapacitor model (dashed lines) and the state equations (dashed-dotted lines). (P. Roblin, S. C. Kang and W. R. Liou, *IEEE Transactions of Electron Devices*, Vol. 38, No. 8, pp. 1706–1718, August 1991. ©1991 IEEE.)

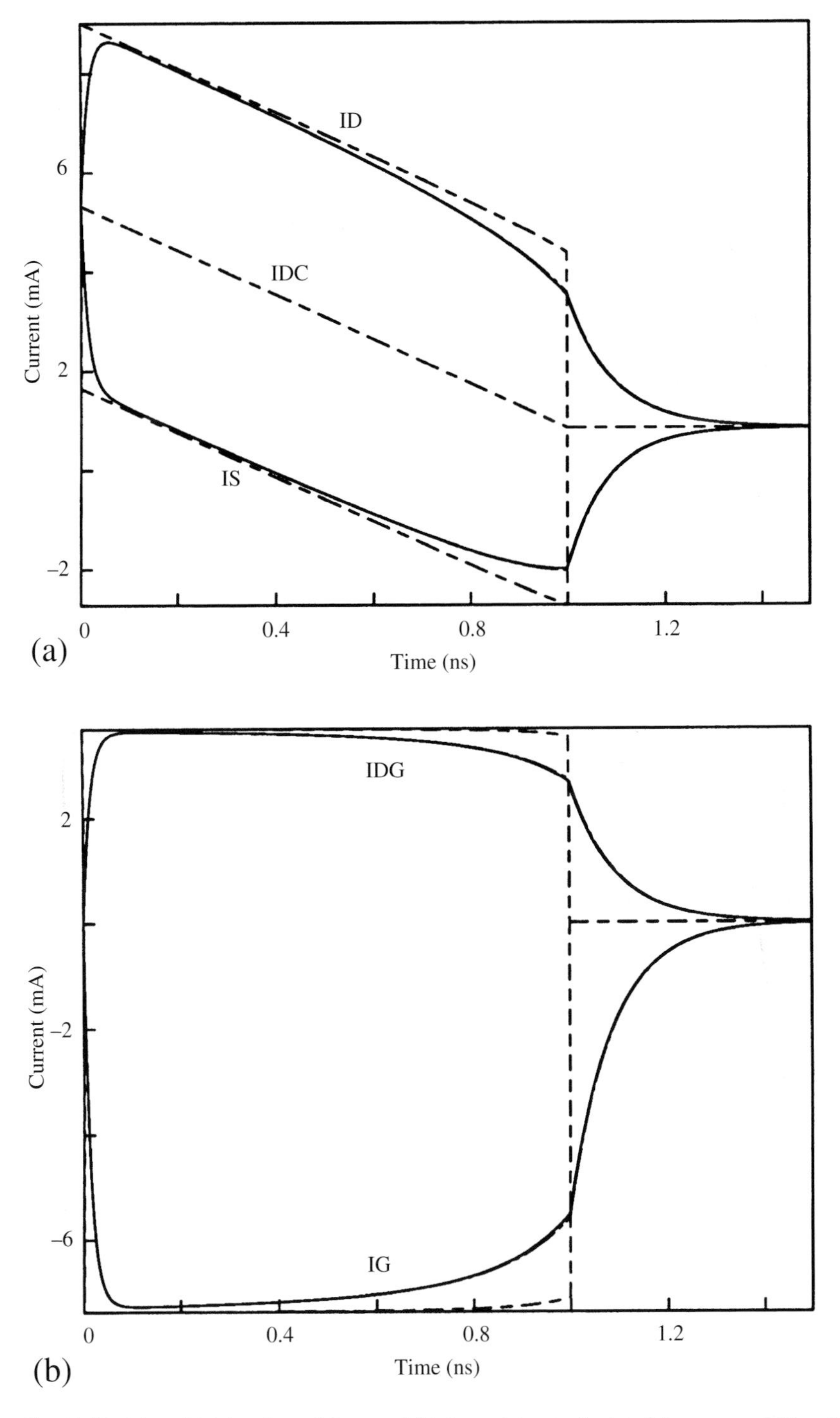

Fig. 11.13. Plot of (a) i_D, i_S and i_{DC} and (b) i_G and i_{DG} calculated for $v_{DS} = 1$ V and v_{GS} varying from 10 to 2 V in 1 ns using the *RC* model (full lines), the transcapacitor model (dashed lines) and the state equations (dashed-dotted lines). (P. Roblin, S. C. Kang and W. R. Liou, *IEEE Transactions of Electron Devices*, Vol. 38, No. 8, pp. 1706–1718, August 1991. ©1991 IEEE.)

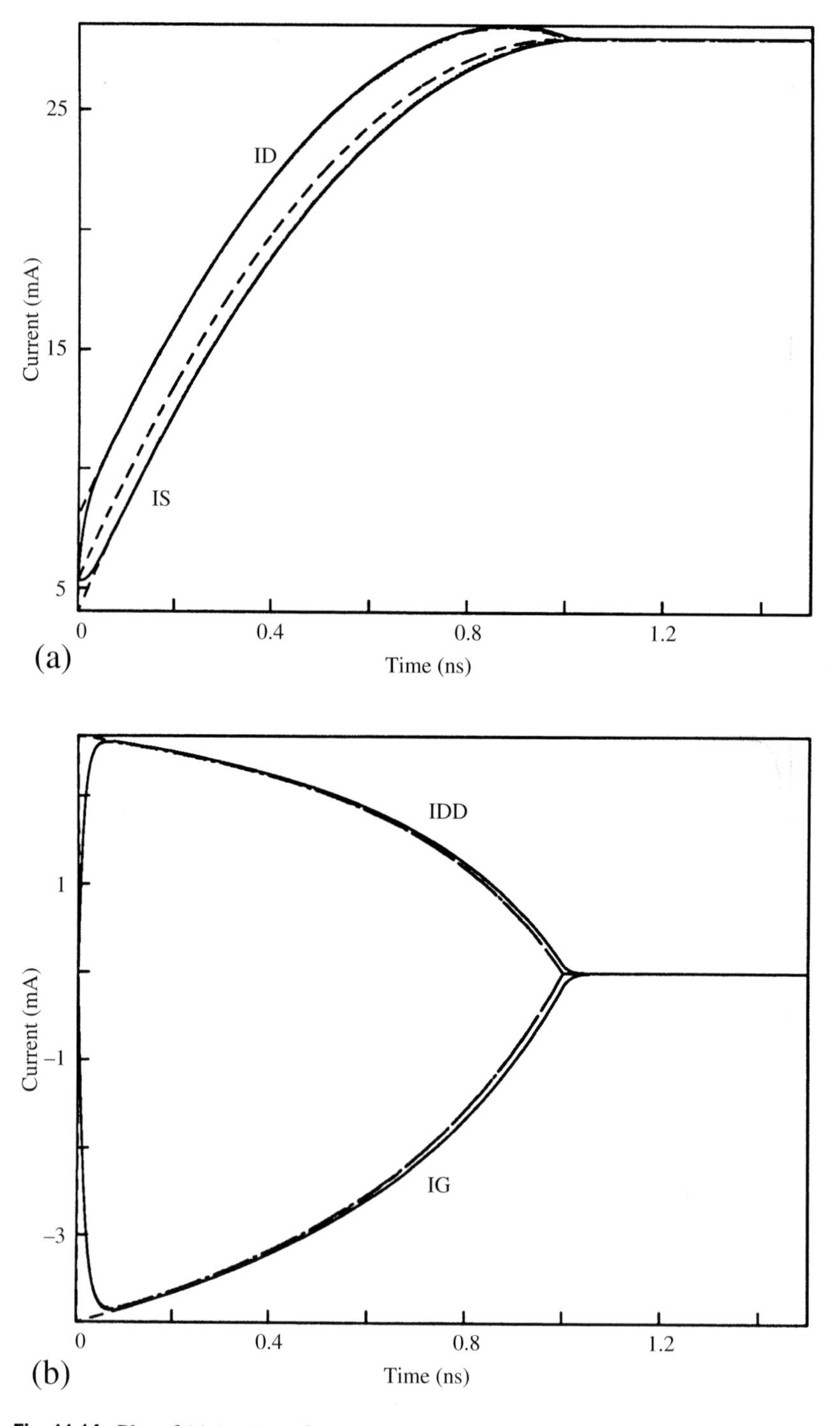

Fig. 11.14. Plot of (a) i_D, i_S and i_{DC} and (b) i_G and i_{DD} calculated for $v_{GS} = 10$ V and v_{DS} varying from 1 to 10 V in 1 ns using the *RC* model (full lines), the transcapacitor model (dashed lines) and the state equations (dashed-dotted lines). (P. Roblin, S. C. Kang and W. R. Liou, *IEEE Transactions of Electron Devices*, Vol. 38, No. 8, pp. 1706–1718, August 1991. ©1991 IEEE.)

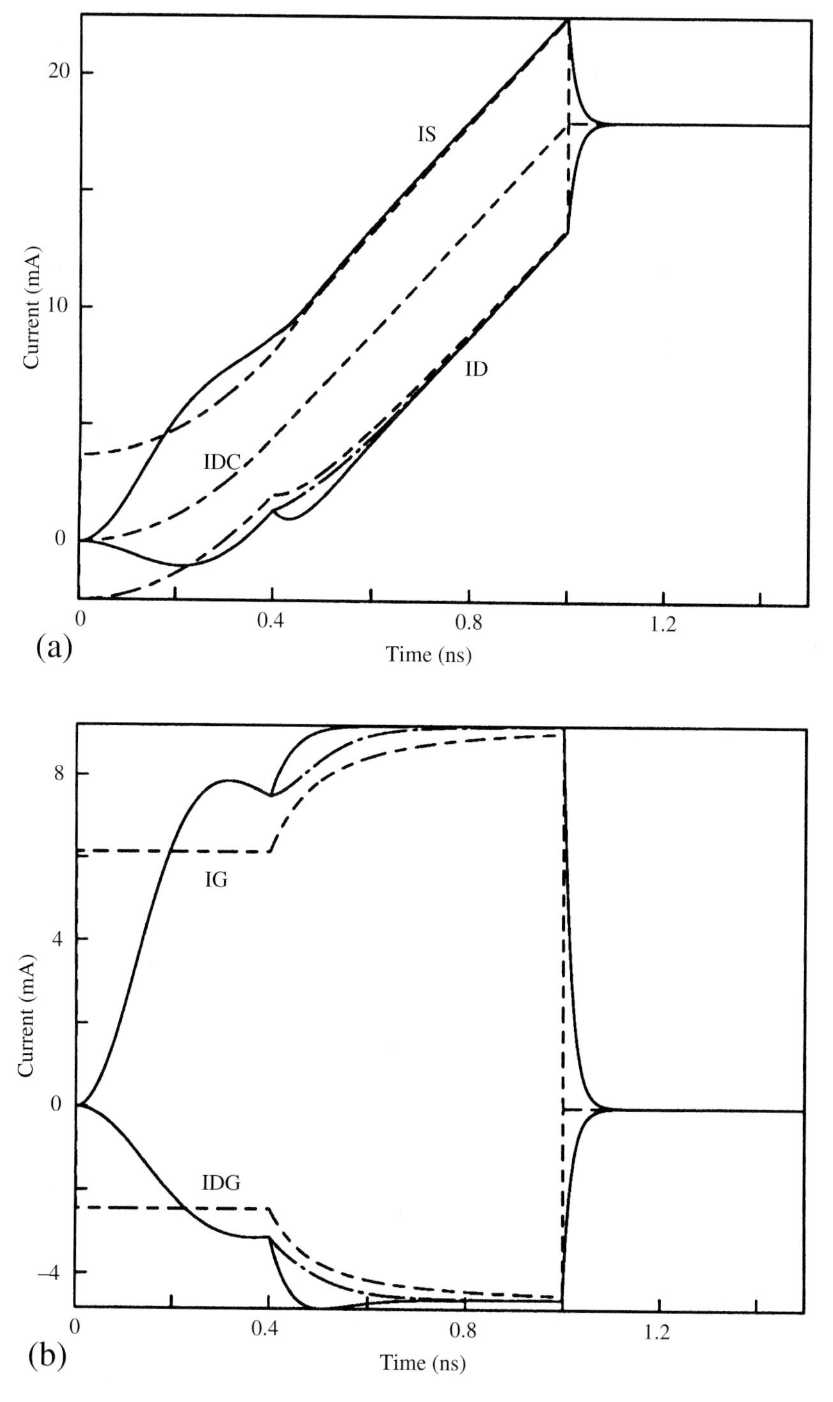

Fig. 11.15. Plot of (a) i_D, i_S and i_{DC} and (b) i_G and i_{DG} calculated for $v_{DS} = 4$ V and v_{GS} varying from 0.0001 to 10 V in 1 ns using the *RC* model (full lines), the transcapacitor model (dashed lines) and the state equations (dashed-dotted lines). (P. Roblin, S. C. Kang and W. R. Liou, *IEEE Transactions of Electron Devices*, Vol. 38, No. 8, pp. 1706–1718, August 1991. ©1991 IEEE.)

differential equations:

$$\left.\begin{aligned}
i_{GG} &= C_{gg}(v_{GS}, v_{DS})\frac{dv_{GS}}{dt} - \frac{d}{dt}\left[C_{gg}(v_{GS}, v_{DS})R_{gg}(v_{GS}, v_{DS})i_{GG}\right],\\
i_{GD} &= C_{gd}(v_{GS}, v_{DS})\frac{dv_{DS}}{dt} - \frac{d}{dt}\left[C_{gd}(v_{GS}, v_{DS})R_{gd}(v_{GS}, v_{DS})i_{GD}\right],\\
i_{DG} &= C_{dg}(v_{GS}, v_{DS})\frac{dv_{GS}}{dt} - \frac{d}{dt}\left[C_{dg}(v_{GS}, v_{DS})R_{dg}(v_{GS}, v_{DS})i_{DG}\right],\\
i_{DD} &= C_{dd}(v_{GS}, v_{DS})\frac{dv_{DS}}{dt} - \frac{d}{dt}\left[C_{dd}(v_{GS}, v_{DS})R_{dd}(v_{GS}, v_{DS})i_{DD}\right].
\end{aligned}\right\} \quad (11.37)$$

The response of the intrinsic MOSFET to the gate and drain voltage ramps as predicted with these new differential equations is shown in Figures 11.12–11.15 using dashed-dotted lines. It is not possible to distinguish this modified *RC* model (dashed-dotted lines) from the *RC* model (full lines) except in Figure 11.15(a) and (b), where a smoother response in agreement with the numerical simulation [17] results when the MOSFET enters the triode mode.

The modified large-signal model still predicts a negative drain current in Figure 11.15. As is explained by Mancini *et al.* [17], the drain current cannot be negative for large drain voltages. Indeed, for large drain voltages, when the device is biased in the saturation region, a fraction of the applied drain voltage is dropped in the built-in potential barrier in the drain region. The resulting increase in the potential barrier at the drain prevents the electrons from diffusing from the drain to the channel to charge the channel. A negative drain current charging the depleted channel is, however, possible in the triode mode (Figures 11.12 and 11.13) since in this case the built-in potential barrier is not increased by the drain voltage (see for example [18]). The boundary conditions of the simple wave-equation used here (see Section 11.2) do not take diffusion into account and cannot therefore predict this effect. Note that large built-in potentials at the drain arise only when the device is biased in saturation (pinch-off). Therefore both the small- and large-signal *RC* models proposed here should be correct for the unsaturated MOSFET and moderately saturated (long-channel) MOSFET. A more complicated equivalent circuit is required for the saturated MOSFET (see next chapter). Note that the use of improved boundary conditions to drive the state equations presented here might not be sufficient by itself to avoid the negative drain current in Figure 11.15. Indeed, the failure to reproduce the exact response (no negative current) in Figure 11.14 also originates from the fact that for small gate voltages V_G the frequency f_0 becomes very small and the channel of the MOSFET quickly behaves like a transmission line. Indeed, the (*RC*) state equations derived above cannot be used for excitation with a frequency component much in excess of f_0 (or $f_{5\%}$). Note, however, that the non-quasi-static state equations generate a current response (dashed-dotted line) far superior to the quasi-static model (dashed line).

11.3.1 Charge conservation

Charge conservation is an important issue in circuit simulation. The charge ΔQ transferred to a device through a terminal X in a time Δt by an in-going current $i_X(t)$ is simply given by

$$\Delta q_X = \int_0^{\Delta t} i_X(t)\, dt.$$

Global charge conservation in the FET model results from Kirchhoff's current law $i_S = i_G + i_D$, as is verified by integration over time. This global charge conservation does not, however, prevent the gate and channel of a large-signal FET model from continuously accumulating charge over time [19]. Such an accumulation of charge in the channel is inconsistent with the assumption of DC I–V characteristics that are not history-dependent. Furthermore, it is known that such unphysical charge accumulation adversely affects the external circuits in a circuit simulator [22].

The gate (or channel) charge Q_G in the MOSFET in the steady state is given by

$$\begin{aligned} Q_G(V_{GS}, V_{DS}) &= W_g \int_0^{L_g} q N_S(x)\, dx \\ &= W_g C_G L_g \frac{2}{3} \frac{(V_{GS} - V_T - V_{DS})^3 - (V_{GS} - V_T)^3}{(V_{GS} - V_T - V_{DS})^2 - (V_{GS} - V_T)^2}. \end{aligned}$$

The variation of the gate charge predicted by the DC model from the steady-state bias condition 1 to steady-state bias condition 2 is

$$\Delta Q_G(1,2) = Q_G[V_{GS}(2), V_{DS}(2)] - Q_G[V_{GS}(1), V_{DS}(1)].$$

Let us verify that the FET state equations (11.37) predict a variation of gate and channel charge which is compatible with the DC model. The instantaneous charge transferred to the gate Δq_G (which is also the charge accumulated in the FET channel) is

$$\begin{aligned} \Delta q_G(t_1, t_2) &= \int_{t_1}^{t_2} i_G\, dt = \int_{t_1}^{t_2} (i_{GG} + i_{GD})\, dt \\ &= \Delta Q_G(t_1, t_2) - \int_{t_1}^{t_2} \left[\frac{d}{dt}(R_{gg} C_{gg} i_{GG}) + \frac{d}{dt}(R_{gd} C_{gd} i_{GD}) \right] dt, \end{aligned}$$

where $\Delta Q_G(t_1, t_2)$ is

$$\Delta Q_G(t_1, t_2) = \int_{t_1}^{t_2} \left[C_{gg} \frac{dv_{GS}(t)}{dt} + C_{gd} \frac{dv_{DS}(t)}{dt} \right] dt. \tag{11.38}$$

The variation of the instantaneous gate charge is then

$$\begin{aligned} \Delta q_G(t_1, t_2) = \Delta Q_G(t_1, t_2) &- \tau_{gg}[v_{GS}(t_2), v_{DS}(t_2)] i_{GG}(t_2) \\ &+ \tau_{gg}[v_{GS}(t_1), v_{DS}(t_1)] i_{GG}(t_1) - \tau_{gd}[v_{GS}(t_2), v_{DS}(t_2)] i_{GD}(t_2) \\ &+ \tau_{gd}[v_{GS}(t_1), v_{DS}(t_1)] i_{GD}(t_1). \end{aligned}$$

If the device is in the steady state at times t_1 and t_2, i_{GG} and i_{GD} must be zero at these times, and we have $\Delta q_G(t_1, t_2) = \Delta Q_G(t_1, t_2)$.

One can easily verify that the capacitor C_{gg} and C_{gd} can be obtained from the gate (or channel) charge Q_G by

$$C_{gg}(v_{GS}, v_{DS}) = \frac{\partial Q_G(v_{GS}, v_{DS})}{\partial v_{GS}},$$

$$C_{gd}(v_{GS}, v_{DS}) = \frac{\partial Q_G(v_{GS}, v_{DS})}{\partial v_{DS}}.$$

Since in the unsaturated MOSFET ($0 \leq k < 1$) the gate charge Q_G admits continuous partial derivatives, its time derivative is then given by

$$\frac{dQ_G}{dt} = C_{gg}(v_{GS}, v_{DS})\frac{dv_{GS}}{dt} + C_{gd}(v_{GS}, v_{DS})\frac{dv_{DS}}{dt}.$$

$\Delta Q_G(t_1, t_2)$ as defined by Equation (11.38) can now be written

$$\Delta Q_G(t_1, t_2) = \int_{t_1}^{t_2} \frac{dQ_G}{dt}\, dt = Q_G[v_{GS}(t_2), v_{DS}(t_2)] - Q_G[v_{GS}(t_1), v_{DS}(t_1)],$$

which is path-independent. Therefore $\Delta q_G(t_1, t_2)$ is equal to $\Delta Q_G(1, 2)$ if the FET is in the steady-state biasing conditions 1 and 2 at times t_1 and t_2, respectively.

The modified large-signal model using the alternative differential equation topology of [10] enforces the desired conservation of charge for the unsaturated FET. Charge conservation is also enforced in the saturated MOSFET ($v_{DS}(t) > v_{GS}(t) - V_T$). Indeed, the saturated MOSFET follows the same state equations since we use $k = 1$ to calculate the RC elements in saturation. Using $k = 1$ is equivalent to applying an effective drain voltage $v_{DS}(t) = v_{GS}(t) - V_T$. However, as we would expect in an ideal pinched-off MOSFET, this effective time-varying drain-to-source voltage $v_{DS}(t)$ does not induce any charging currents in the saturated FET since we have $C_{gd}(k = 1) = C_{dd}(k = 1) = 0$.

11.3.2 Charge conservation in circuit simulators

Note that the capacitors C_{dg} and C_{dd} can be obtained from the partial derivatives of a charge Q_D:

$$C_{dg}(v_{GS}, v_{DS}) = \frac{\partial Q_D(v_{GS}, v_{DS})}{\partial v_{GS}},$$

$$C_{dd}(v_{GS}, v_{DS}) = \frac{\partial Q_D(v_{GS}, v_{DS})}{\partial v_{DS}},$$

where Q_D is the portion of the gate (or channel) charge Q_G associated with the drain and is given by

$$Q_D(v_{GS}, v_{DS}) = C_G W_g L_g$$

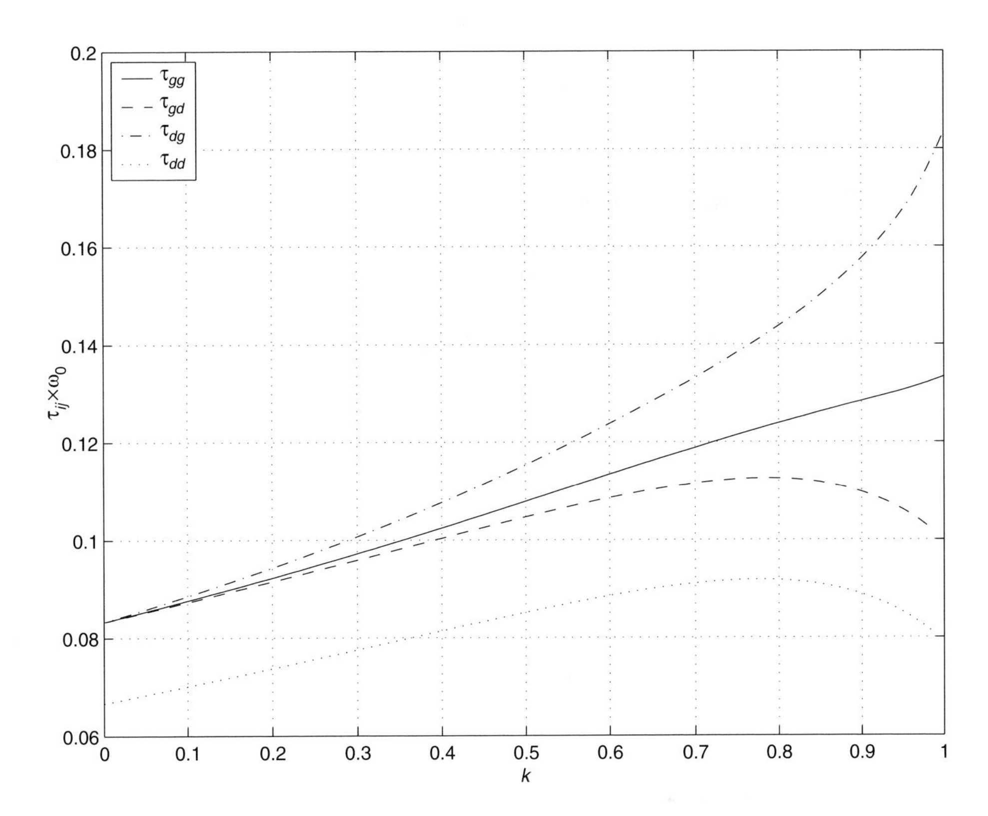

Fig. 11.16. Plot of τ_{gg}, τ_{gd}, τ_{dg}, and τ_{dd} normalized by $1/\omega_0$ versus the biasing parameter k.

$$\times \frac{2}{15} \frac{-15(v_{GS} - V_T)^3 + 25(v_{GS} - V_T)^2 v_{DS} - 15(v_{GS} - V_T)v_{DS}^2 + 3v_{DS}^3}{4(v_{GS} - V_T)^2 - 4(v_{GS} - V_T)v_{DS} + v_{DS}^2}.$$

These identities cannot be used here to reduce the number of differential equations. The large-signal model introduced here therefore requires four differential equations making use of the four time constants τ_{gg}, τ_{gd}, τ_{dg} and τ_{dd} shown in Figure 11.16. By setting $\tau_{gg} = \tau_{gd} = \tau_G$ and $\tau_{dg} = \tau_{dd} = \tau_D$, one can then easily verify that the four differential equations reduce to the following two differential equations:

$$\left.\begin{aligned}(i_{GG} + i_{GD}) &= \frac{dQ_G}{dt} - \frac{d}{dt}[\tau_G(i_{GG} + i_{GD})],\\ (i_{DG} + i_{DD}) &= \frac{dQ_D}{dt} - \frac{d}{dt}[\tau_D(i_{DG} + i_{DD})].\end{aligned}\right\} \quad (11.39)$$

The use of four differential equations instead of two is expected to increase the frequency range of the model. This is demonstrated for the small-signal parameters in Figures 11.8(a)–(d) where the first order y parameters (EQ) resulting from the four differential Equations (11.37) are seen to be valid for $k > 0.1$ to a frequency $f_{5\%}$, 4–12 times that of the first order iterative y parameters (B1) resulting from the two differential equations 11.39. However, the less accurate two-time-constant (τ_G and τ_D) model may be advantageous as it is a charge picture. Indeed, charge conservation is also a numerical issue as non-conservation of charge can also result in practice in models intrinsically conserving charge from numerical errors associated with the calculation of derivatives when voltages are used as state variables in circuit simulators. This problem has been recognized and prompted the development of charge-based quasi-static MODFET models for use in circuit simulators. Such charge-based models are now the default model in microwave circuit simulators.

In a charge-based model the total gate, drain and source currents are given by:

$$\left.\begin{aligned} i_G &= i_{disp,G}, \\ i_D &= I_D(v_{GS}, v_{DS}) + i_{disp,D}, \\ i_S &= i_D + i_G, \end{aligned}\right\} \tag{11.40}$$

where $i_{disp,X}$ is the displacement current associated with the charge element Q_X and is given by

$$i_{disp,X} = \frac{dQ_X(v_{GS}, v_{DS})}{dt} - \frac{d}{dt}[\tau_X(v_{GS}, v_{DS})i_{disp,X}]. \tag{11.41}$$

Note that τ_X is the non-quasi-static charge-redistribution time constant associated with the charge-element Q_X.

Figure 11.17(a) shows a possible topology for implementing this charged-based large-signal model. This charge model which relies on two non-quasi-static time constants (τ_G and τ_D) will be used in Chapter 13 for building an electrothermal FET model. An alternative and equivalent topology (for $\tau_G = \tau_D = \tau_S$) is also shown in Figure 11.17(b). In this alternate topology the current I' and charges Q' are given by:

$$\left.\begin{aligned} I'_D(v_{GS}, v_{GD}) &= I_D(v_{GS}, v_{DS}) \\ Q'_D(v_{GS}, v_{GD}) &= Q_D(v_{GS}, v_{DS}) \\ Q'_S(v_{GS}, v_{GD}) &= -Q_G(v_{GS}, v_{DS}) - Q_D(v_{GS}, v_{DS}). \end{aligned}\right\} \tag{11.42}$$

Note that following [24] a fully symmetric topology is obtained by switching from common-source state-variables (v_{GS}, v_{DS}) to common-gate state-variables (v_{GS}, v_{GD}). This alternate topology is particularly recommended to handle charge conservation when v_{DS} switches from positive to negative voltages.

11.4 Parasitics, extrinsic MODFET and parameter extraction

So far we have mostly discussed the modeling of the intrinsic MODFET. The extrinsic MODFET model is obtained by adding the parasitics equivalent circuit to the intrinsic MODFET model. Typical parasitic equivalent circuits used for the small- and large-signal models are shown in Figure 11.18(a) and (b). C_{gdf} and C_{gsf} are the fringe

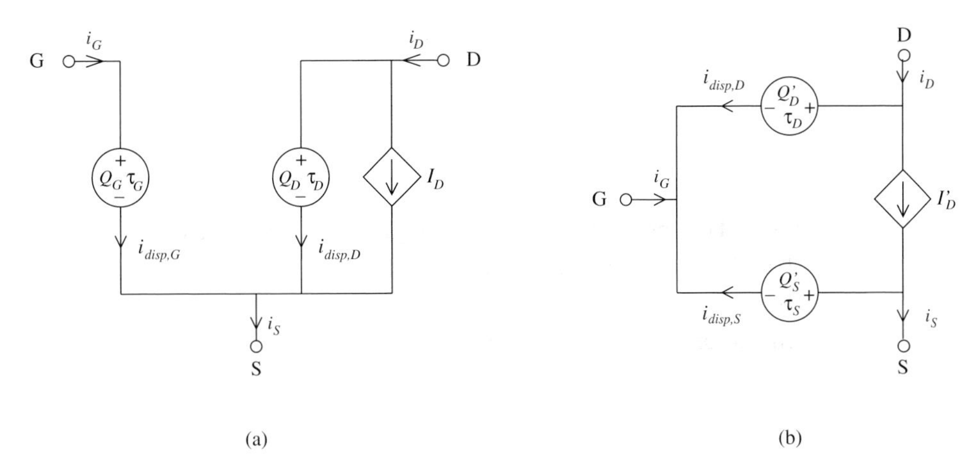

Fig. 11.17. Equivalent circuits for the charge-based large-signal models of the long-channel MODFET. The topology shown in (a) relies on two non-quasi-static time constants τ_G and τ_D. The alternative topology shown in (b) is recommended to enforce charge conservation when v_{DS} switches from positive to negative voltages.

capacitors of the gate. The Schottky diode of the gate contributes the two diodes (or resistance) between gate-and-source and gate-and-drain. R_S, R_D and R_G are the source, drain and gate resistances. C_{pd} and C_{pg} are the pad capacitors and L_S, L_D and L_G are the bond inductances if any. This is the conventional topology used for MESFETs and MODFETs.

Physical models can also be developed for the parasitics. A small-signal analysis of the distributed effect arising from the gate width predicts that the maximum power gain (MAG) of the transistor will decrease if the gate width W_g is larger than $\lambda/12$, where λ is the wavelength of the applied AC signal in the semiconductor. For gate widths smaller than $\lambda/12$, the distributed effects along the gate width can be accurately represented by the resistor R_G in series with the gate. Note that it is also sometimes necessary to account for the frequency dependence of the source and drain resistances [20]. However, the frequency dependence of the parasitics is usually only important at high frequencies. Finally, the large *RC* time delay associated with the variation of charge of the deep donors in the MODFET capacitor (see Chapter 8) can also contribute to the dispersion (variation with frequency) of the transconductance g_m observed at low frequency. Indeed, the DC transconductance $g_m(I\text{–}V)$ calculated from the measured *I–V* characteristics is usually smaller than the $g_m(RF)$ extracted from the RF data using an equivalent circuit. Similarly, the drain conductance $g_d(I\text{–}V)$ and $g_d(RF)$ can be found to be different. This topic will be further explored in detail in Chapter 13.

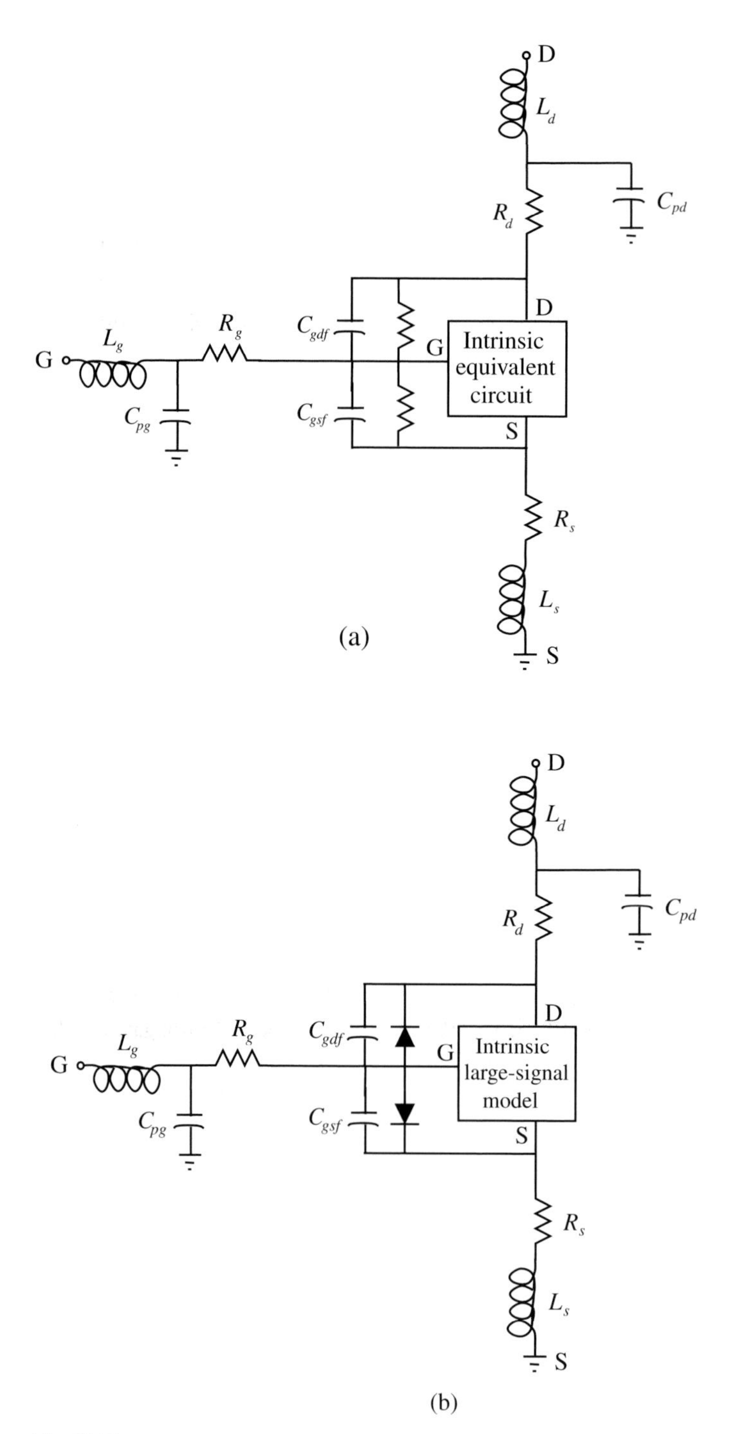

Fig. 11.18. Typical parasitic equivalent circuits used for (a) the small- and (b) the large-signal MODFET models.

11.5 Conclusion

In this chapter we have developed and solved the long-channel MOSFET/MODFET wave-equation. Optimal non-quasi-static small-signal equivalent-circuit and large-signal models were derived.

The non-quasi-static large-signal model introduced in this chapter relied upon a new charge-conserving circuit element consisting of a charge and a time constant. It is interesting to note that this new circuit element also finds application in the accurate modeling of reverse recovery in diodes as has been demonstrated by Yang *et al.* [23].

The calculations in this chapter were based on the long-channel MOSFET/MODFET wave-equation which does not account for velocity saturation effects and is therefore only applicable to the long-channel MODFET. In the next chapter we will generalize our results by studying the velocity-saturated MODFET wave-equation which is applicable to the short-channel MODFET.

11.6 Bibliography

11.6.1 Recommended reading

Y. Tsividis, *Operation and Modeling of the MOS Transistor*, McGraw Hill, New York, 1988.

W. Liu, *Fundamentals of III–V Devices, HBT, MESFETs and HFETs/HEMTs*, Wiley Interscience, New York, 1999.

11.6.2 References

[1] A. J. Tessmer, P. C. Chao, K. H. G. Duh, P. Ho, M. Y. Kao, S. M. J. Liu, P. M Smith, J. M. Ballingall, A. A. Jabra and T. H. Yu, 'Very high performance 0.15 μm gate-length InAlAs/InGaAs/InP lattice-matched HEMTs,' *Proceedings of the 12th Cornell Conference on Advanced concepts in High-Speed Semiconductor Devices and Circuits,* Cornell University, Ithaca, NY, 1989.

[2] J. M. Rollett, 'Stability and power-gain invariants of linear twoports,' *IRE Transactions on Circuit Theory,* Vol. CT-9, No. 3 pp. 29–32, March 1962.

[3] S. J. Mason, 'Power gain in feedback amplifiers,' *IRE Transactions on Circuit Theory,* Vol. CT-1, No. 2, pp. 20–25, June 1954.

[4] G. Gonzalez, *Microwave Transistor Amplifiers Analysis and Design,* Prentice Hall, Englewood Cliffs, New Jersey, 1984.

[5] M. Bagheri and Y. Tsividis 'A small signal dc-to-high-frequency nonquasistatic model for the four-terminal MOSFET valid in all regions of operation,' *IEEE Transactions on Electron Devices*, Vol. ED-32, No. 11, pp. 2383–2391, November 1985.

[6] M. Bagheri, 'An improved MODFET microwave analysis,' *IEEE Transactions on Electron Devices,* Vol. ED-35, No. 7, p. 1147, 1988.

[7] P. Roblin, S. C. Kang, A. Ketterson and H. Morkoç, 'Analysis of MODFET microwave characteristics,' *IEEE Transactions on Electron Devices* Vol. ED-34, No. 9, pp. 1919–1928, September 1987.

[8] P. Roblin, S. C. Kang, A. Ketterson and H. Morkoç, correction to 'Analysis of MODFET microwave characteristics,' *IEEE Transactions on Electron Devices,* Vol. ED-37, No. 3, p. 827, March 1990.

[9] P. Roblin, S. C. Kang and H. Morkoç, 'Analytic solution of the velocity-saturated MOSFET/MODFET wave-equation and its application to the prediction of the microwave characteristics of MODFETs,' *IEEE Transactions on Electron Devices*, Vol. ED-37, No. 7, pp. 1608–1622, July 1990.

[10] K.-W. Chai and J. J. Paulos, 'Unified nonquasi-static modeling of the long-channel four-terminal MOSFET for large and small-signal analyses in all operating regimes,' *IEEE Transactions on Electron Devices*, Vol. 36. No. 11, pp. 2513–2520, November 1989.

[11] A. Van Der Ziel and J. W. Ero, 'Small-signal high-frequency theory of field-effect transistors,' *IEEE Transactions on Electron Devices,* Vol. ED-11, pp. 128–135, April 1964.

[12] V. Ziel and E. N. Wu, 'High-frequency admittance of high electron mobility transistors (HEMTs),' *Solid State Electronics*, Vol. 26, pp. 753–754, 1983.

[13] J. A. Van Nielen, 'A simple and accurate approximation to the high-frequency characteristics of insulated-gate field-effect transistors,' *Solid State Electronics,* Vol. 12, pp. 826–829, 1969.

[14] J. R. Burns, 'High-frequency characteristics of the insulated gate field-effect transistor,' *RCA Review*, Vol. 28, pp. 385–418, September 1967.

[15] J. R. Burns, 'Large-signal transit-time effects in the MOS transistor,' *RCA Review*, Vol. 30, pp. 15–35, March 1969.

[16] S. Y. Oh, D. E. Ward and R. W. Dutton, 'Transient analysis of MOS transistors,' *IEEE Transactions on Electron Devices,* Vol. ED-27, pp. 1571–1578, August 1980.

[17] P. Mancini, C. Turchetti and G. Masetti, 'A nonquasi-static analysis of the transient behavior of the long-channel MOST valid in all regions of operation,' *IEEE Transactions on Electron Devices,* Vol ED-34, pp. 325–344, February 1987.

[18] Y. M. Kim and P. Roblin, 'Two-dimensional charge control model for the MODFETs,' *IEEE Transactions on Electron Devices*, Vol. ED-33, No. 11, pp. 1644–1651, 1986.

[19] R. Daniel, Boeing Electronics Laboratory, private communication, May 1990.

[20] P. Roblin, L. Rice and S. Bibyk, 'Non-linear parasitics in MODFETs and the MODFET IV characteristics,' *IEEE Transactions on Electron Devices*, Vol. ED-35, No. 8, pp. 1207–1214, 1988.

[21] S. C. Kang, 'Analysis of MODFET microwave characteristics,' MSc Thesis, The Ohio State University, 1979.

[22] D. E. Root and B. Hughes, 'Principles of nonlinear active device modeling for circuit simulation,' *32nd ARFTG Conference, Tempe, Az,* IEEE, New York, 1988.

[23] A. T. Yang, Y. Liu and J. T. Yao, 'An efficient nonquasi-static diode model for circuit simulation,' *IEEE Transactions on CAD*, Vol. TCAD-13, pp.231–239, February 1994.

[24] K. Yhland, N. Rorsman, M. Garcia and M. Merkel, 'A symmetrical nonlinear HFET/MESFET model suitable for intermodulation analysis of amplifiers and resistive mixers', *IEEE Transactions on Microwave Theory and Techniques*, Vol. 48, No. 1, January 2000.

11.7 Problems

11.1 Consider the feedback circuit used to unilateralize a two-port device shown in Figure 11.2. In this problem the two-port device is represented by its z small-signal parameters.

(a) Calculate the ratio N of the transformer and the reactance X needed to give $z'_{12} = 0$.

(b) Express the parameters z'_{ij} of the unilateralized two-port device in terms of the parameters z_{ij} of the original two-port device.

11.2 Consider the MOSFET wave-equation

$$\frac{\partial^2 V^2(x,t)}{\partial x^2} = \frac{2}{\mu}\frac{\partial V(x,t)}{\partial t}.$$

For the three-terminal MOSFET the voltage V is simply $v_{GC} - V_T$. The channel current $I(x,t)$ is given by

$$I(x,t) = W_g C_G \mu V(x,t)\frac{\partial V(x,t)}{\partial x}.$$

(a) Verify that for small-signal excitation the MOSFET wave-equation reduces to

$$\frac{d^2}{dx^2}\left[V_0(x)v(x,\omega)\right] = j\frac{\omega}{\mu}v(x,\omega)$$

and that the AC channel current $i(x,\omega)$ is given by

$$i(x,\omega) = W_g C_G \mu \frac{d}{dx}\left[V_0(x)v(x,\omega)\right],$$

where the DC potential $V_0(x)$ is given by

$$V_0(x) = V_{GC}(x) - V_T = (V_{GS} - V_T)\left[1 + (k^2 - 2k)\frac{x}{L_g}\right]^{1/2},$$

with $k = V_{DS}/(V_{GS} - V_T)$.

(b) Verify directly *from the MOSFET wave-equation derived in part (a) and not from its solution* that we can predict that the exact small-signal y parameters obtained from the MOSFET wave-equation can be written in terms of dimensionless parameters:

$$\frac{y_{ij}}{g_0} = f_{ij}\left(k, \frac{\omega}{\omega_0}\right),$$

with g_0 the channel conductance:

$$g_0 = \frac{\mu C_G W_g (V_{GS} - V_T)}{L_g},$$

and with ω_0 a normalization frequency given by

$$\omega_0 = 2\pi f_0 = \mu\frac{(V_{GS} - V_T)}{L_g^2}.$$

Hint: Introduce a normalized position $x' = x/L_g$ and a normalized frequency $\omega_n = \omega/\omega_0$. Note also that the small-signal wave-equation is a linear second order differential equation and therefore admits a solution of the form:

$$v(x') = C_1 v_-(x', \omega_n, k) + C_2 v_+(x', \omega_n, k).$$

11.3 (a) Derive the channel charge Q_G in the MOSFET given by

$$Q_G = W_g \int_0^{L_g} qN_S(x)\,dx,$$

where $qN_S(x)$ is the DC 2DEG channel charge per unit area at position x.

(b) Verify the following identities:

$$C_{gg}(v_{GS}, v_{DS}) = \frac{\partial Q_G(v_{GS}, v_{DS})}{\partial v_{GS}},$$

$$C_{gd}(v_{GS}, v_{DS}) = \frac{\partial Q_G(v_{GS}, v_{DS})}{\partial v_{DS}},$$

(c) Derive Q_D the portion of the channel charge Q_G associated with the drain and given by

$$Q_D = W_g \int_0^{L_g} \frac{x}{L_g} qN_S(x)\,dx,$$

where qN_S is the DC 2DEG channel charge per unit area at position x.

(d) Verify the following identities:

$$C_{dg}(v_{GS}, v_{DS}) = \frac{\partial Q_D(v_{GS}, v_{DS})}{\partial v_{GS}},$$

$$C_{dd}(v_{GS}, v_{DS}) = \frac{\partial Q_D(v_{GS}, v_{DS})}{\partial v_{DS}}.$$

(e) Verify that the gate charge satisfies

$$Q_G(v_{GS}, v_{DS}) = C_{gg}(v_{GS}, v_{DS}) \times (v_{GS} - V_T) + C_{gd}(v_{GS}, v_{DS}) \times v_{DS}.$$

(f) Demonstrate that the relation above originates from the fact that the charge is a linear homogeneous function as it satisfies

$$Q_G(\lambda(v_{GS} - V_T), \lambda v_{DS}) = \lambda \times Q_G(v_{GS} - V_T, v_{DS}).$$

11.4 Consider the *RC* integrator circuit shown in Figure 11.19. The resistance R is implemented using the N-MOSFETs M1 and M2. The ideal opamp 2 and the FETs M3 and M4 act simply as a current mirror which sets the current I_3 equal to I_4 because M3 and M4 have the same source, drain and gate voltages. The capacitor current is then $I_C = I_1 - I_2$.

(a) Verify that we have $I_C = G_C V_{IN}$ and calculate G_C. Use the three-terminal MOSFET model given by Equation (10.9). M1 and M2 are the same FETs but have two different gate voltages.

(b) This configuration permits one to obtain a very high $R_C C$ integration time constant with a transistor of reasonable size $G_C \ll |y_{dd}|$. How does one select V_{g1} and V_{g2} to obtain very large $R_C = 1/G_C$?

(c) Typically a large resistance is desired and a gate length of 100 μm is used for M1 and M2. Calculate $R_C = 1/G_C$. Assume that the gate voltages V_{g1} and V_{g2} are 4 and 5 V; the threshold voltage is 1 V; the gate width is $W_g = 10$ μm; the mobility is 0.0265 $m^2/(V\ s)$; the oxide thickness is 400 Å; the oxide dielectric constant is 3.5×10^{-11} SI; and the source and drain resistances are negligible.

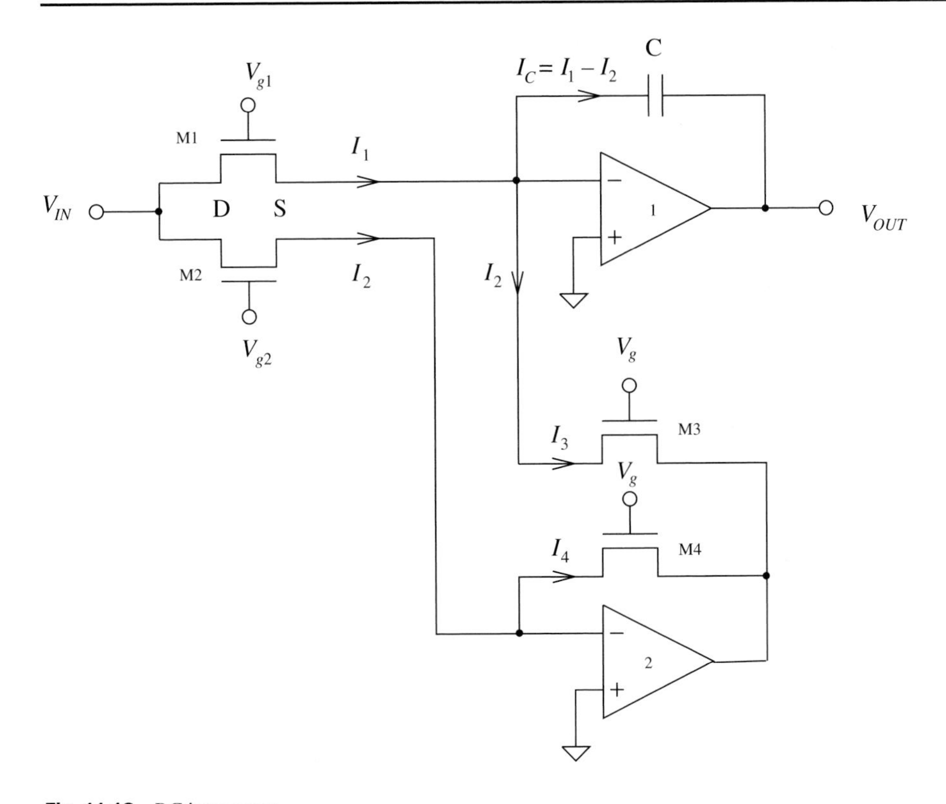

Fig. 11.19. *RC* integrator.

(d) For a 100 μm gate-length distributed effects might become important at low frequencies. Calculate for each MOSFET, the frequencies f_0 given by Equation (11.33) at which distributed effects dominate (see Section 11.2.4).

(e) Calculate the small-signal admittance y_c given by $i_c = y_c(\omega)v_{in}$, where i_c is the current through the capacitor. Express y_c in terms of the y_{ij} parameters of Section 11.2.5. Hint: The gate and source voltages are an AC ground. Note also that $i_s = i_g + i_d$ (see Figure 11.1(b)).

(f) Verify that the same results are obtained for parts (e) and (a) for DC.

(g) Plot $|G_C/y_c|$ versus frequency from 0 to 100 MHz. Assume that the DC part of the input voltage V_{IN} is 0. Note that the g_m of M1 and M2 is 0 when $V_{IN} = 0$. Describe your result.

12 Small- and large-signal AC models for the short-channel MODFET

Time is but the stream I go fishing in.

Walden, Beacon Press 1854, HENRY DAVID THOREAU

12.1 Introduction

The small-signal MOSFET wave-equation introduced in the previous chapter for the intrinsic MOSFET holds only for the region of the channel for which the gradual-channel approximation (GCA) holds. In saturation, it becomes necessary to account for the contribution of the built-in potential in the drain region. We have seen in Chapter 10 that in short-channel MODFETs (modulation doped field-effect transistors), velocity saturation was also taking place in the drain region in saturation. A new wave-equation accounting for space-charge-limited transport therefore needs to be solved in this region. A more complex equivalent circuit will then result for the short-channel MODFET in which the equivalent circuit introduced in Chapter 11 for the MOSFET wave-equation will just be a subcircuit.

Once we have derived the velocity-saturated MODFET wave-equation we will use it in this chapter to develop both a small- and a large-signal model for the short-channel MODFET. The long- and short-channel model topologies will then be compared and their respective merit established.

12.2 Small-signal model for the short-channel MOSFET

12.2.1 The velocity-saturated MOSFET wave-equation

In the short-channel MODFET model introduced in Chapter 12, the FET channel is divided into the GCA and saturation regions of length $X_S = L_g - \ell$ and ℓ, respectively. In the saturation region, the electron velocity is assumed to saturate (to a value v_S)

while the GCA is failing. The channel potential in the saturation region is then assumed to be supported uniquely by the electron distribution in the channel.

In the GCA region, the GCA holds and the 2DEG (two-dimensional electron gas) concentration n_S is controlled by the gate-to-channel potential:

$$qn_S(x,t) = C_G\left[v_{GS}(t) - v_{CS}(x,t) - V_T\right], \tag{12.1}$$

where C_G is the 2DEG gate capacitance. In the saturation region, two-dimensional field effects dominate and the channel potential v_{CS} can be approximately obtained by solving the Poisson equation along the 2DEG channel only (see Chapter 10):

$$\frac{d^2 v_{CS}(x,t)}{dx^2} = \frac{qn_S(x,t)}{d_S\epsilon_1} = \alpha I(x,t), \tag{12.2}$$

where $\alpha = 1/(\epsilon_1 v_S W_g d_S)$. Following the DC model (Chapter 10), we assume that the boundary between the GCA and saturation regions occurs when the channel field $-dv_{CS}/dx$ reaches the critical field $-F_c$.

Let us now establish the wave-equation for this short-channel model. The wave-equation for the GCA region was derived in the previous chapter. The gate length L_g needs to be replaced by X_S and k by $k_s = V_{CS}(X_S)/(V_{GS} - V_T)$, where $V_{CS}(X_S)$ is the DC channel-to-source potential across the entire GCA region.

The channel current in the saturation region can be expressed as

$$I(x,t) = I_D + i(x)\exp(j\omega t) = qW_g n_S(x,t)v_S \tag{12.3}$$

and the continuity equation in the channel as

$$\frac{\partial I(x,t)}{\partial x} = qW_g\frac{\partial v_S n_S(x,t)}{\partial x} = -qW_g\frac{\partial n_S(x,t)}{\partial t} = -\frac{1}{v_S}\frac{\partial I(x,t)}{\partial t}. \tag{12.4}$$

Extracting the AC part from Equation (12.4) and retaining the first order terms yields

$$\frac{di(x)}{dx} = -j\frac{\omega}{v_S}i(x). \tag{12.5}$$

In the saturation region the AC current is related to the AC voltage by the Poisson Equation (12.2). Decomposing Equation (12.2) into DC and AC parts yields the following relationship between the AC voltage $v_{gc}(x)$ and current i:

$$\frac{d^2 v_{gc}(x)}{dx^2} = -\alpha i(x). \tag{12.6}$$

Equations (12.5) and (12.6) make up the wave-equation for the saturation region.

The solution of the wave-equation across the entire channel requires a set of boundary conditions to be enforced at $x = 0$ and $x = L_g$ and at the boundary between the GCA and saturation region. Like for the unsaturated case, the boundary conditions used at $x = 0$ and $x = L_g$ for the common source configuration are

$$v_{gc}(0) = v_{gs}, \tag{12.7}$$

$$v_{gc}(L_g) = v_{gs} - v_{ds}. \tag{12.8}$$

At the GCA/saturation boundary we shall enforce the continuity of the 2DEG carrier concentration, the channel electric field and channel potential, the electron velocity and the current. These are naturally enforced by the continuity of the AC voltage v_{gc} and AC current i at the boundary.

Note that according to our saturation picture, the channel electric field at the floating boundary between the GCA and the saturation region is the DC (constant) critical field F_c. The AC channel field is therefore null at the boundary, and the GCA/saturation boundary must move when AC voltages are applied at the device's terminals so as to maintain a zero AC channel field. In the small-signal analysis, the total (DC + AC) position of the GCA/saturation boundary is written

$$x_S(t) = X_S + x_s \exp(j\omega t), \tag{12.9}$$

where X_S is the DC position and x_s the AC motion of the boundary. Let us now derive the relationship between the AC motion x_s of the GCA/saturation boundary and the GCA AC field v'_{gc}. The total (DC + AC) channel field at the floating boundary x_S is the spatial derivative of the total potential v_{gc} at this boundary:

$$v'_{GC}(x_S) = V'_{GC}(x_S) + v'_{gc}(x_S)\exp(j\omega t). \tag{12.10}$$

Then neglecting second order terms the AC electric field at the floating boundary is

$$v'_{gc}(x_S) = V''_{GC}(X_S)x_s + v'_{gc}(X_S). \tag{12.11}$$

Setting the AC electric field at the floating boundary to zero yields the boundary motion x_s as a function of v'_{gc}:

$$x_s = -\frac{1}{V''_{GC}(X_S)} v'_{gc}(X_S), \tag{12.12}$$

where one can easily calculate $V''_{GC}(X_S)$ to be given by:

$$V''_{GC}(X_S) = -\frac{k_s^2(1-\frac{1}{2}k_s)^2}{(1-k_s)^3}\frac{V_{GS}-V_T}{X_S}. \tag{12.13}$$

The solution of the wave-equation across the entire intrinsic MODFET relies on the continuity of the AC voltage and AC current at the floating boundary. It is therefore necessary to calculate the AC voltage at the floating boundary and to account for the motion of the GCA/saturation boundary. Let us derive the modified GCA channel potential obtained at the floating boundary. The total (DC + AC) channel potential at the floating boundary is given by

$$v_{GC}(x_S, t) = V_{GC}(x_S) + v_{gc}(x_S)\exp(j\omega t), \tag{12.14}$$

where v_{gc} is the AC potential obtained by solving (see previous chapter):

$$Z^2\frac{\partial^2 v_{gc}(s', Z)}{dZ^2} + Z\frac{dv_{gc}(s', Z)}{dZ} - \left(\frac{1}{4} + s'Z^{3/2}\right)v_{gc}(s', Z) = 0, \tag{12.15}$$

where $s' = s/\omega_{0k}$.

Expanding Equation (12.14) with a Taylor series around the DC boundary position X_S for small variations x_s of the boundary position yields the AC voltage $v_{gc}(x_S)$ at the floating boundary x_S (second order terms are neglected)

$$v_{gc}(x_S) = V'_{GC}(X_S)x_s + v_{gc}(X_S) \tag{12.16}$$

$$= -F_c x_s + v_{gc}(X_S). \tag{12.17}$$

The potential drop across the saturation region is also modified by the motion of the boundary which modulates the width of the saturation region. Integrating the Poisson equation

$$\frac{d^2[v_{GC}(x)]}{dx^2} = -\alpha I(x,t) = -\alpha[I_D + i(x)\exp(j\omega t)] \tag{12.18}$$

across the time-varying saturation region yields the AC potential at $x = L_g$

$$v_{gc}(L_g) = (F_c + \alpha I_D l)\, x_s + v_{gc}(x_S) + \Delta v_{gc}(l), \tag{12.19}$$

where $l = (L_g - X_S)$ is the DC width of the saturation region and $\Delta v_{gc}(l)$ is the AC potential $v_{gc}(x)$ obtained by solving the Poisson Equation (12.6) for a fixed saturation region width l, and zero AC potential, $v_{gc}(X_S) = 0$, and zero AC field, $v'_{gc}(X_S) = 0$, at X_S. Substituting Equation (12.17) into Equation (12.19) gives

$$v_{gc}(L_g) = \alpha I_D l x_s + \Delta v_{gc}(l) + v_{gc}(X_S). \tag{12.20}$$

One observes that the contribution of the motion x_s of the GCA/saturation boundary is to add the AC potential term $\alpha I_D l x_s$.

Finally, note that the AC current at the floating boundary x_S is to first order equal to the AC current at the fixed boundary X_S. This originates in the fact that the DC current I_D is constant along the channel.

12.2.2 Exact solution of the velocity-saturated MOSFET wave-equation

The solution of the wave-equation in the GCA region is the same as in Section 11.2 except that L_g must be replaced by X_S and k by $k_s = V_{CS}(X_S)/(V_{GS} - V_T)$:

$$v_{gc}(Z,s) = C_1 I_{2/3}(y) + C_2 I_{-2/3}(y), \tag{12.21}$$

$$i(Z,s) = G_{dos} s'^{1/2} Z^{1/4}[C_1 I_{-1/3}(y) + C_2 I_{1/3}(y)], \tag{12.22}$$

with C_1 and C_2 two arbitrary constants and

$$G_{dos} = \frac{\mu C_G (V_{GS} - V_T)(2k_s - k_s^2)}{X_s},$$

$$y = \frac{4}{3} s'^{1/2} Z^{3/4},$$

$$Z = 1 - (2k_s - k_s^2)\frac{x}{X_S},$$

$$s' = s/\omega_{0k},$$

$$\omega_{0k} = \frac{\mu(V_{GS} - V_T)(2k_s - k_s^2)^2}{X_S^2}.$$

The wave-equation in the saturation region is readily solved. The current wave is obtained by integrating the continuity Equation (12.5)

$$i(x') = i(X_S)\exp\left(-j\frac{\omega}{v_S}x'\right). \tag{12.23}$$

The voltage wave is obtained by integrating the Poisson Equation (12.6). Using the zero AC field and zero voltage boundary conditions, $\Delta v(\ell)$ the voltage drop across the saturation region is found to be

$$\Delta v_{gc}(\ell) = \left\{\alpha\left(\frac{v_S}{\omega}\right)^2\left[\exp\left(-j\frac{\omega}{v_S}\ell\right) - 1\right] + j\alpha\frac{v_S}{\omega}\ell\right\} i(X_S). \tag{12.24}$$

Like for the unsaturated wave-equation, the boundary condition at $x = 0$ is

$$v_{gc}(0) = A_{21}C_1 + A_{22}C_2 = v_{gs},$$

with the same A_{21} and A_{22} coefficients:

$$A_{21} = I_{2/3}\left(\frac{4}{3}s'^{1/2}\right),$$

$$A_{22} = I_{-2/3}\left(\frac{4}{3}s'^{1/2}\right),$$

and like for the unsaturated wave-equation, the boundary condition at $x = L_g$ is

$$\begin{aligned} v_{gc}(L_g) &= \alpha I_D l x_s + \Delta v_{gc}(l) + v_{gc}(X_S) \\ &= A_{11}C_1 + A_{12}C_2 = v_{gs} - v_{ds}, \end{aligned}$$

where the new coefficients A_{11} and A_{12} are after a few manipulations now found to be

$$\begin{aligned} A_{11} = I_{2/3}(y_s) &+ G_{dos}s'^{1/2}Z_s^{1/4}\left\{\alpha\left(\frac{v_S}{\omega}\right)^2\left[\exp\left(-j\frac{\omega}{v_S}\ell\right) - 1\right] + j\alpha\frac{v_S}{\omega}\ell\right\} \\ &\times I_{-1/3}(y_s) - \alpha I_D\ell\frac{(1-k_s)^3 X_S s'^{1/2} Z_s^{1/4}}{k_s\left(1 - \frac{1}{2}k_s\right)V_{out}}[I_{-1/3}(y_s) + I_{5/3}(y_s)], \end{aligned}$$

$$\begin{aligned} A_{12} = I_{-2/3}(y_s) &+ G_{dos}s'^{1/2}Z^{1/4}\left\{\alpha\left(\frac{v_S}{\omega}\right)^2\left[\exp\left(-j\frac{\omega}{v_S}\ell\right) - 1\right] + j\alpha\frac{v_S}{\omega}\ell\right\} \\ &\times I_{1/3}(y_s) - \alpha I_D\ell\frac{(1-k_s)^3 X_S s'^{1/2} Z_s^{-1/4}}{k_s\left(1 - \frac{1}{2}k_s\right)V_{out}}[I_{1/3}(y_s) + I_{-5/3}(y_s)] \end{aligned}$$

with $Z_s = (1 - k_s)^2$.

The remaining calculation of the y-parameters then proceeds like for the unsaturated MOSFET wave-equation. The resulting y-parameters are:

$$y_{dd} = G_{dos}s'^{1/2}Z_s^{1/4}\exp\left(-j\frac{\omega}{v_S}\ell\right)[C_{1ds}I_{-1/3}(y_s) + C_{2ds}I_{1/3}(y_s)],$$

$$y_{gd} = G_{dos}s'^{1/2}\left\{C_{1ds}I_{-1/3}\left(\frac{4}{3}s'^{1/2}\right) + C_{2ds}I_{1/3}\left(\frac{4}{3}s'^{1/2}\right)\right.$$
$$\left. - \; Z_s^{1/4}\exp\left(-j\frac{\omega}{v_S}\ell\right)[C_{1ds}I_{-1/3}(y_s) + C_{2ds}I_{1/3}(y_s)]\right\},$$

$$y_{dg} = G_{dos}s'^{1/2}Z_s^{1/4}\exp\left(-j\frac{\omega}{v_S}\ell\right)[C_{1gs}I_{-1/3}(y_s) + C_{2gs}I_{1/3}(y_s)],$$

$$y_{gg} = G_{dos}s'^{1/2}\left\{C_{1gs}I_{-1/3}\left(\frac{4}{3}s'^{1/2}\right) + C_{2gs}I_{1/3}\left(\frac{4}{3}s'^{1/2}\right)\right.$$
$$\left. - \; Z_s^{1/4}\exp\left(-j\frac{\omega}{v_S}\ell\right)\left[C_{1gs}I_{-1/3}(y_s) + C_{2gs}I_{1/3}(y_s)\right]\right\}.$$

These y-parameters are of the same form as for the unsaturated y-parameters except for the introduction of the multiplicative term $\exp[-j(\omega/v_S)\ell]$ which accounts for the phase shifting of the channel current in the saturation region of length ℓ (see Equation (12.23)). The coefficients C_{1gs}, C_{2gs}, C_{1ds} and C_{2ds} have the same definitions as previously:

$$C_{1gs} = \frac{(A_{22} - A_{12})}{\Delta},$$
$$C_{2gs} = \frac{(A_{11} - A_{21})}{\Delta},$$
$$C_{1ds} = -\frac{A_{22}}{\Delta},$$
$$C_{2ds} = -\frac{A_{21}}{\Delta},$$
$$\Delta = A_{11}A_{22} - A_{12}A_{21}.$$

12.2.3 Equivalent circuit of the velocity-saturated MOSFET wave-equation

Like the exact solution of the unsaturated MOSFET wave-equation the exact solution of the velocity-saturated MOSFET wave-equation is impenetrable. Hopefully, an exact equivalent circuit can be readily developed. This exact equivalent circuit is obtained by rewriting the resulting y-parameters $y_{ij}(sat)$ in terms of the y-parameters of the GCA region $y_{ij}(GCA)$ of reduced gate length $X_S = L_g - \ell$. The procedure is left as an exercise (see Problem 12.1 or [5]). The following expressions are obtained

$$y_{gg}(sat) = y_{gg}(GCA) + y_{gd}(GCA)\delta_s \tag{12.25}$$
$$+ \frac{y_{dg}(GCA) + y_{dd}(GCA)\delta_s}{1 + y_{dd}(GCA)Z_S(\omega)}[1 - \exp(-j\omega\tau_{SAT}) - Z_S(\omega)y_{gd}(GCA)],$$

$$y_{gd}(sat) = y_{gd}(GCA)\gamma_s + \frac{y_{dd}(GCA)\gamma_s}{1 + y_{dd}(GCA)Z_S(\omega)} \times [1 - \exp(-j\omega\tau_{SAT}) - Z_S(\omega)y_{gd}(GCA)],$$

$$y_{dg}(sat) = \frac{y_{dg}(GCA) + y_{dd}(GCA)\gamma_s}{1 + y_{dd}(GCA)Z_S(\omega)} \exp(-j\omega\tau_{SAT}),$$

$$y_{dd}(sat) = \frac{y_{dd}(GCA)\delta_s}{1 + y_{dd}(GCA)Z_S(\omega)} \exp(-j\omega\tau_{SAT}),$$

where $\tau_{SAT} = v_S/\ell$ is the transit time in the saturation region, $Z_S(\omega)$ a generalized impedance

$$Z_S(\omega) = \gamma_s \left\{ I_D \ell B\alpha - \left(\frac{v_S}{\omega}\right)^2 \left[\exp\left(-j\frac{\omega}{v_S}\ell\right) - 1\right] - j\alpha\frac{v_S}{\omega}\ell \right\}$$

and δ_s and γ_s two constants given by

$$\gamma_s = 1 - \delta_s = \frac{1}{1 + \alpha I_D \ell A}$$

with

$$A = \frac{2X_S(1 - k_s)}{(2k_s - k_s^2)(V_{GS} - V_T)} = \frac{1}{F_c},$$

$$B = \frac{4X_S(1 - k_s)^2}{G_{dos}(2k_s - k_s^2)^2(V_{GS} - V_T)} = \frac{1}{\beta F_c^2}.$$

These y-parameters can be represented by the equivalent circuit given in Figure 12.1, where the impedance $Z_S(\omega)$ is approximated by a first order RC network providing the correct second order frequency power series expansion:

$$Z_S(\omega) = R_1 + \frac{R_2}{1 + j\omega C_s R_2}, \tag{12.26}$$

with

$$R_{s1} = \frac{\alpha I_{DC} \ell B - \frac{1}{6}\alpha\ell^2}{1 + \alpha I_D \ell A},$$

$$R_{s2} = \frac{2\alpha\ell^2}{3(1 + \alpha I_D \ell A)},$$

$$C_s = \frac{3}{8}\tau_{SAT}\frac{(1 + \alpha I_D \ell A)}{\alpha\ell^2},$$

using $\alpha = 1/(\epsilon_1 v_S W_g d_S)$ [2].

The resulting equivalent circuit provides an optimal first order non-quasi-static equivalent circuit admitting the correct second order frequency power expansion as well as a graceful degradation. This is demonstrated in Figure 12.2 for an intrinsic

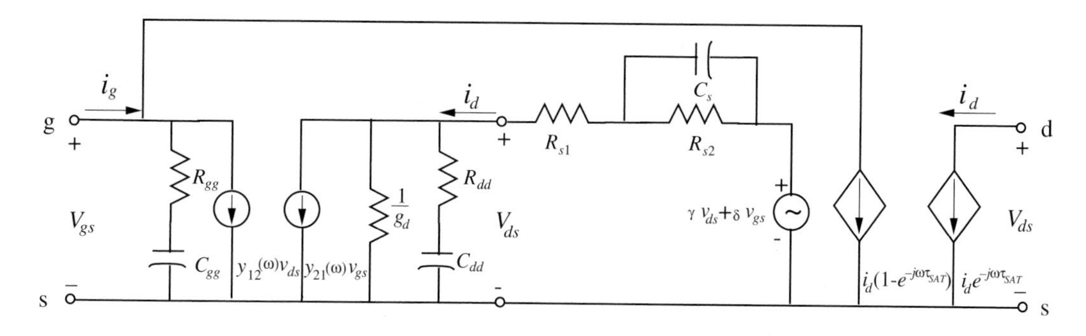

$$y_{12}(\omega) = \frac{j\omega C_{gd}}{1+j\omega R_{gd} C_{gd}}$$

$$y_{21}(\omega) = g_m + \frac{j\omega C_{dg}}{1+j\omega R_{dg} C_{dg}}$$

Fig. 12.1. First order non-quasi-static equivalent circuit II for the short-channel MODFET.

Table 12.1. *Device parameters for the intrinsic short-channel MODFET.*

Parameters		Value
L_g	Gate length (μm)	1
W_g	Gate width (μm)	290
μ	Mobility (cm^2/V s)	4400
v_S	Saturation velocity (m/s)	3.45×10^5
V_T	Threshold voltage (V)	−0.3
d	Gate to channel spacing (Å)	430
d_S	Channel width in saturation (Å)	1500
ϵ_1	Channel dielectric constant	$13.1\epsilon_0$
ϵ_2	Gate dielectric constant	$12.2\epsilon_0$

MODFET with the parameters given in Table 12.1 and for an intrinsic bias of $V_{DS} =$ 3 V and $V_{GS} = 0$ V. The phase and amplitude of y_{dg} versus frequency calculated using this first order *RC* equivalent circuit (dashed-dotted line, EQUI), the exact solution (full line, EXACT), and the frequency power series approximation (dashed line, POWER) are compared in Figure 12.2(a) and (b). The optimal first order *RC* model (EQUI) is seen to hold to a much higher frequency than the frequency power series approximation (POWER).

The equivalent circuit described above gives a natural interpretation of the impact of saturation upon the high-frequency response of the short-channel MODFET. The GCA region with its *RC* topology is easily recognized. The phase shift τ_{SAT} in the saturation region is implemented by a current source acting as a time delay element.

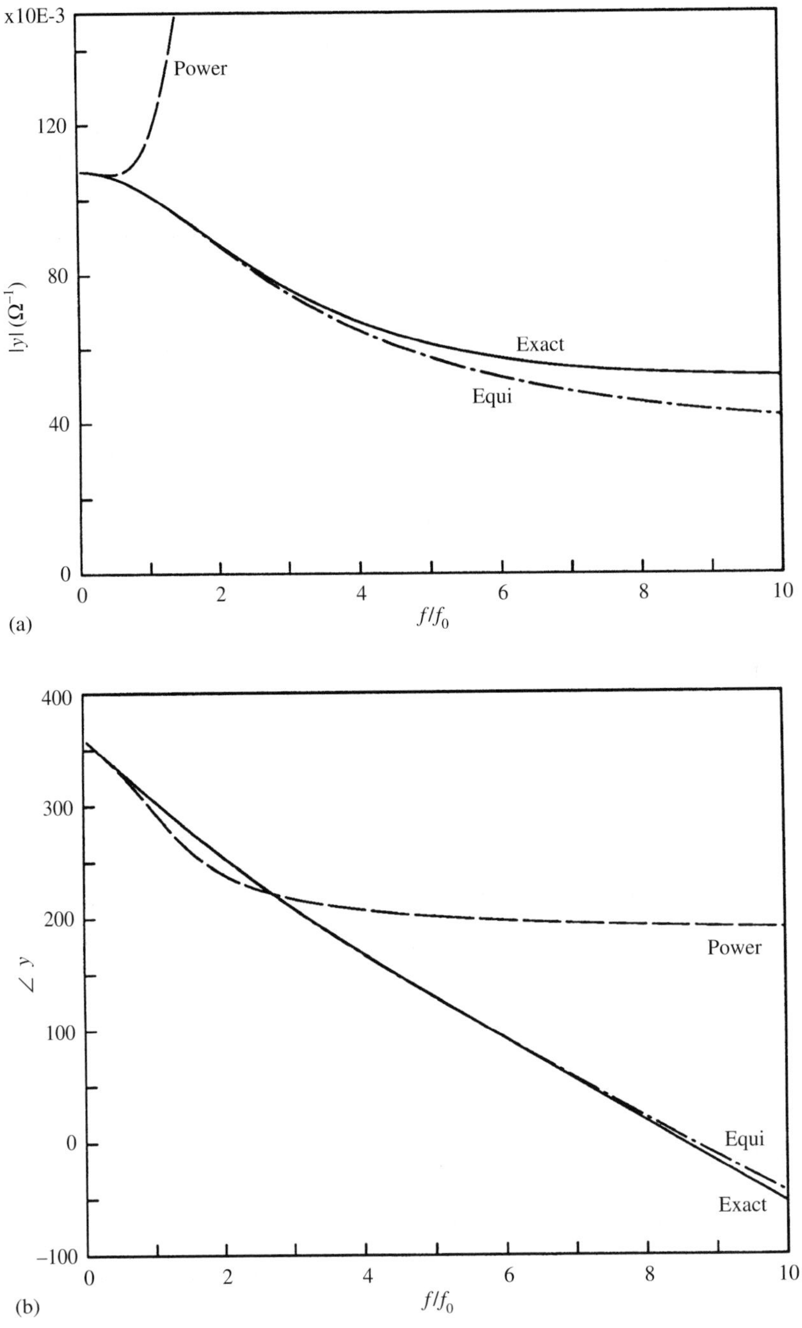

Fig. 12.2. Comparison of (a) the amplitude and (b) the phase of y_{dg} for $V_{DS} = 3$ V and $V_{GS} = 0$ V, obtained with the *RC* equivalent circuit (dashed-dotted line, EQUI), the exact solution (full line, EXACT), and the frequency power series (dashed line, POWER). (P. Roblin, S. C. Kang and W. R. Liou, *IEEE Transactions on Electron Devices*, Vol. 38, No. 8, pp. 1706–1718, August 1991. ©1991 IEEE.)

This phase shift also induces an additional gate current since the gate images the AC channel charge in the saturation region (space-charge neutrality). The impedance $Z_s(\omega)$ accounts for the voltage drop in the saturation region (real part $R_{s1} + R_{s2}$) and charge modulation in the saturation region (imaginary part). Finally, the factors γ_s and δ_s account for the effect of channel-length modulation. It is interesting to note that the complicated solution of the velocity-saturated MOSFET wave-equation admits such a simple (and exact if the exact $y_{ij}(GCA)$ and $Z_s(\omega)$ are used) representation in terms of an equivalent circuit.

12.2.4 High-frequency performance of the short-channel MODFET

The principal features of this non-quasi-static small-signal AC model for the short-channel MOSFET/MODFET are to account for velocity-saturation and channel-length modulation which greatly modify the performance of short-channel MODFETs. This is verified in Figure 12.3(a) where the unity current-gain cut-off frequency f_T plotted versus gate length L_g is seen to vary as $1/L_g^2$ for large gate lengths and as $1/L_g$ for small gate lengths assuming a fixed effective saturation velocity. This result was predicted using a simpler approach in Section 11.1.

In Figure 12.3(b) the unity current-gain cut-off frequency f_T is plotted versus gate length $\alpha = (V_{GS} - V_T)/F_c L_g$. Note that the transition from long to short channel is taking place for $\alpha = 1$ as was predicted in Chapter 12. For the saturation velocity and mobility of Table 12.1 and $V_{DS} = 1$ V the corner point $\alpha = 1$ corresponds to a gate length of 1 and 1.66 μm for $V_{GS} = 0$ and 0.2 V, respectively.

If the effective saturation velocity v_S were to increase with decreasing gate length L_g one would have in the submicron regime a $1/L_g^{\gamma}$ law with $1 \le \gamma \le 2$. However, Rohdin [6] has demonstrated that despite the expected occurence of velocity overshoot the effective saturation velocity is essentially independent of gate length for MODFETs with gate lengths varying from 0.9 to 0.3 μm. His analysis is based on the systematic reverse modeling of a large number of FETs on different wafers. In Table 12.1 a constant saturation velocity of 1.85×10^5 m/s is used.

The new AC-model is seen to predict in Figure 12.4 that for small gate-length MODFETs the decrease of the *extrinsic* unilateral gain with frequency switches from 20 dB per decade to 40 dB per decade for frequencies larger than the extrinsic f_{MAX} [7] when the parasitics are accounted for.

High-frequency analysis [8] of the wave-equation reveals the existence of unilateral power gain resonances in the *intrinsic* MODFET at the frequencies approximately given by $f_n \simeq (n + \frac{1}{2})/2\tau_s$ with τ_s the bias-dependent transit time through the saturation region. Steady-state power gain is therefore conceptually possible in the intrinsic MODFET at frequencies above its (first) maximum frequency of oscillation $f_{MAX}(int)$. The same frequency analysis for the *extrinsic* MODFET predicts, however, that realistic lossy parasitics will suppress these unilateral power gain resonances.

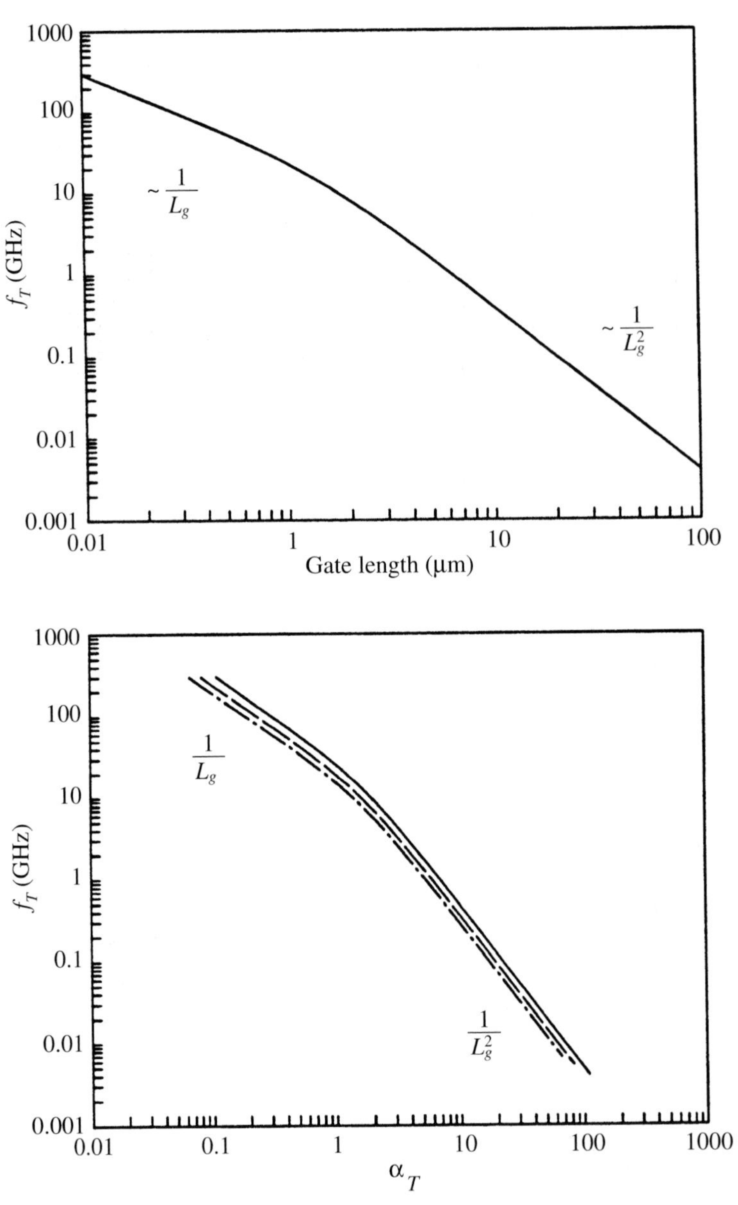

Fig. 12.3. Unity current-gain cut-off frequency plotted versus: (a) L_g and (b) $\alpha_T = (V_{GS} - V_T)/F_c L_g$ for an intrinsic MODFET with $V_{GS} = 0$ V (dashed-dotted line), 0.1 V (dashed line) and 0.2 V (full line) and $V_{DS} = 1$ V. (P. Roblin and S. C. Kang, *IEEE Transactions on Electron Devices*, Vol. 39, No. 6, pp. 1490–1495, June 1992. ©1992 IEEE.)

The parasitics (fringe capacitors and source, drain and gate resistors) which cap the intrinsic device are seen to have a very important effect on the performance of the device. Parasitics are further discussed at the end of this chapter and in Chapters 13–17.

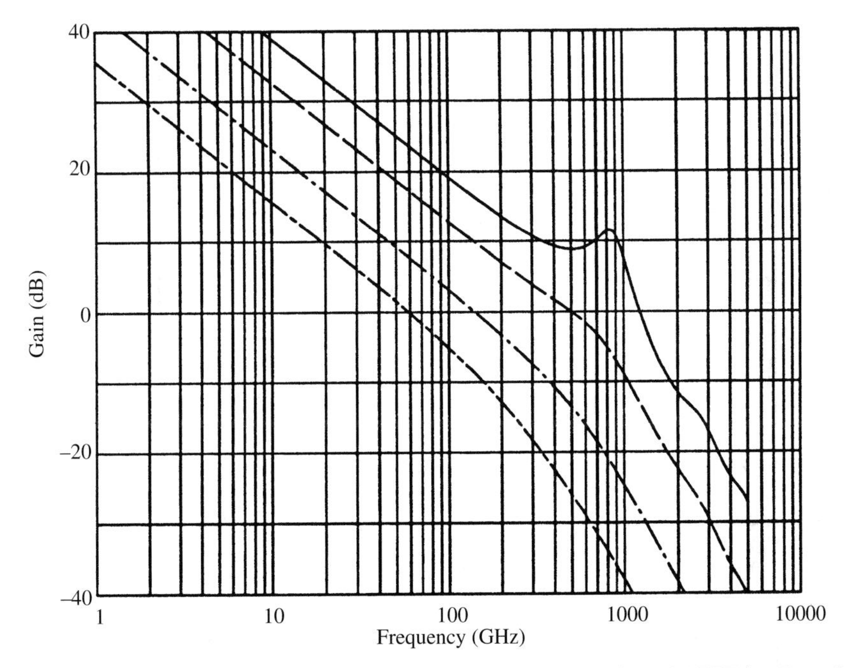

Fig. 12.4. Unilateral power gain versus frequency for the extrinsic MODFET for $R_d = 0.01$ (full line), 0.1 (dashed dotted line), 1 (dashed-dashed line) and 5 (dashed-line) Ω. (P. Roblin and S. C. Kang, *IEEE Transactions on Electron Devices*, Vol. 39, No. 6, pp. 1490–1495, June 1992. ©1992 IEEE.)

12.2.5 Alternate equivalent circuit for the short-channel MODFET

Let us compare the graceful degradation of the novel non-quasi-static equivalent circuit II shown in Figure 12.1 which was developed for the saturated short-channel MODFET, with that of the conventional non-quasi-static equivalent shown in Figure 11.9. For this purpose, we compare in Figure 12.5 the amplitude of y_{dg} as calculated using the exact wave-equation solution (full line), the non-quasi-static equivalent-circuit Model A of Figure 11.9 with y_3 given by Equation (12.28) (dotted-dashed line) and the novel non-quasi-static equivalent circuit II of Figure 12.1 (dashed line).

Recall that the y-parameters are given by

$$y_{gg} = y_1 + y_2,$$

$$y_{gd} = -y_2,$$

$$y_{dg} = y_3 - y_2,$$

$$y_{dd} = y_4 + y_2.$$

In the long-channel Model B, y_1, y_2, y_3 and y_4 are given by

$$y_i = g_i + \frac{j\omega C_i}{1 + j\omega R_i C_i}. \tag{12.27}$$

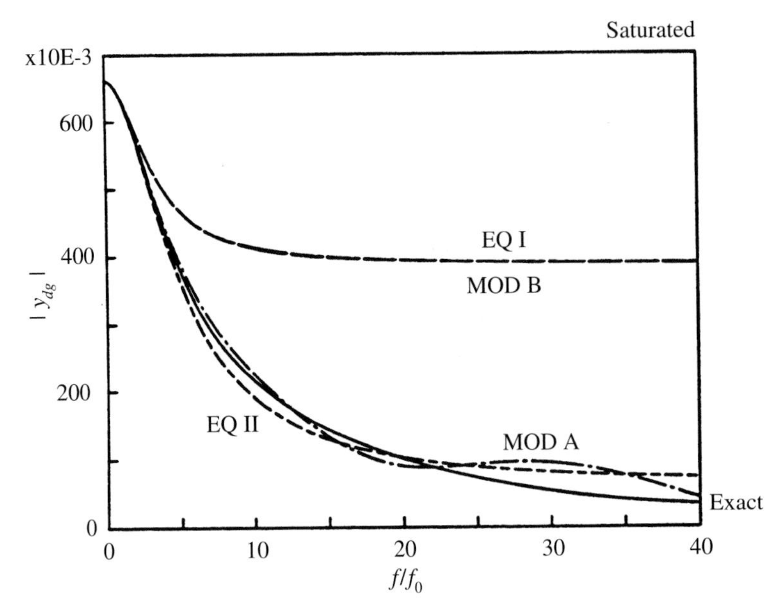

Fig. 12.5. Comparison of $|y_{dg}|$ obtained from the conventional (A and B) and novel equivalent circuits (I and II) for the ideal short-channel intrinsic MODFET biased in saturation. (P. Roblin, Proceedings of the IEEE Catalog No. 93THO333-3, Vol. II, pp. 359–372, 1993. ©1993 IEEE.)

Alternatively in the short-channel Model A, y_3 is selected to be

$$y_3 = \frac{g_m e^{j\omega\tau_{SAT}}}{1 + j\omega\tau_3}. \tag{12.28}$$

Also shown are the long-channel equivalent circuit I of Figure 11.6 and Model B of Figure 11.9 with y_3 given by Equation (12.27) which do not provide a good degradation for the short-channel MODFET since they do not account for the drain delay τ_{SAT}.

Apparently, both equivalent-circuit Model A and equivalent-circuit II fit perfectly the exact MODFET wave-equation solution at low frequencies as intended and exhibit a graceful degradation at high frequencies. However, the equivalent circuit of Model A is sometimes forced to rely on negative elements for its output capacitors. This problem cannot be suppressed by using an inductor shunted by a resistor in series with g_d. We shall therefore rely in the next sections on the more physical non-quasi-static equivalent circuit II shown in Figure 12.1 to develop a charge-conserving large-signal model for short-channel MODFETs.

12.3 Large-signal model for the short-channel MOSFET

This section is concerned with the development of a large-signal model for the velocity-saturated MOSFET wave-equation.

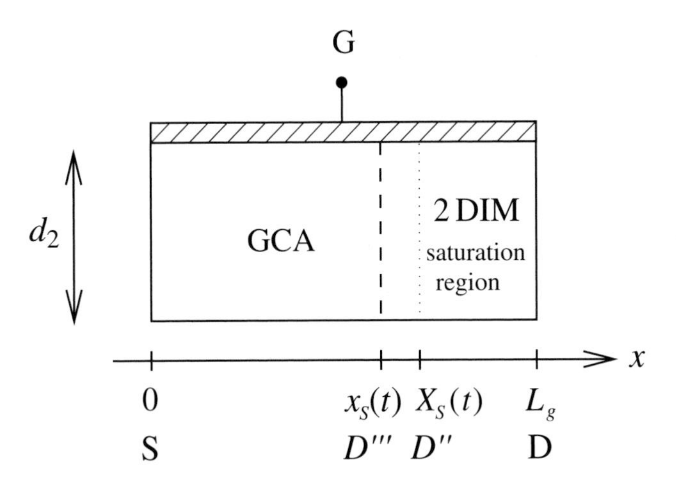

Fig. 12.6. Location of the DC saturation boundary.

We will start by developing a first order non-quasi-static model for the velocity-saturated MOSFET wave-equation. We then introduce the large-signal model before verifying that under small-signal excitation it reduces to the small-signal model. A charge base simplified version of the model will be presented next.

12.3.1 First-order non-quasi-static approximation

The velocity-saturated MOSFET model divides the FET into two region: the GCA region and the velocity saturation region as shown in Figure 12.6. The DC (static) position of the boundary between the GCA region and the saturation region located at X_S is referred as D'. In the presence of AC signals the GCA/saturation boundary varies with time. The instantaneous position of this boundary located at $x_s(t)$ is referred as D'''. In the presence of AC excitation, one can still introduce the DC position D' if we define it as as the time-average value of $x_s(t)$. In the presence of non-periodic transient excitation D' is not physically defined. For the development of the large-signal model we will therefore need to introduce a third quantity, the position D'' which is the low-frequency instantaneous value of the position D''' when displacement currents are negligible. D'' is therefore positioned at $x_s(\omega \simeq 0, t) = X_S(t)$. In small-signal analysis the position D' is constant with time whereas the positions D'' and D''' are time-dependent.

In the mathematical expressions given in the remaining part of this chapter it is necessary to specify which internal node, D', D'' or D''', is used for the voltage dependence when switching from the small-signal to the large-signal model. The following symbolic notation is then adopted to reduce the weight of the mathematical expressions:

$$\left.\begin{aligned}(D') &= (V_{GS}, V_{D'S}),\\ (D'') &= (v_{GS}, v_{D''S}),\\ (D''') &= (v_{GS}, v_{D'''S}).\end{aligned}\right\} \tag{12.29}$$

Note that the position of the GCA/saturation boundary $X_S(D'')$ is given in the velocity-saturated MOSFET model by

$$X_S(D'') = \frac{(v_{GS} - V_T)}{2F_c} \frac{k(2-k)}{1-k}, \tag{12.30}$$

where k is given by

$$k = \frac{v_{D''S}}{v_{GS} - V_T}.$$

The exact solution of the velocity-saturated MOSFET wave-equation leads to the following equations for the small-signal currents:

$$i_g = i_{g,GCA} + i_{g,SAT}$$
$$i_d = i_{d'} \exp[-j\omega\tau_{SAT}(D')],$$

with

$$i_{g,SAT} = i_{d'}\{1 - \exp[-j\omega\tau_{SAT}(D')]\},$$
$$i_{g,GCA} = y^0_{gg}(D')\, v_{gs} + y^0_{gd}(D')\, v_{d's},$$
$$i_{d'} = y^0_{dg}(D')\, v_{gs} + y^0_{dd}(D')\, v_{d's},$$

where $y^0_{ij}(D')$ are the long-channel y-parameters of the GCA region with gate length $X_S(D')$. The AC voltage $v_{d's}$ at the internal drain node D' is related to the external drain voltage v_{ds} by (see Equation 12.20)

$$v_{ds} = v_{d's} - x_s(\alpha I_{DC}\ell) - \Delta v_{gc}, \tag{12.31}$$

where Δv_{gc} is given by (see Equation 12.24)

$$\Delta v_{gc} = \left\{\alpha\left(\frac{v_S}{\omega}\right)^2 [\exp(-j\omega\tau_{SAT}) - 1] + j\alpha\left(\frac{v_S}{\omega}\right)\ell\right\} i_{d'}. \tag{12.32}$$

Note that in Equations (12.31) and (12.32) $\ell = L_g - X_S$ is the width in the saturation region, $\tau_{SAT} = \ell/v_S$ is the traversal time of the saturation region and $\alpha = 1/(\epsilon_1 v_S W_g d_S)$ is a constant specific to the saturation model used [2]. Finally, x_s is the AC motion of the GCA/saturation region boundary which is given by (see [5])

$$x_s = -Bi'_d + A(v_{gs} - v_{d's}).$$

Unnecessarily complicated expressions for A and B were reported in [1]. When X_S (see Equation (12.30)) is substituted in them, one simply obtains $A = 1/F_c$ and $B = 1/(\beta F_c^2)$ so that we have

$$x_s = -\frac{1}{\beta F_c^2} i'_d + \frac{1}{F_c}(v_{gs} - v_{d's}). \tag{12.33}$$

These formulas permit us to rederive the exact solution of the velocity-saturated MOSFET wave-equation if the exact $y_{ij}^0(D')$ parameters of the GCA region (expressed in terms of Bessel functions) are used.

For the development of the large-signal model we shall replace the $y_{ij}^0(D')$ parameters of the GCA region by their first order optimal non-quasi-static RC approximation:

$$y_{ij}^0(D') = g_{ij}^0(D') + \frac{j\omega C_{ij}(D')}{1 + j\omega\tau_{ij}(D')},$$

with $\tau_{ij}(D') = R_{ij}(D')C_{ij}(D')$.

The expressions for g_{ij}^0, C_{ij}, R_{ij} and τ_{ij} are those given in Section 11.2.5 with the gate length L_g replaced by X_S (see Equation (12.30)). For example we have

$$g_m^0 = k\beta\frac{V_{GS} - V_T}{X_S} = \beta F_c \frac{2(1-k)}{2-k},$$

$$g_d^0 = (1-k)\beta\frac{V_{GS} - V_T}{X_S} = \beta F_c \frac{2(1-k)^2}{(2-k)k}.$$

Since a first order non-quasi-static approximation is used for the $y_{ij}^0(D')$ parameters, a first order non-quasi-static approximation should also be used for $-\Delta v_{gc}$. To obtain this we expand $-\Delta v_{gc}$ in a frequency power series

$$-\Delta v_{gc} = i_{d'}\left[\frac{1}{2}\alpha\ell^2 - j\omega\left(\frac{1}{6}\alpha\ell^2\tau_{SAT}\right) - \omega^2\left(\frac{\alpha\tau_{SAT}^2\ell^2}{24}\right)\right].$$

We shall use a model consisting of a resistor $R_{SAT,1}$ in series with a negative inductor $-L_{SAT}$ shunted by a negative resistor $-R_{SAT,2}$ to represent Δv_{SAT}:

$$-\Delta v_{gc} = R_{SAT,1}\, i_{d'} - \frac{j\omega L_{SAT}}{1 + j\omega\dfrac{L_{SAT}}{R_{SAT,2}}} i_{d'} = Z_{SAT}(\omega) i_{d'}.$$

The values of $R_{SAT,1}$, L_{SAT} and $R_{SAT,2}$ which generate the required frequency power series expansion of Δv_{SAT} are then:

$$R_{SAT,1}(D') = \frac{1}{2}\alpha\ell^2,$$

$$L_{SAT}(D') = \frac{1}{6}\alpha\,\ell^2\tau_{SAT},$$

$$\frac{L_{SAT}(D')}{R_{SAT,2}(D')} = \frac{1}{4}\tau_{SAT}.$$

Note that the fact that we selected an effective inductor $(-L_{SAT})$ which is negative is not a problem. An equivalent circuit $R_{SAT,1}||(R'_{SAT,2} + 1/(j\omega C_{SAT}))$ could be used with a series resistance $R'_{SAT,2}$ and a capacitor C_{SAT} which are positive. The motivation for the use of a negative inductor is the simplicity of the relaxation time constant $\tau_{SAT}/4$ and the inductance L_{SAT}. Note, however, that the time delay $L_{SAT}/R_{SAT,2}$ is positive so that the model $-R_{SAT,2}||(-L_{SAT})$ is indeed deterministic.

Similarly we need to develop a first order non-quasi-static model for the saturation part of the gate current:

$$\begin{aligned} i_{g,SAT} &\simeq i_{d'}\,[1 - \exp(-j\omega\tau_{SAT})] \\ &\simeq i_{d'}\, j\omega\tau_{SAT}\,(1 - 1/2\, j\omega\tau_{SAT}) \\ &= \frac{j\omega\tau_{SAT}}{1 + j\omega\frac{1}{2}\,\tau_{SAT}}\, i_{d'}. \end{aligned} \tag{12.34}$$

Thus i_d and $i_{d'}$ are related by

$$i_d = i_{d'} - i_{g,SAT} = i_{d'}\left[\frac{1 - j\omega\frac{1}{2}\tau_{SAT}}{1 + j\omega\frac{1}{2}\tau_{SAT}}\right],$$

where $v_{d'''s}$ is defined as

$$v_{d'''s} = F_c x_s + v_{d's}.$$

These first order non-quasi-static expressions will be used later to develop a new equivalent circuit.

12.3.2 Small-signal equivalent circuit for the D'' internal node

The derivation of the large-signal model from the small-signal one presented above is not immediate. The reason is that the small-signal wave-equation relies on the AC drain voltage measured at the point D' which is the time-average DC (constant) position of the GCA/saturation boundary. As stated above the point D' is not defined in transient large-signal analysis, and the large-signal model must rely instead on the GCA drain voltage at location D'' which is the instantaneous DC (and therefore time-dependent) position of the GCA/saturation boundary given by $X_S(t)$. A change from D' to D'' is required and is performed below.

First we shall evaluate the AC position x_s of the boundary D''' which in turn will permit us to change the reference position from D' to D''. Note that for small-signal excitation, the AC boundary position x_s is related to the instantaneous and DC boundary position $x_S(t)$ by

$$x_S(t) = X_S(D') + x_s\,\exp(j\omega t).$$

An expression for x_s can be derived by comparing the drain currents at D' and D'''. The drain current at the position D' is

$$i_{d'} = y^0_{dg}(D')\, v_{g's} + y^0_{dd}(D')\, v_{d's}. \tag{12.35}$$

The drain current at the position D''' is

$$\begin{aligned} i_{d'''} &= W_g C_g\, \mu F_c(v_{gs} - v_{d'''s}) \\ &= \beta F_c(v_{gs} - v_{d'''s}) \end{aligned} \tag{12.36}$$

with $\beta = W_g C_G \mu$. Equation (12.36) holds simply because by definition the electron velocity at D''' is $v_S = \mu F_c$ and the channel charge at D''' is still controlled by the potential $v_{gd'''}$ (GCA approximation).

Now we know that the potentials $v_{d'''s}$ and $v_{d's}$ are related by

$$v_{d'''s} = F_c x_s + v_{d's}. \tag{12.37}$$

Because the DC current I_{DC} is constant along the channel, the currents $i_{d'}$, $i_{d'''}$ (and also $i_{d''}$) are equal (to first order) despite the AC motion of the boundary. Setting Equations (12.35) and (12.36) equal and using Equation (12.37) leads to

$$-\beta\, F_c^2\, x_s = (y^0_{dg}(D') - \beta F_c)\, v_{gs} + (y^0_{dd}(D') + \beta F_c)\, v_{d's}.$$

Separating x_s into its low-frequency (DC) part $x_{s,dc}$ and its frequency-dependent (AC) part $x_{s,ac}$,

$$x_s = x_{s,dc} + x_{s,ac}(\omega),$$

we have

$$x_{s,dc} = \left[g^0_m(D') - \beta F_c\right] v_{gs} + \left[g^0_d(D') + \beta F_c\right] v_{d's}, \tag{12.38}$$

$$-\beta F_c^2\, x_{s,ac} = \left[\frac{j\omega C_{dg}(D')}{1 + j\omega\tau_{dg}(D')}\right] v_{gs} + \left[\frac{j\omega C_{dd}(D')}{1 + j\omega\tau_{dd}(D')}\right] v_{d's}. \tag{12.39}$$

Now we can use the DC motion $x_{s,dc}$ to derive the expression relating the drain potential $v_{d''s}$ at point D'' to the drain potential $v_{d's}$ at the point D':

$$v_{d''s} = v_{d's} + F_c x_{s,dc}. \tag{12.40}$$

This permits us to evaluate $x_{s,dc}$ in terms of v_{gs} and $v_{d''s}$. Substituting Equation (12.40) into Equation (12.38) we find

$$x_{s,dc} = \left[\frac{g^0_m(D') - \beta F_c}{g^0_d(D') F_c}\right] v_{gs} + \left[\frac{g^0_d(D') + \beta F_c}{g^0_d(D') F_c}\right] v_{d''s}. \tag{12.41}$$

One can further derive the following identities

$$\left.\begin{aligned} \frac{g_m^0(D') - \beta F_c}{g_d^0(D')F_c} &= \frac{\partial X_S(D')}{\partial V_{GS}} = -\frac{1}{F_c}\frac{k^2}{2(1-k)^2}, \\ \frac{g_d^0(D') + \beta F_c}{g_d^0(D')F_c} &= \frac{\partial X_S(D')}{\partial V_{D'S}} = \frac{1}{F_c}\frac{2-2k+k^2}{2(1-k)^2}. \end{aligned}\right\} \tag{12.42}$$

We are now in position to perform the change of reference point from D' to D'' using Equations (12.40) and (12.41). Note that as far as the quantities $y_{ij}^0(D')$ and $X_S(D')$ are concerned a change of variable from (D') to (D'') introduces a second order contribution which is therefore neglected in a first order small-signal analysis.

First we can rewrite the gate current in terms of $v_{d''s}$:

$$\begin{aligned} i_{g,GCA} &= \left(y_{gg}^0 - F_c\, y_{gd}^0 \frac{\partial X_S}{\partial V_{GS}}\right) v_{gs} + \left(y_{gd}^0 - F_c\, y_{gd}^0 \frac{\partial X_S}{\partial V_{D'S}}\right) v_{d''s} \\ &= y_{gg}^1 v_{gs} + y_{gd}^1 v_{d''s}, \end{aligned} \tag{12.43}$$

which defines y_{gg}^1 and y_{gd}^1. Similarly the drain current rewritten in terms of $v_{d''s}$ is now:

$$\begin{aligned} i_{d''} &= \left(y_{dg}^0 - F_c y_{dd}^0 \frac{\partial X_S}{\partial V_{GS}}\right) v_{gs} + \left(y_{dd}^0 - F_c y_{dd}^0 \frac{\partial X_S}{\partial V_{D'S}}\right) v_{d''s} \\ &= y_{dg}^1 v_{gs} + y_{dd}^1 v_{d''s}, \end{aligned} \tag{12.44}$$

which defines y_{dg}^1 and y_{dd}^1. Finally we can rewrite Equation (12.39)

$$\begin{aligned} -\beta F_c^2\, x_{s,ac} = &\left[\frac{j\omega C_{dg}(D')}{1 + j\omega\tau_{dg}(D')} - F_c \frac{\partial X_S(D')}{\partial V_{GS}} \frac{j\omega C_{dd}(D')}{1 + j\omega\tau_{dd}(D')}\right] v_{gs} \\ &+ \left[\frac{j\omega C_{dd}(D')}{1 + j\omega\tau_{dd}(D')} - F_c \frac{\partial X_S(D')}{\partial V_{GS}} \frac{j\omega C_{dd}(D')}{1 + j\omega\tau_{dd}(D')}\right] v_{d''s}. \end{aligned}$$

The last equation is identified as the displacement current (AC) component of the AC drain current $i_{d''}$ given by Equation (12.44). Indeed we can separate the drain current $i_{d'''} = i_{d''} = i_{d'}$ into its low-frequency (dc) part $i_{d'',dc}$ and its frequency-dependent (ac) part $i_{d'',ac}$:

$$i_{d''} = i_{d'',dc} + i_{d'',ac}(\omega).$$

One can then write the following identities:

$$i_{d'',dc} = \beta F_c(v_{gs} - v_{d''s}) = \beta F_c(v_{gs} - v_{d's} - F_c x_{s,dc}), \tag{12.45}$$

$$i_{d'',ac} = -\beta F_c^2 x_{s,ac}. \tag{12.46}$$

Equation (12.45) holds because we have

$$g_m^0 - F_c g_d^0 \frac{\partial X_S}{\partial V_{GS}} = -\left(g_d^0 - F_c g_d^0 \frac{\partial X_S}{\partial V_{D'S}}\right) = \beta F_c.$$

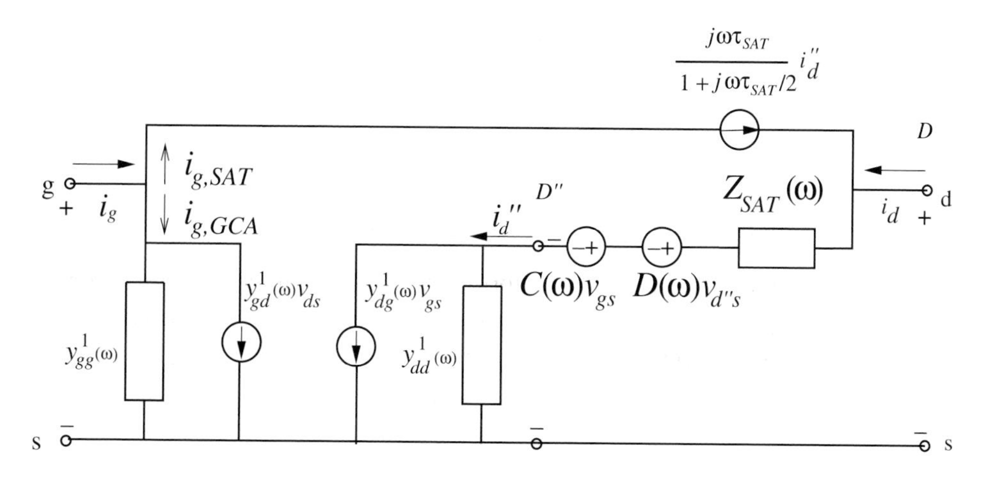

Fig. 12.7. Alternative small-signal topology for the short-channel MOSFET.

Equations (12.45) and (12.46) could have been derived directly from Equation (12.33). This self-consistent result therefore demonstrates that the change of reference from D' to D'' was performed correctly.

To complete the derivation of this small-signal equivalent circuit which uses D'' as the internal node, we need to give an expression directly relating v_{ds} and $v_{d''s}$:

$$\begin{aligned} v_{ds} &= v_{d''s} - F_c x_{s,dc} - \alpha I_{DC}\ell\, x_s + Z_{SAT}(\omega) i_{d''} \\ &= v_{d''s} + C(\omega) v_{gs} + D(\omega) v_{d''s} + Z_{SAT}(\omega) i_{d''}, \end{aligned} \tag{12.47}$$

where we define C and D to be

$$\left.\begin{aligned} C &= -(F_c + \alpha I_{DC}\ell)\frac{\partial X_S(D')}{\partial V_{GS}} + \frac{\alpha I_{DC}\ell}{\beta F_c^2}[y^1_{dg}(\omega) - y^1_{dg}(\omega = 0)], \\ D &= -(F_c + \alpha I_{DC}\ell)\frac{\partial X_S(D')}{\partial V_{D'S}} + \frac{\alpha I_{DC}\ell}{\beta F_c^2}[y^1_{dd}(\omega) - y^1_{dd}(\omega = 0)]. \end{aligned}\right\} \tag{12.48}$$

The new small-signal equivalent circuit using D'' instead of D' for the internal node is shown in Figure 12.7.

12.3.3 Large-signal model

Having derived the small-signal using D'' as the internal node we are now in position to derive the large-signal model for the velocity-saturated MOSFET wave-equation.

This large-signal model must satisfy the following three conditions: (1) its small signal must reduce to that of the first order non-quasi-static model of the velocity-saturated MOSFET wave-equation; (2) it must be expressed in terms of a first order state-equation; and (3) it must conserve charge.

For clarity we first present the state equations (Condition (2)) of this large-signal model and in the process demonstrate Condition (2). We will then verify that

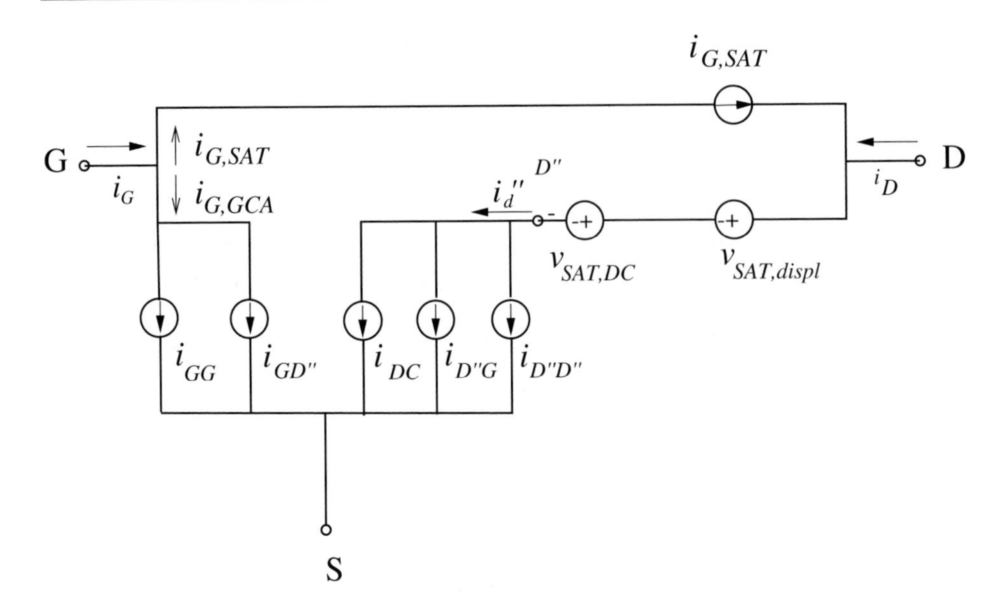

Fig. 12.8. Large-signal topology for the short-channel MOSFET.

Condition (1) is enforced when this is not immediate. We postpone the verification of Condition (3) to the next two sections.

The general topology of the large-signal model is shown in Figure 12.8. This large-signal model makes use of six state equations which are defined below. The gate current is given by

$$i_G = i_{GG} + i_{GD''} + i_{G,SAT},$$

with

$$i_{GG} = C_{gg}(D'')\frac{dv_{GS}}{dt} - \frac{d}{dt}\left[\tau_{gg}(D'')i_{GG}\right], \tag{12.49}$$

$$i_{GD''} = -F_c C_{gd}(D'')\frac{\partial X_S(D'')}{\partial V_{GS}}\frac{dv_{GS}}{dt} + \left[C_{gd}(D'') - F_c C_{gd}(D'')\frac{\partial X_S(D'')}{\partial V_{D''S}}\right] \times \frac{dv_{D''S}}{dt} - \frac{d}{dt}[\tau_{gd}(D'')i_{GD''}], \tag{12.50}$$

$$i_{G,SAT} = C_{g,SAT}(D'')\left(\frac{dv_{GS}}{dt} - \frac{dv_{D'''S}}{dt}\right) - \frac{d}{dt}\left[\frac{1}{2}\tau_{SAT}(D'')i_{G,SAT}\right], \tag{12.51}$$

where $C_{g,SAT}$, the capacitance associated with the saturation region, is

$$C_{g,SAT}(D'') = [L_g - X_S(D'')]W_g C_G.$$

Note that the state equation for $i_{G,SAT}$ is driven among other things by the voltage $v_{D'''S}$ which is related to $v_{D''S}$ by the following definition:

$$v_{D'''S} = v_{D''S} + F_c\, x_{S,displ},$$

where $x_{S,displ}$ is the displacement portion of the instantaneous GCA/saturation boundary:

$$x_S(t) = X_S(D'') + x_{S,displ}$$

Therefore $x_{S,displ}$ becomes a force term like $v_{D''S}$ and v_{GS} in the state equation for $i_{G,SAT}$. The following relation directly relates $x_{S,displ}$ to the displacement part of the drain current (to be introduced below) by the following relation:

$$-\beta F_C^2 x_{S,displ} = i_{D'',displ} = i_{D''G} + i_{D''D''}. \qquad (12.52)$$

The drain current at the node D is given by

$$i_D = i_{D''} - i_{G,SAT},$$

with $i_{D''}$ the drain current at position D''. The drain current $i_{D''}$ is given by

$$i_{D''} = I_{DC}(D'') + i_{D'',displ} = I_{DC}(D'') + i_{D''G} + i_{D''D''},$$

where I_{DC} is the DC current at D'' given by

$$I_{DC}(D'') = \beta F_c(v_{GS} - V_T - v_{D''S})$$

and where $i_{D''G}$ and $i_{D''D''}$ are the displacement drain currents which satisfy the following state equations:

$$i_{D''G} = C_{dd}(D'')\frac{dv_{GS}}{dt} - \frac{d}{dt}\left[\tau_{dg}(D'')i_{D''G}\right], \qquad (12.53)$$

$$i_{D''D''} = -F_c C_{dd}(D'')\frac{\partial X_S(D'')}{\partial V_{GS}}\frac{dv_{GS}}{dt} + \left[C_{dd}(D'') - F_c C_{dd}(D'')\frac{\partial X_S(D'')}{\partial V_{D''S}}\right]\frac{dv_{D''S}}{dt} - \frac{d}{dt}(\tau_{gd}(D'')i_{D''D''}). \qquad (12.54)$$

The internal drain voltage $v_{D''S}$ is obtained from the applied voltage v_{DS} using

$$v_{DS} = v_{D''S} + v_{SAT,DC} + v_{SAT,displ},$$

where $v_{SAT,DC}$ is the instantaneous voltage drop across the saturation region between D'' and D:

$$v_{SAT,DC} = \frac{1}{2}\alpha\, i_{D''}[L_g - x_S(D'')]^2 + F_c[L_g + X_S(D'')].$$

Note that $v_{SAT,DC}$ indirectly involves some displacement terms because it is expressed in terms of the instantaneous drain current $i_{D''}(t)$ and the instantaneous GCA/saturation boundary position $x_S(D'')$.

$v_{SAT,displ}$ is the displacement part of the saturation voltage that is associated with the high-frequency response of the saturation region, and is obtained from the following state equation:

$$v_{SAT,displ} = L_{SAT}(D'')\frac{di_{D''}}{dt} - \frac{d}{dt}\left[\frac{1}{4}\tau_{SAT}(D'')\,v_{SAT,displ}\right],$$

with

$$L_{SAT} = \frac{1}{6}\alpha\,[L_g - X_S(D'')]^2\tau_{SAT}(D'').$$

Note that now six state equations are required for the short-channel FET model instead of the four state equations for the long-channel FET model. As before, four of the state equations are used to account for the GCA region. One state equation is used to account for the drain current delay and its impact on the gate current. The final state equation is used for the motion of the voltage drop across the saturation region.

We need to verify that this large-signal model admits a small-signal model which reduces to that of the first order non-quasi-static model of the velocity-saturated MOSFET wave-equation. This verification of Condition (1) is for the most part immediate.

Equation (12.43) is the small-signal version of the large-signal Equations (12.49) and (12.50) if we define

$$i_{GG} + i_{GD''} = i_{g,GCA}\exp(j\omega t).$$

Equation (12.44) is the small-signal version of the large-signal Equations (12.49) and (12.50) if we define

$$i_{D'',displ} = i_{D''G} + i_{D''D''} = i_{d'',ac}\exp(j\omega t).$$

We can also demonstrate that the large-signal Equation (12.51) for $i_{G,SAT}$ reduces to the small-signal Equation (12.34) for $i_{g,SAT}$. Letting $i_{G,SAT} = i_{g,SAT}\exp(j\omega t)$ in Equation (12.51) we have (using $\tau_{SAT} = (L_g - X_S)/v_s$)

$$i_{g,SAT} = (L_g - X_S)\frac{W_g C_g j\omega(v_{gs} - v_{d'''s})}{1 + j\omega\frac{1}{2}\tau_{SAT}} = \frac{j\omega\tau_{SAT}}{1 + j\omega\frac{1}{2}\tau_{SAT}}\,i_{d'''},$$

which is the same as Equation (12.34) since we have $i_{d'} = i_{d'''}$. Note that X_S and τ_{SAT} are dependent on D'' or equivalently D' for small-signal analysis.

Equation (12.46) is the small-signal equation resulting from the large-signal Equation (12.52) relating $x_{s,displ}$ to $i_{D'',displ}$ if we replace $x_{s,displ}$ by

$$x_{S,displ} = x_{s,ac}\exp(j\omega t).$$

We have now verified that the proposed large-signal model indeed reduces for small-signal analysis to the velocity-saturated MOSFET wave-equation small-signal model.

12.3.4 Charge-based representation

In the next section we will verify that the large-signal model presented above enforces charge conservation. However the practical implementation of charge conservation in a circuit simulator is also a numerical issue. It is well recognized that the use of charges as state variables instead of voltages allows the elimination of the non-conservation of charge which arises from the approximate numerical integration techniques used by the circuit simulator. The use of charge also reduces the number of state equations. Unfortunately the introduction of charges in a non-quasi-static model requires the use of only one [3,1] or two [4] relaxation-time constants instead of the four possible with the use of capacitors.

Let us first introduce the definition of the various charges. The charge in the saturation region Q_{SAT} and its normalized version Q'_{SAT} are defined from the DC channel charge N_S by

$$
\begin{aligned}
Q_{SAT} &= qW_g \int_{X_S}^{L_g} N_S(X_S)\,dx = (L_g - X_S)qN_S(X_S) \\
&= [L_g - X_S(D'')]W_g C_G(v_{GS} - V_T - v_{D''S}) \\
&= (L_g - X_S)Q'_{SAT}(D'').
\end{aligned}
$$

The gate charge Q_G and its normalized version Q'_G are (see previous chapter)

$$
\begin{aligned}
Q_G &= X_S Q'_G \\
&= X_S C_G W_g (v_{GS} - V_T)\frac{2}{3} \times \frac{-3 + 3k - k^2}{k - 2},
\end{aligned}
$$

where k is given by

$$
k = \frac{v_{D''S}}{v_{GS} - V_T}.
$$

The drain charge Q_D and its normalized version Q'_D are (see previous Chapter)

$$
\begin{aligned}
Q_D &= X_S Q'_D \\
&= X_S C_G W_g (v_{GS} - V_T)\frac{2}{15} \times \frac{-15 + 25k - 15k^2 + 3k^2}{4 - 4k + k^2}.
\end{aligned}
$$

The resulting new state equations are presented below. The gate current is given by

$$
i_G = i_{G,GCA} + i_{G,SAT}
$$

$$
i_{G,GCA} = X_S(D'')\frac{dQ_G(D'')}{dt} - F_c C_{gd}(D'')\frac{dX_S(D'')}{dt} - \frac{d}{dt}[\tau_G(D'')i_{G,GCA}] \tag{12.55}
$$

$$
i_{G,SAT} = [L_g - X_S(D'')]\frac{dQ'_{SAT}(D''')}{dt} - \frac{d}{dt}\left[\frac{1}{2}\,\tau_{SAT}(D''')\,i_{G,SAT}\right]. \tag{12.56}
$$

Finally the displacement drain current $i_{D'',displ}$ is now given by

$$i_{D'',displ} = X_S(D'')\frac{dQ_D(D'')}{dt} - F_c C'_{dd}(D'')\frac{dX_S(D'')}{dt} - \frac{d}{dt}[\tau_D(D'')i_{D'',displ}]. \tag{12.57}$$

12.3.5 Charge conservation

We need now to verify that both of these large-signal models conserve charge. Following the approach used in Chapter 11, this is equivalent to verifying that the instantaneous charge transfer to the gate is path-independent:

$$\Delta q_G(t_1, t_2) = \int_{t_1}^{t_2} i_G\, dt = Q_{G,tot}[V_{GS}(2), V_{DS}(2)] - Q_{G,tot}[V_{GS}(1), V_{GS}(1)]$$

if the device is in a steady state at times t_1 and t_2. Note that we can rewrite

$$X_S\frac{dQ'_G}{dt} = \frac{d}{dt}(X_S Q'_G) - Q'_G\frac{dX_S}{dt}$$

$$(L_g - X_s)\frac{dQ'_{SAT}}{dt} = \frac{d}{dt}[(L_g - X_s)Q'_{SAT}] + Q'_{SAT}\frac{dX_s}{dt}.$$

Note also that the following term vanishes:

$$\int_{t_1}^{t_2}(-Q'_G - F_c C_{gd} + Q'_{SAT})\frac{dX_S}{dt}\, dt = 0$$

because of the identity (Problem 12.2)

$$-Q'_G + Q'_{SAT} = F_c C_{gd} = (v_{GS} - V_T)\frac{1}{3}C_G W_g\frac{(3-k)k}{k-2}. \tag{12.58}$$

This leads to

$$\begin{aligned}\Delta q_G(t_1, t_2) &= \int_{t_1}^{t_2}(i_{G,GCA} + i_{G,SAT})\, dt \\ &= \big[Q_G(V_{GS}(2), V_{D''S}(2)) + Q_{SAT}(V_{GS}(2), V_{D''S}(2)) \\ &\quad - Q_G(V_{GS}(1), V_{D''S}(1)) + Q_{SAT}(V_{DS}(1), V_{D''S}(1))\big] \\ &= Q_{G,tot}(V_{GS}(2), V_{DS}(2)) - Q_{G,tot}(V_{GS}(1), V_{DS}(1)),\end{aligned}$$

which is path-independent as required. This derivation of charge conservation holds for both the charge-based (two-τ) and the capacitor-based (four-τ) models.

12.3.6 Model topology

In this section we will conclude our theoretical study of the MOSFET wave-equation with the derivation of a large-signal model for the velocity-saturated MOSFET wave-equation of short-channel MOSFET/MODFET. The large-signal model presented is fairly complicated. Yet the voltage dependences of the y parameters are too simplistic due to the model assumptions of constant capacitor, mobility and saturation velocity. Consequently the increased complexity does not by itself provide an improved fit to the voltage dependence observed in real devices.

The major motivation of this work, however, is to establish as rigorously as possible the large-signal model topology associated with the velocity-saturated MOSFET wave-equation. The topologies of models are of importance because they determine the dispersion (frequency dependence) of the model. Topology does not specify the voltage dependence of the current sources, capacitors or charges, which must be obtained separately. For example, we shall see in the next chapter that the tensor product B-spline can be used in a table model to accurately represent, at a moderate numerical cost, the complex voltage dependence of the model elements while enforcing continuity of the derivatives up to the order desired. With the use of such accurate tools to represent the voltage dependences, the limit in device modeling becomes the topology which establishes the dispersion using circuit-elements which are frequency-independent.

The topology obtained from the velocity-saturated MOSFET wave-equation is more complex than that of the long-channel MOSFET wave-equation. Two additional state equations are introduced and an internal node is required. This originates from the rather complex self-consistent motion of the GCA/saturation boundary.

The large-signal equation of $i_{G,GCA}$ and $i_{D,displ}$ can be rewritten in terms of the voltage D' as

$$i_{G,GGA} = X_s(D'')\,\frac{dQ'_G(D')}{dt} - \frac{d}{dt}\left[\tau_G(D'')\,i_{G,GCA}\right],$$
$$i_{D'',displ} = X_s(D'')\,\frac{dQ'_D(D')}{dt} - \frac{d}{dt}\left[\tau_D(D'')\,i_{D,displ}\right],$$

if we introduce the non-physical voltage $v_{D'S}$:

$$v_{D'S} = v_{D''S} - F_c X_S(D'').$$

We can represent the charge-based large-signal topology shown in Figure 12.9 using Q charge elements with the current given by

$$i_j = X_s(D'')\,\frac{dQ'_j(D')}{dt} - \frac{d}{dt}[\tau_j(D'')\,i_j]$$

and using an L circuit element with current given by

$$v_j = L_j(D'')\,\frac{di_j}{dt} - \frac{d}{dt}\,[\tau_j(D'')\,v_j].$$

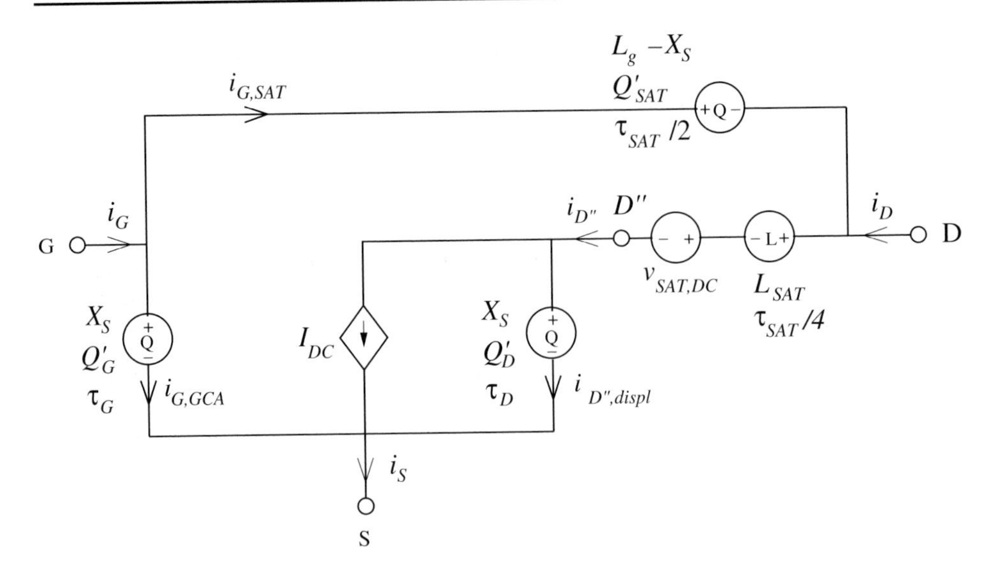

Fig. 12.9. Charge-based large-signal topology.

The topology derived from the short-channel MOSFET wave-equation introduces the displacement current $i_{G,SAT}$ which is supported by the charge Q_{SAT}. This topology clearly indicates that even though in the saturation region the gate-to-channel voltage no longer controls the channel charge (the GCA approximation is not applicable), the charge in the velocity-saturated part of the channel remains imaged in the gate inducing the new additional displacement current $i_{G,SAT}$ in the gate terminal.

12.4 Conclusion

The velocity-saturated MODFET wave-equation analyzed in this chapter and the large-signal model developed from it have provided us with a physical description of the frequency dispersions and partitioning of the charges and of their voltage-dependences in the intrinsic short-channel MODFET. The reader is referred to [2] for an example of its application to the physical modeling of MODFETs and to Chapter 15 for further in-depth analyses.

To conclude Chapter 12, let us note that, as was found in Figure 12.5, the non-quasi-static large-signal model topology (MOD B) developed in Chapter 11 for the long-channel FET is applicable to the short-channel FET up to f_0 which is on the order of f_{max}. Indeed, the new frequency dispersion effects introduced by the non-quasi-static model topology developed for the short-channel FET are usually above f_{max}. Therefore the simpler long-channel model topology established in Chapter 11 is usually sufficient for fitting the frequency response of FETs for frequencies below f_{max} and will be adopted in Chapter 13 for the development of more complex model topologies accounting for electrothermal effects and other low-frequency dispersion effects.

12.5 Bibliography

[1] P. Roblin, S. C. Kang and W. R. Liou, 'Improved small-signal equivalent circuit model and large-signal state-equations for the MOSFET/MODFET wave equation', *IEEE Transactions on Electron Devices*, Vol. 38, No. 8, pp. 1706–1718, August 1991.

[2] P. Roblin, S. C. Kang and H. Morkoç, 'Analytic solution of the velocity-saturated MOSFET/MODFET wave-equation and its application to the prediction of the MODFET microwave characteristics', *IEEE Transactions on Electron Devices* Vol. ED-37, No. 7, pp. 1608–1623, July 1990.

[3] K.-W. Chai and J. J. Paulos, 'Unified nonquasi-static modeling of the long-channel four-terminal MOSFET for large and small-signal analyses in all operating regimes', *Transactions on Electron Devices* Vol. 36. No. 11, pp. 2513–2520, November 1989.

[4] R. R. Daniels, A. T. Yang and J. P. Harrang, 'A universal large/signal signal 3-terminal FET model using a nonquasi-static charge-based approach', *IEEE Transactions on Electron Devices*, Vol. 40, No. 10, pp. 1723–1729, October 1993.

[5] S. C. Kang, 'Small signal AC model for the velocity-saturated MODFET and the prediction of the microwave characteristics of MODFETs', PhD Thesis, The Ohio State University, 1991.

[6] H. Rohdin, 'Reverse modeling of E/D logic sub-micron MODFETs and prediction of maximum extrinsic MODFET current cutoff frequency', *IEEE Transactions on Electron Devices*, Vol. ED-37, No. 4, pp. 920–934, April 1990.

[7] P. Roblin, S. C. Kang and H. Morkoç, 'Microwave characteristics of the MODFET and the velocity-saturated MOSFET wave-equation', *Proceedings of the 1990 International Symposium on Circuits and Systems*, Vol. 2, pp. 1501–1504, May 1990.

[8] P. Roblin and S. C. Kang, 'Unilateral power gain resonances and roll-off with frequency for the velocity-saturated MOSFET wave-equation', *IEEE Transactions on Electron Devices*, Vol. 39, No. 6, pp. 1490–1495, June 1992.

12.6 Problems

12.1 Derive the equivalent circuit of Figure 12.1. Note that the exact equivalent circuit is obtained by rewriting the resulting y-parameters $y_{ij}(sat)$ in terms of the y-parameters of the GCA region $y_{ij}(GCA)$ of reduced gate length $X_S = L_g - \ell$.

12.2 Verify the charge conservation identity given by Equation (12.58).

13 DC and microwave electrothermal modeling of FETs

Patrick Roblin and Siraj Akhtar

The purpose of models is not to fit the data but to sharpen the questions.

11th R. A. Fisher Memorial Lectures, Royal Society, 20 April 1983, SAMUEL KARLIN

13.1 Introduction

The design and simulation of microwave circuits with a circuit simulator requires the availability of fast and accurate models for all components in the circuit. From a design perspective, the reliability of the simulations of microwave circuits is usually limited by the device models that are available in the computer simulation tools used. Accurate and computationally efficient device models, that are easy to extract from measured data and can easily be incorporated into a circuit simulator, are therefore needed to improve and speed up the design of high-frequency circuits.

Although a model can come close, it can never exactly reproduce the performance of a device. Hence it is important to realize that the modeling effort may need to be tailored towards the kinds of circuits being simulated. This can be very important given the fact that not only accuracy, but also speed and convergence are important factors that need to be addressed.

In this chapter, we will present the methodology for the development of a universal field-effect transistor (FET) model for the DC, thermal and microwave modeling of three-terminal FET devices using some of the results derived in Chapter 11. Universal models are models that are applicable to a wide range of technologies. In this chapter we focus on the application of the universal microwave FET models to: (1) SOI (silicon on insulator) MOSFETs [1] for low-power RFIC integrating RF and digital circuitry on a single chip, and (2) LD MOSFETs [2] (laterally diffused MOS) for high-power, linear amplification. These devices are of particular interest as they present some interesting challenges due to their large self-heating (SOI and LD) and the presence of kinks in their DC I–V characteristics (SOI).

Of particular interest is the use of these models for RF/microwave power amplifier design and simulation and associated self-heating and self-biasing effects. The discussion in this chapter is, however, general enough that the methods presented can easily be applied to other devices, technologies and circuits. The reader is referred to [3,4,5] for the application of universal FET models to GaAs FETs and high-electron-mobility transistors (HEMTs).

13.2 Modeling for power amplifier design

For low-cost, low-power, mobile applications (cellular phones and pagers), monolithic microwave integrated circuits (MMICs) and RF CMOS are used to completely integrate the subcomponents of a transceiver system. These components include power amplifiers (PAs), low-noise amplifiers (LNAs), mixers and voltage controlled oscillators (VCOs). Because power consumption is the driving element in such designs, class C and higher amplifiers are used. These amplifiers give high power-added efficiency (PAE) (50–60%) but tend to be very non-linear [6]. The non-linearity is acceptable since constant envelope modulation is used.

For RF CMOS technologies, PAs for transmitters are designed using 'low-frequency techniques' whereby the aspect ratio of the transistor is varied until the desired gain is achieved. For matching purposes, the output resistance of the transistor is transformed to 50 Ω and the capacitance may be canceled by an inductor. For receivers, noise performance is of greater concern than maximizing gain. In such designs where low-cost mass production is essential, it would be very impracticable to extract a device model from measured data for each transistor. In general, the approach has been to try to use available transistor models (such as BISIM3). Once a good design, relatively insensitive to the transistor characteristics, has been achieved, the circuit can be mass produced. As long as device parameter fluctuations are small enough, the circuit performance will be acceptable due to the low power of the signals.

For PAs that are used in expensive and large infrastructure applications (cellular base-stations and high-power transmitters), maximizing output power and linearity is highly important since a number of channels are simultaneously combined and amplified before being sent to the antenna. Since both linearity and PAE are important in such applications, class A or AB biasing is preferred. This inherently implies that DC power consumption will be high (PAE from 12% to 40%) and hence the temperature of the PA will significantly increase, necessitating the need for cooling units. The high power and RF current densities imply that packaged devices and microstrip technology has to be used in such applications. 'Microwave' design techniques that rely on measuring the high-power large-signal S parameters of the transistor at the desired bias point and frequency and then building an input and output matching circuit are used for these amplifiers. Load pull techniques can also be used to obtain data for building matching networks giving constant power gain. Because maximizing gain is very difficult in light of the high power and very low inherent device impedance, the matching circuits must be individually tuned to optimize performance [7].

Extensive circuit simulations are often not performed for such amplifiers due to the unavailability of models that can accurately predict important PA characteristics such as gain harmonics, PAE, two-tone intermodulation distortion (for FDMA AMPS cellular systems), spectral regrowth and adjacent channel power (for CDMA) over a wide bias, temperature and frequency range.* Hence such device models are essential to streamline the design of PAs and to provide a virtual testbed for the designer to carry out all sorts of tests on the amplifier and to optimize the performance accordingly.

Such a model will have to be extracted from measured device data. The substantial variation in transistor parameters calls for a scheme whereby the model extracted for one device can be fitted to another device with the smallest number of additional measurements.

13.3 Physical versus table-based models

SPICE-based FET models, such as versions of BISIM, use physics-based equations to model the device [8].† In order to account for any additional physical effect, a new equation is added. This leads to an increasing number of different, and somewhat complex, equations that have to be accounted for by the model.

In the table-modeling approach, a generic set of equations is used to interpolate between model parameters. This can lead to a reduction in complexity of the equations, while allowing for charge conservation [9,3] and continuity of derivatives, provided that the original set of generic equations can handle this. Table modeling begins with the selection of the right model topology for the device. The disadvantage of the table-modeling approach is that usually the model parameters extracted do not have physical meanings and are simply chosen to give the best fit.

Physics-based models have found increasing use for bipolar type devices, while FET modeling, particularly at high power, is dominated by table-modeling approaches. Yet perhaps, it is the combination of these two methods that will yield the best results. In this chapter, emphasis is given to explaining table-based approaches for device modeling that incorporate physical constraints.

* FDMA = frequency division multiple access; AMPS = advanced mobile phone service; CDMA = code division multiple access.

† SPICE = simulation program for integrated circuits emphasis.

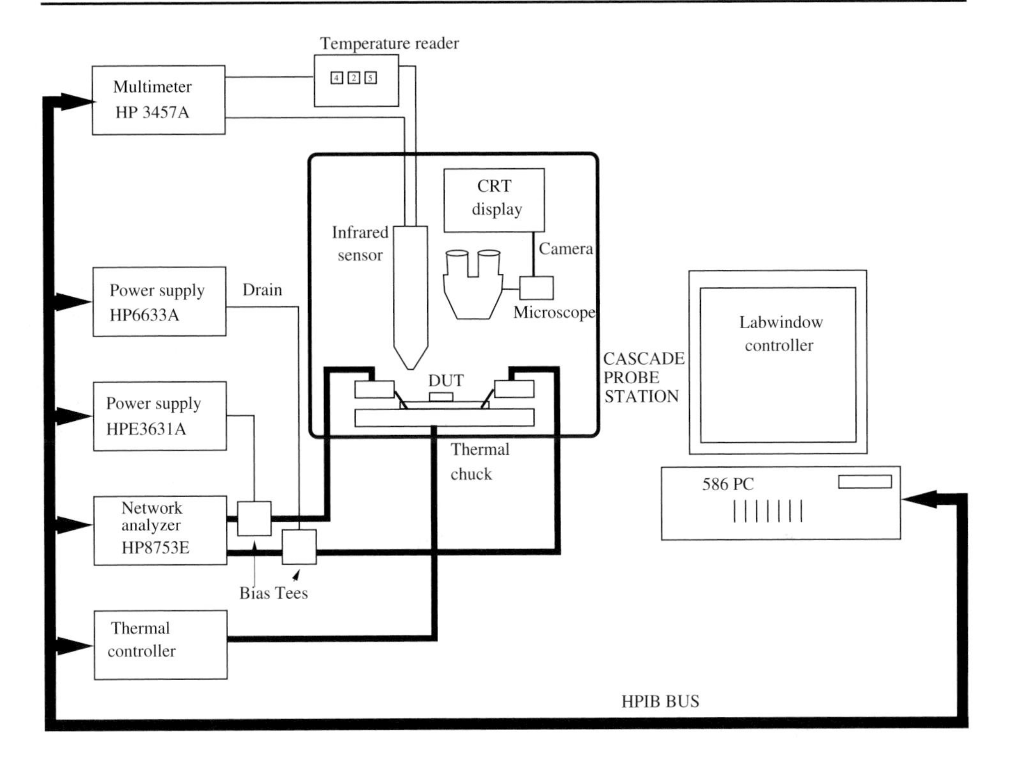

Fig. 13.1. Electrothermal and microwave measurement setup.

13.4 Device characterization

For modeling to be successful, accurate data acquisition is vital. This includes not only current and voltage information, but also microwave and temperature data. The latter is critical for electrothermal modeling, which needs a setup allowing for both substrate heating and cooling. Figure 13.1 shows the setup for an automated data acquisition system under computer control.

Device probing is the best way to acquire accurate data for on-chip devices. However, for large devices that are formed by the parallel combination of many small devices, current crowding and the inherent low impedance of the device makes it difficult to simply combine the extracted model for a unit device. For large packaged devices, DC measurements can be very difficult due to thermal instability, while microwave measurements can be challenging due to the low impedance of the device. It is therefore necessary to ensure that the biasing methodology used for both measurement and circuit design are the same.

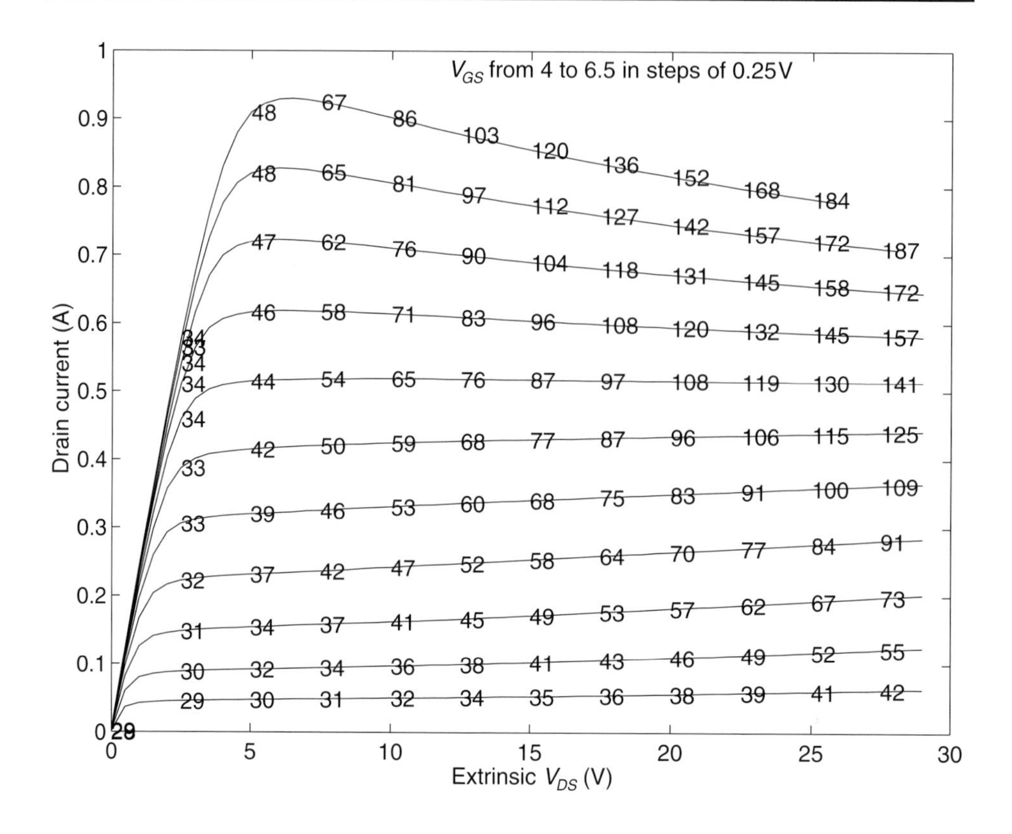

Fig. 13.2. Measured I–V–T of an LD MOSFET for a constant substrate temperature of 29 °C.

13.4.1 DC $I-V-T$

Figure 13.2 shows the measured I–V–T characteristics (noted $I_{D,DC}(V_{GS}, V_{DS}, T_{sub})$) of an LD MOSFET. Superimposed on the figure are the measured device temperature, T_{dev}, at the given bias point. In a measurement such as this, the substrate temperature, T_{sub}, is kept constant. It is clear that there can be a substantial build up of temperature in the device. This high temperature leads to a reduction in the electron mobility in the channel, which gives rise to the negative drain conductance at high bias. This phenomenon is known as self-heating and is one of the causes of low-frequency dispersions in the S parameters of a device.

13.4.2 Pulsed $I-V$ characteristics

Pulsed I–V characteristics, also referred to as transient I–V characteristics, are another type of measurement that is very useful for modeling purposes [10,11,12,13]. In pulsed I–V characteristics, the device is biased at a fixed point that establishes the

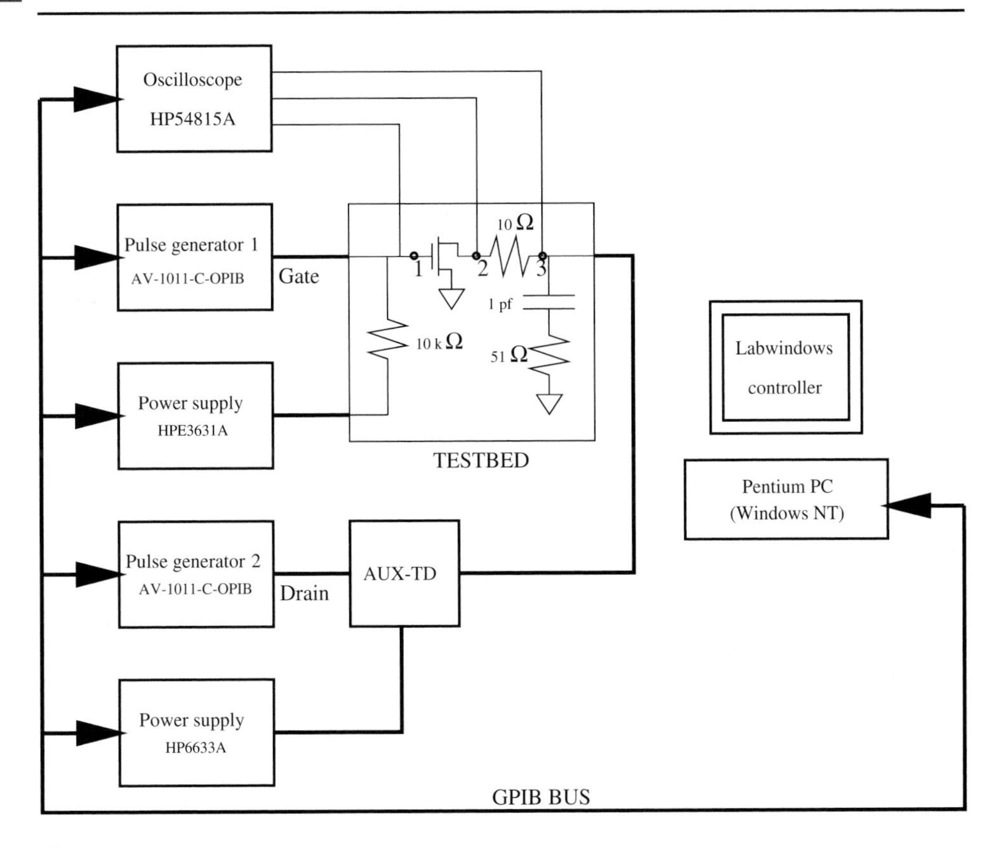

Fig. 13.3. Pulsed I–V measurement setup.

operating point and device temperature (V_{GS}, V_{DS}, I_{DS}, T_{dev}). Pulses of short duration and low duty rate are applied to reach new transient biases (v_{GS}, v_{DS}, i_{DS}, T_{dev}) at which the drain current is recorded. The pulse duration and duty rate are short enough that the device does not have time to respond to low-frequency effects such as traps, parasitic bipolar transistor and thermal effects. The pulse duration should not be too short to allow the FET capacitances to finish updating their charge. Typically the pulse duration is on the order of a microsecond and the duty rate is smaller than 1%. Hence the device operation is isothermal. The pulsed I–V characteristics, which are denoted $I_{D,tran}(v_{GS}, v_{DS}, V_{GS}, V_{DS}, T_{dev})$, provide us with the effective I–V characteristics followed by the transistor at high frequencies when its DC operating point is (V_{GS}, V_{DS}, T_{dev}).

Pulsed I–V test systems can be commercially acquired. However, it is possible to construct such a system using pulsers and an oscilloscope. Figure 13.3 shows such a setup. The trajectory of the voltages and current showing the starting DC bias point is depicted in Figure 13.4 for a measured pulsed I–V characteristic.

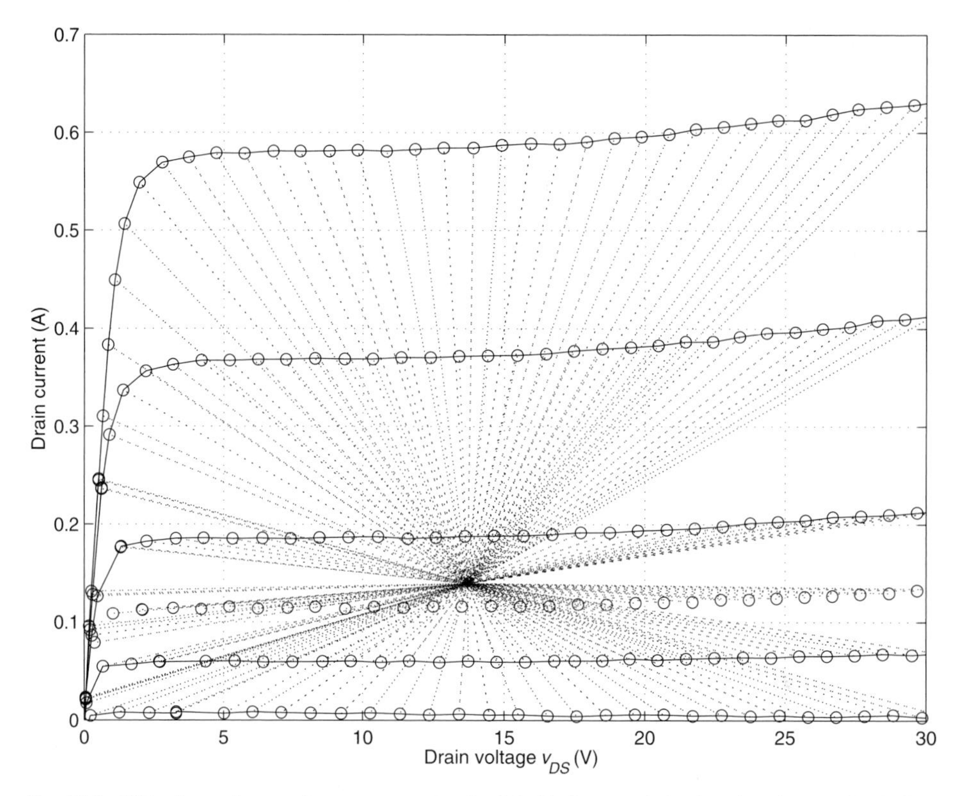

Fig. 13.4. Effective trajectory for a measured pulsed I–V characteristic showing the starting DC bias point. (Pulsed I–V characteristics for $V_{GS} = 4.75$ V and $V_{DS} = 13.7$ V and $T_{sub} = 35$ °C.)

13.4.3 Isothermal $I-V$ characteristics

While pulsed I–V characteristics bypass low-frequency dispersions as a result of both traps, parasitic bipolar transistor and thermal effects, isothermal I–V characteristics bypass only thermal effects. Isothermal I–V characteristics have a constant device temperature yet include the effects of traps and other low-frequency dispersion phenomena. The term isothermal I–V characteristics can be misleading since pulsed I–V characteristics are also isothermal in nature. However, as mentioned earlier, pulsed I–V characteristics are more than just isothermal.

Figure 13.5 shows a pulsed I–V characteristic compared with an isothermal I–V one for a SOI MOSFET. Although both curves have the same average T_{dev}, it is clear that the curves are not the same. Notice that the two I–V characteristics agree at the bias point, indicated by a star, used for measuring the pulsed I–V characteristic. Indeed, at the DC bias point the transient ($I_{D,tran}$) and the isothermal ($I_{D,iso}$) I–V characteristics are equal:

$$I_{D,tran}(v_{GS} = V_{GS}, v_{DS} = V_{DS}, V_{GS}, V_{DS}, T_{dev}) = I_{D,iso}(V_{GS}, V_{DS}, T_{dev}).$$

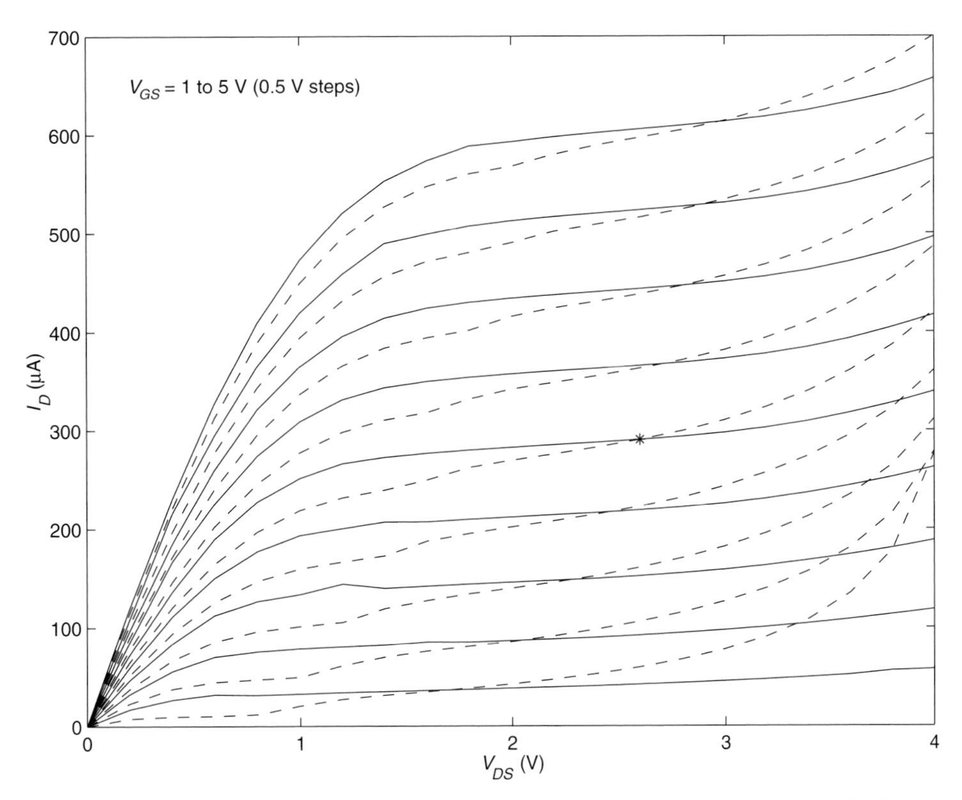

Fig. 13.5. Comparison of pulsed (full line) with isothermal (closed line) I–V characteristics for an SOI MOSFET. The star shows the bias point used to measure the transient data. (S. Akhtar and P. Roblin, *Analog Integrated Circuits and Signal Processing*, Vol. 25, No. 2, pp. 115–131, November 2000.)

The isothermal I–V characteristics of the SOI MOSFET feature a characteristic kink in the drain current associated with the activation in the SOI MOSFET of the parasitic bipolar transistor by the holes generated by impact ionization (see Figure 13.6) at high drain voltages. No such kinks are present in the transient I–V characteristic as the parasitic bipolar transistor does not respond to fast variations of the drain and gate voltages.

Notice further in Figure 13.5 that neither of the two I–V characteristics shows the self-heating-induced negative drain conductance. We shall see in the model section that under DC operation the DC ($I_{D,DC}$) and isothermal ($I_{D,iso}$) I–V drain currents are related by:

$$I_{D,DC}(V_{GS}, V_{DS}, T_{sub}) = I_{D,iso}(V_{GS}, V_{DS}, T_{dev}) \qquad (13.1)$$
$$\text{with} \quad T_{dev} = T_{sub} + R_{th}(T_{sub}) \times P_{avg},$$

where T_{sub} is the substrate temperature, R_{th} is the effective thermal resistance and P_{avg} is the average power dissipated by the FET. It is to be noted that T_{dev} is the average

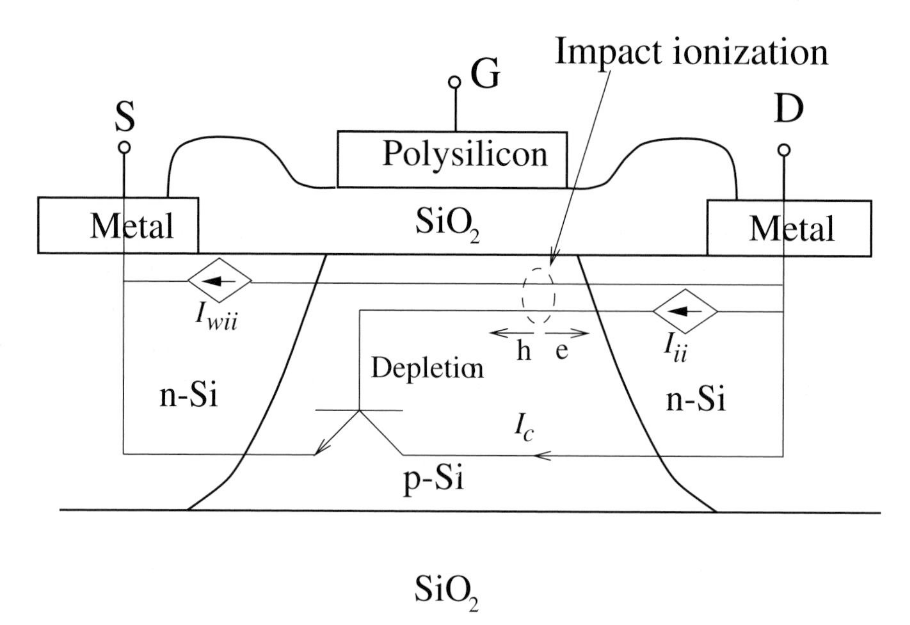

Fig. 13.6. SOI MOSFET layout showing the parasitic bipolar transistor which is activated by the hole (h) current generated by impact ionization on the drain side of the channel.

device temperature in the channel. For large devices a large surface temperature gradient might be present among the gate fingers and a distributed thermal network becomes necessary.

Isothermal I–V characteristics can be measured directly by using an efficient automated data acquisition system that searches for all bias points with the desired T_{dev} as the substrate temperature is varied [14]. Isothermal I–V characteristics can also be extracted directly from a full data set of I_{DS} and T_{dev}, obtained by varying V_{GS}, V_{DS} and T_{sub}. Such an approach allows for the isothermal I–V characteristic at any T_{dev} to be mathematically computed [14].

Both pulsed and isothermal I–V characteristics are important in understanding the behavior of RF circuits such as PAs. In such a circuit, the isothermal I–V characteristic establishes the device bias point, while the pulsed I–V ones are an indication of the device response under RF drive. As the RF power output increases, non-linearities will cause the device operating point and temperature to change. Hence a new isothermal I–V characteristic (one with a different T_{dev}) establishes the new bias point and a different pulsed I–V characteristic predicts the RF behavior.

While isothermal I–V characteristics can be readily obtained using the technique mentioned previously, pulsed I–V characteristics cannot, in general, be repeatedly measured for each bias point. Hence a model must be able to predict pulsed I–V characteristics. This can be accomplished by using the microwave characteristics of the device.

13.5 Small-signal modeling

13.5.1 Microwave data acquisition

The measurement techniques for isothermal I–V characteristics already discussed can be used to acquire isothermal microwave data for a device. This is an acceptable method if the DC bias dependence of the isothermal charge can be neglected. Another approach is to determine the pulsed I–V characteristic followed by pulsed RF measurements. Commercial test systems capable of performing these measurements are available.

In order to deembed the impact of the network analyzer and the measurement test bed, thorough microwave calibrations must be performed. The most ubiquitous calibration scheme used to remove errors from measured S parameter data is SOLT (short, open, load and thru). However, because of the difficulty of fabricating a good load standard, SOLT is not very desirable. The TRL (thru, reflect and line) technique is better since it does not rely on known standard loads but rather uses simple connections to allow for the error boxes to be characterized completely [15]. The 'thru' is a direct connection between the two ports at the desired reference plane. The 'reflect' uses a load with a large reflection coefficient, such as a short or an open. It is not necessary to know the reflection coefficient since it is extracted during the TRL calibration procedure. This then gives an additional verification of the accuracy of the TRL calibration performed. In the 'line' connection, a length of matched transmission line is connected between the two ports. The length of the line need not be known and it need not be lossless, since these parameters will be determined by the TRL procedure [16]. However, the length of the lines used determines the frequency band of validity for the TRL calibration. Further loss has been shown to increase the calibration accuracy, hence a lossy line can provide a usable calibration over a broader band.

In order to span a large frequency band, a multi-line method can be used such that shorter lines are used for higher frequencies and longer lines for lower frequencies.

13.5.2 Small-signal topology

Table modeling begins with the selection of a model topology that fits the bias-dependent small-signal S parameters of the device. Figure 13.7 shows the extrinsic version of the small-signal model topology which was introduced in Chapter 11 for three-terminal FETs.

The y parameters of the intrinsic FET model can be written as:

$$Y_{ij} = g_{ij} + \frac{j\omega C_{ij}}{1 + j\omega\tau_{ij}}, \qquad (13.2)$$

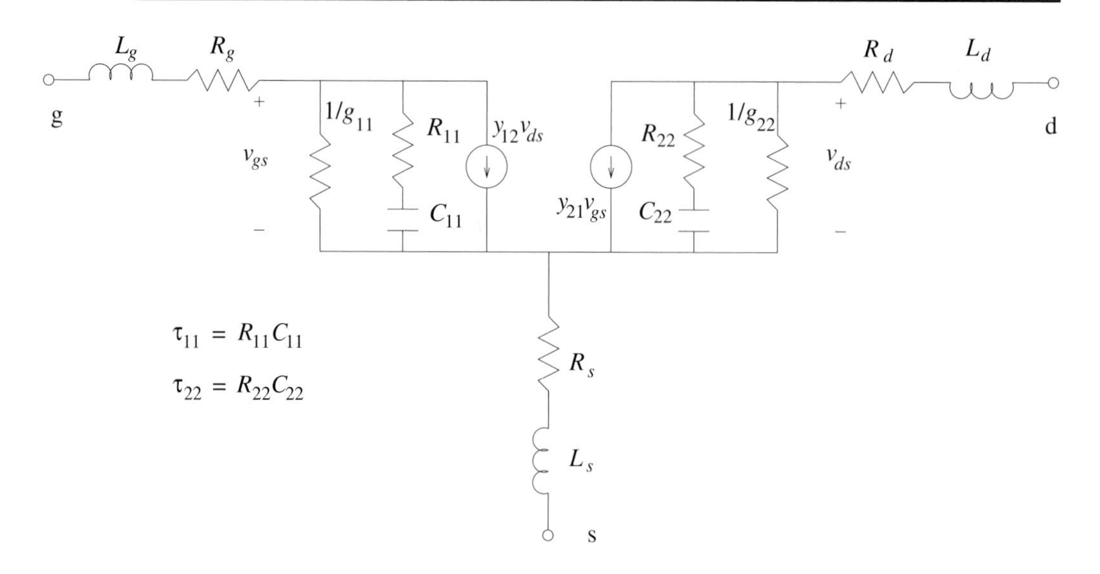

Fig. 13.7. A small-signal RC equivalent-circuit model.

where C_{ij}, g_{ij} and τ_{ij} are the bias-dependent capacitance (transcapacitance), conductance (transconductance) and non-quasi-static time constants, respectively. The non-quasi-static time constants τ_{ij} introduce a redistribution time for the charge, which becomes important at frequencies approaching f_{max} [17].

13.5.3 Parasitic deembedding

Although the intrinsic model is quite simple, the extrinsic model is complicated by a transformation from y to z parameters. Also deembedding of the parasitics associated with the FET and the package does not have an easy solution because of the fact that a DC gate current does not flow in many FETs. In the presence of resistive and inductive parasitics at the terminals, one verifies [18] that the extrinsic z_{ij} for the non-quasi-static model selected assumes a frequency dependence similar to that of the quasi-static model [19,20]:

$$\mathrm{Re}(z_{extr,ij}) = R_{ij} + \frac{A_{ij}}{\omega^2 + B_r^2}, \tag{13.3}$$

$$\frac{1}{\omega}\mathrm{Im}(z_{extr,ij}) = L_{ij} + \frac{B_{ij}}{\omega^2 + B_r^2} + \frac{G_{ij}}{\omega^2\left(\omega^2 + B_r^2\right)}, \tag{13.4}$$

where the various coefficients R_{ij} are now given by

$$R_{12} = R_s + \delta R_{12} \qquad \text{with} \quad \delta R_{12} = \frac{C_{12}\tau_D}{A},$$

Table 13.1. $z_{extr,ij}$ *fit parameters in terms of the model parameters.*

$L_{11} = L_g + L_s$	$L_{22} = L_d + L_s$
$L_{12} = L_{21} = L_s$	$G_{22} = G_{12} = 0$
$B = C_{11}g_d - C_{12}g_m$	$B_r = \frac{B}{A}$
$A_{12} = B_r B_{12}$	$B_{12} = -B_r \delta R_{12} - \frac{C_{12}}{A}$
$A_{22} = B_r B_{22}$	$B_{22} = \frac{C_{11}}{A^2}(B\tau_D + A)$
$I_{p11} = (\tau_G + \tau_D)g_d + C_{22}$	$I_{p21} = -(\tau_G + \tau_D)g_m - C_{21}$
$A_{11} = \frac{g_d}{A} + B_r B_{11}$	$B_{11} = -B_r \delta R_{11} + \frac{I_{p11}}{A}$
$A_{21} = -\frac{g_m}{A} + B_r B_{21}$	$B_{21} = -B_r \delta R_{21} + \frac{I_{p21}}{A}$
$G_{11} = \frac{-Bg_d}{A^2}$	$G_{21} = \frac{Bg_m}{A^2}$

$$R_{22} = R_s + R_d + \delta R_{22} \quad \text{with} \quad \delta R_{22} = -\frac{C_{11}\tau_D}{A},$$

$$R_{11} = R_s + R_g + \delta R_{11} \quad \text{with} \quad \delta R_{11} = -\frac{\tau_G \tau_D g_d + C_{22}\tau_G}{A},$$

$$R_{21} = R_s + \delta R_{21} \quad \text{with} \quad \delta R_{21} = \frac{\tau_G \tau_D g_m + C_{21}\tau_G}{A},$$

with $A = C_{12}C_{21} - C_{11}C_{22} + \tau_D(g_m C_{12} - g_d C_{11})$ and δR_{ij} the non-quasi-static corrections. The remaining parameters are given in Table 13.1.

The measured z-parameter data can be mathematically fitted to Equations (13.3) and (13.4) using least squares methods. The common denominator across all the z parameters can be used to obtain parametric curves between two extrinsic z parameters [21]. The intercept of these linear regressions gives the R_{ij} and L_{ij} coefficients in terms of R_{11} and L_{12}, respectively. Next the independent coefficients R_{11}, A_{ij} and B_r^2 can be obtained from a simultaneous least square fit of the $\text{Re}(z_{ij})$ after rewriting Equation (13.3) in terms of a fraction. Finally G_{21} and L_{12} can be obtained from a simultaneous fit of all the $\text{Im}(z_{ij})$ with the use of the theoretical model constraints: $A_{ij} = \pm g_{ij}/A + B_r B_{ij}$. Figure 13.8 shows an obtained fit for the S parameters of an LD MOSFET.

The model parameters (parasitics plus intrinsic elements) can be expressed in terms of the fitting parameters of Equations (13.3) and (13.4) by inverting their original relationships [18]. One obtains the formulas given in Table 13.5.3 for calculating the extrinsic and intrinsic model parameters in terms of the known fit parameters A_{ij}, R_{ij}, L_{ij}, B_r^2 and G_{ij} [19] and the unknown source resistance R_s. For the two-τ non-quasi-static model considered there exists therefore a continuum of solutions as

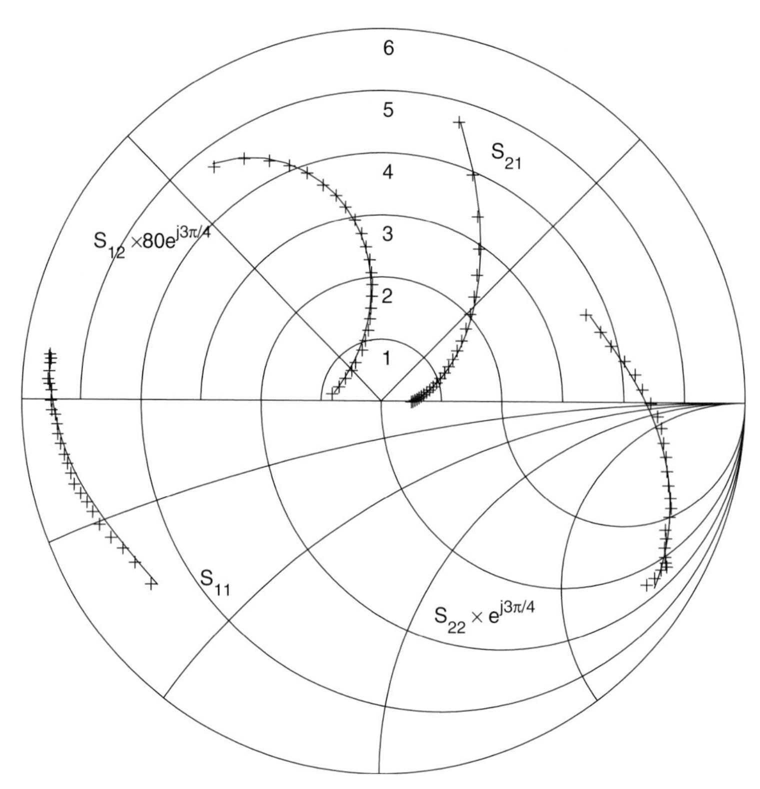

Fig. 13.8. Comparison between small-signal fitted S parameters (full lines) and measured data (plus signs) for $V_{GS} = 4.5$ V, $V_{DS} = 20$ V and T_{dev} of 90 °C for an LD MOSFET. (P. Roblin, S. Akhtar and J. Strahler, *IEEE Microwave and Guided Wave Letters*, Vol. 10, No. 8, pp. 322–324, August 2000. ©2000 IEEE.)

a function of R_s which yield the exact same fit to the z_{ij} or S_{ij} parameters. Note however, that the capacitances C_{11} and C_{12}, the extrinsic g_D and g_M, and the parasitic inductances L_s, L_g and L_d are independent of R_s and hence are uniquely extracted.

A range analysis, such that all extracted parameters have the correct sign, can be used to limit the final solution. Assuming that extrinsic parameters are bias-independent, a multi-bias analysis can be used to pin the values of the parasitics. Figure 13.9 shows such an analysis whereby the parasitic drain and gate resistances are plotted as functions of the source resistance. Points of convergence in the curves indicate the final solution.

Once parasitic elements have been deembedded, the intrinsic data can be fitted to extract the small-signal parameters. It must be noted that the g_m (g_{21}, transconductance) and g_d (g_{22}, drain conductance) obtained may be very different from the DC $g_m(I{-}V)$ and $g_d(I{-}V)$ values. This will be the case if there is low-frequency dispersion in the measured microwave data. The extracted parameters are the AC $g_m(RF)$ and $g_d(RF)$.

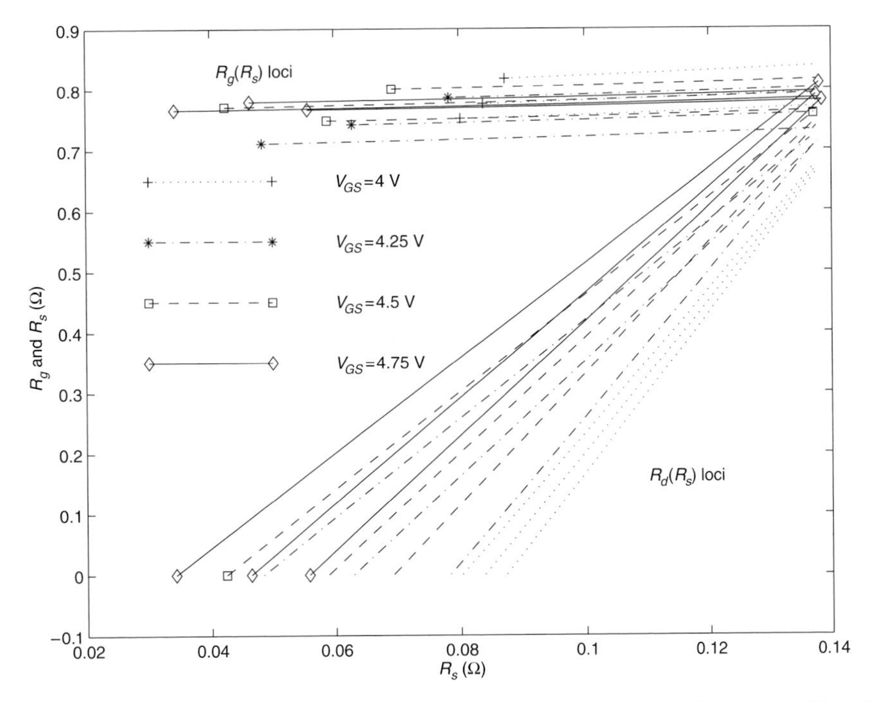

Fig. 13.9. Loci of R_d and R_g as functions of R_s for various gate and drain voltages. Note that the loci are limited to values of R_s giving the correct physical sign for all parameters. (P. Roblin, S. Akhtar and J. Strahler, *IEEE Microwave and Guided Wave Letters*, Vol. 10, No. 8, pp. 322–324, August 2000. ©2000 IEEE.)

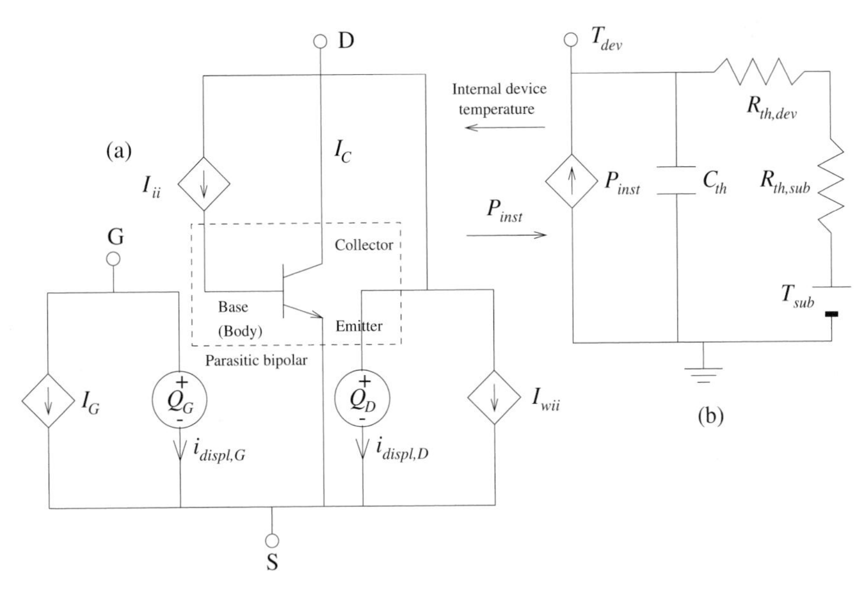

Fig. 13.10. Self-biasing model topology to fit both DC and RF and electrical network representing the thermal network model. (S. Akhtar and P. Roblin, *Analog Integrated Circuits and Signal Processing*, Vol. 25, No. 2, pp. 115–131, November 2000.)

Table 13.2. *Model parameters in terms of the $z_{extr,ij}$ fit parameters.*

$L_s = L_{12} = L_{21}$	$L_g = L_{11} - L_s$
$L_d = L_{22} - L_s$	$B_r = \dfrac{B_{11}B_r^2 - G_{11}}{A_{11}}$
$C_{11} = \left(-\dfrac{G_{11}}{B_r^2} + \dfrac{A_{12}}{A_{22}}\dfrac{G_{21}}{B_r^2}\right)^{-1}$	$C_{12} = \left(-\dfrac{G_{21}}{B_r^2} + \dfrac{A_{22}}{A_{12}}\dfrac{G_{11}}{B_r^2}\right)^{-1}$
$g_M = \dfrac{-G_{21}}{G_{11}(R_{22} + A_{22}/B_r^2) - G_{21}(R_{12} + A_{12}/B_r^2)}$	$g_D = \dfrac{G_{11}}{G_{11}(R_{22} + A_{22}/B_r^2) - G_{21}(R_{12} + A_{12}/B_r^2)}$
$R_d = \left(-1 + \dfrac{A_{22}}{A_{12}}\right) R_s + \left(R_{22} - R_{12}\dfrac{A_{22}}{A_{12}}\right)$	$\tau_D = \dfrac{R_s - R_{12}}{B_r\left(R_{12} - R_s + A_{12}/B_r^2\right)}$
$c_{11p} = -\dfrac{A_{22}}{A_{12}}R_s + \left(\dfrac{A_{22}}{B_r^2} + R_{12}\dfrac{A_{22}}{A_{12}}\right)$	$c_{12p} = R_s - R_{12} - \dfrac{A_{12}}{B_r^2}$
$B = -\dfrac{B_r^2}{c_{11p}G_{11} + c_{12p}G_{21}}$	$R_{p21} = (R_s - R_{21})\dfrac{B}{B_r}$
$g_m = \dfrac{BG_{21}}{B_r^2}$	$g_d = -\dfrac{BG_{11}}{B_r^2}$
$I_{p21} = B_{21}\dfrac{B}{B_r} + B(R_{21} - R_s)$	$\tau_G = -\dfrac{I_{p21}}{2g_m} \pm \left[\left(\dfrac{I_{p21}}{2g_m}\right)^2 + \dfrac{R_{p21}}{g_m}\right]^{1/2}$
$C_{21} = -[I_{p21} + (\tau_D + \tau_G)g_m]$	$C_{22} = -\dfrac{B}{C_{11}}\left(\dfrac{1}{B_r} - \dfrac{C_{12}C_{21}}{B} + \tau_D\right)$
$R_{p11} = \tau_G\tau_D g_d + \tau_G C_{22}$	$R_g = R_{11} - R_s + \dfrac{R_{p11}B_r}{B}$

13.6 Large-signal modeling

While the small-signal model describes the device behavior under low power, a large-signal model is needed to predict device performance for higher-power operation. Obviously the large-signal model used must reduce to the small-signal model under low power.

13.6.1 Model formulation

Figure 13.10 shows an intrinsic large-signal model topology for a three-terminal FET type device. This large-signal model features three distinct subcircuits. The first subcircuit consists of the non-quasi-static FET model derived in Chapter 11 except that

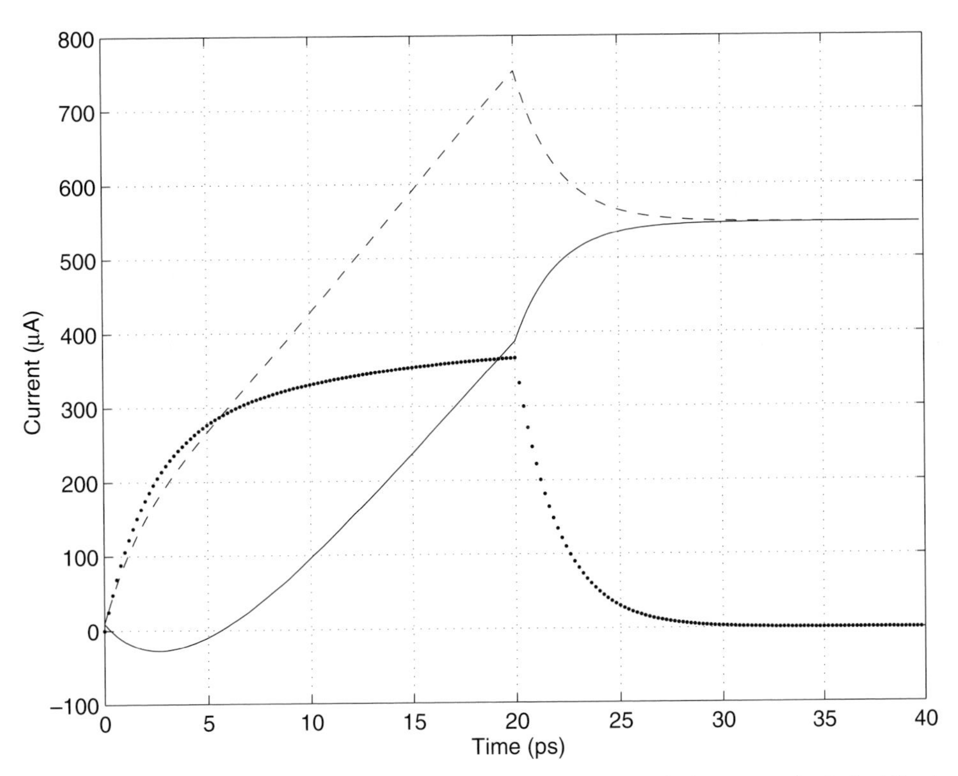

Fig. 13.11. Non-quasi-static transient simulation. The full line is the drain current, the dashed line is the source current, and the dotted line is the gate current.

the drain current is denoted I_{wii} (without impact ionization). The second subcircuit is the parasitic bipolar transistor driven by the impact ionization current I_{ii}. The parasitic transistor is responsible for introducing kinks in the I–V characteristics of SOI MOSFETs and an associated low-frequency dispersion effect. The third subcircuit is the thermal network which is driven by the power dissipated by the FET and accounts for the self-heating effect.

Let us start with the non-quasi-static FET model. In order to better understand the contribution of the various elements of the large-signal model, Figure 13.11 shows the simulated output currents of an SOI MOSFET for an input gate ramp. A voltage ramp from 0.5 to 5.0 V with a 20 ps rise time is applied to the gate while the drain is held at 1.5 V and the source is grounded. In Figure 13.11, if only the large-signal current element existed in the model, both the drain and source current components would be the same. But the charge element introduces a transient gate current which causes the drain and source current to be different. The non-quasi-static times constants give rise to the smooth relaxation in the currents at the points where the slope of the applied voltage changes.

The large-signal representations can be obtained from the extracted small-signal model parameters using the following equations:

$$\left.\begin{aligned}
&I_{D,trans}(v_{GS}, v_{DS}, V_{GS}, V_{DS}, T_{dev}) = \\
&\quad \int_{V_{GS}}^{v_{GS}} g_{21}(V'_{GS}, V_{DS}, T_{dev})\, dV'_{GS} + \int_{V_{DS}}^{v_{DS}} g_{22}(v_{GS}, V'_{DS}, T_{dev})\, dV'_{DS}, \\
&Q_G(v_{GS}, v_{DS}, V_{GS}, V_{DS}, T_{dev}) = \\
&\quad \int_{V_{GS}}^{v_{GS}} C_{11}(V'_{GS}, V_{DS}, T_{dev})\, dV'_{GS} + \int_{V_{DS}}^{v_{DS}} C_{12}(v_{GS}, V'_{DS}, T_{dev})\, dV'_{DS}, \\
&Q_D(v_{GS}, v_{DS}, V_{GS}, V_{DS}, T_{dev}) = \\
&\quad \int_{V_{GS}}^{v_{GS}} C_{21}(V'_{GS}, V_{DS}, T_{dev})\, dV'_{GS} + \int_{V_{DS}}^{v_{DS}} C_{22}(v_{GS}, V'_{DS}, T_{dev})\, dV'_{DS},
\end{aligned}\right\} \quad (13.5)$$

with $I_{D,trans}(V_{GS}, V_{DS}, V_{GS}, V_{DS}, T_{dev}) \simeq I_D(V_{GS}, V_{DS}, T_{sub})$. Note that we will neglect the possible DC bias dependence (V_{GS}, V_{DS}) of the isothermal gate charge Q_G and drain charge Q_D. We will also neglect the dependence of $I_{D,trans}$ on the internal body to source voltage, V_{BS}. Such is the case in LD MOS technology where the source is connected to the body and ground through a sinker diffusion step (see Figure 13.12). In partially depleted SOI MOSFETs, even in the absence of internal body ties, this can be verified to be a good assumption (see Figure 13.18 for a verification). However, for other technologies where traps play an important role, this assumption can be the cause of discrepancies between measured and simulated results.

We shall assume that I_G can be set to zero, which is the case for LD MOSFETs and SOI MOSFETs, but not for other FETs such as MESFETs and HEMTs. The total non-quasi-static current flowing through the charge elements Q_G or Q_D of Figure 13.7 can be written as (using i for G or D):

$$i_{disp,i}(t) = \frac{dQ_i(v_{GS}, v_{DS}, T_{dev})}{dt} - \frac{d}{dt}\left[\tau_i(v_{GS}, v_{DS}, T_{dev}) i_{disp,i}\right], \quad (13.6)$$

where τ_i is the non-quasi-static charge redistribution time associated with the particular charge element, Q_i [17].

The total gate and drain currents can then be expressed as:

$$\left.\begin{aligned}
&i_G(t) = i_{disp,G}(t), \\
&i_D(t) = I_{D,trans}(t) + i_{disp,D}(t).
\end{aligned}\right\} \quad (13.7)$$

13.6.2 Tensor product B-splines

In order to extract the large-signal model parameters, the small-signal parameter data (AC gs and Cs) need to be fitted over the entire bias range and integrated.

This is where table methods for modeling give an advantage by using a generic set of equations to represent important model parameters. The tensor product B-spline

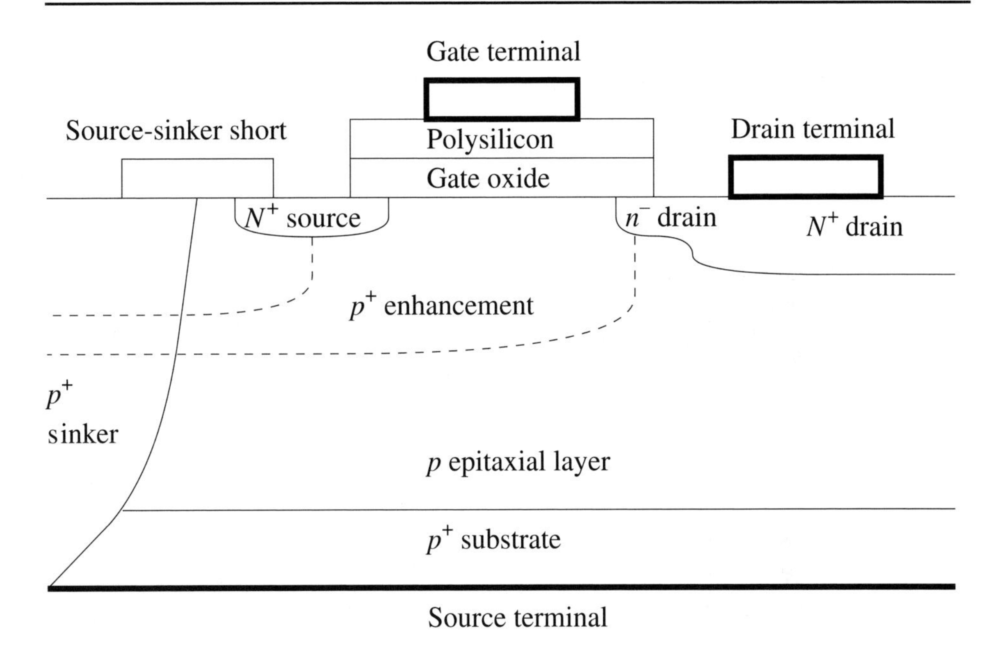

Fig. 13.12. Cross-sectional view of an LD MOSFET. The left-hand p^+ source sinker is used to bring the source contact to the bottom of the transistor. The sinker acts at the same time as a resistive metallic body tie, greatly reducing the floating body and associated parasitic bipolar effects.

(TPS) is one such set of equations [22,23,4]. TPS can be used to represent a bivariate function as follows [24]:

$$S(V_{GS}, V_{DS}) = \sum_{i=1}^{m} \sum_{j=1}^{n} a_{ij}\, B_{i,k_{V_{GS}},t_{V_{GS}}}(V_{GS})\, B_{j,k_{V_{DS}},t_{V_{DS}}}(V_{DS}), \tag{13.8}$$

where $k_{V_{GS}}$ and $k_{V_{DS}}$ are the B-spline orders in the V_{GS} and V_{DS} directions, respectively, $t_{V_{GS}}$ and $t_{V_{DS}}$ are the knot sequences in the V_{GS} and V_{DS} directions, respectively, $B_{i,k_{V_{GS}},t_{V_{GS}}}$ and $B_{j,k_{V_{DS}},t_{V_{DS}}}$ are one-dimensional B-spline polynomial functions, and a_{ij} are the TPS coefficients that need to be determined.

Storage requirements for the TPS method are very reasonable, since for $m = 20$ and $n = 20$, only 400 a_{ij} coefficients are required. For $k_{VG} = 4$ and $k_{VD} = 4$, only 16 coefficients are required to compute functional and derivative values at any particular bias point, due to the variation diminishing property of B-splines. Hence computation proceeds rapidly. By using B-splines of order 4, second order continuity is guaranteed. Also, mixed derivatives match, ensuring that the TPS method will be charge conserving. The TPS method can be used in a least squares approach to extract the a_{ij} coefficients from small-signal gradient and functional data [25]. Using the least squares technique leads to a unique solution, hence avoiding the need to go through time-consuming parameter optimization algorithms.

The TPS method can also be extended to represent multi-variate functions. For

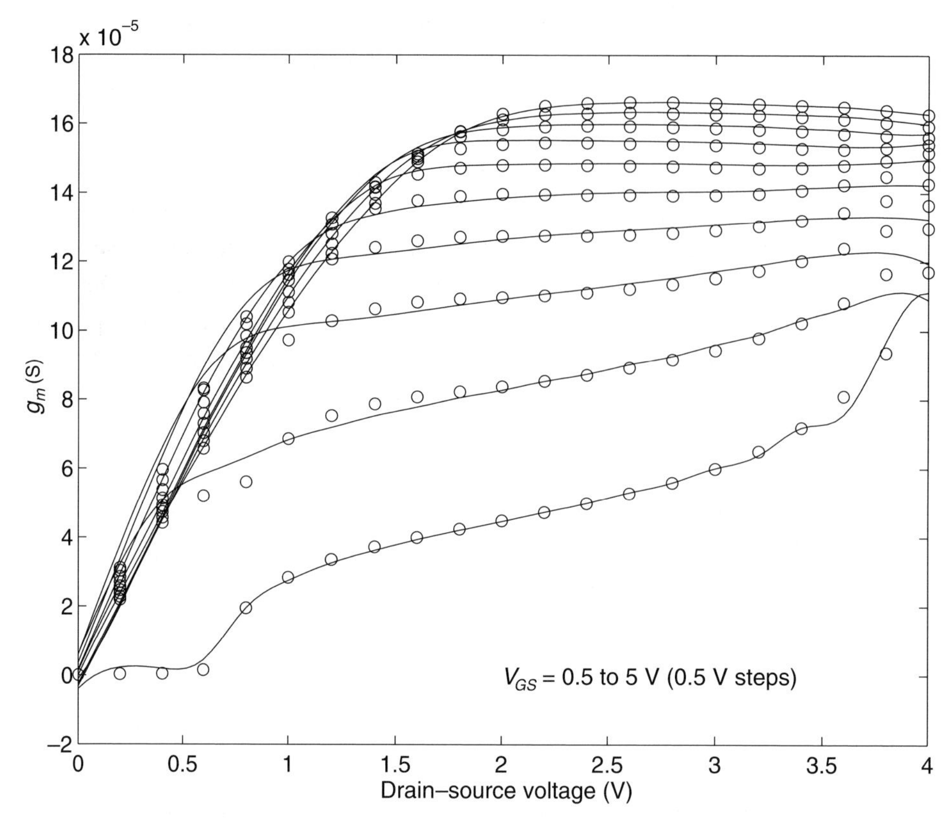

Fig. 13.13. Comparison of AC $g_m(RF)$ obtained from TPS (full lines), with those obtained from the small-signal fit (circles). (S. Akhtar and P. Roblin, *Analog Integrated Circuits and Signal Processing*, Vol. 25, No. 2, pp. 115–131, November 2000.)

example, temperature can be included as follows:

$$S(V_{GS}, V_{DS}, T_{dev}) = \sum_{i=1}^{m}\sum_{j=1}^{n}\sum_{l=1}^{p} a_{ijk} B_{i,k_{GS},t_{GS}}(V_{GS}) B_{j,k_{DS},t_{DS}}(V_{DS}) B_{l,k_{dev},t_{dev}}(T_{dev}).$$

Such a formulation can be used to extract isothermal I–V characteristics from a full data set of I_{DS} and T_{dev}, obtained by varying V_{GS}, V_{DS}, and T_{sub}.

13.6.3 $I-V$ characteristics

TPS can be used to perform the integration of the AC $g_m(RF)$ and $g_d(RF)$ together with a DC bias forcing condition. Figure 13.13 shows the AC $g_m(RF)$ obtained using TPS (full lines) compared with that extracted from the small-signal fit (circles). This integration yields a pulsed I–V characteristic. The forcing condition is necessary because pulsed I–V characteristics are bias-point-dependent and hence there are as many pulsed I–V characteristics as bias points. However, from a modeling point of view, it is not feasible to redo a complete integration for each new bias point. This is especially true for non-linear circuits where the bias point can easily depart from its initial value and hence is dynamically set.

In order to bridge the gap between the isothermal I–V characteristics (which predict the device biasing) and transient I–V characteristics, a parasitic bipolar transistor with a floating base driven by the impact ionization current is used in the model in Figure 13.10 to handle this low-frequency dispersion [26]. Other approaches use a low-pass filter [27] to map this dispersion. The use of a parasitic bipolar is based on physical considerations, since FETs contain a parasitic bipolar junction transistor (BJT). However, depending upon the technology being used, the role of this parasitic can be either substantial (in partially depleted floating body SOI MOSFETs) or minimal (in LD MOSFETs due to the source sinker).

An RF I–V characteristic can be defined as one with derivatives agreeing with the AC $g_m(RF)$ and $g_d(RF)$, while at the same time agreeing with the isothermal I–V characteristic in low-drain-voltage regions. This is based upon observations that low-frequency dispersions are more profound at higher drain voltages. For an SOI MOSFET, the onset of the kink can be used to define the low-drain-voltage region [28].

The model can then predict the appropriate transient I–V characteristics $I_{D,trans}(v_{GS}, v_{DS}, V_{GS}, V_{DS}, T_{dev})$, from the isothermal, $I_{D,iso}(v_{GS}, v_{DS}, T_{dev})$, and the RF, $I_{D,RF}(v_{GS}, v_{DS}, T_{dev})$, I–V characteristics for any bias V_{GS}, V_{DS}. TPS can be used to represent both the isothermal and RF I–V characteristics as shown for an SOI device in Figure 13.14.

13.6.4 Parasitic bipolar topologies

Model topologies for the internal parasitic bipolar transistor are presented in Figure 13.15. For a wide range of drain to source voltages, the internal BJT operates in the active mode, for which model (a) in Figure 13.15 can be used to represent the BJT. For this model I_{wii} and I_{ii} can be expressed in terms of $I_{D,iso}(v_{GS}, v_{DS}, T_{dev})$ and $I_{D,RF}(v_{GS}, v_{DS}, T_{dev})$ as follows:

$$
\begin{aligned}
I_{wii}(v_{GS}, v_{DS}, T_{dev}) = {} & \frac{1}{\alpha}\left[I_{D,RF}(v_{GS}, v_{DS}, T_{dev}) - I_{D,iso}(v_{GS}, v_{DS}, T_{dev})\right] \\
& + I_{D,iso}(v_{GS}, v_{DS}, T_{dev}),
\end{aligned}
\tag{13.9}
$$

$$
I_{ii}(v_{GS}, v_{DS}, T_{dev}) = \frac{1-\alpha}{\alpha}\left[I_{D,iso}(v_{GS}, v_{DS}, T_{dev}) - I_{D,RF}(v_{GS}, v_{DS}, T_{dev})\right].
\tag{13.10}
$$

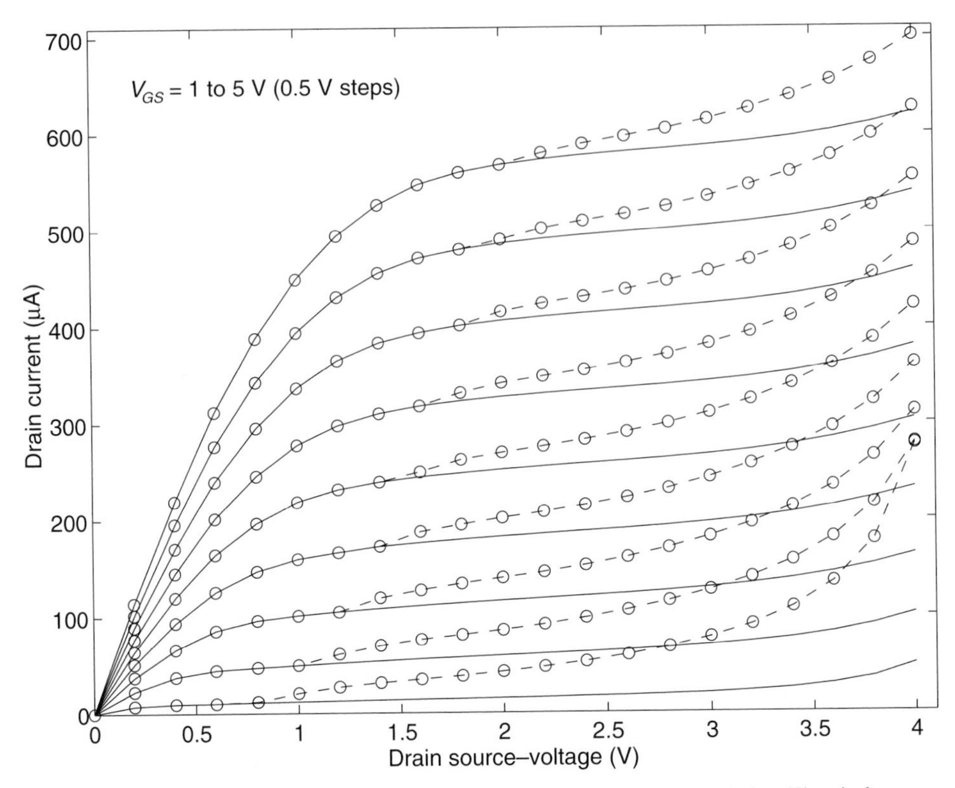

Fig. 13.14. The fitted RF (full lines) and DC (dashed-dashed) I–V characteristics. The circles are DC data. (S. Akhtar and P. Roblin, *Analog Integrated Circuits and Signal Processing*, Vol. 25, No. 2, pp. 115–131, November 2000.)

The transient I–V characteristic associated with this model can then be written as:

$$I_{D,trans}(v_{GS}, v_{DS}, V_{GS}, V_{DS}, T_{dev}) = I_{wii}(v_{GS}, v_{DS}, T_{dev}) + I_{ii}(v_{GS}, v_{DS}, T_{dev}) + \frac{\alpha}{1-\alpha} I_{ii}(V_{GS}, V_{DS}, T_{dev}), \quad (13.11)$$

where V_{GS}, V_{DS} and T_{dev} are the DC bias point around which the transient I–V characteristic is generated.

From Equation (13.11), it is clear that $I_{D,trans}(v_{GS} = 0, v_{DS} = 0, V_{GS}, V_{DS}, T_{dev})$ does not equal zero, yet this is what is measured in reality [29]. The reason for this is that model (a) does not hold when the parasitic BJT goes into cut-off ($v_{DS} < 0.5$ V). In order to correct for this small deviation for V_{DS} below 0.5 V, model (b) in Figure 13.15 can be used. This general model for the bipolar transistor holds for the entire bias range.

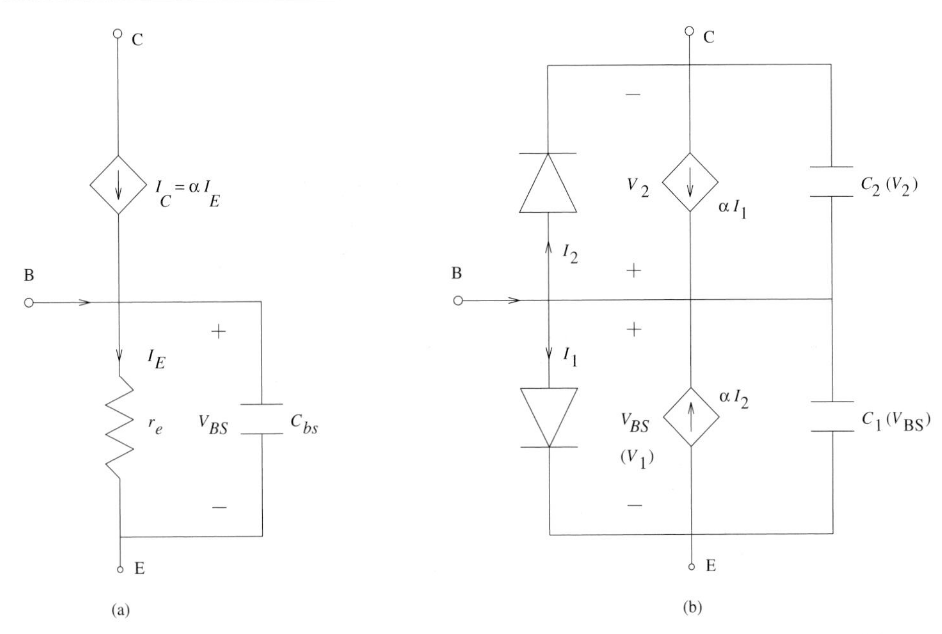

Fig. 13.15. Parasitic bipolar model topologies: (a) a simple model not valid in cut-off (b) valid over the entire bias range. (S. Akhtar and P. Roblin, *Analog Integrated Circuits and Signal Processing*, Vol. 25, No. 2, pp. 115–131, November 2000.)

13.6.5 Charge

The isothermal gate charge, Q_G, can be extracted by fitting C_{11} and C_{12} to the derivatives of the TPSs. Figure 13.16(a) shows the extracted charge. For comparison, the gate charge obtained from a physical device simulator is given in Figure 13.16(b). Excellent agreement is observed between the recovered and measured gate charges. Likewise, the drain charge, Q_D, can be extracted by fitting C_{21} and C_{22} to the derivatives of the TPSs. The non-quasi-static time constants, τ_{11-22}, can be fitted using TPSs in order to provide interpolation between data points.

13.7 Electrothermal modeling

In power devices where device temperatures can soar, thermal modeling and its feedback on the device current and charge, are important to correctly predict circuit behavior.

Figure 13.10 shows a simple thermal network topology which calculates the steady-state isothermal temperature of an FET as a function of the power dissipated by the FET. The computed device temperature is used to compute the corresponding isothermal I–V characteristic and charge. Based on this thermal topology, the thermal

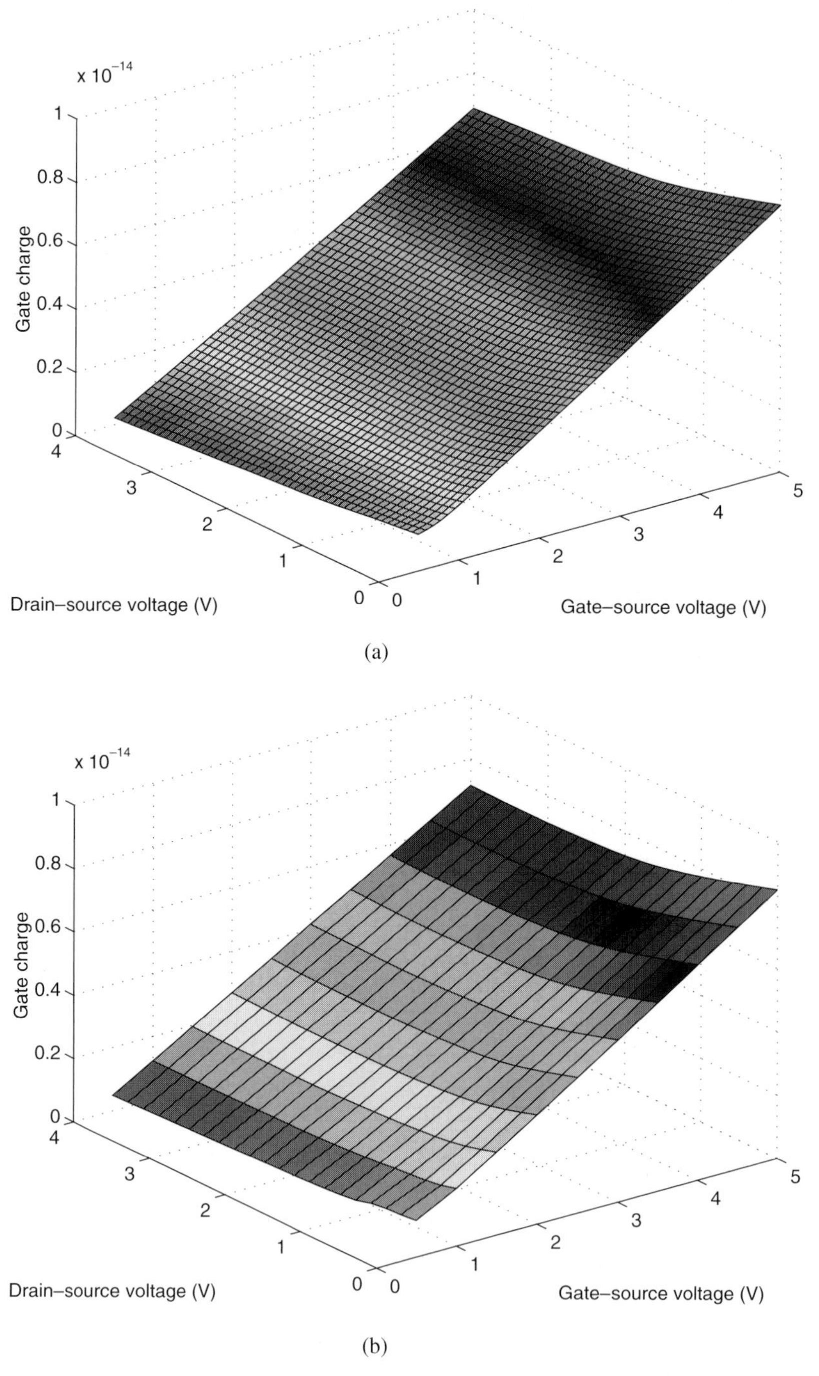

Fig. 13.16. Comparison of gate charge obtained from: (a) TPS and (b) a physical device simulator. (S. Akhtar and P. Roblin, *Analog Integrated Circuits and Signal Processing*, Vol. 25, No. 2, pp. 115–131, November 2000.)

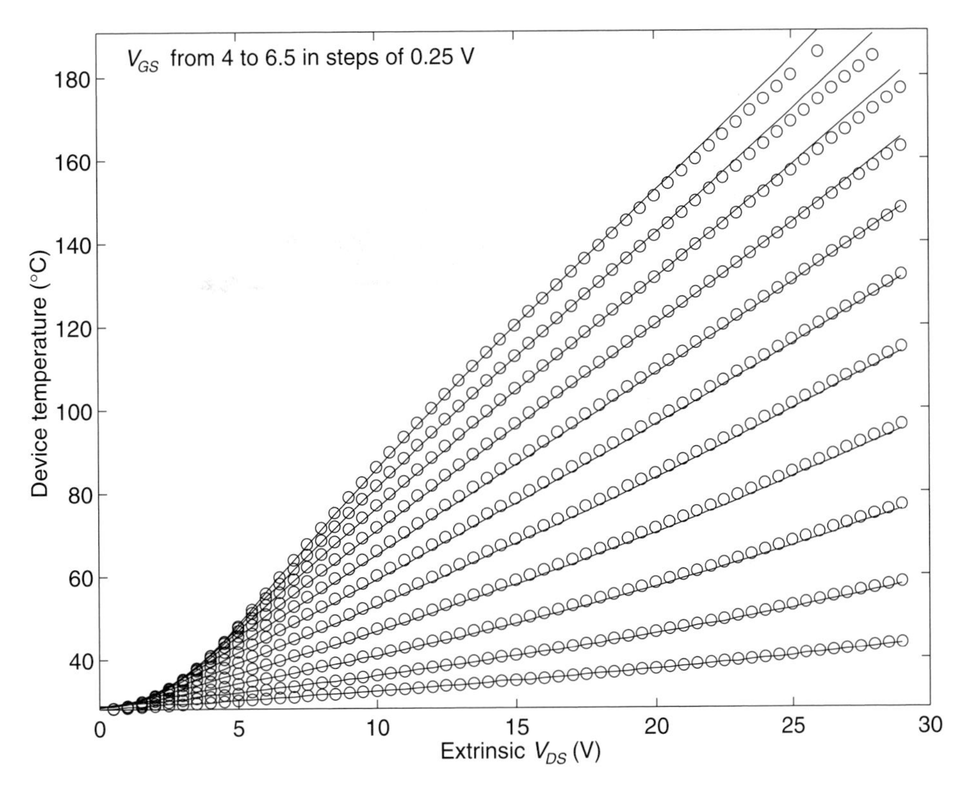

Fig. 13.17. Prediction of device temperature for an LDMOSFET using a single R_{th} for a T_{sub} of 29 °C.

response is given by:

$$P_{inst}(t) = C_{th}\frac{dT_{dev}(t)}{dt} + \frac{T_{dev}(t) - T_{sub}}{R_{th,dev}(T_{sub}) + R_{th,sub}}$$
$$= v_{DS}(t)i_D(t) + v_{GS}(t)i_G(t), \tag{13.12}$$

where $P_{inst}(t)$ is the instantaneous device power dissipated by the FET, $R_{th,dev}(T_{sub})$ is the device thermal resistance, $R_{th,sub}(T_{sub})$ is the substrate thermal resistance, and C_{th} is the device thermal capacitance. The contribution of the thermal capacitance is to average the instantaneous power dissipation in the FET. Hence for steady-state operation, the device temperature is linearly related to the power dissipated.

The simple thermal subcircuit, using a single R_{th}, is able to predict well the entire thermal map of the device. Figure 13.17 shows the measured device temperature (circles) for an LD MOSFET, while the device temperature predicted by the model is given by the full lines. Although a single R_{th} can do an excellent job of predicting the steady-state response, to obtain better accuracy for the transient thermal response, a distributed thermal topology would have to be used [30].

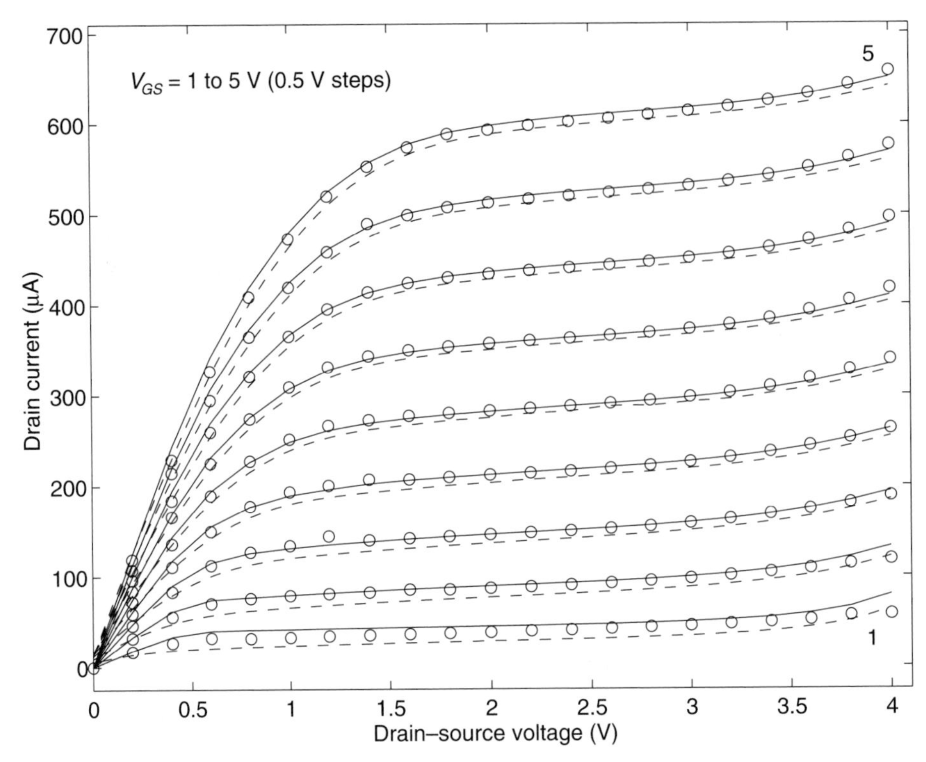

Fig. 13.18. Transient I–V characteristics recovered from AC $g_m(RF)$ and $g_d(RF)$ integration and bias point force (dashed lines), transient I–V characteristics generated by the self-biasing microwave model (full lines), and measured transient I–V characteristics (for V_{GS} = 3.0 V and V_{DS} = 2.6 V) (circles). (S. Akhtar and P. Roblin, *Analog Integrated Circuits and Signal Processing*, Vol. 25, No. 2, pp. 115–131, November 2000.)

13.8 Circuit simulations

Any model developed must be easily implementable in a microwave circuit simulator environment. Many commercial simulators allow for implementation of compiled user-defined models by writing external code. Simulators that allow for harmonic balance simulations are well suited for designing and optimizing microwave circuits. Once a model has been implemented, the simulator becomes a virtual test bed.

13.8.1 Pulsed I–V characteristics

The model's ability to predict pulsed I–V characteristics is highlighted in Figure 13.18. In this simulation, a fixed DC bias point of V_{GS} = 3 V and V_{DS} = 2.6 V is used. The simulated pulsed I–V characteristic is given by the full lines, while that measured is given by circles. The figure also includes an I–V characteristic recovered by the direct integration of the AC $g_m(RF)$ and $g_d(RF)$ plus a bias point force (dashed

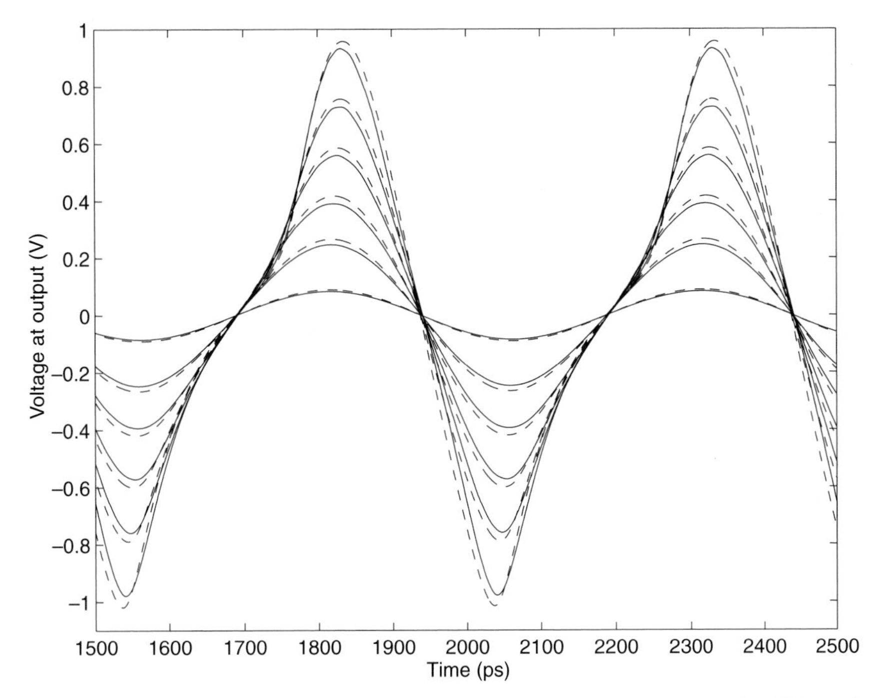

Fig. 13.19. Simulated (full lines) output voltage across 50 Ω load resistance for different input power levels versus measured data (dashed lines). (S. Akhtar and P. Roblin, *Analog Integrated Circuits and Signal Processing*, Vol. 25, No. 2, pp. 115–131, November 2000.)

lines). The model using isothermal and RF I–V characteristics is actually able to do a better job and can predict pulsed I–V characteristics for any bias point.

13.8.2 Power amplifier

Figures 13.19–13.21 compare the simulated response with the measured data of a power amplifier circuit using a single SOI MOSFET biased in class A with a simple output matching circuit at a fundamental frequency of 2 GHz. Figure 13.19 compares the output voltage across a 50 Ω load resistance for various input power levels obtained on a circuit simulator (full lines) with measured data (dashed lines). Figure 13.20 compares the simulated output power and power gain (left-hand axis), and power-added efficiency (right-hand axis), for various input power levels P_{in} (dBm) (full lines) with measured data (dashed lines with circles). Finally, Figure 13.21 compares the first three output power harmonics for various input power levels obtained on a circuit simulator (full lines) with measured data (dashed lines with circles). For input power levels greater than the dashed vertical line, the range of the model is exceeded. In this region, the model can use linear extrapolation, which leads to a graceful degradation in the model predictions.

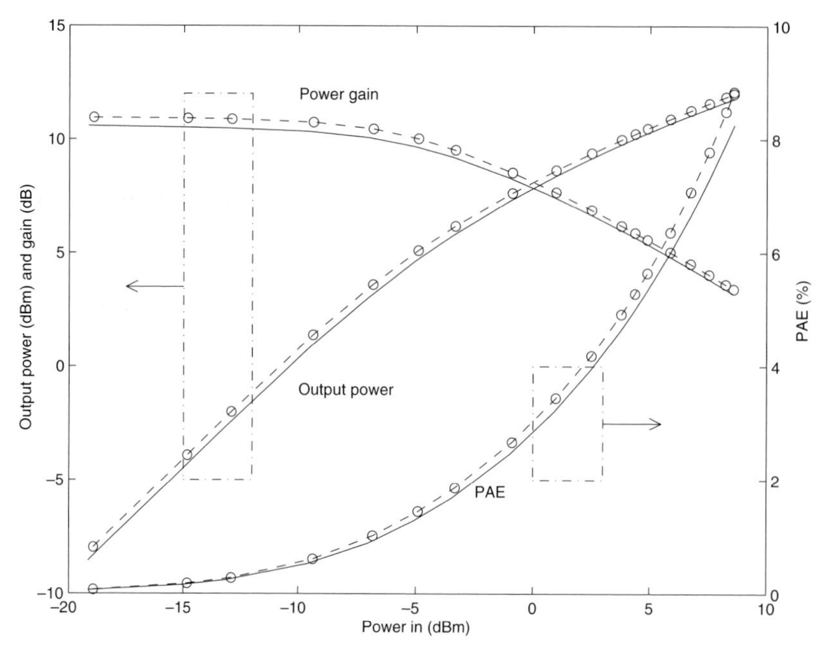

Fig. 13.20. Simulated (full lines) output power, power gain (left-hand axis) and power-added efficiency (right-hand axis) versus measured data (dashed lines). (S. Akhtar and P. Roblin, *Analog Integrated Circuits and Signal Processing*, Vol. 25, No. 2, pp. 115–131, November 2000.)

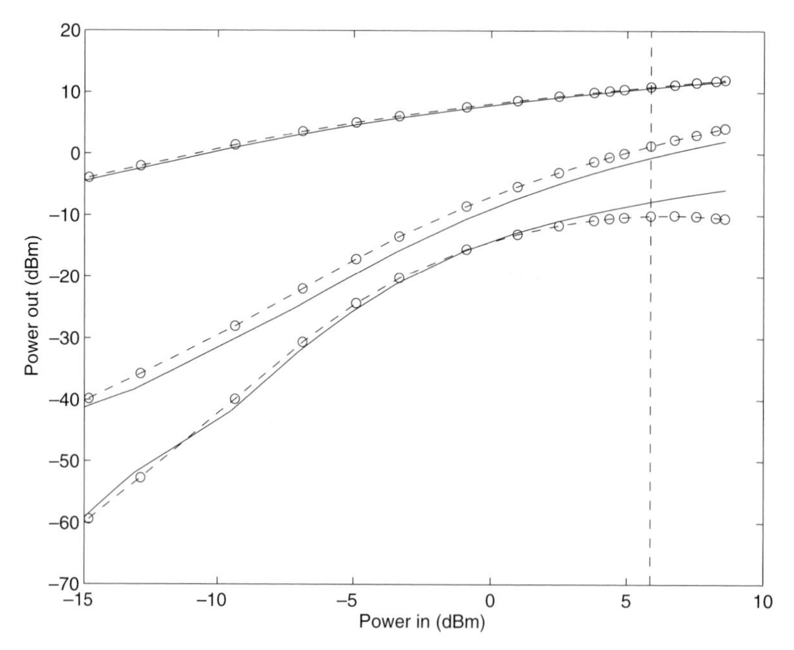

Fig. 13.21. Simulated (full lines) first three output power harmonics versus measured data (dashed lines). Model range is limited to input power levels less than dashed vertical line. (S. Akhtar and P. Roblin, *Analog Integrated Circuits and Signal Processing*, Vol. 25, No. 2, pp. 115–131, November 2000.)

13.9 Conclusion

This chapter has examined approaches to the microwave modeling of three-terminal FET devices using table methods that incorporate physical constraints. The techniques, however, can easily be applied to other types of FETs and bipolar devices, provided that the correct topology is first developed.

Device modeling is a very interdisciplinary study. It demands a thorough knowledge of not only device physics, but also measurement and characterization techniques, computational algorithms, programming aspects, and circuit design, simulation and measurement. Combination of this knowledge makes for better device modeling. This chapter has attempted to highlight all of these areas.

13.10 Bibliography

[1] J.-P. Colinge, *Silicon-on-Insulator Technology: Materials to VLSI*, Kluwer, Norwell, MA, 1991.

[2] J. Pritiskutch and B. Hanson, 'Understanding LDMOS device fundamentals,' *Microwaves & RF*, Vol. 37, No. 8, pp. 114–116, August 1996.

[3] D. E. Root and B. Hughes, 'Principles of nonlinear active device modeling for circuit simulation,' *Proceedings 32nd ARFTG Conference, Tempe, AZ*, Piscataway, New Jersey, pp. 3–26, 1988.

[4] R. R. Daniels, A. T. Yang and J. P. Harrang, 'A universal large/small signal 3-terminal FET model using a non-quasi-static charge-based approach,' *IEEE Transactions on Electron Devices*, Vol. 40, No. 10, pp. 1723–1729, October 1993.

[5] J. L. B. Walker, *High Power GaAs FET Amplifiers*, Artech House, Norwood, MA, 1993.

[6] B. Razavi, *RF Microelectronics*, Prentice Hall, Upper Saddle River, New Jersey, 1998.

[7] T. Ruttan, 'Designing amplifiers for wireless systems,' *Microwaves & RF*, Vol. 37, No. 1, pp. 89–100, January 1998.

[8] P. Antognetti and G. Massobrio, *Semiconductor Device Modeling with SPICE*, McGraw-Hill, New York, 1988.

[9] D. E. Ward, Charge-based modeling of capacitance in MOS transistors, Doctoral Dissertation, Stanford University, June 1981.

[10] K. A. Jenkins and J. Y.-C. Sun, 'Measurement of I–V curves of silicon-on-insulator (SOI) MOSFETs without self-heating,' *IEEE Electron Device Letters*, Vol. 16, No. 4, pp. 145–147, April 1995.

[11] A. Wei, M. J. Sherony and D. A. Antoniadis, 'Transient behavior of the kink effect in partially-depleted SOI MOSFETs,' *IEEE Electron Device Letters*, Vol. 16, No. 11, pp. 494–496, November 1995.

[12] J. Gautier and J. Y.-C. Sun, 'On the transient operation of partially depleted SOI nMOSFET's,' *IEEE Electron Device Letters*, Vol. 16, No. 11, pp. 497–499, November 1995.

[13] J.-P. Tayssier, P. Bouysse, Z. Ouarch, D. Barataud, T. Peyretaillade and R. Quere, '40-GHz/150-ns versatile pulsed measurement system for microwave transistor isothermal characterization,' *IEEE Transactions on Microwave Theory and Techniques*, Vol. 46, No. 12, pp. 2043–2052, December 1998.

[14] S. Akhtar and P. Roblin, 'Electro-thermal and iso-thermal characterizations of power RF silicon LDMOSFETs,' *To appear in IEEE Transactions on Electron Devices.*

[15] G. Engen, and C. Hoer, 'Thru-reflect-line: An improved technique for calibrating the dual six-port automatic network analyzer,' *IEEE Transactions on Microwave Theory and Techniques*, Vol. MTT-27, No. 12, pp. 987–993, December 1979.

[16] D. Pozar, *Microwave Engineering*, Addison Wesley, Reading, MA, 1997.

[17] P. Roblin, S. C. Kang and W. R. Liou, 'Improved small-signal equivalent circuit model and large-signal state equations for the MOSFET/MODFET wave equation,' *IEEE Transactions on Electron Devices*, Vol. 38, No. 8, pp. 1706–1718, August 1991.

[18] P. Roblin, S. Akhtar, and J. Strahler, 'New non-quasi-static theory for extracting small-signal parameters applied to LDMOSFETs,' *IEEE Microwave and Guided Wave Letters*, Vol. 10, No. 8, pp. 322–324, August 2000.

[19] S. L. Lee, H. K. Yu, C. S. Kim, J. G. Koo and K. S. Nam, 'A novel approach to extracting small-signal model parameters of silicon MOSFETs,' *IEEE Microwave Guided Wave Letters,* Vol. No. 7, No. 3, March 1997.

[20] J. P. Raskin, G. Dambrine and R. Gillon, 'Direct extraction of the series equivalent circuit parameters for the small-signal model of SOI MOSFETs,' *IEEE Microwave Guided Wave Letters,* Vol. 7, No. 12, pp 408-410, December 1997.

[21] J.-P. Raskin, R. Gillon, J. Chen, D. Vanhoenacker-Janvier and J.-P. Colinge, 'Accurate SOI MOSFET characterization at microwave frequencies for device performance optimization and analog modeling,' *IEEE Transactions on Microwave Theory and Techniques*, Vol. 45, No. 5, pp. 1017–1025, May 1998.

[22] G. Bischoff and J. P. Krusius, 'Technology independent device modeling for simulation of integrated circuits for FET technologies,' *IEEE Transactions on Computer-Aided Design*, Vol. CAD-4, No. 1, pp. 99–110, January 1985.

[23] E. J. Prendergast, P. Lloyd and H. Dirks, 'The extraction of terminal charges from two-dimensional device simulations of MOS transistors,' *The International Journal for Computation and Mathematics in Electrical and Electronic Engineering*, Vol. 6, No. 2, pp. 107–114, 1987.

[24] C. de Boor, *A Practical Guide to Splines*, Springer-Verlag, New York, 1978.

[25] S. Akhtar, Highly accurate large-signal microwave-model for SOI-MOSFETs, MSc, The Ohio State University, February 1996.

[26] D. Suh and J. G. Fossum, 'A physical charge-based model for non-fully depleted SOI MOSFETs and its use in assessing floating-body effects in SOI CMOS circuits,' *IEEE Transactions on Electronic Devices*, Vol. 42, No. 4, p. 728, April 1995.

[27] D. E. Root, S. Fan and J. Meyer, 'Technology independent large signal non quasi-static FET models by direct construction from automatically characterized device data,' *Proceedings of the 21st European Microwave Conference, Stuttgart, Germany*, Piscataway, New Jersey, 1991, pp. 927–932.

[28] M. H. Somerville, J. A. del Alamo and W. Hoke, 'Direct correlation between impact ionization and the kink effect in InAlAs/InGaAs HEMTs,' *IEEE Electron Device Letters*, Vol. 17, No. 10, pp. 473–475, October 1996.

[29] K. Jeon, Y. Kwon and S. Hong, 'A frequency dispersion model of GaAs MESFET for large-signal applications,' *IEEE Microwave and Guided Wave Letters*, Vol. 7, No. 3, pp. 78–80, March 1997.

[30] A. L. Caviglia, Development and modeling of microwave SOI MOSFETs, Doctoral Dissertation, The University of Maryland, 1995.

13.11 Problems

13.1 (a) Rederive the expressions given in Table 13.1.

(b) Rederive the extraction formula given in Table 13.2.

13.2 Demonstrate that model (b) in Figure 13.14 leads to the experimentally verified condition of

$$I_{D,trans}(v_{GS} = 0, v_{DS} = 0, V_{GS}, V_{DS}, T_{dev}) = 0.$$

13.3 Using a resistor R of arbitrary value and in series with a regular quasi-static charge element $i = dQ_{QS}/dt$, develop a simple circuit which permits the implementation in a circuit simulator of the non-quasi-static charge element of charge $Q_i(v_{GS}, v_{DS})$ and time constant $\tau_i(v_{GS}, v_{DS})$ satisfying the equation:

$$i_{disp,i}(t) = \frac{dQ_i(v_{GS}, v_{DS})}{dt} - \frac{d}{dt}\left[\tau_i(v_{GS}, v_{DS})i_{disp,i}\right].$$

Hint: Use a quasi-static charge which is a function of three voltages $Q_{QS}(v_{GS}, v_{DS}, v_R)$ where v_R is the voltage across the resistor R.

14 Analytical DC analysis of short-gate MODFETs

The most creative theories are often imaginative visions imposed on facts.

STEPHEN J. GOULD

14.1 Introduction

Of the various three-terminal devices proposed or demonstrated over the last couple of decades, the modulation doped field-effect transistor (MODFET) (or high-electron-mobility transistor (HEMT)) and the heterojunction bipolar transistor (HBT) (Chapters 2, 18 and 19) are the most successful. The two-terminal resonant tunneling diode (RTD) (Chapters 4 and 6) is also of great interest because of the extremely compact low-power high-speed digital circuitry it makes possible when integrated, for instance, with HEMTs (see e.g. [1]). Table 14.1 shows, for several transistor technologies, representative values (1999) for circuit frequency range, device cut-off frequencies, off-state breakdown voltage, maximum output power, power-added efficiency and noise figures with associated gain. The SiGe HBT is very attractive because of the potentially low cost of manufacturing and cut-off frequencies high enough for most wireless applications. For applications in a similar frequency range that require larger microwave output power, the GaAs-based HBT is an even better candidate. This device also shows very low $1/f$ noise, which translates into low oscillator phase noise. In addition to its speed, one advantage of the InP-based HBT is its surface properties which allow smaller devices. When special processing techniques are employed such down-scaling can result in very impressive power-gain cut-off frequency and circuit performance [2, 3].

When high frequencies and low noise are required the device of choice is typically a III–V Schottky-barrier-gate field-effect transistor FET (SBGFET), several examples of which appear in Table 14.1. GaAs-based metal–semiconductor field-effect transistors (MESFETs) and pseudomorphic HEMTs (PHEMTs) with gate lengths approaching 0.1 μm can be produced at low cost (e.g. [4]). From a speed standpoint the most advanced III–V SBGFETs are sub-0.1-μm InP-based HEMTs. These have record

Table 14.1. *Indicative comparison of various high-speed technologies based on recent (⟶ 2000) conferences and abstracts.*

Technology	Circuit freq. range (GHz)	f_t (GHz)	f_{max} (GHz)	BV_{off} (V)	P_{out} (freq, bias, size)	PAE (%)	NF_{min}/G_a (dB)
Si CMOS 0.1 μm NMOS	→ 3–4	55 > 150					0.51/18.5 (2 GHz) 2.4/5.5 (10 GHz)
Si BJT		26–50 100 (R&D)	40–80	3–7			
SiGe HBT	1–30 40 GHz VCO	40–60 130 (R&D)	50–80 160? (R&D)	3–6	630 mW (2 GHz)	80	0.8/13 (2 GHz) 0.9/- (10 GHz, R&D)
GaAs HBT	0.9–64	60 170 (R&D)	100 224 (R&D)	10–30	4.33 W/mm (20 GHz,10.5 V) 0.77 W/mm (3.4 V)	66 61	0.83/16.9 (2 GHz) 1.1/11 (4 GHz) 1.7/10 (18 GHz)
InP HBT (mostly DHBT)	2–94	60–180 250 (SHBT, R&D)	90–200 800 (SHBT, R&D)	5–20	2.7 W/mm (10 GHz, SHBT) 7.5 W/mm (10 GHz) 4.9 W/mm (18 GHz) 1.9 W/mm (30 GHz)	43 54 36	0.46/- (2 GHz) 2/- (10 GHz) 3.3/- (18 GHz)

Table 14.1. *Continued.*

Technology	Circuit freq. range (GHz)	f_t (GHz)	f_{max} (GHz)	BV_{off} (V)	P_{out} (freq, bias, size)	PAE (%)	NF_{min}/G_a (dB)
0.4 μm GaAs MESFET (HP)	−26	31–23	55	12	0.65 W/mm (18 GHz,6 V)		
0.25 μm GaAs PHEMT	−50	64	> 120	10	0.65 W/mm (40 GHz,5 V)	> 30	1.3/- (18 GHz)
0.1–0.15 μm GaAs PHEMT	12–94	90–120	150–290	 8 9 5	2.8 W (77 GHz,5 V) 0.72 W/mm (44 GHz) 0.60 W/mm (60 GHz) 0.40 W/mm (94 GHz)	48 26 29 13	0.38/10.5 (12 GHz) 1.5/- (60 GHz) 2.1 (94 GHz)
0.2 μm InP HEMT	50–94	110	3–400?		0.45 W/mm (60 GHz,4 V)	26	
0.1 μm InP HEMT	12–213	160–210 340 (R&D)	3–400?	5 6	0.40 W/mm (40 GHz,2.5 V) 0.21 W/mm (60 GHz,2 V) 0.34 W/mm (60 GHz,3 V) 0.53 W/mm (60 GHz,4 V)	 42 39	0.35/16.8 (12 GHz) 0.8/8.9 (60 GHz) 1.2/7.2 (94 GHz)
0.15–0.25 μm (Al)GaN HEMT	4–20	73	140	→ 100	470 mW (8 GHz,150 μm) 3 W (4 GHz, 2 mm,L_g = 1 μm)	46 30	1.2/- (10 GHz)

current-gain cut-off frequencies [5, 6] and noise figures (e.g. [7]). Applications include: wireless millimeter-wave communications [8]; fiber-radio personal communication systems [9]; automobile collision avoidance radar; and low-noise receivers for communication by optical fiber and satellites (DBS) [10]. Because of their impressive performance, important applications, and interesting device physics, these FETs are a main topic of Chapters 14–17. These chapters build on the theory in Chapters 8, 9 and 11, but take the very practical perspective of an engineer interested in designing a state-of-the-art MODFET technology and bringing it to the market, or using such a technology to design high-performance circuits and systems. Since much interesting and basic physics is encountered on the way, this perspective is not as limiting as it may seem at the outset. Compared to earlier treatises (e.g. [11]) the present focus is on short-gate devices where previously neglected effects can no longer be avoided. Despite the increased complexity, we maintain an analytical modeling approach.

Faced with the task of designing, optimizing and fabricating high-frequency III–V MODFETs, with integration into an MMIC (monolithic microwave integrated circuit) process often the ultimate goal, models yielding physical insight are very useful. They should be able to help answer the question 'Is what we're producing what it could or should be?' This is a question that both a process engineer and a circuit designer might have. Commercially available two-dimensional numerical models, intended to encompass all pertinent physics, some covered in the preceding chapters, can in principle do this job. Indeed, there are physical effects, such as those related to reliability (Section 14.7), which ultimately require a numerical solution. However, because of the time it takes to learn the intricacies and idiosyncrasies of such programs, the time it takes to run them, and the hurdles that are often encountered (convergence problem, bugs etc.), this approach can be cumbersome and expensive. Also, there are often important effects that occur in a 'real life' fabrication environment that are overlooked. In this respect, not even fundamental numerical models can be used blindly as 'black-box' tools. Analytically tractable models, developed with physically meaningful parameters, can be of great help in at least two respects. First, they allow quick mapping of the parameter space and thus provide guidance in process development. Second, they can actually be quite accurate; sometimes accurate enough to allow deeper physical insight. In fact, the analytical modeling approach that we take has helped uncover, or elucidate, two effects of practical and fundamental importance, namely the on-state breakdown in InP-type MODFETs (Section 14.6.5), and the interfacial gate resistance in SBGFETs (Chapter 16).

The final element in a high-performance, uniform, reproducible and reliable IC process is an interface to the circuit designer. This interface is the CAD model. This must be accurate and computationally efficient. The first of these required features disqualifies most analytical models, including the ones to be developed here. The second feature disqualifies the fundamental numerical models. The most practical and efficient solution to this dilemma is to use fit- or measurement-based CAD models which store in lookup tables the bias-dependent currents and charges, measured and extracted on fabricated devices [12, 13]. Such extractions can also be made on solutions by a complete numerical physics-based model [14]. This allows designers to evaluate circuit and system performance before significant investment is made in process development. Despite being inadequate for providing a CAD tool, the analysis that we are about to embark on has helped to clarify how to do a CAD model extraction for a more scalable result (Section 16.3). Without scalability, a time-consuming model extraction has to be made on each device used in the circuit design, a very unpractical and expensive scenario.

14.2 Background to the FET DC modeling approach

A realistic DC model for the MODFET is required in order to address important large-signal issues. It also lays the groundwork for the small-signal equivalent circuit and Y parameters (Chapter 15) which are of primary importance for a high-speed device. With a deeper understanding of the gate resistance (Chapter 16), one can then model the high-frequency figures of merit (Chapter 17). The physics of the intrinsic part of the device will be captured in experimentally validated terms that are less complex than many of those used in preceding chapters. However, we now include important extrinsic effects that cannot be neglected for the short-gate lengths of interest.

Figure 14.1 depicts a 0.1-μm recessed-gate SBGFET, and defines many of the parameters and quantities to be used and referred to in Chapters 14–17. The source and drain electrodes should be ohmic contacts that let electrons in and out of the channel with a small resistance R_C for each contact. The source- and drain-side caps have been removed laterally $L_{us} = L_{ud}$, preferably by a vertically selective recess etch as discussed in Section 17.7. The rectifying Schottky-barrier gate electrode has the dimension L_g (length) in the direction (y) of electron flow between the source and drain, and W_g (width) in the perpendicular direction (z) along the gate finger. The gate is spaced in the x direction to the effective center of the channel by the distance d_{gc}. For gate bias V_G less positive than the Schottky-barrier height Φ_B, the gate controls the concentration of electrons available for conduction in the channel. A positive voltage V_D on the drain causes a flow of electrons from the source to the drain. V_D generates a lateral channel drift field $F_c = dV_c/dy$ which gives the electrons their drift velocity $v = \mu F_c$, where μ is the electron mobility. We use F for electric field to distinguish it from energy (E), and, because of the negative charge $-q$ on the electron, it is convenient to define $F_c = +dV_c/dy$, rather than using the conventional $-dV_c/dy$ definition. As shown in Section 9.3.1, the mobility is given by

$$\mu = \frac{q\langle\tau_m\rangle}{m^*} \tag{14.1}$$

where $\langle\tau_m\rangle$ is average momentum relaxation time, and m^* is the electron effective

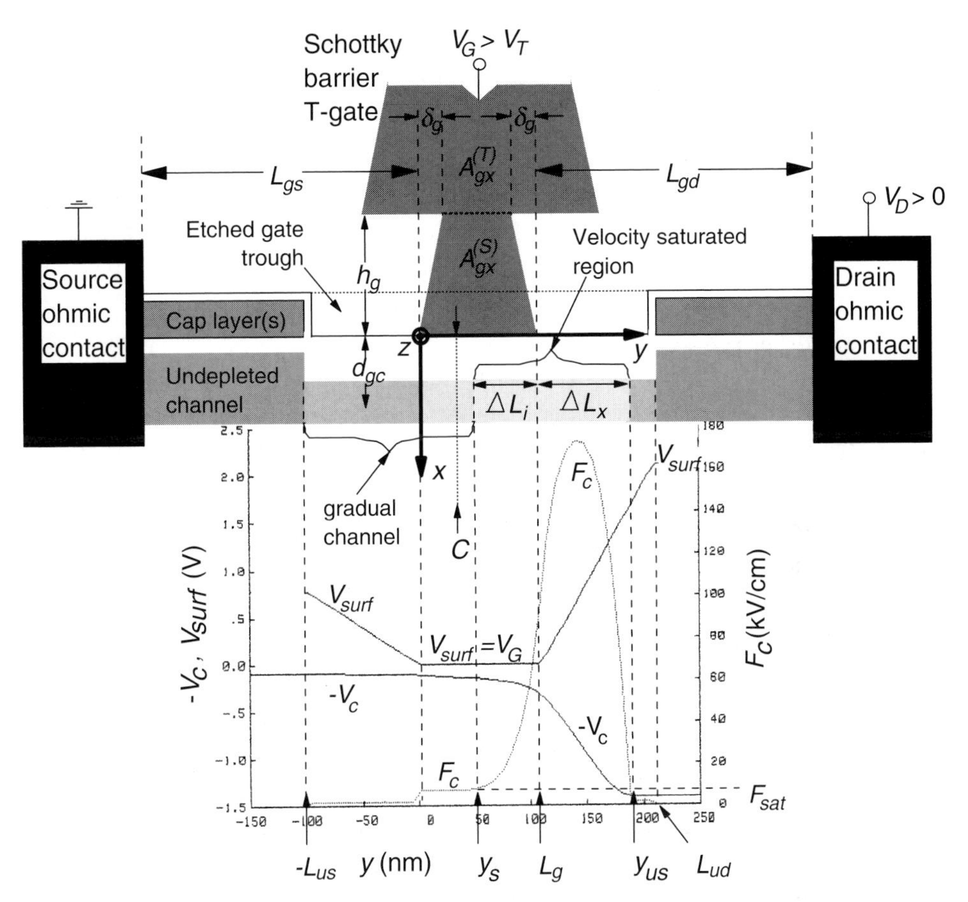

Fig. 14.1. 0.1-μm recessed-gate SBGFET with channel and surface potential, and channel drift field, at $V_G = 0$ V ($\rightarrow$ maximum transconductance) and $V_D = 1.5$ V. The source will be assumed to be at ground. Depleted regions are indicated qualitatively by white background.

mass in the channel. m^* is significantly smaller than the free electron mass m_0 (e.g. $m^* = 0.044m_0$ in $In_{0.53}Ga_{0.47}As$).

Because the surface potential is held constant by the gate, while the channel potential V_c is increasing towards the drain, the channel is increasingly reversed-biased and pinched down as the electrons approach the drain-side edge of the gate, more so the higher the drain voltage. The DC drain or output conductance $g_{ds0} = \partial I_D/\partial V_D$ decreases (indicated by the $1 - k$ term in Y_{dd} in Section 11.2.4). At the same time, the DC transconductance $g_{m0} = \partial I_D/\partial V_G$, a most important parameter for gain and high-frequency performance, increases (indicated by the k term in Y_{dg}). Diffusion is neglected since this does not change the total current significantly [15]. Current continuity in the low-field source-side part of the gated channel is then maintained by the electrons moving faster to compensate for the reduced concentration as the drain is approached. The increased velocity is sustained by an increase in the lateral drift field. For a sufficiently large V_D, the field near the drain-side edge of the gate becomes too large for the electron velocity to continue to increase in proportion. Instead, the electron velocity becomes effectively saturated at a value denoted v_{sat}. The magnitude of v_{sat} correlates in magnitude with the mobility [16]. For current continuity to be maintained, the electron channel sheet concentration n_s becomes constant. Thus, velocity saturation prevents classical pinch-*off* from occuring. The large increase in the lateral drift field in this region helps support (through Gauss's law) the channel charge. This shift of charge support from the gate to the drain-induced field is enhanced by spreading of the electrons as they gain energy and escape the epitaxially defined channel (Section 10.3.2, [17]).

This simple picture is adequate for explaining the essential features of even very short gate (< 0.1 μm) FETs. It thus provides for rather good understanding and optimization of the device. From a practical standpoint, this is convenient, but it is also quite surprising since electron transport over short distances in high fields is considerably more complicated than assumed here. The electron velocity is more correctly described as being tied to the local electron energy rather than to the local field, as discussed in Chapter 9. The net gain in energy over some distance depends on the local electric field and the scattering rate. When analyzing and optimizing the device, one should then solve Equation (9.17) to account for the fact that it takes time for an electron (even in a constant field) to gain, or lose, enough energy to reach its steady-state velocity associated with this field. To account accurately for transient velocity one should really account for two-valley transport (Section 9.5), and also solve for momentum balance [18]. Even with today's computer power, these important effects generally have to be neglected. In the practical commercial device modeling software packages that the authors have experience with, they certainly are, as are most quantum mechanical effects. In an FET, such as that depicted in Figure 14.1, the field varies rapidly. Thus, an electron is unlikely to have its velocity coincide with the steady-state velocity associated with the local field at any point; its velocity is always either 'undershooting' or 'overshooting'. Despite these fundamental complications, the simple field-based picture is consistent with a large body of experimental work (e.g. [19, 20, 5, 6]).

14.3 Brief semiconductor materials history for SBGFETs

Device engineers look for semiconductor channels with large electron mobility, saturation velocity, and full-channel sheet concentration n_{s0}. In addition to being beneficial for the intrinsic and extrinsic small-signal performance, these increase the maximum drain current:

$$I_D^{(max)} = q W_g n_{s0} v_{sat}. \tag{14.2}$$

A large current is good for driving interconnect capacitances in digital applications, increases the signal output power, and improves the large-signal linearity of the device. The first step towards better μ and v_{sat} in FET materials was to reduce m^* by fanning out in the periodic table from the column-IV element silicon to compounds of column-III and -V materials. In undoped form, these materials have peak electron velocities that are 2–3 times higher than the $\sim 1 \times 10^7$ cm/s in silicon. GaAs was first to be tried, in doped form, for the fabrication of MESFETs [21]. InP MESFETs were first fabricated by Barrera and Archer [22], and some interesting transport-induced difference between MESFETs on the two binary compounds were observed [23]. With the ionized impurity scattering in doped material, which reduces $\langle \tau_m \rangle$, a III–V MESFET does not take full advantage of the potential transport properties. The remedy for this originated in the pioneering work by Dingle *et al.* [24] who, by growing the wide-bandgap III–V alloy AlGaAs epitaxially on GaAs, and doping only the AlGaAs, managed to separate the donors from the mobile electrons. The electrons energetically favor the undoped GaAs where, by the attraction to the remote donors, they form a two-dimensional electron gas (2DEG) adjacent to the AlGaAs/GaAs heterojunction. The method was named modulation doping, soon thereafter leading to the term MODFET for a FET made on such an epitaxial structure. The most popular alternative name is HEMT, for high electron mobility transistor. The introduction of In in the channel reduces the bandgap and improves n_{s0} significantly. It also reduces m^* even further. GaInAs channels with up to 25% In mole fractions, with ~ 125 Å thickness can be grown on GaAs substrates. FETs made on such material are called pseudomorphic MODFETs, or PHEMTs. These FETs exhibit an excellent combination of speed and output power. Lattice mismatch of GaInAs to GaAs prevents In mole fractions larger than $\sim 30\%$ on GaAs, unless special growth techniques are employed (Section 17.7). With InP substrate, In mole fractions can approach 100% for sufficiently thin layers [25]. In this material system, the wide-bandgap electron supply layer typically is AlInAs, although InP can also be employed (e.g. [26]). The lattice-matched (and most common) In mole fractions for GaInAs and AlInAs on InP are 53% and 52%, respectively. Because of their very high v_{sat} ($\sim 3 \times 10^7$ cm/s) and n_{s0} ($\sim 3 \times 10^{12}$ cm^{-2}), FETs in this material system (and variations thereof) have shown the best noise and speed performance of any transistor (e.g. [5]). High-speed MODFETs, including several applications, are reviewed in the tutorial article by Nguyen *et al.* [27].

14.4 2DEG gate charge control in a heavily dual pulse-doped MODFET structure

Figure 14.2(a) illustrates, with the important parameters defined, a typical conduction-band diagram $E_C(x)$ along the cut-line C under the gate in Figure 14.1. At this stage of the analysis $V_D = 0$, and it does not matter where under the gate ($0 < y < L_g$) the cut is made. As in Chapter 8, we are interested in the gate control of the 2DEG concentration, i.e. the $n_s(V_G)$ curve in Figure 14.2(b). For increased generality, we now consider a double-heterojunction double-sided pulse-doped MODFET structure. Pulse doping [28] is particularly important in the AlInAs/GaInAs system because of the rather low Schottky-barrier height of AlInAs ($\Phi_B \approx 0.7$ eV). An undoped top layer thickness $d_t = 50$–200 Å is typically used. The additional pulse below the channel [29] is beneficial in increasing the maximum 2DEG concentration n_{s0}.

In the vicinity of the optimum operating point ($n_s = n_{si} \sim n_{s0}/2$) $n_s(V_g)$ is approximately linear, as will be assumed in the AC modeling (Chapters 11, 12, 15, 17):

$$n_s(V_G) = \frac{\varepsilon(V_G - V_T)}{q d_{gc}} \tag{14.3}$$

(cf. Equation (8.8)). V_T is the threshold voltage (Figure 14.2(b)), defined by linear extrapolation from the optimum inflection-point gate bias $V_G = V_{Gm}$, and can be shown to be given by (Problem 14.1)

$$V_T = \frac{\Phi_B - \Delta E_C + E_{F0}}{q} - \frac{q N_{d1} d_{p1}(d_t + d_{p1}/2)}{\varepsilon} - \frac{q N_{d2} d_{p2}(d_t + d_{p1} + d_{s1})}{\varepsilon}. \tag{14.4}$$

For dual doped structures with sufficiently large back-side pulses (in particular for the extreme case of an 'inverted' structure where $N_{d1} = 0$) there is no meaningful inflection point in $n_s(V_G)$, and no obviously optimum gate bias. For such structures, the threshold voltage has to be defined differently. Inverted, or almost inverted, structures are rather uncommon, however, for at least two reasons. The most fundamental is that the back-side heterojunction generally is rougher, leading to lower mobility of electrons residing in its vicinity. It is also more difficult to make contact to the 2DEG (cf. Section 14.6.2) when there is no or little doping in the front. Thus, we will not consider such structures here.

The effective gate-to-channel distance in Equation (14.3) is given by

$$d_{gc} = d_t + d_{p1} + d_{s1} + \Delta d. \tag{14.5}$$

The two parameters E_{F0} and Δd result from the Fermi level (Figure 14.2(a)) moving up in the well as this is filled with electrons (cf. Problem 8.1(b)). In analytical treatments the linear relationship

$$E_F(n_s) = E_{F0} + \left(\frac{q^2 \Delta d}{\varepsilon}\right) n_s \tag{14.6}$$

is typically assumed, although one could for better accuracy also include a quadratic term in n_s. The term E_{F0} is typically rather small, and the constant of proportionality between E_F and n_s has been written in a form that includes what turns out in

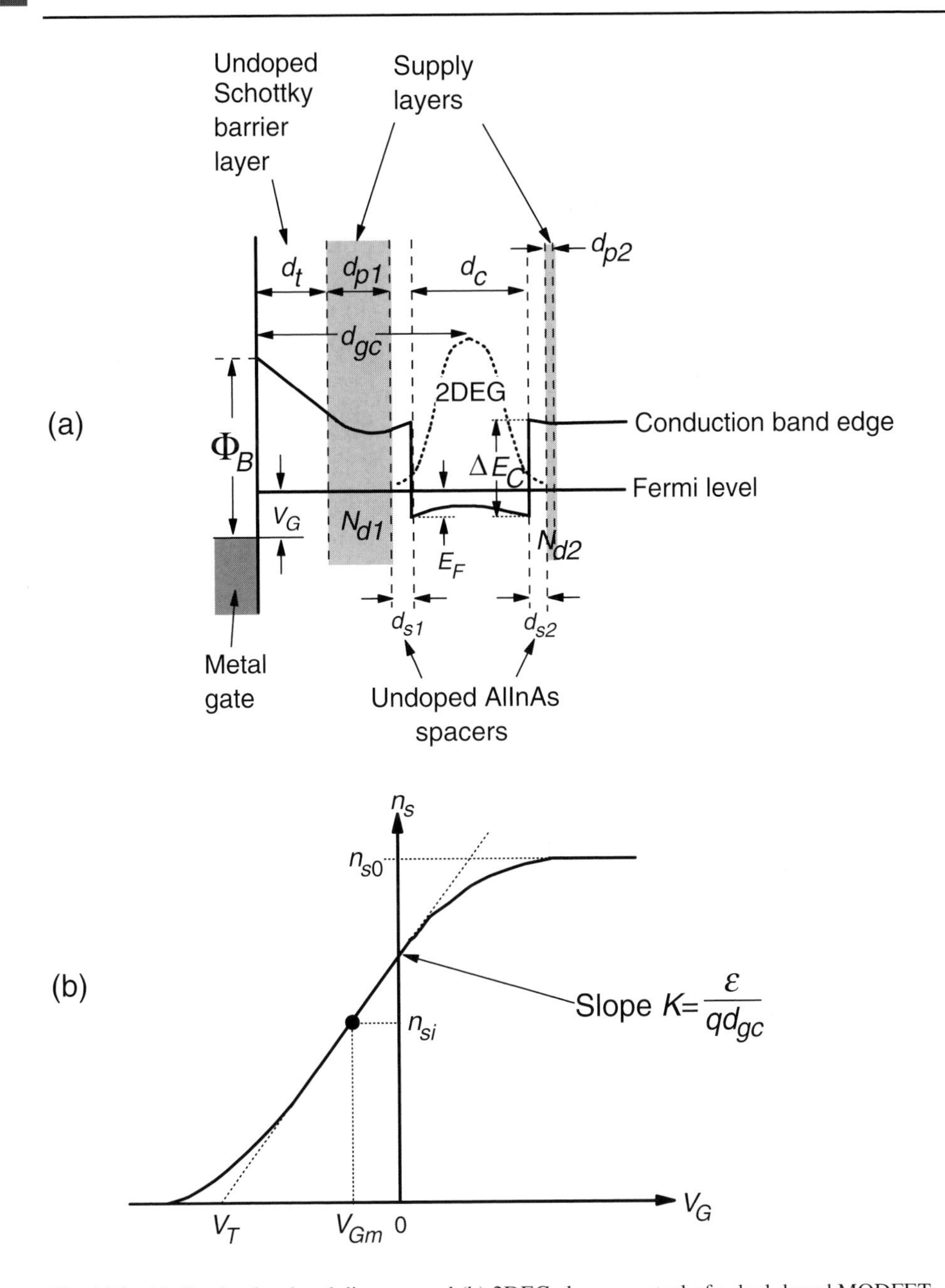

Fig. 14.2. (a) Conduction-band diagram and (b) 2DEG charge control of a dual doped MODFET.

Equation (14.5) to be the effective distance Δd from the top heterojunction to the center of the 2DEG.

In a MESFET ($\Delta E_C = 0$), the channel electrons move among the charged donors, and Coulomb scattering reduces the electron velocity compared to a MODFET. Because of the absence of a confining well, the electron concentration in a MESFET channel remains essentially constant, and the gate modulates the channel width (and thus d_{gc}). In a well-designed MODFET, the gate modulates the channel electron concentration, while the width of the channel, and thus d_{gc}, is relatively V_G-independent in a rather wide range. In terms introduced by Foisy *et al.* [30], such a MODFET exhibits 'efficient charge modulation', in the sense that little parasitic charge (Q_{SL}) in the top supply layer is being induced and modulated by the gate voltage near $V_G = V_{Gm}$. One accomplishes this by designing for large n_{s0}. A well-designed MODFET with large n_{s0} is somewhat similar to a Si MOSFET, with the top wide-band electron supply layer playing the role of the gate oxide. Of course, as V_G approaches Φ_B/q the MODFET gate will start to conduct non-negligible current (Section 8.6), and prior to this, as n_s approaches n_{s0}, parasitic charge in the form of neutralized donors and free electrons will be induced in the supply layer [30]. Typically the concentration of free electrons is small, and with the low electron mobility in the doped wide-bandgap supply layer, the contribution to the current of the so-called parasitic MESFET [31] can be neglected.

For single-sided doping, the maximum 2DEG concentration is approximately [32]

$$n_{s0} = \left(N_{d1}^2(d_{s1} + \Delta d)^2 + 2\varepsilon N_{d1}\Delta E_C^{(eff)}/q^2\right)^{1/2} - N_{d1}(d_{s1} + \Delta d), \tag{14.7a}$$

where the effective conduction-band offset is given by

$$\Delta E_C^{(eff)}(N_{d1}) = \Delta E_C + E_{FC}(N_{d1}) - E_x(N_{d1}) - E_{F0}. \tag{14.7b}$$

This explicit expression follows from Equation (8.6) by inserting the approximate $E_F(n_s)$ expression in Equation (14.6)*. Equation (14.7) is derived with the conduction band in the supply layer pulled down by a positive gate bias, to essentially replicate the semiinfinite uniformly doped situation in Figure 8.2. $E_{FC}(N_{d1})$ in Equation (14.7b) is the position of the Fermi level relative to the conduction-band edge far from the heterojunction, i.e. it is $-\delta$ in Figure 8.2. With symmetric double-sided doping ($N_{d2} = N_{d1}$), n_{s0} is still given by Equation (14.7), after the following substitutions in Equation (14.7a): $N_{d1} \to 2N_{d1}$, $\Delta d \to 2\Delta d$, and $\Delta E_C^{(eff)} \to 2\Delta E_C^{(eff)}$ (Problem 14.2).

Doping concentrations in typical pulses are well into the 10^{18} cm^{-3} range, and can be as high as $\sim 10^{19}$ cm^{-3} in thin pulses. At these levels the semiconductor is degenerate, a situation that actually starts at a rather low doping level $N_d \sim N_C$ where

$$N_C = 2\left(\frac{m^* kT}{2\pi\hbar^2}\right)^{3/2} \sim 5 \times 10^{17}\text{cm}^{-3} \tag{14.8}$$

is the effective density of states in the conduction band. Assuming isolated donor levels, a calculation of electron concentrations n in bulk AlInAs (with $m^* = 0.085m_0$) doped with Si to [Si] $\sim 10^{19}$ cm^{-3} results in $n \sim 10^{18}$ cm^{-3}, while, in fact, Hall measurements show that n is very close to [Si]. The reason for the discrepancy

* The same insertion into Equation (8.9) is what makes the threshold voltage an explicit function of the structural parameters (Equation (14.4); Problems 8.1(b) and 14.1).

is that the donor levels are not isolated, but form a band which merges with the conduction band to form a metal-like conduction band, as discussed in Section 2.2.2. In a degenerately doped semiconductor E_{FC} is approximately given by

$$E_{FC}\left(N_{d1}\right) = kT \ln \left\{ a\left(\frac{N_{d1}}{N_C}\right)^2 + \left[a^2\left(\frac{N_{d1}}{N_C}\right)^4 + \left(\frac{N_{d1}}{N_C}\right)^2\right]^{1/2} \right\}. \tag{14.9}$$

This expression comes from a modified Ehrenberg approximation [33] for the electron concentration:

$$n = N_C \frac{\exp(E_{FC}/kT)}{\left[1 + 2a\exp(E_{FC}/kT)\right]^{1/2}}, \qquad a = 2^{-3/2}, \tag{14.10}$$

This approximation for $n(E_{FC})$ has four useful properties. First, it approaches the correct Boltzmann expression for $-E_{FC} > kT$. Second, a Taylor expansion in $\exp(E_{FC}/kT)$ has the right first order correction. Third, as opposed to the original Ehrenberg approximation, it does not saturate for $E_{FC} \gg kT$. This correctly allows more electrons in the conduction band as E_{FC} increases, albeit at a higher rate than the 'metallic' free-electron-gas limit

$$n = \frac{1}{3\pi^2}\left(\frac{2m^* E_{FC}}{\hbar^2}\right)^{3/2} \tag{14.11}$$

[34], which would be a better approximation in this limit. Fourth, Equation (14.10) is easily inverted to give E_{FC} as a function of $n = N_D$ (Equation (14.9)). To allow analytical treatment, it is assumed that the conduction band maintains its intrinsic effective mass. This will introduce some error into the analysis [35, 36] which we cannot account for analytically.

As the degenerate conduction band forms, the energy of each electron is reduced by the attractive interaction with its 'exchange hole' of charge $+q$ associated with the ionic background. This interaction is denoted E_x (> 0) in Equation (14.7b), and can be estimated by (Problem 14.3)

$$E_x\left(N_{d1}\right) = \frac{3}{2}\frac{q^2}{4\pi\varepsilon}\left(N_{d1}\right)^{1/3}. \tag{14.12}$$

This expression is quite consistent with careful treatments of the shift of the conduction-band edge (e.g. [37]), as well as with Equation (2.2). The exchange interaction considered here is similar to that which causes the image potential as an electron is removed from a metal. The difference is that in this case the hole remains behind on the image plane just inside the metal surface, and flattens out to a disk as the electron is further removed [38]. The reason for the correction $-E_x$ in $\Delta E_C^{(eff)}$ is illustrated in Figure 14.3(a). Equation (14.7) was derived assuming non-degenerate electron concentration ($E_{FC} < 0$, $|E_{FC}| > kT$) in the neutral supply layer, and using the depletion approximation between this and the heterojunction. First of all,

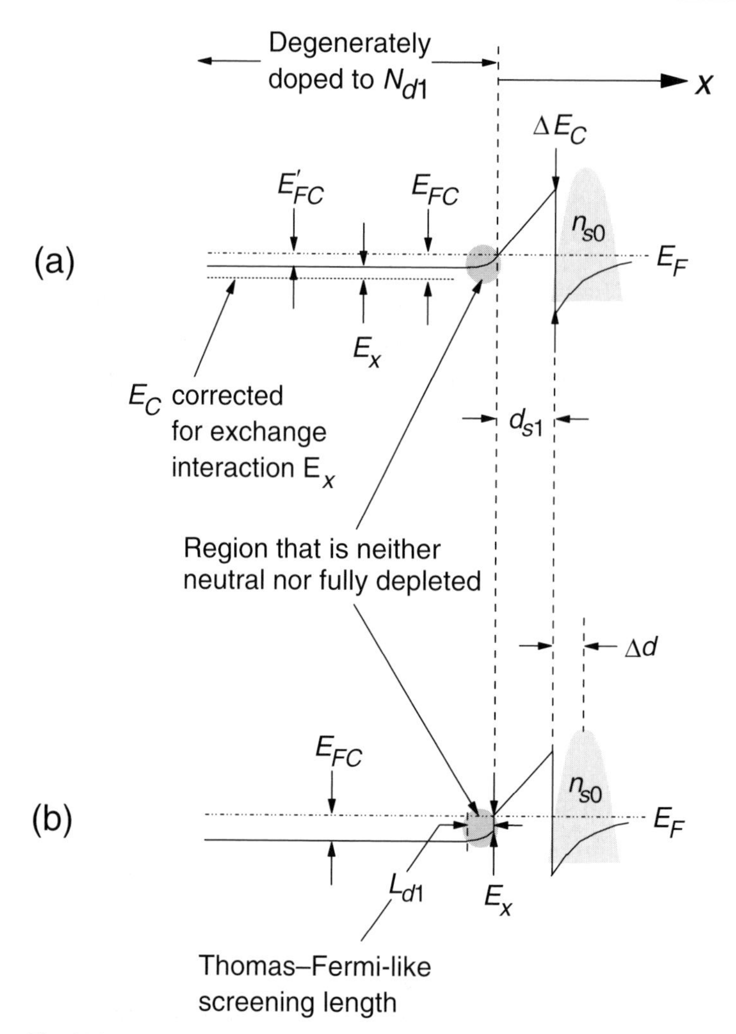

Fig. 14.3. Qualitative conduction-band diagrams for two ways of calculating the maximum 2DEG concentration with degenerate donor concentration. (a) Standard depletion approximation using the uncorrected (full curve) conduction-band diagram. E_x is included post-analysis by using the electrostatic-potential-based $E'_{FC} = E_{FC} - E_x$ in the expression for n_{s0}. (b) Partial metal-like depletion. E_x is included approximately in the analysis by lowering the conduction band in the doped layer ($x < 0$) uniformly, including the partially depleted screening layer.

using the depletion approximation in the regime indicated with the shaded circle in Figure 14.3(a) is questionable, and to check the validity of Equation (14.7), we will take an alternative approach based on Figure 14.3(b). Secondly, E_{FC}, which goes into the potential problem that yields n_{s0}, is the uncorrected value, denoted E'_{FC} in Figure 14.3(a), but this is in fact equal to the actual E_{FC}in the bulk of the supply layer minus E_x, as the figure illustrates. The exchange interaction in the neutral supply layer pulls E_C down by E_x, but must bring E_F with it since n must remain essentially equal to N_{d1}. Since E_F is constant throughout the structure it is brought down also where the 2DEG is, and thus n_s suffers. Many-body effects, such as the exchange interaction, in the 2DEG [39] are assumed to be contained in E_{F0} and Δd.

The alternative expression for n_{s0} is based on Figure 14.3(b), and assumes degenerate doping as used in most designs. We now assume a metallic free electron gas (Equation (14.11)) in the neutral supply layer, and a constant exchange interaction E_x for $x < 0$ where the degenerate doping resides. This results in a small conduction-band discontinuity at $x = 0$, which tends to confine the supply-layer electrons, and it is assumed that the spacer layer ($0 < x < d_{s1}$) is depleted ($n = 0$). The supply layer, however, is only partially depleted, and the depletion drops exponentially away from $x = 0$. The situation is analogous to Thomas–Fermi screening at the surface of a metal. Poisson's equation is solved in $x < 0$, where the charge density is $q(N_{d1} - n(x))$. The conduction-band edge deviates only moderately from its bulk value at $x = -\infty$, and we thus expand Equation (14.11) only to first order in E_{FC}. From the field at $x = 0$ we arrive at

$$n_{s0} = \frac{\varepsilon \Delta E_C^{(eff)}}{q^2 \left(L_{d1} + d_{s1} + \Delta d\right)}, \tag{14.13a}$$

where

$$L_{d1} = \frac{\hbar}{q} \left(\frac{\varepsilon}{m^*}\right)^{1/2} \left(\frac{\pi^4}{3N_{d1}}\right)^{1/6} \tag{14.13b}$$

is the characteristic Thomas–Fermi screening length for the supply layer. Because of the small (1/6) exponent, L_{d1} is only weakly dependent on N_{d1}. A typical value of L_{d1} is 30 Å. The spacer layer (d_{s1}), inserted to reduce Coulomb scattering, reduces n_{s0} somewhat, but can be kept sufficiently small compared to the unavoidable $L_{d1} + \Delta d$ (~90 Å), not to have a large effect. $d_{s1} = 20$ Å is a typical value. Similarly to Equation 14.7 above, Equation (14.13) can also, after the exchanges $\Delta d \rightarrow 2\Delta d$ and $\Delta E_C^{(eff)} \rightarrow 2\Delta E_C^{(eff)}$ in Equation (14.13a), be used to calculate the maximum improvement in n_{s0} with double-sided doping. Figure 14.4 shows n_{s0} calculated by the two methods, for both single- and double-sided doping. Somewhat surprisingly, the two agree very well for large doping levels, where one would expect Equation (14.7) to fail. Mathematically, this is because the two expressions have the same $\varepsilon \Delta E_C^{(eff)}/[q^2(d_{s1} + \Delta d)]$ limit as $N_{d1} \rightarrow \infty$. At low doping levels, where Equation (14.7) becomes increasingly accurate, the two methods deviate. Thus, Equation (14.7) is quite accurate for all doping levels. The sought-after large n_{s0} requires a large effective conduction-band discontinuity and a large donor concentration. Given typical specifications for threshold voltage and gate leakage, the latter requires the use of pulse-doping.

One major reason for the success of MODFETs is that the concentration of heterojunction interface states that could trap electrons can be made negligible [40]. However, early MODFETs were plagued by another trap phenomenon, one associated

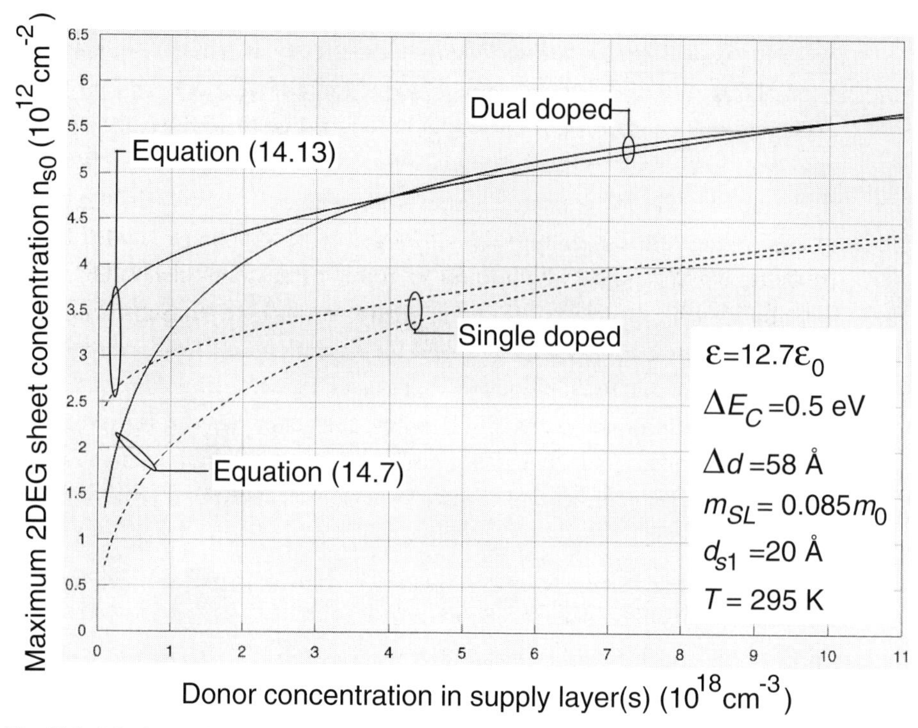

Fig. 14.4. Maximum 2DEG sheet concentration versus donor concentration for single- and dual-doped $Al_{0.48}In_{0.52}As$ and $Ga_{0.53}In_{0.47}As$ channels, calculated in the two ways described in the text (and Figure 14.3). SL = supply layer.

with the $Al_xGa_{1-x}As$ supply layer donor species (typically Si). The trap was denoted DX (see Section 2.2.2) to indicate a donor complex [41], and shows up for $x > 0.2$. Later work (see [17] for a good review) indicates that the DX center is due to the substitutional donor itself. This is supported by the theory of Chadi and Chang [42]. For $x < 0.2$, the DX center is resonant with the $Al_xGa_{1-x}As$ conduction band, and is typically not seen. For larger x it enters the wider bandgap and becomes a deep trap with large emission and capture barriers associated with large lattice relaxation. This in effect makes $E_{FC}(N_d)$ a relatively significant negative number, thus reducing n_{s0}. It also leads to low-frequency dispersion in the AC parameters, and other 'misbehaviors' [40]. In digital circuits the result can be loss and distortion of pulses due to time-dependent threshold shifts [43, 44]. The DX center concentration can be reduced by concentrating the Si, as much as possible, into a plane [45], using Se as the dopant [46], or by using InGaP as the supply-layer material (e.g. [47]). Another solution [48] employs a short-period ($\sim$40 Å) AlAs/GaAs superlattice as the supply layer, with the Si doping only in the GaAs. Even with a large average Al mole fraction, resulting in a large effective ΔE_C, DX centers can be eliminated with this method. The most practical and therefore common solution for GaAs-based

MODFETs, however, has become to simply reduce the Al mole fraction in the supply layer to below $\sim 20\%$. This unfortunately reduces the conduction-band discontinuity to the GaAs channel, with continued low n_{s0} as a result. Because of the large n_{s0} you can get in a MESFET (which goes a long way in compensating for the lower mobility), this made it hard for early MODFETs to compete. The PHEMT and the InP-based HEMT with their high-mobility InGaAs channel and large ΔE_Cs came to the rescue, allowing MODFETs to significantly outperform MESFETs. AlInAs exhibits no DX-related problems for Al mole fractions less than 60% [49].

Fully numerical self-consistent calculation solving both Schrödinger's and Poisson's equations (e.g. [50, 39, 27]), now available commercially, can be used to determine Δd and E_{F0} for a particular material system. This is done by comparing the linear part of the numerical $n_s(V_G)$ curve to Equations (14.3–5). As an example of particular interest, $\Delta d \approx 58$ Å for the $Ga_{0.47}In_{0.53}As/Al_{0.48}In_{0.52}As$ system. These calculations also show that, in order to avoid reducing the electron concentration in the high-mobility channel below the value predicted by Equation (14.7), d_c should not be much smaller than $2\Delta d$. This is not surprising given the interpretation of Δd above: the center of the 2DEG needs to be spaced from the bottom heterojunction by at least the same amount Δd as it is from the top. This avoids having the electron wavefunction excessively spilling over into the wide-bandgap lower-mobility material. E_{F0} is roughly proportional to d_c^{-2} (like the energy levels in a square potential), and can typically be neglected. For double-sided doping, $d_c \sim 2\Delta d$ is a good choice. If d_c is much larger, the channel separates into two parallel channels, with non-optimum composite gate-modulation characteristics, and potentially larger output conductance. With double-sided doping, the back-side of the channel starts filling up with electrons at a more negative gate voltage than indicated by the inflection-point-based threshold voltage (Equation (14.4)). The shift is $-qN_{d2}d_{p2}d_c/\varepsilon$, and is another reason not to use a thicker channel than necessary.

The analytical expressions for V_T and n_{s0} are quite accurate and useful, and are, together with Δd (which is essentially a channel/heterojunction material parameter), in many respects sufficient for a design. Still, it is advisable to do some cross-checking with a full numerical Poisson–Schrödinger calculation. A numerical calculation is necessary if it is important to know accurately the gradual non-linear approaches of n_s to 0 and n_{s0} (Figure 14.2(b)). The result of the numerical simulation can be used to best pick the parameters α and V_{Gm} in the analytical representation (cf. Problem 10.3)

$$n_s(V_G) = n_{s0}\left[\alpha + (1-\alpha)\tanh\left(\frac{V_G - V_{Gm}}{V_1}\right)\right] \tag{14.14}$$

for the non-linear charge control, where $(V_{Gm}, n_s(V_{Gm}) = \alpha n_{s0})$ is the inflection point. Equation (14.14) allows analytical integration in the gradual channel of the FET, which is one step in the development of the $I_D(V_G, V_D)$ model (Section 14.6). In addition, the analytical non-linear charge-control representation in Equation (14.14) has the good features of producing the experimentally observed smooth bell-shaped $g_{m0}(V_G)$ curve, and the correct finite $I_D^{(max)}$ (Equation (14.2)). α must be chosen < 0.5 for the predicted $I_D(V_G, V_D)$ to pinch off. A purely analytical design could start with layer thicknesses and doping levels being picked to give desired V_T and d_{gc}. Then α would be chosen and n_{s0} calculated (Equation (14.7)). V_{Gm} is given by

$$V_{Gm} = V_T + \alpha \frac{q n_{s0} d_{gc}}{\varepsilon}, \tag{14.15}$$

and the remaining parameter V_1 by

$$V_1 = (1 - \alpha) \frac{q n_{s0} d_{gc}}{\varepsilon}. \tag{14.16}$$

For better accuracy, the choice of parameters in Equation (14.14) can be boiled down to a least-squares fit (with V_1 as the one floating parameter) to a numerically calculated $n_s(V_G)$ [51].

14.5 An analytically manageable 2DEG transport model

As outlined in the introduction, the transport model is based on the two-piece velocity-field curve in Figure 14.5. For fields below a critical value F_{sat}, where the *steady-state* electron velocity peaks (III–V materials) or saturates (silicon), one can assume a linear relationship $v = \mu F$ between the velocity and the local electric field. For larger fields, one can assume, for the prediction of the basic performance of even ultra-short FETs, that the velocity remains equal to an effective saturation velocity, independent of any further increase in the field. For III–V materials, empirically, the effective saturation velocity essentially coincides with the peak velocity. The fact that the ultimate scattering limited velocity v'_{sat} (Figure 14.5) is no larger than in silicon appears to be of little consequence. Thus, velocity overshoot does, in some respect, appear to show up as a measurable effect, but, surprisingly, the effective saturation velocity does not increase as the gate length is reduced, and does not exceed the peak velocity by much. This experimentally based notion of a gate-length-independent effective saturation velocity has also been suggested by Monte Carlo simulations [52], although these simulations tend to overestimate the experimental value.

Considering the constancy of v_{sat}, it may at first seem odd to include a gate-length dependence in the low-field mobility. Nevertheless, a significant (almost an order of magnitude) reduction in effective mobility, as the gate length is reduced from 1 to 0.1 μm, has been inferred from the same Monte Carlo simulations [52] that suggested a constant effective saturation velocity. Velocity saturation in III–V materials used for FETs requires acceleration of the electrons in the fast central Γ valley to an energy $E_\Gamma^{(sat)}$ where significant scattering to slower satellite valleys can take place. This led Foisy *et al.* [30] to point out that the drain voltage V_{Dsat} at onset of velocity

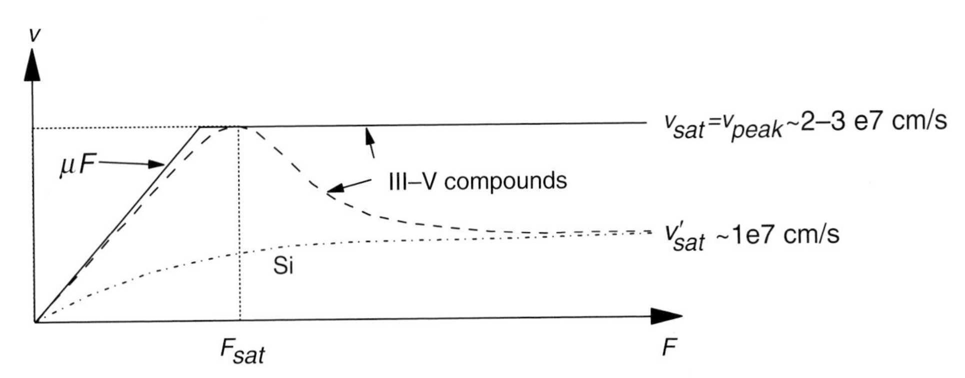

Fig. 14.5. Qualitative electron velocity-field curves for Si and III–V compounds. The solid two-piece linear approximation, possibly with a correction for the moderate non-linearity at low fields, is appropriate for analytical modeling of III–V FETs.

saturation cannot be lower than $E_{\Gamma}^{(sat)}/q^*$. At the same time, however, as $L_g \to 0$, V_{Dsat} approaches $F_{sat}L_g$. A gate-length-dependent phenomenological mobility

$$\mu(L_g) = \mu_0\left[1 - \exp\left(-\frac{L_g}{L_\mu}\right)\right], \tag{14.17a}$$

where we have defined

$$L_\mu = \frac{E_{\Gamma}^{(sat)}}{qF_{sat0}}, \tag{14.17b}$$

accounts for the finite V_{Dsat} as$L_g \to 0$, and correctly approaches the bulk mobility μ_0 as $L_g \to \infty$. The Γ–L separation is ~0.5 eV in $In_{0.53}Ga_{0.47}As$ and ~0.3 eV in GaAs, and the critical field is ~4.9 and ~4.0 kV/cm, respectively [53]. This corresponds to a characteristic length L_μ of 1.3 and 0.6 μm for $In_{0.53}Ga_{0.47}As$ and GaAs, respectively. These values are in reasonable accordance with the 0.8–1.1 μm that the gate-length-dependent mobility in [52] corresponds to. However, for our analytical modeling, we are really after the average energy the electrons have gained at the point they travel at their peak average velocity ($v_{peak} = v_{sat}$ in Figure 14.5). Monte Carlo simulations of GaAs transport [54] show this energy to be ~55 meV. This leads to L_μ = 1670 Å, a value that is suggestively close to the electron mean free path in the GaAs Γ valley [55].

* This was done in the context of inefficient charge modulation. The parasitic supply-layer charge (Q_{SL}) dominates the loss of efficiency for $V_{Gm} < V_G < \Phi_B$, while the 'excess' gradual-channel charge ΔQ_{GC} (the 'slow' extra electron charge in the gradual channel necessary to satisfy current continuity) dominates for $V_T < V_G < V_{Gm}$. Near V_{Gm} the two conspire to reduce the maximum modulation efficiency (ME) to less than 100%, thus reducing the apparent v_{sat}. The onset of Q_{SL} can be moved out to higher V_G by increasing n_{s0}, thus increasing the modulation efficiency. However, the minimum V_{Dsat} prevents pushing the onset of ΔQ_{GC} to arbitrarily low V_G. The effect was analyzed at DC for 1-μm devices [30]. It does not seem to prevent extracting a physical v_{sat} at microwave frequencies on AlGaAs/GaAs and AlGaAs/InGaAs MODFETs with 0.07–0.7 μm gate length, biased for maximum current-gain cut-off frequency and with parasitics properly removed [20, 88]. AlInAs/InGaAs MODFETs, with their larger ΔE_C, are even less affected.

The velocity-field curve up to F_{sat} is not quite linear, as indicated in Figure 14.5. A field dependence in the mobility can be included within the analytical framework with the velocity-field curve

$$v(F) = \frac{\mu(L_g)F}{1 + F/F_1(L_g)}, \qquad F \leq F_{sat}(L_g) \tag{14.18a}$$

$$= v_{sat}, \qquad F \geq F_{sat}(L_g), \tag{14.18b}$$

where

$$F_{sat}(L_g) = \frac{F_{sat0}}{1 - \exp(-L_g/L_\mu)}, \tag{14.18c}$$

and

$$F_1(L_g) = \frac{F_{sat}(L_g)}{(\mu_0 F_{sat0})/v_{sat}) - 1}. \tag{14.18d}$$

Assuming the simplified linear charge control in Equation (14.3), one can derive

$$V_{Dsat} = \frac{V_G - V_T + F_{sat}L_g}{1 - \Gamma} - \left[\left(\frac{V_G - V_T + F_{sat}L_g}{1 - \Gamma}\right)^2 - \frac{2(V_G - V_T)F_{sat}L_g}{1 - \Gamma}\right]^{1/2}. \tag{14.19a}$$

for the intrinsic drain voltage at onset of velocity saturation voltage. Γ is given by

$$\Gamma = \frac{F_{sat}(L_g)}{F_1(L_g)} = \frac{F_{sat0}}{F_1(\infty)} = \frac{\mu_0 F_{sat0}}{v_{sat}} - 1, \tag{14.19b}$$

and is independent of L_g. The drain saturation voltage approaches the classical gradual-channel value $V_G - V_T$ as $L_g \to \infty$. As $L_g \to 0$, V_{Dsat} approaches $F_{sat}L_g$, which in turn approaches $E_\Gamma^{(sat)}/q$ as intended.

14.6 Quasi-two-dimensional model for electrostatics and I–V characteristics

With the models chosen for gate charge control $n_s(V_G)$ and transport $v(F)$, we can now move on to calculate the channel potential $V_c(y)$ and lateral channel drift field $F_c(y) = dV_c/dy$, as well as the drain current $I_D(V_G, V_D)$. The equation that yields this information is that for the continuity of channel current:

$$I_D = qW_g \left\{ n_s \left(V_{surf}(y) - V_c(y)\right) + \frac{\varepsilon\delta_c}{q}\frac{d^2V_c}{dy^2} \right\} v\left(\frac{dV_c}{dy}\right), \tag{14.20}$$

where diffusion and gate current have been neglected. The application of a drain bias $V_D > 0$ has affected the 2DEG charge control, i.e. the factor in braces, in three ways.

First, the channel potential (equal to 0 in the one-dimensional gate charge control $n_s(V_G)$) is now the position-dependent $V_c(y)$. As $V_c(y)$ increases under the gate towards the drain, n_s is reduced, i.e. the channel is pinched down. Second, with a drain bias large enough to velocity-saturate the device, one has to consider the charge control outside the gate region ($0 < y < L_g$). This is particularly important for the cutting-edge short-gate FETs, where the laterally etched access regions adjacent to the gate become increasingly important to the performance. Thus V_G is replaced with a position-dependent effective gating voltage $V_{surf}(y)$ (Figure 14.1). Under the gate ($0 < y < L_g$), $V_{surf}(y) = V_G$ as before, while outside the gate $V_{surf}(y)$ will approach a rather large positive value at the edges of the gate trough, corresponding to the large n_s in the capped regions ($y < -L_{us}$ and $y > L_g + L_{ud}$ in Figure 14.1). Third, the channel charge will be supported, not only by the vertical field (the one-dimensional $n_s(V_{surf} - V_c)$), but also by the lateral drift field. Assuming a position-independent effective channel thickness δ_c in the saturation region $y_s < y < y_{us}$, the second term in the braces of Equation (14.20) models this effect. This is what gives the model its two-dimensional nature. A more accurate version of this term was developed by Foisy [17].

14.6.1 The low-field gradual channel

Equation (14.20) is quite a complicated equation for the channel potential V_c. For the analytical approach taken, we benefit from the choices of $n_s(V_G)$ (Equation (14.14)) and $v(F_c)$ (Equation (14.18)). For the low-field gradual-channel region ($0 < y < y_s$), Equation (14.20) is simplified by neglecting the lateral charge support (second term in the braces). Integration of Equation (14.20) in the gradual channel, and rearrangement of terms and factors, yield (see Problem 10.3)

$$y_s = y_s(V_G, I_D) = L_g - \Delta L_i$$
$$= \frac{qW_g n_{s0}\mu V_1}{I_D}\left\{\alpha\frac{V_c(y_s) - V_c(0)}{V_1} + (1-\alpha)\ln\left(\frac{\cosh\left(\frac{V_G - V_c(0) - V_{Gm}}{V_1}\right)}{\cosh\left(\frac{V_G - V_c(y_s) - V_{Gm}}{V_1}\right)}\right)\right\} - \frac{V_c(y_s) - V_c(0)}{F_1} \qquad (14.21a)$$

for the end of the gradual channel, i.e. the location for onset of velocity saturation. The channel potential at this point, $V_c(y_s)$, is obtained from the condition

$I_D = qW_g n_s(V_G - V_c(y_s))v_{sat}$ for onset of velocity saturation:

$$V_c(y_s) = V_G - V_{Gm} - V_1 \tanh^{-1}\left(\frac{\dfrac{I_D}{I_D^{(max)}} - \alpha}{1-\alpha}\right) \equiv V_{sat}. \tag{14.21b}$$

The practically unavoidable parasitic voltage at the intrinsic source ($y = 0$) is expressed as

$$V_c(0) = V_{Slin} + V_{Sugt}(I_D, L_{us}) \equiv V_0, \tag{14.21c}$$

where

$$V_{Slin} = V_c(-L_{us}) = \left[\frac{R_C + R(L_{gs} - L_{us})}{W_g}\right] I_D \tag{14.21d}$$

is the linear part of the parasitic voltage on the source side. V_{Slin} is due to two simple linear resistors. R_C is the contact resistance normalized to unit gate width, and R is the sheet resistance of the full unetched epitaxial structure. Similarly, the linear part of the drain-side parasitic voltage is given by

$$V_{Dlin} = V_D - V_c(L_g + L_{ud}) = \left[\frac{R_C + R(L_{gd} - L_{ud})}{W_g}\right] I_D. \tag{14.22}$$

V_{Sugt} is non-linear in I_D and is the voltage that drops over the source-side part of the ungated etched trough. The two parameters V_{sat} (Equation (14.21b)) and V_0 (Equation (14.21c)) have been defined for later use. The DC source resistance is $R_S = V_c(0)/I_D$. The AC source resistance, appropriate for the small-signal high-frequency performance (Chapters 15–17), is given by $R_s = \partial V_c(0)/\partial I_D$. The measurement of these and the prediction of R_C are discussed in a separate subsection below.

V_{Sugt} limits the drain current that can be passed through the device. There are two non-linear effects at work. The primary is that the potential of the free etched surface is modulated very ineffectively by the adjacent gate. Thus, as the gate voltage becomes sufficiently large, the current is limited by the carrier concentration that the free surface depletion allows. A free (unmetallized) III–V surface tends to have similar surface potential and band-bending as when metallized. This has to do with similar mid-gap defects being generated by metallization, oxidation, or deposition of typical passivating dielectric films. These mid-gap defects pin the Fermi level at the surface. Some of these effects will be discussed in Chapter 16. The free surface can be particularly detrimental to the current for enhancement-mode FETs, i.e. FETs with positive threshold voltage. A second non-linear effect in V_{Sugt} has to be considered to account for the small drain current that can flow even in the extreme case of the free surface pinching off the channel in $-L_{us} < y < 0$. A finite I_D

results from the lateral charge support in the presence of a drain bias. The effect is qualitatively the same as that modeled by the second term in the square brackets of Equation (14.20), but is quantitatively typically much smaller on the source side. To find an analytical solution for V_{Sugt}, we simplify Equation (14.20). First, the y-dependent $V_{surf}(y) - V_c(y)$ is replaced with an effective constant surface-to-channel voltage V_{Gugt}. Second, the two-piece velocity-field curve (Equation (14.18) is replaced with an effective one-piece Si-like expression [56]. We are then left to solve the non-linear differential equation

$$I_d = qW_g \left[n_s(V_{Gugt}) + \frac{\varepsilon\delta_c}{q}\frac{d^2V_c}{dy^2} \right] \frac{\mu_{ugt}\dfrac{dV_c}{dy}}{1 + \dfrac{\mu_{ugt}}{v_{ugt}}\dfrac{dV_c}{dy}} \tag{14.23}$$

for $V_c(y)$ in $-L_{us} < y < 0$. An approximate solution can be found by Taylor expanding V_c to second order in $y + L_{us}$. The result is

$$V_{Sugt}(I_D, L_{us}) = \left[F_2 + \left(F_2^2 + F_3^2\right)^{1/2} \right] L_{us}, \tag{14.24a}$$

where we have defined

$$F_2 = \frac{1}{2}\left\{ \frac{RI_D}{W_g} + \frac{\left[I_D - qW_g n_s(V_{Gugt})v_{ugt}\right] L_{us}}{2\varepsilon W_g \delta_c v_{ugt}} \right\} \tag{14.24b}$$

and

$$F_3^2 = \frac{I_D L_{us}}{2\varepsilon W_g \delta_c \mu_{ugt}}. \tag{14.24c}$$

In the non-velocity-saturated part of L_{ud} on the drain side, we assume an average field F_{ugt} that is the same as in L_{us}, i.e.

$$F_{ugt} = \frac{V_{Sugt}}{L_{us}}. \tag{14.25}$$

Within the analytical framework, the treatment of the low-field parts of the ungated trough is rather crude because of the complex non-linear effects accounted (approximately) for. The problems of the gradual channel under the gate (Equation (14.21), and the velocity-saturated part in the following subsection, are actually solvable with less questionable assumptions. Since F_{ugt} is relatively small under typical conditions, errors introduced by the approximate analysis should not be critical. The approach can quantitatively account for the maximum current of FETs being considerably smaller than the hypothetical full channel current $I_D^{(max)}$ (Equation (14.2)). One central parameter that controls this is the effective gating voltage V_{Gugt}, since $n_s(V_{Gugt})$ is the channel electron concentration allowed by the surface. As an example, for a MODFET with $N_{d1}d_{p1} + N_{d2}d_{p2} = 4.5 \times 10^{12}$ cm^{-2}, with a 25-Å Pt gate and silicon nitride passivation, we would choose $V_{Gugt} = +0.37$ V. Of this, 0.15 V comes from the ungated passivated AlInAs having approximately this amount less band bending than when gated [57], and 0.22 V comes from the Pt gate sintering ~ 35 Å into the AlInAs Schottky-barrier layer (Section 17.7), thereby shifting the effective threshold voltage associated with the adjacent free AlInAs surface by −0.22 V [58].

The other central parameter controlling the maximum current allowed by the free surface is the effective saturation velocity v_{ugt}. When L_{us} is sufficiently short, one would choose v_{ugt} equal to the standard saturation velocity v_{sat} used in the high-field drain-side region. For sufficiently large L_{us}, however, v_{ugt} approaches the significantly lower high-field stationary saturated velocity v'_{sat} (6.5×10^6 cm/s for $In_{0.53}Ga_{0.47}As$). One could account for this, for instance, by assuming an exponential drop in v_{ugt} with L_{us}:

$$v_{ugt} = v_{sat} - (v_{sat} - v'_{sat})\left[1 - \exp(-L_{us}/L_{ugt})\right]. \tag{14.26}$$

Based on Monte Carlo simulations of high-field steady-state 2DEG transport in GaAs [54] the mean free path between Γ–L scattering events ranges from negligible at very high fields to 0.5 μm at fields close to that corresponding to onset of velocity saturation. Based on this one might choose an intermediate value $L_{ugt} = 0.25$ μm for the characteristic length in Equation (14.26). This is comparable to the characteristic length L_μ used for the gate-length-dependent mobility (Equation 14.17)). The final parameter to choose is the effective mobility μ_{ugt}. Without further consideration, this could simply be set to μ_0. Alternatively, it could be considered an adjustable parameter.

14.6.2 Source, drain and contact resistances

Source and drain resistances can be measured rather accurately with DC methods (e.g. [59–65]. The following measurement recipe has proved to be quite accurate for the DC source resistance:

$$R_S = \left.\frac{\partial V_G}{\partial I_D}\right|_{I_G=const>0,\,V_D<V_{Dsat}}. \tag{14.27}$$

As indicated, the measurement should be taken in the linear drain-bias regime, with a constant forward current sent through the gate. This avoids contributions to the measured R_S from the gate metallization (Chapter 16) and diode resistances. A fraction of the channel resistance will contribute to the measured R_S, but this component can be minimized by using as large a gate current as is deemed safe ($V_G = V_{Gmax}$). The exponential voltage dependence of the gate current will crowd the current towards the source. A good test for whether a channel resistance correction is necessary is to check whether the slope of $V_G(I_D)$ drops significantly if a larger I_G is used. Typically, this is not the case, but if it is, some of the corrections in the references given above may be useful. R_D can be measured by reversing source and

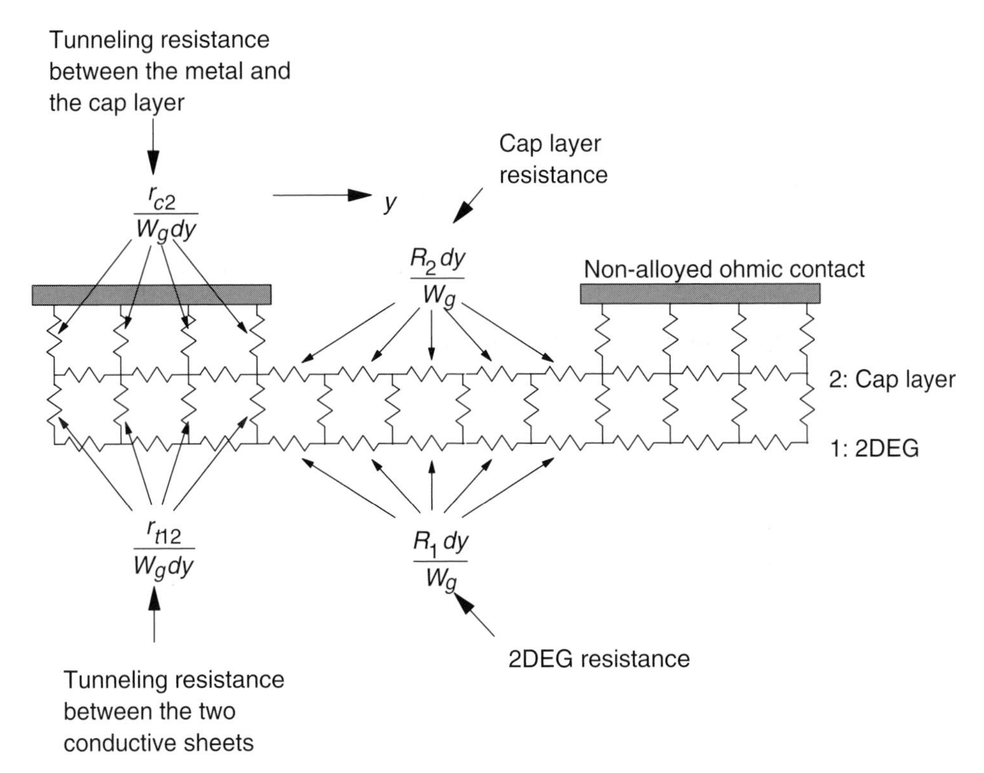

Fig. 14.6. Resistive network representing an unetched MODFET structure with non-alloyed ohmic contacts.

drain, or by other methods described in [59–65]. One way of estimating the AC source and drain resistances at a particular I_D, is to multiply the DC values (Equation (14.27)) with $R_{ds}(I_D)/R_{ds}(0)$, where R_{ds} is the small-signal drain-source resistance measured as

$$R_{ds} = \left.\frac{\partial V_D}{\partial I_D}\right|_{V_G = V_{Gmax}}. \tag{14.28}$$

If the cap layers are depleted, the unetched parts between the source and the drain can also contribute to the non-linearity in source and drain resistance.

Non-alloyed tunneling contacts (see also Section 17.7) are preferable to alloyed contacts if the contact resistance R_C can be kept low enough. Representing the contacts and epitaxial structure by the resistive network in Figure 14.6, it is shown in Problem 14.4(a) that R_C of a non-alloyed contact is given by

$$R_C = \frac{R}{W_g k} \frac{\dfrac{k}{\kappa} + \left(\dfrac{\kappa}{k}\right)^2}{1 - \left(\dfrac{\kappa}{k}\right)^2}. \tag{14.29a}$$

R is the sheet resistance of the full unetched epitaxial structure, and results from the parallel combination of the 2DEG (R_1) and the cap layer(s) (R_2):

$$R = \frac{R_1 R_2}{R_1 + R_2}; \tag{14.29b}$$

k and κ are exponential decay constants given by

$$k^2 = \frac{R_1 + R_2}{r_{t12}}, \tag{14.29c}$$

and

$$\kappa^2 = \frac{1}{2}\left(\frac{R_1}{r_{t12}} + \frac{R_2}{r_{t12}} + \frac{R_2}{r_{c2}}\right) - \frac{1}{2}\left[\left(\frac{R_1}{r_{t12}} + \frac{R_2}{r_{r12}} + \frac{R_2}{r_{c2}}\right)^2 - \frac{4R_1R_2}{r_{t12}r_{c2}}\right]^{1/2}. \tag{14.29d}$$

It is possible to make r_{c2}, the specific metal contact resistance to the top layer 2, small by using a doped narrow-bandgap material. However, if r_{t12}, the specific tunneling resistance from the cap layer to the channel, becomes large, R_C becomes unacceptable. This is illustrated by the full curves in Figure 14.7, which show the dependence of R_C on r_{c2}, with r_{t12} as a parameter, assuming typical sheet resistances. As r_{t12} approaches 10^{-6} Ω cm^2, R_C starts to become unacceptable. This situation typically arises with a larger Al mole fraction in the Schottky-barrier layer. One is then forced to use an alloyed ohmic contact for which there is typically less understanding and predictability of R_C. Alloyed ohmic contacts are also necessary when a depleted cap is used. In this case $R_2 = \infty$, and the tunneling resistance directly to layer 1 is prohibitively large. The analysis leading to Equation (14.29) is similar to that in [66]. The diagonal dashed line in Figure 14.7 corresponds to the standard expression $R_C = (r_{c2}R)^{1/2}$ for a single channel [67]. As required, Equation (14.29) approaches this case when $r_{t12}\to 0$. The horizontal lines in Figure 14.7 correspond to the hypothetical case of contacting layer 1 directly with the metal, with a specific contact resistance $r_{c1} = r_{t12}$. The derivation of this R_C is similar to that of Equation (14.29). The result is included in Figure 14.7 to test another limit of Equation (14.29), i.e. $r_{c2}\to 0$, corresponding to the metal essentially replacing layer 2 under the contact. The two R_C's do indeed coincide in this limit. It is interesting that R_C given by Equation (14.29) has a minimum below this limit (Problem 14.4(b)).

For the AC case there is a possibility of dispersion because of the capacitive coupling [68] between the metal and layer 2, and layers 1 and 2. One can predict the effect of this by replacing $1/r_{t12}$ in Equation (14.29) with $1/r_{t12} + j\omega C_{12}$, where C_{12} is the specific capacitance between layer 1 and 2, and similarly for $1/r_{c2}$.

14.6.3 The high-field velocity-saturated region

Equation (14.21) gives explicitly the position $y = y_s$ for onset of velocity saturation, and the potential $V_c(y_s)$ there, as functions of drain current I_D and gate voltage

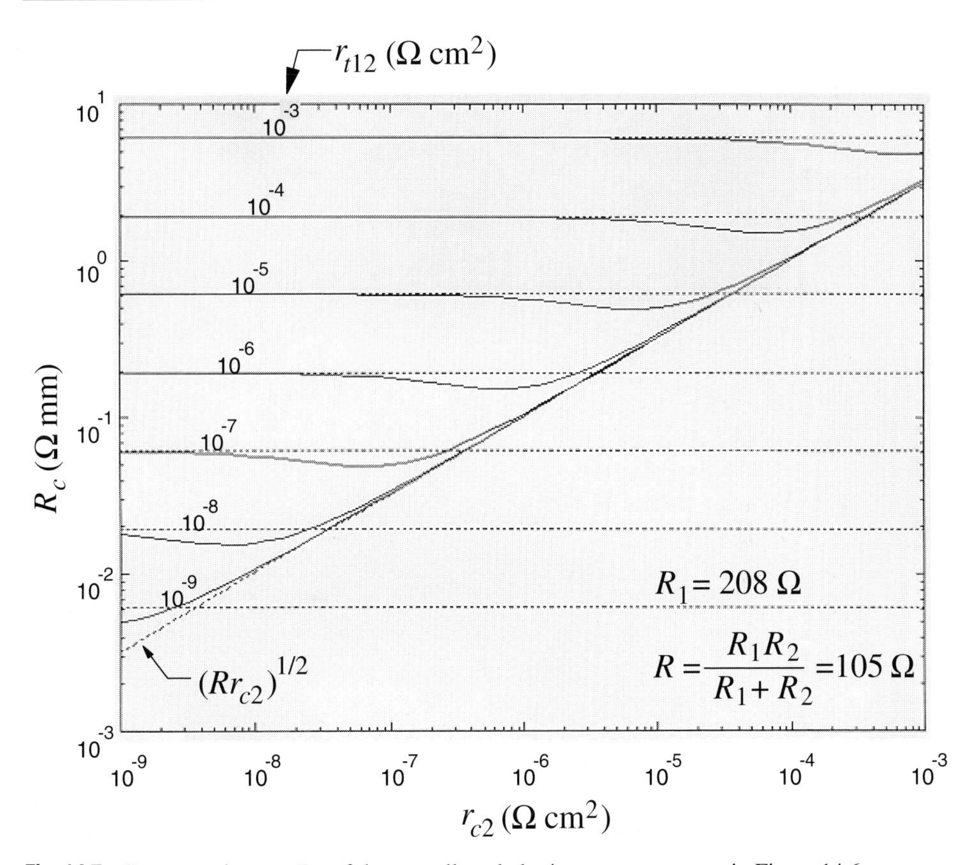

Fig. 14.7. Contact resistance R_C of the non-alloyed ohmic contact structure in Figure 14.6 as a function of the specific contact resistance r_{c2} to the top layer 2, with the specific tunneling resistance r_{t12} between the two conductive layers as the parameter. The dashed diagonal line is the standard contact resistance $(Rr_{c2})^{1/2}$ to a single channel with sheet resistance R and specific contact resistance r_{c2}. This case corresponds to zero r_{t12}. The horizontal lines correspond to the hypothetical case of contacting the first (bottom 2DEG) layer directly with the metal, assuming a specific contact resistance $r_{c1} = r_{t12}$. This case approaches R_C when r_{c2} approaches zero.

V_G. The rest of the problem boils down to calculating analytically and explicitly the drain voltage $V_D = V_D(V_G, I_D) = V_c(L_g + L_{gd}) + R_C I_D$. With a root-finding algorithm, one can alternatively calculate I_D as a function of applied bias, i.e. $I_D = I_D(V_G, V_D)$. Either way, one must calculate $V_c(y)$ for $y_s < y < L_g + L_{gd}$. In the velocity-saturated region Equation (14.20) is simplified by the electron velocity being independent of the field. The remaining complexity is the non-linear characteristic of $n_s(\cdot)$ introduced in Equation (14.14) to model the gradual pinch-off and saturation. These important aspects of MODFET charge control, however, are accounted for in the I–V characteristics by the gradual channel, i.e. Equation (14.21). In order to solve for $V_c(y)$ for $y > y_s$ analytically, we approximate Equation (14.20) in the saturated region by

$$I_d = qW_g n_s \left(V_{surf}(y) - V_c(y) + a^2 \frac{d^2 V_c}{dy^2} \right) v_{sat}, \tag{14.30}$$

where

$$a^2 = \frac{\varepsilon \delta_c}{q \dfrac{dn_s}{d(V_{surf} - V_c)}(y_s)} = \frac{\varepsilon \delta_c V_1}{q(1-\alpha) n_{s0}} \cosh^2 \left(\frac{V_G - V_{Gm} - V_c(y_s)}{V_1} \right). \tag{14.31}$$

Since $I_d = qW_g n_s (V_{surf}(y_s) - V_c(y_s)) v_{sat}$, we can rewrite Equation (14.30) as the linear second order differential equation

$$\frac{d^2 V_c}{dy^2} = \frac{[V_{surf}(y_s) - V_c(y_s)] - [V_{surf}(y) - V_c(y)]}{a^2} \tag{14.32}$$

for the channel potential. As the channel is further pinched down under the gate beyond the point of velocity saturation, i.e. when the numerator in the right-hand side of Equation (14.32) is positive, the channel field will continue to increase. Beyond the gate, as V_{surf} increases, the field will eventually peak (at $y = y_{max}$), and start decreasing. At a point $y = y_{us} = L_g + \Delta L_x$ further towards the drain, the transport in the channel will again become linear (subscript *us* for *u*n-*s*aturation). The channel field, which would be essentially constant between the source and drain without a gate (barring Gunn-domain formation (Section 9.6)), becomes instead strongly peaked a small distance $y_{max} - L_g$ beyond the drain-side edge of the gate. The rate of change in the field is determined by $V_{surf}(y)$ and the characteristic distance a. With a given by Equation (14.31), Equation (14.30) is only an approximation based on a first order series expansion of Equation (14.20). As $V_c(y)$ increases rapidly beyond y_s, higher orders would in principle have to be accounted for. This would, however, preclude an analytical solution of $V_c(y)$. Instead, we maintain the simple soluble form of Equation (14.32), but adjust the choice of a based on the following considerations. For sufficiently large drain bias, $V_c(y)$ will indeed increase well beyond the small deviation that was the basis for Equations (14.30–14.31), and at some point $y = y_{dc} < L_g$ under the gate, V_c will reach a value $V_c(y_{dc})$ that makes $n_s(V_G - V_c(y_{dc}))$ equal to zero. From the standpoint of one-dimensional ($V_D = 0$) gate charge control, this corresponds to pinch-off of the channel, which, with the gate charge control assumed (Equation (14.14)), occurs at the gate bias

$$V_{Goff} = V_G - V_c(y_{dc}) = V_{Gm} - V_1 \tanh^{-1}\left(\frac{\alpha}{1-\alpha} \right). \tag{14.33}$$

In the actual two-dimensional case ($V_D > 0$), the situation corresponds to total transfer of charge support from the gate to the drain. For $y > y_{dc}$, when the drain-voltage-induced lateral field fully supports the channel charge, the channel potential is governed by:

$$\frac{d^2V_c}{dy^2} = \frac{I_D}{W\varepsilon\delta_c v_{sat}}. \tag{14.34}$$

For $y_s < y < y_{dc}$ we integrate Equation (14.32), and pick a so that, in addition to V_c and dV_c/dy, also dV_c^2/dy^2 is continuous at $y = y_{dc}$. The result is an alternative expression for a^2 that is a function of the gate voltage and drain current:

$$a^2(V_G, I_D) = \left[\frac{V_G - V_{Goff} - V_c(y_s)}{I_D}\varepsilon W_g v_{sat}\right]\delta_c. \tag{14.35}$$

This ensures that Equations (14.20) and (14.30) are equivalent not only close to $y = y_s$ where dV_c^2/dy^2 is negligible, but also at $y = y_{dc}$ where dV_c^2/dy^2 totally dominates. a^2 given by Equation (14.35) reflects a mixture of gate and drain charge control in the regime $y_s < y < y_{dc}$, i.e. between onset of velocity saturation and local loss of gate charge control. The gate control is evident in the appearance in square brackets of an effective gate-to-channel spacing. The drain control occurs because of the finite channel thickness δ_c. This mixture becomes even clearer with the simple linear charge control in Equation (14.3), when a^2 becomes $d_{gc}\delta_c$, using either of the Equations (14.31) or (14.35). Also with the realistic non-linear charge control in Equation (14.14) the two expressions for a^2 give very similar values.

Once gate charge support is lost, $V_c(y)$ is governed by Equation (14.34) up to the point $y = y_{rc} > L_g$ where the effective forward bias of the free surface reintroduces the mixed charge control. Analytical solutions for the different points of interest (y_s, y_{dc}, y_{rc}, y_{max} and y_{us}) require a manageable $V_{surf}(y)$. As mentioned, for typical III–V semiconductors, a free surface and a metallized surface have a similar Fermi-level pinning position, and thus a similar depletion width. However, along a free surface, the potential can vary. For analytical modeling we assume that the effect of the free surface adjacent to the gate is that of a linearly varying gating potential, illustrated in Figure 14.1, similar to [69, 70]. In reality, the electrostatic potential, obeying Poisson's equation, will vary in a more complicated way along the surface, with a larger slope near the gate and cap. For the present level of modeling the assumption of a linearly varying V_{surf} is as complex as can be handled analytically, and we thus assume, in each of the three regions, that

$$V_{surf}(y) = V_G + F_{surf}(y - y_0). \tag{14.36}$$

F_{surf} and y_0 are region-dependent. Under the gate, $F_{surf} = 0$, while in the unmetallized part of the trough F_{surf} and y_0 are given by

$$F_{surf} = -\frac{V_{Slin} + V_{Gcap} - V_G}{L_{us}} \equiv -F_{surfS}, \quad y_0 = 0, \tag{14.37a}$$
$$\text{for} \quad -L_{us} \le y \le 0,$$

and

$$F_{surf} = \frac{V_D - V_{Dlin} + V_{Gcap} - V_G}{L_{ud}} \equiv F_{surfD}, \quad y_0 = L_g, \qquad (14.37b)$$
$$\text{for} \quad L_g \leq y \leq L_g + L_{ud}.$$

V_{Gcap} is an effective gate voltage (free to float above the channel potential) that would fill the channel under the unetched cap to its appropriate carrier concentration (Equation (14.14)).

To capture its general behavior also at lower currents, the channel thickness in the saturated region is assumed to have the following dependence on I_D:

$$\delta_c(I_D) = \frac{\gamma I_D / W_g}{1 + \frac{\gamma I_D / W_g}{\delta_c^{(max)}}}. \qquad (14.38)$$

$\delta_c^{(max)}$ can be used to adjust the output conductance of the FET in its active on-state region (cf. Section 10.3.2). However, as will be shown in Section 15.4, realistic output conductance can be predicted by considering the capacitive coupling between drain and source. Thus, we typically leave $\delta_c^{(max)}$ at the nominal epitaxial channel layer thickness d_c. In the other limit, as threshold is approached and I_D is reduced, Equation (14.38) allows the channel thickness to be reduced in a way consistent with the punch-through voltage V_{Dpt}, the drain voltage necessary to force a current through a nominally off channel ($V_G < V_{Goff}$). This requires that the channel charge is supported solely by the drain-induced field. I_D is zero before the onset of punch-through, so the parasitic voltages are negligible. V_{Dpt} is typically large enough for it to be assumed that the entire channel $0 < y < L_g + L_{ud}$ is velocity-saturated. Equation (14.34) can then be used and integrated to yield

$$V_{Dpt} = \frac{(L_g + L_{ud})^2}{2\varepsilon \gamma v_{sat}}. \qquad (14.39)$$

In short-gate devices, punch-through is likely to occur before off-state breakdown due to impact ionization or gate tunneling current. Thus, the off-state breakdown voltage measured at a low I_D (~ 1 mA/mm) [71, 72] can often be a reasonable estimate of V_{Dpt}, from which γ can be chosen.

14.6.4 Impact ionization in the channel and gate tunneling

With the field and potential in the channel, the breakdown characteristics can be estimated. Two mechanisms are of particular interest: impact ionization in the channel and tunneling of electrons from the gate to the channel. Off-state breakdown in InP-type HEMTs has been proposed to be due to a combination of the two [73]. Numerical models (e.g. [74]) are required to include these effects self-consistently.

With an analytical model one can study breakdown effects by doing first order calculations using the unperturbed electrostatics.

The drain current with impact ionization in the channel is

$$I'_D = M_n I_D, \tag{14.40}$$

where M_n is the electron multiplication factor. This is given by

$$M_n = \frac{1}{1 - I_{np}}, \tag{14.41a}$$

where

$$I_{np} = \int_{y_s}^{y_{us}} dy\, \alpha_n(y) \exp\left(-\int_{y_s}^{y} dy' \left[\alpha_n(y') - \alpha_p(y')\right]\right). \tag{14.41b}$$

This expression comes from considering in one dimension the generation rate $(\alpha_n J_n + \alpha_p J_p)/q$ of electron–hole pairs, the individual continuity equations for the electrons and holes and the constancy of the total current density $J = J_n + J_p$ [75]. With impact ionization based on the local field F, the electron ionization coefficient is often expressed as [76]

$$\alpha_n(F) = \alpha_{n0} \exp\left(-\frac{F_n}{F}\right), \tag{14.42}$$

where α_{n0} is a constant. The exponential term originates from a Boltzmann factor, where the activation energy is the threshold energy E_{Tn} for an electron to generate an electron–hole pair, and the temperature is the temperature T_n of an electron accelerated in the high field F to its saturation velocity. Assuming a constant mean free path l_n for the energy relaxation, Equation (14.42) results, and the characteristic field F_n is given by [76]

$$F_n = \frac{3E_{Tn}}{2ql_n}. \tag{14.43}$$

By considering momentum and energy balance of the impact event, one derives

$$E_{Tn} = \frac{2m_e + m_h}{m_e + m_h} E_G, \tag{14.44}$$

for the threshold energy [76]. With the effective mass m_e of electrons typically being significantly smaller than that for holes (m_h), E_{Tn} exceeds the energy bandgap E_G of the channel only by a rather small amount.

Avalanche breakdown occurs when $I_{np} = 1$, but long before then the presence of impact ionization in the channel can have profound effects on the device. One example is the kink phenomenon in InP-type HEMTs, which causes the DC drain current to be depressed for moderate drain biases. Although there may be several alternative origins for the kink phenomenon (e.g. [77]), a commonly occuring kink in these devices is due to a subtle interplay between the unmetallized part of the gate trough and impact ionization [78]. The depression in I_D is caused by negatively charged surface states, primarily in $-L_{us} < y < 0$. In terms of our analytical model, this corresponds to a low V_{Gugt} and a high $V_c(0)$ (Section 14.6.1). However, because of holes generated by impact ionization in the velocity-saturated region, starting at a rather low V_D ($\sim E_{Tn}/q \sim 0.8$ V), the negative surface charge is reduced, allowing a larger I_D to flow. The quantitative treatment of this phenomenon requires full numerical methods and a more fundamental energy-based viewpoint of the impact ionization [78].

The most straightforward estimate of M_n is one based on the unperturbed field, and is simplified further if one can assume that the hole ionization coefficient is given by $\alpha_p = k_p \alpha_n$. The double integral I_{np} will then only involve the single integral

$$I_n = \int_{y_s}^{y_{us}} \alpha_n(y)\, dy, \tag{14.45}$$

and will be given by

$$I_{np} = \frac{1 - \exp(-(1 - k_p) I_n)}{1 - k_p}. \tag{14.46}$$

For $Ga_{0.47}In_{0.53}As$ channels, $\alpha_p = k_p \alpha_n$ is less good an assumption with Urquhart's [79] $\alpha_n(F)$ and $\alpha_p(F)$, compared to Pearsall's [80] and Osaka's [81]. Since the large fields (> 300 kV/cm) of most importance force us to use the $\alpha_n(F)$ and $\alpha_p(F)$ expressions beyond the fields used for their measurements, it is not always clear what the best choice for k_p is. $k_p = 0.5$ has been used [82], but with the later experimental data [79] $k_p = 1$ may be a better choice.

Having calculated $F_c(y)$ and $V_c(y)$, one can also estimate the tunneling reverse gate current I_{Gt}. Using the low-temperature limit of the reverse I–V characteristics derived by Padovani and Stratton [83], and assuming a triangular barrier defined by the Schottky-barrier height Φ_B and the electric field $F_g(y)$ at the gate–semiconductor interface (again with opposite sign to the convention), one arrives at (e.g. [84])

$$I_{Gt} = \frac{q^3 m_M W_g}{8\pi h m_{SBL} \Phi_B} \int_0^{L_g} dy\, F_g^2(y) \exp\left(-\frac{4(2m_{SBL})^{1/2} \Phi_B^{3/2}}{3q\hbar F_g(y)} \right), \tag{14.47}$$

where m_M and m_{SBL} are the electron mass in the metal and Schottky-barrier layer, respectively. In the region $0 < y < y_s$, where the lateral field is neglected, $F_g(y)$ is given approximately by

$$F_g(y) = \frac{q\left(N_{d1} d_{p1} + N_{d2} d_{p2}\right)}{\varepsilon} - \frac{V_G - V_T - V_c(y)}{d_{gc}}. \tag{14.48}$$

This comes out as a byproduct as one derives Equation (14.4) for the threshold voltage. For $y_s < y < L_g$ one cannot neglect the lateral field. Since the field of interest is now at the gate ($x = 0$), while the boundary conditions for the field are known in the channel ($x = d_{gc}$), we face the full two-dimensional problem in $0 < x < d_{gc}$,

$y_s < y < L_g$. This is the Grebene–Ghandhi [85] problem solved in Section 10.3.1. The solution to the potential V is thus the sum of a y-independent (since the doping is y-independent) particular solution $V_p(x)$, and a homogeneous solution $V_h(x, y)$ satisfying Laplace's equation. The one-dimensional gradual-channel solution, with the channel potential set to $V_c(y_s)$, is the natural choice for $V_p(x)$, since this leads to the simplest boundary conditions for $V_h(x, y)$. Two of the boundary conditions, $V_h(x, y_s) = 0$ and $V_h(0, y) = 0$, are satisfied by the solution

$$V_h(x, y) = B \sin(\beta x) \sinh(\beta (y - y_s)). \tag{14.49a}$$

β and B are constants determined by the boundary conditions $\partial V_h/\partial y(d_{gc}, y_s) = F_{sat}$ and $\partial V_h/\partial y(d_{gc}, L_g) = F_c(L_g)$ to be

$$\beta = \frac{\cosh^{-1}\left(\dfrac{F_c(L_g)}{F_{sat}}\right)}{L_g - y_s} \tag{14.49b}$$

and

$$B = \frac{F_{sat}}{\beta \sin(\beta d_{gc})}, \tag{14.49c}$$

respectively. In the saturated region $y_s < y < L_g$, the field at the gate thus becomes

$$F_g(y) = \frac{q(N_{d1}d_{p1} + N_{d2}d_{p2})}{\varepsilon} - \frac{V_G - V_T - V_c(y_s)}{d_{gc}} + F_{sat}\frac{\sinh(\beta(y - y_s))}{\sin(\beta d_{gc})}. \tag{14.50}$$

The integrals in Equations (14.45) and (47) are calculated most easily with Simpson's formula.

In the saturated channel ($x = d_{gc}$, $y_s < y < L_g$), the solution $V_p(x) + V_h(x, y)$ to the full two-dimensional problem is identical to the solution to the one-dimensional problem for the channel potential (Equation (14.32)) with $V_{surf}(y) = V_G$. The two parameters a and β are related by $a\beta = 1$. The simple one-dimensional second order linear differential equation (Equation (14.32)) for the channel potential was developed primarily to allow straightforward, albeit approximate, incorporation of the regions external to the gate. With the choice of a in Equation (14.35), the approach has the additional benefit of effectively (and approximately) incorporating the effect of higher order terms in the potential, i.e. additional terms of the form in Equation (14.49a) with increasing β's. With the choice of a linear $V_{surf}(y)$, the solution to Equation (14.32), and the determination of y_{dc}, y_{max}, y_{rc}, y_{us} etc., are straightforward compared to solving a two-dimensional electrostatic problem.

14.6.5 Application examples and some large-signal issues

The first application example of the DC model is the potential and field profiles shown in Figure 14.1 for an InP-type MODFET biased at $V_D = 1.5$ V, near maximum transconductance. After the field reaches F_{sat} at y_s, it increases exponentially under the gate. Beyond the gate, V_{surf} is free to increase to its value $V_D - V_{Dlin} + V_{Gcap}$ (cf. Equation (14.37b)) at the drain-side edge of the gate trough. The field no longer increases exponentially, but instead peaks at $y = y_{max}$, and then drops down to a point where the channel transport becomes linear again. The surface-induced high-field domain differs from the common view of saturation in a MESFET, dominated by a transport-induced stationary Gunn domain beyond the gate [86]. Fully numerical MODFET modeling [87] appears to validate the approach. In velocity-saturated regions with mixed charge control, Equation (14.32) has the solution

$$V_c(y) = V_c(y_s) + \left[V_G - V_{surf}(y_s)\right] + F_{surf}(y - y_0) + V_+ \exp(y/a) + V_- \exp(-y/a), \quad (14.51)$$

where V_+ and V_- are given by the known channel potential and field at the beginning of the region. The use of a linear $V_{surf}(y)$ allows us to find any point y of interest associated with a particular field, or derivative of the field. y_{dc}, for instance, is most conveniently found by the condition $\left(d^2V_c/dy^2\right)(y_{dc}) = I_D/(W\varepsilon\delta_c v_{sat})$, which, per Equation (14.34), is a condition for the channel charge being fully supported by the lateral field. Starting at $y = y_{dc}$, $V_c(y)$ is determined by Equation (14.34) which is trivial to solve. The point $y = y_{rc}$, where the surface gating voltage has risen enough to re-introduce the mixed charge control, is determined by $V_{surf}(y_{rc}) - V_c(y_{rc}) = V_{Goff}$. For $y > y_{rc}$, $V_c(y)$ is once again governed by Equation (14.32), with Equation (14.51) as the formal solution. The point $y = y_{max}$ of maximum field $\left(\left(d^2V_c/dy^2\right)(y_{max}) = 0\right)$, and $y = y_{us}$ of 'unsaturation' $((dV_c/dy)(y_{us}) = F_{sat})$ are easily found. The evolution of the interesting points as V_D increases is shown in Figure 14.8 for a PHEMT [88].

Typically velocity saturation sets in under the gate, and $V_{surf}(y_s) - V_G$ in Equation (14.51) is zero. However, for sufficiently large drain voltage (and short gates) the entire channel can become velocity-saturated, and the point of onset moves into the source side of the ungated trough as shown in Figure 14.8 for $V_D \sim 2$ V. This is the onstate version of the punch-through situation discussed above, and results in a larger output conductance since the channel is now modulated by the free surface with a varying gating voltage. Only in this situation does the source-side $V_{surf}(y)$ (Equation (14.37a)) enter into the calculation. Punch-through starts when $V_c(0)$ (Equation (14.21c)) becomes larger than $V_c(y_s)$ (Equation (14.21b)), i.e. when $V_0 > V_{sat}$. Under the gate, analytical treatment of linear transport is possible because of the constant $V_{surf} = V_G$, which allows direct integration of Equation (14.20), leading to Equation (14.21a). As we saw when developing expressions for V_{Sugt} and F_{Sugt}, this is not the case in $-L_{us} < y < 0$, because of the varying $V_{surf}(y)$. For the present case with $y_s < 0$, we assume that the effective field F_{ugt} (Equations (14.24)–(14.25)) can still be used for $-L_{us} < y < y_s$. The saturation point y_s is determined by

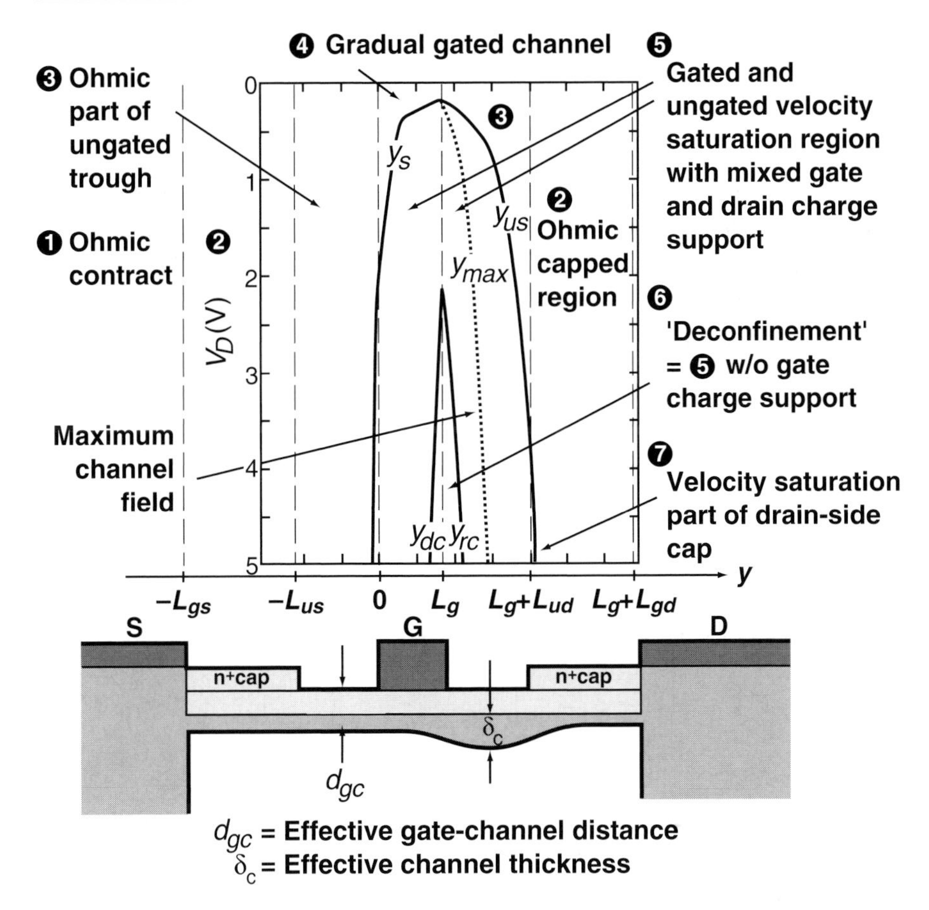

Fig. 14.8. Example of the evolution of the different regions in a MODFET as the drain bias is increased.

$I_D = qW_g n_s \left(V_{surf}(y_s) - V_c(y_s)\right) v_{sat}$, where $V_c(y_s) = V_{Slin} + F_{ugt}(y_s + L_{us})$. This results in

$$y_s = -\left(\frac{V_0 - V_{sat}}{F_{ugt} + F_{surfS}}\right), \quad \text{when } V_0 > V_{sat}. \tag{14.52}$$

Further towards $y = -L_{us}$, the free surface rapidly 'unpinches' the channel, preventing $y_s(V_D)$ from moving far into negative-y territory, and making the use of F_{ugt} in $-L_{us} < y < y_s$ a reasonable simplification.

Figure 14.8 also shows that for sufficiently large drain biases y_{us} will move out into the drain-side cap region. In this typically highly conductive (sheet resistance R) region, F_{surf} is assumed equal to its value RI_D/W_g in the linear part. Thus,

$$V_{surf}(y) = V_D - V_{Dlin} + V_{Gcap} + \frac{RI_D}{W_g}\left[y - (L_g + L_{ud})\right], \tag{14.53}$$
$$L_g + L_{ud} < y < L_g + L_{gd},$$

in this fourth region of potential velocity saturation.

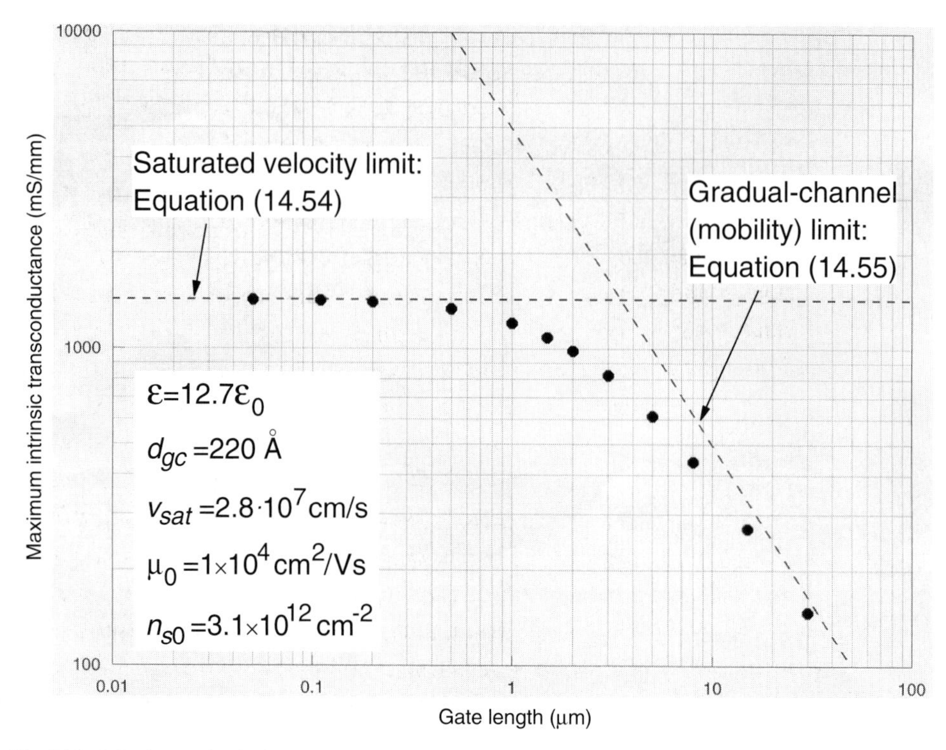

Fig. 14.9. Maximum intrinsic transconductance with parameters representative of InP-type MODFETs versus gate length. Filled circles are calculated with the full DC model. Dashed lines are the two limits.

Figure 14.9 shows calculated maximum intrinsic transconductance versus gate length for a typical InAlAs/InGaAs MODFET structure. To this end R_C, R, L_{us}, V_{Gugt}, v_{ugt}, etc. were picked to produce negligible parasitic voltage $V_c(0)$. For the short-gate lengths of primary interest in this chapter, the intrinsic DC transconductance in saturation approaches the limit*

$$g_{m0,sat}^{(i)} = \frac{\varepsilon W_g v_{sat}}{d_{gc}}, \tag{14.54}$$

which is directly proportional to the saturation velocity, and independent of the gate length and mobility. This expression will be used for the small-signal analysis in Chapters 15 and 17. For longer gates the transconductance will begin to be degraded

* The notation introduced here for small-signal parameters is heavy on sub- and superscripts, and may appear cumbersome. However, it is important to be aware of whether the parameter is intrinsic (i) or extrinsic (x), whether it is the DC (0) or AC (ω) value, and finally whether the parameter is associated with the gradual channel (gc) or the velocity-saturated region (sat). Capacitances and gate leakages are generally understood to be intrinsic parameters, i.e. there are no parasitic resistors hidden inside of them. In the case of transconductance and output conductance, however, it is not always clear whether their quoted values are for the intrinsic or extrinsic case. Thus, we are careful to carry along the appropriate identifying superscript.

by the low-field, essentially resistive, part of the device. It eventually approaches Shockley's [89] original gradual-channel limit, where

$$g_{m0,gc}^{(i)} = \frac{qW_g n_s(V_G)\mu_0}{L_g} \tag{14.55}$$

(Problem 14.5), even though the channel never totally pinches off, and the field remains finite [51]. In this limit, the maximum transconductance does not correspond to the maximum slope of $n_s(V_G)$, but rather its maximum value n_{s0}. Note that $(g_{m0,gc}^{(i)})^{-1}$ is the open ($V_D = 0$) channel resistance. Since the same v_{sat} was used throughout the gate-length range in Figure 14.9, it illustrates that a gate-length dependence in the transconductance (e.g. [90]) is not necessarily a sign of velocity overshoot. It is interesting that, based on $g_{m0}^{(i)} - L_g$ curves which do not saturate as $L_g \rightarrow 0$, experimental observation of velocity overshoot in Si NMOSFETs has been claimed (e.g. [91]). The room temperature saturation velocity, even at $L_g = 50$ nm, however, is only $\sim 30\%$ above the steady-state v'_{sat}.

Figure 14.10(a) shows the modeled drain current characteristic for an AlInAs/GaInAs MODFET with the 0.1-μm geometry in Figure 14.1. Because of the short gate and lateral charge support the channel also conducts for $V_G \leq V_{Goff}$, particularly above the ~ 3 V punch-through voltage that results from our choice of γ. $M_n I_D$ is shown in dashed lines. The impact ionization contributes *directly* to the drain current only moderately, by adding to the output conductance beyond a V_G-dependent V_D. Recall, however, that a small amount of impact ionization at lower V_D's can affect the drain current and output conductance *indirectly* through the kink effect [78]. A small amount of impact ionization (much smaller than that which would correspond to bulk avalanche breakdown) is also responsible for the on-state breakdown (= burn-out) voltage for InP-type MODFETs. Because of the low level of impact ionization current relative to the primary drain current, one cannot generally distinguish an upturn in I_D before the FET breaks down. The presence of impact ionization is, however, clearly evident as a negative gate current, corresponding to holes being collected by the gate. Figure 14.10(b) shows burn-out biases ($BV_{ds}^{(on)}$, I_D) for 0.1-μm $In_{0.53}Ga_{0.47}As$-channel MODFETs with two gate widths (2×11 and 2×75 μm) across a wafer, at various current levels. Superimposed from Figure 14.10(a) are the top $I_D(V_D)$ curve, and three curves of constant predicted secondary current $(M_n - 1)I_D$. Added are five loci of constant DC power dissipation, and two potential load-lines for power amplifier design. Several things are noteworthy. First, unlike the situation for MESFETs [92] and PHEMTs [93], the burn-out biases do not follow a constant-power curve. Instead, they track rather closely the locus of $(M_n - 1)I_D = 10$ mA/mm, calculated using the approximate method described above, with published field-dependent impact ionization coefficients [79], and $k_p = 1$. Relative to the 22-μm FETs, the 150-μm zapping points tend to deviate somewhat towards a constant power locus, but not in a major way. One might interpret this as

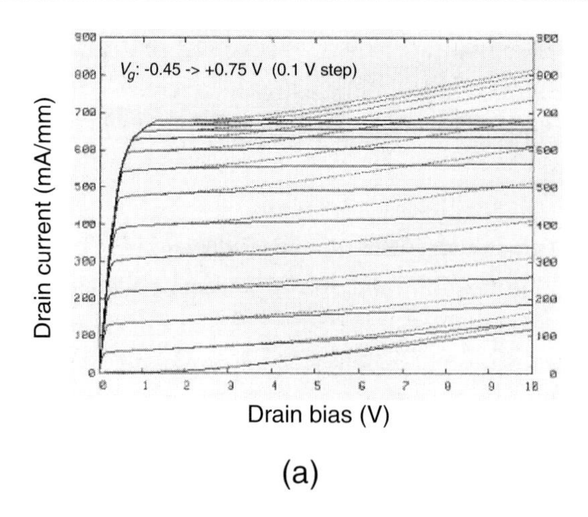

(a)

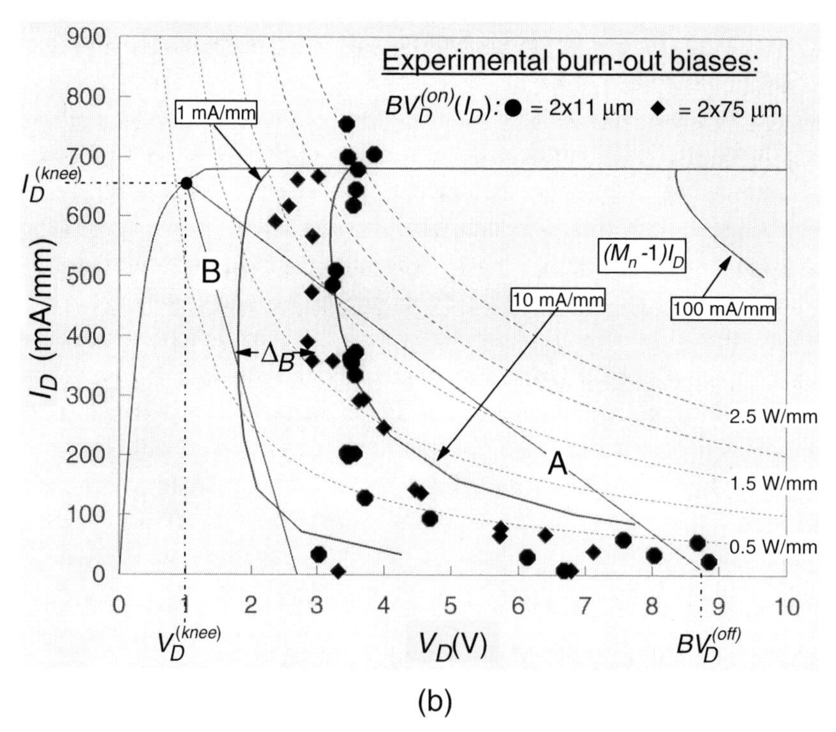

(b)

Fig. 14.10. (a) Calculated extrinsic I–V characteristics with impact ionization included (dashed curves) of the FET in Figure 14.1 with typical InP-type MODFET parameters. (b) Experimental burn-out (or on-state breakdown) points on actual FETs, with top I–V and impact ionization curves from (a) superimposed. Constant power dissipation (dashed curves) and two load-lines are also indicated.

a small thermal burn-out component, caused by the increase in the impact ionization for $In_{0.53}Ga_{0.47}As$ with temperature [94]. Typically one expects impact ionization to be reduced with temperature because of increased scattering. For narrow-bandgap material, however, the reduction in bandgap with temperature [75] can reverse the trend. Overall, the adherence of $BV_{ds}^{(on)}(I_D)$ to the impact ionization curves is striking, particularly the increase in $BV_{ds}^{(on)}$ for the smaller FETs at the highest currents. This is due to the reduced channel field as the channel becomes more uniformly filled. Since the field appears exponentially in the impact ionization coefficients this overwhelms the moderate increase in primary drain current. Near off-state, it is the increase in primary current that causes the fast drop in $BV_{ds}^{(on)}(I_D)$.

As a consequence of these results, the often quoted *off*-state breakdown voltage $BV_{ds}^{(off)}$ is not that relevant for these InP-type FETs. With load-line A in Figure 14.10(b) the classical back-of-the-envelope estimate for the maximum output power available from a device is

$$P_{out}^{(max)} = \frac{1}{8} I_D^{(knee)} \left(BV_D^{(off)} - V_D^{(knee)} \right) \tag{14.56}$$

for class-A operation. With the device biased at the midpoint of the load-line, near maximum transconductance, class-A allows the highest bandwidth (e.g. [27])*. Although the predicted millimeter-wave output power would be impressive with load-line A in Figure 14.10(b), the FET would be biased in (and the voltage would swing further into) the region of destructive breakdown. For reliable operation one should back off to load-line B, which accounts for the smaller on-state breakdown voltage with some margin Δ_B. The breakdown voltage has a flat minimum near I_D of maximum transconductance ($\sim I_D^{(max)}/2$). This minimum, $BV_D^{(on,min)}$, becomes a more conservative and significant measure of breakdown for high-speed narrow-bandgap FETs, and the maximum output power available from such a device is better expressed as

$$P_{out}^{(max)} = \frac{1}{4} I_D^{(knee)} \left(V_D^{(bias)} - V_D^{(knee)} \right), \tag{14.57a}$$

where

$$V_D^{(bias)} = BV_D^{(on,min)} - \Delta_B. \tag{14.57b}$$

Note that it is the *knee* current that counts†. Although there is a rather small difference in the values of $I_D^{(knee)}$ and $I_D^{(max)}$ in the analytical DC model, the difference can be significant in real cases where there is a large output conductance. This situation can, for instance, occur for very short gates, or in FETs with a large In mole fraction in the channel.

* It is, however, not the most efficient class due to the large quiescent power P_{DC} being dissipated. As the name implies, the power-added efficiency (PAE) listed in Table 14.1 is defined as $PAE = (P_{out} - P_{in})/P_{DC}$, where P_{in} is the input microwave power [27]. The PAE will peak at a smaller input power than what is required to reach maximum output power. The maximum conceivable PAE in class A is 50%, and would correspond to the hypothetical situation of $V_D^{(knee)} = 0$.

† We have assumed a typical class-A design where the knee point in the drain characteristics is one 'anchor' for the load-line. For a device with high maximum current, and low on-state breakdown voltage, a shallower load-line that does not reach the knee point, but allows a larger voltage swing, may have to be considered.

The simple DC expressions for maximum output power can be quite accurate even at high frequencies (e.g. [58]), provided that the device has reasonably low trap-related dispersion, is not too non-linear and sees the proper load at the output. Finding the optimum load admittance $Y_L^{(opt)}$ experimentally involves so-called load-pull measurements, where a plethora of loads are presented to the device and the optimum one is picked. With a good large-signal CAD model, simulation should be able to produce the same result. The problem is inherently large-signal and non-linear, and requires one of the two approaches. Nevertheless, a good first guess is

$$\mathrm{Re}\left(Y_L^{(opt)}\right) = \frac{I_D^{(knee)}}{2\left(V_D^{(bias)} - V_D^{(knee)}\right)} \tag{14.58a}$$

$$\mathrm{Im}\left(Y_L^{(opt)}\right) = -2\pi f_0 C_{22}^{(x)}, \tag{14.58b}$$

where f_0 is the center of the frequency band of operation, and $C_{22}^{(x)}$ is the extrinsic small-signal output capacitance of the FET at the bias point. This and other small-signal properties of the device are the topics of Chapter 15.

14.7 Reliability

Determining the reliable operating regime is much more complex a process than the discussion of Figure 14.10(b) may have suggested. It is an important task that must be done for any process destined for manufacturing. In addition to a repeatable uniform process and a representative CAD model, the end user needs to know that the devices are reliable in the specified gate and drain voltage ranges. Establishing reliability involves measures both DC stress at elevated temperatures (HTOL = high-temperature operating lifetime), and large-signal RF stress (RFOL), often performed at room temperature. The bottom-line outcome of the stressing is an extrapolated mean time to failure (*MTTF*). An *MTTF* of 10^6 hour at 125 or 150° C channel temperature (T_0) is typically required. 10^6 hours (= 114 years) may seem excessive, particularly with today's product lifetime, but it is a number with a history, and does leave room for the ~50% fraction of devices that fail before the mean time. The following is a very brief discussion centered around InP-type MODFETs. For a thorough review of testing methods and degradation mechanisms, an interested reader can consult the rich literature on reliability (e.g. [95–97]).

Reliability work is difficult and time-consuming. Most often, it is not just a matter of establishing the *MTTF*, but also of improving it. Fully closing the loop between the two is a major undertaking, and can only be done for a major process upgrade. Determining the *MTTF* from HTOL is a long process that has to be done with great care. One has to start with sufficiently many devices to make the *MTTF* meaningful.

The devices have to be packaged, and 'infant deaths' prudently removed from the population. The degradation is accelerated by elevating the channel temperature T well above T_0. The channel temperature is determined by liquid crystal measurements, or thermal modeling of its rise over the base temperature. At least three T's should be used, so that a linear fit to $\log_{10}(MTTF)$ vs $1/T$ can be made with some confidence, and the activation energy determined:

$$E_A = \frac{k}{\log_{10}(e)} \frac{d \log_{10}(MTTF)}{d(1/T)}. \tag{14.59}$$

The choice of the T's is a tradeoff, and a judgement call. T has to be large enough so that the degradation is sufficiently accelerated to be determined, but not so large that the dominant failure mechanism differs from the one at specified operating temperatures. The sought-after $MTTF$ is found by extrapolation from an elevated temperature T_e:

$$MTTF\,(T_0) = MTTF\,(T_e) \exp\left(\frac{E_A}{k}\left(\frac{1}{T_0} - \frac{1}{T_e}\right)\right). \tag{14.60}$$

A typical HTOL failure criterion for FETs is a 10–15% drop in maximum extrinsic transconductance or I_{DSS} $(= I_D(V_G = 0, V_D > V_{Dsat} + (R_S + R_D)I_D))$. The use of I_{DSS} as a failure criterion hails from the days when threshold voltages were very negative (by today's standards), and I_{DSS} was essentially equal to the maximum drain current. Today, for a typical depletion-mode ($V_T < 0$) device, the maximum transconductance occurs near $V_G = 0$. Thus, I_{DSS} will depend on V_T, $g^{(i)}_{m0,sat}$, $g^{(i)}_{ds0,sat}$, R_S, and R_D, and therefore on the large number of underlying physical parameters that we have encountered. The number of basic parameters and device regions that go into the maximum drain current are fewer, making it more telling and fundamental. For an enhancement-mode device ($V_T > 0$), the choice is even more clear. As one learns more about the mechanisms of degradation, other failure criteria, such as a maximum increase in R_D (e.g. [98–100]), may have to be considered.

With regard to improving the reliability, there is not much quantitative help from modeling, although the type of analysis that established the correlation between impact ionization and burn-out voltage in Figure 14.10(b) can be helpful in guiding the process development. It also suggests loci of similar reliability, which is useful considering that the rather involved process outlined in the last paragraph was for a single bias. The modeling is also useful in prescribing acceptable load-lines, both for power amplifier design and RFOL. The physics of degradation, however, is mostly too complex even for the most sophisticated models available. For the case of primary interest here, $Ga_{0.47}In_{0.53}As/Al_{0.48}In_{0.52}As$ MODFETs, it is clear that impact ionization plays an important role (e.g. [99, 100]), particularly with its positive temperature coefficient [94]. Impact ionization by itself, however, is not necessarily a bad thing. After all, there are devices (avalanche photodiodes) installed in communication systems that rely on it. The search continues for the mechanism that ultimately causes the degradation (possibly a recombination process). In addition, there are many other potential degradation mechanisms at elevated temperatures and/or high fields. These include gate sinking (e.g. [101, 98]), degradation (oxidation) of the AlInAs surface (e.g. [102, 103]), indiffusion (potentially field-enhanced) of foreign atoms that compensate the donors (e.g. [104]), hot-electron-induced degradation of the passivation–semiconductor interface (e.g. [105]), and degradation of the ohmic contacts (e.g. [106]). Complicating the task of drawing general conclusions is that different mechanisms appear to dominate in ostensibly similar processes and epitaxial structures from different labs. Examples of lessons learned in one laboratory can be found in [107].

14.8 Conclusion

This chapter has begun to introduce the reader to some advanced topics on MODFETs. The goal has been to thoroughly analyze important practical issues for DC and large-signal operation, while maintaining an analytical approach. We have been able to gain insight into and model charge control and transport in degenerately dual-doped structures. We have managed to take the modeling far enough to predict breakdown characteristics that affect output power and reliability. In the following three chapters we will continue the analytical approach, but will shift the focus to the high-frequency performance.

14.9 Bibliography

[1] A. Seabaugh, B. Brar, T. Broekaert, F. Morris, P. van der Wagt and G. Frazier, 'Resonant-tunneling mixed-signal circuit technology', *Solid State Electronics*, Vol. 43, pp. 1355–1365, 1999.

[2] U. Bhattacharya, M.J. Mondry, G. Hurtz, I.-H. Tan, R. Pullela, M. Reddy, J. Guthrie, M.J.W. Rodwell and J.E. Bowers, 'Transferred substrate Schottky-collector heterojunction bipolar transistors: First results and scaling laws for high f_{max}', *IEEE Electron Device Letters*, Vol. 16, pp. 357–359, 1995.

[3] M. Rodwell, Q. Lee, D. Mensa, J. Guthrie, S.C. Martin, R.P. Smith, R. Pullela, B. Agarwal, S. Jaganathan, T. Mathew and S. Long, 'Transferred-substrate HBT integrated circuits', *Solid State Electronics*, Vol. 43, pp. 1489–1495, 1999.

[4] J.-E. Müller, A. Bangert, T. Grave, M. Kärner, H. Riechert, A. Schäfer , H. Siweris, L. Schleicher, H. Tischer, L. Verweyen, W. Kellner and T. Meier, 'A GaAs HEMT MMIC chip set for automotive radar systems fabricated by optical stepper lithography', *Technical Digest of the GaAs IC Symposium*, IEEE, Piscataway, pp. 189–192, 1996.

[5] L.D. Nguyen, A.S. Brown, M.A. Thompson and L.M. Jelloian, '50-nm self-aligned-gate pseudomorphic AlInAs/GaInAs high electron mobility transistors', *IEEE Transactions on Electron Devices*, Vol. 39, pp. 2007–2014, 1992.

[6] T. Suemitsu, T. Ishii, H. Yokoyama, Y. Umeda, T. Enoki, Y. Ishii and T. Tamamura, '30-nm-gate InAlAs/InGaAs HEMT's lattice-matched to InP substrates', *Technical Digest of the International Electron Devices Meeting*, IEEE, Piscataway, pp. 223–226, 1998.

[7] U.K. Mishra and J.B. Shealy, 'InP-based Hemts: Status and potential', *Proceedings of the International Conference on Indium Phosphide and Related Materials*, IEEE, Piscataway, pp. 14–17, 1994.

[8] R.L. Van Tuyl, 'Unlicensed millimeter wave communications: A new opportunity for MMIC technology at 60 GHz', *Technical Digest of the GaAs IC Symposium*, IEEE, Piscataway, pp.3–5, 1996.

[9] D. Polifko and H. Ogawa, 'The merging of photonic and microwave technologies', *Microwave Journal*, Vol. 35, pp. 75–78, 1992.

[10] S. Yajima and A. Niedzwiecki, 'Direct broadcast satellite applications', *Hewlett-Packard Journal*, Vol. 49, pp. 52–55, 1998.

[11] H. Rohdin, 'Analytical forward and reverse modeling of MODFETs'. In *Heterojunction Transistors and Small Size Effects in Devices,* ed. M. Willander, Studentlitteratur/Chartwell-Bratt, Lund, pp. 107–194, 1992.

[12] G. Bischoff and J.P. Krusius, 'Technology independent device modeling for simulation of integrated circuits for FET technologies', *IEEE Transactions on Computer-Aided Design*, Vol. 4, pp. 99–110, 1985.

[13] D.E. Root, S. Fan and J. Meyer, 'Technology independent large signal non quasi-static FET models by direct construction from automatically characterized device data', *Proceedings of the European Microwave Conference*, pp. 927–932, 1991.

[14] W.M. Coughran, Jr, W. Fichtner and E. Grosse, 'Extracting transistor charges from device simulations by gradient fitting', *IEEE Transactions on Computer-Aided Design*, Vol. 8, pp. 380–394, 1989.

[15] Y.M. Kim and P. Roblin, 'Two-dimensional charge-control model for MODFETs', *IEEE Transactions on Electron Devices*, Vol. 33, pp. 1644–1651, 1986.

[16] J. Xu and M. Shur, 'Velocity-field dependence in GaAs', *IEEE Transactions on Electron Devices*, Vol. 34, pp. 1831–1832, 1987.

[17] M.C. Foisy, A physical model for the bias dependence of the modulation-doped field-effect transistor's high-frequency performance. PhD Thesis, Cornell University, 1990.

[18] P.A. Sandborn, A. Rao and P.A. Blakey, 'An assessment of approximate nonstationary charge transport models used for GaAs device modeling', *IEEE Transactions on Electron Devices*, Vol. 36, pp. 1244–1253, 1989.

[19] N. Moll, M.R. Hueschen and A. Fischer-Colbrie, 'Pulse-doped AlGaAs/InGaAs pseudomorphic MODFETs', *IEEE Transactions on Electron Devices*, Vol. 35, pp. 879–886, 1988.

[20] H. Rohdin, 'Reverse modeling of E/D logic submicrometer MODFETs and prediction of maximum extrinsic MODFET current gain cut-off frequency', *IEEE Transactions on Electron Devices*, Vol. 37, pp. 920–934, 1990.

[21] W.W. Hooper and W.I. Lehrer, 'An epitaxial GaAs field-effect transistor', *Proceedings of the IEEE*, Vol. 55, pp. 1237–1238, 1967.

[22] J.S. Barrera and R.J. Archer, 'InP Schottky-gate field-effect transistor', *IEEE Transactions on Electron Devices*, Vol. 22, pp. 1023–1030, 1975.

[23] R.W.H. Engelmann and C.A. Liechti, 'Bias dependence of GaAs and InP MESFET parameters', *IEEE Transactions on Electron Devices*, Vol. 24, pp. 1288–1296, 1977.

[24] R. Dingle, H.L. Stormer, A.C. Gossard and W. Wiegmann, 'Electron mobilities in modulation-doped semiconductor heterojunction superlattices', *Applied Physics Letters*, Vol. 33, pp. 665–667, 1978.

[25] A.S. Brown, A.E. Schmitz, L.D. Nguyen, J.A. Henige and L.E. Larson, 'The growth of high performance $In_xGa_{1-x}As(.52 < x < .9)$-$Al_{.48}In_{.52}As$ high electron mobility transistors by MBE', *Proceedings of the International Conference on Indium Phosphide and Related Materials*, IEEE, Piscataway, pp. 263–266, 1994.

[26] A. Mesquida Küsters, A. Kohl, S. Brittner, V. Sommer and K. Heime, 'Effect of indium mole fraction on charge control, DC and RF performance of single quantum-well InP/$In_xGa_{1-x}As$/InP ($0.53 \leq x \leq 0.81$) HEMTs', *Proceedings of the International Conference on Indium Phosphide and Related Materials*, IEEE, Piscataway, pp. 323–326, 1994.

[27] L.D. Nguyen, L.E. Larson and U.K. Mishra, 'Ultra-high-speed modulation-doped field-effect transistors: A tutorial review', *Proceedings of the IEEE*, Vol. 80, pp. 494–518, 1992.

[28] M. Hueschen, N. Moll, E. Gowen and J. Miller, 'Pulse doped MODFETs', *Technical Digest of the International Electron Devices Meeting*, IEEE, Piscataway, pp. 348–351, 1984.

[29] M. Hueschen, N. Moll, and A. Fischer-Colbrie, 'High-current GaAs/AlGaAs MODFETs with f_T over 80 GHz', *Technical Digest of the International Electron Devices Meeting*, IEEE, Piscataway, pp. 596-599, 1987.

[30] M.C. Foisy, P.J. Tasker, B. Hughes and L.F. Eastman, 'The role of inefficient charge modulation in limiting the current-gain cut-off frequency of the MODFET', *IEEE Transactions on Electron Devices*, Vol. 35, pp. 871–878, 1988.

[31] K. Lee, M.S. Shur, T.J. Drummond and H. Morkoç, 'Parasitic MESFET in (Al,Ga)As/GaAs modulation-doped FET's and MODFET characterization', *IEEE Transactions on Electron Devices*, Vol. 31, pp. 29–35, 1984.

[32] K. Lee, M. Shur, T.J. Drummond and H. Morkoç, 'Electron density of the two-dimensional gas in modulation doped layers', *Journal of Applied Physics* , Vol. 54, pp. 2093–2096, 1983.

[33] R.A. Smith, *Semiconductors*, Cambridge University Press, Cambridge, 1978.

[34] C. Kittel, *Solid State Physics*, 5th ed., John Wiley & Sons, New York, 1976.

[35] S.R. Dhariwal, V.N. Ojha and G.P. Srivastava, 'On the shifting and broadening of impurity bands and their contribution to the effective electrical bandgap narrowing in moderately doped semiconductors', *IEEE Transactions on Electron Devices*, Vol. 32, pp. 44–48, 1985.

[36] Lj.D. Živanov and M.B. Živanov, 'Determination of average effective masses of majority carriers as function of impurity concentrations for heavily doped GaAs', *Proceedings of the International Semiconductor Conference*, IEEE, Piscataway, pp. 103–106, 1995.

[37] S.C. Jain, J.M. McGregor and D.J. Roulston, 'Band-gap narrowing in novel III–V semiconductors', *Journal of Applied Physics*, Vol. 68, pp. 3747–3749, 1990.

[38] P.A. Serena, J.M. Soler, N. Garcia, 'Self-consistent image potential in a metal surface', *Physical Review B*, Vol. 34, pp. 6767–6769, 1986.

[39] F. Stern and S. Das Sarma, 'Electron energy levels in GaAs-$Ga_{1-x}Al_xAs$ heterojunctions', *Physical Review B*, Vol. 30, pp. 840–848, 1984.

[40] N. Moll, 'HFETs: A study in developmental device physics', in *Compound Semiconductor Transistors: Physics and Technology*, ed. S. Tiwari, IEEE, Piscataway, pp. 3–20, 1993.

[41] D.V. Lang, R.A. Logan and M. Jaros, 'Trapping characteristics and a donor-complex (DX) model for the persistent-photoconductivity trapping center in Te-doped $Al_xGa_{1-x}As$', *Physical Review B*, Vol. 19, pp. 1015–1030, 1979.

[42] D.J. Chadi and K.J. Chang, 'Theory of atomic and electronic structure of DX centers in GaAs and $Al_xGa_{1-x}As$ alloys', *Physical Review Letters*, Vol. 61, pp. 873–876, 1988.

[43] R.T. Kaneshiro, C.P. Kocot, R.P. Jaeger, J.S. Kofol B.J.F. Lin, E. Littau, H. Luechinger and H.G. Rohdin, 'Anomalous nanosecond transient component in a GaAs MODFET technology', *IEEE Electron Device Letters*, Vol. 9, pp. 250–252, 1988.

[44] B.J.F. Lin, H. Luechinger, C.P. Kocot, E. Littau, C. Stout, B. McFarland, H. Rohdin, J.S. Kofol, R.P. Jaeger and D.E. Mars, 'A 1-μm MODFET process yielding MUX and DMUX circuits operating at 4.5 Gb/s', *Technical Digest of the GaAs IC Symposium*, IEEE, Piscataway, pp. 143–146, 1988.

[45] B. Etienne and V. Thierry-Mieg, 'Reduction in the concentration of DX centers in Si-doped GaAlAs using the planar doping technique', *Applied Physics Letters*, Vol. 52, p. 1237, 1988.

[46] T. Yokoyama, M. Suzuki, T. Maeda, T. Ishikawa, T. Mimura and M. Abe, 'Se-doped AlGaAs/GaAs HEMT's for stable low-temperature operation', *IEEE Electron Device Letters*, Vol. 11, pp. 197–199, 1990.

[47] A.A. Aziz and M. Missous, 'InGaP/InGaAs/GaAs pseudomorphic HEMT grown by solid source MBE', *Proceedings of the High-Performance Electron Devices for Microwave and Optoelectronic Applications Workshop*, pp. 145–152, 1996.

[48] T. Baba, T. Mizutani and M. Ogawa, 'AlAs/n-GaAs superlattice and its application to high-quality two-dimensional electron gas systems', *Journal of Applied Physics*, Vol. 59, pp. 526–532, 1986.

[49] H. Sari and H.H. Wieder, 'DX centers in $In_xAl_{1-x}As$', *Journal of Applied Physics*, Vol. 85, pp. 3380–3382, 1999.

[50] B. Vinter, 'Subbands and charge control in a two-dimensional electron gas field-effect transistor', *Applied Physics Letters*, Vol. 44, pp. 307–309, 1984.

[51] H. Rohdin and P. Roblin, 'A MODFET dc model with improved pinchoff and saturation characteristics', *IEEE Transactions on Electron Devices*, Vol. 33, pp. 664–672, 1986.

[52] G.U. Jensen, B. Lund, T.A. Fjeldly and M. Shur, 'Monte Carlo simulation of short-channel heterostructure field-effect transistors', *IEEE Transactions on Electron Devices*, Vol. 38, pp. 840–851, 1991.

[53] M.V. Fischetti, 'Monte Carlo simulation of transport in technologically significant semiconductors of the diamond and zinc-blende structures – Part I: Homogeneous transport', *IEEE Transactions on Electron Devices*, Vol. 38, pp. 634–649, 1991.

[54] T. Wang, private communication, 1985.

[55] V. Choudhry and V.K. Arora, 'Field dependence of drift velocity and electron population in satellite valleys in gallium arsenide', *Electronics Letters*, Vol. 22, pp. 271–272, 1986.

[56] D.R. Greenberg and J.A. del Alamo, 'Velocity saturation in the extrinsic device: A fundamental limit in HFETs', *IEEE Transactions on Electron Devices*, Vol. 41, pp. 1334–1339, 1994.

[57] M. Arps, H.-G. Bach, W. Passenberg, A. Umbach and W. Schlaak, 'Influence of SiN_x passivation on the surface potential of GaInAs and AlInAs in HEMT layer structures', *Proceedings of the International Conference on Indium Phosphide and Related Materials*, IEEE, Piscataway, pp. 308–311, 1996.

[58] H. Rohdin, A. Nagy, V. Robbins, C.-Y. Su, A.S. Wakita, J. Seeger, T. Hwang, P. Chye, P.E. Gregory, S.R. Bahl, F.G. Kellert, L.G. Studebaker, D.C D'Avanzo and S. Johnsen, '0.1-μm gate-length AlInAs/GaInAs/GaAs MODFET MMIC process for applications in high-speed wireless communication', *Hewlett-Packard Journal*, Vol. 49, pp. 37–38, 1998.

[59] P. Urien and D. Delagebeaudeuf, 'New method for determining the series resistances in a MESFET or TEGFET', *Electronics Letters*, Vol. 19, pp. 702–703, 1983.

[60] K. Lee, M. Shur, A.J. Valois, G.Y. Robinson, X.C. Zhu, and A. Van der, Ziel, 'A new technique for characterization of the 'end' resistance in modulation-doped FETs', *IEEE Transactions on Electron Devices*, Vol. 31, pp. 1394–1398, 1984.

[61] R.P. Holmstrom, W.L. Bloss and J.Y. Chi, 'A gate probe method of determining parasitic resistance in MESFETs', *IEEE Electron Device Letters*, Vol. 7, pp. 410–412, 1986.

[62] S.-M.J. Liu, S.-T. Fu, M. Thurairaj and M.B. Das, 'Determination of source and drain series resistances of ultra-short gate-length MODFET's', *IEEE Electron Device Letters*, Vol. 10, pp. 85–87, 1989.

[63] W.J. Azzam and J.A. del Alamo, 'An all-electrical floating-gate transmission line model technique for measuring source resistance in heterostructure field-effect transistors', *IEEE Transactions on Electron Devices*, Vol. 37, pp. 2105–2107, 1990.

[64] Y. Zhu, Y. Ishimaru and M. Shimizu, 'Direct determination of source, drain and channel resistances of HEMTs', *Electronics Letters*, Vol. 31, pp. 318–320, 1995.

[65] G.T. Cibuzar, 'Effects of gate recess etching on source resistance', *IEEE Transactions on Electron Devices*, Vol. 42, pp. 1195–1196, 1995.

[66] M.D. Feuer, 'Two-layer model for source resistance in selectively doped heterojunction transistors', *IEEE Transactions on Electron Devices*, Vol. 32, pp. 7–11, 1985.

[67] R.E. Williams, *Gallium Arsenide Processing Techniques*, Artech House, Dedham, 1984.

[68] P. Roblin, L. Rice, S.B. Bibyk and H. Morkoç, 'Nonlinear parasitics in MODFET's and MODFET I–V characteristics', *IEEE Transactions on Electron Devices*, Vol. 35, pp. 1207–1214, 1988.

[69] T. Hariu, K. Takahashi and Y. Shibata, 'New modeling of GaAs MESFETs', *IEEE Transactions on Electron Devices*, Vol. 30, pp. 1743–1749, 1983.

[70] R.B. Darling, 'Generalized gradual channel modeling of field-effect transistors', *IEEE Transactions on Electron Devices*, Vol. 35, pp. 2302–2314, 1988.

[71] S.R. Bahl and J.A. del Alamo, 'A new drain-current injection technique for the measurement of off-state breakdown voltage in FETs', *IEEE Transactions on Electron Devices*, Vol. 40, pp. 1558–1560, 1993.

[72] H. Rohdin, A. Nagy, V. Robbins, C.-Y. Su, C. Madden, A. Wakita, J. Raggio and J. Seeger, 'Low-noise, high-speed $Ga_{.47}In_{.53}As/Al_{.48}In_{.52}As$ 0.1-μm MODFETs and high-gain/bandwidth three-stage amplifier fabricated on GaAs substrate', *Proceedings of the International Conference on Indium Phosphide and Related Materials*, IEEE, Piscataway, pp. 73–76, 1995.

[73] S.R. Bahl and J.A. del Alamo, 'Physics of breakdown in InAlAs/n$^+$-InGaAs heterostructure field-effect transistors, *IEEE Transactions on Electron Devices*, Vol. 41, pp. 2268–2275, 1994.

[74] K.W. Eisenbeiser, J.R. East and G.I. Haddad, 'Theoretical analysis of the breakdown voltage in pseudomorphic HFETs', *IEEE Transactions on Electron Devices*, Vol. 43, pp. 1778–1787, 1996.

[75] S.M. Sze, *Physics of Semiconductor Devices*, 2nd Ed., John Wiley & Sons, New York, 1981.

[76] K. Seeger *Semiconductor Physics*, Springer-Verlag, New York, 1973.

[77] A.S. Brown, U.K. Mishra, C.S. Chou, C.E. Hooper, M.A. Melendes, M. Thompson, L.E. Larson, S.E. Rosenbaum and M.J. Delaney, 'AlInAs-GaInAs HEMT's utilizing low-temperature AlInAs buffers grown by MBE', *IEEE Electron Device Letters*, Vol. 10, pp. 565–567, 1989.

[78] T. Suemitsu, T. Enoki, N. Sano, M. Tomizawa and Y. Ishii, 'An analysis of the kink phenomena in InAlAs/InGaAs HEMT's using two-dimensional device simulation', *IEEE Transactions on Electron Devices*, Vol. 45, pp. 2390–2399, 1998.

[79] J. Urquhart, D.J. Robbins, R.I. Taylor and A.J. Moseley, 'Impact ionisation coefficients in $In_{0.53}Ga_{0.47}As$', *Semiconductor Science and Technology*, Vol. 5, pp. 789–791, 1990.

[80] T.P. Pearsall, 'Impact ionization rates for electrons and holes in $Ga_{0.47}In_{0.53}As$', *Applied Physics Letters*, Vol. 36, pp. 218–220, 1980.

[81] F. Osaka, T. Mikawa and T. Kaneda, 'Impact ionization coefficients of electrons and holes in (100)-oriented $Ga_{1-x}In_xAs_yP_{1-y}$', *IEEE Journal of Quantum Electronics*, Vol. 21, pp. 1326–1338, 1985.

[82] G.-G. Zhou, A. Fischer-Colbrie J. Miller, Y.-C. Pao, B. Hughes, L. Studebaker and J.S. Harris, 'High output conductance of InAlAs/InGaAs/InP MODFET due to weak impact ionization in the InGaAs channel', *Technical Digest of the International Electron Devices Meeting*, IEEE, Piscataway, pp. 247–250, 1991.

[83] F.A. Padovani and R. Stratton, 'Field and thermionic-field emission in Schottky barriers', *Solid State Electronics*, Vol. 9, pp. 695–707, 1966.

[84] M.H. Somerville and J.A. del Alamo, 'A model for tunneling-limited breakdown in high-power HEMTs', *Technical Digest of the International Electron Devices Meeting*, IEEE, Piscataway, pp. 35–38, 1996.

[85] A.B. Grebene and S.K. Ghandhi, 'General theory for pinched operation of the junction-gate FET', *Solid State Electronics*, Vol. 12, pp. 573–589, 1969.

[86] C.A. Liechti, 'Microwave field-effect transistors – 1976', *IEEE Transactions on Microwave Theory and Techniques*, Vol. 24, pp. 279–300, 1976.

[87] Y. Awano, 'New transverse-domain formation mechanism in a quarter-micrometre-gate HEMT', *Electronics Letters*, Vol. 24, pp. 1315–1317, 1988.

[88] H. Rohdin and A. Nagy, 'A 150 GHz sub-0.1-μm E/D MODFET MSI process, *Technical Digest of the International Electron Devices Meeting*, IEEE, Piscataway, pp. 327–330, 1992.

[89] W. Shockley, 'A unipolar 'field-effect' transistor', *Proceedings of the IRE*, pp. 1365–1376, 1952.

[90] P.R. de la Houssaye, D.R. Allee, Y.-C. Pao, D.G. Schlom, J.S. Harris and R.F.W. Pease, 'Electron saturation velocity variation in InGaAs and GaAs channel MODFET's for gate lengths to 550 Å', *IEEE Electron Device Letters*, Vol. 9, pp. 148–150, 1988.

[91] G.A. Sai-Halasz, M.R. Wordeman, D.P. Kern, S. Rishton and E. Ganin, 'High transconductance and velocity overshoot in NMOS devices at the 0.1-μm gate-length level', *IEEE Electron Device Letters*, Vol. 9, pp. 464–466, 1988.

[92] V.A. Vashchenko, N.A. Kozlov, V.F. Sinkevitch, J.B. Martynov and A.S. Tager, 'Simulation of the GaAs MESFET burnout', *Proceedings of the 25th European Microwave Conference*, pp. 229–233, 1995.

[93] J.A. del Alamo and M.H. Somerville, 'Breakdown in millimeter-wave power InP HEMTs: A comparison with PHEMTs', *Technical Digest of the GaAs IC Symposium*, IEEE, Piscataway, pp. 7–10, 1998.

[94] G. Meneghesso, A. Mion, A. Neviani, M. Matloubian, J. Brown, M. Hafizi, T. Liu, C. Canali, M. Pavesi, M. Manfredi and E. Zanoni, 'Effects of channel quantization and temperature on off-state and on-state breakdown in composite channel and conventional InP-based HEMTs', *Technical Digest of the International Electron Devices Meeting*, IEEE, Piscataway, pp. 43–46, 1996.

[95] M.J. Howes and D.V. Morgan (eds.), *Reliability and Degradation. Semiconductor Devices and Circuits*, John Wiley & Sons, Chichester, 1981.

[96] A. Christou (ed.), *Reliability of Gallium Arsenide MMICs*, John Wiley & Sons, Chichester, 1992.

[97] M. Ohring, *Reliability and Failure of Electronic Materials and Devices*, Academic Press, San Diego, 1998.

[98] T. Enoki, H. Ito and Y. Ishii, 'Reliability study on InAlAs/InGaAs HEMT's with an InP recess-etch stopper and refractory gate metal', *Solid State Electronics*, Vol. 41, pp. 1651–1656, 1997.

[99] A.S. Wakita, H. Rohdin, C.-Y. Su, N. Moll, A. Nagy and V.M. Robbins, 'Drain resistance degradation under high fields in AlInAs/GaInAs MODFETs', *Proceedings of the International Conference on Indium Phosphide and Related Materials*, IEEE, Piscataway, pp. 376–379, 1997.

[100] H. Rohdin, C.-Y. Su, N. Moll, A. Wakita, A. Nagy, V. Robbins and M. Kauffman, 'Semi-analytical analysis for optimization of 0.1-μm InGaAs-channel MODFETs with emphasis on on-state breakdown and reliability', *Proceedings of the International Conference on Indium Phosphide and Related Materials*, IEEE, Piscataway, pp. 357–360, 1997.

[101] M. Kuzuhara, K. Onda, A. Fujihara, E. Mizuki, Y. Hori and H. Miyamoto, 'InP-based heterojunction FET processing for high-reliability millimeter-wave application', *Proceedings of the International Conference on Indium Phosphide and Related Materials*, IEEE, Piscataway, pp. 809–812, 1995.

[102] Y. Ashizawa, C. Nozaki, T. Noda, A. Sasaki and S. Fujita, 'Surface related degradation of InP-based HEMTs during thermal stress', *Solid State Electronics*, Vol. 38, pp. 1627–1630, 1995.

[103] K. Weigel, M. Warth, K. Hirche and K. Heime, 'Degradation effects and stabilization of InAlAs/InGaAs-HFETs', *Proceedings of the International Conference on Indium Phosphide and Related Materials*, IEEE, Piscataway, pp. 662–665, 1996.

[104] T. Sonoda, Y. Yamamoto, N. Hayafuji, H. Yoshida, H. Sasaki, T. Kitano, S. Takamiya and M. Ostubo, 'Manufacturability and reliability of InP HEMTs', *Solid State Electronics*, Vol. 41, pp. 1621–1628, 1997.

[105] R. Menozzi, M. Borgarino, Y. Baeyens, M. Van Hove and F. Fantini, 'On the effect of hot electrons on the DC and RF characteristics of lattice-matched InAlAs/InGaAs/InP HEMTs', *IEEE Microwave Guided Wave Letters*, Vol. 7, pp. 3–5, 1997.

[106] M. Hafizi and M.J. Delaney, 'Reliability of InP-based HBT's and HEMT's: Experiments, failure mechanisms, and statistics', *Proceedings of the International Conference on Indium Phosphide and Related Materials*, IEEE, Piscataway, pp. 299–302, 1994.

[107] H. Rohdin, A. Wakita, A. Nagy, V. Robbins, N. Moll and C.-Y. Su, 'A 0.1-μm MHEMT millimeter-wave IC technology designed for manufacturability', *Solid State Electronics*, Vol. 43, pp. 1645–1654, 1999.

14.10 Problems

14.1 Derive the threshold voltage expression (Equation (14.4)) for the dual-doped MODFET. Assume that the field is zero below the bottom pulse, and that near the inflection point in $n_s(V_G)$ where $n_s \sim n_{s0}/2$, there is negligible change in the potential across the 2DEG.

14.2 (a) Show that Equation (14.7) applies to the symmetrically dual-doped case after the substitutions $N_{d1} \rightarrow 2N_{d1}$, $\Delta d \rightarrow 2\Delta d$, and $\Delta E_C^{(eff)} \rightarrow 2\Delta E_C^{(eff)}$.

(b) Discuss the design of the backside pulse (N_{d2} and d_{p2}) in a practical MODFET.

14.3 Derive Equation (14.12) for the electron exchange interaction. Hints: Pick a manageable geometry for the 'exchange hole', i.e. the surrounding fixed ionic background that can be 'assigned' to the electron. Calculate the Coulomb interaction energy.

14.4 (a) From the resistive network in Figure 14.6 develop the expression for the resistance of a non-alloyed ohmic contact (Equation (14.29).

(b) Why is there a minimum in R_c vs r_{c2} in Figure 14.7?

14.5 Derive Equation (14.55) for the intrinsic transconductance in the limit of long gate lengths. Note that $n_s(V_G)$ is not assumed to be linear. Hint: For long gate lengths, saturation can be assumed to coincide with pinch-off at the drain-side edge of the gated channel.

15 Small-signal AC analysis of the short-gate velocity-saturated MODFET

That must be wonderful, I don't understand it at all.

MOLIÈRE

15.1 Introduction

This chapter develops the small-signal equivalent circuit and Y parameters for a short-gate high-speed MODFET. We start with the intrinsic device, and discuss in detail the effects of the velocity-saturated region. Finally, we incorporate extrinsic elements, in particular the access resistances. Of these, the source and drain resistances were discussed in Section 14.6.2. The gate resistance is a considerably more complex component, with multiple physical origins, and distributed incorporation into the equivalent circuit. Detailed discussion of this is deferred to Chapter 16.

15.2 Equivalent circuit for the intrinsic device

We are interested in a bias near optimum for small-signal performance. This means biasing the gate where the charge control is essentially linear (as assumed in Chapter 11). This occurs near the inflection point V_{Gm} in the $n_s(V_G)$ curve in Figure 14.2(b), and produces maximum intrinsic capacitance,

$$C_0 = qKW_gL_g, \tag{15.1}$$

where

$$K = \left(\frac{dn_s}{dV_G}\right)_{max} = \frac{\varepsilon}{qd_{gc}}. \tag{15.2}$$

This bias point minimizes the degrading effects of parasitic capacitances which do not modulate the two-dimensional electron gas (2DEG). The associated open-channel

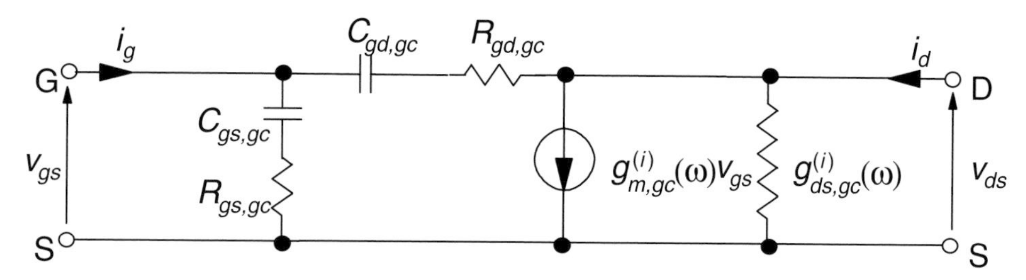

Fig. 15.1. Equivalent circuit for the linear MODFET analyzed in Section 11.2. It is similar to those in Figure 11.9, and its topology is more conventional than the original one in Figure 11.6. Six of the eight elements that describe the AC characteristics (Equations (15.7), (15.8), (15.13)–(15.16)) are easily identifiable components in the equivalent circuit. The remaining two are the time delays that go into the frequency dependence of the output conductance and transconductance (Equations (15.9), (15.10)). The controlling voltage for the transconductance is shown explicitly.

conductance is

$$g_0 = \frac{1}{R_0} = \frac{q\mu n_{si} W_g}{L_g}, \tag{15.3}$$

where R_0 is the open-channel ($V_D = 0$) resistance at $V_G = V_{Gm}$ and n_{si} is the 2DEG inflection-point carrier concentration. With the non-linear gate charge control in Equation (14.14), $n_{si} = \alpha n_{s0}$. The optimum drain bias for speed produces a moderate saturation which balances the benefits of high electron velocity under the gate with the detrimental effect of the external delay associated with ΔL_x (Section 15.4). A significant fraction of the channel under the gate ($0 < y < y_s$ in Figure 14.1) will typically still be linear, and we base the development of the equivalent circuit on the linear-case topology in Figure 15.1. Even before the modifications we will make to account for velocity saturation, this is quite similar to the final topology, used universally by experimentalists and designers, even deep in saturation. As shown in Problem 15.1 the elements in Figure 15.1 are determined by the coefficients $I_{ij}(k)$ and $R_{ij}(k)$ derived in Section 11.2. These are functions of the dimensionless bias parameter

$$k = \frac{V_D}{V_G - V_T}. \tag{15.4}$$

One finds:

$$g^{(i)}_{m,gc}(\omega) = g^{(i)}_{m0,gc} \exp(-j\omega\tau^{(TL)}_{d,gc}) \frac{\sin(\omega\tau^{(s)}_{d,gc})}{\omega\tau^{(s)}_{d,gc}}, \tag{15.5}$$

$$g^{(i)}_{ds,gc}(\omega) = g^{(i)}_{ds0,gc} \exp(-j\omega\tau^{(TL)}_{d,gc}) \frac{\sin(\omega\tau^{(s)}_{d,gc})}{\omega\tau^{(s)}_{d,gc}}, \tag{15.6}$$

$$g^{(i)}_{m0,gc} = g_0 k, \tag{15.7}$$

$$g_{ds0,gc}^{(i)} = g_0(1-k), \tag{15.8}$$

$$\tau_{d,gc}^{(TL)} = \tau_{d,gc}^{(1)}, \tag{15.9}$$

$$\tau_{d,gc}^{(s)} = \sqrt{3}\left[\left(\tau_{d,gc}^{(2)}\right)^2 - \left(\tau_{d,gc}^{(1)}\right)^2\right]^{1/2}, \tag{15.10}$$

$$\tau_{d,gc}^{(1)} = \frac{C_0}{g_0}\frac{I_{dg}(k) - I_{gd}(k)}{k}, \tag{15.11}$$

$$\tau_{d,gc}^{(2)} = \frac{C_0}{g_0}\left[2\frac{R_{dg}(k) - R_{gd}(k)}{k}\right]^{1/2}, \tag{15.12}$$

$$C_{gs,gc} = C_0\left[I_{gg}(k) - I_{gd}(k)\right], \tag{15.13}$$

$$C_{gd,gc} = C_0 I_{gd}(k), \tag{15.14}$$

$$R_{gs,gc} = R_0\frac{R_{gg}(k) - R_{gd}(k)}{\left[I_{gg}(k) - I_{gd}(k)\right]^2}, \tag{15.15}$$

$$R_{gd,gc} = R_0\frac{R_{gd}(k)}{I_{gd}^2}. \tag{15.16}$$

The subscript gc stands for the gradual channel to distinguish these parameters from those to be introduced for the velocity-saturated part of the channel. $\tau_{d,gc}^{(1)}$ and $\tau_{d,gc}^{(2)}$ are the delays in the drain current appearing in the expansion to second order in $j\omega$ of $g_{x,gc}^{(i)}(\omega)$ $(x = m, ds)$:

$$g_{x,gc}^{(i)}(\omega) = g_{x0,gc}^{(i)}\left[1 - j\omega\tau_{d,gc}^{(1)} - \tfrac{1}{2}\left(\omega\tau_{d,gc}^{(2)}\right)^2\right]. \tag{15.17}$$

In Equations (15.5) and (15.6) we have replaced $\tau_{d,gc}^{(1)}$ and $\tau_{d,gc}^{(2)}$ with the drain delays $\tau_{d,gc}^{(TL)}$ and $\tau_{d,gc}^{(s)}$, respectively, and rewritten $g_{x,gc}^{(i)}(\omega)$ in a form which has the correct second-order Taylor expansion of Equation (15.17), but which does not blow up as $\omega \to \infty$. The functional form in Equations (15.5)–(15.6) degrades 'gracefully' [1]. At the present stage in the analysis, the choice of functional form may appear arbitrary. Considering the gradual-channel case, where the charge control resembles a parallel-plate capacitor with some series resistance, it may have been more natural to use a factor $1/\left(1 + j\omega\tau_{d,gc}^{(RC)}\right)$ instead of $\sin\left(\omega\tau_{d,gc}^{(s)}\right)/\left(\omega\tau_{d,gc}^{(s)}\right)$. The latter functional form, however, is one we will encounter for the velocity-saturated case, and we might as well use it here too. It accounts for the degradation in magnitude, while the transmission-line-like factor $\exp\left(-j\omega\tau_{d,gc}^{(TL)}\right)$ accounts for the increased phase delay for higher frequencies. Note that, as developed in Chapter 11 and in this section, the controlling voltage for the transconductance is the entire intrinsic gate voltage, i.e. not just the fraction that drops across $C_{gs,gc}$. This is the physical way of approaching the problem, since we do not know *a priori* the values for $C_{gs,gc}$ and $R_{gs,gc}$. Compared to more phenomenological models where the controlling voltage is assumed to be across $C_{gs,gc}$, the drain delays will be longer.

It is interesting that the frequency dependences of $g_{ds,gc}^{(i)}$ and $g_{m,gc}^{(i)}$ are identical. The gradual-channel current responds to an AC stimulus from the gate very similarly to one from the drain. It is only the k-dependent magnitude of the response that is different. (At the 'half-way' point ($k = \frac{1}{2}$) between open channel (k, $V_D = 0$) and classical pinch-off ($k = 1$, $V_D = V_G - V_T$), the magnitudes are also the same.) Note that the output impedance of the linear FET appears inductive: $1/g_{ds,gc}^{(i)}(\omega) \approx 1/g_{ds0,gc}^{(i)} + j\omega\tau_{d,gc}^{(TL)}/g_{ds0,gc}^{(i)}$.

In Figure 15.2 we have plotted, in the normalized form of Chapter 11, the six non-linear fundamental parameters in Equations (15.9), (15.10) and (15.13)–(15.16) against the dimensionless saturation parameter k. (With their simple linear k dependence, $g_{m0,gc}^{(i)}$ and $g_{ds0,sat}^{(i)}$ were not included.) In addition, we have plotted the two RC gate delays

$$\tau_{gs,gc}^{(RC)} = R_{gs,gc}C_{gs,gc} \tag{15.18}$$

and

$$\tau_{gd,gc}^{(RC)} = R_{gd,gc}C_{gd,gc}, \tag{15.19}$$

also in normalized form. Like the two drain delays, the two RC delays are quite similar in magnitude. Despite the strong k dependence of $R_{gd,gc}$ and $C_{gd,gc}$ individually, their product is rather independent of k even up to $k = 1$. Similarly, although $C_{gs,gc}$ increases, and $C_{gd,gc}$ decreases, as k increases, the sum of the two is practically constant up to $k \sim 0.5$. With finite saturation velocity, preventing classical pinch-off, k is less than 1 for all gate lengths ([2]; Section 14.6.5)). The DC model in Chapter 10 can be used to calculate $k = V_c(y_s)/(V_G - V_T)$ associated with the gradual channel. The value is indeed less than 1, but will depend somewhat on L_g, V_G, and the value we assign to $E_\Gamma^{(sat)}$ (Section 14.5). Given the relative insensitivity to k of the time constants and the total gradual-channel capacitance, it is quite reasonable for the short gates of interest to evaluate the parameters at $k = 0$.

In saturation, several effects occur that force us to modify the equivalent circuit. Before going into the more fundamental quantitative analysis in Sections 15.3 and 15.4, we take the qualitative approach illustrated in Figure 15.3. Referring again to Figure 14.1, the high-field region extends by ΔL_i towards the source, and by ΔL_x towards the drain. The partially depleted high-field region $\Delta L_i + \Delta L_x$ separates the large concentration of electrons in the ohmic gradual channel from the similar situation at the drain. Thus a drain–source capacitance develops. ΔL_x also increases the charge separation between the input (gate) and the output (drain). This will physically redistribute $C_{gd,gc}$ ($\sim C_0/2$) largely to an internal node S' which is separated only by a small resistance (associated with the gradual channel) from the source S, but by a

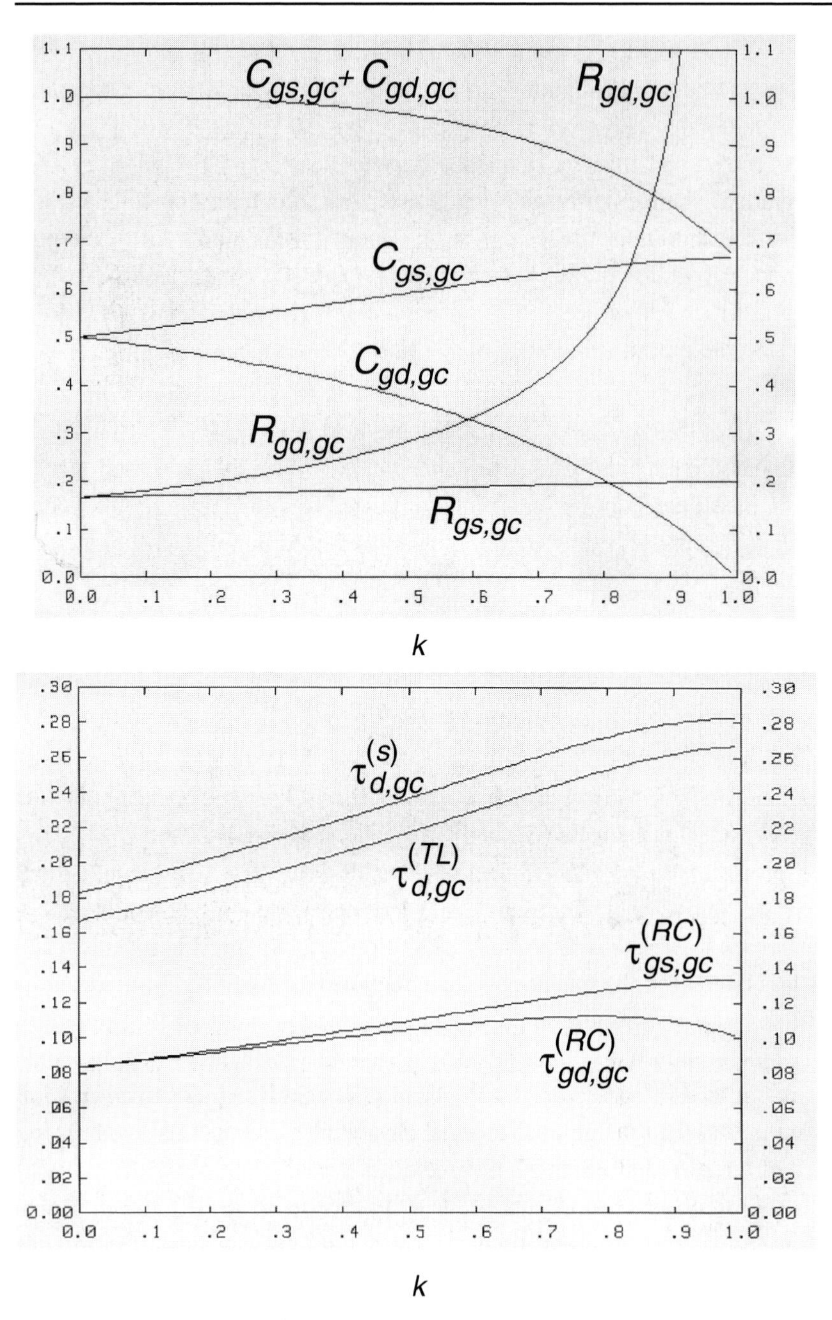

Fig. 15.2. The eight fundamental equivalent circuit elements, two RC gate delays, and the total gate capacitance for a linear MODFET vs the bias parameter k. All parameters are normalized as in Chapter 11.

large resistance (associated with the velocity-saturated region) to the drain D. There will still be a capacitance from the gate to the drain, but this can be considerably

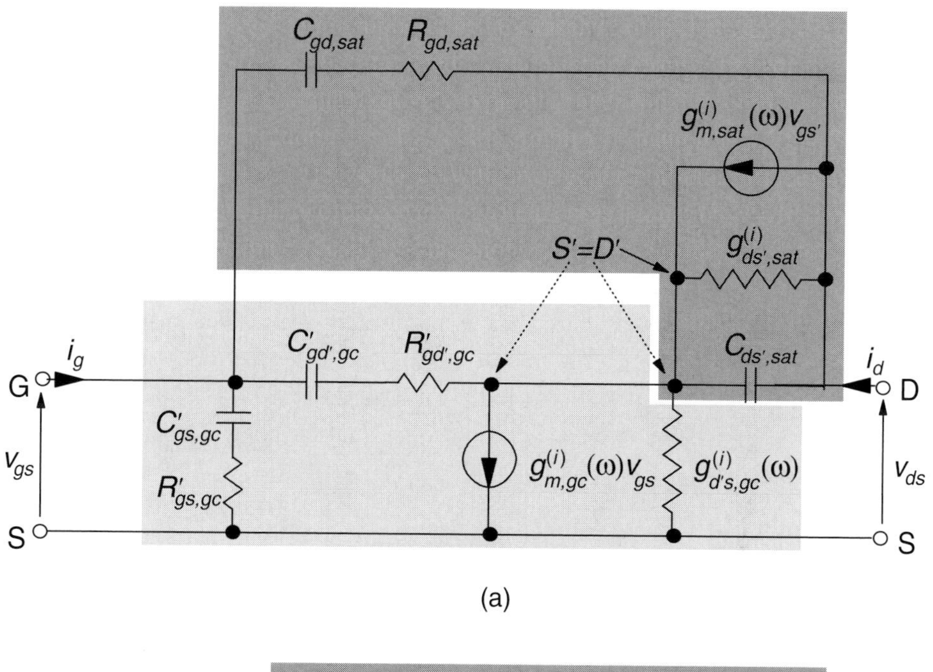

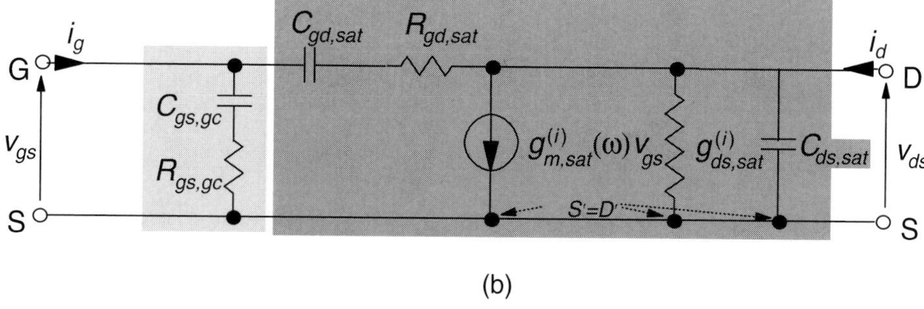

Fig. 15.3. Heuristic development of the equivalent circuit for an intrinsic short-gate velocity-saturated MODFET. (a) The gradual part of the channel ($y < y_s$ in Figure 14.1) is still described by the equivalent circuit in Figure 15.1. This part is indicated by the lighter shaded area. The prime attached to the R's and C's merely indicates that the length of this region is no longer L_g, but $L_g - \Delta L_i$. Elements in the darker shaded area are added, as discussed in the text, to model the velocity-saturated part of the channel ($y > y_s$). The border ($y = y_s$) between the two regions is associated with the node denoted alternatively by S′ and D′, depending on whether the element is connected to it from the drain (D) or source (S) side, respectively. (b) A simplified equivalent circuit resulting from letting the low-voltage S′/D′ node 'collapse' to the source node (ground), as discussed in the text. This happens to be the standard equivalent circuit also for longer-gate FETs, even though the S′/D′-collapse is less warranted then.

smaller because of the reverse bias. For the low voltage at S' (corresponding to a small k), particularly for the short gates of interest here, the conductance $g^{(i)}_{d's,gc}$ is very large, and can be approximated by a short. What little resistance is overlooked by doing this can appear as a small component in the extrinsic source resistance R_s (Section 14.6.2).

The transconductance $g_{m,gc}^{(i)}$ can be neglected beside $g_{d's,gc}^{(i)}$. The equivalent circuit for the intrinsic MODFET will thus 'collapse' into that shown in Figure 15.3(b). The two RC branches on the input side in Figure 15.3(a) can be combined into the one-series combination $R_{gs,gc} - C_{gs,gc}$ in Figure 15.3(b), where now

$$C_{gs,gc} = C_0 \frac{L_g - \Delta L_i}{L_g} = qKW_g \left(L_g - \Delta L_i\right). \tag{15.20}$$

As indicated by the appended subscripts *gc* and *sat* in Figure 15.3(b), the input of the saturated FET will be dominated by the source-side gradual channel, while the output and feedback characteristics will be dominated by the drain-side velocity-saturated region.

The gate delay associated with the gradual channel of a short-gate MODFET is estimated as

$$\tau_{g,gc}^{(RC)} = R_{gs,gc} C_{gs,gc}\big|_{k=0} = R_{gd,gc} C_{gd,gc}\big|_{k=0} = \frac{1}{12}\tau_{gc}, \tag{15.21}$$

where the time constant τ_{gc} is associated with the gradual channel:

$$\tau_{gc} = \frac{\varepsilon \left(L_g - \Delta L_i\right)^2}{q\mu n_{si} d_{gc}}. \tag{15.22}$$

The series charging resistance associated with the gate-source capacitance then becomes

$$R_{gs,gc} = \frac{\tau_{g,gc}^{(RC)}}{C_{gs,gc}} = \frac{1}{12} \frac{L_g - \Delta L_i}{q\mu n_{si} W_g}. \tag{15.23}$$

The delays in the transconductance and output conductance, also evaluated at $k = 0$, are

$$\tau_{d,gc}^{(TL)} = \frac{1}{6}\tau_{gc} \tag{15.24}$$

and

$$\tau_{d,gc}^{(s)} = \frac{1}{\sqrt{30}}\tau_{gc}. \tag{15.25}$$

For the gate lengths of interest, the transconductance $g_{m0,sat}^{(i)}$ is given by its saturated-velocity limit, i.e. by Equation (14.54). Unlike the transconductance, however, the output conductance depends strongly on the gate length. Empirically, with the FETs biased for maximum speed, i.e. moderately into saturation with ΔL_x small, the relationship is approximately inverse:

$$g_{ds0,sat}^{(i)} = g_{ds0,sat}^{(sq,i)} \frac{W_g}{L_g}. \tag{15.26}$$

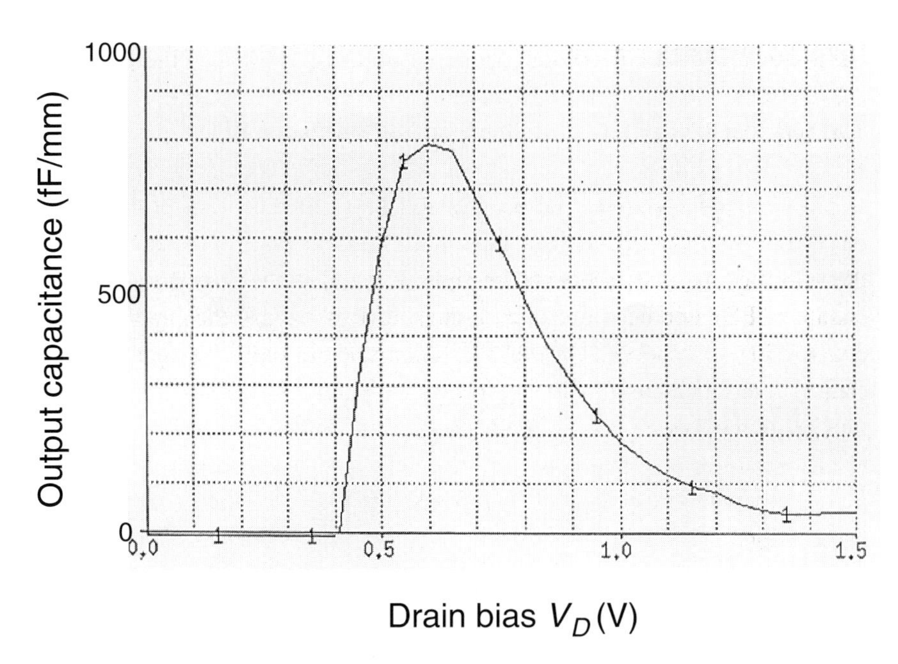

Fig. 15.4. Measured extrinsic output capacitance of a 0.1-μm InGaAs/AlInAs MODFET near maximum transconductance versus drain bias.

Some observations [3] indicate that the factor $g_{ds0,sat}^{(sq,i)}$ is not a constant, but instead is proportional to d_{gc}. This has some intuitive appeal since the output conductance would then explicitly improve with the gate aspect ratio L_g/d_{gc}, consistent with 'conventional wisdom'. For larger ΔL_i and ΔL_x, $g_{ds0,sat}^{(i)}$ is reduced to an extent which depends in a rather complex way on geometry and transport. The output admittance is $g_{ds0,sat}^{(i)} + j\omega C_{ds,sat}$, where the drain–source capacitance $C_{ds,sat}$ can be quite large once it appears, but typically drops as V_D (and thus $\Delta L_i + \Delta L_x$) is further increased. Such a case is shown in Figure 15.4. Note that the capacitance is negative for $V_D < 0.4$ V, before velocity saturation has set in. This is due to the inductive nature of $g_{ds,gc}^{(i)}$. The gate-drain feedback capacitance $C_{gd,sat}$ in Figure 15.3 is the remnant of the 'original' $C_{gd,gc}$ which does not terminate on the gradual channel, but on channel points external to the gate.

It is clear that the simple picture used so far will not be able to predict $C_{gd,sat}$, or explain the intricate behavior of $g_{ds0,sat}^{(i)}$ and $C_{ds,sat}$. A semiempirical, and essentially one-dimensional model was useful for predicting $g_{m0,sat}^{(i)}$ and $C_{gs,gc}$, and the associated time constants and delays, but the other three components are determined by the two-dimensional nature of the high-field velocity-saturated drain region. This can certainly warrant fully numerical methods, but we will be able to learn quite a bit in the next two sections by developing, with some simplifying assumptions, a semianalytical two-dimensional model.

15.3 Displacement currents

The heuristic approach to developing an equivalent circuit for the velocity-saturated MODFET in the last section could not take us all the way. It does, however, suggest a convenient division of the device into two parts. One consists of the metal contacts, any heavily doped cap layers and the low-field ohmic parts of the channel. The other is the velocity-saturated part of the channel. If we view the first part as a set of metallic conductors, this division becomes very fruitful as we can then apply some powerful electrostatic approaches. As the theory for vacuum tubes was developed it became clear that terminal currents to conductors with applied voltages and charged particles moving in the space between them are the sum of two components. The first is due to the capacitive coupling between the conductors. The second current component is induced by the moving charges, in our case the velocity-saturated electrons. This will be dealt with in the following section. We first address the capacitance problem.

We apply the method of conformal mapping [4] to the simplified geometry in Figure 15.5(a), which we let represent the intrinsic MODFET. We will use it to estimate the charges induced on two grounded FET terminals by applying $V_0 = 1$ V on the third (either the drain or gate). This immediately yields estimates for the capacitances of interest. We map the space inside the path in the z-plane (Figure 15.5(a)) onto the upper half of the w-plane (Figure 15.5(b),(c)). The counter-clockwise path, hugging the conductors in the z-plane, should map onto the entire real axis in the w-plane, so that the conductor with voltage V_0 maps onto the negative part, and the grounded conductors map onto the positive part. The resulting potential in the upper w-halfplane ($0 \leq \arg(w) \leq \pi$) is then given by (Problem 15.2(b))

$$V_w(w) = V_0 \frac{\arg(w)}{\pi}. \tag{15.27}$$

The potential of ultimate interest is that in the z-plane. This is given by $V_z(z) = V_w(w(z))$ (Problem 15.2(c)), and the problem thus boils down to determining the transform $w(z)$. Of particular use for us is the expression

$$Q_{ij} = -\frac{\varepsilon W_g V_0}{\pi} \ln\left(\frac{x_{j2}}{x_{j1}}\right) \tag{15.28}$$

(Problem 15.2(d)) for the charge induced between points z_{j1} and z_{j2} on a grounded conductor j when one of the other conductors i is biased at a voltage V_0, while all the other conductors are grounded. x_{j1} and x_{j2} are the points on the real w axis corresponding to z_{j1} and z_{j2}.

For the path in the z-plane in Figure 15.5(a) we can use the Schwartz–Christoffel transform for polygons [4]:

$$\frac{dz}{dw} = A \prod_j \left(w - b_j\right)^{-\varphi_j/\pi}. \tag{15.29}$$

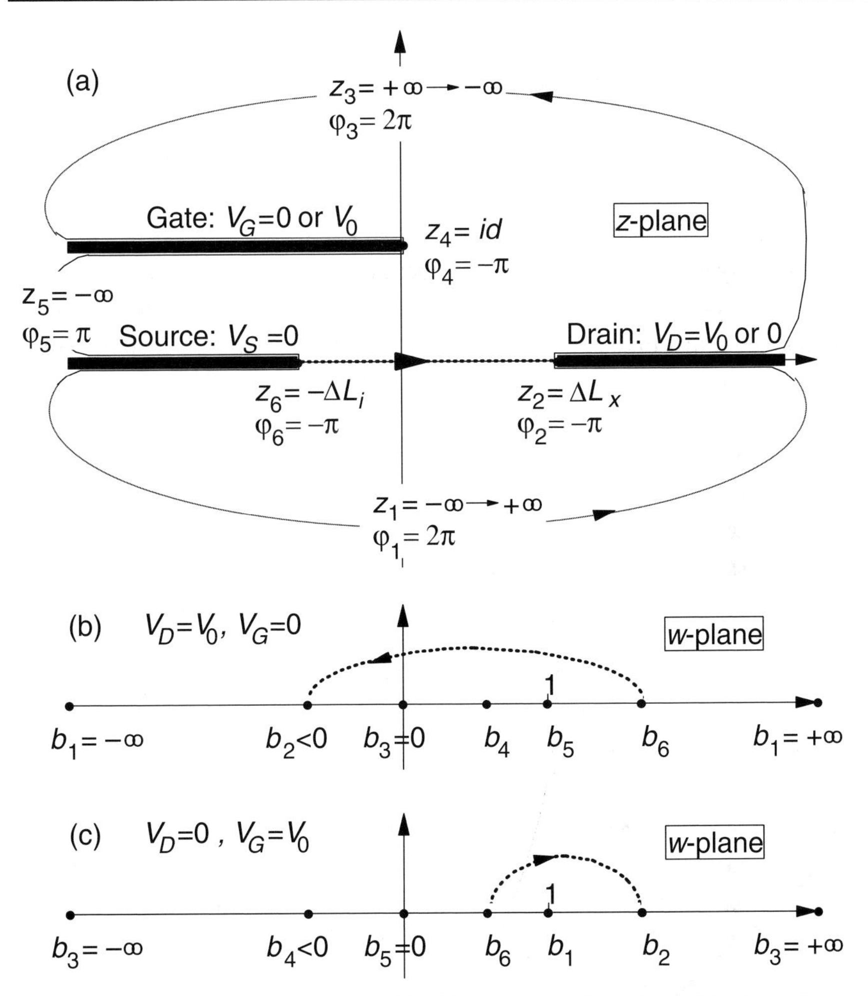

Fig. 15.5. Conformal mapping of a simplified FET geometry from the z-plane (a) to the upper half of the w-plane for two different biases: (b) $V_D = V_0$, $V_G = 0$; (c) $V_D = 0$, $V_G = V_0$. The electron path across the velocity-saturated region has been indicated in both the z- and w-planes.

The b_j's are the points on the real w axis that correspond to the corners of the polygon(s) in the z-plane, and the φ_j's are the corresponding left-turn angles as one follows the circumference of the polygon in the counterclockwise manner. A is a constant to be determined. The z points that separate conductors at zero potential from those at V_0 should map onto either $w = \pm\infty$ or $w = 0$. Of these, only the one at $w = b = 0$ enters into the product in Equation (15.29). To make the number of unknowns equal to the number of equations, we have to pick a value for one of the b_j's. The rest of the b_j's, together with A and an integration constant, are determined by solving the equations resulting from plugging $w = b_j$ into the $z(w)$ expression.

We are interested in charges induced on the gate and source by a voltage on the drain, as well as the charges induced on the source and drain by a voltage on the gate. The two bias conditions are $V_D = V_0$, $V_G = 0$ and $V_G = V_0$, $V_D = 0$, respectively, as indicated in Figure 15.5 and Table 15.1. Figure 15.5 (b) and (c) define the transformations for the two cases, and Table 15.1 summarizes all the information necessary to calculate semianalytical estimates for the four charges. The capacitances of interest are trivially related to these charges:

$$C_{gs,gc} = \frac{Q_{gs}}{V_0}, \tag{15.30a}$$

$$C_{ds,sat} = \frac{Q_{ds}}{V_0}, \tag{15.30b}$$

and

$$C_{gd,sat} = \frac{Q_{gd}}{V_0}. \tag{15.30c}$$

The two non-linear coupled equations $f_1(\cdot) = 0$ and $f_2(\cdot) = 0$ in Table 15.1 are not solvable by analytical means. The root (b_4, b_6) or (b_6, b_2), respectively, is best found with Newton's method. With the bounds and initial guesses given in Table 15.1, this is a very fast and robust process. When a sequence of various ΔL_i, ΔL_x or d_{gc} are of interest, the best choice of initial guess is obviously the last solution. The problem is considerably less numerically involved than a full two-dimensional numerical solution involving typically thousands of grid points.

We have made a couple of simplifications in Figure 15.5(a). First, contrary to the case of an actual FET where the gate is surrounded on top by either air or a finite passivating oxide or nitride layer, we consider the three conductors to be embedded in the semiconductor. Conformal mapping with a position-dependent dielectric constant would appear impossible (except in special coplanar situations [5]), since the whole concept is based on analytical functions where the real and imaginary parts satisfy the two-dimensional Laplace equation. Second, we treat the conductors as infinitely thin semiinfinite sheets. We do this so we can determine the conformal mapping analytically. Making turns in the z-plane by angles that are multiples of π leads to a nicely integrable dz/dw expression. Neglecting the finite thickness of the channel does not introduce much error. Neglecting the actual thickness of the gate is, however, a significant simplification, which we have to account for. The amount of correction will depend on the geometry of the actual gate and the passivation.

When we calculate the capacitances of interest, we only add up the charge to the appropriate position on the semiinfinite conductor (Equation (15.28)). For instance, to calculate the charge Q_{dg} induced on the gate by a drain voltage, we add the contributions from $z = id$ to $z = -L_g + id$ on both sides of the infinitesimally thin gate sheet. Similarly, when we calculate the charge Q_{gs} induced on the source by a voltage on the (semiinfinite) gate, we add the contributions from z=- ΔL_i to $z = -L_g$ on both sides of the infinitesimally thin source sheet. This would correctly account for the parallel-plate part of C_{gs} under the gate, as well as fringing capacitance

Table 15.1. *A recipe, using the conformal mapping in Figure 15.5, for estimating charges in a MODFET.*

Induced charges of interest (refer to Equation (15.28))	$\boldsymbol{Q_{ds}}$ **and** $\boldsymbol{Q_{dg}}$	$\boldsymbol{Q_{gs}}$ **and** $\boldsymbol{Q_{gd}}$
Voltages on the conductors in Figure 15.5(a)	$V_D = V_0, \quad V_G = 0$	$V_G = V_0, \quad V_D = 0$
Choice of fixed b	$b_5 = 1$ (Figure 15.5(b))	$b_1 = 1$ (Figure 15.5(c))
Choice of independent b variables (real and > 0)	b_4, b_6 (Figure 15.5(b))	b_6, b_2 (Figure 15.5(c))
$z(w)$ transformation	$z(w) = A\left[w - \frac{p}{w} + q\ln\left(\frac{w}{w-1}\right) + B\right]$	$z(w) = A\left[w - \frac{p}{w-1} + r\ln(w) + B\right]$
Dependent variables: all real and fully determined by the two independent variables	$b_2 = 1 - b_4 - b_6 \quad (< 0)$ $p = b_2 b_4 b_6 \quad (< 0)$ $q = -(b_4 + b_6)(b_2 + b_4 b_6) \quad (> 0)$ $A = -\frac{d_{gc}}{\pi q} \quad (< 0)$ $B = -b_4 + \frac{p}{b_4} - q\ln\left(\frac{b_4}{1-b_4}\right)$	$b_4 = \frac{2-b_6-b_2}{1-b_6 b_2} \quad (< 0)$ $r = -b_2 b_4 b_6 \quad (> 0)$ $p = 2r - 1 + b_4 b_6 + b_2(b_4 + b_6) \quad (< 0)$ $A = \frac{d_{gc}}{\pi r} \quad (> 0)$ $B = -b_4 - \frac{p}{1-b_4} - r\ln(-b_4)$
Non-linear coupled equations for the two independent variables	$f_1(b_4, b_6) = b_6 - \frac{p}{b_6} + q\ln\left(\frac{b_6}{b_6-1}\right) + B + \frac{\Delta L_i}{A} = 0$ $f_2(b_4, b_6) = b_2 - \frac{p}{b_2} + q\ln\left(\frac{b_2}{b_2-1}\right) + B - \frac{\Delta L_x}{A} = 0$	$f_1(b_6, b_2) = b_6 + \frac{p}{1-b_6} + r\ln(b_6) + B + \frac{\Delta L_i}{A} = 0$ $f_2(b_6, b_2) = b_2 - \frac{p}{b_2-1} + r\ln(b_2) + B - \frac{\Delta L_x}{A} = 0$

Table 15.1. *Continued.*

Induced charges of interest (refer to Equation (15.28))	Q_{ds} and Q_{dg}	Q_{gs} and Q_{gd}
Bounds for the independent variables	$0 < b_4 < 1$ $b_6 > 1$	$0 < b_6 < 1$, $2 - b_6 < b_2 < 1/b_6$
Good first guess for $\Delta L_i \approx \Delta L_x \approx d_{gc}$	$b_4 = 0.21$ $b_6 = 1.46$	$b_6 = 0.32$ $b_2 = 2.49$
Non-linear equation with two real roots x_{j1} and x_{j2} that determine the induced charge Q_{ij} (Equation 15.28).	$f(x) = x - \frac{p}{x} + q \ln \left\| \frac{x}{x-1} \right\| + B - \frac{a}{A} = 0$ a = real part of the z coordinate of the point furthest left on the gate or source in Figure 15.5(a) that should contribute to the charge	$f(x) = x - \frac{p}{x-1} + r \ln(x) + B - \frac{a}{A} = 0$ a = real part of the z coordinate of the point furthest left (right) on the gate (drain) in Figure 15.5(a) that should contribute to the charge
a, and the upper and lower limits for x_{j1} and x_{j2}, for the four charges of interest. L_{gs}, L_g, ΔL_x and other geometrical parameters: See Figure 14.1. $\Delta L_g^{(s)}$ and ΔL_d: See Section 15.3.	Q_{ds}: $a = -(L_g + L_{gs})$ $1 + \exp\left(\frac{1}{q}\left(\frac{-a}{A} + 1 - \frac{p}{b_6} + B\right)\right) < x_{j1} < b_6$ $b_6 < x_{j2} < \frac{a}{A} - B$ Q_{dg}: $a = -L_g$ $0 < x_{j1} < b_4$ $b_4 < x_{j2} < 1 - b_4 \exp\left(\frac{1}{q}\left(\frac{-a}{A} + b_4 - p + B\right)\right)$	Q_{gs}: $a = -L_g - \Delta L_g^{(s)}$ $\exp\left(\frac{1}{r}\left(\frac{a}{A} - b_6 - p - B\right)\right) < x_{j1} < b_6$ $b_6 < x_{j2} < 1 - \frac{p}{\frac{a}{A} - 1 - B}$ Q_{gd}: $a = \Delta L_x + \Delta L_d$ $1 - \frac{p}{\frac{a}{A} - 1 - B} < x_{j1} < b_2$ $b_2 < x_{j2} < \frac{a}{A} - r \ln(b_2) - B$

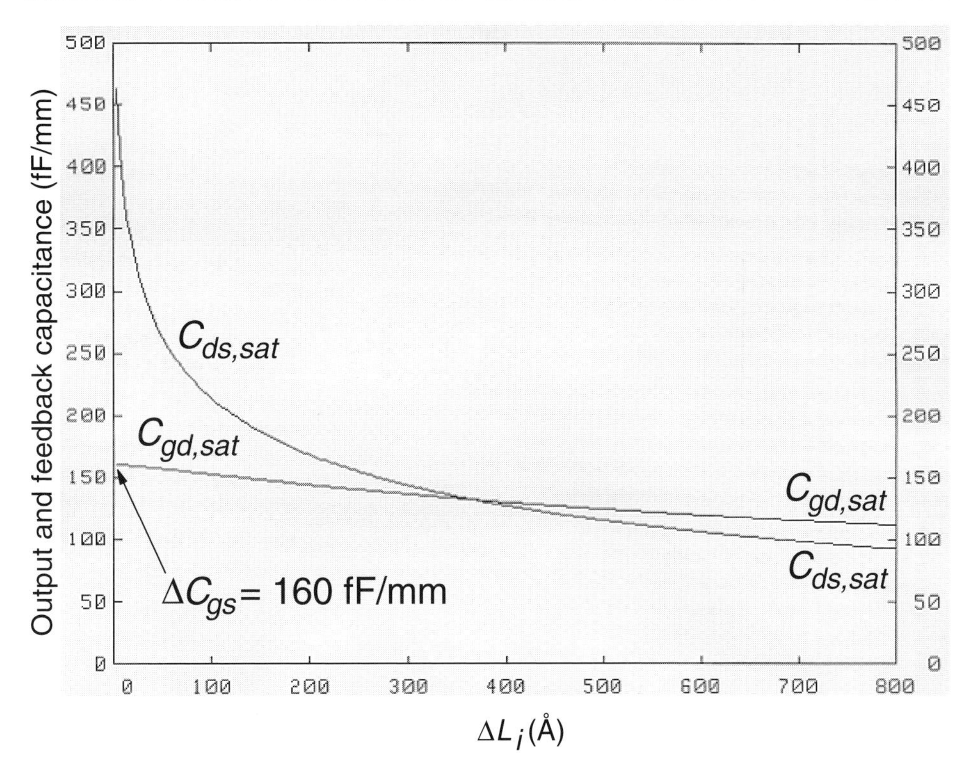

Fig. 15.6. Semianalytical conformal-mapping solution for the FET output and feedback capacitance. The geometry and parameter values are those defined in Figure 15.5 and Table 15.2.

at the right-hand edge of the source ($z = -\Delta L_i$). However, it would miss the fringing capacitance, denoted ΔC_{gs}, at the left-hand edge ($z = -L_g$) of the actual finite gate. Rather than simply adding ΔC_{gs} to Q_{gs}/V_0 calculated to $z = -L_g$, we instead, as indicated in Table 15.1, calculate Q_{gs} to $z = -L_g - \Delta L_g^{(s)}$, where

$$\Delta L_g^{(s)} = \frac{\Delta C_{gs} d_{gc}}{\varepsilon W_g} \tag{15.31a}$$

is the effective fringing length. This allows ΔL_i to approach L_g (and actually exceed it) without sudden unphysical changes in C_{gs}.

For the gate-drain feedback capacitance $C_{gd,sat}$ we use Q_{gd}/V_0 (Equation (15.30c)) rather than Q_{dg}/V_0, or some combination of the two. The reason is that the feedback capacitance should not be a strong function of L_g (and Q_{dg} is). Q_{gd} is the charge induced on the drain from $z = \Delta L_x$ to $z = \Delta L_x + \Delta L_d$ (again on both sides of the drain conductor), where ΔL_d is an adjustable parameter to allow our simplified geometry to yield realistic feedback capacitance. The source-side fringing capacitance ΔC_{gs} in Equation (15.31a) can be calculated as

$$\Delta C_{gs} = C_{gd,sat}(\Delta L_i \to 0, \Delta L_x \to 0). \tag{15.31b}$$

Figure 15.6 shows $C_{ds,sat}$ and $C_{gd,sat}$ versus ΔL_i for the representative high-speed

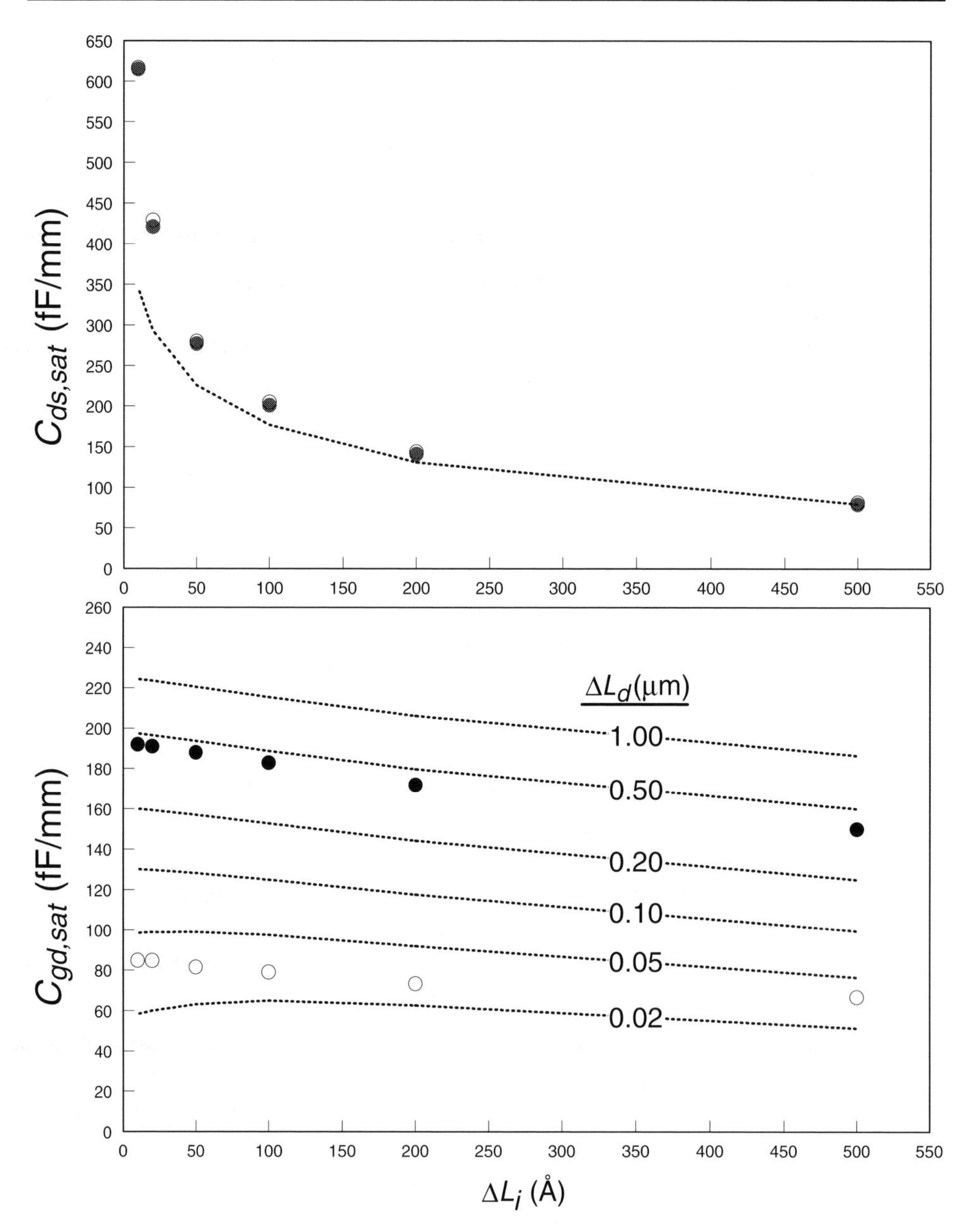

Fig. 15.7. Output and feedback capacitance versus the internal extent of the velocity-saturated region. Circles: full two-dimensional numerical solution for the geometry in Figure 15.8. Filled circles: with nitride passivation. Open circles: no passivation. Dashed line: semianalytical solution in Figure 15.6 for the simplified example FET (Figure 15.5 and Table 15.2) with different ΔL_d.

MODFET defined in Table 15.2. We will use these parameters throughout this section. Note the rapid drop in $C_{ds,sat}$, as already indicated experimentally in Figure 15.4. $C_{gd,sat}$ varies much less rapidly despite ΔL_x increasing 50% faster than ΔL_i. This is

Table 15.2. *MODFET parameters used in Sections 15.3 and 15.4.*

Type	Parameter	Value	Ref.
Semiconductor material	Relative dielectric constant	$\varepsilon = 13.18\varepsilon_0$	Equations (15.2) and (14.54)
	Effective saturation velocity	$v_{sat} = 2.8 \times 10^7$ cm/s	Equation (14.54)
Cross-sectional two-dimensional geometry	Gate-channel spacing	$d_{gc} = 250$ Å	Equations (15.2) and (14.54)
	Gate length	$L_g = 100$ nm	Equation (15.1)
	Source-side extent of lateral recess	$L_{us} = 100$ nm	Equation (15.54)
	Gate-source spacing	$L_{gs} = 1$ μm	Figure 14.1
	Effective length of drain for C_{gd}	$\Delta L_d = 0.2$ μm	Section 15.3
	Source-side intrinsic fringing capacitance	$\Delta C_{gs} = 160$ fF/mm	Equation (15.31) Figure 15.6
Bias-dependent	Extent of velocity saturation external to gate (set by V_D and V_G)	ΔL_x = varied	Figure 14.1 Equation (15.54)
	Ratio of extents of external and internal velocity saturation	$\Delta L_x/\Delta L_i = 1.5$	Figure 14.1 Equation (15.54)

because, as ΔL_i and ΔL_x increase, and $C_{ds,sat}$ drops rapidly, a larger fraction of the field lines emanating from the drain will terminate on the gate. This will to some extent negate the reduction in $C_{gd,sat}$ due to the wider depletion (ΔL_x) between the gate and drain. Figure 15.7 compares semianalytical calculations with fully numerical two-dimensional solutions of the Laplace equation using Silvaco International's ATLAS simulator. The FET-like structure is shown in Figure 15.8. It is similar to that in Figure 14.1, and is a realistic structure with a gate of finite thickness, and conductors representing heavily doped cap layers. The thickness for the 2DEG was chosen to be 50 Å. Figure 15.7 shows results with and without nitride passivation. With the finite gate, this makes a big difference in $C_{gd,sat}$. However, it has little effect on $C_{ds,sat}$, since the associated field lines are to a much larger extent confined to the semiconductor. This, and the huge initial value, should make $C_{ds,sat}$ a good indicator of the onset of velocity saturation. Because of the finite thickness used for the channel in Figure 15.8, $C_{ds,sat}$ calculated with the fully numerical approach is larger than that calculated with the semianalytical solution, particularly as $\Delta L_i + \Delta L_x$ gets small. Although the values differ from the full two-dimensional solution, the simplified semianalytical solution predicts the general shape of the curves well, i.e. the relative constancy of $C_{gd,sat}$, and the rapid variation of $C_{ds,sat}$. By choosing ΔL_d judiciously we can represent the shape and magnitude of $C_{gd,sat}(\Delta L_i)$ quite well.

Figure 15.9 shows the gate-source capacitance $C_{gs,gc}$ versus ΔL_i for the example FET. The value calculated with the conformal-mapping method (curve (b); Equation (15.30a)) is larger than that calculated using the simple formula (curve a;

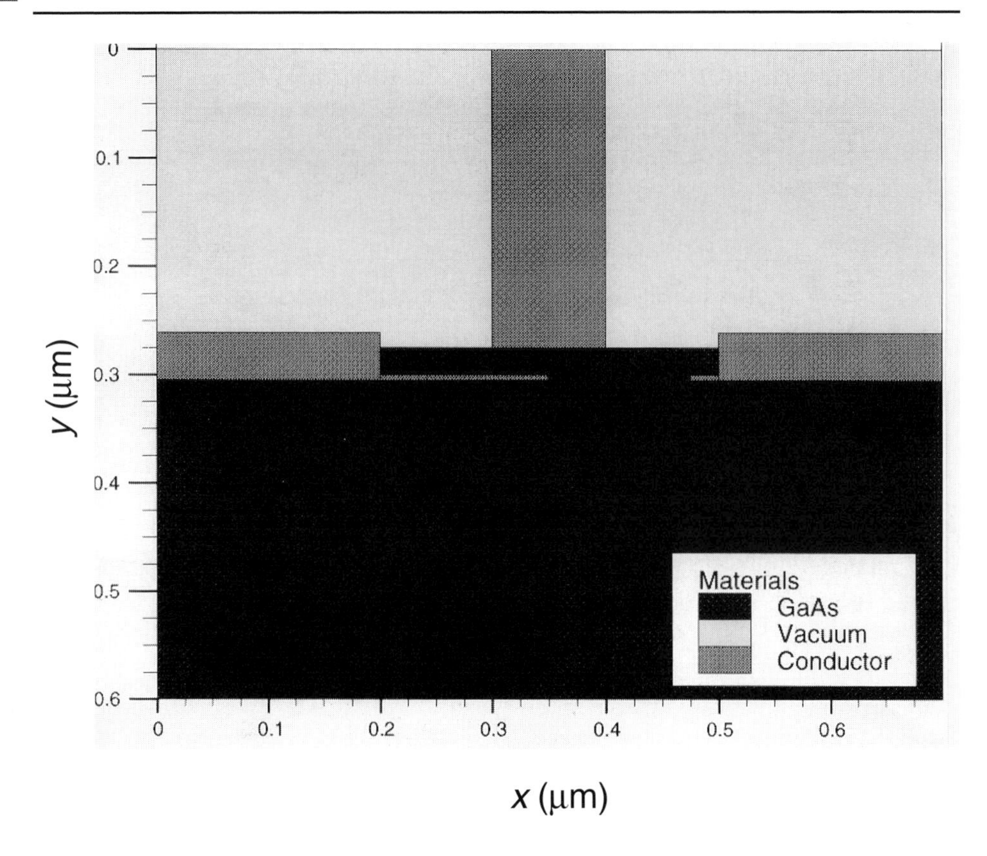

Fig. 15.8. Realistic FET-like geometry for full two-dimensional solution of Laplace's equation.

Equation (15.20)) because of the fringing fields. For ΔL_i less than $\sim d_{gc}$ the fringing effect is only on the source side (ΔC_{gs}), extending the length of the modulated channel by $\sim \Delta L_g^{(s)}$. For larger ΔL_i, fringing occurs at the drain-side edge of the gradual channel, and the slope of the curve (b) correctly approaches that predicted by Equation (15.20). Curve (c) will be discussed in Section 15.4.

With the approximate semianalytical method for calculating the intrinsic capacitances, we can now estimate the displacement part of the gate and drain currents in saturation:

$$i_{g,sat}^{(displ)} = \frac{j\omega C_{gs,gc}}{1 + j\omega\tau_{g,gc}^{(RC)}} v_g + \frac{j\omega C_{gd,sat}}{1 + j\omega\tau_{g,gc}^{(RC)}} \left(v_g - v_d\right), \tag{15.32}$$

$$i_{d,sat}^{(displ)} = \frac{j\omega C_{ds,sat}}{1 + j\omega\tau_{g,gc}^{(RC)}} v_d + \frac{j\omega C_{gd,sat}}{1 + j\omega\tau_{g,gc}^{(RC)}} \left(v_d - v_g\right). \tag{15.33}$$

We have reintroduced the charging time constant (Equations (15.21), (15.22)) to approximately account for the fact that the channel conductor is not a perfect metal, but rather the gradual part of the semiconductor 2DEG. Figure 15.2 showed that the charging time constants for the gate-source and the gate-drain capacitances are quite

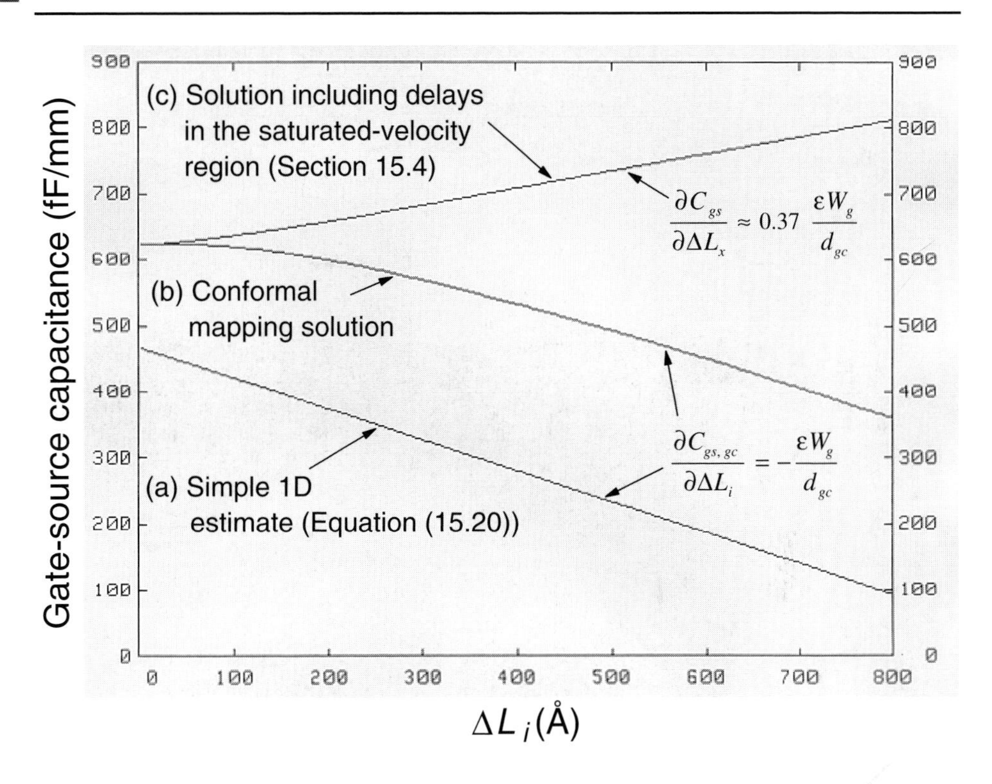

Fig. 15.9. Solutions for the gate-source capacitance for the example FET.

similar, even up to the hypothetical situation of total pinch-off ($k = 1$), where the field would become infinite. We have assumed in Equations (15.32) and (15.33) that this continues to be the case also in the less extreme case of velocity saturation, and that it also applies to the intrinsic drain–source capacitance. Considering the fringing modulation, a more conservative estimate for not only $\tau_{g,gc}^{(RC)}$ (Equations (15.21) and (15.22)), but also $R_{gs,gc}$, $\tau_{d,gc}^{(TL)}$ and $\tau_{d,gc}^{(s)}$ (Equations (15.23)–(15.25)), results if we replace L_g with $L_g + \Delta L_g^{(s)}$. In effect $\Delta L_g^{(s)}$ is the increase in the electrical gate length on the source-side due to fringing fields. For the example MODFET in this section, $\Delta L_g^{(s)} = 34$ nm. This is non-negligible compared to the 100-nm gate (as is the similar effect on the drain-side associated with $C_{gd,sat}$). As indicated by Figure 15.7, the actual effect could be larger or smaller, depending on the passivation. ΔC_{gs} is one reason to keep the aspect ratio L_g/d_{gc} large. In Chapter 17, where we push for performance, we will be a bit more aggressive in our choice of d_{gc}.

15.4 Conduction-induced currents and delays

We now move to address the second component in the drain and gate currents, i.e. the one induced by the velocity-saturated electrons. For the drain current we

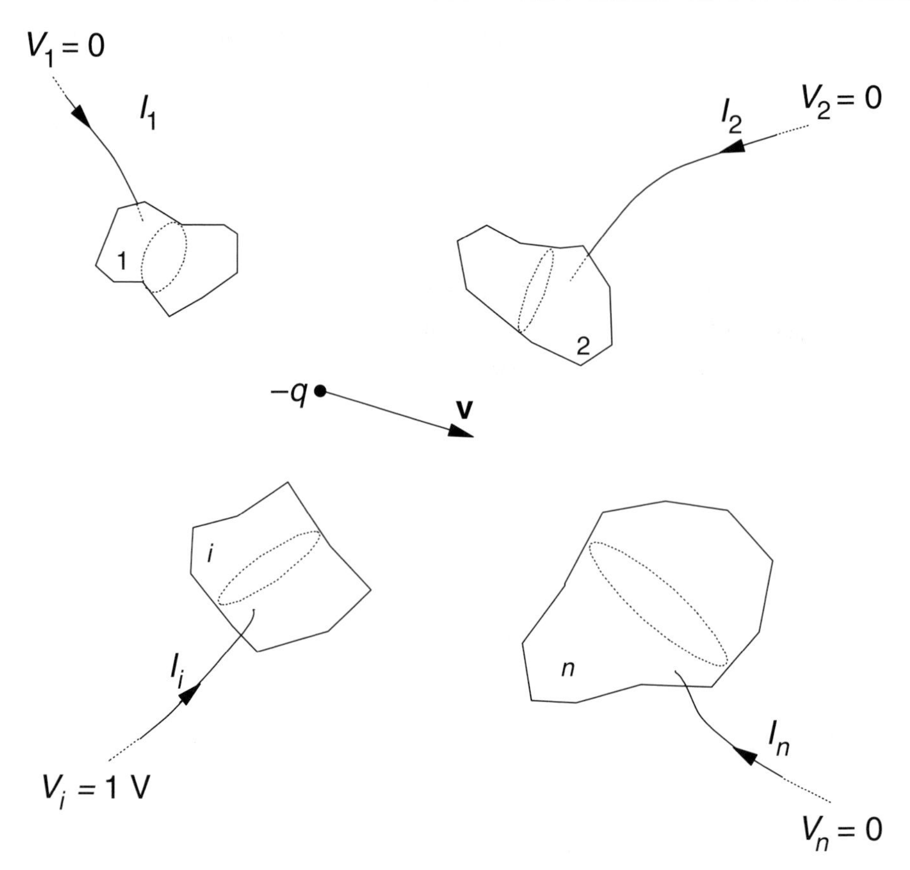

Fig. 15.10. General three-dimensional geometry and biasing for deriving the currents induced in the wires to arbitrarily shaped conductors (in this case conductor i) by a moving electron. In addition to the electron there can be any distribution of fixed charge in the space not occupied by the n conductors. This space can have position-dependent dielectric constant.

certainly expect two conductive components due to the transconductance and the output conductance. We will, however, also find a less obvious additional capacitive component in the gate current, as indicated by curve (c) in Figure 15.9. This important result will explain why the FET current-gain cut-off frequency does not improve further, but actually drops, as the drain bias increases deeper into saturation. The drain delays that cause this were analyzed mathematically by Moll [6]. The analysis is very general and instructive, and it will be included in the following.

Currents induced by moving charges were first analyzed by Shockley [7] and Ramo [8], for the case of a uniform dielectric between the conductors, no fixed charge in the dielectric, and constant voltages on the conductors. The result has more recently been shown to be valid also for arbitrary position-dependent dielectric constant and fixed charge, and with time-dependent voltages on the conductors [9]. We are interested in the currents induced on the gate and drain by the electrons traveling in the velocity-saturated region from the end of the gradual channel at $y = y_s$ (Figure 14.1) to the point $y = y_{us}$ of 'unsaturation' (Section 14.6). Induced currents in essentially one-dimensional devices such as HBTs and pin mesa photodetectors are rather straightforward [10, 11]. In an FET the problem is more complicated because of the two-dimensional electrostatics, something we began to deal with in the last section. Before becoming FET-specific, however, we can learn some important general lessons about currents induced by moving electrons in a three-dimensional geometry. We make use of the following alternative form of Green's theorem introduced by Kim *et al.* [9]:

$$\int d\mathbf{S} \cdot (\Phi\varepsilon\nabla\Psi - \Psi\varepsilon\nabla\Phi) = \int dV\,(\Phi\nabla\cdot(\varepsilon\nabla\Psi) - \Psi\nabla\cdot(\varepsilon\nabla\Phi)). \tag{15.34}$$

As with the more familiar forms of Green's theorem, this is easily derived from the divergence theorem. We refer to Figure 15.10, which depicts an arbitrary three-dimensional geometry of conductors embedded in a medium with arbitrary dielectric constant $\varepsilon(\mathbf{r})$ and fixed space-charge density $\rho(\mathbf{r})$. Equation (15.34) allows us to derive the general expression for the current I_i induced in the wire to conductor i by the electron charge $-q$ at $\mathbf{r}_0$, moving at velocity $\mathbf{v}$ between the conductors. We do this by choosing Φ in Equation (15.34) to be the electrostatic potential that results from setting $q = 0$ and $\rho = 0$, and applying a voltage V_i to conductor i. We will attach an identifying subscript i to Φ, and divide by V_i to make Φ_i dimensionless (normalized to 1 V). It satisfies

$$\nabla\cdot(\varepsilon\nabla\Phi_i) = 0 \tag{15.35}$$

in the space between the conductors. This is the problem we addressed in the last subsection for the FET, generalized to three dimensions. The solution yields the capacitances C_{ij} between conductor i and the other conductors ($j \neq i$). Ψ is chosen to be the electrostatic potential (not normalized) that results from the moving electron, and the fixed space charge, with all conductor voltages set to zero. Ψ satisfies Poisson's equation:

$$\nabla\cdot(\varepsilon\nabla\Psi) = -\rho + q\delta\,(\mathbf{r} - \mathbf{r}_0). \tag{15.36}$$

$V_i\Phi_i + \Psi$ is then the solution to the entire linear electrostatic problem illustrated in Figure 15.10, of which our simplified model of an FET is a special case. As mentioned, the first term yields currents due to direct capacitive coupling between the conductors. The second term will yield the additional currents induced by the moving charges. Performing the integrations in Equation (15.34), remembering that $\Phi_i = \Psi = 0$ on all the conductor surfaces, with the exception $\Phi_i = 1$ on the ith conductor's surface (S_i), we arrive at

$$Q_i = \int_{S_i} d\mathbf{S}\cdot(\varepsilon\nabla\Psi) = -\int dV\,\Phi_i\rho + q\Phi_i(\mathbf{r}_0), \tag{15.37}$$

where Q_i is the charge induced by the moving electron on conductor i of interest. Since ρ represents fixed charge, the current induced to the conductor is

$$I_i = \frac{dQ_i}{dt} = q\frac{d\Phi_i(\mathbf{r}_0)}{d\mathbf{r}_0} \cdot \frac{d\mathbf{r}_0}{dt} = q\nabla\Phi_i \cdot \mathbf{v}. \tag{15.38}$$

$\nabla\Phi_i$ has the dimension of m^{-1}, and depends only on the geometry of the conductors. If the electron charge $-q$ moves from conductor j to conductor $i \neq j$, the time integral of the current induced to conductor i should equal q, and it does:

$$\begin{aligned}\int_{t_j}^{t_i} dt\, I_i &= q\int_{t_j}^{t_i} dt\, \nabla\Phi_i \cdot \mathbf{v} = q\int_{\mathbf{r}_j}^{\mathbf{r}_i} d\mathbf{r} \cdot \nabla\Phi_i \\ &= q\left[\Phi_i(\mathbf{r}_i) - \Phi_i\left(\mathbf{r}_j\right)\right] = q(1-0) = q.\end{aligned} \tag{15.39}$$

In this classical picture the current is delivered to conductor i during the entire time the electron travels; i.e. the terminal current is not a delta-function at the time the electron arrives at the conductor. This view is appropriate in devices in which many electrons are involved, and is of importance for the proper predictions of instantaneous currents from Monte Carlo simulations where 'superparticles' consisting of many electrons are considered [12]. One can similarly show that when a charged particle travels from conductor $j \neq i$ to $k \neq i$, the time integral of the current induced to conductor i is zero. However, during the transit, there will be a time-dependent current (with no DC component). Note also that, in the electrostatic approximation used [9], Kirchhoff's current law is still valid, i.e.

$$\sum_i I_i = 0 \tag{15.40}$$

(Problem 15.3). Except for a rare occasion when a three-dimensional time-domain solution of Maxwell's equations is included (e.g. [13]), all FET simulations use the electrostatic approximation, i.e. only Poisson's equation is solved self-consistently with the solid-state transport equations.

In the particular case we are interested in, the velocity-saturated MODFET, we continue to assume the simplified conductor geometry in Figure 15.5. We are interested in the velocity-saturated electrons transiting in the z-plane from point z_6 to point z_2. We will assume that they follow a straightline trajectory, i.e. that they do not leave the channel. The corresponding paths in the w-plane are indicated qualitatively in Figure 15.5(b) and (c). y will be the coordinate along the real z axis, consistent with Figure 14.1, but now with $y = 0$ at the drain-side edge of the gate. During the transit, the electrons will induce currents in the drain and gate according to Equation (15.38). With a constant saturation velocity (v_{sat}) in the channel, the time dependence of the induced currents will trace out the y dependence of the 'field' $\nabla\Phi_i$*. Figure 15.11

* It should be reiterated that this field has the unit of inverse distance, and is only a function of the conductor geometry. It should not be confused with the actual total electric field in the device.

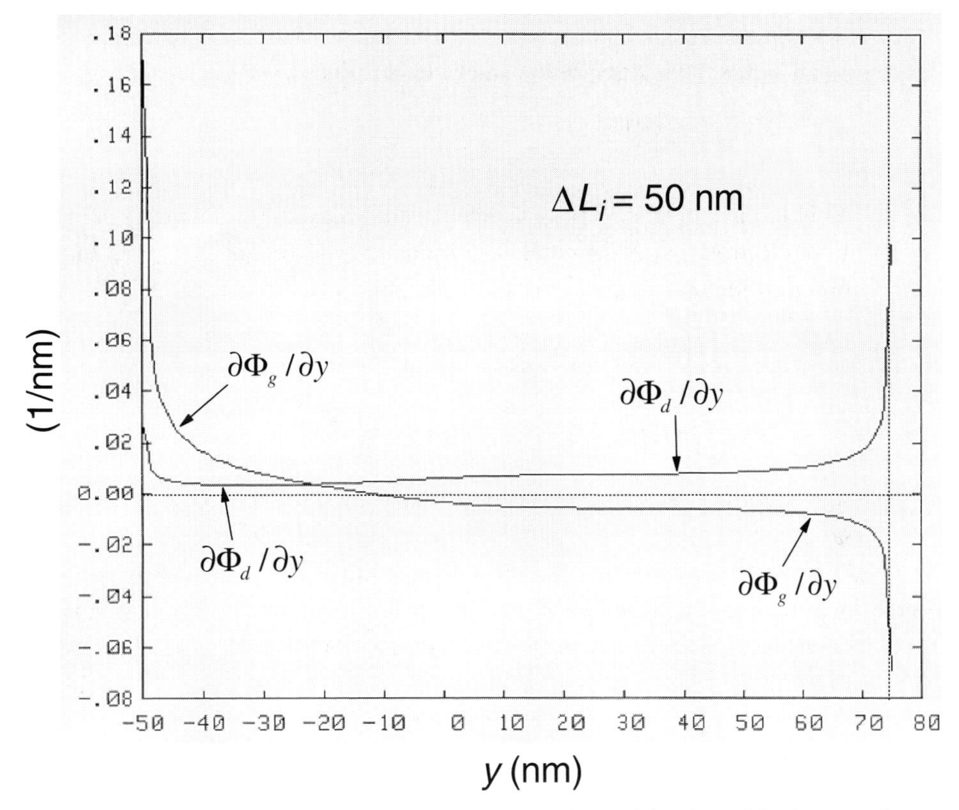

Fig. 15.11. The gradients of Φ_g and Φ_d along the saturated part of the channel for the example FET. The time dependences of the currents induced by an electron traversing with constant saturation velocity have the same shapes.

shows $\partial\Phi_g/\partial y$ and $\partial\Phi_d/\partial y$ for the example FET, with $\Delta L_i = 50$ nm. There is no DC component in the gate current, i.e. the integral of $\partial\Phi_g/\partial y$ is zero. Although currents are indeed induced during the entire transit of the electrons, there will be some 'spikiness' at the beginning and end due to the large slopes in Φ_g and Φ_d. Finding the solutions for Φ_g and Φ_d is straightforward, having earlier solved for the transformation $z(w)$ for the two bias cases in Table 15.1. It is a matter of finding the $w(y)$ trajectories from b_6 to b_2 in Figure 15.5 (b) and (c), which boils down to solving the two simultaneous equations $\text{Re}[z(w)] = y$ and $\text{Im}[z(w)] = 0$. As in the last section, this is preferably done by Newton's method. Solving from $y = -\Delta L_i$ to $y = \Delta L_x$ as we did in Figure 15.11, the obvious initial guess for each new y is the solution for the last y. At $y = -\Delta L_i$ the solution is $b_6 + j0$. With $w(y)$ determined, $\Phi_g(y)$ and $\Phi_d(y)$ are given by the right-hand side of Equation (15.27).

We add up the current from all the electrons in the saturated part of the channel:

$$I_i(t) = qW_g v_{sat} \int_{y_s(t)}^{y_{us}(t)} dy \frac{\partial\Phi_i\left(y_s(t), y_{us}(t), y\right)}{\partial y} n_s'(y,t), \tag{15.41}$$

where n_s' is the 2DEG concentration in the saturated region $y_s < y < y_{us}$. We consider small-signal modulation of the gate voltage:

$$V_g(t) = V_G + v_g \exp(j\omega t), \tag{15.42}$$

and analogously for the drain voltage $V_d(t)$. The voltage $V_s(t)$ at the point y_s of onset of velocity saturation is given by

$$V_s(t) = V_S + (\gamma_g v_g - \gamma_d v_d) \exp(j\omega t), \tag{15.43}$$

where

$$\gamma_g = \frac{\partial V_S}{\partial V_G} \tag{15.44}$$

and

$$\gamma_d = -\frac{\partial V_S}{\partial V_D}. \tag{15.45}$$

In the present context, we neglect the phase in γ_g and γ_d (Section 12.3.2). Both γ_g and γ_d are larger than zero, and can be estimated by the DC model in Chapter 14. The 2DEG carrier concentration at the saturation point y_s is

$$n_{ss}'(t) = n_s\left(V_g(t) - V_s(t)\right) = n_{si} + K\left[(1-\gamma_g)v_g + \gamma_d v_d\right] \exp(j\omega t), \tag{15.46}$$

where K is given by Equation (15.2). In the limit of small signals, the carrier concentration further into the saturated region $y_s < y < y_{us}$ is

$$n_s'(y,t) = n_{ss}'\left(t - \frac{y - y_{s0}}{v_{sat}}\right), \tag{15.47}$$

where y_{s0} is the DC value of $y_s(t)$. The parameters in the small-signal time-dependence of y_s and y_{us} (corresponding to γ_g and γ_d for V_s in Equation (15.43)) will not enter into I_i after we 'filter out' the $j(2\omega)t$ harmonic terms by replacing the time-dependent integration limits in Equation (15.41) with their DC values y_{s0} and y_{us0}. With these geometrical parameters now fixed we only need to maintain the y dependence in Φ_i*. We arrive at

$$\begin{aligned} I_i = q W_g v_{sat} \{ & n_{si} [\Phi_i(y_{us}) - \Phi_i(y_s)] \\ & + K\left[(1-\gamma_g)v_g + \gamma_d v_d\right] F_i(\omega) \exp(j\omega t)\} \end{aligned} \tag{15.48}$$

for the current induced to conductor i by the electrons in the velocity-saturated region. The term in first square brackets yields a DC current. For the induced gate current ($i = g$), $\Phi_i(y_{us}) = \Phi_i(y_s) = 0$, and this term is zero. For the induced drain current ($i = d$), however, $\Phi_i(y_{us}) = 1$, and this term yields the DC drain current $q W_g n_{si} v_{sat}$.

* We have altogether left out the x dependence (Figure 14.1), since we are considering straight electron trajectories in the channel.

The second term in the braces is the small-signal modulation of n_s, which yields an AC current in both cases. $F_i(\omega)$ in this term is given by

$$F_i(\omega) = \int_{y_{s0}}^{y_{us0}} dy \frac{\partial \Phi_i}{\partial y} \exp\left(-j\omega \frac{y - y_{s0}}{v_{sat}}\right). \tag{15.49}$$

In a one-dimensional geometry with a velocity-saturated drift region of extent d (applicable for instance in the base–collector depletion region in a bipolar transistor or the i-region in a pin diode), $F_i(\omega)$ becomes the factor [14]

$$F_i^{(1D)} = \exp(-j\omega\tau) \frac{\sin(\omega\tau)}{\omega\tau}, \tag{15.50}$$

where τ is *half* of the transit time d/v_{sat}. With the 2D geometry we face for our FET structure, we will, analogously to the gradual-channel case in Section 15.2, expand $F_i(\omega)$ to second order in $j\omega$, and rewrite the result in a form which degrades gracefully as $\omega \to \infty$. We use the $\sin(\omega\tau)/\omega\tau$ frequency dependence suggested by the one-dimensional case, and arrive at

$$i_{d,sat}^{(cond)} = \left(g_{m0,sat}^{(i)} v_g + g_{ds0,sat}^{(i)} v_d\right) \exp\left(-j\omega\tau_{d,sat}^{(TL)}\right) \frac{\sin(\omega\tau_{d,sat}^{(s)})}{\omega\tau_{d,sat}^{(s)}} \tag{15.51}$$

for the conduction-induced drain current. This belatedly explains why we chose this functional form in the gradual-channel case above (Equations (15.5) and (15.6)). The transconductance and output conductance continue to have the same frequency dependence. Their DC values are

$$g_{m0,sat}^{(i)} = qW_g K(1 - \gamma_g) v_{sat}, \tag{15.52}$$

$$g_{ds0,sat}^{(i)} = qW_g K \gamma_d v_{sat}. \tag{15.53}$$

γ_g contains the gate-length dependence in the transconductance, introduced by the gradual channel and illustrated in Figure 14.9. It approaches 0 for short gates where $g_{m0,sat}^{(i)}$ is given by Equation (14.54). γ_d is also small, as predicted by the DC model in Chapter 14. The actual output conductance is typically significantly larger than predicted by Equation (15.53). In Chapter 10 it was shown that an increase in the effective channel thickness increases $g_{ds0,sat}^{(i)}$. The channel thicknesses that are required to explain experimentally observed output conductances are often much larger than the epitaxial thickness of the channel (e.g. [15]). One physical effect that has been proposed to explain this observation is the spreading into the buffer that can occur as electrons are heated in the high fields [16]. This would mean that complexities introduced by non-stationary transport can no longer be ignored if we expect to understand and predict the elusive parameter $g_{ds0,sat}^{(i)}$, including its empirical inverse gate-length dependence (Equation (15.26)). This would force us to abandon the analytical approach, and resign to a fully numerical alternative. However, there is a less complex effect that we can evaluate by taking the same semianalytical approach as in Section 15.3, and this is the direct modulation of the gradual-channel 2DEG concentration by the drain voltage. When we calculated $C_{ds,sat}$ we allowed the drain voltage to modulate charge all the way back to the source ohmic contact ($a = -(L_g + L_{gs})$ for Q_{ds} in Table 15.1). In the case of $g^{(i)}_{ds0,sat}$ however, we are only interested in the drain voltage modulation of the part of the 2DEG where the electrons have a significant (but unsaturated) drift velocity. Thus, we should go no further back than to $y = -L_{us}$ in Figure 14.1. Between $y = -L_{us}$ and $y = y_s$ in Figure 14.1, the electrons accelerate from essentially zero drift velocity to v_{sat}. In our simple estimate we will assume that the electrons in this low-field source side of the channel all travel at the 'average' velocity $v_{sat}/2$. With a recalculated drain–source capacitance, this time setting $a = -(L_g + L_{us})$ and denoting the result $C'_{ds,sat}$, the output conductance due solely to two-dimensional electrostatics becomes

$$g^{(i)}_{ds0,sat} = \frac{C'_{ds,sat}\,(v_{sat}/2)}{L_g + L_{us} - \Delta L_i} = g^{(i)}_{m0,sat}\frac{1}{2\pi}\frac{d_{gc}}{L_g + L_{us} - \Delta L_i}\ln\left(\frac{x'_{s2}}{x'_{s1}}\right). \tag{15.54}$$

Thus, considering only drain-induced charge modulation as the cause of output conductance, the parameter γ_d is given approximately by

$$\gamma_d = \frac{1}{2\pi}\frac{d_{gc}}{L_g + L_{us} - \Delta L_i}\ln\left(\frac{x'_{s2}}{x'_{s1}}\right). \tag{15.55}$$

The first equality in Equation (15.54) is directly analogous to the expression for transconductance, with the capacitance in the numerator describing the total charge being modulated, and the denominator being the extent of the channel involved in the modulation. x'_{s1} and x'_{s2} are calculated as prescribed for Q_{ds} in Table 15.1, with $a = -(L_g + L_{us})$. Note the appearance of an inverse dependence on gate length as suggested by experiments. With L_{us}, ΔL_i and the logarithmic factor, the overall geometrical dependence is more complicated. This is illustrated in Figure 15.12 for the example MODFET. Figure 15.12(a) shows the predicted dependence of $g^{(i)}_{ds0,sat}$ on ΔL_i. The upper limit is obviously not infinity as suggested here, but rather $\sim g_0$ (Equation (15.3)). The predicted gate-length dependence in Figure 15.12(b) is less steep than inverse, but the overall predictions are quite in agreement with experiments, as is the reduction in $g^{(i)}_{ds0,sat}$ with increasing L_{us} that Equation (15.54) predicts. The simple transport model thus continues to be quite adequate in most respects.

The transmission-line delay in $i^{(cond)}_{d,sat}$ is given by

$$\tau^{(TL)}_{d,sat} = \tau^{(1)}_{d,sat} = \frac{1}{v_{sat}}\int_{y_{s0}}^{y_{us0}} dy\,[1 - \Phi_d(y)]. \tag{15.56}$$

The delay $\tau^{(s)}_{d,sat}$ responsible for reducing the magnitude of $i^{(cond)}_{d,sat}$ is given by

$$\tau^{(s)}_{d,sat} = \sqrt{3}\left[\left(\tau^{(2)}_{d,sat}\right)^2 - \left(\tau^{(1)}_{d,sat}\right)^2\right]^{1/2}, \tag{15.57}$$

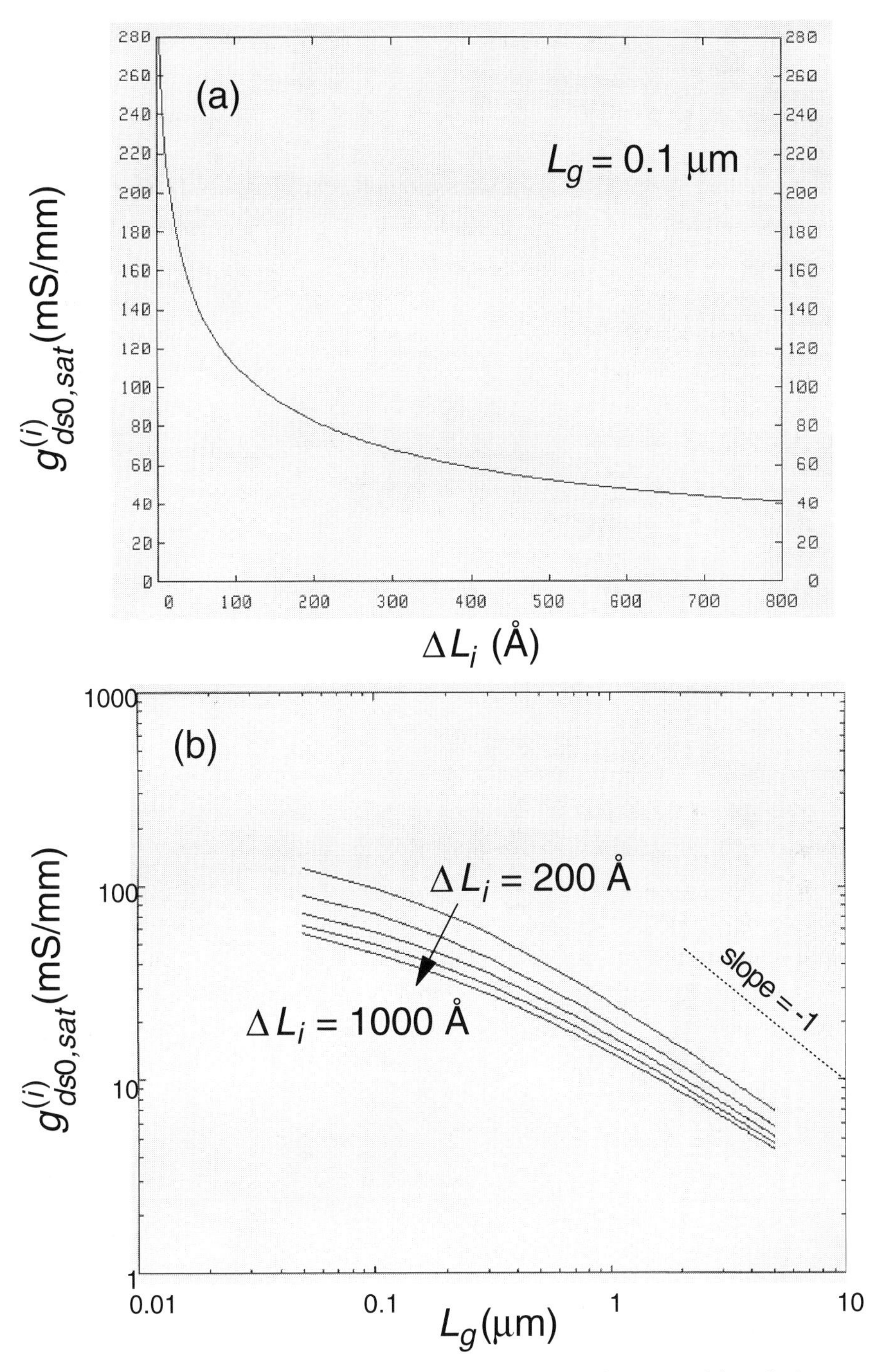

Fig. 15.12. Output conductance of the example FET, versus (a) the extent of the velocity-saturated region, and (b) the gate length with the extent of the velocity-saturated region as a parameter.

where

$$\tau_{d,sat}^{(2)} = \frac{1}{v_{sat}} \left\{ 2 \int_{y_{s0}}^{y_{us0}} dy \, (y - y_{s0}) \left[1 - \Phi_d(y)\right] \right\}^{1/2}. \tag{15.58}$$

$\tau_{d,sat}^{(1)}$ and $\tau_{d,sat}^{(2)}$ are the delays appearing in the expansion of $F_d(\omega)$:

$$F_d(\omega) = 1 - j\omega\tau_{d,sat}^{(1)} - \frac{1}{2}\left(\omega\tau_{d,sat}^{(2)}\right)^2 + \cdots, \tag{15.59}$$

analogously to Equation (15.17).

For the conduction-induced gate current we arrive at

$$i_{g,sat}^{(cond)} = \frac{j\omega g_{m0,sat}^{(i)} \tau_{g,sat}^{(1)}}{1 + j\omega\tau_{g,sat}^{(RC)}} v_g + \frac{j\omega g_{ds0,sat}^{(i)} \tau_{g,sat}^{(1)}}{1 + j\omega\tau_{g,sat}^{(RC)}} v_d, \tag{15.60}$$

where $\tau_{g,sat}^{(1)}$ is analogous to $\tau_{d,sat}^{(1)}$:

$$\tau_{g,sat}^{(1)} = \frac{1}{v_{sat}} \int_{y_{s0}}^{y_{us0}} dy \, \Phi_g(y), \tag{15.61}$$

and the gate RC delay is

$$\tau_{g,sat}^{(RC)} = \frac{1}{v_{sat}^2 \tau_{g,sat}^{(1)}} \int_{y_{s0}}^{y_{us0}} dy \, (y - y_{s0}) \, \Phi_g(y). \tag{15.62}$$

The total gate and drain currents are given by the sum of their respective displacements (Equations (15.32) and (15.33)) and conduction-induced components (Equations (15.51) and (15.60)). Expressed in terms of the intrinsic Y parameters, the result is (with the notation in Chapter 11, and a superscript (i) to emphasize that this is for the intrinsic device):

$$Y_{gg}^{(i)} = \frac{j\omega\left(C_{gs,gc} + C_{gd,sat}\right)}{1 + j\omega\tau_{g,gc}^{(RC)}} + \frac{j\omega g_{m0,sat}^{(i)} \tau_{g,sat}^{(1)}}{1 + j\omega\tau_{g,sat}^{(RC)}}, \tag{15.63a}$$

$$Y_{gd}^{(i)} = -\frac{j\omega C_{gd,sat}}{1 + j\omega\tau_{g,gc}^{(RC)}} + \frac{j\omega g_{ds0,sat}^{(i)} \tau_{g,sat}^{(1)}}{1 + j\omega\tau_{g,sat}^{(RC)}}, \tag{15.63b}$$

$$Y_{dg}^{(i)} = g_{m0,sat}^{(i)} e^{-j\omega\tau_{d,sat}^{(TL)}} \frac{\sin(\omega\tau_{d,sat}^{(s)})}{\omega\tau_{d,sat}^{(s)}} - \frac{j\omega C_{gd,sat}}{1 + j\omega\tau_{g,gc}^{(RC)}}, \tag{15.63c}$$

$$Y_{dd}^{(i)} = g_{ds0,sat}^{(i)} e^{-j\omega\tau_{d,sat}^{(TL)}} \frac{\sin(\omega\tau_{d,sat}^{(s)})}{\omega\tau_{d,sat}^{(s)}} + \frac{j\omega\left(C_{ds,sat} + C_{gd,sat}\right)}{1 + j\omega\tau_{g,gc}^{(RC)}}. \tag{15.63d}$$

The effects of the conduction-induced delays on the capacitances measured/extracted at sufficiently low frequencies are: (1) an increase in the gate-source capacitance by $\left(g_{m0,sat}^{(i)} + g_{ds0,sat}^{(i)}\right) \tau_{g,sat}^{(1)}$; (2) a decrease in the gate-drain feedback capacitance by

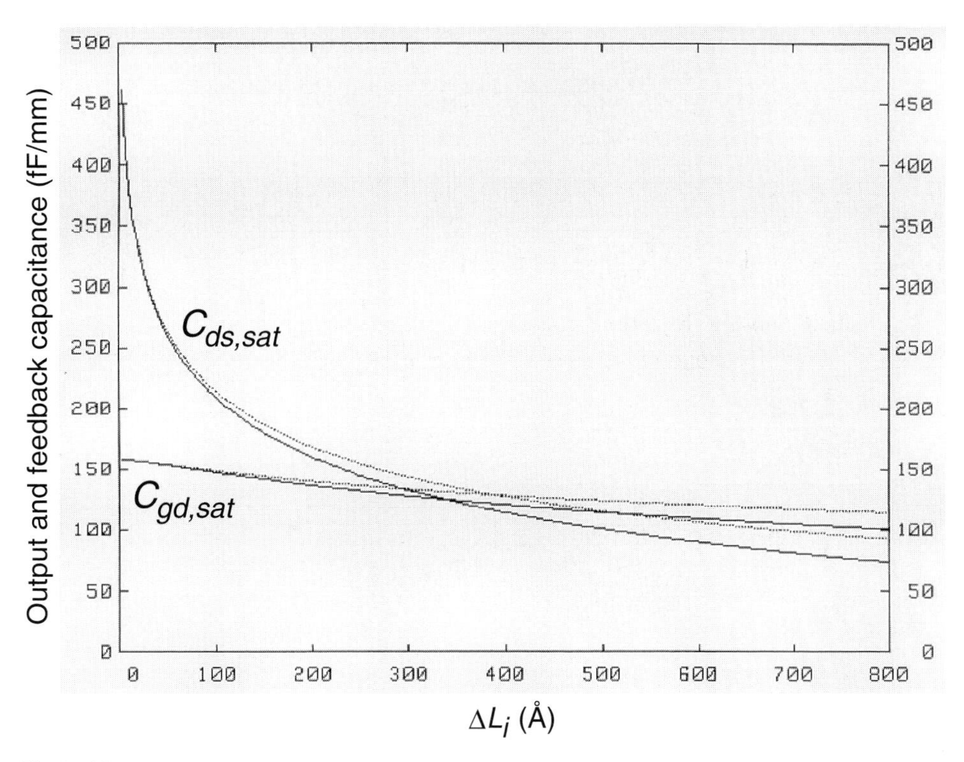

Fig. 15.13. Output and feedback capacitance for the example FET, with (full line) and without (dotted line; = Figure 15.6) the delays introduced by the velocity-saturated region.

$g_{ds0,sat}^{(i)}\tau_{g,sat}^{(1)}$; and (3) a decrease in the output capacitance by $g_{ds0,sat}^{(i)}\tau_{d,sat}^{(TL)}$. Curve (c) in Figure 15.9 illustrates the first effect for the example FET (Table 15.2). The rapid drop in $C_{gs,gc}$ is more than compensated for by $\left(g_{m0,sat}^{(i)} + g_{ds0,sat}^{(i)}\right)\tau_{g,sat}^{(1)}$. This is expected, since the capacitance $\varepsilon W_g \Delta L_i/d_{gc}$ subtracted out in $C_{gs,gc}$ (Equation (15.20)) has to reappear at some point, since, after all, the charge in the saturated region is modulated. The capacitive 'price' for modulating the charge *external* to the gate is, however, less than $\varepsilon W_g \Delta L_x/d_{gc}$, as indicated by the factor 0.37 in Figure 15.9. The effect on $C_{gd,sat}$ and $C_{ds,sat}$ is less dramatic, at least in an absolute sense, and is shown in Figure 15.13. The velocity-saturation region has redistributed some capacitance from C_{gd} to C_{gs}.

The final effect of the conduction-induced delays is on the transconductance, which will be degraded in magnitude ($\tau_{d,sat}^{(s)}$) and in phase ($\tau_{d,sat}^{(TL)}$). Figure 15.14 shows the drain and gate delays introduced by the velocity-saturated region for the example FET. $\tau_{g,sat}^{(1)}$ will always be smaller than $\tau_{d,sat}^{(1)}$; in marginal saturation significantly so. This is because Φ_g is always $< 1 - \Phi_d$ as illustrated in Figure 15.15. For typical bias, such as the 1.5 V drain bias in Figure 14.1, ΔL_i and ΔL_x are on the order of 0.1 μm. The delays in Figure 15.14 are then more than two orders of magnitude larger than the time constants associated with the gradual part of the channel. It is evident that the velocity-saturated part of the device will play the dominant role in the small-signal

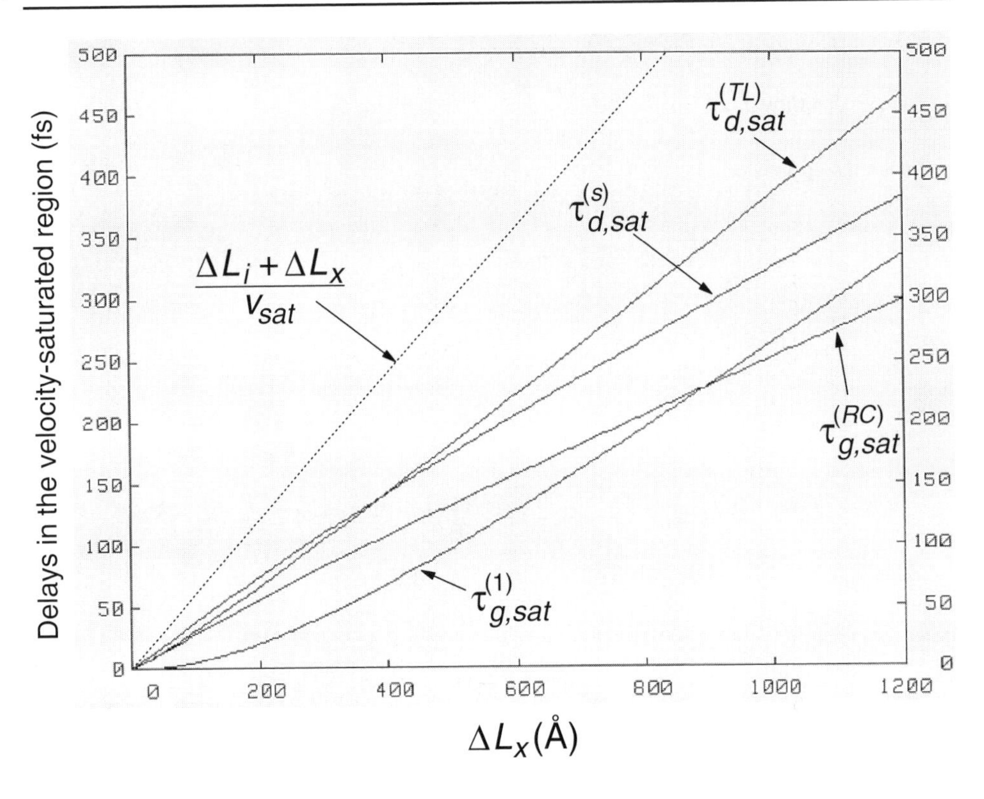

Fig. 15.14. Delays in the example FET introduced by the velocity-saturated region.

frequency performance of a device biased for optimum speed. The major effect of the gradual channel is its contribution to the gate-source capacitance.

Efforts have been made to express the rather complex two-dimensional geometrical dependence of the capacitance and delay components in an FET in simple linear terms of ΔL_i and ΔL_x. This can work to some extent for $\tau^{(TL)}_{d,sat}$ and $\tau^{(s)}_{d,sat}$, because $1 - \Phi_d$ will always go from 1 at $y = -\Delta L_i$ to 0 at $y = \Delta L_x$ (Figure 15.15), with moderate changes in shape as ΔL_i are ΔL_x are varied. Even so, it is difficult to assign coefficients with wide applicability. This is even more the case for C_{gs}. With the simple numerical analysis involved, we are better off doing an approximate two-dimensional analysis using the conformal-mapping method. The phenomenological dimensionless coefficient $\alpha = \left(d_{gc}/\varepsilon W_g\right)\left(\partial C_{gs}/\partial \Delta L_x\right)$ (defined similarly as in [17]) only approaches a constant ($\sim$0.4) rather deep in saturation as illustrated in Figure 15.9. In weaker saturation, α is better approximated by 0. The difficulty in assigning a universal constant value to α stems from the complex dependence of the shape and maximum value of Φ_g (Figure 15.15) on ΔL_i and ΔL_x. For $\Delta L_i + \Delta L_x$ small compared to d_{gc}, the equipotentials of Φ_g are not able to significantly penetrate the $\Delta L_i + \Delta L_x$ gap, and $\Phi_g^{(max)}$ will be depressed. The good news is that this particular feature contributes to making the non-trivial development

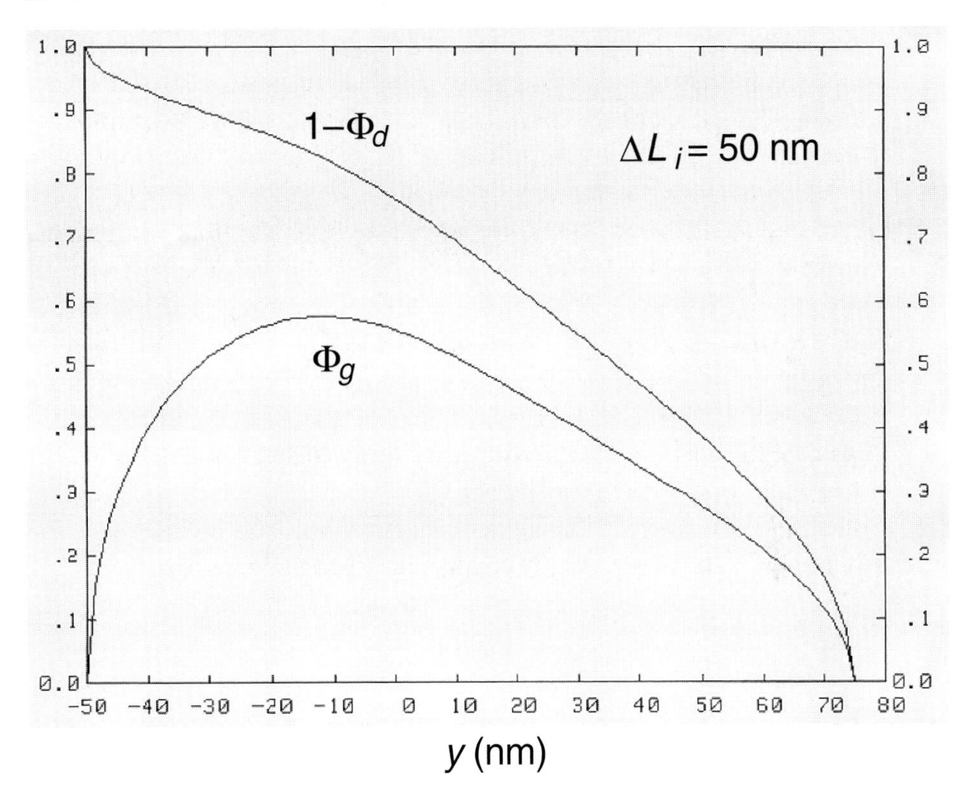

Fig. 15.15. Φ_g and $1 - \Phi_d$ along the saturated part of the channel for the example FET.

task of scaling the gate length into the deep submicron regime (Section 17.7) worth while. But it also shows that the operating drain voltage may have to be scaled down for reasons other than breakdown and reliability. An ideal FET material system for scaling down L_g would be one with good transport properties, and a sufficiently large channel bandgap to allow scaling down also of L_{ud} (Figure 14.1), without introducing unacceptable breakdown voltage. This would limit ΔL_x, and allow more maintained high-frequency performance as V_d is increased deep into saturation. A good candidate to achieve this situation is GaN/AlGaN. Of course, one has to balance the benefits of a lower gate-source capacitance with the drawbacks of a larger feedback capacitance. The two-dimensional analysis in this section, augmented by the parasitic elements introduced in the Section 15.5 and Chapter 16, will allow us to do this with minimal numerical analysis.

15.5 *Y* parameters and equivalent circuit for the extrinsic device

We are now able to estimate rather accurately all the components in the equivalent circuit in Figure 15.3(b). This circuit was developed phenomenologically from the simpler gradual-channel case, and has the topology used almost universally by device and circuit designers. We now need to include the unavoidable parasitic elements. Figure 15.16(a) shows the conventional equivalent circuit for the extrinsic MODFET, including the parasitic access resistances. C_{gs}, C_{gd} and C_{ds} are the total capacitances resulting from the displacement-current and conduction-induced components. There can be some additional gate capacitance due to field lines from the actual thicker gate to the ohmic metallization (or a heavily doped cap) through a passivation layer. We can use ΔL_d to adjust this (Figures 15.7 and 15.8). Alternatively, or in addition, we include corrective components $C_{gs}^{(f)}$ (to the source) and $C_{gd}^{(f)}$ (to the drain). These allow us to account for any asymmetry in the gate. If the dielectric constant is sufficiently small the corrections could be negative, as shown for the air case in Figure 15.7. Either way, the correction can be approximated rather well by a constant. Similarly, we allow for a drain–source corrective fringing capacitance $C_{ds}^{(f)}$, which is expected to be quite small. These corrective fringing capacitances do not show up as separate components in Figure 15.16. We will simply consider them additive parts of the total C_{gs}, C_{gd} and C_{ds}. In terms of the physical quantities introduced in the analysis, the capacitances are given by

$$C_{gs} = C_{gs,gc} + \left(g_{m0,sat}^{(i)} + g_{ds0,sat}^{(i)}\right)\tau_{g,sat}^{(1)} + C_{gs}^{(f)}, \tag{15.64a}$$

$$C_{gd} = C_{gd,sat} - g_{ds0,sat}^{(i)}\tau_{g,sat}^{(1)} + C_{gd}^{(f)}, \tag{15.64b}$$

$$C_{ds} = C_{ds,sat} - g_{ds0,sat}^{(i)}\tau_{d,sat}^{(1)} + C_{ds}^{(f)}. \tag{15.64c}$$

We have also included the two components g_{gs} and g_{gd} associated with the gate diode parallel conductance*. These should be small and of secondary importance near the optimum bias point for small-signal high-frequency operation, but do affect gain and the noise figure at low frequencies (Chapter 17). We have also included a gate resistance R_g.

R_{gs}, R_{gd} and τ in the conventional equivalent circuit are merely fitting parameters, unable to account for the full dispersion. For device optimization (Chapter 17) we instead will make use of the full physics-based expressions for the Y parameters. The 'true' equivalent circuit with elements containing the full and rather complex dispersion of the Y parameters is shown in Figure 15.16(b). Here we have augmented the intrinsic Y parameters in Equation (15.63) with the fringe capacitance corrections and gate conductances, as indicated by the prime in the (i') superscript:

$$Y_{11}^{(i')} = Y_{gg}^{(i)} + g_{gs} + g_{gd} + j\omega\left(C_{gs}^{(f)} + C_{gd}^{(f)}\right), \tag{15.65a}$$

$$Y_{12}^{(i')} = Y_{gd}^{(i)} - g_{gd} - j\omega C_{gd}^{(f)}, \tag{15.65b}$$

* Like most of the components (but unlike the transconductance and output conductance) g_{gs} and g_{gd} are understood to be intrinsic parameters, i.e. to be 'inside' the access resistances.

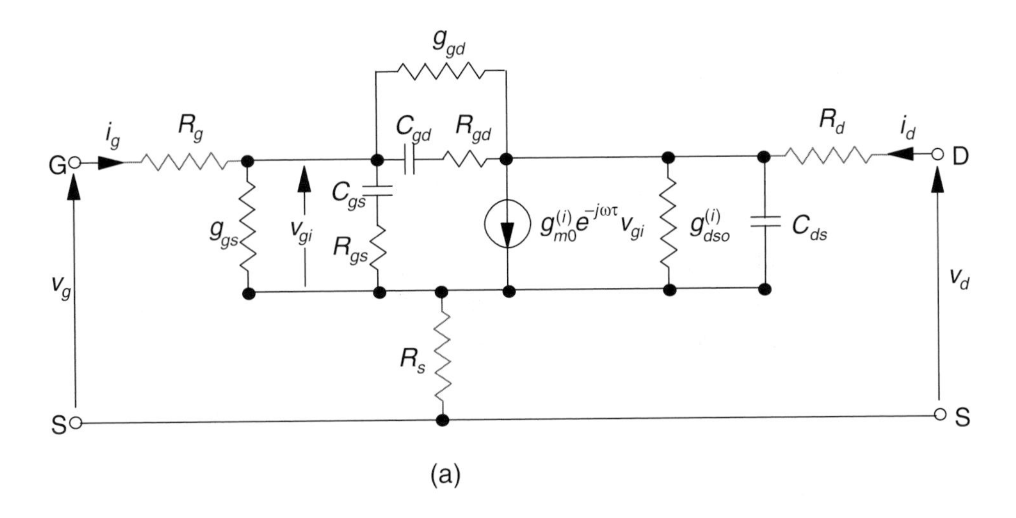

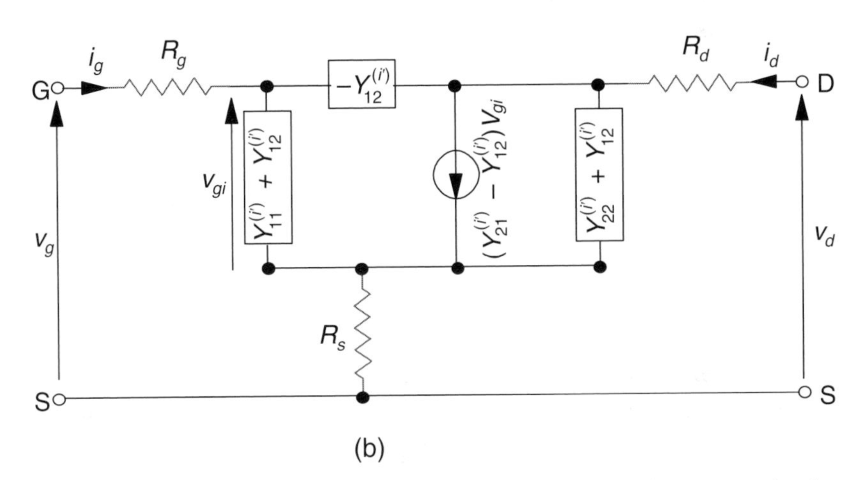

Fig. 15.16. MODFET equivalent circuits with extrinsic elements included. (a) conventional topology based on Figure 15.3(b); (b) using elements that incorporate the physically based Y parameters (Equations (15.63) and (15.65)).

$$Y_{21}^{(i')} = Y_{dg}^{(i)} - g_{gd} - j\omega C_{gd}^{(f)}, \tag{15.65c}$$

$$Y_{22}^{(i')} = Y_{dd}^{(i)} + g_{gd} + j\omega\left(C_{gd}^{(f)} + C_{ds}^{(f)}\right). \tag{15.65d}$$

These are conveniently used as the components in a matrix $\underline{Y}^{(i')}$, and we therefore now use standard numeric indices, as opposed to alphabetical. The Y-matrix $\underline{Y}^{(x')}$ for the extrinsic case with non-zero parasitic resistances is determined by accounting for the parasitic gate, drain, and source voltage drops $R_g i_g$, $R_d i_d$, and $R_s(i_g + i_d)$, respectively*. The current flowing in the source resistance is the sum of the gate and

* The prime in the superscript (x') indicates that we have one more effect to account for before we arrive at the final extrinsic Y parameters. This will be done in Section 16.8.

drain currents since Kirchhoff's current law is still valid. The delays appearing in the analysis, even those we have referred to as 'transmission-line' delays, are introduced by transport and electrostatic charge control, not by electromagnetic wave propagation. Thus, they do not show up as phase delays between, say, the source and drain current. We get for the extrinsic Y-parameter matrix:

$$\underline{Y}^{(x')} = \underline{A}^{-1}\underline{Y}^{(i')}, \tag{15.66a}$$

where

$$\underline{A} = \begin{bmatrix} 1 + R_s\left(Y_{11}^{(i')} + Y_{12}^{(i')}\right) + R_g Y_{11}^{(i')} & R_s\left(Y_{11}^{(i')} + Y_{12}^{(i')}\right) + R_d Y_{12}^{(i')} \\ R_s\left(Y_{21}^{(i')} + Y_{22}^{(i')}\right) + R_g Y_{21}^{(i')} & 1 + R_s\left(Y_{21}^{(i')} + Y_{22}^{(i')}\right) + R_d Y_{22}^{(i')} \end{bmatrix}. \tag{15.66b}$$

The inversion of a 2×2 matrix is a quickly executed command in modern programming environments. Alternatively, the following is a more explicit way of expressing the extrinsic Y parameters:

$$Y_{11}^{(x')} = F\left(Y_{11}^{(i')} + D\left(R_s + R_d\right)\right), \tag{15.67a}$$

$$Y_{12}^{(x')} = F\left(Y_{12}^{(i')} - DR_s\right), \tag{15.67b}$$

$$Y_{21}^{(x')} = F\left(Y_{21}^{(i')} - DR_s\right), \tag{15.67c}$$

$$Y_{22}^{(x')} = F\left(Y_{22}^{(i')} + D\left(R_s + R_g\right)\right), \tag{15.67d}$$

where

$$D = Y_{11}^{(i')}Y_{22}^{(i')} - Y_{12}^{(i')}Y_{21}^{(i')} \tag{15.67e}$$

and

$$F^{-1} = 1 + R_s(Y_{11}^{(i')} + Y_{12}^{(i')} + Y_{21}^{(i')} + Y_{22}^{(i')}) + R_d Y_{22}^{(i')} + R_g Y_{11}^{(i')} + R_{sdg}^2 D. \tag{15.67f}$$

In Equation (15.67f) we have used

$$R_{sdg}^2 = R_s R_d + R_s R_g + R_d R_g. \tag{15.67g}$$

We are often interested in moderate frequencies, where a Taylor expansion to $(j\omega)^2$, suffices, and we can neglect g_{gs} and g_{gd}. The Y parameters with fringe capacitance corrections but no parasitic resistances included (Equation (15.65)) are then given by

$$Y_{11}^{(i')} = j\omega C_{11}^{(i')} + \omega^2 B_{11}^{(i')}: \tag{15.68a}$$

$$C_{11}^{(i')} = C_{gs,gc} + C_{gd,sat} + g_{m0,sat}^{(i)}\tau_{g,sat}^{(1)} + C_{gs}^{(f)} + C_{gd}^{(f)} = C_{gs} + C_{gd}, \tag{15.68b}$$

$$B_{11}^{(i')} = \left(C_{gs,gc} + C_{gd,sat}\right)\tau_{g,gc}^{(RC)} + g_{m0,sat}^{(i)}\tau_{g,sat}^{(1)}\tau_{g,sat}^{(RC)}; \tag{15.68c}$$

$$Y_{12}^{(i')} = -j\omega C_{12}^{(i')} - \omega^2 B_{12}^{(i')} : \tag{15.68d}$$

$$C_{12}^{(i')} = C_{gd,sat} - g_{ds0,sat}^{(i)} \tau_{g,sat}^{(1)} + C_{gd}^{(f)} = C_{gd}, \tag{15.68e}$$

$$B_{12}^{(i')} = C_{gd,sat} \tau_{g,gc}^{(RC)} - g_{ds0,sat}^{(i)} \tau_{g,sat}^{(1)} \tau_{g,sat}^{(RC)}; \tag{15.68f}$$

$$Y_{21}^{(i')} = g_{m0,sat}^{(i)} - j\omega C_{21}^{(i')} - \omega^2 B_{21}^{(i')} : \tag{15.68g}$$

$$C_{21}^{(i')} = g_{m0,sat}^{(i)} \tau_{d,sat}^{(1)} + C_{gd,sat} + C_{gd}^{(f)}, \tag{15.68h}$$

$$B_{21}^{(i')} = \tfrac{1}{2} g_{m0,sat}^{(i)} \left(\tau_{d,sat}^{(2)}\right)^2 + C_{gd,sat} \tau_{g,gc}^{(RC)}; \tag{15.68i}$$

$$Y_{22}^{(i')} = g_{ds0,sat}^{(i)} + j\omega C_{22}^{(i')} + \omega^2 B_{22}^{(i')} : \tag{15.68j}$$

$$C_{22}^{(i')} = C_{ds,sat} + C_{gd,sat} - g_{ds0,sat}^{(i)} \tau_{d,sat}^{(1)} + C_{ds}^{(f)} + C_{gd}^{(f)}, \tag{15.68k}$$

$$B_{22}^{(i')} = \left(C_{ds,sat} + C_{gd,sat}\right) \tau_{g,gc}^{(RC)} - \tfrac{1}{2} g_{ds0,sat}^{(i)} \left(\tau_{d,sat}^{(2)}\right)^2. \tag{15.68l}$$

The two constants D and F become:

$$D = j\omega D_1 + \omega^2 D_2 : \tag{15.69a}$$

$$\begin{aligned} D_1 &= g_{ds0,sat}^{(i)} C_{11}^{(i')} + g_{m0,sat}^{(i)} C_{12}^{(i')} \\ &= g_{ds0,sat}^{(i)} \left(C_{gs} + C_{gd}\right) + g_{m0,sat}^{(i)} C_{gd}, \end{aligned} \tag{15.69b}$$

$$D_2 = g_{ds0,sat}^{(i)} B_{11}^{(i')} + g_{m0,sat}^{(i)} B_{12}^{(i')} - C_{11}^{(i')} C_{22}^{(i')} + C_{12}^{(i')} C_{21}^{(i')}; \tag{15.69c}$$

$$F = F_0 \left(1 - j\omega F_1 - \omega^2 F_2\right): \tag{15.69d}$$

$$F_0 = \frac{1}{1 + R_s\left(g_{m0,sat}^{(i)} + g_{ds0,sat}^{(i)}\right) + R_d g_{ds0,sat}^{(i)}}, \tag{15.69e}$$

$$\begin{aligned} F_1 = F_0 \Big(& R_s \left(C_{11}^{(i')} - C_{12}^{(i')} - C_{21}^{(i')} + C_{22}^{(i')}\right) \\ & + R_d C_{22}^{(i')} + R_g C_{11}^{(i')} + R_{sdg}^2 D_1 \Big); \end{aligned} \tag{15.69f}$$

$$\begin{aligned} F_2 = F_0 \Big(& R_s \left(B_{11}^{(i')} - B_{12}^{(i')} - B_{21}^{(i')} + B_{22}^{(i')}\right) \\ & + R_d B_{22}^{(i')} + R_g B_{11}^{(i')} + R_{sdg}^2 D_2 \Big) + F_1^2. \end{aligned} \tag{15.69g}$$

The extrinsic Y parameters expanded to $(j\omega)^2$ are given by:

$$Y_{11}^{(x')} = j\omega C_{11}^{(x')} + \omega^2 B_{11}^{(x')} : \tag{15.70a}$$

$$C_{11}^{(x')} = F_0 \left(C_{11}^{(i')} + (R_s + R_d) D_1\right), \tag{15.70b}$$

$$B_{11}^{(x')} = F_0 \left(B_{11}^{(i')} + (R_s + R_d) D_2\right) + F_1 C_{11}^{(x')}; \tag{15.70c}$$

$$Y_{12}^{(x')} = -j\omega C_{12}^{(x')} - \omega^2 B_{12}^{(x')} : \tag{15.70d}$$

$$C_{12}^{(x')} = F_0\left(C_{12}^{(i')} + R_s D_1\right), \tag{15.70e}$$

$$B_{12}^{(x')} = F_0\left(B_{12}^{(i')} + R_s D_2\right) + F_1 C_{12}^{(x')}; \tag{15.70f}$$

$$Y_{21}^{(x')} = g_{m0,sat}^{(x)} - j\omega C_{21}^{(x')} - \omega^2 B_{21}^{(x')} : \tag{15.70g}$$

$$g_{m0,sat}^{(x)} = F_0 g_{m0,sat}^{(i)}, \tag{15.70h}$$

$$C_{21}^{(x')} = F_0\left(C_{21}^{(i')} + R_s D_1 + F_1 g_{m0,sat}^{(i)}\right), \tag{15.70i}$$

$$B_{21}^{(x')} = F_0\left(B_{21}^{(i')} + R_s D_2 + F_2 g_{m0,sat}^{(i)} + F_1\left(C_{21}^{(i')} + R_s D_1\right)\right); \tag{15.70j}$$

$$Y_{22}^{(x')} = g_{ds0,sat}^{(x)} + j\omega C_{22}^{(x')} + \omega^2 B_{22}^{(x')} : \tag{15.70k}$$

$$g_{ds0,sat}^{(x)} = F_0 g_{ds0,sat}^{(i)}, \tag{15.70l}$$

$$C_{22}^{(x')} = F_0\left(C_{22}^{(i')} + \left(R_s + R_g\right) D_1 - F_1 g_{ds0,sat}^{(i)}\right), \tag{15.70m}$$

$$B_{22}^{(x')} = F_0\left(B_{22}^{(i')} + \left(R_s + R_g\right) D_2 - F_2 g_{ds0,sat}^{(i)} + F_1\left(C_{22}^{(i')} + \left(R_s + R_g\right) D_1\right)\right). \tag{15.70n}$$

Note that the factor F_0 (< 1; Equation (15.69e)) degrades the extrinsic transconductance and output conductance identically (Equations (15.70h), (15.70l)). F_0 is also a factor in the extrinsic capacitances (e.g. Equation (15.70b)), but an additional term reduces the effect.

15.6 Conclusion

This chapter built on the treatment in Chapter 11 in developing expressions for calculating the Y parameters for the extrinsic short-gate MODFET. We developed analytical tools for estimating the capacitances and the output conductance in saturation, as well as the effect of delays on the small-signal terminal currents. Before we are ready to make full use of these methods and tools in design and optimization, we must, however, take a closer look at what determines the gate resistance R_g. This parameter significantly impacts gain and noise, and its origins, as we will learn in the next chapter, are quite rich in nature and content.

15.7 Bibliography

[1] M. Bagheri, 'An improved MODFET microwave analysis', *IEEE Transactions on Electron Devices*, Vol. 35, pp. 1147–1149, 1988.

[2] H. Rohdin and P. Roblin, 'A MODFET dc model with improved pinchoff and saturation characteristics', *IEEE Transactions on Electron Devices*, Vol. 33, pp. 664–672, 1986.

[3] J. Braunstein, P.J. Tasker, A. Hülsmann, K. Köhler, W. Bronner, and M. Schlechtweg, 'G_{ds} and f_T analysis of pseudomorphic MODFETs with gate lengths down to 0.1 μm', *Proceedings of the International Symposium on Gallium Arsenide and Related Compounds*, Insitute of Physics, pp. 41–45, 1994.

[4] P.M. Morse and H. Feshbach, *Methods of Theoretical Physics*, Pt I, McGraw-Hill, New York, 1953.

[5] R.E. Collin, *Foundations for Microwave Engineering*, 2nd ed., McGraw-Hill, New York, 1992.

[6] N. Moll, 'Delay and current due to charged particles in generalized space-charge regions', unpublished manuscript, 1991.

[7] W. Shockley, *Journal of Applied Physics*, Vol. 9, p. 635, 1938.

[8] S. Ramo, 'Currents induced by electron motion', *Proceedings of the IRE*, Vol. 27, pp. 584–585, 1939.

[9] H. Kim, H.S. Min, T.W. Tang and Y.J. Park, 'An extended proof of the Ramo–Schockley theorem', *Solid State Electronics*, Vol. 34, pp. 1251–1253, 1991.

[10] S.E. Laux and W. Lee, 'Collector signal delay in the presence of velocity overshoot', *IEEE Electron Device Letters*, Vol. 11, pp. 174–176, 1990.

[11] J.E. Bowers and C.A. Burrus, Jr, 'Ultrawide-band long-wavelength p-i-n photodetectors', *Journal of Lightwave Technology*, Vol. 5, pp. 1339–1350, 1987.

[12] S. Babiker, A. Asenov, N. Cameron, S.P. Beaumont and J.R. Barker, 'Complete Monte Carlo RF analysis of 'real' short-channel compound FET's', *IEEE Transactions on Electron Devices*, Vol. 45, pp. 1644–1652, 1998.

[13] S.M.S. Imtiaz and S.M. El-Ghazaly, 'Performance of MODFET and MESFET: A comparative study including equivalent circuits using combined electromagnetic and solid-state simulator', *IEEE Transactions on Microwave Theory and Techniques*, Vol. 46, pp. 923–931, 1998.

[14] J.M. Early, 'P–N–I–P and N–P–I–N junction transistor triodes', *Bell Systems Technical Journal*, Vol. 33, pp. 517–533, 1954.

[15] H. Rohdin and A. Nagy, 'A 150 GHz sub-0.1-μm E/D MODFET MSI process', *Technical Digest of the International Electron Devices Meeting*, IEEE, Piscataway, pp. 327–330, 1992.

[16] M.C. Foisy; 'A physical model for the bias dependence of the modulation-doped field-effect transistor's high-frequency performance', PhD. Thesis, Cornell University, 1990.

[17] P.H. Ladbrooke, 'Reverse modelling of GaAs MESFETs and HEMTs', *GEC Journal of Research*, Vol. 6, pp. 1–9, 1988.

15.8 Problems

15.1 Translate the $R_{ij}(k)$ and $I_{ij}(k)$ coefficients, derived in Section 11.2 from the long-channel MOSFET wave-equation, into the equivalent circuit in Figure 15.1; i.e., derive Equations (15.11)–(15.16).

15.2 Consider the two-dimensional electrostatic problem consisting of several conductors of arbitrary shape embedded in a uniform dielectric. Use Cartesian coordinates x and y to describe the geometry. All conductors are grounded except for one which is held at potential V_0. Assume that the conductors are 'almost touching'; i.e. that, as you completely trace their circumferences without lifting the pen, only relatively short dual parallel-line segments are necessary to bridge the gaps between the conductors. Now think of x and y as the coordinates in the complex z-plane ($z = x + iy$), and the potential $V(x, y)$ as the real part $V_z(z)$ of a complex potential $V_z(z) + iU_z(z)$. Assume that you have an analytical function $w(z)$ that maps the area outside the conductors in the z-plane onto the upper half-plane in the complex w-plane, such that the segment that traces the conductor held at potential V_0 maps onto the entire negative real w axis. This leads to a simple geometry for the complex potential $V_w(w) + iU_w(w)$ in the w-plane.

(a) Show that the equipotentials of $U_z(z)$ are field lines of $V_z(z)$. Hint: Recall Cauchy–Riemann's equation for analytical functions.

(b) Show that $V_w(w)$ is given by the simple expression in Equation (15.27). Hint: Recall that a function is the electrostatic potential of the problem if it satisfies Laplace's equation and the boundary conditions.

(c) Show that $V_z(z)$ is given by $V_w(w(z))$, i.e., that the potential in the w-plane can be directly used to calculate the more complicated potential in the z-plane once we have determined $w(z)$.

(d) Derive the charge formula (Equation (15.28)). Hint: Do the charge integration in the w-plane.

15.3 By direct application of Equation (15.38), show that Kirchhoff's current law (Equation (15.40)) holds in the 'electrostatic approximation'. Hint: What is the solution to Laplace's equation when all conductors are biased to 1 V?

16 Gate resistance and the Schottky-barrier interface

Make everything as simple as possible, but not simpler.

ALBERT EINSTEIN

16.1 Introduction

The gate resistance R_g (Figure 15.16) has long been recognized as a very important parasitic parameter that can be difficult to reduce to an acceptable value. R_g degrades the noise figure and power gain. For the field-effect transistor (FET) depicted in Figure 14.1, the gate and drain voltages (and source ground) are applied to metal pads outside the active FET area. The gate, drain and source metallizations carry the currents laterally in the yz-plane onto the active FET area, and deliver the currents in an essentially uniform fashion in the x direction to the semiconductor. For the source and drain this is typically not a problem because of their larger extension in the y direction, and the thick interconnect metallizations available (neither is shown in Figure 14.1, which only depicts the central core of the device). The source and drain resistances (R_s and R_d) are thus not limited by the metallization resistance, but by the contact and semiconductor components discussed in Section 14.6.2. The situation is different for the gate because of its much smaller extension in the y direction. Consequently, in order to reduce the gate resistance for submicron gates, much effort has gone into developing T-gate processes (Figure 14.1; Section 17.7). This chapter will show that there is much more to the gate resistance, and therefore that there are additional ways to reduce it. In particular, we will discuss an interfacial component which, in its purest form, is intimately tied to the mechanism responsible for Schottky-barrier formation.

16.2 Components in the input resistance

The gate metallization resistance contributes to R_g in a distributed way, which reduces the effect to a third of the end-to-end gate finger resistance [1, 2]:

$$R_{ga} = \frac{r_{ga} W_g}{3N_f^2}. \tag{16.1a}$$

We have introduced the subscript a in this component to indicate *a*ccess resistance along the gate finger. r_{ga} is the normalized end-to-end gate metallization resistance given by

$$r_{ga} = \frac{\rho}{A_{gx}}. \tag{16.1b}$$

ρ is the gate metal bulk resistivity. $A_{gx} = A_{gx}^{(T)} + A_{gx}^{(S)}$ is the gate cross-sectional area (Figure 14.1), enhanced by the fat top of the T ($A_{gx}^{(T)}$) over the thin stem ($A_{gx}^{(S)}$). As before, W_g is the total gate width. N_f is the number of parallel fingers (of width W_g/N_f) that make up the gate. Increasing N_f is obviously a very effective way of reducing R_{ga}.

The skin effect will introduce frequency dependence in the AC gate metallization access resistance. Using an expression for the skin effect that is a good approximation for isolated metal strips [3], we can express the frequency dependence of the gate metallization access resistance as:

$$r_{ga}^{(ac)}(f) = r_{ga}\left(1 + \frac{f}{f_{se}}\right)^{1/2}. \tag{16.2a}$$

f_{se} is a characteristic frequency for onset of significant skin effect:

$$f_{se} = \beta_{se}\frac{r_{ga}}{\mu_0}. \tag{16.2b}$$

$\mu_0 = 4\pi \times 10^{-7}$ V s/(A m) is the vacuum permeability (*not* mobility in this case), and β_{se} is a geometric factor, approximately equal to 3.5 for a square cross-section. For a typical $r_{ga} = 150$ Ω/mm, f_{se} is 420 GHz. It has been shown with three-dimensional numerical modeling that the gate skin effect in a realistic 0.1-μm MODFET T-gate structure is well described by Equation (16.2), and is quite negligible [4].

The other resistive components on the input side of the FET are the charging resistances R_{gs} and R_{gd} in Figure 15.16(a). With $V_D > 0$, R_{gd} can be much larger than R_{gs}, as we saw in Figure 15.2 for the linear MODFET. Nevertheless, it is most often omitted in the equivalent circuit. This is because R_{gd} is difficult to extract, which is to say that it is still too small to have an effect on the measured Y parameters. One reason for this is that, while R_{gd} is large, what really counts is the time constant $R_{gd}C_{gd}$, and this is, as we saw for the linear MODFET in Figure 15.2, actually smaller than $R_{gs}C_{gs}$. An extrinsic gate-drain fringe capacitance could reduce the importance of R_{gd} even further. Thus, the important intrinsic resistive component is R_{gs}, given by $R_{gs,gc}$ in Equation (15.23), and often referred to as R_i. Introducing

$$R_{sq} = \frac{1}{q\mu n_{so}} \tag{16.3}$$

for the sheet resistance of a full two-dimensional electron gas (2DEG), Equation (15.23) implies that

$$R_{gs} \approx \frac{1}{12} R_{sq} \left(\frac{L_g + \Delta L_g^{(s)}}{W_g} \right) \left(\frac{I_D^{(max)}}{I_D} \right) = \frac{\left(L_g + \Delta L_g^{(s)}\right) v_{sat}}{12\mu I_D}, \tag{16.4}$$

before onset of deep velocity saturation ($\Delta L_i \sim 0$). We have included the source-side fringing effect (Equation (15.31)). Deeper in saturation, as $\tau_{g,sat}^{(1)}$ and $\tau_{g,sat}^{(RC)}$ (Section 15.4) become significant, R_{gs} will increase beyond the estimate in Equation (16.4), and at the same time the effective mobility μ can be reduced below its long-gate 'bulk' value μ_0 (Section 14.5).

Equivalent circuit extraction methods have difficulty separating R_g from R_{gs}. However, the *total* gate-side input resistance $R_g + R_{gs}$ is well captured by these methods [5]. Experimental values of this sum are typically significantly larger than predicted by Equations (16.1) and (16.4), suggesting an additional unaccounted component in the input resistance [4]. There are very few regions in the device, not already accounted for in the preceding sections, that could be physically acceptable origins for the 'missing' resistance. One is the metal–semiconductor interface, which could lead to an additional gate resistance

$$R_{gi} = \frac{r_{gi}}{W_g L_g}, \tag{16.5}$$

where r_{gi} is a normalized (or specific) interfacial gate resistance. The other is the resistance associated with current flow in the x direction in the stem of the gate (Figure 14.1). Including so-called gate 'necking' ([6]; Section 17.7), described by the geometrical parameters δ_g, h_g, and L_g in Figure 14.1, the gate stem resistance is given by

$$R_{g,stem} = \frac{\rho h_g}{W_g L_g} \left[-\frac{L_g}{2\delta_g} \ln \left(1 - \frac{2\delta_g}{L_g} \right) \right]. \tag{16.6}$$

The W_g scaling of the two components is the same. With no necking ($\delta_g = 0$), the L_g scaling is also the same. The constant of proportionality (effectively r_{gi}) would then be ρh_g. For a state-of-the-art 0.1-μm e-beam T-gate process (Section 17.7), $\rho h_g \sim 4 \times 10^{-11}$ $\Omega\,\text{cm}^2$. If we, instead of bulk resistivity, use a thin film value for the stem resistance, the estimate might increase to $\sim 2 \times 10^{-10}$ Ω cm^2. The use of high-resistance refractory gate metallization can bring the value up by two orders of magnitude [7]. With necking, the correction factor in square brackets in

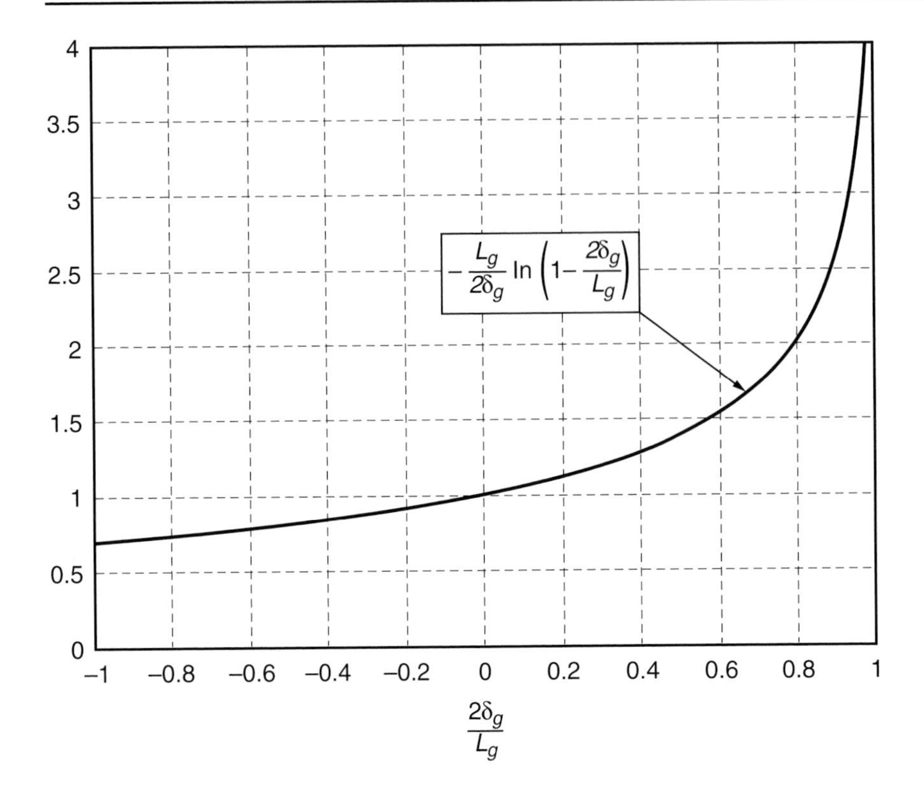

Fig. 16.1. Correction factor in the stem resistance (Equation (16.6)) due to necking.

Equation (16.6) will deviate from unity as shown in Figure 16.1. Necking, i.e., $\delta_g > 0$, results from a gate process in which the gate cut (in resist or dielectric) has steep walls. The more advantageous situation, $\delta_g \leq 0$, results from oxide spacer processes (Section 17.7), or from resist processes where the resist walls slope outwards. Either way, Figure 16.1 shows that the correction factor is rather moderate, except near $\delta_g = L_g/2$ where the wide top of the gate is completely dislodged from the stem. In this extreme case the gate resistance goes back to being properly described by R_{ga}, i.e., proportional to W_g, but with a large r_{ga} corresponding to the cross-sectional area $A_{gx}^{(S)}$ of the now triangular stem.

16.3 Measurement and scaling of the gate resistance

As implied by the discussion, determining the origins of the gate resistance involves measuring/extracting its value versus gate width and frequency. In order to accurately determine the components that make up the gate resistance, several gate widths, in

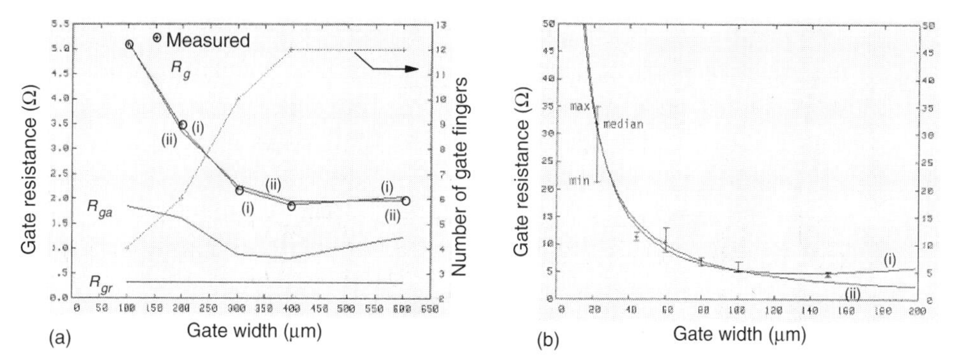

Fig. 16.2. (a) Gate resistance and number of gate fingers for multi-finger 0.2-μm PHEMTs versus total gate width [8]. (i) Full fit → $r_{ga} = 885\ \Omega$/mm, $r_{gi} = 5.3 \times 10^{-7}\ \Omega\,\text{cm}^2$. (ii) Fit with 614 Ω/mm estimate of DC end-to-end $r_{ga} \rightarrow r_{gi} = 5.9 \times 10^{-7}\ \Omega\,\text{cm}^2$. R_{ga} and R_{gr} are the gate metallization access resistance and the residual fixed resistance, respectively, for fit (i). (b) Gate resistance of two-finger 0.1-μm AlInAs/GaInAs MODFETs versus total gate width. (i) Full fit → $r_{ga} = 560\ \Omega$/mm, $r_{gi} = 9.2 \times 10^{-7}\ \Omega\,\text{cm}^2$. (ii) Fit with 82 Ω/mm end-to-end DC measurement of $r_{ga} \rightarrow r_{gi} = 8.1 \times 10^{-7}\ \Omega\,\text{cm}^2$. (H. Rohdin, N. Moll, C. -Y. Su and G. Lee, *IEEE Transactions on Electron Devices*, Vol. 45, pp. 2407–2416, 1998.)

a wide range, should be used. For fixed frequency, the measured values can then be fitted with confidence to the equation

$$R_g(W_g) = R_{gr} + \frac{r_{ga} W_g}{3N_f^2} + \frac{r_{gi}}{W_g L_g}. \tag{16.7}$$

The first term allows for a possible fixed offset related to probing, layout and/or calibration (Section 17.2). The other two terms were introduced above, and the associated constants $r_{ga}/(3N_f^2)$ and r_{gi}/L_g are determined from the fit. Ideally, one should, in addition, do the measurement and fitting with FETs of various gate lengths. This would be difficult for a couple of reasons. First, for longer gates, R_{gs} may start to become significant even at $V_D = 0$, where R_g, R_s and R_d measurements are made. Since it is difficult to separate R_{gs} from R_g, the result could be distorted. Second, and more serious, the task would require several well-defined gate lengths of submicron dimension, all with the same metal–semiconductor interface properties. Most processes do not have this flexibility.

If the extracted r_{ga} at high frequencies is significantly larger than the DC value $r_{ga}^{(DC)}$ measured on special end-to-end structures, or estimated by Equation (16.1b), necking could be an issue. This should then be further investigated with a cross-section SEM (scanning electron microscope), and, if necessary, a process fix developed. If, on the other hand, r_{ga} is on the order of $r_{ga}^{(DC)}$, and/or r_{gi} is larger than the estimate for the stem resistance, the metal–semiconductor interface is a likely origin of the 'excess' gate resistance.

Figure 16.2 shows two examples of the gate resistance of MODFETs plotted versus gate width. They are good examples because of both their similarities and their differences. They are similar in that neither process suffers from necking, and in that the results extracted are based on several gate widths. This promotes accurate separation of the three terms in R_g. In both cases the measurements were done with $V_D = 0$ V. The $R_g(W_g)$ data in Figure 16.2(a) were taken on 0.2-μm power pseudomorphic high-electron-mobility transistors (PHEMTs) with various number of fingers [8]. These gates were intentionally trapezoidal (no T) with rather large access resistance r_{ga}. The extraction was done with the gate forward biased (drawing a significant current) using the technique of Dambrine *et al.* [9]. A low frequency (~ 1 GHz) gives the best accuracy for this method. In Figure 16.2(b) are $R_g(W_g)$ data for two-finger 0.1-μm symmetric-T-gate InP-type MODFETs (Section 17.7). The following simple formula for estimating the gate resistance from the measured Y parameters was used:

$$R_g = \mathrm{Re}\left(\frac{1}{Y_{11}}\right) - \frac{1}{4\,\mathrm{Re}\,(Y_{22})}. \tag{16.8}$$

The resistance of the gated channel is already small because of the short gate. By biasing the gate for full channel occupation, without significant DC gate leakage, it is further reduced, and is neglected beside the R_s and R_d. For a symmetric device ($R_s = R_d$), the second term subtracts out the resistance made up of R_s and R_d in parallel. Under the full-channel zero-V_D condition, R_{gs} (Equation (16.4)) is also particularly small. R_g is evaluated by Equation (16.8) at sufficiently high frequencies where the measure becomes unaffected by g_{gs} and g_{gd}, and is frequency independent. With g_{gs}, g_{gd}, R_{gs}, R_{gd} and $(g_{ds})^{-1}$ negligible, Equation (16.8) is easily derived from the equivalent circuit in Figure 15.16(a). Despite the differences in material, gate process and extraction technique, the main features of the two sets of data in Figure 16.2 are very similar. Both exhibit a prominent inverse gate-width scaling of R_g with excellent fit to Equation (16.7). The values for r_{gi} are similar (5–$9\times10^{-7}\ \Omega\,\mathrm{cm}^2$), and are much larger than the stem resistance. In the first case there is no separate stem, and in the second case the stem resistance is very small ($\sim 10^{-10}\ \Omega\,\mathrm{cm}^2$). In neither case is there a sign of frequency dependence. Less accurately estimated values for r_{gi}, based on fixed gate-width data in the literature, fall in a range 5×10^{-8}–$10^{-6}\ \Omega\,\mathrm{cm}^2$ [4]. Another interesting data point is provided by the complete Monte Carlo RF analysis by Babiker *et al.* [10] of a 0.12-μm PHEMT biased at $V_D = 1.5$ V, and their comparison with measurements on an actual device with the same nominal structure. In this case, with the device biased in saturation, the predicted R_{gs} (R_i) is ~ 4 times larger than predicted by Equation (16.4), presumably due to delays and effective mobility degradation. Even so, the measured value is even larger, and the discrepancy corresponds to $r_{gi} = 2\times 10^{-7}\ \Omega\,\mathrm{cm}^2$.

The interfacial gate resistance has several interesting and important consequences for the device AC performance and optimization. Some of these will be covered in Chapter 17. In addition, there is a very important practical consequence for device model extraction. As an example we consider the Agilent EEsof Root FET Model, although the lesson is easily implemented in other models. When the measurement-based large-signal device model [11] is extracted for an FET under test, the first step is to determine the parasitic resistances from S parameter data (Section 17.2) with $V_D = 0$ and $V_G > 0$ ('cold-FET configuration') as in the first example above (Figure 16.2(a)). Once these have been calculated, and the S parameters of the FET have been measured in the entire active biasing regime, the non-linear voltage-controlled charge and current functions of the three intrinsic nodes can be determined. The result, an accurate large-signal non-quasi-static table-based model of the device, can then be used for reliable circuit design. Although several gate widths were used in the two examples above, this is not always practical. Instead, the extraction is typically done on a representative FET. During the design of a circuit using various gate widths, the simulator assumes the proportional gate width scaling of Equation (16.1). The software determines the constant of proportionality $r_{ga}/(3N_f^2)$ during the extraction. Since the gate resistance is actually dominated by a term which scales *inversely* with W_g, it is clear that this can lead to large scaling errors, and inaccurate circuit modeling. The way around this problem for the Agilent EEsof Root FET Model is to replace the total extracted R_g with a separately measured or calculated (and typically much smaller) R_{ga} in the 'Model Variable Table'. An internal time constant model parameter will then approximately take care of the remaining dominant W_g^{-1}-term R_{gi}. The time constant, which is determined by the extraction algorithm, accounts approximately for the internal delay $R_{gs}C_{gs}$. By replacing R_g with R_{ga}, we have then essentially set $R_{gs} = R_{gi}$. This redistribution of the input resistance may not be correct physically, but it does lead to correct W_g scaling, and thus a more accurate and reliable circuit simulation. The different gate-length scaling of R_{gs} and R_{gi} (Equations (16.4) and (16.5)) is not an issue since, as mentioned earlier in this section, a particular process typically only produces one submicron gate length. Even if several gate lengths were available, there are no gate-length scalings established for the other parameters in the model, and separate models for each gate length would have to be extracted.

The interfacial gate resistance scales as a contact resistance, and is in this respect reminiscent of JFET behavior. There the effect is well understood in terms of a standard ohmic contact resistance, such as r_{c2} in Figure 14.6. The origin of an interfacial component in the gate resistance of an SBGFET (SBG = Schottky-barrier gate) would have to be very different, and needs to be clarified for a more complete understanding of FET operation. We proceed by reviewing Schottky-barrier formation (Section 16.4), a field that remains a topic of research, and some controversy. We then analyze what goes on electronically at the metal–semiconductor interface as we modulate the gate voltage at high frequencies. The analysis is done in two steps. First, we investigate the frequency dependence of the interfacial admittance and show how this is related to the measured r_{gi} (Section 16.5). Second, as an underlying physical mechanism, we will analyze tunneling between the gate and localized states in the semiconductor near the interface (Sections 16.6 and 16.7). A reader more interested in the bottom line, and satisfied with a cursory treatment of the physical origin of the interfacial gate resistance, can skip to Section 16.8, where we go through the modifications to the equivalent circuit and Y parameters introduced by the two gate resistance components.

16.4 Interfacial gate resistance and Schottky barriers

The evidence encountered points to an interfacial origin for R_{gi}. As more and more data have been collected, and techniques for minimizing contamination and oxide formation at the metal–semiconductor interface implemented (Section 17.7), it appears, for InP- and GaAs-based SBGFETs, that there is a minimum reproducibly achievable value $r_{gi}^{(min)} \sim 10^{-7}\ \Omega\,\mathrm{cm}^2$ for the specific interfacial gate resistance. The question then arises whether this is a fundamental limit. If so, it is expected to be closely related to Schottky-barrier formation. Interestingly, the most widely used and analytically tractable model for Schottky barriers specifically invokes an interfacial layer between the metal and semiconductor. Cowley and Sze [12] did this to explain the fundamental barrier height Φ_{B0} in Si, GaP, GaAs and CdS, for a variety of metals. There were two requirements put on the interfacial layer. First, it should have a thickness on the order of a few atomic layers. It would then be transparent to electrons as the metal and semiconductor were conceptually brought together, allowing the Fermi levels of the metal and semiconductor surface states to freely line up. Second, the interfacial layer had to withstand a potential across it. This was necessary to accomplish the primary goal of the model, i.e. to account for the deviation of experimental Φ_{B0}'s from the classical Schottky limit $\Phi_M - \mathrm{X}_{SC}$, where Φ_M is the metal work function and X_{SC} the semiconductor affinity. An interfacial layer of 4–5 Å thickness, with vacuum electronic properties, was assumed. Section 16.5 extends this static model to the microwave and millimeter-wave frequencies of interest. A tunneling analysis (Sections 16.6 and 16.7) quantifies the transparency of the interfacial layer. It turns out that this layer presents an AC resistance that cannot be ignored, and is in fact very close in magnitude to the experimentally observed r_{gi}. We have to realize, however, that although it can explain the static characteristics of a wide variety of barriers with a few physically reasonable macroscopic parameters, and can provide a remarkably good prediction of r_{gi}, the Cowley–Sze picture is not a precise representation of physical reality. The AC tunneling behavior is much more sensitive to microscopic details than the barrier height is, and it becomes important to understand more precisely the actual physical situation that the Cowley–Sze interfacial layer represents. We therefore now briefly review alternative, more microscopically specific, physical Schottky-barrier models from the literature.

First-principles models for Schottky-barrier formation typically assume an ideal barrier in the sense that the metal and semiconductor atoms are nearest neighbors with no interpenetration of materials. In the most ideal case (e.g. [13]), metal electron wave-functions penetrate into the semiconductor, where they populate part of a continuum of gap states, which are referred to as MIGSs (metal-induced gap states). Despite their name, MIGSs are semiconductor Bloch states, like those in the conduction and valence bands, except that they are in the forbidden gap between these. Thus, they have complex wave-vectors, and decay away from the surface. The Fermi level is pinned near a 'canonical' energy, where the gap states switch from valence to conduction-band character, a point that corresponds to local charge neutrality [14]. This picture of Schottky-barrier formation is elegant but controversial. Its failure, upon closer inspection [15–18] to predict the generally observed essentially metal-independent Fermi level pinning suggests that the problem is more complicated. The most likely complication is that defects, with associated energy levels pinning the Fermi level inside the forbidden gap, form in the semiconductor close to the interface. There are many possible alternative microscopic origins of these defects, including anti-sites (for III–V compounds) and vacancies [16, 18], new chemical compounds and changes in atomic geometry [15]. The relative importance of these can be affected by such factors as morphology, stoichiometry, surface reconstruction, surface preparation and metal reactivity [19]. One prominent defect model, developed by Spicer and coworkers [20, 21], relies on semiconductor native defects, and is in its final refined form [21] referred to as the advanced unified defect model (AUDM). Another proposed complication is the formation of a thin bond-disordered layer in the semiconductor near the interface, resulting in a continuum of disorder-induced gap states (DIGS) [22, 23]. Similar to the MIGS case, the Fermi level is proposed to be pinned near a 'neutral level', which depends only on the bulk semiconductor band structure. Some combination of these [24] and other mechanisms [25, 26] may be involved in determining the barrier height.

While the details of why the mid-gap states come to exist differ radically in all these pictures, the existence of a dipole layer between the metal and semiconductor is not controversial. Thus in some sense, the Cowley–Sze model can be adapted to any physical reality, whether it corresponds to a barrier dominated by MIGSs, defects or bond-disorder, through judicious choice of the properties of the interfacial layer and mid-gap states. For some interfacial conditions the Cowley–Sze model would not just be a convenient construct for analytical modeling, but also a good physical representation. In addition to the case of a thin native oxide, surface reconstruction of GaAs has been proposed to result in true MIS-like Schottky barriers [27]. Sections 16.5 and 16.6 use the Cowley–Sze picture to lay the theoretical groundwork for the existence of the interfacial gate resistance, without tying it to any particular physical interpretation. Section 16.7 examines the interfacial gate resistance for different choices of interfacial layer properties, appropriate to different physical models.

16.5 Admittance analysis of a Schottky barrier with semiconductor surface states

Figure 16.3 is based on Figure 15 in Sze's Chapter 5 [28], and uses similar notation. Cowley and Sze's solution of the static band lineup problem lead to the following important expression for the fundamental (i.e. without image force lowering) barrier height:

$$\Phi_{B0} = \gamma\,(\Phi_M - \mathrm{X}_{SC}) + (1-\gamma)\left(E_g - E_0\right). \tag{16.9}$$

E_0 is the 'neutral level' measured from the top of the valence band. This is the energy below which the surface states must be filled for the surface to be charge neutral. The other important parameter in Equation (16.9), not defined in Figure 16.3, is γ, given by

$$\gamma = \frac{1}{1 + q^2 D_S d_i/\varepsilon_i}. \tag{16.10}$$

The density of states D_S was assumed to be constant in Cowley and Sze's analysis. With a finite interfacial-layer thickness d_i, γ decreases from 1 to 0 as D_S increases from zero to infinity. The $D_S = 0$ limit thus yields the classical ideal Schottky expression $\Phi_{B0} = \Phi_M - \mathrm{X}_{SC}$. The other limit $D_S = \infty$ yields perfect pinning: $\Phi_{B0} = E_g - E_0$. The neutral level E_0 corresponds to the canonical pinning position in the MIGS and DIGS models. Equations (16.9–10) express the reduced sensitivity $\partial\Phi_{B0}/\partial\Phi_M = \gamma$ of the Schottky-barrier height to metal work function in the presence of surface states and an interfacial layer. The experimental value for GaAs is $\gamma = 0.074 \pm 0.05$ [12].

The Cowley–Sze model has been extended to analyze I–V and low-frequency C–V characteristics [29–33]. This subsection extends the model to show quantitatively how an interfacial layer can produce a component in the parasitic gate resistance of SBGFETs. It examines the effect of modulating the metal (gate) voltage at microwave and millimeter-wave frequencies. At these frequencies, and normal bias, recombination of bulk semiconductor carriers at the surface state can be neglected. This makes the admittance analysis less complicated than in some of the I–V and C–V references above.

Section 16.7 considers situations where the interfacial layer is not a vacuum, and we therefore introduce an affinity X_i, a conduction-band edge E_{Ci} and a valence band edge E_{Vi} for the interfacial layer. A non-stationary voltage V is applied to the metal. This results in a varying electron quasi-Fermi level E_{FSC} in the semiconductor (SC), and a Fermi level split $E_{Fi} = E_{FS} - E_{FM}$ across the thin interfacial layer, where E_{FS} and E_{FM} are the Fermi levels for the surface states and the metal, respectively. The built-in voltage V_{bi} in the semiconductor will also change, and we denote this modified

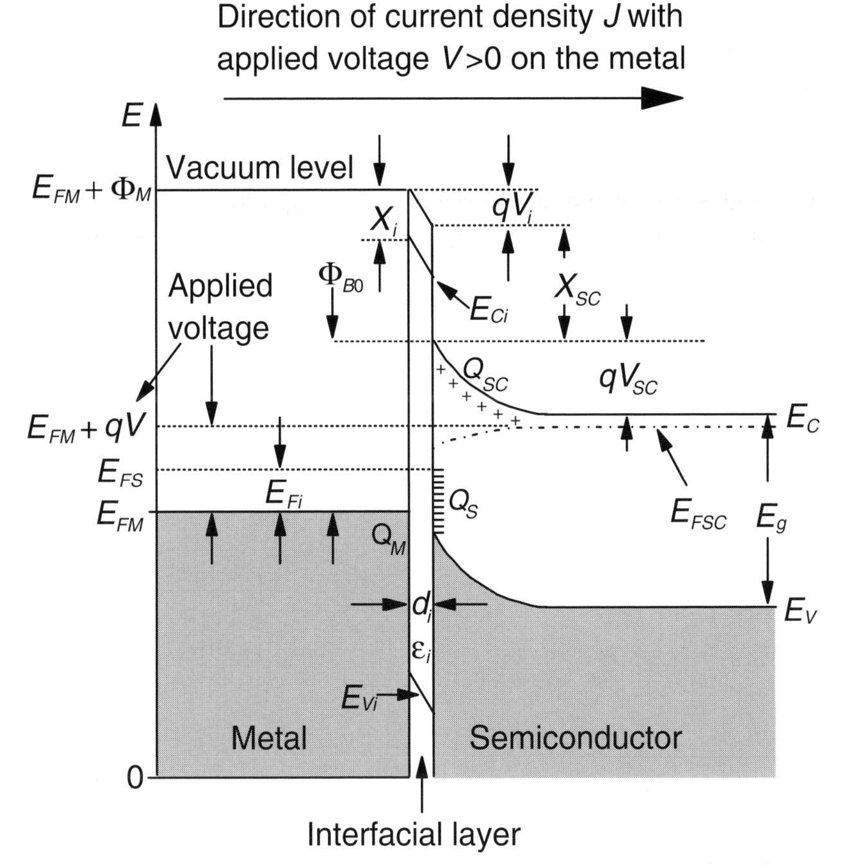

Fig. 16.3. Cowley–Sze's [12] energy-band diagram for a metal–semiconductor (n-type) contact with an interfacial layer. The electron energy E is referenced to the bottom of the metal conduction band. The voltage drops V_i and V_{SC}, and the Fermi energy split, E_{Fi}, are positive as drawn. (H. Rohdin, N. Moll, A.M. Bratkovsky and C.-Y. Su, *Physical Review B*, Vol. 59, pp. 13 102–13 113, 1999.)

voltage V_{SC}. Ideally, $V_{SC} = V_{bi} - V$, but even at DC this is generally not the case [32]. We are interested in small-signal AC variations, prefixed by δ, in a bias and frequency regime not dominated by DC conduction across the semiconductor barrier (i.e. g_{gs} and g_{gd} in Figure 15.16(a) are negligible). There will be a deviation from ideality due to the voltage division between the semiconductor and the interfacial layer. The AC driving force, δV, results in similarly denoted AC variations in the 'primary' variables E_{Fi}, V_i and V_{SC}. Variations in Q_{SC}, Q_M, Q_S and J can be expressed in terms of these. The other parameters in Figure 16.3 are fixed, independent of V.

The voltage division is expressed by

$$\delta V = -\delta V_i - \delta V_{SC}. \tag{16.11}$$

Charge conservation requires that

$$\delta Q_{SC} + \delta Q_M + \delta Q_S = 0, \tag{16.12}$$

where

$$\delta Q_{SC} = c_D(V_{SC})\delta V_{SC}, \tag{16.13}$$

$$\delta Q_M = -c_i \delta V_i \tag{16.14}$$

and

$$\delta Q_S = -c_S(\delta V_i + \delta E_{Fi}/q). \tag{16.15}$$

$c_D(V_{SC})$ is the doping- and structure-dependent semiconductor depletion capacitance per unit area. The other two normalized capacitances are

$$c_i = \frac{\varepsilon_i}{d_i} \tag{16.16}$$

for the interfacial layer, and

$$c_S = q^2 D_S \tag{16.17}$$

for the surface states, where D_S is the density of surface states at the bias position of E_{FS}. Cowley and Sze assumed a constant D_S. Some defect models involve sharp peaking of $D_S(E)$ at the pinning position [20]. The MIGS model suggests a rather uniform distribution with an increase in $D_S(E)$ near the valence and conduction-band edges. In the DIGS model the Fermi level is pinned near a pronounced minimum in $D_S(E)$. These details do not affect the small-signal AC analysis. For small δE_{Fi}, the tunneling current density δJ_t can be expressed in terms of a linear resistance r_{it}, the interfacial tunneling resistance. δJ_t is a real (in-phase) conductive current in the interfacial layer and the metal; i.e. unlike the parallel displacement current, it does not contribute to the metal charge Q_M. Like the displacement current, however, δJ_t is an AC current, since in the high-frequency operating regime of interest here (g_{gs}, g_{gd} negligible) there is negligible DC charging (or discharging) of the surface states. δJ_t is proportional to the small AC 'unpinning' of the Fermi level, described by δE_{Fi}:

$$\delta J_t = \frac{d\delta Q_S}{dt} = j\omega\delta Q_S = \frac{\delta E_{Fi}/q}{r_{it}}. \tag{16.18}$$

The preceding equations lead to solutions, first for the voltage division:

$$\frac{\delta V_{SC}}{\delta V} = -\frac{1}{1+\dfrac{c_D}{c_i+\dfrac{c_S}{1+j\omega c_S r_{it}}}}, \tag{16.19}$$

and ultimately for the admittance per unit area:

$$y(\omega) = \frac{\delta J}{\delta V} = j\omega c_D\left(\frac{-\delta V_{SC}}{\delta V}\right) = \frac{j\omega c_D}{1+\dfrac{c_D}{c_i+\dfrac{c_S}{1+j\omega c_S r_{it}}}}. \tag{16.20}$$

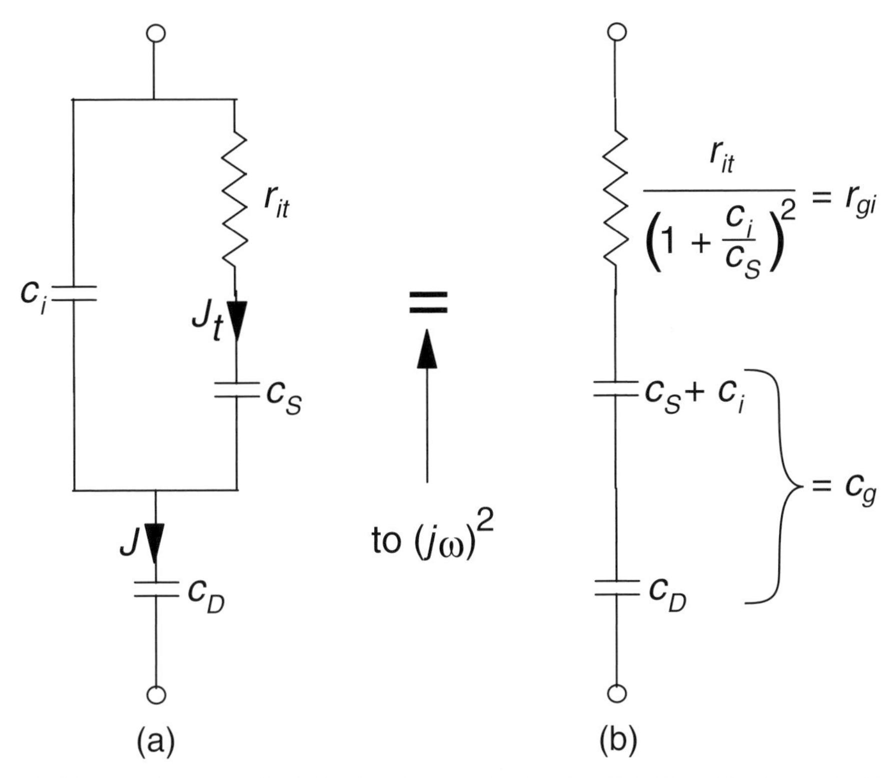

Fig. 16.4. (a) Equivalent circuit for the admittance in Equation (16.20), and (b) the standard FET equivalent circuit, corresponding to Equation (16.21). The two are equivalent to second order in $j\omega$. (H. Rohdin, N. Moll, A.M. Bratkovsky and C.-Y. Su, *Physical Review B*, Vol. 59, pp. 13 102–13 113, 1999.)

For the total AC current density δJ in Equation (16.20), we used the expression $j\omega c_D(-\delta V_{SC})$, which is valid in the depletion region. The expression $\delta J_t + j\omega c_i(-\delta V_i)$, valid in the metal and interfacial layer, leads to the same result, as it should. The denominator in Equation (16.19) is the AC ideality factor associated with a non-ideal ($c_i < \infty$) interface. Not surprisingly, the result in Equation (16.20) is very similar to that of Terman's [34] MOS admittance analysis. Terman's 'energy loss mechanism', associated with charging and discharging the surface states through the semiconductor, becomes in the present case the loss due to tunneling through the interfacial layer.

The equivalent circuit for the admittance in Equation (16.20) is shown in Figure 16.4(a). We also show the circuit in Figure 16.4(b) because it corresponds to the zero-drain-bias SBGFET equivalent circuit (without the source and drain access resistances) used to extract the gate resistance (Equation (16.8)). It has the simple one-pole admittance

$$y(\omega) = \frac{j\omega c_g}{1 + j\omega r_{gi} c_g}. \tag{16.21}$$

With

$$r_{gi} = \frac{r_{it}}{\left(1 + \frac{c_i}{c_S}\right)^2} \tag{16.22}$$

and

$$c_g = \frac{c_D}{1 + \left(\frac{c_D}{c_i + c_S}\right)} \tag{16.23}$$

the circuit in Figure 16.4(b) has the same admittance as that in Figure 16.4(a), to second order in $j\omega$ (Problem 16.1). The phenomenological, experimentally inferred, interfacial gate resistance r_{gi} has thereby been identified in terms of the underlying physical tunneling resistance r_{it}, the interfacial layer capacitance c_i and capacitance c_s associated with the surface states. The normalized SBGFET gate capacitance c_g is identified as c_d and $c_i + c_s$ in series. The denominator in Equation (16.23) is the DC limit of the AC ideality factor discussed above. The value is close to unity for a physically ideal Schottky barrier. For GaAs one can demonstrate this by assuming, as suggested by Cowley and Sze [12], interfacial-layer parameters $d_i = 5$ Å and $\epsilon_i = \epsilon_0$, and using their expression for γ (Equation (16.10)). With $\gamma = 0.074$, one calculates $D_S = 1.38 \times 10^{14}$ cm^{-2}/eV. This results in $c_s = 22$ μF/cm^2 and $c_i = 2.0$ μF/cm^2 *. With a typical depletion capacitance $c_D = 0.45$ μF/cm^2, corresponding to a 25-nm gate-channel spacing, the DC limit of the ideality factor is 1.02. With the same parameters, Figure 16.5 shows the frequency dependence of y (Equation (16.20)), in terms of the equivalent series capacitance $-1/[\omega\,\mathrm{Im}(1/y)]$, and the r_{gi} estimate $\mathrm{Re}(1/y)$. $r_{it} = 5.18 \times 10^{-7}$ Ω cm^2 is the result of the theory in Section 16.6, with $d_i = 5$ Å. It leads to $r_{gi} = 4.37 \times 10^{-7}$ Ω cm^2, which is remarkably close to the experimentally observed $r_{gi}^{(min)}$. Up to about 50 GHz the equivalent series resistance and capacitance equal those given by Equations (16.22) and (16.23). At frequencies > 100 GHz significant deviations develop. For the lower end of Cowley–Sze's interfacial-layer thickness range (4 Å) the dispersion is much lower.

16.6 Theory for the interfacial tunneling resistance

The results in the last section indicate that the minimum interfacial gate resistance is consistent with metal-to-surface-state tunneling, and does not go away at typical frequencies of interest. The next step is to calculate the basic underlying tunneling resistance r_{it}. With appropriate approximations, this becomes an analytically manageable example of applied quantum mechanics, with interesting ties to both theoretical

* It is worth noting that this value for c_i is quite similar to values extracted on metal–Si diodes by fitting interfacial-layer theory to experimental DC I–V curves [31].

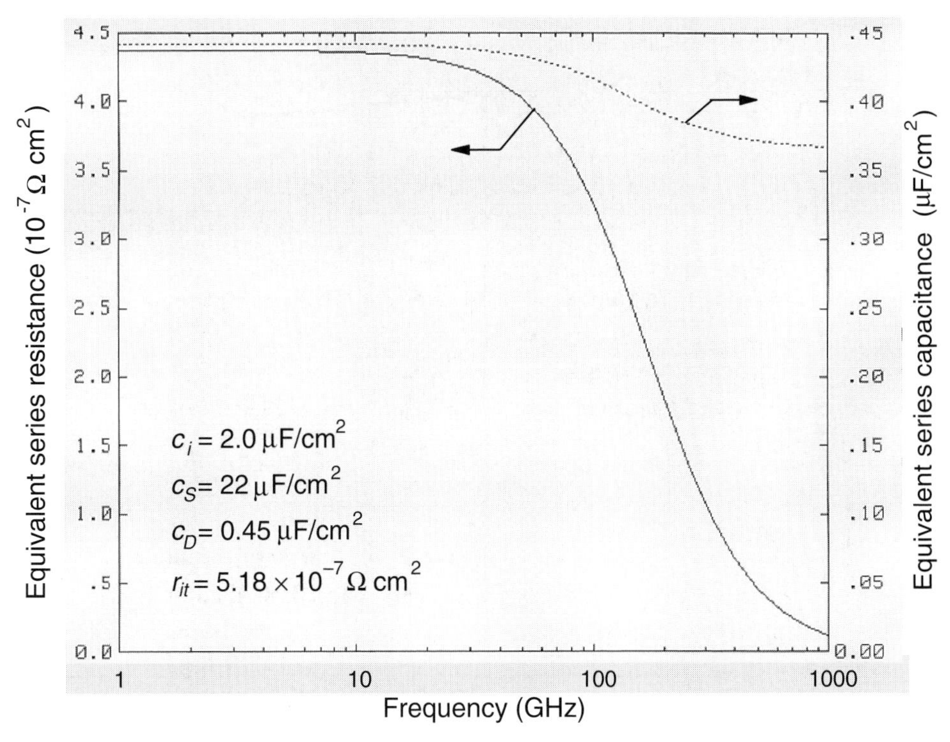

Fig. 16.5. Frequency dependence of the equivalent series RC circuit elements for the admittance in Equation (16.20), with a typical value for the interfacial gate resistance ($r_{gi} \sim r_{gi}^{(min)}$).

and experimental solid-state physics, and with important consequences for technology and design. For clarity, we divide the analysis into subsections, starting with general tunneling theory. Ultimately (Section 16.7), we can make quantitative predictions in the context of the various Schottky-barrier pictures discussed in Section 16.4.

16.6.1 General formalism for tunneling between metal and surface states

The tunneling problem to be solved here is illustrated in Figure 16.6. An expression for tunneling current can be derived from Fermi's golden rule, i.e. the expression for the probability per unit time of transition from an initial state i to a final state state f (e.g. [35, 36]):

$$P_{if} = \frac{2\pi}{\hbar} \left|M_{if}\right|^2 \rho_f f_i \left(1 - f_f\right). \tag{16.24}$$

M_{if} is the matrix element for the transition, ρ_f is the density of final states, and f_i and f_f are the probabilities of occupation of state i and f, respectively. We apply this to the metal–semiconductor system ($i \to S$, $f \to M$) of interest. Energy E (referenced to the bottom of the metal conduction band), transverse momentum $\mathbf{k}_{xy}$ and spin are assumed to be conserved. The densities of states are thus calculated for fixed $\mathbf{k}_{xy}$

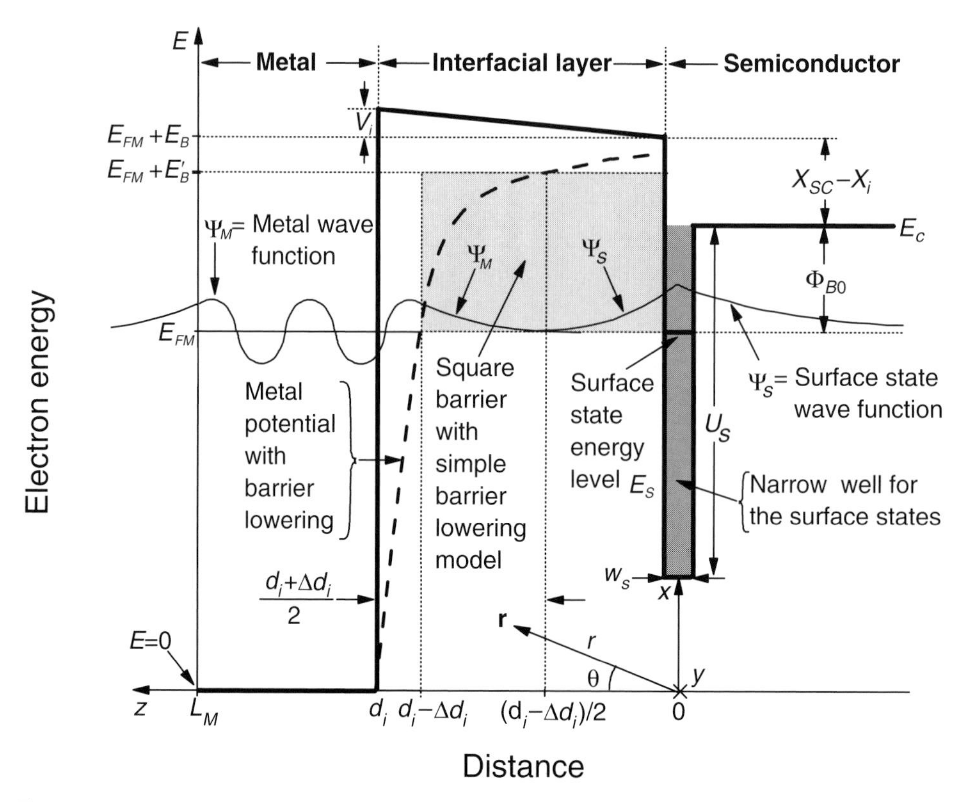

Fig. 16.6. Detailed energy-band diagram for the metal-to-surface-state tunneling problem. (H. Rohdin, N. Moll, A.M. Bratkovsky and C.-Y. Su, *Physical Review B*, Vol. 59, pp. 13 102–13 113, 1999.)

and E, spin not included. The occupation probabilities for electrons are given by the Fermi–Dirac distribution function, denoted by $f(E)$. As is often done in tunneling analyses, we will in the end assume zero temperature so that $f(E)$ becomes an easily manageable step function (going from 1 to 0 at the Fermi energy). For the tunneling current, one integrates $\rho_S P_{SM} - \rho_M P_{MS}$ over E, sums over all $\mathbf{k}_{xy}$, and multiplies by $2q$ for electrons with two possible initial spins [36]. The sum over $\mathbf{k}_{xy}$ becomes an integral:

$$\sum_{k_{xy}} = \frac{A_{xy}}{(2\pi)^2} \int d^2k_{xy}, \tag{16.25}$$

where A_{xy} is the cross-sectional area in the xy-plane, and $A_{xy}/(2\pi)^2$ is the density of states in two-dimensional k space. With the Fermi level raised by E_{Fi} on the semiconductor side by the applied bias V (Figure 16.3), the tunneling current density becomes:

$$J_t = \frac{q}{\pi\hbar}\int d^2k_{xy}\int dE\,[f(E-E_{Fi})-f(E)]\,\rho_S(E-E_{Fi})$$
$$\times\,\rho_M\left(E,k_{xy}\right)\left|M_{SM}\left(E,k_{xy}\right)\right|^2. \tag{16.26}$$

With cosine wave-function solutions (Figure 16.6) and associated (non-periodic) boundary conditions, the one-dimensional density-of-states in the metal is given by

$$\rho_M\left(E,k_{xy}\right) = \frac{L_M}{\pi}\left(\frac{\partial E}{\partial k_z}\right)^{-1} = \frac{L_M m_M}{\pi\hbar^2 k_z\left(E,k_{xy}\right)}, \tag{16.27}$$

where L_M is the thickness of the metal, k_z is the component along the z axis of the wave-vector, and m_M is the electron effective mass in the metal. Assuming isotropic effective mass, k_z is given by

$$k_z^2\left(E,k_{xy}\right) = \frac{2m_M E}{\hbar^2} - k_{xy}^2. \tag{16.28}$$

The density of surface states is

$$\rho_S(E) = \sum_{s'}\delta\left(E-E_{S'}\right), \tag{16.29}$$

where the $E_{S'}$'s are the energies of the surface states. We will solve for the tunneling current from a single localized surface state (energy E_S) to the metal, and then perform the sum, which (similar to the sum in k_{xy} space above) can be expressed as an integral:

$$\sum_{S'} = A_{xy}\int dE_{S'}\,D_S\left(E_{S'}\right). \tag{16.30}$$

The integral should be taken over the E_{Fi} gap in Figure 16.3. For small excursions from equilibrium, the sum becomes the factor $A_{xy}D_S E_{Fi}$.

The matrix element for electrons tunneling from the semiconductor surface state to the metal is calculated by Bardeen's method [35]:

$$M_{SM}\left(E,k_{xy}\right) = -\frac{\hbar^2}{2m_i}\int_{S_i} d\mathbf{S}\cdot\left(\Psi_S^*\nabla\Psi_M - \Psi_M\nabla\Psi_S^*\right). \tag{16.31}$$

m_i is the electron effective mass in the interfacial tunneling barrier. Ψ_M and Ψ_S are solutions with the same energy to two simplified problems: (1) metal (M) + semi-infinite barrier (B), and (2) semi-infinite barrier (B) + surface state (S) + semi-infinite semiconductor (SC), respectively. The perturbation to this situation occurs by thinning the barrier B to an interfacial layer i so that the two exponential tails start overlapping. The parts of the tails that extend into the other region are assumed to remain exponential. S_i is a simple, but arbitrarily-shaped, surface that completely separates the surface state from the metal, and lies in the interfacial-layer barrier between the metal and semiconductor. There is a similarity between Equation (16.31) and the standard expression for probability current in quantum mechanics. The original full proof of Equation (16.31) was given by Bardeen [35], from the many-particle point of view, for the case of constant band structure. Harrison [36] built on this, took an independent–particle point of view, and allowed a spatially-varying band structure. We offer a simplified derivation, which is similar to one given by Duke [37], but tailored to our particular problem. The matrix element is given by

$$\begin{aligned} M_{SM} &= \int_{M+i+S+SC} dV\, \Psi_S^* (H - E)\Psi_M \\ &= \int_{SC+S+\delta i} dV \left[\Psi_S^* (H - E)\, \Psi_M - \Psi_M\, (H - E)\, \Psi_S^*\right]. \end{aligned} \tag{16.32}$$

The first volume integral, taken over the entire space of the problem, is the well-known result from perturbation theory in quantum mechanics. The operator is the difference between the total (H) and unperturbed ($H^{(0)}$) Hamiltonian. We have used $H^{(0)}\Psi_M = E\Psi_M$, where E is the energy of our two states. It is assumed that Ψ_M is a solution to the Schrödinger equation in the metal and in the barrier, but not in the semiconductor since it is exponential (as opposed to oscillatory) there [35]. The converse is assumed for Ψ_S. This means that we only have to integrate over the semiconductor and the attractive core of the surface state. But we can also extend the volume of integration to any fraction δi of the barrier region between the surface state and the metal, where both Ψ_M and Ψ_S are assumed to be good solutions. This is what we do in the second step in Equation (16.32). However, in this volume, since Ψ_S is a good solution there, we can subtract $\Psi_M(H - E)\Psi_S*$ ($= 0$) from the original integrand. The two terms involving E simply cancel, and we are left with the two Hamiltonian terms. For a position-dependent effective mass, the Hamiltonian is [38]

$$H = -\nabla \cdot \frac{\hbar^2}{2m(\mathbf{r})}\nabla + V(\mathbf{r}), \tag{16.33}$$

where the mass can no longer be brought entirely outside the kinetic energy operator. Although we ultimately assume constant masses for the metal and the surface state, we have to have an $\mathbf{r}$ dependence in the effective mass in Equation (16.33) since the integration volume includes part of the tunneling barrier where the effective mass can be different. Inserting the generalized Hamiltonian into Equation (16.32), the two terms with $V(\mathbf{r})$ as a factor will cancel, and we are left with

$$\begin{aligned} M_{SM} &= -\frac{\hbar^2}{2}\int_{SC+S+\delta i} dV \left[\Psi_S^* \nabla \cdot \left(\frac{1}{m}\nabla\Psi_M\right) - \Psi_M \nabla \cdot \left(\frac{1}{m}\nabla\Psi_S^*\right)\right] \\ &= -\frac{\hbar^2}{2}\int_{S_i} d\mathbf{S} \cdot \left[\Psi_S^* \left(\frac{1}{m}\nabla\Psi_M\right) - \Psi_M \left(\frac{1}{m}\nabla\Psi_S^*\right)\right]. \end{aligned} \tag{16.34}$$

The last step makes use of the alternative form of Green's theorem in Equation (15.34). The integration surface S_i should lie outside the attractive core of the surface state. We can make it consist of two infinitely large parts, one with a disk-like shape in the interfacial layer between the semiconductor and the metal, the other a hemispherical shell in the semiconductor, infinitely removed from the surface state. With the exponential drops in the wave-functions, the latter contributes nothing. Since the contributing integration surface is in the interfacial barrier, where we assign a constant effective mass m_i, Equation (16.31) follows.

16.6.2 Interfacial tunneling barrier

Before we can determine the two wave-functions in Equation (16.31), we must define the barrier. Considering that we are interested in very thin layers with only approximately known characteristics, details in the barrier shape are neglected, as are changes to the barrier induced by the AC voltage. We express the tunneling barrier in terms of quantities on the semiconductor side in Figure 16.6:

$$E_B = \Phi_{B0} + X_{SC} - X_i. \tag{16.35}$$

Using (with our notation) Blakemore's [39] asymptotically correct expression

$$\frac{(E_{FM} + E_B)^2}{E_{FM} + E_B + \dfrac{q^2}{16\pi\varepsilon_i^{(\infty)}(d_i - z)}} \tag{16.36}$$

for the barrier lowered by the image force (dashed curve in Figure 16.6), the range of the metal potential is extended beyond the background ionic core by

$$\Delta d_i = \frac{q^2}{16\pi\varepsilon_i^{(\infty)} E_B} \frac{E_{FM}}{E_{FM} + E_B}, \tag{16.37}$$

at $E = E_{FM}$. In the simple free-electron-gas model, E_{FM} is expressed as

$$E_{FM} = \frac{\hbar^2 k_{FM}^2}{2m_M}, \tag{16.38}$$

where the Fermi sphere radius is given by:

$$k_{FM} = \left(3\pi^2 n_M\right)^{1/3} \approx 1.2\ \text{Å}^{-1} \tag{16.39}$$

[40]. n_M is the concentration of metal carriers, assumed to be equal to the atomic concentration ($\sim 6 \times 10^{22}$ cm^{-3}). $\epsilon_i^{(\infty)}$ is the high-frequency dielectric constant appropriate for tunneling [41]. The tunneling distance for electrons at the Fermi level will be reduced from the geometrical interfacial-layer thickness d_i to

$$d_i' = d_i - \Delta d_i. \tag{16.40}$$

A representative choice for m_M appears to be one half of the free electron mass m_0 [42]. The corresponding Δd_i is 0.5 Å for a metal–vacuum interface, and 0.3 Å for a metal–GaAs interface. Additional effects that occur for real atomic metals are either too small, or too complex [43], to include in a simple barrier model. The assumption of a constant barrier (and barrier lowering) is reasonable [44], particularly considering uncertainties in parameter values. It allows direct integration of Equation (16.26).

To estimate the lowered barrier

$$E'_B = E_B - \Delta E_B, \tag{16.41}$$

we again apply Equation (16.36), and use, as a uniformly lowered barrier E'_B, the value calculated at the midpoint $z = (d_i - \Delta d_i)/2$. Thus, the barrier lowering is estimated as

$$\Delta E_B = E_{FM} + E_B - \frac{(E_{FM} + E_B)^2}{E_{FM} + E_B + \dfrac{q^2}{8\pi\varepsilon_i^{(\infty)}(d_i + \Delta d_i)}}. \tag{16.42}$$

16.6.3 Metal wave-function tail and tunneling effective mass

The wave-function and the associated probability current density should be conserved across boundaries [36]. With different effective masses, we determine the metal wave-function in the tunneling region by matching Ψ_M and $m^{-1}\partial\Psi_M/\partial z$ at the effective metal-barrier interface $z = d'_i = d_i - \Delta d_i$ (Figure 16.6). The metal wave-function tail in the tunneling layer is then given by (Problem 16.2):

$$\Psi_M = \left(\frac{2}{A_{xy}L_M}\right)^{1/2} \frac{m_i k_z}{\left(m_M^2\eta_i^2 + m_i^2k_z^2\right)^{1/2}} \exp\left(-\eta_i(d'_i - z)\right) \exp\left(i\mathbf{k}_{xy}\cdot\mathbf{r}_{xy}\right). \tag{16.43}$$

$\mathbf{k}_{xy}$ and k_z are related by Equation (16.28), with $E = E_{FM}$ (Equation (16.38)), i.e.,

$$k_{xy}^2 + k_z^2 = k_{FM}^2. \tag{16.44}$$

η_i is the decay factor in the tunneling barrier:

$$\eta_i(k_{xy}) = \frac{1}{\hbar}\left(2m_iE'_B + \hbar^2k_{xy}^2\right)^{1/2}. \tag{16.45}$$

$\mathbf{r}_{xy}$ is the transverse component of $\mathbf{r}$ in Figure 16.6, i.e. the projection of $\mathbf{r}$ onto the xy-plane. If the tunneling barrier is a Cowley–Sze interfacial layer with vacuum properties, the choice for m_i is obvious ($= m_0$). If it is the semiconductor itself, or an oxide, the choice is less obvious. The effective masses for conduction-band electrons (m_e) and valence-band holes (m_h) may be known, but we are dealing with tunneling in the forbidden gap. A simple continuation of the valence and conduction bands into the forbidden bandgap is illustrated in Figure 16.7. The two bands are joined together smoothly by simply adding k^{-2} for the two cases. The decay factor for a state at energy

E_S can then be estimated*. If we identify $E_C - E_S$ as the tunneling barrier (to which we apply the lowering discussed in the previous section) the tunneling effective mass is

$$m_i = \frac{m_{ei} m_{hi} \left(E_S - E_{Vi}\right)}{m_{hi} \left(E_S - E_{Vi}\right) + m_{ei} \left(E_{Ci} - E_S\right)}, \qquad (16.46)$$

where we have attached a subscript i for interfacial layer. We used the partitioning of $k^2(E_S)$ into the transverse propagating component k_{xy}^2 and the decaying component $-\eta_i^2$. As a more fundamental alternative to the phenomenological continuation of the bands in Figure 16.7, we can use a result of a $\mathbf{k} \cdot \mathbf{p}$ theory [45]. For a 'small gap' isotropic semiconductor Kane derived the following cubic secular equation

$$(E - E_C)(E - E_V)(E - E_V + \Delta) - \frac{\hbar^2 k^2}{2m_0} E_P \left(E - E_V + \frac{2}{3}\Delta\right) = 0 \qquad (16.47)$$

for calculating $E(k)$ of the conduction band ($E = E_C$ at Γ), light-hole band ($E = E_V$ at Γ) and split-off band ($E = E_V - \Delta$ at Γ). Δ is the valence-band spin-orbit split at Γ (0.34 eV for GaAs), and E_P is an interaction energy ($\approx$ 23 eV for III–V semiconductors). Higher order k terms have been neglected. Rather than calculating $E(k)$, we are interested in $k^2(E_S) = k_{xy}^2 - \eta_i^2$. Inserting this into Equation (16.47), a comparison with Equation (16.45) (again with $E_B = E_C - E_S$ as the tunneling barrier) yields

$$m_i = m_0 \frac{(E_S - E_{Vi})(E_S - E_{Vi} + \Delta)}{E_P (E_S - E_{Vi} + 2\Delta/3)} \qquad (16.48)$$

for the tunneling effective mass in the interfacial layer. For GaAs, the $k^2(E)$ relation resulting from the simple continuation in Figure 16.7 is essentially the same as that given by the 'small gap' $\mathbf{k} \cdot \mathbf{p}$ result, if m_{hi} in Equation (16.46) is set to the light-hole mass. For a GaAs interfacial-layer tunneling barrier, with $E_B = E_C - E_S = 0.85$ eV, the tunneling effective mass is $0.031 m_0$. This is a very small value, about one half of the conduction-band effective mass ($m_e = 0.067 m_0$ [46]). It corresponds to a 6.3 Å characteristic distance $1/2\eta_i(0)$ for the radial exponential drop in electronic charge of the individual state. This value compares well with the 2.8-Å one-dimensional decay length of MIGS (d_{MIGS}) [13]. The smaller value for the latter is expected since it results from averaging the electronic charge parallel to the interface.

In case of an oxide interfacial layer, we still attempt to apply Equation (16.48), although the term 'small-gap' would not seem to apply in this case. For a typical oxide barrier ($E_{Ci} - E_S = 4$ eV, $E_S - E_{Vi} = 5$ eV, $E_g = 9$ eV, $X_i = 1$ eV), assuming that the interaction energy E_P is the same as for III–V semiconductors (23 eV), and that the spin-orbit splitting Δ is negligible, one calculates $m_i = 0.22 m_0$. This is quite close to the $0.29 m_0$ used by Stratton [41] for Al_2O_3.

* Point B in Figure 16.7, which has the shortest decay length, is the branch point in the complex band structure thought to be the 'canonical' pinning position in the MIGS model.

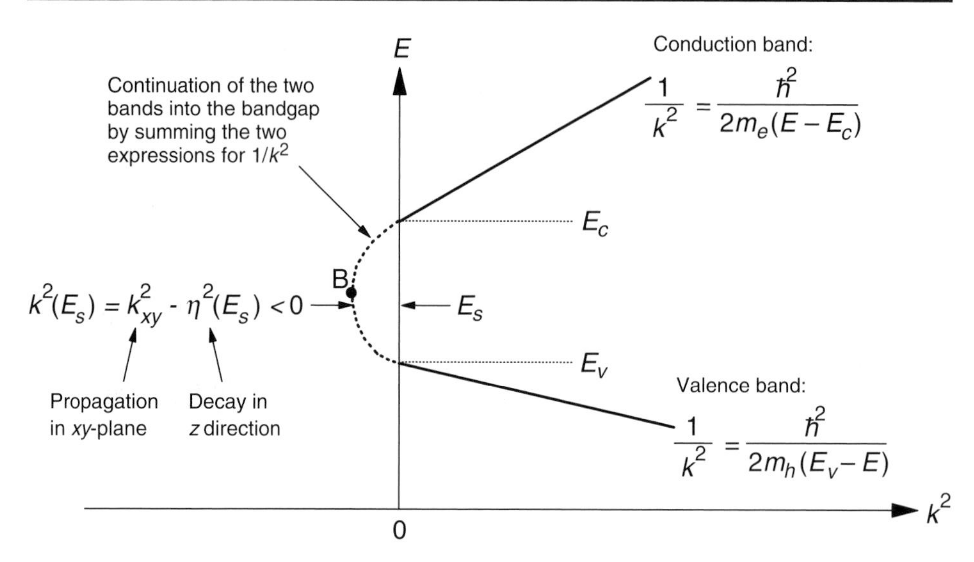

Fig. 16.7. Continuation of the conduction and valence bands of an oxide or semiconductor tunneling barrier into its forbidden bandgap. This can be used for estimating the decay constant of the metal tail in the z direction. B is the point in the complex band structure with the fastest decay.

16.6.4 Surface-state wave-function

The surface states are assumed to be localized. We will take the integration surface S_i in Equation (16.31) to be a hemispherical shell, with radius a_S, around the surface state, outside the attractive core, plus the $z = 0+$ plane outside the radius a_S. We refer to Figure 16.6, but initially consider the mid-gap state of energy E_S to be a semiconductor bulk state. For this, we assume a spherically symmetric square-well potential [47] of short-range $w_s/2$. This approximates the potential felt by an electron from an atomic core, screened by deeper-lying electrons. The potential is E_C for $r > w_s/2$. The ground state will be a spherically symmetric s-type state. We let a_S approach zero in order to integrate Equation (16.31) analytically. This, in effect, requires that we assume a three-dimensional delta-function potential, which has no bound excited states. In this limit, the wave-function has the simple form

$$\Psi_S = \left(\frac{\eta_{si}}{2\pi}\right)^{1/2} \frac{\exp(-\eta r)}{r}, \tag{16.49}$$

where η is a decay constant, and η_{si} is determined by normalization. For the spherically symmetric delta potential, $\eta_{si} = \eta$. For a bulk semiconductor defect state $\eta = \eta_s$, where

$$\eta_s = \frac{1}{\hbar}\left[2m_S\left(E_C - E_S\right)\right]^{1/2} \tag{16.50}$$

The effective mass m_S in this equation is given by Equation (16.48) (or Equation (16.46)), without the i subscript. As good as the predictions of the decay constant and effective mass seem to be with the $\mathbf{k} \cdot \mathbf{p}$-based approach, it should be noted that it is only approximate when dealing with deep localized levels. This is because wave-functions associated with such levels can have components with **k**s significantly larger than the $\mathbf{k} \cdot \mathbf{p}$ theory can account for.

When the state is a surface state, it is no longer spherically symmetrical, since it is adjacent to a potential barrier $> E_C - E_S$. The wave-function then decays more slowly into the semiconductor than it does into the tunneling barrier. We cannot solve this quantum mechanical problem analytically, and we instead account for the spherical asymmetry approximately. We simply assume that the surface state decays with $\eta = \eta_s$ into the semiconductor ($\pi/2 < \theta \leq \pi$), and with $\eta = \eta_i(0) \equiv \eta_{i0}$ into the barrier ($0 \leq \theta < \pi/2$)*. The resulting normalization constant in Equation (16.49) will be given by

$$\eta_{si} = \frac{2\eta_s \eta_{i0}}{\eta_s + \eta_{i0}} \tag{16.51}$$

16.6.5 Tunneling resistance and capture cross-section

With the metal and surface-state wave-functions in Subsections 16.6.3 and 16.6.4 the tunneling matrix element M_{SM} (Equation (16.31)) becomes:

$$M_{SM} = -\hbar^2 \left(\frac{4\pi \eta_{si}}{A_{xy} L_M} \right)^{1/2} \frac{k_{FM}}{\left[(m_M \eta_{i0})^2 + (m_i k_{FM})^2 \right]^{1/2}} \exp(-d_i' \eta_i(k_{xy})), \tag{16.52}$$

where, by evaluating the pre-factor at $\mathbf{k}_{xy} = \mathbf{0}$, we have made use of the fact that the strongest dependence on $\mathbf{k}_{xy}$ occurs in the exponential factor. This approximation allows us to perform the integral in Equation (16.26) analytically. We get for the tunnel current density

$$J_t = \frac{2q\hbar D_S E_{Fi}}{m_i d_i'^2} \frac{(m_M \eta_{si})(m_i k_{FM})}{(m_M \eta_{i0})^2 + (m_i k_{FM})^2} \left(1 + 2d_i' \eta_{i0}\right) \exp\left(-2d_i' \eta_{i0}\right). \tag{16.53}$$

The expression for the gate tunneling resistance (Equation (16.18)) becomes

$$r_{it} = \frac{\delta E_{Fi}/q}{\delta J_t} = \frac{m_i d_i'^2}{2q^2 \hbar D_S} \frac{(m_M \eta_{i0})^2 + (m_i k_{FM})^2}{(m_M \eta_{si})(m_i k_{FM})} \frac{\exp\left(2d_i' \eta_{i0}\right)}{1 + 2d_i' \eta_{i0}}. \tag{16.54}$$

The accuracy of this solution is limited to a large degree by how well the actual surface-state potential is approximated by the (θ-dependent) delta-function. A more realistic potential with a non-zero range $w_s/2$ (Figure 16.6) would lead to a smaller η_{si} than predicted by Equation (16.51). This would reduce the wave-function tail outside the attractive core of the defect potential, where the integral in Equation (16.31) is taken. This would increase the tunneling resistance.

Freeman and Dahlke [29] developed a similar theory for tunneling through the insulator between metal and surface states in an MOS structure. Theirs differs

* This approximation leads to minor differences compared to the more gradual θ dependence assumed in [42].

from the present analysis mainly in the quantum mechanical representation of the surface state. Freeman and Dahlke treated the surface states as a 2DEG confined in the interface plane by a one-dimensional delta potential. The effects of lateral localization were accounted for by a tunneling capture cross-section σ_T introduced phenomenologically, independent of the tunneling problem. In applying the theory to problems of interest, a value for σ_T has to be chosen. In the theory developed in this section, the wave-function parameters, which are the result of the microscopic model for the surface state, lead directly to an analytical solution for the tunneling current. There is no need for an independent cross-section. Nevertheless, capture cross-section is a very useful quantity. Together with the energy level, it describes many of the experimentally observed properties of a localized state. There is a large number of published experimental cross-sections for bulk traps (e.g. [48]). By requiring that the two theories give the same result, we have an opportunity to compare a simple theoretical estimate of the tunneling capture cross-section with experimental bulk values. This can give us a feeling for the accuracy to be expected from the model. With some minor modifications to Freeman and Dahlke's model, one can derive an alternative expression for r_{it} [42] which differs from Equation (16.54) only in an additional factor $2\pi/(\sigma_T \eta_{i0}^2)$. The two expressions for r_{it}, derived under otherwise identical assumptions, allow us then to identify

$$\sigma_T = \frac{2\pi}{\eta_{i0}^2}. \tag{16.55}$$

This is the effective tunneling cross-section that corresponds to the model for the surface state in Section 16.6.4. Equation (16.55) expresses the interesting point that the tunneling cross-section depends on the barrier. For non-spherical ($\eta_{i0} > \eta_s$) tunneling cases, the cross-section can be significantly smaller than the cross-section $\sigma_T^{(B)}$ for the spherically symmetric bulk case ($\eta_{i0} = \eta_s$). With GaAs parameters, for a bulk trap located at 0.85 eV ($\approx \Phi_{B0}$) below the conduction band we get $\sigma_T^{(B)} = 1.0 \times 10^{-13}$ cm^2. The experimental range for such traps is from about 10^{-14} to 10^{-11} cm^2 [48]. Neglecting experimental errors, the wide range is due to the different, and basically unknown, bulk impurity potentials. If one assumes that the uncertainty in surface-state potential is equally large, one conservatively concludes that the predictions of r_{it} in the next section could have error bars of ± 2 orders of magnitude.

16.7 Application to various Schottky-barrier models

We now apply the theory developed in Sections 16.5 and 16.6 by adapting the parameters to represent alternative pictures of Schottky-barrier formation. The most critical tunneling parameters describe the interfacial layer and its interface with the semiconductor. These parameters are varied, as shown in Table 16.1, and explained in

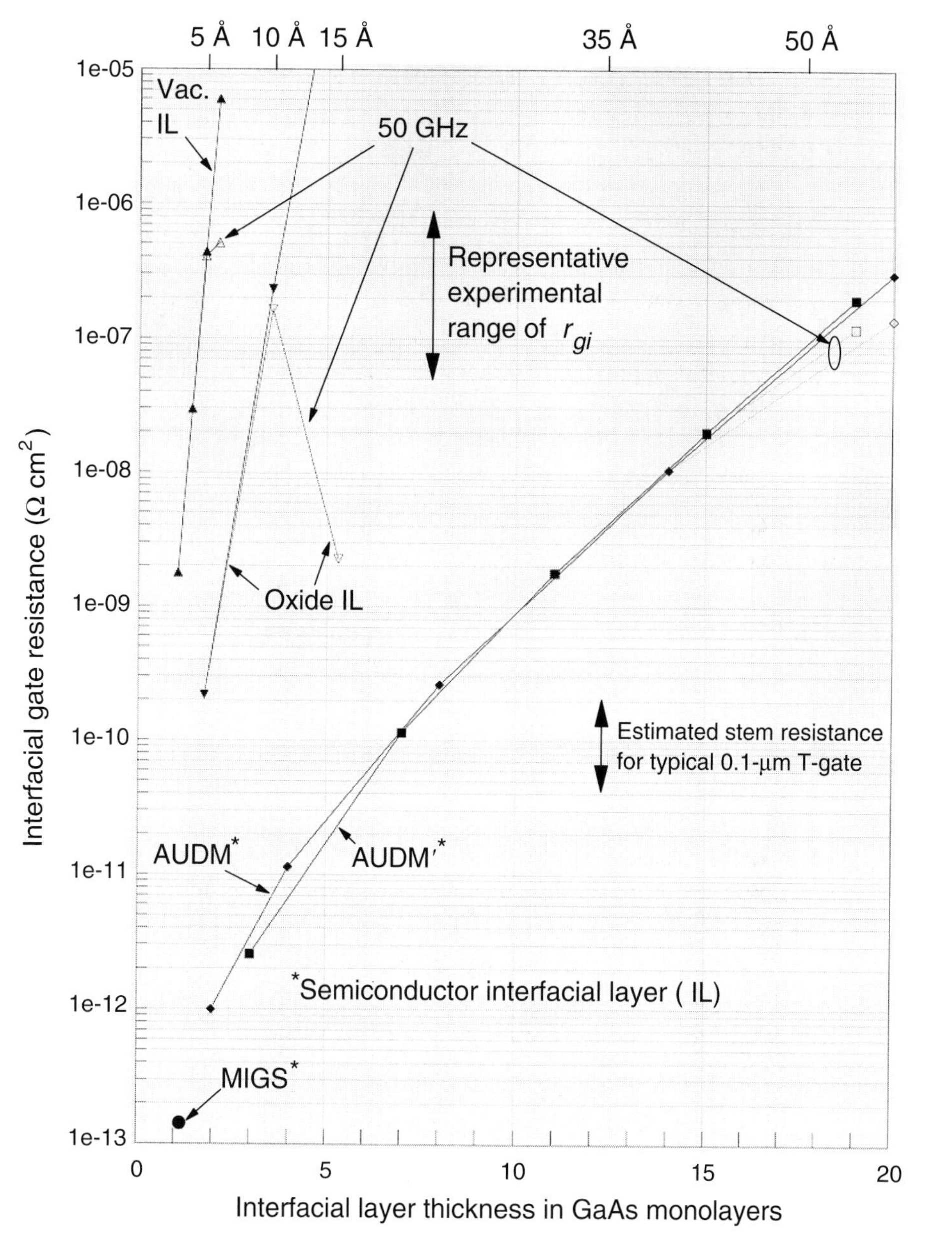

Fig. 16.8. Calculated interfacial gate resistance at DC (solid symbols) and 50 GHz (open symbols) versus interfacial-layer (IL) thickness for the five cases defined in Table 16.1. (H. Rohdin, N. Moll, A.M. Bratkovsky and C.-Y. Su, *Physical Review B*, Vol. 59, pp. 13 102–13 113, 1999.)

footnotes. The remaining parameters, those describing the metal and semiconductor, are fixed as shown in the first footnote. The resulting interfacial gate resistance at DC and at 50 GHz is shown in Figure 16.8.

Cowley–Sze's 4–5 Å vacuum picture leads to the observed values for the minimum

Table 16.1. *Interface parameters for the tunneling calculation, corresponding to four different models for Schottky-barrier formation on GaAs*[a]. *(H. Rohdin, N. Moll, A.M. Bratkovsky and C.-Y. Su, Physical Review B, Vol. 59, pp. 13 102–13 113, 1999.)*

Model	Vacuum IL[c]	Oxide IL[d]	MIGS[e]	AUDM[f]	AUDM′[g]
X_i (eV)	0	0.95	4.12	4.07	4.07
m_i/m_0	1	0.29	0.031	0.031	0.031
ϵ_i/ϵ_0	1	9	13.1	13.1	13.1
$\epsilon_i^{(\infty)}/\epsilon_0$ [b]	1	2.99	13.3	13.3	13.3
d_i	3–6 Å	5–15 Å	$d_{TF} + d_{MIGS} = 3.3$ Å	$(1–20)\times d_{ML}$	$(3–19)\times d_{ML}$
$\gamma = \frac{\partial \Phi_{B0}}{\partial \Phi_M}$	0.074	0.074	$\frac{1}{1 + q^2 D_S \left(\frac{d_{TF}}{\varepsilon_0} + \frac{d_{MIGS}}{\varepsilon_{SC}} \right)} = 0.13$		0.074
D_S	$\frac{\varepsilon_i}{q^2 d_i}\left(\frac{1}{\gamma} - 1\right)$	$\frac{\varepsilon_i}{q^2 d_i}\left(\frac{1}{\gamma} - 1\right)$	5×10^{14} cm^{-2}/eV	2.5×10^{14} cm^{-2}/eV	$\frac{\varepsilon_i}{q^2 d_i}\left[\cosh^{-1}\left(\frac{1}{\gamma}\right)\right]^2$

[a] Fixed parameters: $m_M = 0.5 m_0$, $n_M = 6 \times 10^{22}$ cm^{-3}; $E_g = 1.424$ eV, $X_{SC} = 4.07$ eV, $\Phi_{B0} = 0.85$ eV, $m_e = 0.067 m_0$, $m_{lh} = 0.087 m_0$, $d_{ML} = 2.83$ Å, $c_D = 0.45$ μF/cm^2.

[b] The square of the index of refraction.

[c] The Cowley–Sze [12] vacuum interfacial-layer (IL) model. D_S is calculated using the experimental γ and Equation (16.10).

[d] First alternative: The vacuum is replaced, more realistically, with an unintentional oxide. Reasonable candidates are Ga and Al oxides. We use the effective mass and dielectric properties of Al_2O_3 [41] and the affinity of SiO_2. A thickness range thought to be reasonable for an unintentional oxide is chosen.

[e] Second alternative: Based on numerical *ab initio* modeling of ideal intimate contact on GaAs(110) [13]. The tunneling barrier is primarily in the semiconductor. The values for X_i, m_i, ϵ_i and $\epsilon_i^{(\infty)}$ are thus those for GaAs; X_i is increased by 0.05 eV to compensate for the slightly lower 0.8 eV Schottky-barrier height calculated in [13]. The density of MIGS used is that calculated in [13].

[f] Third alternative: Based on numerical *ab initio* modeling of intimate contact on GaAs(110) with *bulk-like defects* in monolayers near the interface [18]. The barrier is in the semiconductor. The value for D_S is the density of states calculated in [18] for the case of all the defects located in one monolayer, but is then distributed over the number of monolayers that the defects are spread over.

[g] Fourth alternative: Differs from the third alternative in that it invokes the experimental γ to determine D_S. Hasegawa's [22] expression (Equation (16.57)) for γ with bulk defects (volume density of states D_V) in a thin surface layer (of thickness d_i) is used. D_S is then calculated as $D_V d_i$, and spread over the number of monolayers as in the AUDM case.

interfacial gate resistance, and the observed lack of dispersion. The sensitivity to thickness is, however, particularly strong here because of the large intrinsic barrier (the metal work function) and interfacial-layer effective mass (= the free electron mass). The effect of including barrier lowering (and narrowing) is also particularly large, because of the low dielectric constant. The strong exponential dependence on interfacial-layer thickness is evident in Figure 16.8, and also in Figure 16.9 where we let d_i and D_S vary independently. Figure 16.9 shows r_{gi} and r_{it} versus surface-state density for an interfacial-layer thickness range extended 1 Å above and below that suggested by Cowley and Sze. In the 10^{13}–10^{15} cm^{-2}/eV D_S range, which should cover reasonable experimental conditions, r_{gi} varies less than the underlying r_{it} because of Equation (16.22). In fact, r_{gi} has a maximum at $D_S = \epsilon_i/q^2 d_i$ before it starts to approach zero as D_S approaches zero. In this ideal limit, the FET-degrading gate-resistance parameter r_{gi} approaches zero, even as r_{it}, for a finite d_i, approaches infinity.

A moderately thick oxide can also lead to predictions consistent with experiments as seen for the 10-Å case in Figure 16.8. Thicker oxides, however, quickly approach the MOS case, where there is a very large (ideally infinite) tunneling resistance, which does not degrade the FET performance since it is bypassed at very low frequencies. The large low-frequency value (off-scale at 1.3×10^{-4} $\Omega\,\text{cm}^2$) and dispersion for the 15-Å case in Figure 16.8 are inconsistent with experimental observations on normal FETs. However, the larger average and experimental spread in r_{gi} sometimes seen during process development [4] could well be due to an interfacial oxide and possibly organic residues. A III–V surface can be sensitive to even a small controlled exposure to oxygen [20], and real device wafers in a fabrication environment can get a significant amount of uncontrolled exposure. Measures can be taken to minimize the detrimental effect resulting from this, as discussed in Section 17.7. An effective cleanup dip just prior to gate metal deposition, and a sintered Pt gate typically lead to $r_{gi} \sim 10^{-7}$ $\Omega\,\text{cm}^2$ with little dispersion up to 50 GHz. The $PtAs_2$–semiconductor interface resulting from the sintering may be as close to an ideal intimate metal–semiconductor Schottky contact as one can expect to get in a practical processing environment.

The MIGS case in Table 16.1 and Figure 16.8 represents the most ideal Schottky barrier physically conceivable. In this case there is no physical tunneling barrier. If, however, we still represent the situation with an interfacial layer, as was done in [13] to estimate γ (Equation (16.10)), we can get an upper conceivable r_{gi} limit. d_i/ϵ_i was replaced by $d_{TF}/\epsilon_0 + d_{MIGS}/\epsilon_{SC}$, where d_{TF} (= 0.5 Å) is the Thomas–Fermi metal screening length [40]. This results in $\gamma = 0.13$ for GaAs, which is in the upper part of the experimental range. For our upper-limit tunneling calculation we choose $d_i = d_{TF} + d_{MIGS}$ and $\epsilon_i = \epsilon_{SC}$. The predicted value for r_{gi} in Figure 16.8 is still exceedingly small, $\sim$6 orders of magnitude lower than experimental observations, and $\sim$3 orders of magnitude smaller than the estimate for the stem resistance.

From the preceding results and discussion it appears that there remains, in SG-

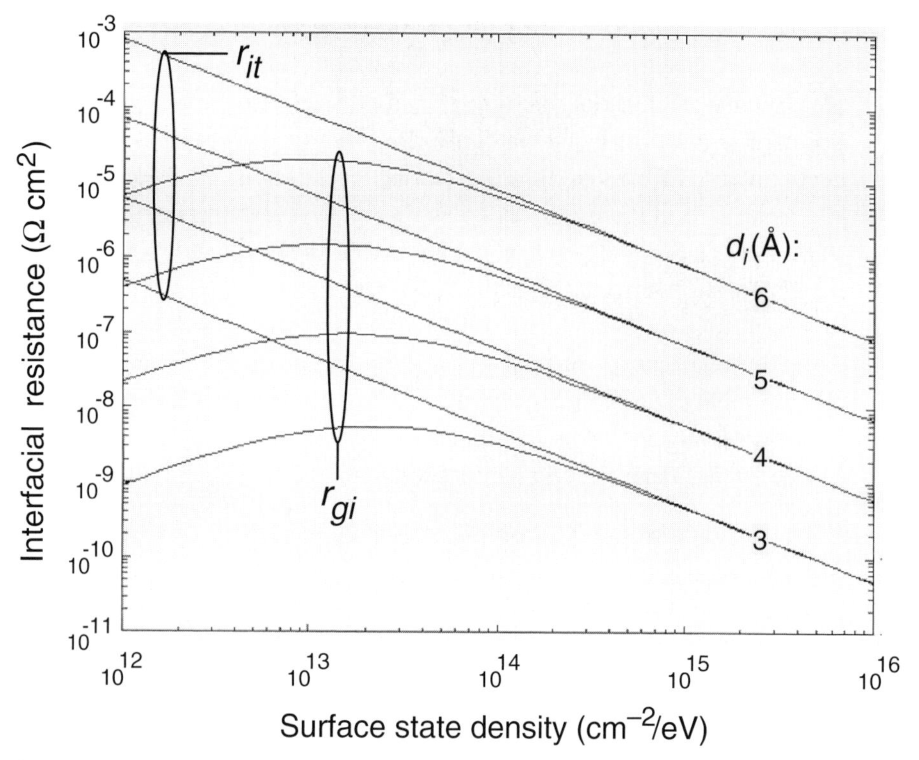

Fig. 16.9. Calculated interfacial tunneling and gate resistance versus surface-state density for a 3–6 Å range of interfacial-layer thickness. Cowley–Sze's [12] vacuum picture is assumed with parameters given in Table 16.1 (except that the dependence of D_S on d_i is abandoned). (H. Rohdin, N. Moll, A.M. Bratkovsky and C.-Y. Su, *Physical Review B*, Vol. 59, pp. 13 102–13 113, 1999.)

BFETs produced in practical fabrication environments, even with sintered gates, a significant interfacial tunneling barrier. The gate resistance is too large to be reconciled with an ideal defect-free intimate metal–semiconductor contact, but it can be accounted for by a thin oxide, or a vacuum interfacial layer. The oxide picture is troublesome, however, in the context of sintered gates. The vacuum picture is unsatisfactory in that it is physically somewhat unrealistic. There is, however, a third alternative, denoted AUDM in Table 16.1 and Figure 16.8. Here, the barrier is composed of metal in intimate contact with the semiconductor, like in the MIGS case, but with defects in the semiconductor near (and not just at) the interface. This is an important distinction, since such defects will act as terminal states for tunneling from the metal, with a non-zero barrier in the semiconductor itself. The original importance of the defects, and their location away from the surface, was that this picture explains the experimentally observed insensitivity of the Schottky-barrier height on GaAs to the choice of metal, i.e. the small γ. This was shown by van Schilfgaarde and Newman [18], using a first-principles numerical model. They also showed [17, 18] that the defect-free MIGS model cannot explain this important feature. This may be

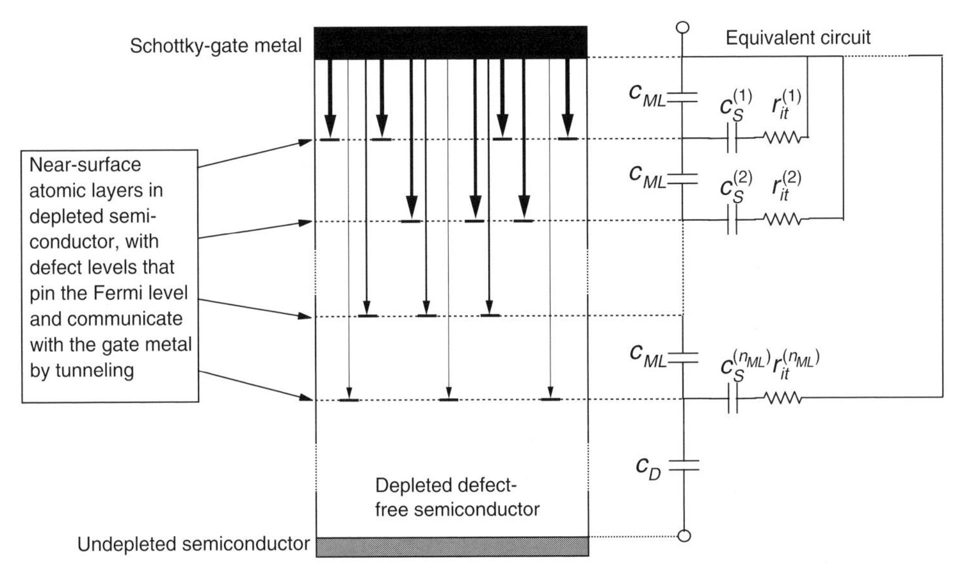

Fig. 16.10. Physical illustration and equivalent circuit for the case of tunneling between the metal gate and defects in the semiconductor near the interface. The metal and semiconductor are in intimate contact. The defects are located in equidistant monolayers of the semiconductor, and are thus no longer surface states. (H. Rohdin, N. Moll, A.M. Bratkovsky and C.-Y. Su, *Physical Review B*, Vol. 59, pp. 13 102–13 113, 1999.)

surprising considering the rather good prediction of γ based on the interfacial-layer analysis above. However, other analyses have resulted in similar criticism of the MIGS model [15, 16]. Given the intimate contact, it may well be that an interfacial-layer representation fails in the MIGS case. The interfacial gate resistance would then indeed be even lower than the 10^{-13} $\Omega\,\text{cm}^2$ in Figure 16.8.

The physical picture of [18] is based in spirit on the advanced unified defect model [21], which is the reason AUDM is used to denote this case. A likely defect to be involved in Schottky-barrier formation on GaAs, and the one used in [18], is the As_{Ga} anti-site, which is believed to be the same as the deep donor EL2. The Schottky-barrier height and EL2 bulk binding energy are very similar. It is worth noting that the capture cross-section of 10^{-13} cm^2 that we effectively use (Equation (16.55)) is relatively close to the experimental values of 10^{-14}–10^{-12} cm^2 for this level [48]. In [18], all the EL2 defects were located in a monolayer at various distances from the interface. It was shown that, for the experimentally observed pinning, the defects need to be located in the second monolayer from the surface, or deeper. In reality, the defects will be spread over several monolayers. To account in the tunneling analysis for the spatial distribution of the defect states, we generalize the dispersion analysis of Section 16.5 as illustrated in Figure 16.10. The normalized admittance, with tunneling to n_{ML}

layers, is given by the following set of expressions (Problem 16.3):

$$y^{(n_{ML})} = \left(\frac{1}{j\omega c_D} + \frac{z_{n_{ML}}}{1 + q_{n_{ML}}} \right)^{-1}, \tag{16.56a}$$

$$z_k = r_{it}^{(k)} + \frac{1}{j\omega c_S^{(k)}}, \quad k = 1, 2, \ldots, n_{ML} \tag{16.56b}$$

$$q_k = \frac{p_k}{1 + \dfrac{p_{k-1}}{1 + q_{k-1}}}, \quad q_0 \neq -1, \tag{16.56c}$$

$$p_k = j\omega c_{ML} z_k, \quad p_0 = 0, \tag{16.56d}$$

where $r_{it}^{(k)}$ is the tunneling resistance to layer k, $c_S^{(k)} = q^2 D_S^{(k)}$ is the 'surface' state capacitance associated with layer k, and $c_{ML} = \epsilon_{SC}/d_{ML}$ is the capacitance for a semiconductor monolayer thickness d_{ML}. In the present case we assume that all $D_S^{(k)}$ are equal, and are given by D_S/n_{ML}, where D_S is determined from [18]. The earlier Equation (16.20) for tunneling to one layer of states is recovered by setting $n_{ML} = 1$ and $c_{ML} = c_i$. We ignore that the top capacitor in Figure 16.10 is somewhat larger than the others because of the barrier narrowing ($\Delta d_i \approx 0.3$ Å $\approx 0.1 \times d_{ML}$). The result for bulk-like defects spread over n_{ML} monolayers, with n_{ML} between 2 and 20, is shown in Figure 16.8.

A variation on the AUDM is denoted AUDM′ in Table 16.1 and Figure 16.8. Instead of using the theoretical D_S from [18], we use the experimental γ. To determine the experimentally based D_S, we use the analytical connection [22] between γ and a volume (bulk) defect density-of-states D_V distributed over a thickness d_i in the semiconductor. With $D_S = D_V d_i$, this connection is (Problem 16.4)

$$\gamma = \frac{1}{\cosh\left(\left(\dfrac{q^2 D_S d_i}{\varepsilon_{SC}}\right)^{1/2}\right)}. \tag{16.57}$$

This expression was derived in the context of the DIGS model. However, it is not limited to a particular view of the microscopic origin of the defects. It is simply the general bulk-defect analog of the surface-state case in Equation (16.10). It is interesting how similar the results in Figure 16.8 are for the theoretically- and experimentally-based D_S. For both the AUDM and the AUDM′ cases, the predicted interfacial gate resistance is quite close to the experimental values for defect depths of 15–20 monolayers. This is a quite reasonable finding. Based on a striking correlation between the Fermi level pinning position at metal–semiconductor interfaces and in irradiated bulk material, Walukiewicz [49] has developed a more detailed defect model for Schottky-barrier formation than the ones above, which relies on amphoteric native defects* in a surface layer of similar thickness.

* V_{Ga} and $As_{Ga} + V_{As}$ in a balance that depends on doping.

It is worth reemphasizing that the curves of Figure 16.8 are subject to a variety of uncertainties which may move the curves up or down by an order of magnitude. As mentioned, an actual deep level may confine more of the wave-function to the attractive core potential, leading to a larger value of r_{it}. The effective cross-section $\sigma_T^{(B)}$ is close enough to experimental values for EL2 that this source of variation should not be more than an order of magnitude. The dependence on D_S is qualitatively similar to that shown in Figure 16.9 for the Cowley–Sze vacuum picture. r_{gi} peaks quite near $D_S = 10^{14}$ cm^{-2}/eV. An order-of-magnitude variation in r_{gi} would require more than an order of magnitude change in D_S, and could only decrease r_{gi}. The sensitivity to a variation in the metal effective mass is also rather weak; in a typical case a factor of 2 change from the nominal $m_M = 0.5m_0$ changes r_{gi} only by 25%. However, if the tunneling effective mass were larger or smaller than the original estimate of $0.031m_0$ the r_{gi} curve would move up or down considerably. Figure 16.11 shows the effect for the case of a semiconductor barrier with bulk-like defects spread over the first 10 monolayers (28 Å). A moderate increase of m_i to $0.06m_0$ increases the predicted r_{gi} from 10^{-9} $\Omega\,\text{cm}^2$ to 10^{-8} $\Omega\,\text{cm}^2$, with negligible increase in dispersion. In addition to explaining Fermi level pinning during Schottky-barrier formation, the defect model can thus get quite close to predicting the experimentally observed interfacial gate resistance. Deep penetration of defects leads to larger dispersion, as seen in Figure 16.8. This may be one reason that one can occasionally observe FETs with large r_{gi} and dispersion.

16.8 Summary and modifications to the equivalent circuit and Y-parameters

Sections 16.3–16.7 have covered in some depth a generally overlooked component in the gate resistance R_g of SBGFETs. We found that the always 'larger-than-expected' R_g is caused by a component R_{gi} which scales *inversely* with gate width. We interpret R_{gi} as a metal–semiconductor *interfacial* gate resistance. The dominance of R_{gi} profoundly affects model scaling and device optimization. For GaAs- and InP-based SBGFETs there appears to exist a smallest practically achievable normalized interfacial gate resistance r_{gi} on the order of 10^{-7} $\Omega\,\text{cm}^2$. Using the Cowley–Sze representation of a Schottky barrier, we showed that this lower practical bound is conceptually and quantitatively well explained by electron tunneling between metal and semiconductor surface states. Physical models for Schottky-barrier formation involving near-surface crystal imperfections are also quite consistent with the experimental r_{gi} on practical III–V FETs. The dispersion in the measured r_{gi} is insignificant up to the maximum frequencies used in typical measurements (26–50 GHz). In fact, it appears to be lower than predicted in Sections 16.5 and 16.7. Realizing that several effects could not be included in the analytical treatment, it then seems prudent to

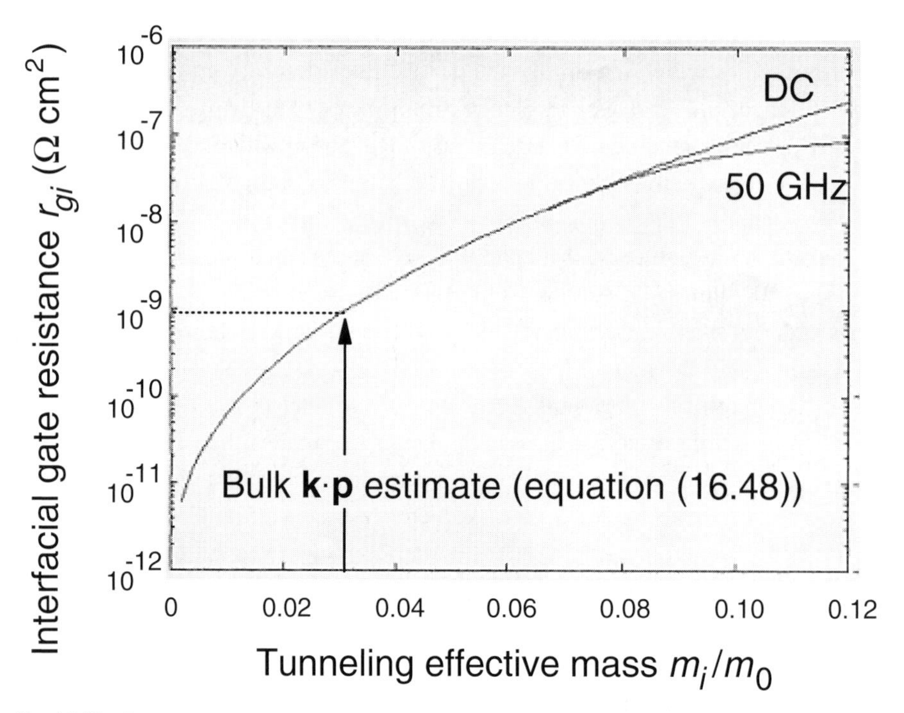

Fig. 16.11. Calculated interfacial gate resistance at DC and 50 GHz versus tunneling effective mass m_i for bulk-like defects distributed over ten monolayers. (H. Rohdin, N. Moll, A.M. Bratkovsky and C.-Y. Su, *Physical Review B*, Vol. 59, pp. 13 102–13 113, 1999.)

neglect dispersion altogether, and thus to identify R_g in Figure 15.16 with the simple resistance R_{gi}. In particular, one of the effects neglected is tunneling or hopping [30] between defect states. Another is the possibility of a resistive interfacial layer, in addition to the defects in the semiconductor. This could be a very thin oxide or some other residue, a layer with intermixing of atoms from the metal and semiconductor, or the surface reconstruction layer proposed by Freeouf *et al.* [27]. These neglected effects could well make the interfacial gate resistance more DC-like, with little dispersion. The similar r_{gi}s extracted with the two different methods in Figure 16.2 would suggest such a picture, since the data in Figure 16.2(a) were extracted under near-DC conditions in heavy forward bias, while the data in Figure 16.2(b) were extracted at high frequency in a bias regime where the DC conduction is negligible. Electron conduction current across the Schottky barrier thus appears to experience a resistance of similar magnitude to that observed at microwave frequencies. This supports the standard placement of the two diode conductances g_{gs} and g_{gd} 'inside' R_g ($= R_{gi}$) in Figure 15.16(a). The apparent similarity between the DC and microwave interfacial gate resistances is not at all obvious. Prediction of the former would be significantly more complicated than our analysis at microwave frequencies, and maybe not as worthwhile, considering the main applications for these FETs. It would appear

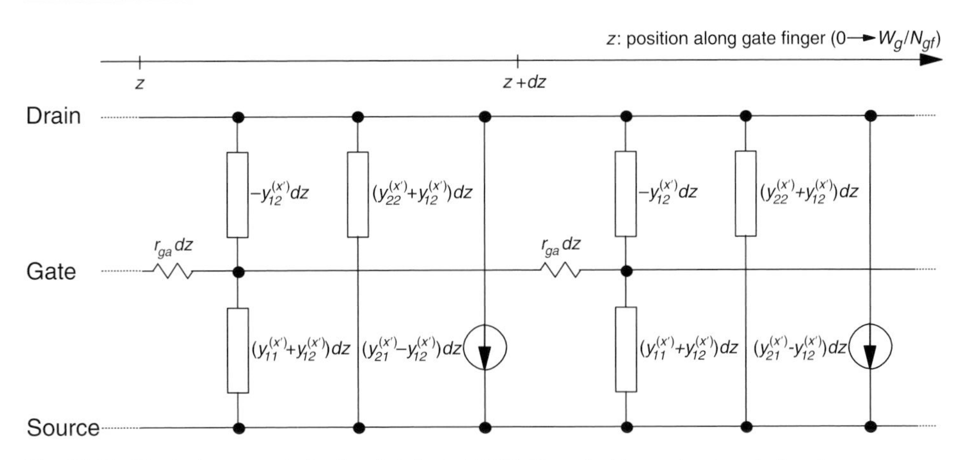

Fig. 16.12. Equivalent circuit for the distributed FET. The admittance (rectangles) and transconductance elements have been expressed in the Y parameters calculated with Equation (15.66), normalized to unit gate width.

that the gate resistance is best extracted at the high frequencies intended for the device operation.

The reason why we have not yet included the metallization access resistance R_{ga} in R_g is that the distributing effect of r_{ga} should be included more carefully [1] than suggested by Equation (16.1). We have to consider small width-segments (dz) of the FET, each with the equivalent circuit in Figure 15.16. The gate voltage is fed laterally along the resistive gate finger (z direction). The equivalent circuit of this situation is shown in Figure 16.12, where two things should be noted. First, as discussed in the introductory part of this section, we neglect distributed resistance associated with the source and drain. Second, the lower case used for the Y parameters indicates that these are the values derived from Equation (15.66), but then normalized to unit gate width. This means, for instance, that R_g in Equation (15.66b) is not R_{gi}, but r_{gi}/L_g. The full extrinsic Y parameters for the distributed FET (allowing us to finally drop the $'$ in the superscript) are (Problem 16.5(a)):

$$Y_{ij}^{(x)} = Y_{ij}^{(x')} F_{ga}, \quad ij \neq 22, \tag{16.58a}$$

$$Y_{22}^{(x)} = Y_{22}^{(x')} \left[1 - \frac{Y_{12}^{(x')} Y_{21}^{(x')}}{Y_{11}^{(x')} Y_{22}^{(x')}} \left(1 - F_{ga} \right) \right], \tag{16.58b}$$

where $Y_{ij}^{(x')} = y_{ij}^{(x')} W_g$, i.e. the Y-parameters for an FET of width W_g with $r_{ga} = 0$, and

$$F_{ga} = \frac{\tanh\left(\left[Y_{11}^{(x')} \left(r_{ga} W_g / N_f^2 \right) \right]^{1/2} \right)}{\left[Y_{11}^{(x')} \left(r_{ga} W_g / N_f^2 \right) \right]^{1/2}} \tag{16.58c}$$

is a factor containing all the distributed effects due to $r_{ga} W_g > 0$.

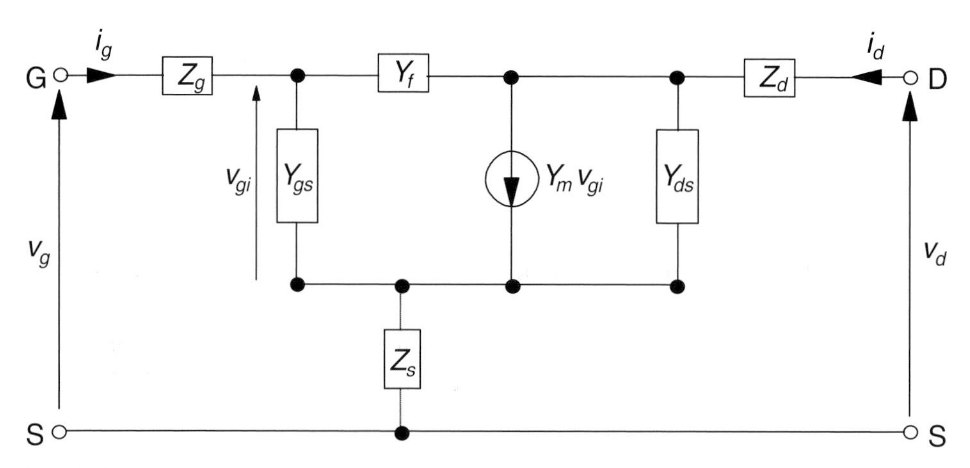

Fig. 16.13. Lumped equivalent circuit for the full distributed FET. The elements are given by Equations (16.59)–(16.61).

Alternatively, the effect of r_{ga} can be represented by a gate impedance Z_{ga} in series with the input to the two-port described by $Y_{ij}^{(x')}$, i.e the device with $r_{ga} = 0$. It is easy to show (Problem 16.5(b), (c)) that Z_{ga} is given by

$$Z_{ga} = \frac{1}{Y_{11}^{(x')}} \left(\frac{1}{F_{ga}} - 1 \right) \xrightarrow[r_{ga} W_g \to 0]{} \frac{r_{ga} W_g}{3N_f^2}. \tag{16.59}$$

As indicated, this has the limiting value given without a derivation in Equation (16.1a). We can represent the distributed FET with the lumped equivalent circuit in Figure 16.13. As in Figure 15.16, the 'almost-intrinsic' Y parameters in Equation (15.65) have been represented by the three admittances:

$$Y_{gs} = Y_{11}^{(i')} + Y_{12}^{(i')}, \tag{16.60a}$$

$$Y_f = -Y_{12}^{(i')}, \tag{16.60b}$$

$$Y_{ds} = Y_{22}^{(i')} + Y_{12}^{(i')} \tag{16.60c}$$

and the transadmittance

$$Y_m = Y_{21}^{(i')} - Y_{12}^{(i')}, \tag{16.60d}$$

where we have now dropped the somewhat cumbersome superscript. The parasitic elements not contained in Equation (15.65) have been lumped into the three series impedances

$$Z_g = R_{gi} + Z_{ga} + j\omega L_{gx}, \tag{16.61a}$$

$$Z_s = R_s + j\omega L_{sx} \tag{16.61b}$$

and

$$Z_d = R_d + j\omega L_{dx}, \tag{16.61c}$$

where we allow for fixed (W_g-independent) series inductances external to the device. These would typically come about from bonding wires. Representing these with lumped inductors is only valid for sufficiently low frequencies. At higher frequencies, or if the additional series impedances are of a different nature, the $j\omega L$'s have to be replaced with the appropriate frequency-dependent impedances.

Given that Z_{ga} (Equation (16.59) is not independent of the other components in the equivalent circuit, the lumped equivalent circuit in Figure 16.13 is not helpful in predicting the Y parameters. One still has to use Equation (16.58), where a source (or drain) inductance is easily included by simply replacing R_s with $R_s + j\omega L_{sx}$ (or R_d with $R_d + j\omega L_{dx}$). A gate inductance is accounted for by doing another F_{ga}-like correction (Equations (16.58a), (16.58b)), this time to $Y_{ij}^{(x)}$, and using the correction factor

$$F_{gx} = \frac{1}{1 + j\omega L_{gx} Y_{11}^{(x)}}. \tag{16.62}$$

The expression for F_{gx} becomes obvious by inspecting Equation (16.59). If we wish to account for the skin effect (Equation (16.2)) and an inductance l_{ga} per unit length of the gate finger, we can replace r_{ga} in Equation (16.58c) with $r_{ga}^{(ac)} + j\omega l_{ga}$. The inductance can be estimated as

$$l_{ga} = \frac{\varepsilon_{ga} W_g}{c^2 \left(C_{gs,gc} + C_{gd,sat} + C_{gs}^{(f)} + C_{gd}^{(f)}\right)}, \tag{16.63}$$

where c is the speed of light in vacuum, and ϵ_{ga} is an average of the relative dielectric constants of the semiconductor and passivation. The main usefulness of the lumped equivalent circuit in Figure 16.13 is in noise calculations, as we will see in Section 17.6.

One very important III–V FET application is power amplifiers in wireless communication. Designing and optimizing for large microwave or millimeter-wave output power with good efficiency is a bigger challenge than suggested by the cursory discussion in Section 14.6.5*. The total gate width can be many millimeters, and the gate then has to be divided up into many parallel fingers. The input signal has to be distributed to these in a way that does not lead to unacceptable phase difference between fingers near the center of the device, and fingers furthest away at the edges. Devices extending less than about 1/16 of a wavelength can be modeled well with lumped elements (e.g. [50]). For large-power FETs, however, electromagnetic wave

* For long battery life, hand-held single-supply applications require enhancement-mode devices ($V_T > 0$) with very low drain-current leakage in the off-state ($V_G = 0$). The concurrent requirements of high $I_D^{(max)}$ and $g_m^{(max)}$, and the finite Φ_B, present the device and process designers with their own set of challenges.

propagation delays† must be properly accounted for with distributed elements [50]. With aggressive shrinking of the device, the drain resistance and inductance may have to be accounted for in the same distributed way as the gate access resistance. This leads to more complicated expressions than Equation (16.58) which may also include the mutual inductance between gate and drain [51]. For power FETs the ground is typically ideally distributed on the back-side of the wafer, and one can account for the inductance and resistance of source ground vias with lumped elements (cf. Equation (16.61b)).

16.9 Conclusion

This chapter has covered in great detail the last building block in our analytical model for the high-frequency performance of MODFETs. The analysis of the gate resistance has brought us to physical depths not normally reached with analytical device modeling. In particular, we have had to review and evaluate, from a practical device standpoint, various models of Schottky-barrier formation, a continuing topic of basic research since the 1930s. In Section 16.8 we put everything together for a physics-based small-signal model applicable to real extrinsic FETs. We are thus ready for a real optimization example. This is one of the topics of the next chapter.

16.10 Bibliography

[1] P. Wolf, 'Microwave properties of Schottky-barrier field-effect transistors', *IBM Journal of Research and Development*, Vol. 9, pp. 125–141, 1970.

[2] H. Fukui, 'Determination of the basic device parameters of a GaAs MESFET', *Bell Systems Technical Journal*, Vol. 58, pp. 771–797, 1979.

[3] R. Faraji-Dana and Y. Chow, 'Edge condition of the field and A.C. resistance of a rectangular strip conductor', *IEE Proceedings, Pt. H*, Vol. 137, pp. 133–140, 1990.

[4] H. Rohdin, N. Moll, C.-Y. Su, and G. Lee, 'Interfacial gate resistance in Schottky-barrier-gate field-effect transistors', *IEEE Transactions on Electron Devices*, Vol. 45, pp. 2407–2416, 1998.

[5] A.D. Patterson, V.F. Fusco, J.J. McKeown and J.A.C. Stewart, 'A systematic optimization strategy for microwave device modelling', *IEEE Transactions on Microwave Theory and Techniques*, Vol. 41, pp. 395–405, 1993.

[6] P.C. Chao, 'Gate processes for GaAs PHEMTs. Short course', *International Conference on Gallium Arsenide Manufacturing Technology*. 1998.

[7] K. Onodera, K. Nishimura, S. Aoyama, S. Sugitani, Y. Yamane and M. Hirano, 'Extremely

† Unlike the transport-induced delays discussed in Section 15.3, these delays do result in Kirchhoff's current law breaking down.

low-noise performance of GaAs MESFET's with wide-head T-shaped gate', *IEEE Transactions on Electron Devices*, Vol. 46, pp. 310–319, 1999.

[8] S.-W. Chen,O. Aina, W. Li, L. Phelps and T. Lee, 'An accurately scaled small-signal model for interdigitated power P-HEMT up to 50 GHz', *IEEE Transactions on Microwave Theory and Techniques*, Vol. 45, pp. 700–703, 1997.

[9] G. Dambrine, A. Cappy, F. Heliodore and E. Playez, 'A new method for determining the FET small-signal equivalent circuit', *IEEE Transactions on Microwave Theory and Techniques*, Vol. 36, pp. 1151–1159, 1988.

[10] S. Babiker, A. Asenov, N. Cameron, S.P. Beaumont and J.R. Barker, 'Complete Monte Carlo RF analysis of 'real' short-channel compound FETs', *IEEE Transactions on Electron Devices*, Vol. 45, pp. 1644–52, 1998.

[11] D.E. Root, S. Fan and J. Meyer, 'Technology independent large signal non quasi-static FET models by direct construction from automatically characterized device data', *Proceedings of the European Microwave Conference*, pp. 927–932, 1991.

[12] A.M. Cowley and S.M. Sze, 'Surface states and barrier height of metal–semiconductor systems', *Journal of Applied Physics*, Vol. 36, pp. 3212–3220, 1965.

[13] S.G. Louie, J.R. Chelikowsky and M.L. Cohen, 'Theory of semiconductor surface states and metal–semiconductor interfaces', *Journal of Vacuum Science and Technology*, Vol. 13, pp. 790–797, 1976.

[14] J. Tersoff, 'Schottky barrier heights and the continuum of gap states', *Physical Review Letters*, Vol. 52, pp. 465–468, 1984.

[15] C.B. Duke and C. Mailhiot, 'A microscopic model of metal–semiconductor contacts', *Journal of Vacuum Science and Technology B*, Vol. 3, pp. 1170–1177, 1985.

[16] W.A. Harrison, 'Theory of band line-ups', *Journal of Vacuum Science and Technology B*, Vol. 3, pp. 1231–1238, 1985.

[17] M. van Schilfgaarde and N. Newman, 'Electronic structure of ideal metal/GaAs contacts', *Physical Review Letters*, Vol. 65, pp. 2728–31, 1990.

[18] M. van Schilfgaarde and N. Newman, 'Electronic structure of ideal and nonideal metal/GaAs contacts', *Journal of Vacuum Science and Technology B*, Vol. 9, pp. 2140–2145, 1991.

[19] L.J. Brillson, 'Metal–semiconductor interfaces', *Surface Science*, Vol. 299/300, pp. 909–927, 1994.

[20] W.E. Spicer, I. Lindau, P. Skeath and C.Y. Su, 'Unified defect model and beyond', *Journal of Vacuum Science and Technology*, Vol. 17, pp. 1019–26, 1980.

[21] W.E. Spicer, Z. Liliental-Weber, E. Weber, N. Newman, T. Kendelewicz, R. Cao, C. McCants, P. Mahowald, K. Miyano and I. Lindau, 'The advanced unified defect model for Schottky barrier formation', *Journal of Vacuum Science and Technology B*, Vol. 6, pp. 1245–1251, 1988.

[22] H. Hasegawa, 'Theory of Schottky barrier formation based on unified disorder induced gap state model, *Proceedings of the 18th International Conference on the Physics of Semiconductors*, World Scientific, Singapore, pp. 291–4, 1986.

[23] H. Hasegawa and H. Ohno, 'Unified disorder induced gap state model for insulator–semiconductor and metal–semiconductor interfaces', *Journal of Vacuum Science and Technology B*, Vol. 4, pp. 1130–1138, 1986.

[24] W. Mönch, 'Metal–semiconductor contacts: electronic properties', *Surface Science*, **299/300**, pp. 928–944, 1994.

[25] R. Ludeke, G. Jezequel and A. Taleb-Ibrahimi, 'Screening and delocalization effects in Schottky barrier formation', *Journal of Vacuum Science and Technology B*, Vol. 6, pp. 1277–1283, 1988.

[26] J.L. Freeouf and J.M. Woodall, 'Schottky barriers: An effective work function model', *Applied Physics Letters*, Vol. 39, pp. 727–729, 1981.

[27] J.L. Freeouf, J.M. Woodall, L.J Brillson and R.E. Viturro, 'Metal/(100)GaAs interface: Case for a metal–insulator–semiconductor-like structure', *Applied Physics Letters*, Vol. 56, pp. 69–71, 1990.

[28] S.M. Sze, *Physics of Semiconductor Devices*, 2nd Ed., John Wiley & Sons, New York, 1981.

[29] L.B. Freeman and W.E. Dahlke, 'Theory of tunneling into interface states', *Solid State Electronics*, Vol. 13, pp. 1483–1503, 1970.

[30] P. Muret and A. Deneuville, 'Capacitance spectroscopy of localized states at metal–semiconductor interfaces. I. Theory', *Journal of Applied Physics*, Vol. 53, pp. 6289–6299, 1982.

[31] Y.-S. Lou, and C.-Y. Wu, 'A self-consistent characterization methodology for Schottky-barrier diodes and ohmic contacts, *IEEE Transactions on Electron Devices*, Vol. 41, pp. 558–566, 1994.

[32] G. Gomila and J.M. Rubi, 'Relation for the nonequilibrium population of the interface states: Effects on the bias dependence of the ideality factor', *Journal of Applied Physics*, Vol. 81, pp. 2674–2681, 1997.

[33] P. Cova, A. Singh and R.A. Masut, 'A self-consistent technique for the analysis of the temperature dependence of current–voltage and capacitance–voltage characteristics of a tunnel metal–insulator–semiconductor structure', *Journal of Applied Physics*, Vol. 82, pp. 5217–5226, 1997.

[34] L.M. Terman, 'An investigation of surface states at a silicon/silicon oxide interface employing metal-oxide-silicon diodes', *Solid State Electronics*, Vol. 5, pp. 285–299, 1962.

[35] J. Bardeen, 'Tunnelling from a many-particle point of view', *Physical Review Letters*, Vol. 6, pp. 57–59, 1961.

[36] W.A. Harrison, 'Tunneling from an independent-particle point of view', *Physical Review*, Vol. 123, pp. 85–89, 1961.

[37] C.B. Duke, *Tunneling in Solids*, Academic Press, New York, 1969.

[38] H. Mizuta and T. Tanoue, *The Physics and Applications of Resonant Tunneling Diodes*, Cambridge University Press, Cambridge, 1995.

[39] J.S. Blakemore, *Solid State Physics*, 2nd Ed., Cambridge University Press, Cambridge, 1985.

[40] C. Kittel, *Solid State Physics*, 5th Ed., John Wiley & Sons, New York, 1976.

[41] R. Stratton, 'Volt–current characteristics for tunneling through insulating films', *The Journal of Physics and Chemistry of Solids*, Vol. 23, pp. 1177–1190, 1962.

[42] H. Rohdin, N. Moll, A.M. Bratkovsky and C.-Y. Su, 'Dispersion and tunneling analysis of the interfacial gate resistance in Schottky barriers', *Physical Review B*, Vol. 59, pp. 13 102–13 113, 1999.

[43] M. Garcia-Hernandez, P.S. Bagus and F. Illas, 'A new analysis of image charge theory', *Surface Science*, Vol. 409, pp. 69–80, 1998.

[44] G. Binnig, N. Garcia, H. Rohrer, J.M. Soler and F. Flores, 'Electron–metal-surface interaction potential with vacuum tunneling: Observation of the image force', *Physical Review B*, Vol. 30, pp. 4816–4818, 1984.

[45] E.O. Kane, 'The $k \cdot p$ method', in *Semiconductors and Semimetals*, ed. R.K. Willardson and A.C. Beer, Vol. 1, pp. 75–100, Academic Press, New York, 1966.

[46] S. Adachi, 'GaAs, AlAs, and $Al_xGa_{1-x}As$: Material parameters for use in research and device applications', *Journal of Applied Physics*, Vol. 58, pp. R1–29. 1985.

[47] L.I. Schiff, *Quantum Mechanics*, 3rd Ed., McGraw-Hill, New York, 1968.

[48] G.M. Martin, A. Mitonneau and A. Mircea, 'Electron traps in bulk and epitaxial GaAs crystals', *Electronics Letters*, Vol. 13, pp. 191–193, 1977.

[49] W. Walukiewicz, 'Mechanism for Schottky-barrier formation: The role of amphoteric native defects', *Journal of Vacuum Science and Technology B*, Vol. 5, pp. 1062–1067, 1987.

[50] S.M. Lardizabal, R.E. Leoni III, R. Mallavarpu, D. Teeter and M. Snow, 'A fast, scalable FET model that accounts for propagation effects', *Technical Digest of the Gallium Arsenide IC Symposium*, IEEE, Piscataway, pp. 135–138, 1999.

[51] S.J. Nash, A. Platzker and W. Struble, 'Distributed small signal model for multi-fingered GaAs PHEMT/MESFET devices', *IEEE MTT-S International Microwave Symposium Digest*, IEEE, Piscataway, pp. 1075–1078, 1996.

16.11 Problems

16.1 Derive the Equations (16.22) and (16.23) for the low-frequency limit of the components in the standard series RC equivalent circuit for the gate-to-channel admittance.

16.2 By properly matching the sinusoidal and exponential parts of the metal wave-functions and their derivatives (see text), derive the normalized wave-function tail in the interfacial layer (Equation 16.43). Hints: Use WKB wave-functions (see any textbook on quantum mechanics) for the z-dimension, and the factor $A_{xy}^{-1/2} \exp(i\mathbf{k}_{xy} \cdot \mathbf{r}_{xy})$ for the unconfined transverse component. (Convince yourself that this factor is normalized.) Allow L_M to be large enough for the interior of the metal to totally dominate the normalization integral in z.

16.3 Derive Equation (16.56) for the admittance resulting from direct tunneling between the gate and defect states located in equidistant layers in the underlying semiconductor. Hints: Use the equivalent circuit part of Figure 16.10 and work downward to the next monolayer. Each time check what the resulting admittance would be if this were the last monolayer containing defects. Take note of the recursive pattern developing, and generalize to an arbitrary number of layers.

16.4 Consider a metal in intimate contact with a semiconductor that has acceptor-like bulk defects distributed evenly to a depth d_i below the interface. The position-dependent charge density associated with the defects is $-qD_V E_{F0}(z)$ $(0 < z < d_i)$, where D_V is the defect density (units = cm^{-3}/eV), and E_{F0} is the energy between the Fermi level and the neutral level ([12, 28]; Section 16.5). Show that the Schottky-barrier height is given by Equation (16.9), and that its sensitivity γ to the metal work function is given by Equation (16.57). Hints:

Solve Poisson's equation, neglecting the field beyond the interfacial layer. A review of the Cowley–Sze analysis for the case of semiconductor surface states separated from the metal by a defect-free insulator [12, 28] might be helpful.

16.5 (a) With the aid of Figure 16.12, derive the full extrinsic Y parameters in Equation (16.58).

(b) Show that the alternative representation using an external element Z_{ga} (Equation (16.59)) produces the same Y parameters.

(c) What is the physical reason for the factor $1/3$ in the low-frequency limit (Equation (16.59))?

17 MODFET high-frequency performance

Few things are harder to put up with than the annoyance of a good example.

MARK TWAIN

17.1 Introduction

As one reason for delving into our rather deep analysis of the device, we stated in the introduction of Chapter 14 that we, as device physicists, process engineers or circuit designers, would like to be in the position to answer the question 'Is what we're producing what it could or should be?' In this chapter we put everything together, so that we can finally predict various important figures-of-merit, and answer that question. The measurements that we compare our AC predictions with will be done at high frequencies, and we start with a brief review of some practical basics in this area. Since the question posed above implies that we are also involved or at least interested in fabricating devices and circuits, we will towards the end take a look at some of the issues involved in this the most essential task. Finally, we will briefly outline the process of reverse modeling which can be useful if the answer to the question above is 'no'.

17.2 Some high-frequency measurement issues

The FET's frequency-dependent performance is most naturally described by the small-signal Y parameters, which relate the AC currents resulting from applied AC voltages:

$$i_g = Y_{11}v_g + Y_{12}v_d \tag{17.1a}$$

$$i_d = Y_{21}v_g + Y_{22}v_d. \tag{17.1b}$$

The Y parameters, however, are not directly measurable at the high frequencies we are interested in. The reason is the wave-nature of the voltages and currents along the transmission-line cables connecting the network analyzer (NWA) and the device under

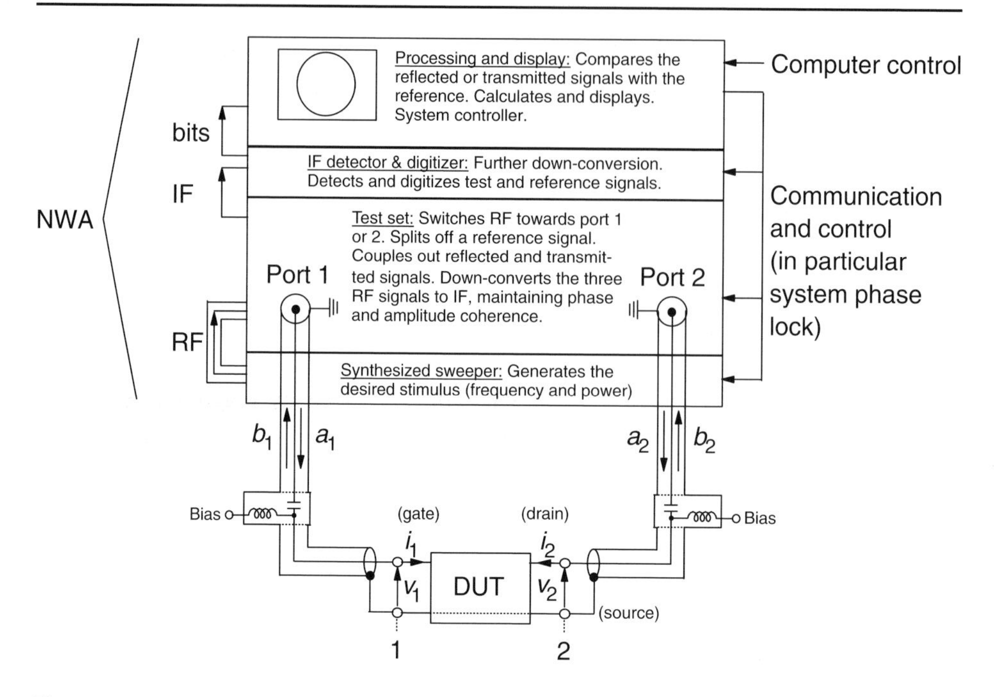

Fig. 17.1. Overview of a network analyzer (NWA) connected to a device under test (DUT).

test (DUT), as depicted in Figure 17.1. We are interested in the Y parameters at the 'planes' 1 and 2 that define the extent of the DUT, but can only do the measurements at ports 1 and 2 of the NWA. In between, the magnitude and phase of the voltage and current waves vary. By connecting and measuring known passive structures in place of the DUT, corrections for the phase shift and loss of the cables can be made, effectively moving the measurement ports to coincide with the device planes. This is done in the calibration menu on the front panel of the NWA. There are three standard calibration structures for each port: a short, an open and a resistive load. There is also a 'thru' line connecting the two ports. The user needs in effect to supply numbers describing the calibration structures, primarily the capacitance of the open, the inductance of the short, the resistance of the load and the delay of the thru. The calibration method requiring a short, an open, a load and a thru is referred to as the SOLT method. Other methods are available. The line-reflect-match (LRM) method has the advantage of not requiring the short and open which are more difficult to model at high frequencies than transmission lines and loads [1].

Given the distributed nature of the measurement system and the wave nature of the voltages and currents, the NWA reports the 'scattering' of power waves [2] by the DUT. The incoming power waves to the two ports of the DUT, launched from either of the two ports of the NWA, are denoted a_1 and a_2, respectively. The power waves emanating from the device, either from reflection or amplification of a_1 or a_2, are denoted b_1 and b_2, respectively. The power waves are related to the forward (away from the NWA; +) and backward (toward the NWA; −) traveling voltage waves, and the characteristic impedance Z_0 of the cables (essentially always 50 Ω), by

$$a_j = \frac{v_j^{(+)}}{Z_0^{1/2}} \tag{17.2a}$$

and

$$b_j = \frac{v_j^{(-)}}{Z_0^{1/2}}, \quad j = 1, 2. \tag{17.2b}$$

These are complex quantities, generally expressed by magnitude and phase. The forward and backward traveling current waves are related to the voltage waves through Z_0:

$$i_j^{(+/-)} = \frac{v_j^{(+/-)}}{Z_0}. \tag{17.3}$$

The total voltage at a point along the transmission line is given by the sum of the forward and backward traveling waves:

$$v_j = v_j^{(+)} + v_j^{(-)}; \tag{17.4}$$

while the net forward traveling current is given by the difference:

$$i_j = i_j^{(+)} - i_j^{(-)}. \tag{17.5}$$

The forward and backward traveling powers are given by

$$p_j^{(+)} = v_j^{(+)} \left(i_j^{(+)}\right)^* = |a_j|^2 \tag{17.6a}$$

and

$$p_j^{(-)} = v_j^{(-)} \left(i_j^{(-)}\right)^* = |b_j|^2, \tag{17.6b}$$

respectively, which is the reason for the naming and definition of a_i and b_i. For small signals, the power waves are linearly related by the scattering (or S) parameters:

$$b_1 = S_{11}a_1 + S_{12}a_2, \tag{17.7a}$$

$$b_2 = S_{21}a_1 + S_{22}a_2. \tag{17.7b}$$

It is these S parameters that are displayed and reported by the NWA. For passive (non-amplifying) devices the magnitudes of the S parameters are less than 1, i.e. they fall inside the unit circle in the complex plane. For an amplifying FET, S_{21} will fall outside. Before a calibration and measurement is started it is advisable to study the (uncorrected) frequency dependence of $|S_{11}|$ and $|S_{22}|$ without a DUT. This could reveal a loose connection, while a smooth gradual drop with frequency indicates that there are no major discontinuities in the 50-Ω environment.

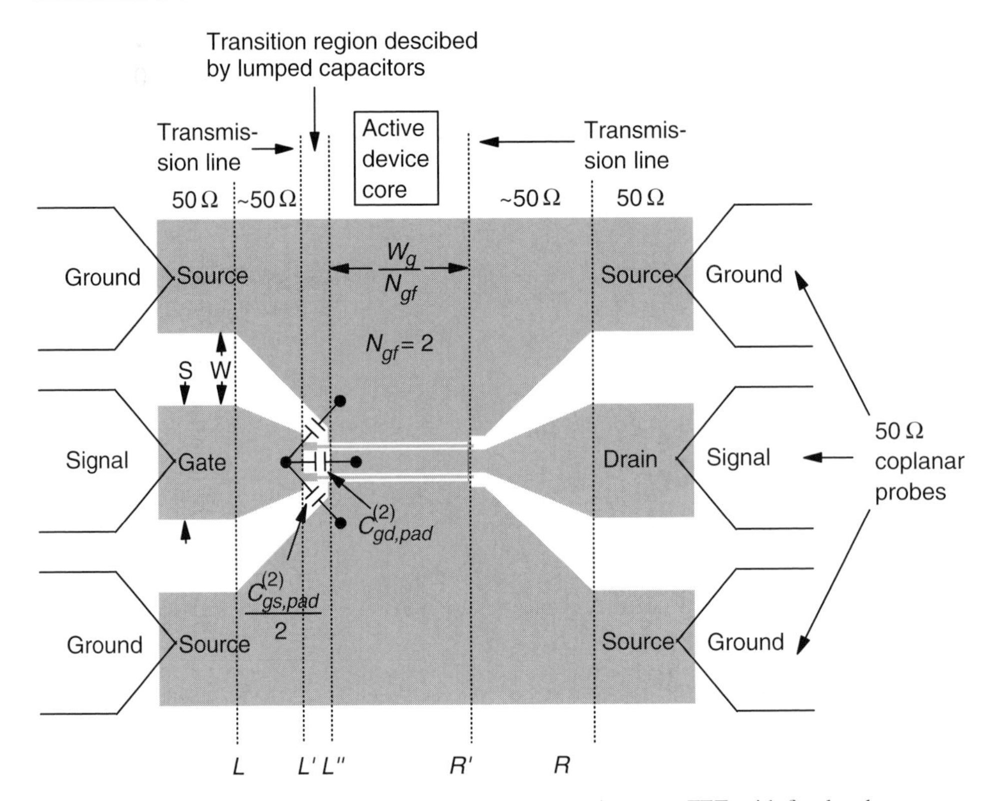

Fig. 17.2. Typical layout of a discrete probable two-finger microwave FET, with fixed pad capacitances indicated.

Equations (17.1)–(17.5) and (17.7) yield the Y parameters in terms of the measured S parameters (Problem 17.1):

$$Y_{11} = \frac{1}{Z_0}\frac{(1-S_{11})(1+S_{22})+S_{12}S_{21}}{(1+S_{11})(1+S_{22})-S_{12}S_{21}}, \tag{17.8a}$$

$$Y_{12} = \frac{1}{Z_0}\frac{-2S_{12}}{(1+S_{11})(1+S_{22})-S_{12}S_{21}}, \tag{17.8b}$$

$$Y_{21} = \frac{1}{Z_0}\frac{-2S_{21}}{(1+S_{11})(1+S_{22})-S_{12}S_{21}}, \tag{17.8c}$$

$$Y_{22} = \frac{1}{Z_0}\frac{(1+S_{11})(1-S_{22})+S_{12}S_{21}}{(1+S_{11})(1+S_{22})-S_{12}S_{21}}. \tag{17.8d}$$

A typical layout for on-wafer testing is shown in Figure 17.2 for a two-finger FET. The pad geometries to the left of L and to the right of R are short segments of an on-wafer 50-Ω coplanar transmission line, long enough to be contacted with coplanar probes connected by cables to the measurement system. All of these components are designed for $Z_0 = 50\ \Omega$ to keep reflections outside the device itself to a minimum.

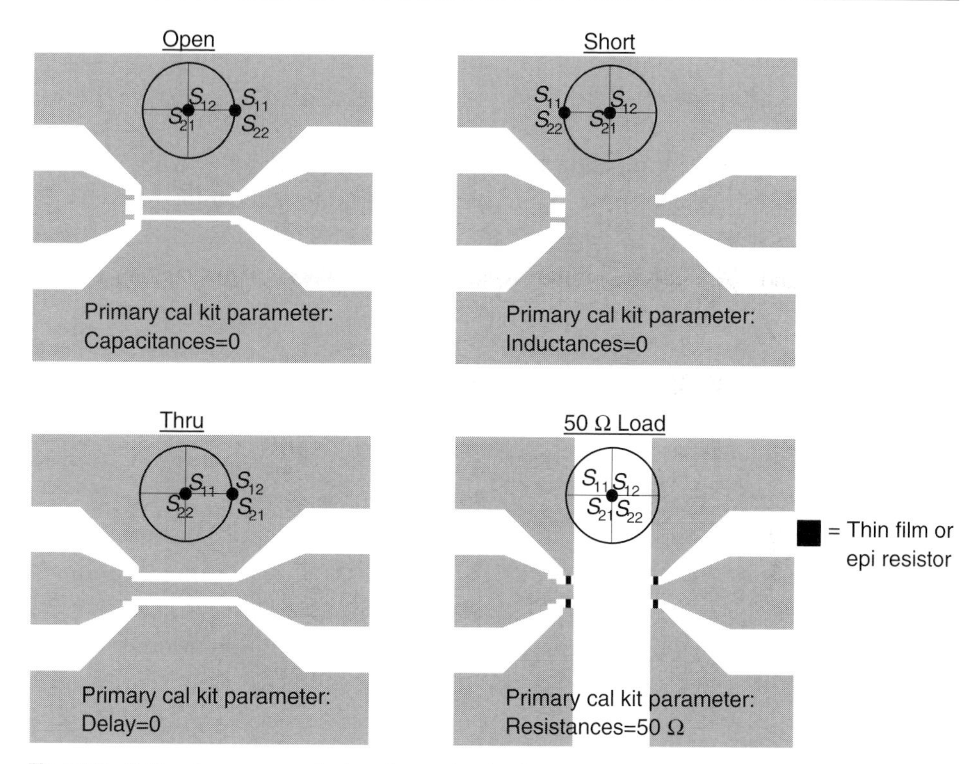

Fig. 17.3. Calibration patterns designed to make the S-parameters measured on the FET in Figure 17.2 correspond closely to the active device core. Capacitances and inductances external to the device will be 'taken out' by the calibration. The primary 'cal kit' parameters should be set as indicated. Indicated in the circles (radius = 1) are the locations in the complex plane of the S parameters when measured on the calibration standards after the calibration is completed.

The loss, phase-shift and reflections in the cables and probes are accounted for by the calibration. Typical on-wafer calibration patterns are shown in Figure 17.3. As illustrated in Figure 17.2, efforts are generally made to keep the 50-Ω environment as close to the device as possible. The approach is to taper gradually the metal feeds between L and L', and between R' and R, keeping the ratio S/W constant. For uniform (non-tapered) coplanar transmission lines, S/W determines the characteristic impedance [3], assuming: (1) negligible thickness of the metal, (2) thick substrate and (3) wide ground planes. The tapering tends to preserve the transmission-line nature of the pad layout up to quite close to the device. The deviation from an ideal uniform transmission line can be accounted for by laying out the calibration patterns similarly (Figure 17.3). On the gate side, in particular, there will be a transition region (L'–L'') that can be described and modeled rather well by fixed (W_g-independent) lumped capacitances, denoted $C^{(2)}_{gs,pad}$ and $C^{(2)}_{gd,pad}$ for the two-finger case in Figure 17.3. It is reasonable to assume that layouts with more than two (but still an even number of) fingers are laid out so that

$$C_{gx,pad}^{(N_{gf})} = \frac{N_{gf}}{2} C_{gx,pad}^{(2)}, \quad x = s, d; \qquad N_{gf} = 4, 6, 8, \ldots. \tag{17.9}$$

There is also a small additional gate inductance associated with L'–L''. A major reason for S-parameter measurement is to determine a model for the core of the device between L'' and R' in Figure 17.2. This is the part for which we have scaling rules for W_g and N_{gf}. Although the measurement is small-signal, a large-signal table-based model for computer-aided circuit design can be generated by doing the small-signal measurements in a wide biasing regime, including deep pinch-off where the currents and intrinsic semiconductor mobile charge are negligible. The Y parameters can then be integrated in bias space to generate bias-dependent total currents and charges for the drain and gate (e.g. [4]). Device engineers are primarily interested in the small-signal characteristics of the core of the device, near optimum channel modulation. The SOLT calibration patterns in Figure 17.3, with the main cal kit parameters indicated, are designed for this purpose, since they in effect move the measurement right up to the L''- and R'-planes. With the LRM calibration method, the same can be accomplished after calibration, by contacting the open and adjusting the port extensions on the NWA so that, as the frequency is swept, the displayed traces of S_{11} and S_{22} fall as close as possible to a dot at $(1, 0)$. Circuit designers cannot avoid the problem of fixed parasitics by simply calibrating them out. Instead, an important task is to determine them. One practical way is to start with a calibration with well-known standards, then measure the S parameters of device-like layout such as those in Figure 17.3, and finally back out fixed capacitances and inductances associated with the device layout. Another approach is to calculate numerically the scattering matrix associated with segments such as L–L'' and R'–R in Figure 17.2, and then include the results in the circuit design. Such problems are three-dimensional, but linear, i.e. the result is independent of the voltages that will appear in the segments. Commercial software packages are available for this task. For very high frequencies, well into the millimeter-wave regime, neither of these approaches may be accurate enough, and a full three-dimensional numerical solution of the entire circuit, including the active devices, may be warranted [5].

17.3 Recap of procedure and parameters for calculating MODFET Y parameters

Chapters 15 and 16 went into some depth to develop physically based expressions for Y_{11}, Y_{12}, Y_{21} and Y_{22}. The intrinsic Y parameters are given by Equation (15.63). These include the important effects of delays introduced by the velocity-saturated region. The delays (Equations (15.56)–(15.58), (15.61) and (15.62)), as well as the three intrinsic capacitances (Equation (15.30)) and the output conductance (Equation (15.54)), can be calculated semianalytically using the conformal-mapping recipe in Table 15.1. The generally small gradual-channel time constant is given by Equations (15.21) and (15.22), with the gate length increased by the source-side fringing modulation (Equation (15.31)). The DC transconductance is given by Equation (14.54). In sufficiently weak saturation, the time constants for the drain current will be those of the gradual channel, although, as discussed in Chapter 15, very quickly the saturated region will start to dominate, as assumed in Equations (15.63c) and (15.63d)). For a smooth transition we make the substitution

$$\tau_{d,sat}^{(x)} \rightarrow \left[\left(\tau_{d,gc}^{(x)}\right)^2 + \left(\tau_{d,sat}^{(x)}\right)^2\right]^{1/2}, \quad x = s, TL. \tag{17.10}$$

in these equations. Equation (15.65) introduced parasitic gate conductances and fringe capacitance corrections, proportional to W_g, into the Y parameters. The parasitic source and drain resistances (Section 14.6.2), as well as the interfacial part of the gate resistance (Sections 16.2, 16.3 and 16.8), scale inversely with gate width W_g, and are accounted for in Equation (15.66). We included in Equations (16.58) the effect of the gate metallization resistance, which is to a small degree increased by the skin effect (Equation (16.2)), and gate metallization inductance (Equation (16.63)).

Table 17.1 lists in an organized way all the parameters that go into predicting the small-signal frequency performance of the MODFET. There are references to where the parameters are introduced and discussed in detail. In the following we will vary some of the parameters and study the effect on performance. In particular, we will optimize the device with respect to gate length and width, and study the effect of the interfacial gate resistance and the extent of the velocity-saturated region. Most of the parameters will be left at their default values. We will assume the semiconductor material parameters in Table 17.1, which are representative for AlInAs/GaInAs MODFETs. Except for the gate length, and a brief look at the effect of reducing the gate–channel spacing significantly (2×), we will leave the cross-sectional geometry at its default, which is representative for a high-frequency device with acceptable gate leakage. We have reduced the default gate–channel spacing by 20% compared to Chapter 15, since we will consider very short gates, and would like to keep the aspect ratio L_g/d_{gc} reasonably large. d_{gc} and ΔL_d will affect ΔC_{gs} as shown in Figure 17.4. We have left the parasitic gate capacitances $C_{gs}^{(f)}$ and $C_{gd}^{(f)}$ at zero since the values for ΔC_{gs} and $C_{gd,sat}$ (Figure 15.6) are quite realistic (i.e. consistent with measurement) by themselves. We allow for a small fixed (i.e. W_g-independent) parasitic gate–source and gate–drain capacitances $C_{gs}^{(N_{gf})}$ and $C_{gd}^{(N_{gf})}$. The default values in Table 17.1 are zero, however, indicating that in most cases we assume that they were calibrated out as discussed above. We do not concern ourselves with a parasitic drain-source capacitance, either a fixed component or one proportional to W_g, since this will not affect either the current or power gains of primary interest to us. The parasitic impedances in Table 17.1 are typical, based on measurements and

Table 17.1. *Parameters determining the MODFET high-frequency performance. Default values are for a typical and/or optimized InP-based device. No tweaking of external impedances ($L_{gx} = L_{dx} = L_{sx} = 0$; Equations (16.61) and (16.62)).*

Type	Parameter	Default value	Ref.
Semiconductor material	(Average) relative dielectric constant	$\varepsilon = 12.7\varepsilon_0$	Equations (15.2) and (14.54)
	Effective saturation velocity	$v_{sat} = 2.8 \times 10^7$ cm/s	Equation (14.54)
	Full channel sheet resistance	$R_{sq} = 200\ \Omega$/sq.	Equation (16.3)
	Characteristic length in $\mu(L_g)$	$L_\mu = 167$ nm	Equation (14.17)
Cross-sectional two-dimensional geometry (Figure 14.1)	Gate–channel spacing	$d_{gc} = 200$ Å	Equations (15.2) and (14.54)
	Gate length	$L_g = 100$ nm	Equation (15.1)
	Source-side extent of lateral recess	$L_{us} = 100$ nm	Equation (15.54)
	Gate–source spacing	$L_{gs} = 1\ \mu$m	Figure 14.1
	Effective length of drain for C_{gd}	$\Delta L_d = 0.2\ \mu$m	Section 15.3
	Source-side intrinsic fringing capacitance	$\Delta C_{gs} = 160$ fF/mm	Equations (15.31), Figure 15.6
Layout geometry	Total gate width	$W_g = 50\ \mu$m	Equations (16.7), (16.58), (16.59)
	Number of parallel gate fingers	$N_{gf} = 2$	Equations (16.7), (16.58)
Parasitic gate admittances ($\propto W_g$)	Gate–source parasitic capacitance correction	$C_{gs}^{(f)} = 0$ fF/mm	Equation (15.65)
	Gate–drain parasitic capacitance correction	$C_{gd}^{(f)} = 0$ fF/mm	Equation (15.65)
	Gate–source parasitic conductance	$g_{gs} = 100\ \mu$S/mm	Equation (15.65)
	Gate–drain parasitic conductance	$g_{gd} = 50\ \mu$S/mm	Equation (15.65)
Parasitic impedances	Source resistance ($\propto W_g^{-1}$)	$R_s = 0.35\ \Omega$ mm	Equations (14.27)–(14.29)
	Drain resistance ($\propto W_g^{-1}$)	$R_d = 0.40\ \Omega$ mm	Equations (14.27)–(14.29)
	Interfacial gate resistance ($\propto W_g^{-1}$)	$r_{gi} = 3 \times 10^{-7}\ \Omega\,\text{cm}^2$	Sections 16.2, 16.3, 16.8, Equations (16.5), (16.7)
	Gate metallization resistance ($\propto W_g$)	$r_{ga} = 100\ \Omega$/mm	Equations (16.1), (16.7), (16.58), (16.59)
	Effective dielectric constant for the inductance of the gate metallization	$\varepsilon_{ga} = 9\varepsilon_0$	Equation (16.63)
	Geometric factor in the expression for the skin effect	$\beta_{se} = 3.5$	Equation (16.2)
Pad parasitics (W_g independent)	Fixed gate–source capacitance for a two-finger FET	$C_{gs,pad}^{(2)} = 0$ fF	Section 17.2 Figure 17.2
	Fixed gate–drain capacitance for a two-finger FET	$C_{gd,pad}^{(2)} = 0$ fF	Section 17.2 Figure 17.2

Table 17.1. *Continued.*

Type	Parameter	Default value	Ref.
Bias dependent	Drain current normalized to maximum channel current (set by V_G)	$I_D/I_D^{(max)} = 0.5$	Equation (16.4)
	Extent of velocity saturation external to gate (determined by V_D and V_G)	$\Delta L_x = 400$ Å	Figure 14.1 Equation (15.54)
	Ratio of extents of external and internal velocity saturation	$\Delta L_x/\Delta L_i = 1.5$	Figure 14.1 Equation (15.54)

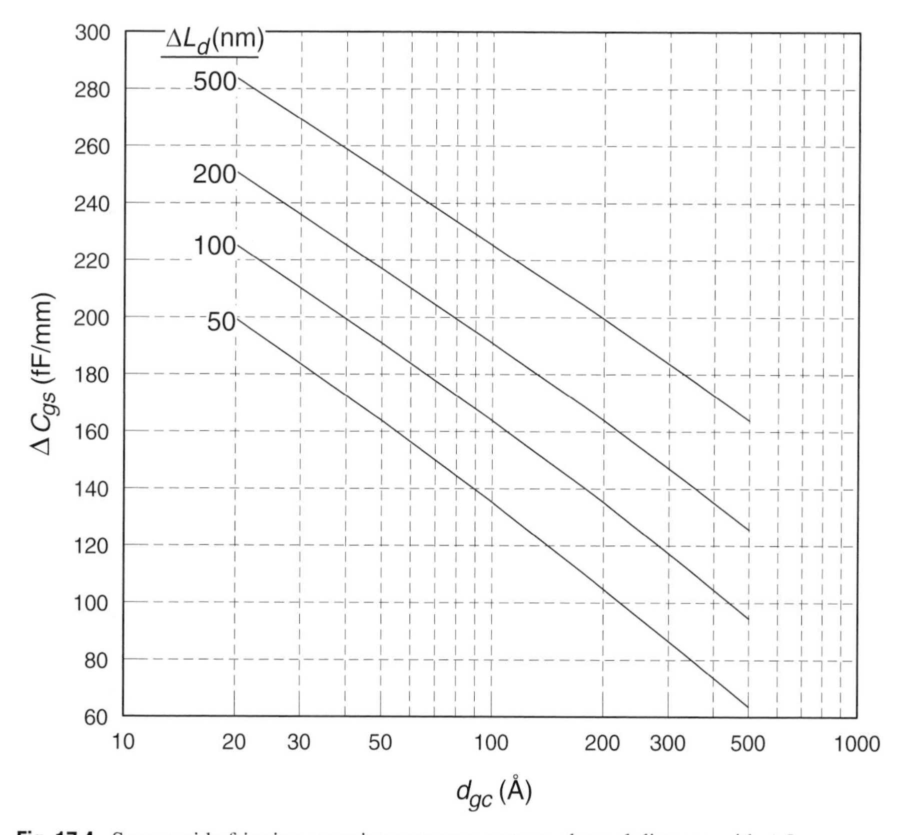

Fig. 17.4. Source-side fringing capacitance versus gate-to-channel distance with ΔL_d as a parameter. ΔC_{gs} is proportional to ε. Here, $\varepsilon = 12.7\varepsilon_0$.

calculations. The parasitic gate conductances g_{gs} and g_{ds} can vary significantly with process and bias. Unless they are abnormally large, g_{gs} and g_{ds} will not affect the performance at high frequencies. The extents ΔL_i and ΔL_x of the velocity-saturated regions can be calculated with the DC model in Chapter 14, but here we simply assume that $\Delta L_x = 1.5\Delta L_i$, as we did in Chapter 15.

17.4 Current gain, optimum power gain and cut-off frequencies

The most widely used measures of performance for a high-frequency device are the cut-off frequencies of various gains; i.e. the frequencies where these gains are unity. The cut-off frequencies are usually significantly larger than can be measured, and extrapolation has to be employed. This has to be done prudently, in order not to misrepresent the actual performance. The frequency dependences of the three most common gains are shown in Figure 17.5 for the default parameters in Table 17.1. The most basic of the gains, involving no matching circuitry at the input or output of the device, is the current gain with the output short-circuited*:

$$G_i \equiv \left|\frac{i_d}{i_g}\right|^2_{v_d=0} = \left|\frac{Y_{21}}{Y_{11}}\right|^2. \tag{17.11}$$

With a normal device and a good calibration, the current gain follows very closely a −20 dB/decade slope up to very high (and usually untestable) frequencies. This is illustrated in Figure 17.5(a). Considering the increased difficulty of getting and maintaining a good calibration at the highest test frequencies, it has become the rule [6] among device and process researchers to estimate the extrinsic cut-off frequency $f_T^{(x)}$, by extrapolating G_i from low-to-intermediate frequencies, using a −20 dB/decade slope. The actual cut-off frequency is larger, but cannot be reliably determined for cutting-edge devices. $f_T^{(x)}$ is important to a device/process engineer, particularly early in the development, since it indicates an upper frequency of potential transistor operation. This is because G_i of the active device core (Figure 17.2) is not severely degraded by the source and drain resistances, and not at all by the gate resistance. Although gate fringe capacitances will degrade $f_T^{(x)}$, it is a rather good measure of the intrinsic device speed, and thus the potential of the technology. The reasons for this become clear as we use the Taylor series expansion of the extrinsic Y parameters in Equations (15.68)–(15.70) to find the proper −20-dB/decade-based expression for $f_T^{(x)}$ [7]:

$$f_T^{(x)} = \frac{g_{m0}^{(i)}}{2\pi\left(C_{gs} + C_{gd} + \left\{(R_s + R_d)\left[g_{ds0}^{(i)}\left(C_{gs} + C_{gd}\right) + g_{m0}^{(i)} C_{gd}\right]\right\}\right)}. \tag{17.12}$$

$C_{gs} + C_{gd}$ is the total gate capacitance given by Equations (15.64a) and (15.64b). Equations (15.70h), (15.70b), (15.68b) and (15.69b) were used to arrive at Equation (17.12). The factor F_0 (Equation (15.69e)) initially appears in both numerator

* In this section we drop the superscript identifying parameters as extrinsic (x) or intrinsic (i), except for the transconductance, output conductance and current-gain cut-off frequency where there is often confusion on this issue. Without (x) or (i), the parameter is understood to be extrinsic, i.e. it includes all parasitic access resistances, fringe capacitances and distributed effects. We also drop the *sat* subscript for the transconductance and output conductance, since it is understood that we are interested only in the case where the device is velocity-saturated. For the capacitances, we must keep the subscripts to avoid confusing, say, $C_{gs,gc}$ with the total gate–source capacitance C_{gs} (Equation (15.64a)).

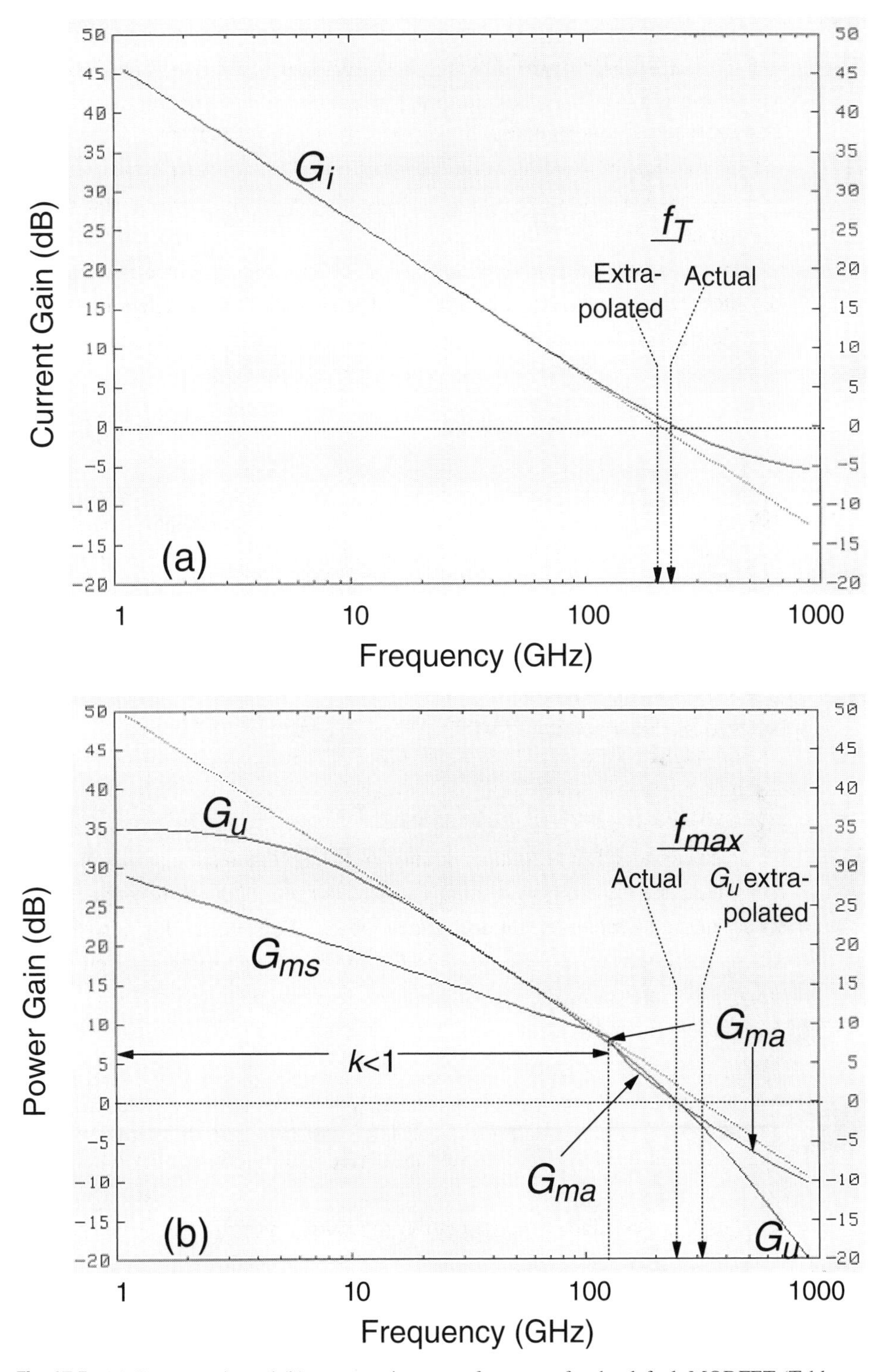

Fig. 17.5. (a) Current gain and (b) power gain versus frequency for the default MODFET (Table 17.1).

and denominator of f_{Tx}. Despite generally being significantly less than unity ($\sim$0.66 for the default FET), and thus significantly reducing the extrinsic transconductance (Equation (15.70h)), F_0 has no direct effect on $f_T^{(x)}$. However, $f_T^{(x)}$ is still reduced below its intrinsic value

$$f_T^{(i)} = \frac{g_{m0}^{(i)}}{2\pi\left(C_{gs} + C_{gd}\right)}, \tag{17.13}$$

by the additional capacitive term $\{\cdots\}$ in the denominator of Equation (17.12). Like F_0, this is introduced by the presence of source and drain resistances. However, the two terms in $g_{m0}^{(i)} C_{gd} + g_{ds0}^{(i)} \left(C_{gs} + C_{gd}\right)$ ($= D_1$; Equation (15.69b)) are each products with one rather small component (C_{gd} and $g_{ds0}^{(i)}$, respectively). The result of this, and the canceling of F_0, is that $f_T^{(x)}$ is less degraded by the source and drain resistances than $g_{m0}^{(x)}$ is.

Once sufficiently high $f_T^{(x)}$ has been demonstrated, and backend processes have been developed, the device can become attractive for MMIC circuit application, provided its power gain, noise figure (Section 17.6) and reliable output power (Sections 14.6.5 and 14.7) are adequate. The small-signal power gain can be optimized at a particular frequency with the right feedback and input/output matching networks, which generally involve segments of coplanar or microstrip transmission lines [8]. For a device embedded in a *lossless* network N (i.e. no resistive components) Mason [9] showed that the quantity

$$G_u = \frac{|Y_{21} - Y_{12}|^2}{4\left[\mathrm{Re}(Y_{11})\,\mathrm{Re}(Y_{22}) - \mathrm{Re}(Y_{12})\,\mathrm{Re}(Y_{21})\right]} \tag{17.14}$$

(U in Section 11.1.3) is independent of N. This includes the situation in Figure 17.2, i.e. the device by itself, in which case N is the null network. Figure 17.5(b) includes $G_u(f)$ using the default device parameters in Table 17.1. For any fixed frequency of interest, one can design N such that the resulting embedded device is unilateral ($Y_{12} = 0$), and input and output resistances are positive ($\Rightarrow \mathrm{Re}(Y_{11}), \mathrm{Re}(Y_{22}) > 0$). Mason identified G_u as the available power gain for this situation. As a result, G_u is referred to as Mason's unilateral power gain. Parasitic capacitors between extrinsic nodes of the device (i.e. 'outside' the parasitic gate, drain and source resistances in Figure 15.16), do not degrade G_u. The reason is that the parasitic admittance of such a capacitance can be exactly canceled by a parallel lossless inductor, at the fixed frequency of interest. The fixed pad capacitances in Table 17.1 and Figure 17.2 are of this nature. If not calibrated out, however, they would degrade G_i. As mentioned in Section 17.3, we do not consider a fixed drain-source pad capacitance, since this would degrade neither G_u nor G_i.

The cut-off frequency of G_u is most often referred to as f_{max} (Section 11.1.3). This quantity, almost exclusively determined experimentally by extrapolation at -20 dB/decade, is a frequently used figure-of-merit, for several reasons. One is that

G_u is defined by Equation (17.14) for all frequencies (like G_i), including typical test frequencies. Except for situations where G_u exhibits resonances [10, 11], this eases the task of extrapolation for the tester. For the circuit designer, G_u and f_{max} are attractive since they indicate the ultimate gain and maximum possible frequency of operation to expect from the device in a well-designed circuit. For the device/process engineer f_{max} is important since it indicates how well parasitics are kept under control. Power gain is more sensitive to feedback capacitance, output conductance and gate leakage than the current gain. In particular, power gain is, unlike current gain, degraded by the gate resistance R_g. At low frequencies G_u is limited by g_{gs} and g_{gd} (Figure 17.5(b))*. For a well-designed device with low parasitics, the extrapolated f_{max} can be significantly larger than $f_T^{(x)}$. As with $f_T^{(x)}$ (Equation (17.12)) we can get an explicit analytical expression for f_{max} extrapolated at −20 dB/decade:

$$f_{max} = \frac{g_{m0}^{(x)}}{4\pi \left(g_{m0}^{(x)} B_{12}^{(x')} + g_{ds0}^{(x)} B_{11}^{(x')} \right)^{1/2}}. \tag{17.15}$$

The four quantities in this expression are given by Equations (15.67g) and (15.68)–(15.70), with $R_g = R_{ga} + R_{gi}$ (Equations (16.1) and (16.5)). Unlike $f_T^{(x)}$ the expression for the extrapolated f_{max} does not collapse into an expression in terms of intrinsic parameters and parasitic elements that would fit on one line. Note in Figure 17.5 that Equation (17.12) conservatively underestimates the generally unmeasurable actual $f_T^{(x)}$, while Equation (17.15) significantly overestimates the actual f_{max}.

It is important to realize two things about G_u. First, it is a narrow-band concept, since, strictly speaking, the lossless network N works as intended only at the fixed frequency it was designed for. Second, as pointed out by Mason, G_u is not necessarily the largest gain obtainable from the device. A lossy network, as would typically be involved for conjugate match [8], can give larger gain. A gain based on simultaneous conjugate match of input and output of the device was introduced by Rollett [12]. This becomes the maximum available gain

$$G_{ma} = \left| \frac{Y_{21}}{Y_{12}} \right| \left[k - (k^2 - 1)^{1/2} \right] \tag{17.16a}$$

(MAG in Section 11.1.2), where k is the 'stability factor'

$$k = \frac{2\,\mathrm{Re}(Y_{11})\,\mathrm{Re}(Y_{22}) - \mathrm{Re}(Y_{12}Y_{21})}{|Y_{12}Y_{21}|}. \tag{17.16b}$$

For $k > 1$, which occurs for sufficiently large frequencies (> 100 GHz for 0.1-μm InP-type MODFETs), the FET is unconditionally stable, i.e. it will not oscillate with any passive loads. For these high frequencies, Equation (17.16) is valid for the maximum available gain. For most applications (< 100 GHz) the device is potentially unstable.

* So is G_i, but the effect, barely noticeable in Figure 17.5(a), sets in at much lower frequencies.

It can be stabilized by input and/or output shunt resistors [13] without affecting Y_{21} or Y_{12}. Thus, for $k \leq 1$ the maximum stable gain is

$$G_{ms} = \left| \frac{Y_{21}}{Y_{12}} \right| . \tag{17.17}$$

G_{ms} and G_{ma} can indeed be larger than G_u, as illustrated by the example in Figure 17.5 in a small frequency range near $k = 1$. Generally G_{ma} is not easily extrapolated since it is only defined at large (usually untestable) frequencies, and there its slope varies significantly. However, as illustrated in Figure 17.5(b), its *actual* cut-off frequency is equal to the *actual* f_{max}. While this has been proven to be fundamentally true [14], the fact that the *actual* $f_T^{(x)}$ in Figure 17.5 is also essentially equal to the actual f_{max}, is a coincidence.

For optimization purposes, focusing, as we will in the following section, on the *actual* power gain cut-off frequency is a good choice for several reasons. First, the two power gains considered reflect how the device could ultimately work in a well-designed system. Second, it avoids misleading extrapolation. Third, by focusing our attention on the power gain *cut-off* frequency, rather than on some fixed frequency range, we avoid having to deal with possible resonances (G_u) and the stability border (G_{ma}/G_{ms}) moving in and out of this range as we vary physical parameters. We are interested in a general 'broad-band' optimization of the device physical parameters. Narrow-band optimization is left to the circuit designer. Fourth, the equality of the actual power gain cut-off frequencies lends an important uniqueness to the common value, which is also referred to as the maximum frequency of oscillation. The actual f_{max} is easily found with a bisectioning root-finding algorithm. With regards to resonances in G_u, it is interesting that these occur much more rarely with the consistent treatment of output capacitance and output conductance in Sections 15.3 and 15.4, as compared to choosing the two independently [10, 11]. Resonances in G_u do still occur, but generally only for very small r_{ga} and r_{gi}.

17.5 Optimization of f_{max}

Figure 17.6 shows the dependence of the cut-off frequencies as the device is taken into velocity saturation, for various gate lengths. $f_T^{(x)}$ peaks near $\Delta L_x = 400$ Å, essentially independently of L_g. Further into saturation $f_T^{(x)}$ drops because the gate–source capacitance increases faster (Figure 15.9) than the gate–drain capacitance drops (Figure 15.6). This is qualitatively consistent with all experimental data. However, the effect is more severe in actual devices than indicated by Figure 17.6. This is because of three related effects. First, the effective electron saturation velocity eventually drops below the peak velocity in Figure 14.5 as the high-field region gets wider and the electrons scatter more. Second, the peak velocity itself is reduced

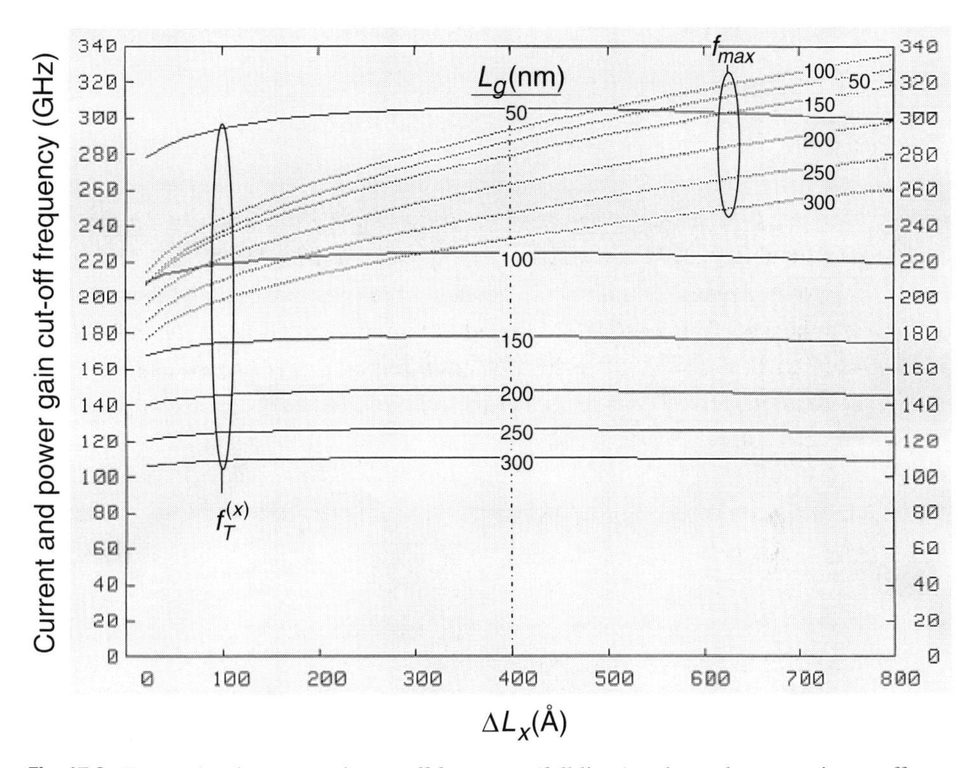

Fig. 17.6. Extrapolated current-gain cut-off frequency (full lines) and actual power gain cut-off frequency (dotted lines) versus the external extent of the velocity-saturated region. The gate length is varied in steps of 50 nm from 50 to 300 nm. Other parameters have default values (Table 17.1). The vertical dashed-dotted line indicates an essentially gate-length-independent flat optimum for $f_T^{(x)}$ at $\Delta L_x = 400$ Å.

due to lattice heating as phonons are emitted in these scattering events. Third, with increased scattering the electrons will not traverse the velocity-saturated region in a simple straightline trajectory. Our assumptions will thus fail deep in saturation, and the delays will be longer than we have predicted. For more accurate predictions in this regime fully numerical models that include energy balance have to be employed. The lower effective saturation velocity will also reduce the DC transconductance. The effect on f_{max} is that in reality it does not continue to increase with ΔL_x as indicated in Figure 17.6, but rather tends to saturate before dropping at a lower rate than $f_T^{(x)}$. Up to peak $f_T^{(x)}$ our semianalytical model is adequate, and we will thus do our optimization at $\Delta L_x = 400$ Å. Another figure-of-merit sometimes quoted for the device (mostly when it is large) is $f_{max}/f_T^{(x)}$. The higher this quotient is, the lower is the relative effect of parasitics. As Figure 17.6 illustrates, it is difficult to keep $f_{max}/f_T^{(x)}$ large as the gate length is shrunk.

Figure 17.7 shows the cut-off frequencies for the MODFET in Table 17.1 versus gate length with the interfacial gate resistance r_{gi} as a parameter. r_{gi} is varied from

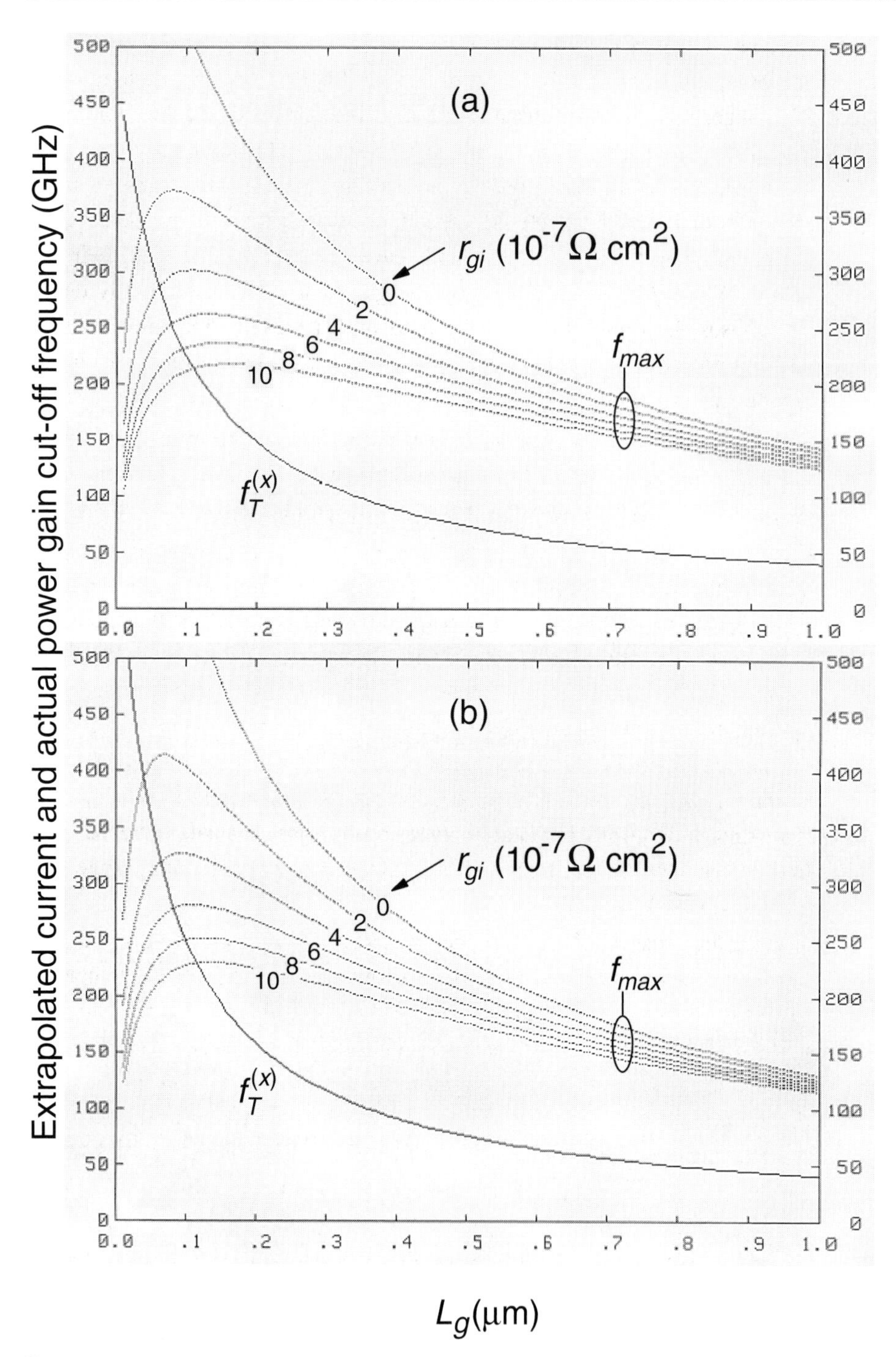

Fig. 17.7. Current gain (full lines) and power gain cut-off frequency (dotted lines) versus gate length for two values of gate-to-channel spacing: (a) $d_{gc} = 200$ Å, (b) $d_{gc} = 100$ Å. The interfacial gate resistance is varied as indicated. Gate access resistance and inductance are set to zero to avoid distributed effects along the gate finger. Other parameters have default values (Table 17.1).

the ideal (but unrealistic) value of zero to the excessively high (but still occasionally occuring) value of $1 \times 10^{-6}\ \Omega\,\text{cm}^2$. The $3 \times 10^{-7}\ \Omega\,\text{cm}^2$ default value in Table 17.1 appears to be the lowest value that is reproducibly achievable in a practical fabrication environment (Sections 16.3 and 16.8). In this first optimization step we are interested in the basic performance of the device, not degraded by distributed effects and fixed layout-related capacitances. We thus set r_{ga} and l_{ga} to zero for the calculations in Figure 17.7, and leave the pad parasitics to their zero default value. While $f_T^{(x)}$ is unaffected by the gate resistance, r_{gi} strongly degrades f_{max}, and causes the optimum gate length to increase as r_{gi} increases. As opposed to earlier optimizations [11], $f_T^{(x)}(L_g)$ does not peak, but continues to increase as L_g is reduced (albeit significantly more slowly than L_g^{-1}). The main reason is that the intrinsic output conductance $g_{ds0}^{(i)}$, degrading $f_T^{(x)}$ according to Equation (17.12), is now calculated as described in Section 15.4, with a resulting gate-length dependence (Figure 15.12(b)) less steep than the L_g^{-1} dependence assumed in the earlier work (Equation (15.26)). The more analytical estimate of $g_{ds0}^{(i)}$ is also the reason why f_{max} is now predicted to peak at shorter gates. Thinning the gate-to-channel distance by a factor of 2 reduces the output conductance and the feedback capacitance (relative to the total gate capacitance). The result in Figure 17.7(b) is a moderate increase in both cut-off frequencies for shorter gates. However, 0.1 μm remains approximately the optimum gate length for general broad-band circuit applications. 0.1-μm gates can be fabricated with good yield by several methods, as discussed in Section 17.7, and are viable for production. Experimental gates as short as 30 nm have been made [15]. The rather simple semianalytical model is remarkably accurate in predicting the cut-off frequencies even for this extreme cutting-edge case (by year 2000 standards).

With the cross-sectional geometry determined, we now look at the effect of gate metallization resistance. Figure 17.8 shows the two cut-off frequencies of interest versus total gate width of the two-finger FET in Table 17.1. Included is the case of a small (0.5 fF) fixed capacitance associated with the pads, one for each of the two gate branches (Figure 17.2). A pad-related capacitance could come about because the calibration does not take it out, or because the system drifts out of calibration. As illustrated by Figure 17.8, f_{max} is unaffected by this, while $f_T^{(x)}$ for small FETs can be significantly affected. If the cause is a calibration drift, $f_T^{(x)}$ of smaller FETs could appear either high or low. With the optimization being based on f_{max}, the width of a gate finger should not be much larger than 25 μm for the default r_{ga}. If a total gate width significantly larger than 50 μm is required for high output power at high frequencies (Section 14.6.5), it has to be made up of more than two fingers. As discussed briefly at the end of Chapter 16, the simultaneous distribution of gate voltage to the fingers of large FETs can become an additional challenge and design issue. From the practical standpoint of process evaluation it is valuable to have a variety of gate-finger widths. Occasionally, as was demonstrated in Chapter 16, a little

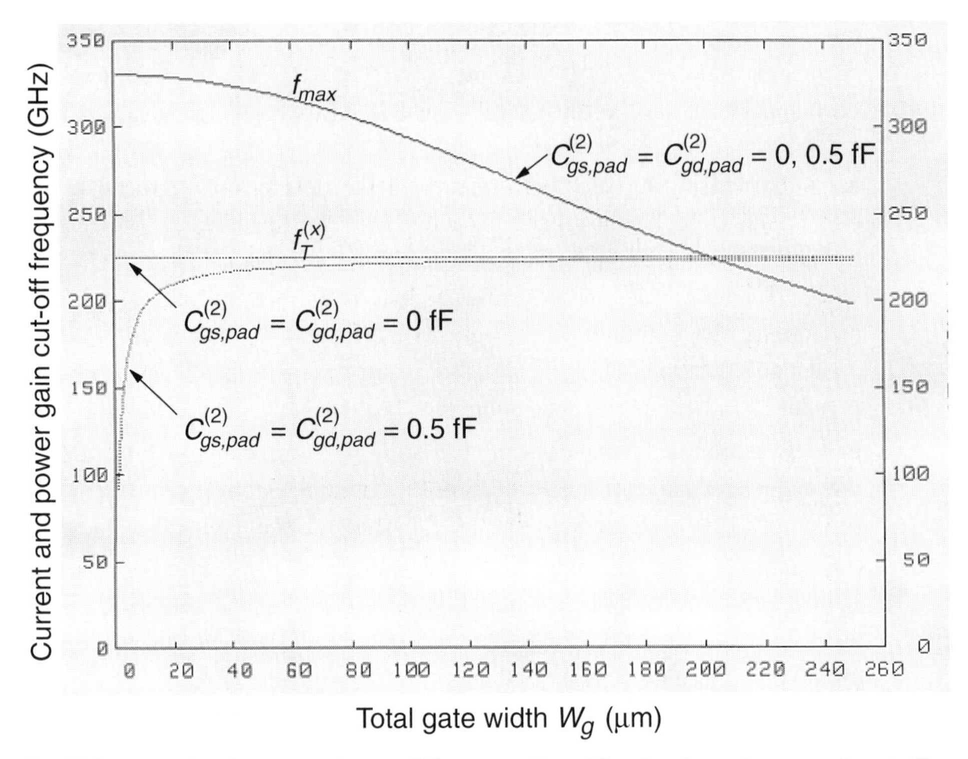

Fig. 17.8. Extrapolated current-gain cut-off frequency (dotted lines) and actual power gain cut-off frequency (solid line) for the two-finger MODFET in Table 17.1 versus total gate width. Two values for pad parasitics were assumed: $C^{(2)}_{gs,pad} = C^{(2)}_{gd,pad} = 0$ and 0.5 fF.

splurging with mask and wafer real estate can lead to a deeper understanding of the device physics.

Figure 17.9 shows how the performance of the optimized device is affected if the components in the gate resistance differ from the assumed default values. The dependence of f_{max} on the standard metallization resistance is rather weak, while a change in the Schottky interface properties can have a large effect. If r_{gi} could be reduced significantly, sub-0.1-μm gates would become worth pursuing for further improvement in the power gain.

17.6 Noise, noise figure and associated gain

In addition to the high-frequency gains discussed in the preceding section and the maximum output power discussed briefly in Section 14.6.5, the noise characteristics of the device are of great importance. While the high-frequency noise tends to be lower in devices with higher cut-off frequencies because of reduced channel resistance (the intrinsic source of thermal noise), the opposite is the case for low-frequency

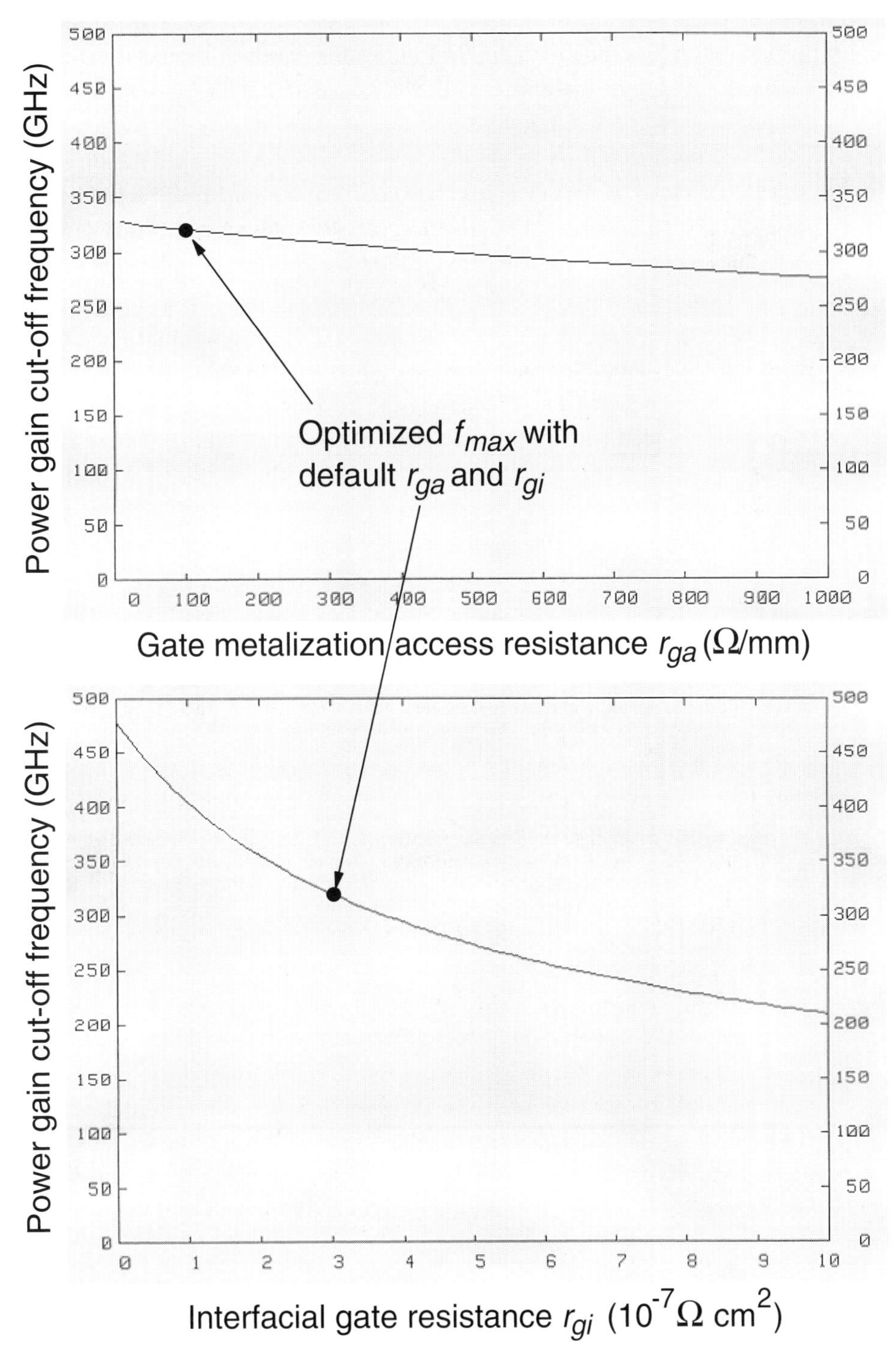

Fig. 17.9. Sensitivity of the power gain cut-off frequency for the optimized $0.1 \times 50\ \mu m^2$ two-finger MODFET in Table 17.1 on variations in the two gate resistance components.

excess noise [16]. Low-frequency noise is particularly difficult to model because of the multiple complex origins and intricate frequency dependence. The dominant origin of low-frequency noise for MESFETs and MODFETs appears to be generation–recombination at traps in a depletion region of the device, but the amount of noise, and even the spectral shape ($1/f^{\alpha}$, $\alpha = 1$–2) is next to impossible to predict with much accuracy. Easier to predict is the upconversion of the low-frequency noise to oscillator phase noise [17]. Heterojunction bipolar transistors (HBTs) with their low low-frequency noise are better suited for oscillator applications. Beyond the corner frequency where the low-frequency noise drowns in the background noise of thermal origin, MODFETs are quieter than HBTs, and are thus well suited for the high-frequency amplification that we have just optimized for.

Also high-frequency noise modeling can be quite complex. There are thermal noise sources associated with all resistive elements in the device, i.e. channel and parasitic access resistances (Figure 17.10(a)). Thermal noise, or Johnson noise, occurs in a resistor because of random electron thermal motion. In a bandwidth B of the frequency domain that we are working in, a resistor R at temperature T can be represented by a noiseless resistor R in series with a phasor noise voltage e_R with a mean-square value given by

$$\overline{|e_R|^2} = \overline{e_R^* e_R} = 4kTRB. \tag{17.18a}$$

This is the famous Nyquist theorem, which is most easily arrived at by connecting the resistor to a capacitor and equating the energy stored in the capacitor due to the noise voltage with $kT/2$ according to the equipartition theorem (e.g. [18]). The thermal noise from the resistor is practically white, i.e. the spectral density $\overline{|e_R|^2}/B$ has no frequency dependence up to 10^{12}–10^{14} Hz [19, 20]. By Norton's theorem, an alternative equivalent representation is the noiseless resistor in parallel with a white noise current source j_R:

$$\overline{|j_R|^2} = \overline{j_R^* j_R} = 4kT\frac{1}{R}B. \tag{17.18b}$$

The voltage (e) and current (j) noise sources in Figure 17.10(a) do in general have a frequency dependence, but only because they are the result of having applied Thevenin's or Norton's theorem to the subnetworks that make up the respective impedance (Z) or admittance (Y). In fact, a linear one-port (like an individual Y and Z in Figure 17.10(a)) with all its resistive components at the same temperature T can be represented by its noiseless impedance Z in series with an AC noise voltage e_Z with

$$\overline{|e_Z|^2} = \overline{e_Z^* e_Z} = 4kT\,\mathrm{Re}\,(Z)\,B, \tag{17.19a}$$

or by its noiseless admittance Y in parallel with an AC noise current j_Y with

$$\overline{|j_Y|^2} = \overline{j_Y^* j_Y} = 4kT\,\mathrm{Re}\,(Y)\,B. \tag{17.19b}$$

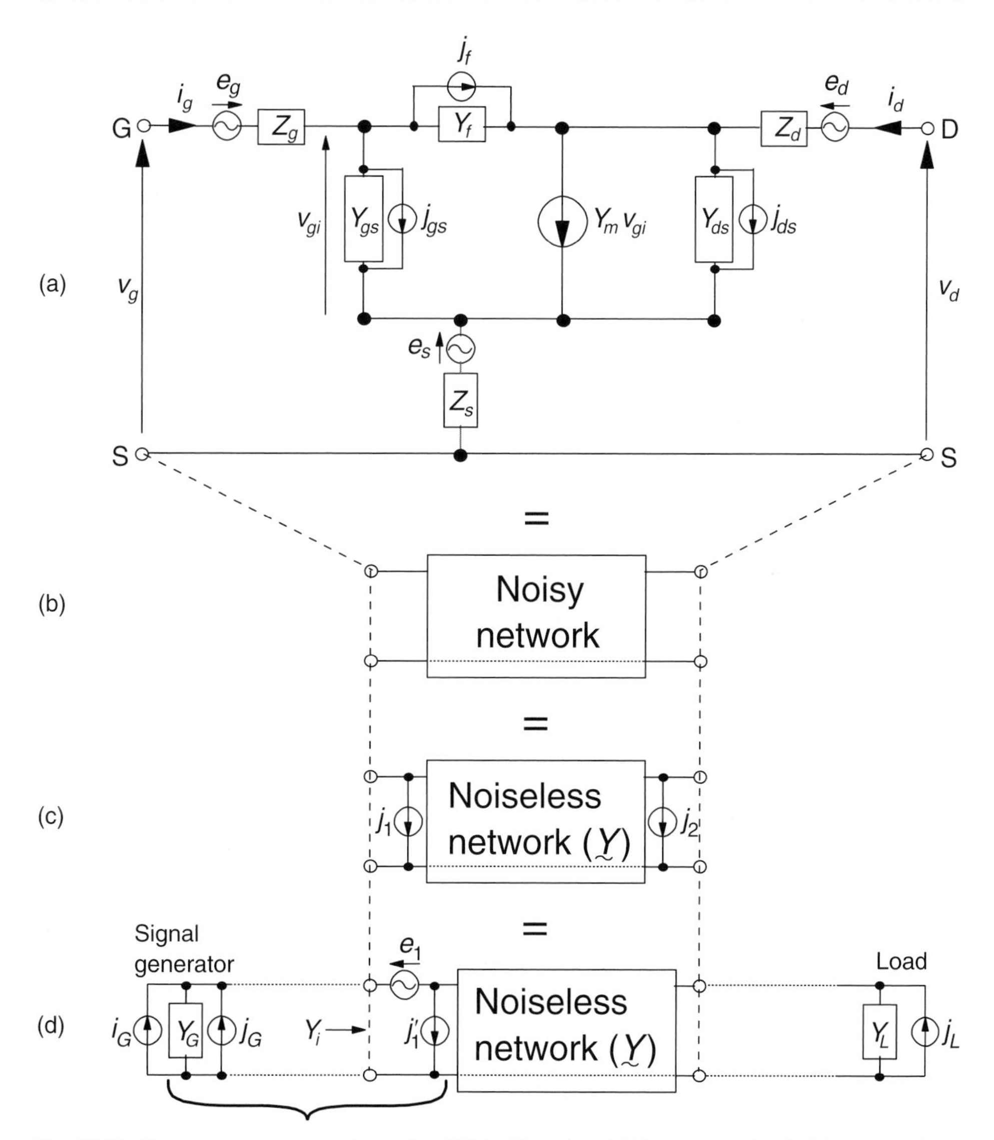

Fig. 17.10. Equivalent representations of an FET with noise. (a) Equivalent circuit (Figure 16.13) with thermal noise sources at their physical locations. (b) Standard two-port box representation with noise sources still inside. (c) Short-circuit noise current sources brought outside a now noiseless FET. (d) Equivalent representation using two noise sources on the input, one voltage, one current. In (d) is indicated how one would use the network for amplifying a signal from a generator with an admittance Y_G. The output load admittance is Y_L and Y_i is the input impedance loading the generator. Noise currents are denoted by j, noise voltages by e.

A frequency dependence in $\mathrm{Re}(Z)$ and $\mathrm{Re}(Y)$ will show up in the associated noise source spectrum. Just like an internal one-port component can be represented by a noiseless element attached to a noise source, so can the entire two-port in Figure 17.10(b), except that two noise sources are required. Determining the terminal noise sources in Figure 17.10(c) or (d) becomes a circuit analysis problem. The two equivalent representations in Figure 17.10(c) or (d) are related by:

$$j_1' = j_1 - \frac{Y_{11}}{Y_{21}} j_2; \tag{17.20a}$$

$$e_1 = -\frac{j_2}{Y_{21}}. \tag{17.20b}$$

The internal sources are uncorrelated, and all cross-terms (e.g. $\overline{j_{gs}^* j_{ds}}$) will be zero. The two terminal sources, however, will be correlated, since each internal source will contribute to both.

The noise figure F is the most common description of the noisiness of an amplifier. It depends on both the individual rms values and the correlation cross-product of the terminal noise sources. F also depends on the admittance $Y_G = G_G + jB_G$ of the generator driving the amplifier (Figure 17.10(d)). The generator is at the temperature T_0, which under most test conditions is near the so-called standard temperature 290 K (typical room temperature), and generates noise accordingly (Equation (17.19)). The maximum noise power that can be delivered to the amplifier, the so-called available noise power, is (Problem 17.2)

$$P_{AN_i} = kT_0B. \tag{17.21a}$$

We have attached a subscript N for noise, and i for input (to the amplifier). The only parameter of the generator that affects its available noise power is its temperature T_0. The available signal input power is given by (Problem 17.2)

$$P_{AS_i} = \frac{|i_G|^2}{4\,\mathrm{Re}\,(Y_G)}. \tag{17.21b}$$

Whether noise or signal, the power delivered to the amplifier is maximum with so-called conjugate matching, i.e. when $Y_i = Y_G^*$. The input noise is amplified, and to this the amplifier adds its own internally generated noise. The noise figure F is defined as the total noise power (in the bandwidth B) coming out of the amplifier divided by the fraction which is purely amplified input noise. Thus,

$$F \equiv \frac{P_{AN_o}}{G_A P_{AN_i}} = \frac{P_{AS_i}/P_{AN_i}}{P_{AS_o}/P_{AN_o}} = \frac{(S/N)_i}{(S/N)_o}, \tag{17.22}$$

where subscript o stands for output from the amplifier. G_A is its so-called available gain defined by

$$G_A = \frac{P_{AS_0}}{P_{AS_i}} \tag{17.23}$$

Defining the gain of an amplifier in terms of available (rather than the actual) input and output powers is a fairer measure of the amplifier, since it also reflects how well the input is matched to the preceding stage. It would not do much good to design an amplifier in which the power absorbed by the input is amplified with a large gain, if only a small fraction of the incoming power is actually absorbed. In the last two steps of Equation (17.22) we have rewritten the noise figure in terms of input and output signal-to-noise ratios. In dB ($F^{(dB)} = 10\log_{10}(F)$), the noise figure becomes the inevitable loss in signal-to-noise ratio from input to output, a number a circuit designer would strive to minimize. This may make it clearer why it is F (as opposed to, say, P_{N_o}) which is typically the object of minimization. For a hypothetical noiseless amplifier $F = 1$. In reality $F > 1$, and the design of a low-noise amplifier (LNA) involves minimizing F. The circuit designer does this, for the device at hand, by choosing an optimum generator admittance Y_G for the frequency band of interest. A device designer is interested in minimizing $F_{min}(f) = F\left(Y_G^{(F)}(f)\right)$, where $Y_G^{(F)}(f)$ is the generator admittance for minimum noise figure at frequency f. For fixed f, F_{min} depends only on device parameters, since these determine the optimum Y_G.

The noise figure of a multi-stage amplifier can be expressed in terms of the noise figures and available gains of the individual stages (Problem 17.3):

$$F_{123\ldots n} = F_1 + \frac{F_2 - 1}{G_{A_1}} + \frac{F_3 - 1}{G_{A_2}G_{A_1}} + \cdots + \frac{F_n - 1}{G_{A_{n-1}} \cdots G_{A_2}G_{A_1}}. \tag{17.24}$$

In the present context, the main importance of Equation (17.24) is to illustrate that the first stage (subscript 1) is the most important for overall low noise, if its gain is sufficiently large. InP-type MODFETs are thus of particular interest as gain elements in the first stage of an amplifying system. At the end of the chain, output power becomes more important. Because of the breakdown issues discussed in Section 14.6.5, pseudomorphic high-electron-mobility transistors (PHEMTs) are better suited here.

17.6.1 The FET noise model by Pucel, Haus and Statz

Pucel, Haus and Statz (PHS) [21] developed a theory for noise in GaAs metal–semiconductor FETs (MESFETs) that has been the basis for much of the later work, including that which led to the famous Fukui equation for F_{min} [22]. The PHS theory neglects several elements in the equivalent circuit (Figure 15.16(a)) that are of secondary importance to the noise figure. The intrinsic elements considered were C_{gs}, R_{gs} and $g_{m0}^{(i)}$. In addition, R_s and R_g, the two most important parasitic resistors to degrade the gain and generate noise, were included. Transit-time effects leading to the ω dependence in Equation (15.63c) were neglected. The first part of the PHS analysis concerned the intrinsic device. j_1 was modeled as induced by the channel noise coupling capacitively to the gate electrode. j_2 is more directly tied to the channel noise, which was assumed to have two origins associated with the two regions under the gate. The gradual channel contributes thermal noise, enhanced by field-dependent electron heating [23]. The velocity-saturated part of the channel contributes what amounts to diffusion noise. This comes about as follows. Using the Einstein relationship $D = kT\mu/q$ for the diffusion constant D, the thermal noise current (Equation (17.18b)) in a segment $(y, y + \Delta y)$ of the FET channel can be reformulated in terms of D:

$$\overline{|j_c(y, \Delta y)|^2} = 4q^2 BD(y)n_s(y)W_g \frac{1}{\Delta y}. \tag{17.25a}$$

Thus the thermal noise picture can be extended into the velocity-saturated region, where the mobility is non-applicable, but where diffusion still takes place. However, to make this picture compatible with neglecting diffusion in the DC analysis (Chapter 14), Statz *et al.* [24] reinterpreted $\overline{|j_c(y, \Delta y)|^2}$ in the saturated region as an effective shot noise current. Shot noise occurs when a current I flows from a cathode as a result of electrons being randomly emitted at an average rate $r = I/q$. The probability of a certain number of electrons being emitted in a time interval is given by the Poisson distribution, which is characterized by having the same average as it has variance. The shot noise rms current is given by

$$\overline{|j_I|^2} = \overline{j_I^* j_I} = 2qIB. \tag{17.25b}$$

In [24] an effective I_{eff} was considered such that $\overline{|j_c(y, \Delta y)|^2} = 2qI_{eff}B$. Noise current pulses occur in parallel with a channel element Δy at an average rate $r = I_{eff}/q = 2Dn_s W_g/\Delta y$. Each event creates a noise dipole which is assumed to drift unchanged, at the saturated velocity, towards the drain while inducing a drain open circuit noise voltage. The analyses leading to expressions for the intrinsic FET equivalent gate and drain noise currents are quite complex. The two extrinsic elements R_s and R_g are then introduced, and the three-term Taylor series

$$F_{min} = 1 + 2K_g^{1/2}\left[\frac{\omega C_{gs}}{g_{m0}^{(i)}}\right]\left(K_r + g_{m0}^{(i)}\left(R_g + R_s\right)\right)^{1/2} + 2K_g\left[\frac{\omega C_{gs}}{g_{m0}^{(i)}}\right]^2 g_{m0}^{(i)}\left(K_c R_{gs} + R_g + R_s\right) \tag{17.26a}$$

for the minimum noise figure was derived for $T = T_0$. A correction factor T/T_0 should be applied to the resistances if $T \neq T_0$. The three new parameters that appear are given by

$$K_g = P\left\{\left[1 - C(R/P)^{1/2}\right]^2 + \left(1 - C^2\right)R/P\right\}, \tag{17.26b}$$

$$K_r = \frac{\left(1 - C^2\right)R}{\left[1 - C(R/P)^{1/2}\right]^2 + \left(1 - C^2\right)R/P}, \tag{17.26c}$$

$$K_c = \frac{1 - C(R/P)^{1/2}}{\left[1 - C(R/P)^{1/2}\right]^2 + \left(1 - C^2\right)R/P}, \tag{17.26d}$$

where the real and dimensionless P, R and C are the basic noise parameters determined by the average powers of the intrinsic gate and drain noise currents, and the correlation between the two:

$$P = \frac{\overline{|j_2|^2}}{4kT_0Bg_{m0}^{(i)}}, \tag{17.26e}$$

$$R = \frac{g_{m0}^{(i)}\overline{|j_1|^2}}{4kT_0B\omega^2C_{gs}^2}, \tag{17.26f}$$

$$jC = \frac{\overline{j_1^* j_2}}{\left(\overline{|j_2|^2}\,\overline{|j_1|^2}\right)^{1/2}}. \tag{17.26g}$$

Note in Equation (17.26a) that, except for R_g and R_s, the equivalent circuit elements are intrinsic, and only three of these appear. In particular, C_{gd} is not included, which means that the item inside the square brackets in Equation (17.26a) is smaller than $f/f_T^{(i)}$. Note also that R_{gs} only appears explicitly in the ω^2 term. It also shows up implicitly to some degree in K_r, but apparently not to the extent that $K_r + g_{m0}^{(i)}\left(R_g + R_s\right)$ can be replaced by the suggestive form $g_{m0}^{(i)}\left(R_{gs} + R_g + R_s\right)$.

17.6.2 The Fukui equation and Pospieszalski's thermal model

It is probably clear from the brief description of the theory in [21, 24] that the basic parameters P, R and C are quite difficult to calculate, and vary with bias in complicated ways. Even so, there are several effects left out, such as external velocity saturation and non-stationary transport leading to electron heating not directly related to the local field. These are important effects for today's deep submicron devices, as are some of the equivalent circuit elements left out of the analysis. Device researchers have in general taken the semiempirical approach introduced by Fukui [22]. He found that the expression

$$F_{min} = 1 + K_f\left(\frac{\omega C_{gs}}{g_{m0}^{(i)}}\right)\left[g_{m0}^{(i)}\left(R_g + R_s\right)\right]^{1/2}, \tag{17.27a}$$

with the fitting parameter $K_f \sim 2.5$, gave good predictions for the minimum noise figure of GaAs MESFETs at ambient temperature $T_a = T_0$, regardless of what bias was required for F_{min}. The Fukui factor K_f has always been considered a materials-related parameter. Its value has been reduced over the years as devices with higher-mobility channels have been developed. The Fukui equation (Equation (17.27a)) is a simplified and truncated form of Equation (17.26a). The theory in [21, 24] would say that

$$K_f = 2K_g^{1/2}. \tag{17.27b}$$

The Fukui equation suggests that R_{gs} should not appear in any form as a noise source. Of course, we know from Chapter 16 that it is hard to separate R_{gs} and R_g, so it is not inconceivable that R_g in Equation (17.27a) should actually be $R_{gs} + R_g$. It seems natural that R_{gs} should appear beside R_g with similar weight. A difference in the weight would arise if R_g and R_{gs} were associated with different temperatures. This viewpoint was introduced by Pospieszalski [25] whose noise model is an interesting alternative to the Fukui and PHS models. Pospieszalski's is a thermal model for the intrinsic device using the two resistive components R_{gs} and $g_{ds0}^{(i)}$ in Figure 15.16(a) as the intrinsic noise sources. These are associated with the temperatures T_g and T_d, respectively. The benefit of the approach is that it contains the basic physics prescribed by Nyquist's thermal noise theorem, while lumping all the complexity, which can never be accounted for with analytical models anyway, into just the two adjustable parameters T_g and T_d. The approach is supported by two experimental observations. First, the two temperatures are frequency-independent, as expected for resistive noise sources. Second, T_g is essentially equal to the ambient temperature T_a, as expected for a component largely associated with a region of the device (the gradual channel) where there is only moderate electron heating. Deeper into velocity saturation, as discussed briefly in Sections 16.2 and 16.3, delays increase the effective R_{gs}, and T_g may at the same time increase. However, low-noise FETs usually are not biased in a regime where this would be a dominant effect. Another reason for T_g to be close to T_a arises when the interfacial gate resistance R_{gi} is accounted for by R_{gs} (Chapter 16), since R_{gi} is expected to be even less affected by electron heating in the channel. T_d is significantly larger than T_a, but quite reasonably so, considering the heating of the electrons in the velocity-saturated part of the channel which dominates the output conductance $g_{ds0}^{(i)}$. We can attempt a very rough estimate of T_d for a typical InP-type MODFET (Figure 14.1). Even for typical low-noise biases ($V_D \sim 1$ V) there will be some impact ionization in the channel (Section 14.6.5). That means that a fraction of the electrons will have gained at least the threshold energy $E_{Tn} \sim 0.8$ eV (Section 14.6.4). For $\Delta L_i + \Delta L_x$ smaller than, or on the order of, the mean free path, this applies to a substantial fraction of the electrons. As the electrons traverse the high-field region, starting out at ~ 0.1 eV, a reasonable estimate for their average energy then is ~ 0.4 eV. Equating this with $3kT_d/2$ leads to $T_d \sim 3100$ K. An alternative estimate, based on electron energies in published energy-based numerical modeling of various HEMTs, suggests T_d in the range 2200–4600 K. The two estimates are consistent with each other and, generally, with published measured noise figures.

Pospieszalski's model is equivalent to that in [21] if $C = (R/P)^{1/2}$, which leads to $K_r = R$, $K_g = P - R$, and $K_c = 1$. In terms of the two noise temperatures, $K_g = \left(g_{ds0}^{(i)} T_d\right) / \left(g_{m0}^{(i)} T_0\right)$ and $K_r = g_{m0}^{(i)} R_{gs} \left(T_g / T_0\right)$. Insertion of these into Equation (17.26a), keeping only the first two terms, produces the modified Fukui equation

$$F_{min} = 1 + K_f \left(\frac{\omega C_{gs}}{g_{m0}^{(i)}}\right) \left[g_{m0}^{(i)} \left(R_{gs}\frac{T_g}{T_0} + R_g\frac{T_a}{T_0} + R_s\frac{T_a}{T_0}\right)\right]^{1/2}. \tag{17.28a}$$

This is richer in content than the original equation for two reasons. First, R_{gs} is included explicitly beside R_s and R_g, as argued intuitively above. Second, K_f can now be estimated from physical electron transport considerations with the formula

$$K_f = 2K_g^{1/2} = 2\left(\frac{g_{ds0}^{(i)} T_d}{g_{m0}^{(i)} T_0}\right)^{1/2}. \tag{17.28b}$$

Note that, although $g_{ds0}^{(i)} \ll g_{m0}^{(i)}$, the fact that $T_d \gg T_0$ makes $g_{ds0}^{(i)}$ an important noise source. With $g_{m0}^{(i)}$ tied to the electron saturation velocity v_{sat} (Equation (14.54)), K_f is indeed affected by material quality. $g_{ds0}^{(i)}$ depends on the gate length and the extent of the velocity-saturation region (Figure 15.12). There is thus also geometry and V_D dependence in K_f. The strongest V_D dependence, more than compensating for the drop in $g_{ds0}^{(i)}$ with V_D, however, comes from T_d, as argued above. Alternative expressions for K_f that include an explicit drain current dependence have been proposed (e.g. [26, 27]). The powerful and physically appealing simplification of introducing T_d and T_g is somewhat analogous to lumping the complexities of high-field transport into the two parameters $\mu(L_g)$ and v_{sat} (Section 14.5). However, compared to these (and T_g), T_d is more variable, and less predictable.

17.6.3 General formalism for noise figure and power gain

We have neglected the effect of the resistive (and therefore noisy) components g_{gs} and g_{gd}. For high-speed FETs used at lower frequencies, e.g. DBS (direct broadcast satellite) applications, the noise contribution from these can become non-negligible [28]. At high frequencies, the neglect of delays, feedback capacitance, etc., also introduces errors. With the lumped equivalent circuit we developed in Chapters 15 and 16 (Figure 16.13), the general thermal noise source expressions in Equation (17.19) and Pospieszalski's thermal model, we can take the noise analysis a bit further. The theory for dealing with noise in linear two-ports was developed by Rothe and Dahlke [29]. This work is the foundation for most transistor noise models. Appendix L of Gonzalez's textbook [20] provides a nice rendition that we will essentially follow. We use the representation in Figure 17.10(d). P_{N_i} in the definition of noise figure (Equation (17.22)) is as always given by Equation (17.21a). P_{N_0} is most easily found by applying Norton's theorem to the bracketed portion of Figure 17.10(d). We then find that

$$F = 1 + \frac{\overline{|j_1' + e_1 Y_G|^2}}{\overline{|j_G|^2}}. \tag{17.29}$$

We split j_1' into two parts, one correlated with e_1 through an admittance Y_c, the other (j_{1u}') uncorrelated:

$$j_1' \equiv Y_c e_1 + j_{1u}'. \tag{17.30a}$$

The correlation admittance is given by

$$Y_c = \frac{\overline{e_1^* j_1'}}{\overline{|e_1|^2}} = G_c + jB_c. \tag{17.30b}$$

We also define an effective noise resistance

$$R_F \equiv \frac{\overline{|e_1|^2}}{4kT_0B} \tag{17.31}$$

and an effective noise conductance

$$G_u \equiv \frac{\overline{|j_{1u}'|^2}}{4kT_0B} = \frac{\overline{|j_1'|^2}}{4kT_0B} - R_F\,|Y_c|^2. \tag{17.32}$$

Y_c, R_F and G_u can be evaluated once we get specific about the device (Section 17.6.4). Inserting Equations (17.30) and (17.31) into Equation (17.29) yields

$$F = 1 + \frac{G_u}{G_G} + \frac{R_F}{G_G}\left[(G_G + G_c)^2 + (B_G + B_c)^2\right]. \tag{17.33}$$

The minimum noise figure requires the generator admittance $Y_G = Y_G^{(F)} = G_G^{(F)} + jB_G^{(F)}$, where

$$B_G^{(F)} = -B_c \tag{17.34a}$$

and

$$G_G^{(F)} = \left(G_c^2 + \frac{G_u}{R_F}\right)^{1/2}. \tag{17.34b}$$

The minimum noise figure is

$$F_{min} = 1 + 2R_F\left(G_G^{(F)} + G_c\right), \tag{17.35}$$

and for non-optimum loads the noise figure is given by

$$F = F_{min} + \frac{R_F}{G_G}\left[\left(G_G - G_G^{(F)}\right)^2 + \left(B_G - B_G^{(F)}\right)^2\right]. \tag{17.36}$$

Loci in the Y_G-plane of constant noise figures ('iso-noise figures') are non-concentric circles. With the effective noise resistance R_F appearing in both terms of Equation (17.36), it is not surprising that a small F_{min} (determined by the device) is important, not only in itself, but also in promoting broad-noise circles [30], i.e. in producing a flatter minimum.

It is clear from Equation (17.24) that, for an overall low noise figure of a system that includes several stages of amplification, the first stage should not only have a low noise figure, but also a high gain. The available gain for the device is given by (Problem 17.4) [31]

$$G_A = \frac{|Y_{21}|^2\,\mathrm{Re}\,(Y_G)}{\mathrm{Re}\,(Y_{22})\,|Y_{11} + Y_G|^2 - \mathrm{Re}\left[Y_{12}Y_{21}\,(Y_{11} + Y_G)^*\right]}. \tag{17.37}$$

The often quoted 'associated gain' is the available power gain at $F = F_{min}$. This is in general not the maximum gain, which occurs at an optimum generator admittance $Y_G^{(G_A)}$ generally different from $Y_G^{(F)}$. The admittance $Y_G^{(G_A)} = G_G^{(G_A)} + jB_G^{(G_A)}$ is most easily found by first solving $\partial G_A^{-1}/\partial B_G = 0$, yielding

$$B_G^{(G_A)} = -\operatorname{Im}(Y_{11}) + \frac{\operatorname{Im}(Y_{12}Y_{21})}{2\operatorname{Re}(Y_{22})}, \quad (17.38a)$$

and then solving $\left(\partial G_A^{-1}/\partial G_G\right)_{B_G = B_G^{(G_A)}} = 0$, yielding

$$G_G^{(G_A)} = \frac{|Y_{12}Y_{21}|}{2\operatorname{Re}(Y_{22})}\left(k^2 - 1\right)^{1/2}, \quad (17.38b)$$

where k is the stability factor (Equation (17.16b). Inserting these optimum values into Equation (17.37) one finds that this equals Rollett's maximum available gain, quoted without derivation in Equation (17.16a). During the derivation of Equation (17.38) it becomes clear that the denominator in Equation (17.37) consists of two second order polynomials in G_G and B_G, with no cross-terms, and with the same coefficient for B_G^2 as for G_G^2. We can thus write

$$\frac{1}{G_A} = \frac{1}{G_{ma}} + \frac{R_{G_A}}{G_G}\left[\left(G_G - G_G^{(G_A)}\right)^2 + \left(B_G - B_G^{(G_A)}\right)^2\right], \quad (17.39a)$$

where we have defined the resistance

$$R_{G_A} = \frac{\operatorname{Re}(Y_{22})}{|Y_{21}|^2}. \quad (17.39b)$$

The form in Equation (17.39a) was used by Fukui [31] because of its interesting similarity with the noise figure (Equation (17.36)). Loci in the Y_G-plane of constant gain ('iso-gains') are also non-concentric circles. Unfortunately, the noise figure is not minimum for the same source impedance that maximizes the gain, and a compromise has to be made. If R_{G_A} and R_F are small, the loss in gain and/or noise figure as a result of the compromise can be kept small.

17.6.4 Noise figure and associated gain of the MODFET

We apply this general theory to the noisy FET in Figure 17.10 (cf. [32]). The first step is to derive the two terminal noise currents in Figure 17.10(c). In addition to j_1 and j_2, there are three unknown currents, namely those flowing through Y_{gs}, Y_f and Y_{ds}. Solving two node current and three loop voltage equations yields expressions for the five unknowns. The elements in the equivalent circuit in Figure 17.10(a) are given by Equations (16.59)–(16.61). The associated noise sources are determined by Equation (17.19). The noise temperature for the sources is assumed to be the ambient temperature T_a, except for Y_{ds}, which we associate with T_d. External capacitors do not contribute to the short-circuit noise currents j_1 and j_2, but do affect the Y parameters as discussed in Section 17.2. This ultimately influences the noise figure to some degree, because of the transformation of (j_1, j_2) to (j_1', e_1) (Equation (17.20)). The initial circuit analysis leads to

$$j_1 = \frac{(1 + A_{22}) I_1 - A_{12} I_2}{\Delta_A}, \tag{17.40a}$$

$$j_2 = \frac{(1 + A_{11}) I_2 - A_{21} I_1}{\Delta_A}, \tag{17.40b}$$

where we have introduced

$$A_{11} = Y_{gs}\left(Z_g + Z_s\right) + Y_f Z_g, \tag{17.41a}$$

$$A_{12} = Y_{gs} Z_s - Y_f Z_d, \tag{17.41b}$$

$$A_{21} = Y_{ds} Z_s + Y_m\left(Z_g + Z_s\right) - Y_f Z_g, \tag{17.41c}$$

$$A_{22} = Y_{ds}\left(Z_d + Z_s\right) + Y_m Z_s + Y_f Z_d \tag{17.41d}$$

and

$$\Delta_A = (1 + A_{11})(1 + A_{22}) - A_{12} A_{21}. \tag{17.41e}$$

These are dimensionless parameters, and functions of the equivalent circuit elements. I_1 and I_2 are 'composite' noise currents:

$$I_1 = j_{gs} + j_f + \left(Y_{gs} + Y_f\right) e_g - Y_{gs} e_s - Y_f e_d, \tag{17.42a}$$

$$I_2 = j_{ds} - j_f + \left(Y_m - Y_f\right) e_g - \left(Y_m + Y_{ds}\right) e_s + \left(Y_{ds} + Y_f\right) e_d. \tag{17.42b}$$

The transformation to input-side noise sources yields

$$e_1 = Z_1 I_1 + Z_2 I_2, \tag{17.43a}$$

$$j_1' = K_1 I_1 + K_2 I_2, \tag{17.43b}$$

where we have introduced

$$Z_1 = \frac{A_{21}}{Y_{21} \Delta_A}, \tag{17.44a}$$

$$Z_2 = -\frac{1 + A_{11}}{Y_{21} \Delta_A} \tag{17.44b}$$

and

$$K_1 = \frac{1 + A_{22} + A_{21} \frac{Y_{11}}{Y_{21}}}{\Delta_A}, \tag{17.45a}$$

$$K_2 = -\frac{A_{12} + (1 + A_{11}) \frac{Y_{11}}{Y_{21}}}{\Delta_A}. \tag{17.45b}$$

Recall that our goal is to calculate R_F, Y_c and G_u (Equations (17.30)–(17.32)). These involve $\overline{|e_1|^2}$, $\overline{\left|j_1'\right|^2}$ and $\overline{e^* j_1'}$, which are given in terms of the quantities introduced:

$$\overline{|e_1|^2} = |Z_1|^2\,\overline{|I_1|^2} + |Z_2|^2\,\overline{|I_2|^2} + 2\,\mathrm{Re}\left[Z_1^* Z_2\left(\overline{I_1^* I_2}\right)\right], \tag{17.46a}$$

$$\overline{\left|j_1'\right|^2} = |K_1|^2\,\overline{|I_1|^2} + |K_2|^2\,\overline{|I_2|^2} + 2\,\mathrm{Re}\left[K_1^* K_2\left(\overline{I_1^* I_2}\right)\right], \tag{17.46b}$$

$$\overline{e^* j_1'} = Z_1^* K_1 \overline{|I_1|^2} + Z_2^* K_2 \overline{|I_2|^2} + Z_1^* K_2\left(\overline{I_1^* I_2}\right) + Z_2^* K_1\left(\overline{I_1^* I_2}\right)^* . \tag{17.46c}$$

Remaining to be evaluated are the three quantities $\overline{|I_1|^2}$, $\overline{|I_2|^2}$ and $\left(\overline{I_1^* I_2}\right)$. Remembering that the individual noise sources on the right-hand side of Equation (17.42) are uncorrelated, this is easily done with the help of Equation (17.19):

$$\overline{|I_1|^2} = 4kT_a B\,\mathrm{Re}\left(Y_{gs} + Y_f + \left|Y_{gs} + Y_f\right|^2 Z_g + \left|Y_{gs}\right|^2 Z_s + \left|Y_f\right|^2 Z_d\right), \tag{17.47a}$$

$$\begin{aligned}\overline{|I_2|^2} = {} & 4kT_a B\,\mathrm{Re}\left[\left(\frac{T_d}{T_a}\right) Y_{ds} + Y_f + \left|Y_m - Y_f\right|^2 Z_g \right.\\ & \left. + |Y_m + Y_{ds}|^2 Z_s + \left|Y_{ds} + Y_f\right|^2 Z_d\right],\end{aligned} \tag{17.47b}$$

$$\begin{aligned}\left(\overline{I_1^* I_2}\right) = {} & 4kT_a B\big[-\mathrm{Re}\left(Y_f\right) + \left(Y_{gs} + Y_f\right)^*\left(Y_m - Y_f\right)\mathrm{Re}\left(Z_g\right)\\ & + Y_{gs}^*\left(Y_m + Y_{ds}\right)\mathrm{Re}\left(Z_s\right) - Y_f^*\left(Y_{ds} + Y_f\right)\mathrm{Re}\left(Z_d\right)\big].\end{aligned} \tag{17.47c}$$

Equations (17.41) and (17.44)–(17.47) allow us to calculate all the noise parameters in Section 17.6.3, and in particular the two central device parameters, F_{min} and the associated gain. Since here we are interested exclusively in the device itself, we do not include external impedances ($L_{sx} = L_{dx} = L_{gx} = 0$), but use the full complex lumped equivalent $Z_g = Z_{ga} + R_{gi}$ (Equations (16.59) and (16.61a)) for the gate 'resistance'. Figure 17.11 shows minimum noise figure and gain versus frequency, for the default MODFET in Table 17.1. The noise parameter values are $T_g = T_0 = 290$ K and $T_d = 3100$ K. In Figure 17.11(a) we adjusted five of the default parameters to make the device in effect have the simplified equivalent circuit in [21]. Thus, we set the drain resistance and gate leakage to zero, and choose values for $C_{gs}^{(f)}$ and $C_{gd}^{(f)}$ that result in zero total drain-source and gate–drain capacitance. We can then compare the full noise model with Equations (17.28) and (17.26a), setting $K_g = \left(g_{ds0}^{(i)} T_d\right) / \left(g_{m0}^{(i)} T_0\right)$, $K_r = g_{m0}^{(i)} R_{gs}\left(T_g/T_0\right)$ and $K_c = 1$. It is clear that a first order equation is insufficient to model F_{min} at high frequencies of interest for InP-type MODFETs. The second order expression, although not strictly applicable*, comes much closer (within ~ 0.05 dB at 50 GHz). Actually, the full model predicts $F_{min}^{(dB)}$ to

* The K_g, K_r and K_c that lead to equivalency between the PHS model and Pospieszalski's are, strictly speaking, only applicable up to first order in frequency [33].

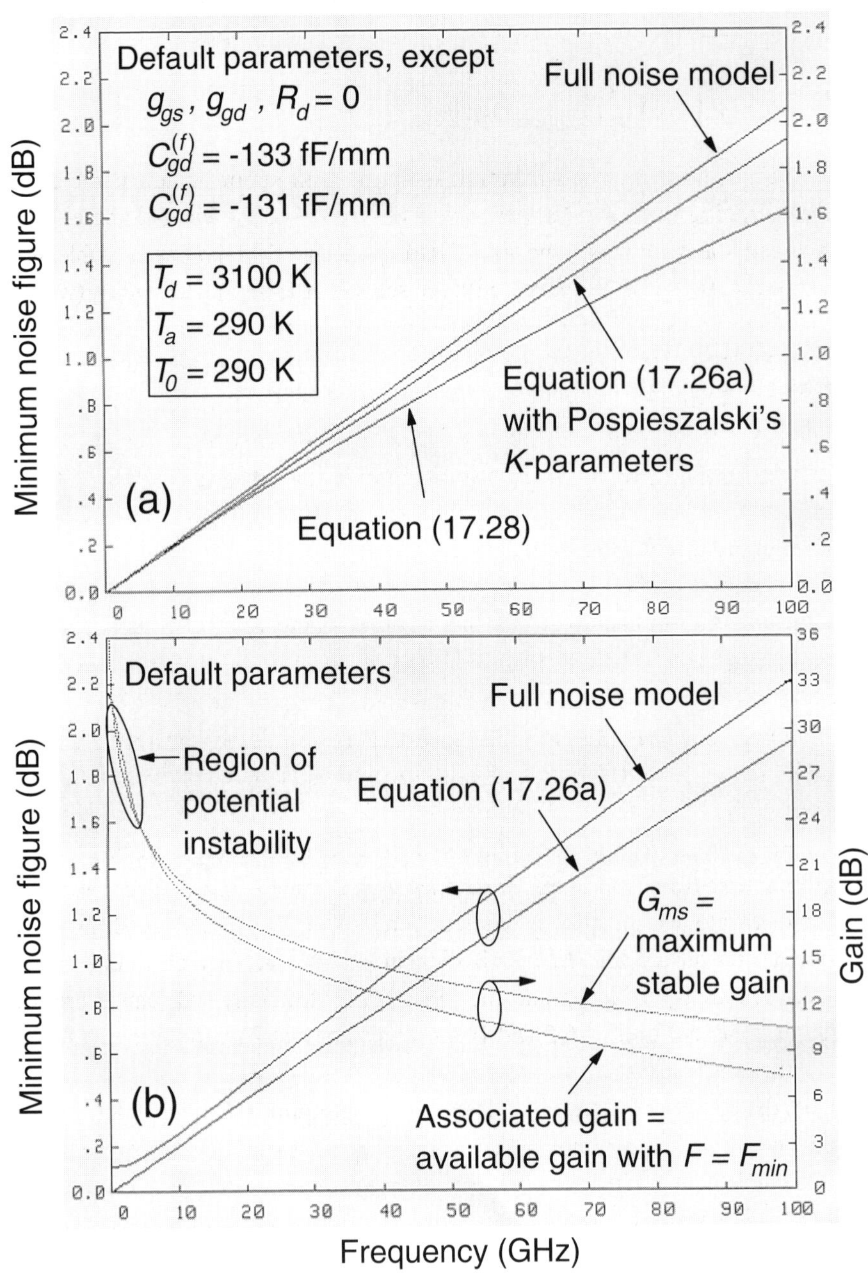

Fig. 17.11. (a) Comparison of the minimum noise figure predicted by Equations (17.26) and (17.28) with that predicted by the full noise model in Sections 17.6.3–17.6.4. The default parameters in Table 17.1 were modified to make the comparison meaningful. (b) Similar, but with the default gate leakage, drain resistance, gate–drain capacitance and drain-source capacitance reinstated. Associated and maximum stable gain are included in (b). The noise temperatures are the same for (a) and (b).

be almost linear with frequency (cf. [34]). Thus a better, more conservative estimate for the minimum noise figure with the simplified equivalent circuit might actually be

$$F_{min} = \exp\left(K_f \left(\frac{\omega C_{gs}}{g_{m0}^{(i)}}\right)\left[g_{m0}^{(i)}\left(R_{gs}\frac{T_g}{T_0} + R_g\frac{T_a}{T_0} + R_s\frac{T_a}{T_0}\right)\right]^{1/2}\right). \tag{17.48}$$

In Figure 17.11(b), we have reinstated the full set of default parameters in Table 17.1, keeping the noise temperatures the same. For this realistic case, we have also included the gain, both the associated gain and the maximum stable gain (Equation (17.17)). The discrepancy with the second order estimate of F_{min} (which does not include drain resistance, gate leakage and gate–drain and drain-source capacitance) is now larger. At low frequencies the discrepancy is mainly due to the gate leakage. At 50 GHz it is mostly due to the neglect of gate–drain feedback capacitance ($\sim$0.13 dB), while R_d contributes to a smaller degree ($\sim$0.04 dB). Figure 17.11(b) also shows the gain traded off for low noise at high frequencies ($\sim$2 dB at 50 GHz). For low-frequency applications, one may have to reduce the gain below the optimum for noise in order to stabilize the device.

With a fixed T_d (and T_g) F_{min}, as predicted with the 'full' noise model, has a minimum with respect to ΔL_x rather close to where $f_T^{(x)}$ peaks in Figure 17.6. Below the optimum ΔL_x, F_{min} increases primarily because $g_{ds0}^{(i)}$ increases (Figure 15.12), while above the optimum, F_{min} increases primarily because the gate–source capacitance increases (Figure 15.9). The noise model is, however, not completely 'full', since there are several effects still not accounted for. The most important of these is the dependence of T_d on bias, and thus on ΔL_x. For InP-type MODFETs $T_d = 3100$ K was chosen, based on the electron energy, to be a representative intermediate value between two rather well-defined limits. The lower limit is the onset of velocity saturation, required for large $g_m^{(i)}/g_{ds}^{(i)}$ and good gain. The upper limit is the onset of impact ionization in the channel, which initially may cause a kink, eventually will degrade the noise, and ultimately will cause breakdown and degradation (Section 14.6.5 and 14.7). It is clear that T_d will increase as V_D increases and the electrons attain higher energies. Another effect not accounted for is the dependence of $g_m^{(i)}$ on ΔL_x. Below the optimum ΔL_x the reduced $g_m^{(i)}$ and the reduced T_d counteract one another in their effect on F_{min}. Above the optimum $\Delta L_x$the continued increase in T_d causes a significant increase in F_{min}. The actual optimum ΔL_x is thus shifted negatively. In addition to its V_D dependence, T_d increases rapidly at a fixed V_D, as the device is turned on by an increasing V_G [35]. This leads to an optimum I_D for noise that is less than that at maximum transconductance, which is the bias point assumed for the small-signal AC modeling. Still, the model in Sections 17.6.3–17.6.4, together with a representative, physically based choice of T_d, provides a tool for the device engineer to optimize for low noise.

If a deeper understanding of noise-related complexities is required, it takes methods (e.g. [36, 37]) which go beyond the analytical modeling approach we have adopted throughout these sections as a practical aid to understanding and optimizing a MODFET.

17.7 Process and manufacturability issues

Figure 17.12(a) shows a three-stage MMIC amplifier using 0.1-μm InP-type HEMTs. In addition to three 100-μm wide four-finger FETs, there are other important components. The circuit is coplanar in the sense that everything is on top of the wafer, including the ground plane. This is a convenient and workable choice for this circuit. A back-side via process and ground plane would allow microstrip transmission lines to be used. In Figure 17.12(a) the coplanar transmission lines feed the signals to and from the chip, and perform interstage matching. Thin film resistors are used as loads in the first two stages, in the feedback loops of the first and last stages, and in the biasing of the gates. Spiral inductors are used in the optimization of the frequency response, a measurement of which is shown in Figure 17.12(c). The design of this high-frequency MODFET-based circuit, with all its additional lumped and distributed components, requires the CAD tools that we have sprinkled references to throughout our treatise of the device itself. The very high gain over a very wide frequency range is made possible by the careful device optimization that the previous sections have dealt with, and serves as a good illustration of why MODFET technology is of great interest. Figure 17.12(b) illustrates the uniformity and yield possible even in a cutting-edge FET technology like this, when sufficient thought and work has gone into the process. This crucial aspect will be discussed briefly in this section. Reliability, the third important component of a successful process, was discussed in Section 14.7.

Figure 17.13 shows the epitaxial structure, grown by molecular beam epitaxy (MBE) (Chapter 1), of a state-of-the art InP-type HEMT. It may not be the most basic or typical structure for this type of device, but the heart of the structure, comprising the channel, spacer, supply and Schottky-barrier layers, is quite typical. As we have seen, these layers determine the basic device characteristics. The rest of the structure is designed with manufacturability in mind. There are many alternative approaches to achieve this, but the example in Figure 17.13 is good in that it illuminates several issues. We are careful to refer to this HEMT as InP-*type*, rather than InP-*based*, since there actually is no InP in this structure. The high-mobility channel material is $Ga_{0.47}In_{0.53}As$, which usually requires InP as a substrate. Relative to GaAs, however, InP substrates are more expensive and brittle, particularly for larger wafer sizes ($\geq 3''$). One can, as illustrated in Figure 17.13, overcome this obstacle by implementing a buffer technology that allows the use of GaAs substrates [39, 40]. The first layer grown is a linearly-graded low-temperature buffer (LGLTB) in which the lattice constant is varied from that of GaAs (5.65 Å) to that of InP (5.89 Å), by gradually replacing Ga in $Al_{0.48}Ga_{0.52}As$ with In. After ~ 1 μm of material growth, the grading is complete.

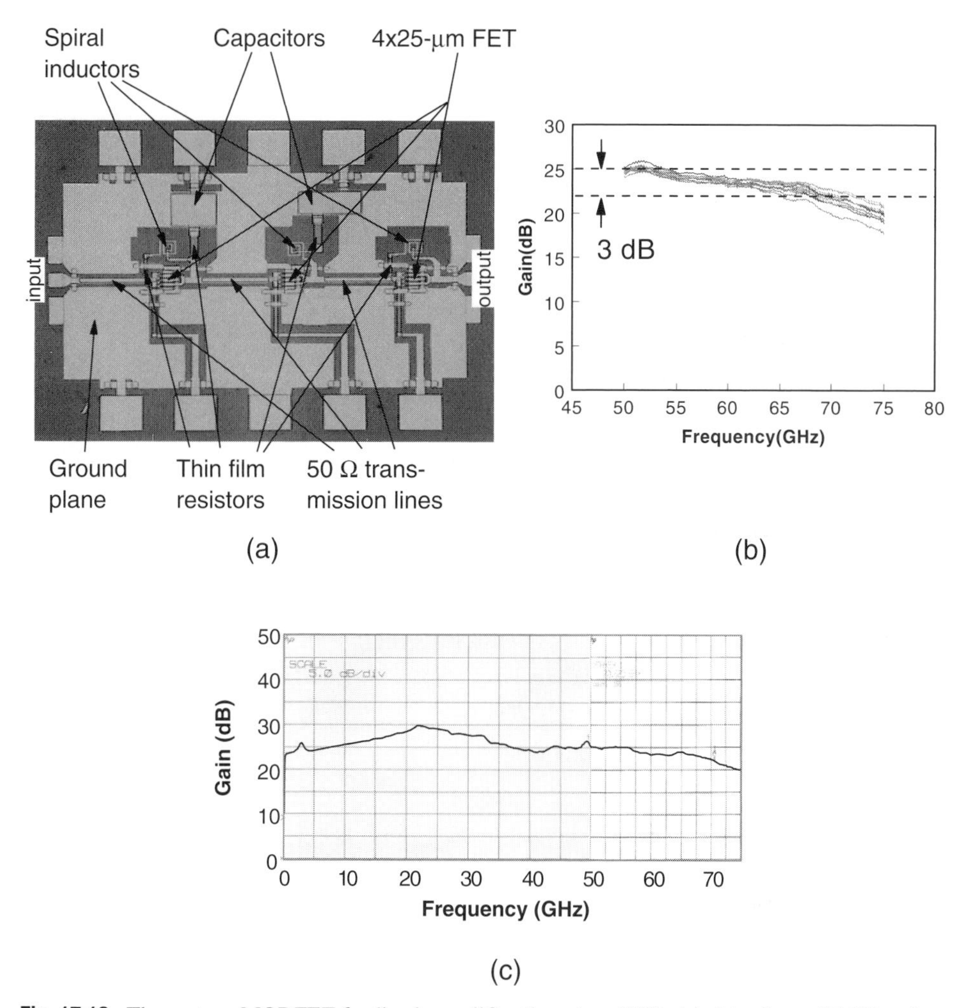

Fig. 17.12. Three-stage MODFET feedback amplifier (based on [38]): (a) chip photo; (b) V-band gain for several amplifiers on a 2″ wafer; (c) wide-band gain measurement of a reoptimized amplifier. ((a) and (b) reproduced with permission from H. Rohdin, A. Wakita, A. Nagy, V. Robbins, N. Moll and C.-Y. Su, *Solid State Electronics*, Vol. 43, pp. 1645–1654, 1999.)

The buffer is completed by a ∼0.25-μm layer of $Al_{0.48}In_{0.52}As$ on which any desired combination of layers lattice-matched to InP can be grown. Misfit dislocations are generated during the grading, but the vast majority remains confined to the LGLTB. The threading dislocation density in the device layers does not reduce the yield of typical circuits of interest below that on wafers utilizing InP as a substrate. Equally important is that the device/circuit performance [41, 10] and reliability [42] do not suffer. The LGLTB adds to the growth time, but subtracts substrate cost and wafer breakage. It provides for a seamless fit of this high-performance InP-type FET process with existing GaAs IC manufacturing infrastructure. This can be an overriding economic advantage, particularly with 6″ GaAs substrates.

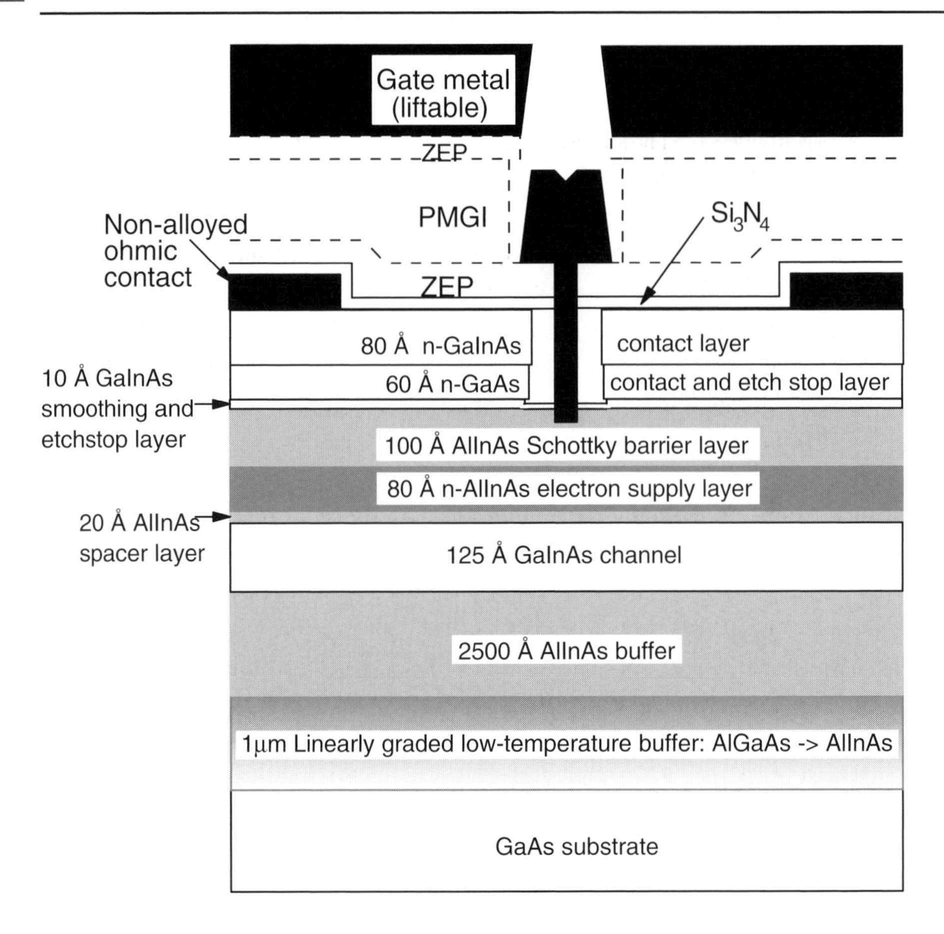

Fig. 17.13. Epitaxial structure and gate definition of a 0.1-μm $In_{0.5}Ga_{0.5}As$-channel MODFET on GaAs substrate. (H. Rohdin, A. Wakita, A. Nagy, V. Robbins, N. Moll and C.-Y. Su, *Solid State Electronics*, Vol. 43, pp. 1645–1654, 1999.)

One interesting aspect of the LGLTB approach is that it allows some freedom in the choice of In mole fraction. The conduction-band offset ΔE_C for the 2DEG is maximum for $\sim 30\%$ In, which could lead to a larger maximum carrier concentration n_{s0} (Section 14.4) and breakdown voltage BV_D (Section 14.6.5) [43]. However, it also makes it harder, if not impossible, to achieve good contact resistance (R_c) with non-alloyed ohmic contacts, because of the $\sim 70\%$ Al in the Schottky-barrier layer. Non-alloyed ohmic contacts are in general preferable to alloyed ones because of their good reproducibility. The control and reproducibility of MBE and metal deposition are typically better than of alloying processes. One of the two purposes of the top two layers in Figure 17.13 is to provide for low parasitic source and drain resistances. Low contact resistance is accomplished by the choice of material (GaInAs and GaAs) and by high Si doping (6×10^{18}–3×10^{19} cm^{-3}). This results in a low barrier for the electrons, and thus low tunneling resistance (Section 14.6.2).

The second purpose of the top two layers in Figure 17.13 is to provide for a well-defined, uniform and reproducible threshold voltage. This is obviously of importance for circuit yield. Threshold control is accomplished by a two-step selective recess process [44]. The first (wet) etch should remove the top GaInAs layer at a rate that is significantly higher than in the underlying GaAs. This allows for sufficient lateral etch of the GaInAs before the GaAs layer is consumed. The amount of lateral etch affects several of the DC parameters encountered in Chapter 14, such as $I_D^{(max)}$ and $BV_D^{(off)}$ [10, 41], as well as $V_D^{(knee)}$ since R_S and R_D will have a non-negligible component associated with the etched higher-resistance regions ([45]; Section 14.6.1). Excessive lateral etch can introduce an unacceptably large 'kink' in the drain I–V characteristics [46], which degrades $I_D^{(knee)}$. There is no lateral etch stop, but with good etching procedures the lateral extent can be kept under repeatable control [44]. The second etch (for the remaining GaAs) can be a dry (plasma) reactive ion etch (RIE) which stops on Al-containing layers [47, 48, 49]. Alternatively, wet chemical etches that stop on the underlying layers with a large ($\sim 50\%$) In mole fraction, can be used [44]. The selectivity of the second etch, whether wet or dry, is the key to threshold control.

The thickness of the GaAs etch layer (started on a thin GaInAs layer [44]) has to be chosen with care. If it is too thick it will not be coherent with the layers below (i.e. dislocations will be generated), and the doping efficiency will be reduced dramatically, preventing a low R_c. If it is too thin, the first etch will consume it and some uncontrollable fraction of the underlying AlInAs Schottky-barrier layer, leading to a non-uniform V_T. Although Matthews and Blakeslee's [50] theory leads to a 40 Å critical thickness for GaAs clad by $Ga_{0.47}In_{0.53}As$, the actual critical thickness for layers in tensile stress can be significantly larger than the theoretical one [51]. The layer structure in Figure 17.13 was designed for standard solid-source MBE. With MOCVD, an approach utilizing an InP etch-stop layer [46] is an excellent alternative.

The T-shaped gate illustrated in Figure 17.13 is defined by: (1) direct e-beam writing in a trilayer resist [52], (2) development of the exposed resist [52], (3) RIE of the bottom thin nitride layer, (4) recess etching (as just discussed), (5) metal evaporation, and (6) 'lifting off' the excess metal sitting on the unexposed resist stack (by immersing the wafer in a solvent). The gate length is defined by the cut in the bottom layer resist. The cut in the top layer resist, which is of the same kind as the bottom, is larger because of a more concentrated developer. The different middle layer resist has after selective development the widest lateral extent, allowing for easy lift-off. Gates on the order of 0.1 μm can be defined with this approach. The one-step exposure self-aligns the top of the T-shaped gate with its stem. The approach relies on the excellent selectivity of the developers. In special single-layer resists gate cuts as short as 30 nm can be defined [15]. In this case an additional optical exposure is used to define the wider top of the gate.

In lithography for very short gates, the aspect ratio in the resist cut (resist height divided by the gate length) can become quite large. Since gate metal is not only

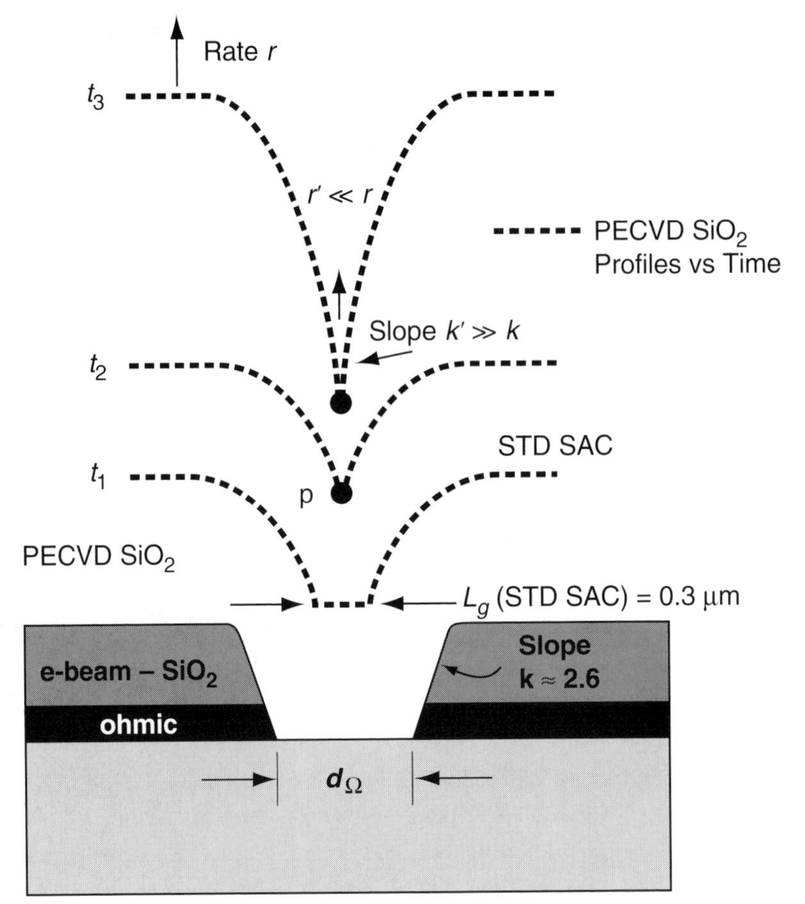

Fig. 17.14. Mechanism for forming a deep narrow notch during conformal PECVD (plasma-enhanced chemical vapor deposition) growth of SiO_2. Prior to this step, ohmic contact metal and SiO_2 were evaporated with the same mask. A self-limiting process develops for $t > t_2$, when the growth rate in the notch is reduced below that in the field [56]. (H. Rohdin and A. Nagy, *Technical Digest of the International Electron Devices Meeting*, IEEE, Piscataway, pp. 327–330, 1992.)

deposited on the semiconductor at the bottom of the cut, but also on the resist edges, the gate mask becomes progressively smaller during the deposition. Starting with a rectangular cut, the resulting bottom stem of the T-gate will be trapezoidal (Figure 14.1). This is the necking phenomenon discussed in Chapter 16. If the aspect ratio is too large, the opening closes completely, thus detaching the then triangular stem from the top of the T. The problem is managed by using a sufficiently thin resist, and can be avoided if the bottom resist layer has sufficiently outward-sloping walls.

The low through-put of direct e-beam writing is a manufacturing and cost issue. Alternative methods for defining ultra-short gates, but based on high through-put optical lithography, have also been developed. These techniques include angle evaporation [53], phase-shifting techniques [54, 55] and self-limiting oxide spacers

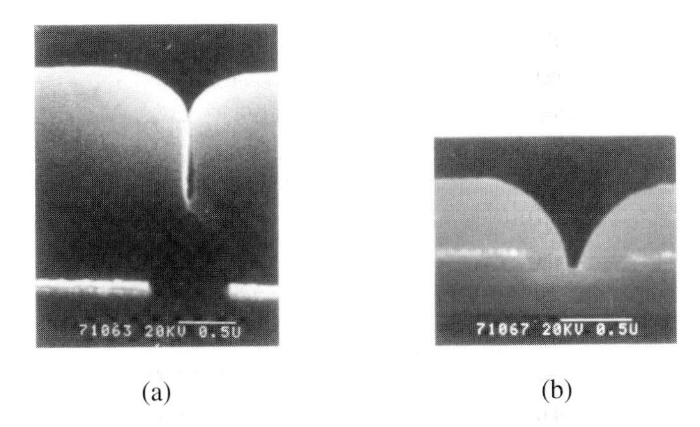

Fig. 17.15. Formation of an ultra-short gate cut by optical lithography and spacer technology: (a) deep notch after completion of the conformal oxide growth illustrated in Figure 17.14; (b) 0.1-μm gate cut created after RIE back-etching [56]. (H. Rohdin and A. Nagy, *Technical Digest of the International Electron Devices Meeting*, IEEE, Piscataway, pp. 327–330, 1992.)

[56]. The last is based on the behavior of conformal growth of oxide between optically defined ohmic contacts. Figure 17.14 shows schematically, and Figure 17.15(a) actually, how a deep notch can develop. After anisotropic RIE back-etch of the oxide, a deep submicron gate cut remains, as shown in Figure 17.15(b). One of the interesting features of this approach is that the uniformity of the final cuts is better than the uniformity of the micrometer-size interohmic spacing. This is a result of the self-limiting feature of the notch formation. This type of approach has been shown to be well suited for production [57]. With the slope of the oxide in Figure 17.15(b), there is no necking problem with this approach.

In a good T-gate process, the gate metallization access resistance r_{ga} (along the dimension into the paper in Figures 17.13–17.15) is kept low ($\leq 100\ \Omega$/mm). However, as we have seen in Chapter 16, this is only one of two components that make up the total gate resistance of the FET. A typically more dominant one is associated with the metal–semiconductor interface. Measures should be taken to improve the quality of this interface as much as possible. Some measures that work well are: (1) using an effective cleanup dip prior to evaporation; (2) minimizing the delay to evaporation; and (3) using a gate metal that can be controllably sintered into the semiconductor, while consuming remaining native oxide and/or contaminants. From the standpoints of a clean interface and of barrier height, a good gate metal appears to be Pt, which has been studied and used by workers in the III–V FET field for quite some time [58]. Because of the ~ 1.4 scaling factor between sintering depth and initial Pt thickness, the Pt thickness must be limited [59, 60, 44].

Other steps necessary to complete the MMIC process include device isolation, nitride deposition for device encapsulation (or passivation) and capacitors, and at least two levels of interconnect metallization. We refer the interested reader to Williams

[61] where these 'back-end' processes and many other process-related topics are discussed.

17.8 Reverse modeling

If the answer to the question that we posed in the introductions to Chapters 14 and 17 is 'no', inverse, or reverse, modeling [62, 63, 64] can be quite useful. As the name implies, this involves using measured device data, and backing out underlying basic physical parameters, the most important intrinsic ones being L_g, v_{sat} and d_{gc}. To get good estimates of these, one preferably measures the Y parameters of the device under conditions that make the theory in Chapters 14 and 15 particularly applicable, and easy to 'invert'. First, intermediate frequencies are used, so that Equations (15.68)–(15.70) are valid. Second, the gate is biased for optimum modulation near maximum $g_{m0}^{(x)}$ and $f_T^{(x)}$, so that Equations (14.54), (15.1) and (15.2) and are applicable, and the degrading effects of parasitic capacitances are kept to a relative minimum. The drain bias should be large enough to velocity saturate the device, but small enough to leave the intrinsic fringing essentially symmetric ($C_{gd,sat} = \Delta C_{gs}$), and $\tau_{g,sat}^{(1)}$ negligible. If we assume also that the extrinsic fringing is symmetric ($C_{gs}^{(f)} = C_{gd}^{(f)}$), and that R_s and R_d can be measured at DC as discussed in Section 14.6.2, Equations (15.68b), (15.68e), (15.69b), (15.69e), (15.70b), (15.70e), (15.70h) and (15.70l) can be used to calculate the intrinsic parallel-plate gate capacitance C_0 (Equations (15.1) and (15.2)), and the intrinsic transconductance (Equation (14.54)). This gives us two equations for the three unknowns L_g, v_{sat} and d_{gc}. The third piece of information can come from either Y_{11} measurements on a FATFET [63], or direct measurement of L_g in an SEM (scanning electron microscope). A FATFET is a FET with larger well-known gate length (typically several micrometers). Using this as a 'reference', however, has two limitations. First, there may be some uncertainty in the gate bias at which to evaluate the gate capacitance [63]. Second, it requires that the recess depth is gate-length-independent. This in turn requires either a very good wet etching technique (e.g. [44], Figure 3), or a selective etch. Neither may be available at the time of troubleshooting by reverse modeling. The SEM approach has the disadvantage of being destructive unless done on a lithographically nominally identical test structure. Its advantage is its accuracy, and, when done on a cleaved FET, it provides an opportunity to check the lateral recess determining L_{us} and L_{ud}.

An alternative way of getting the third piece of information is to rely on lessons learned from earlier reverse modeling. One such important lesson on GaAs-based MODFETs is the constant v_{sat} for gate lengths ranging from 0.07 to 0.7 μm, and the essential equality of v_{sat} and v_{peak} (Figure 14.5) [63, 56]. Based on measured $f_T^{(x)}$ and forward modeling, this also holds true for InP-based MODFETs, even down to 30 nm gates [15].

The experimental threshold voltage that corresponds to the theoretical V_T in Equation (14.4) is measured at a fixed low V_D ($< V_{Dsat}$). The inflection-point tangent of $I_D(V_G)$ is then used to extrapolate to zero I_D at $V_G = V_T'$. The threshold voltage can then be calculated as $V_T = V_T' - V_D/2$. The materials parameters that go into V_T (Φ_B, ΔE_C, E_{Fo}, ε) are rather well established, as is Δd (Equation (14.5)). Thus, with some additional reliable MBE-related information (doping and/or layer thicknesses) the properly measured V_T can also be of help when troubleshooting a process.

We will not delve further into the topic of reverse modeling, and simply refer the interested reader to the references, with the one further comment that, as processes evolve, so can the assumptions used in the reverse modeling. One example is the determination of the source resistance. In the 0.25–0.7 μm gate symmetric self-aligned process studied in [63] extra care was taken when determining R_s. In later 0.1-μm non-self-aligned processes the relative contribution to the measured R_s from the gated channel is negligible, and the DC measurement outlined in Section 14.6.2 is typically sufficient. Another parameter that could excessively degrade the device performance, particularly the gain and noise figure, is the parasitic gate resistance. Its measurement and separation into physical components, were discussed in Section 16.3.

17.9 Conclusion

In this last chapter on MODFETs we have, as an example, applied the analytical theory developed over several chapters to the optimization of an InP-type MODFET. We focused on the actual f_{max} since this cut-off frequency is particularly fundamental. We went on to cover noise in some depth, and included a physically based thermal noise model that includes all equivalent circuit elements. Processing issues that affect not only the performance, our primary focus, but also manufacturability and yield, were briefly discussed by means of examples from industry. Finally, an even briefer discussion of reverse modeling rounded off the section on MODFETs. The last two chapters in the book deal with modeling and material/processing issues of another important device for high-speed applications: the heterojunction bipolar transistor.

17.10 Bibliography

[1] S. Lautzenhiser, A. Davidson, K. Jones, 'Improve accuracy of on-wafer tests via LRM calibration', *Microwaves & RF*, Vol. 29, pp. 105–109, 1990.

[2] K. Kurokawa, 'Power waves and the scattering matrix', *IEEE Transactions on Microwave Theory and Techniques*, Vol. 13, pp. 194–202, 1965.

[3] R.E. Collin, *Foundations for Microwave Engineering*, 2nd Ed., McGraw-Hill, New York, 1992.

[4] D.E. Root, S. Fan and J. Meyer, 'Technology independent large signal non quasi-static FET models by direct construction from automatically characterized device data', *Proceedings of the European Microwave Conference*, pp. 927–932, 1991.

[5] A.P. Freundorfer, 'Millimeter-wave design. Short course', *21st Annual IEEE Gallium Arsenide IC Symposium*, IEEE, Piscataway, 1999.

[6] L.D. Nguyen, P.J. Tasker and W.J. Schaff, 'Comments on "A new low-noise AlGaAs/GaAs 2DEG FET with a surface undoped layer"', *IEEE Transactions on Electron Devices*, Vol. 34, p. 1187, 1987.

[7] P.J. Tasker and B. Hughes, 'Importance of source and drain resistance to the maximum f_T of millimeter-wave MODFETs', *IEEE Electron Device Letters*, Vol. 10, pp. 291–293, 1989.

[8] G.D. Vendelin, *Design of Amplifiers and Oscillators by the S-Parameter Method*, Wiley-Interscience, New York, 1982.

[9] S.J. Mason, 'Power gain in feedback amplifier', *Transactions of the IRE*, Vol. 1, pp. 20–25, 1954.

[10] H. Rohdin, A. Nagy, V. Robbins, C.-Y. Su, A.S. Wakita, J. Seeger, T. Hwang, P. Chye, P.E. Gregory, S.R. Bahl, F.G. Kellert, L.G. Studebaker, D.C. D'Avanzo and S. Johnsen, '0.1-μm gate-length AlInAs/GaInAs/GaAs MODFET MMIC process for applications in high-speed wireless communication', *Hewlett-Packard Journal*, Vol. 49, pp. 37–38, 1998.

[11] H. Rohdin, N. Moll, C.-Y. Su and G. Lee, 'Interfacial gate resistance in Schottky-barrier-gate field-effect transistors', *IEEE Transactions on Electron Devices*, Vol. 45, pp. 2407–2416, 1998.

[12] J.M. Rollett, 'Stability and power-gain invariants of linear twoports', *Transactions of the IRE*, Vol. 9, pp. 29–32, 1962.

[13] P. Wolf, 'Microwave properties of Schottky-barrier field-effect transistors', *IBM Journal of Research and Development*, Vol. 9, pp. 125–141, 1970.

[14] H.-O. Vickes, 'Gain partitioning: a new approach for analyzing the high-frequency performance of compound semiconductor FETs', *IEEE Transactions on Microwave Theory and Techniques*, Vol. 39, pp. 1383–1390, 1991.

[15] T. Suemitsu, T. Ishii, H. Yokoyama, Y. Umeda, T. Enoki, Y. Ishii and T. Tamamura, '30-nm-gate InAlAs/InGaAs HEMT's lattice-matched to InP substrates', *Technical Digest of the International Electron Devices Meeting*, IEEE, Piscataway, pp. 223–226, 1998.

[16] C.-Y. Su, H. Rohdin and C. Stolte, '$1/f$ noise in GaAs MESFETs', *Technical Digest of the International Electron Devices Meeting*, IEEE, Piscataway, pp. 601–604, 1983.

[17] H. Rohdin, C.-Y. Su and C. Stolte, 'A study of the relation between device low-frequency noise and oscillator phase noise for GaAs MESFETs', *IEEE MTT-S International Microwave Symposium Digest*, IEEE, Piscataway, pp. 267–269, 1984.

[18] A. Van der Ziel, *Noise in Measurements*, John Wiley & Sons, New York, 1976.

[19] F. Reif, *Fundamentals of Statistical and Thermal Physics*, McGraw-Hill Kogakusha, Tokyo, 1965.

[20] G. Gonzalez, *Microwave Transistor Amplifiers: Analysis and Design*, Prentice-Hall, Upper Saddle River, 1997.

[21] R.A. Pucel, H.A. Haus and H. Statz, 'Signal and noise properties of gallium arsenide microwave field-effect transistors'. In *Advances in Electronics and Electron Physics*, ed. L. Marton, Vol. 38, Academic Press, New York, pp. 195–265, 1975.

[22] H. Fukui, 'Optimal noise figure of microwave GaAs MESFETs', *IEEE Transactions on Electron Devices*, Vol. 26, pp. 1032–1037, 1979.

[23] W. Beachtold, 'Noise behavior of GaAs field-effect transistors with short gate lengths', *IEEE Transactions on Electron Devices*, Vol. 19, pp. 674–680, 1972.

[24] H. Statz, H.A. Haus and R.A. Pucel, 'Noise characteristics of gallium arsenide field-effect transistors', *IEEE Transactions on Electron Devices*, Vol. 21, pp. 549–562, 1974.

[25] M.W. Pospieszalski, 'Modeling of noise parameters of MESFETs and MODFETs and their frequency and temperature dependence', *IEEE Transactions on Microwave Theory and Techniques*, Vol. 37, pp. 1340–1350, 1989.

[26] D. Delagebeaudeuf, J. Chevrier, M. Laviron and P. Delescluse, 'A new relationship between the Fukui coefficient and optimal current value for low-noise operation of field-effect transistors', *IEEE Electron Device Letters*, Vol. 6, pp. 444–445, 1985.

[27] A. Cappy, A. Vanoverschelde, M. Schortgen, C. Versnaeyen and G. Salmer, 'Noise modeling in submicrometer-gate two-dimensional electron-gas field-effect transistors', *IEEE Transactions on Electron Devices*, Vol. 32, pp. 2787–2796, 1985.

[28] R. Reuter, S. van Waasen and F.J. Tegude, 'A new noise model of HFET with special emphasis on gate-leakage', *IEEE Electron Device Letters*, Vol. 16, pp. 74–76, 1995.

[29] H. Rothe and W. Dahlke, 'Theory of noisy fourpoles', *Proceedings of the IRE*, Vol. 44, pp. 811–818, 1956.

[30] B. Hughes, 'Designing FETs for broad noise circles', *IEEE Transactions on Microwave Theory and Techniques*, Vol. 41, pp. 190–198, 1993.

[31] H. Fukui, 'Available power gain, noise figure and noise measure of two-ports and their graphical representation', *IEEE Transactions on Circuit Theory*, Vol. 13, pp. 137–142, 1966.

[32] P. Heymann, M. Rudolph, H. Prinzler, R. Doerner, L. Klapproth and G. Bock, 'Experimental evaluation of microwave field-effect-transistor noise models', *IEEE Transactions on Microwave Theory and Techniques*, Vol. 36, pp. 156–163, 1999.

[33] M. Pospieszalski, private communication, 2000.

[34] B. Hughes, 'A linear dependence of F_{min} on frequency for FETs', *IEEE Transactions on Microwave Theory and Techniques*, Vol. 41, pp. 979–982, 1993.

[35] M.W. Pospieszalski and A.C. Niedzwiecki, 'FET noise model and on-wafer measurement of noise parameters', *IEEE MTT-S International Microwave Symposium Digest*, IEEE, Piscataway, pp. 1117–1120, 1991.

[36] B. Carnez, A. Cappy, R. Fauquembergue, E. Constant and G. Salmer, 'Noise modeling in submicrometer-gate FET's', *IEEE Transactions on Electron Devices*, Vol. 28, pp. 784–789, 1981.

[37] A. Cappy, 'Noise modeling and measurement techniques', *IEEE Transactions on Microwave Theory and Techniques*, Vol. 36, pp. 1–10, 1988.

[38] C.J. Madden, R.L. Van Tuyl, M.V. Le and L.D. Nguyen, 'A 17 dB gain, 0.1-70 GHz InP HEMT amplifier IC', *Digest of Technical Papers IEEE International Solid-State Circuits Conference*, IEEE, Piscataway, pp. 178–179, 1994.

[39] A. Fischer-Colbrie, G.G. Zhou and G. Hasnain, 'High-quality $In_{.53}Ga_{.47}As/In_{.52}Al_{.48}As$ MODFETs and PINs Grown on GaAs Substrates', *Electronic Materials Conference*, paper A4, 1993.

[40] R.S. Goldman, J. Chen, K.L. Kavanagh, H.H. Weider, V.M. Robbins and J.N. Miller, 'Structural and magnetotransport properties of InGaAs/InAlAs heterostructures grown on linearly-graded Al(InGa)As buffers on GaAs', *Compound Semiconductors 1994. Proceedings of the 21st International Symposium*, Institute of Physics, Bristol, Vol. 141, p. 313, 1995.

[41] H. Rohdin, A. Nagy, V. Robbins, C.-Y. Su, C. Madden, A. Wakita, J. Raggio and J. Seeger, 'Low-noise, high-speed $Ga_{.47}In_{.53}As/Al_{.48}In_{.52}As$ 0.1-μm MODFETs and high-gain/bandwidth three-stage amplifier fabricated on GaAs substrate', *Proceedings of the International Conference on Indium Phosphide and Related Materials*, IEEE, Piscataway, p. 73–76, 1995.

[42] A. Wakita, H. Rohdin, V. Robbins, N. Moll, C.-Y. Su, A. Nagy and D. Basile, 'Low-noise bias reliability of AlInAs/GaInAs modulation-doped field effect transistors with linearly graded low-temperature buffer layers grown on GaAs substrates', *Japanese Journal of Applied Physics*, Vol. 38, pp. 1186–1189, 1999.

[43] A. Cappy, 'Metamorphic InGaAs/AlInAs heterostructure field effect transistors: Layer growth, device processing and performance', *Proceedings of the International Conference on Indium Phosphide and Related Materials*, IEEE, Piscataway, pp. 3–6, 1996.

[44] H. Rohdin, A. Wakita, A. Nagy, V. Robbins, N. Moll and Su, C.-Y. 'A 0.1-μm MHEMT millimeter-wave IC technology designed for manufacturability', *Solid State Electronics*, Vol. 43, pp. 1645–1654, 1999.

[45] G.T. Cibuzar, 'Effects of gate recess etching on source resistance', *IEEE Transactions on Electron Devices*, Vol. 42, pp. 1195–1196, 1995.

[46] T. Enoki, T. Kobayashi and Y. Ishii, 'Device technologies for InP-based HEMTs and their applications to ICs', *Technical Digest of the Gallium Arsenide IC Symposium*, IEEE, Piscataway, pp. 337–339, 1994.

[47] K. Hikosaka, T. Mimura and K. Joshin, 'Selective dry etching of AlGaAs–GaAs heterojunction', *Japanese Journal of Applied Physics*, Vol. 20, pp. L847–L850, 1981.

[48] K.L. Seaward, N.J. Moll and W.F. Stickle, 'The role of aluminum in selective reactive ion etching of GaAs on AlGaAs', *Journal of Vacuum Science and Technology B*, Vol. 6, pp. 1645–1649, 1988.

[49] S. Kuroda, N. Harada, S. Sasa, T. Mimura and M. Abe, 'Selectively dry-etched n^+-GaAs/N-InAlAs/InGaAs HEMT's for LSI', *IEEE Electron Device Letters*, Vol. 11, pp. 230–232, 1990.

[50] J.W. Matthews and A.E. Blakeslee, 'Defects in epitaxial multilayers. I. Misfit dislocations', *Journal of Crystal Growth*, Vol. 27, pp. 118–125, 1974.

[51] P.S. Pizani, T.M. Boschi, F. Lanciotti Jr., J. Groenen, R. Carles, P. Maigné and M. Gendry, 'Alloying effects on the critical layer thickness in $In_xGa_{1-x}As$/InP heterostructures analyzed by Raman scattering', *Applied Physics Letters*, Vol. 72, pp. 436–438, 1998.

[52] A.S. Wakita, C.-Y. Su, H. Rohdin, H.-Y. Liu, A. Lee, J. Seeger and V.M. Robbins, 'Novel high-yield trilayer resist process for 0.1 μm T-gate fabrication', *Journal of Vacuum Science and Technology B*, Vol. 13, pp. 2725–2728, 1995.

[53] N. Moll, M.R. Hueschen and A. Fischer-Colbrie, 'Pulse-doped AlGaAs/InGaAs pseudomorphic MODFETs', *IEEE Transactions on Electron Devices*, Vol. 35, pp. 879–886, 1988.

[54] H.-Y. Liu, C.-Y. Su, N. Farrar and B. Gleason, 'Fabrication of 0.1 μm T-shaped gates by phase-shifting optical lithography', *Proceedings of SPIE*, Vol. 1927, Pt. 1, pp. 42–52, 1993.

[55] J.G. Wang, K.Y. Hur, L.G. Studebaker, B.C. Keppeler and A.T. Quach, '0.15 micron gate AlInAs/GaInAs MHEMT fabricated on GaAs using deep-UV phase-shifting mask lithography', *Technical Digest of the Gallium Arsenide IC Symposium*, IEEE, Piscataway, pp. 74–77, 1997.

[56] H. Rohdin and A. Nagy, 'A 150 GHz sub-0.1-μm E/D MODFET MSI process', *Technical Digest of the International Electron Devices Meeting*, IEEE, Piscataway, pp. 327–330, 1992.

[57] J.-E. Müller, A. Bangert, T. Grave, M. Kärner, H. Riechert, A. Schäfer, H. Siweris, L. Schleicher, H. Tischer, L. Verweyen, W. Kellner and T. Meier, 'A GaAs HEMT MMIC chip set for automotive radar systems fabricated by optical stepper lithography', *Technical Digest of the Gallium Arsenide IC Symposium*, IEEE, Piscataway, pp. 189–192, 1996.

[58] V. Kumar, 'Reaction of sputtered Pt films on GaAs', *The Journal of Physics and Chemistry of Solids*, Vol. 36, pp. 535–541, 1975.

[59] N. Harada, S. Kuroda and K. Hikosaka, 'N-InAlAs/InGaAs HEMT DCFL inverter fabricated using Pt-based gate and photochemical dry etching', *IEICE Transactions on Electronics*, Vol. E75-C, pp. 1165–1171, 1992.

[60] K.J. Chen, T. Enoki, K. Maezawa, K. Arai and M. Yamamoto, 'High-performance enhancement-mode InAlAs/InGaAs HEMT's using non-alloyed ohmic contact and Pt-based buried-gate', *Proceedings of the International Conference on Indium Phosphide and Related Materials*, IEEE, Piscataway, pp. 428–431, 1995.

[61] R.E. Williams, *Gallium Arsenide Processing Techniques*, Artech House, Dedham, 1984.

[62] P.H. Ladbrooke, 'Reverse modelling of GaAs MESFETs and HEMTs', *GEC Journal of Research*, Vol. 6, pp. 1–9, 1988.

[63] H. Rohdin, 'Reverse modeling of E/D logic submicrometer MODFET's and prediction of maximum extrinsic MODFET current gain cut-off frequency, *IEEE Transactions on Electron Devices*, Vol. 37, pp. 920–934, 1990.

[64] S.J. Mahon, 'The modelling and inverse modelling of high electron mobility transistor devices and circuits'. PhD Thesis, University of Sydney, 1992.

17.11 Problems

17.1 Translate the measured S parameters into Y parameters, i.e. derive Equation (17.8).

17.2 Show that the available noise and signal powers from the generator are given by Equation (17.21), and that the maximum power is delivered when generator and amplifier are conjugately matched ($Y_i = Y_G^*$).

17.3 Express the noise figure of a chain of amplifiers in terms of the noise figures and available gains of the individual stages, i.e. derive Equation (17.24). Hints: Remember that the noise figure is always defined (Equation (17.22)) with respect to the generator-independent available noise power (Equation (17.21a)). Start by expressing the noise power generated by the last stage in terms of its gain and noise figure. Then work backward.

17.4 Derive Equation (17.37) for the available gain (Equation (17.23)) of a two-port. Hint: Apply Norton's theorem to the network comprising the two-port and input signal generator.

18 Modeling high-performance HBTs

D.L. Pulfrey

And we must take the current when it serves,
or lose our ventures

King Lear, Act 4, Scene 3, WILLIAM SHAKESPEARE

18.1 Introduction

Epitaxial-layer, bipolar transistors are intrinsically well suited to high-frequency applications because their critical, physical dimensions are mainly in the direction of the semiconductor film growth, which can be controlled on a near-atomic scale. This is in contrast to field-effect transistors (FETs), where the critical dimension of gate length must be determined lithographically.

Among the family of bipolar transistors, heterojunction bipolar transistors (HBTs) are particularly attractive for operation at high frequencies because their employment of a wide-bandgap emitter allows a highly doped base region to be used without compromising the current gain [1]. With a highly doped base, the base width can be reduced while still maintaining an acceptable base resistance. A short base width leads directly to an improved cut-off frequency, f_T, which, when coupled with the lower base resistance, leads to an improved oscillation frequency, f_{max}. These, and other, attributes of HBTs have been reviewed [2].

In this chapter, some important aspects of modeling high-performance HBTs are discussed. The aims are twofold: (i) to gain some insight into the workings of an HBT at the microscopic level; (ii) to use this insight to examine, or develop, analytical expressions which may be useful in the engineering design of high-frequency and high-speed devices.

At the microscopic level, the emphasis is on the collector current density, J_C. This is an important parameter for high-frequency devices because, via the charge-control method, it is involved in the calculation of the overall signal delay time, τ_{EC}, which, in turn, is related to the common-emitter, unity gain, current cut-off frequency, f_T which, in its turn, is related to the unity gain, power cut-off frequency, f_{max}. These two frequencies are the main performance metrics for a high-frequency device. At the engineering level, the emphasis is on providing compact models for J_C, f_T and f_{max}. Brief consideration is also given to the incorporation of the compact model for J_C into a large-signal equivalent circuit suitable for the simulation of HBTs in switching applications.

18.2 Microscopic modeling of HBTs

18.2.1 Introduction

A complete microscopic model of the HBT would invoke the Boltzmann transport equation (BTE) to describe charge flow in the bulk regions of the device, the Schrödinger wave-equation (SWE) to describe charge concentrations and flows at regions of abruptly changing potential, and Poisson's equation (PE) to link internal charge distributions to external voltages. Less demanding models might be realized by, for example, using the drift-diffusion equation (DDE) as an approximate solution of the BTE, using the Jeffreys–Wentzel–Kramers–Brillouin (JWKB) approximation as a solution for the SWE, and not requiring these solutions to be consistent with PE.

In this chapter, as an example of a tractable microscopic model for HBTs, an iterative approach is described which solves directly the one-dimensional BTE in a field-free base, and accounts for quantum-mechanical tunneling at the emitter base junction via the JWKB approximation [3,4]. Limiting the quantum mechanical treatment to tunneling through junctions is reasonable as other quantum aspects, such as bound energy states, are unlikely to be of importance because of the near-negligible depth of any potential notches in the device (see Figure 18.1). Limiting the bulk-transport treatment to the base is reasonable, at least when computing the collector current density, J_C, because this is the zone of the associated minority-carrier flow. Limiting base transport to the field-free case is also reasonable, at least in instances of uniform doping and composition, because the high doping in the base ensures that low-level injection conditions apply. Limiting the problem to one dimension is reasonable, at least for the calculation of J_C, as the quasi-neutral base width, W_B, is short compared to any relevant lateral dimension of a modern HBT.

18.2.2 Direct solution of the BTE

When subjected to the restrictions listed above, the BTE reduces to [4]:

$$\begin{aligned} v_z \frac{df(z,k,\theta)}{dz} &= C_{in}(z,k,\theta) - C_{out}(z,k,\theta) \\ &= C_{in}(z,k,\theta) - \frac{f(z,k,\theta)}{\tau(k)}, \end{aligned} \tag{18.1}$$

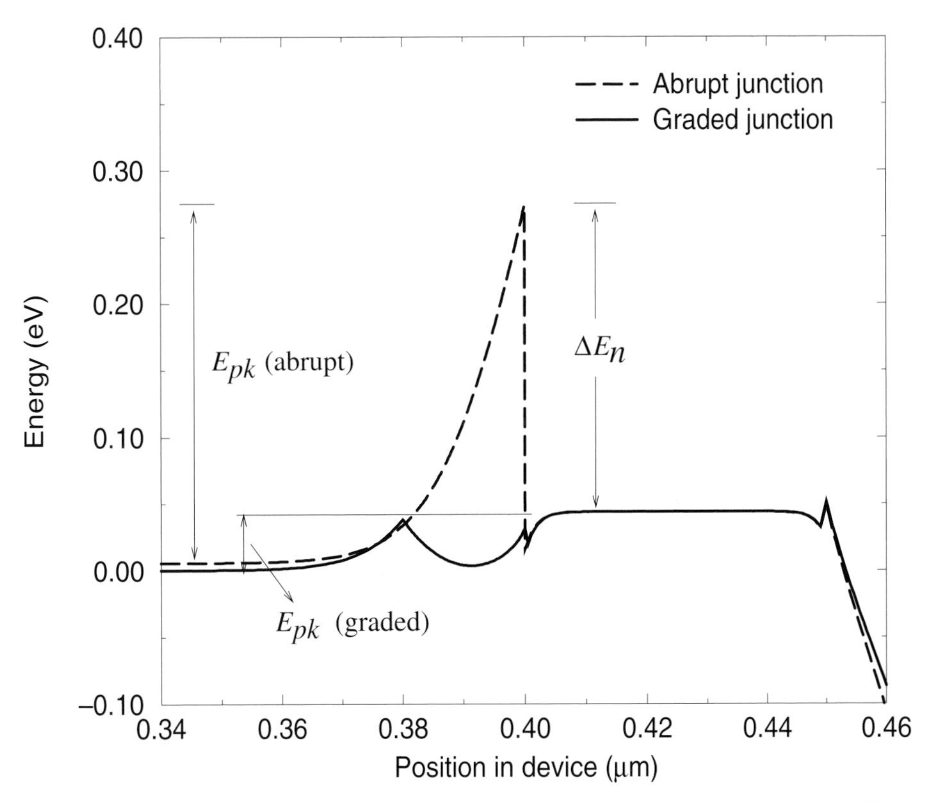

Fig. 18.1. Conduction-band profiles for abrupt- and graded-emitter $Al_xGa_{1-x}As/GaAs$ HBTs operating under similar forward-bias conditions. In the abrupt case $x = 0.3$; in the graded case x is linearly changed over a distance of 0.02 μm. The base region extends from 0.40–0.45 μm. Note the presence of a small barrier at the base–collector junction. This is caused by bandgap narrowing in the base region. This phenomenon is also responsible for the slight spike remaining in the conduction band at the emitter–base junction of the graded device. Note also the two different definitions of the peak barrier height E_{pk}, depending on whether the emitter–base junction is abrupt or graded.

where v_z is the electron velocity, k is the magnitude of the electron wave-vector and is directed at an angle θ to the z axis, C_{in} and C_{out} are the incoming- and outgoing-collision integrals, respectively, and τ is the scattering lifetime. The last three properties can be expressed in the non-degenerate case as, respectively:

$$C_{in}(z, k, \theta) = \frac{1}{(2\pi)^3} \int_{\mathbf{k}'} f(z, k', \theta') S(\mathbf{k}', \mathbf{k})\, d\mathbf{k}',$$

$$C_{out}(z, k, \theta) = \frac{1}{(2\pi)^3} \int_{\mathbf{k}'} f(z, k, \theta) S(\mathbf{k}, \mathbf{k}')\, d\mathbf{k}' = \frac{f(z, k, \theta)}{\tau(k)},$$

where $S(\mathbf{k}, \mathbf{k}')$ is identified with a particular scattering mechanism and describes the rate of transition of carriers from an occupied state with wave-vector $\mathbf{k}$ to an empty state with wave vector $\mathbf{k}'$. The scattering mechanisms considered here are those due to

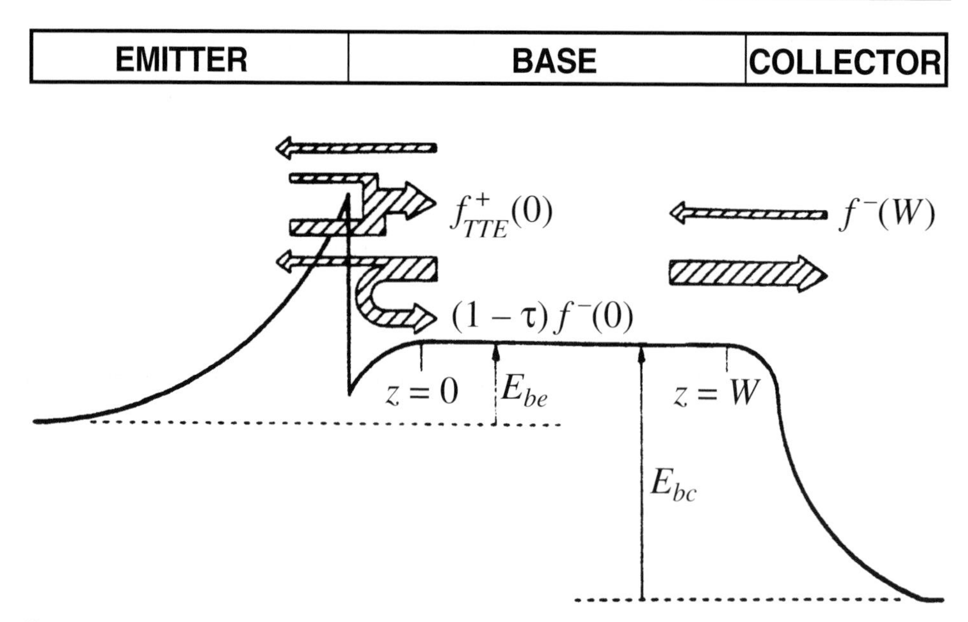

Fig. 18.2. Montage illustrating the injected and reflected electron fluxes at the emitter–base junction, and the collected and back-injected fluxes at the base–collector junction [4]. Note that W in the figure is equivalent to W_B in the text. (A.R. St Denis and D.L. Pulfrey, *Journal of Applied Physics*, Vol. 84, pp. 4959–4965, 1998.)

screened, ionized impurities and polar-optical phonons; these are the most important ones for the $Al_xGa_{1-x}As$/GaAs material system, from which the examples in this chapter are drawn. Equation (18.1) is a first order, ordinary differential equation and, in principle, can be solved using an integrating factor. The difficulty lies in not being able to evaluate the integration constant because f, the distribution function, is not fully specified at any boundary. However, by splitting up f into forward-going (f^+) and negative-going (f^-) parts, the known partial boundary conditions are sufficient to allow an iterative solution to be obtained [4,5]. The basic forms of the two components of the distribution are:

$$f^+(z,k,\theta) = f^+(0,k,\theta)\exp(-z/v_z\tau) + \int_0^z \exp[(z'-z)/v_z\tau]\frac{1}{v_z}C_{in}\,dz', \tag{18.2}$$

$$f^-(z,k,\theta) = f^-(W_B,k,\theta)\exp[(W_B-z)/v_z\tau] + \int_{W_B}^z \exp[(z'-z)/v_z\tau]\frac{1}{v_z}C_{in}\,dz'. \tag{18.3}$$

For an abrupt junction, electrons are injected by thermionic emission over the barrier, and by tunneling through it, giving rise to a distribution function $f^+_{TTE}(0,k,\theta)$. Additionally, electrons backscattered in the base can be reflected into the forward-going ensemble by the potential spike at the junction, giving rise to a distribution function $f^+_{RFL}(0,k,\theta)$. These features are illustrated in Figure 18.2. Thus, the boundary condition at $z = 0$ for the forward-going part of the electron ensemble in an abrupt-junction device is [4]:

$$f^+(0, k, \theta) = f^+_{TTE}(0, k, \theta) + f^+_{RFL}(0, k, \theta), \tag{18.4}$$

where

$$f^+_{TTE}(0, k, \theta) = \frac{n_E}{N_c} \exp(-E_{be}/k_B T) \exp(-E_k/k_B T)\mathcal{T}(k, \theta), \tag{18.5}$$

$$f^+_{RFL}(0, k, \theta) = f^-(0, k, \pi - \theta)[1 - \mathcal{T}(k, \theta)]. \tag{18.6}$$

In the above, n_E is the equilibrium (Maxwellian) electron concentration in the emitter at the distal edge of the emitter–base space-charge region, from which the injected flux originates, N_c is the effective density of states in the conduction band, E_{be} is the potential energy difference shown in Figure 18.2, $k_B T$ is the thermal energy and $\mathcal{T}(k, \theta)$ is the JWKB approximation for the tunneling transmission probability. The electron energy, E_k, is computed here by assuming parabolic bands and spherical constant-energy surfaces, i.e.,

$$E_k = \frac{\hbar^2}{2m^*}(k_x^2 + k_y^2 + k_z^2), \tag{18.7}$$

where m^* is the electron effective mass.

For graded-emitter junctions, the absence of a dominant potential spike (see Figure 18.1) means that $f^+(0, k, \theta)$ can take the form of a simple hemi-Maxwellian.

At the other end of the base, $z = W_B$, a classical homojunction is assumed (see Figure 18.2), regardless of the type of emitter–base junction. The collector is considered to be perfectly absorbing, but is also capable of injecting electrons into the base under appropriate bias conditions. The distribution for this flux is taken to be hemi-Maxwellian and, analogously to the emitter case, is given by

$$f^-(W_B, k, \theta) = \frac{n_C}{N_c} \exp(-E_{bc}/k_B T) \exp(-E_k/k_B T), \tag{18.8}$$

where n_C is the equilibrium electron concentration in the collector, and the barrier E_{bc} is shown in Figure 18.2.

Substituting the appropriate boundary conditions and collision integrals into Equations (18.2) and (18.3) permits an iterative solution for the forward- and backward-going parts of the electron distribution function to be obtained. These can be summed at any position to determine $f(z, k, \theta)$, from which it is straightforward to compute useful parameters such as: carrier concentration, carrier mean velocity and current density.

The results which follow are for $Al_xGa_{1-x}As$/GaAs HBTs, with the mole fraction, x, being 0.3 for abrupt-junction devices. For graded-emitter devices, the barrier for electron flow in the npn devices considered here is determined by the height, and not by the shape, of the conduction-band barrier (see Figure 18.1). Thus it is appropriate to set $x = 0.0$ for the purpose of estimating J_C in a graded-emitter HBT when using the microscopic model. For both graded and abrupt devices, the emitter, base and collector doping densities in cm^{-3} are, respectively, 5×10^{17}, 1×10^{19} and 5×10^{16}. In presenting the results, the base width is often normalized to the the overall mean free path length, l_{sc}, which is computed from the individual energy-dependent scattering lengths for screened, ionized impurity (SII) scattering, and polar-optical phonon (POP) scattering, as follows:

$$l_{sc} = \frac{1}{n} \frac{2}{(2\pi)^3} \int_{\mathbf{k}} \left[\frac{1}{l_{sc,SII}(k)} + \frac{1}{l_{sc,POP}(k)} \right]^{-1} f_{\mathrm{MB}}(\mathbf{k})\, d\mathbf{k}. \tag{18.9}$$

A Maxwell–Boltzmann distribution function, f_{MB}, is used only for the purpose of computing this normalizing value of l_{sc}. Its value here is 46 nm, which is in reasonable agreement with the value estimated from actual time-of-flight measurements on similarly doped material [7].

Components of the distribution function

The components of the distribution function for the case of an abrupt-junction HBT with base width $W_B = 1 l_{sc}$ are shown in Figure 18.3; the distribution function is normalized with respect to $n_E \exp(-E_{pk}/k_B T)/N_c$, where E_{pk} is the peak barrier height at the base–emitter junction (E_{pk}(abrupt) in Figure 18.1); the energy is normalized to the POP energy (0.036 eV); the operating voltages are $V_{BC} = 0$ and $V_{BE} = 0.8\, V_{bi}$, where V_{bi} is the built-in potential at the emitter–base junction. The carriers injected into the base from the emitter form the ballistic component at $z = 0$; the distribution is sharply peaked in energy and strongly focused about the z axis ($\theta = 0$). This distinctive shape of the distribution is characteristic of tunneling through a conduction-band spike. The maximum tunnel flux occurs, irrespective of bias, at an energy which is close to 80% of the peak barrier height [8]; below this energy tunneling is reduced by the increasing thickness of the barrier; above this energy tunneling is reduced by the decreasing electron population. The distribution is focused around the z axis because of the dependence of tunneling on the longitudinal component of energy ($E_z = \hbar^2 k_z^2/2m$).

The ballistic component is reduced by scattering as it transits the base. The corresponding scattered component is shown in row 3 of Figure 18.3; there is a noticeable step at the phonon energy in this distribution at $z = W_B$ due to the loss of carriers by electron–POP interactions. Notice also that this distribution approaches the form of a hemi-Maxwellian, i.e., it is nearly exponential in shape over the full range of the forward angle. This is somewhat surprising in view of the fact that the base has a width of only one scattering length. Similar behavior in short-base Si homojunction devices has been reported by others and discussed in [9]. In Equation (18.2), the first term on the right-hand side represents those carriers that are still traveling ballistically at some point z: this fraction can be written as $\exp[-z/(l_{sc} \cos\theta)]$.

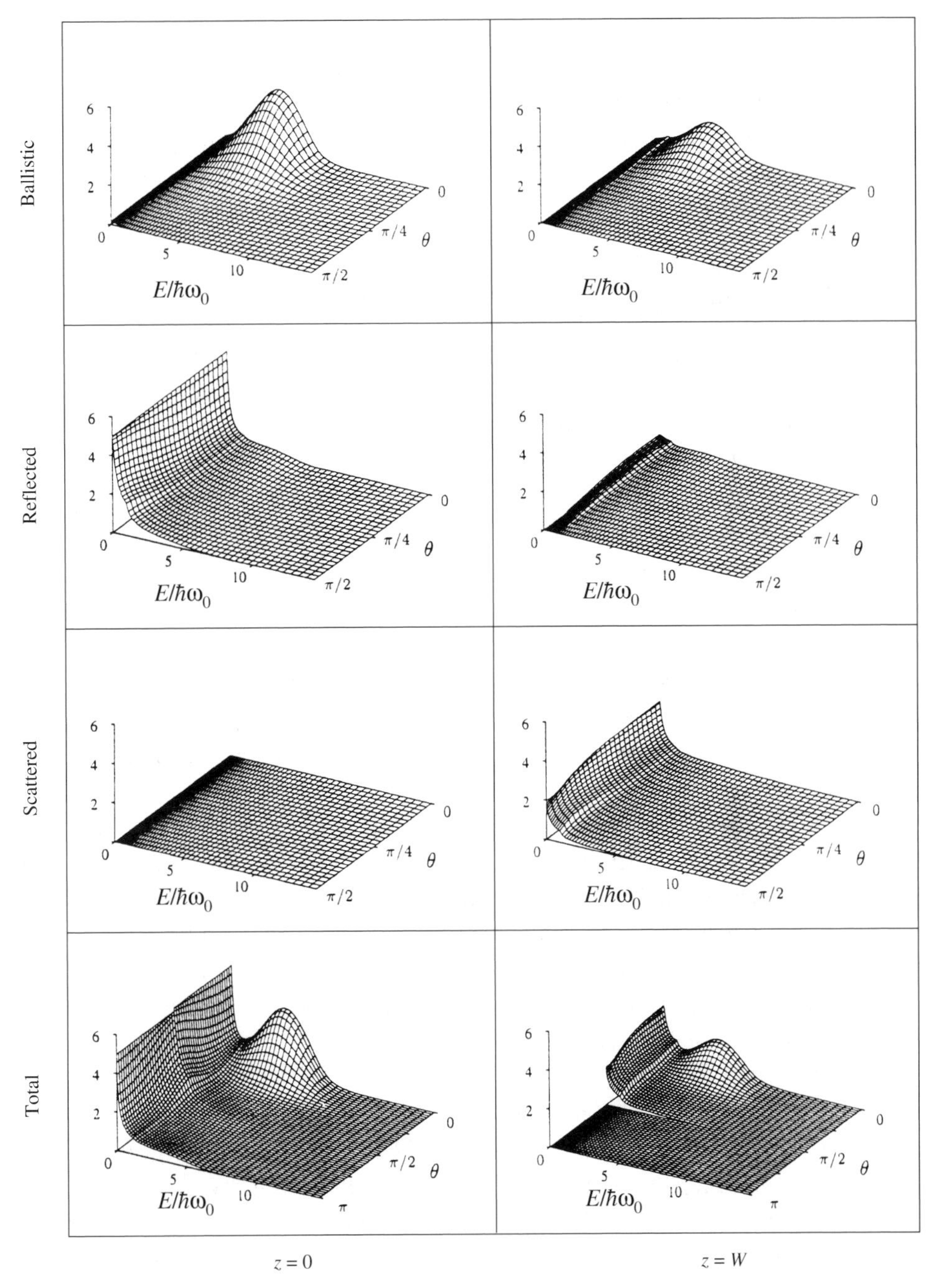

Fig. 18.3. Normalized components of the electron distribution function at the ends of the quasi-neutral base of an abrupt-junction HBT with $V_{BE} = 0.8V_{bi}$ and $V_{BC} = 0$ [4]. For the forward direction, $0 \leq \theta < \pi/2$, and for the backward direction, $\pi/2 < \theta \leq \pi$. Note: $W \equiv W_B$. (A.R. St Denis and D.L. Pulfrey, *Journal of Applied Physics*, Vol. 84, pp. 4959–4965, 1998.)

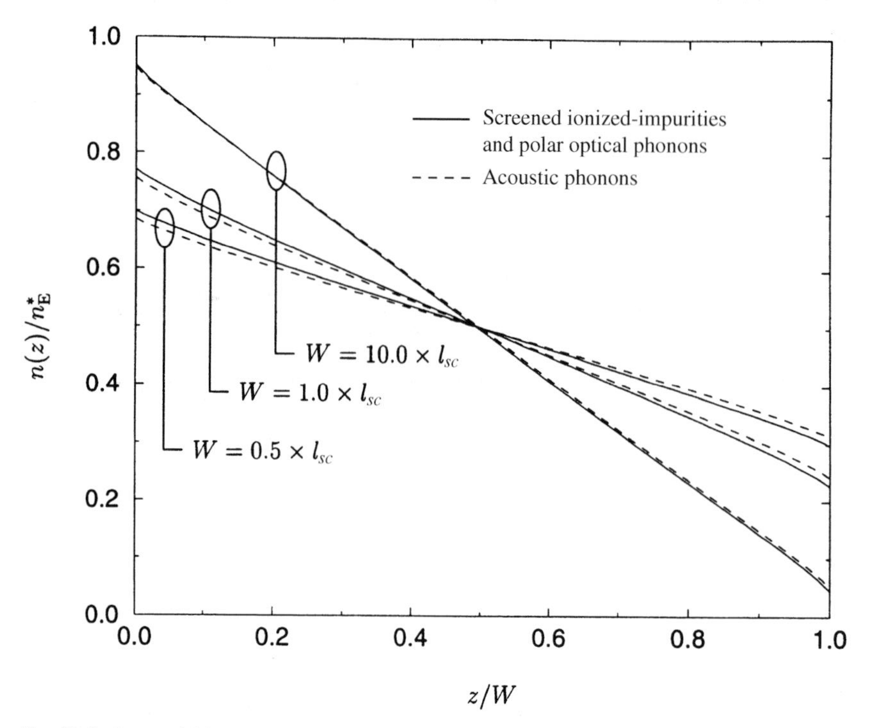

Fig. 18.4. Base width dependence of the carrier profile for two different sets of scattering mechanisms for a graded-emitter HBT under operating conditions of $V_{BE} = 0.8V_{bi}$ and $V_{BC} = 0$ [3]. Note: $W \equiv W_B$. (A.R. St Denis, PhD thesis, University of British Columbia, 1999.)

Therefore, for $W_B = 1l_{sc}$, it is clear that even for the least favorable scattering case of $\cos\theta = 1$, over 60% of the carriers will scatter before exiting the base. Thus, there is significant scattering from the entire, injected distribution, which produces a near-thermalized, scattered component of electrons at the collector end of a short GaAs base. Further scattering, which is experienced by the backscattered electrons, serves to drive the scattered component of the distribution even closer to that of a hemi-Maxwellian. A large fraction of the backscattered electrons at $z = 0$ is reflected by the conduction-band spike, which accounts for the near-hemi-Maxwellian nature of the reflected distribution shown in Figure 18.3.

The bottom row of Figure 18.3 shows both the forward- and backward-directed components of the total distribution. While the backward component builds into a near-hemi-Maxwellian distribution at the emitter end of the quasi-neutral base, the full distribution is far from thermalized at any point in the base.

For the case of the graded-emitter HBT, the components of the distribution are not nearly so interesting. In the absence of a conduction-band spike at the emitter–base interface, the distribution is injected as a near-equilibrium hemi-Maxwellian, and remains in this condition across the base, as any scattering serves only to drive it closer to its equilibrium form.

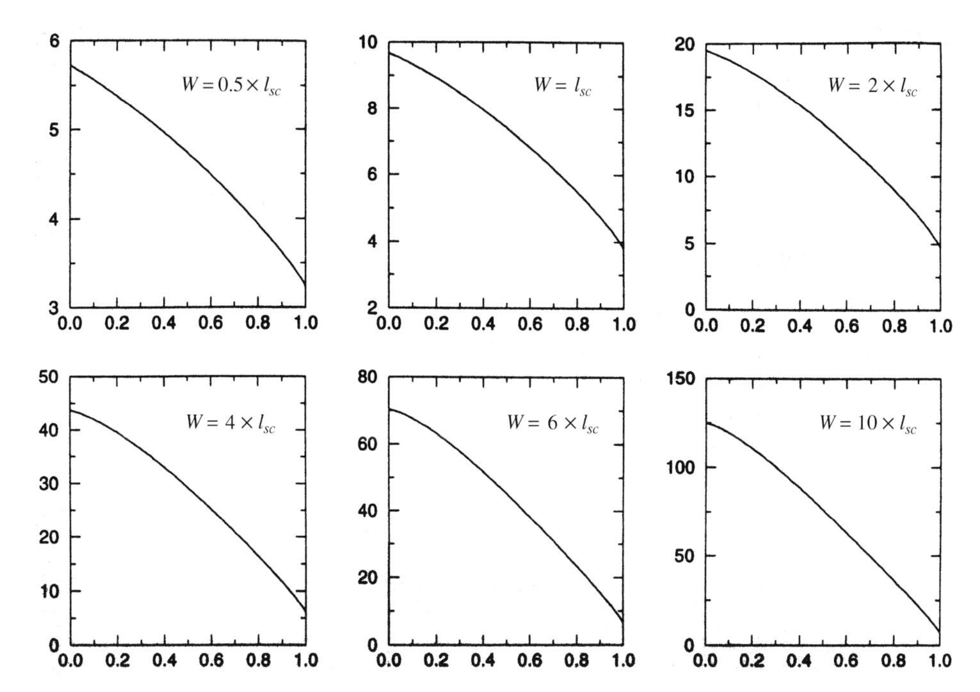

Fig. 18.5. Normalized carrier concentration profiles in an abrupt-junction HBT for various base widths and the same bias conditions as in Figure 18.4. The abscissae are z/W_B and the ordinates are $n(z)/n_E^*$ [3]. Note: $W \equiv W_B$. (A.R. St Denis, PhD thesis, University of British Columbia, 1999.)

Profiles of electron concentration and velocity

The electron concentration profiles for the graded-emitter case and various base widths are shown in Figure 18.4. The concentration is normalized to $n_E^* = n_E \exp(-E_{pk}/k_BT)$, i.e., to twice the value of the concentration in the hemi-Maxwellian distribution injected over the emitter–base potential barrier of height E_{pk} (i.e., E_{pk}(graded) in Figure 18.1). As the base width shrinks, the profiles become progressively non-linear, and the mean gradient decreases as the flow moves from being diffusion-limited to ballistic-limited. This sort of behavior has been documented before for Si-like bipolar transistors [5], for which acoustic phonons provide the dominant scattering mechanism. Figure 18.4 shows that, for the case of a graded-emitter device, the concentration profiles are barely sensitive to the nature of the scattering mechanism. This is a consequence of the forward- and backward-going components of the distribution maintaining near-hemi-Maxwellian forms over the entire base width.

The concentration profiles for the abrupt-junction case are shown in Figure 18.5. The normalization is again to n_E^* but, note that in this case, normalized values in excess of unity occur. This is because of both forward tunneling through the barrier and reflection of backscattered electrons at the interface (see Figure 18.2). The tunneling probability, $\mathcal{T}(k, \theta)$, for the backscattered component, under the conditions of the

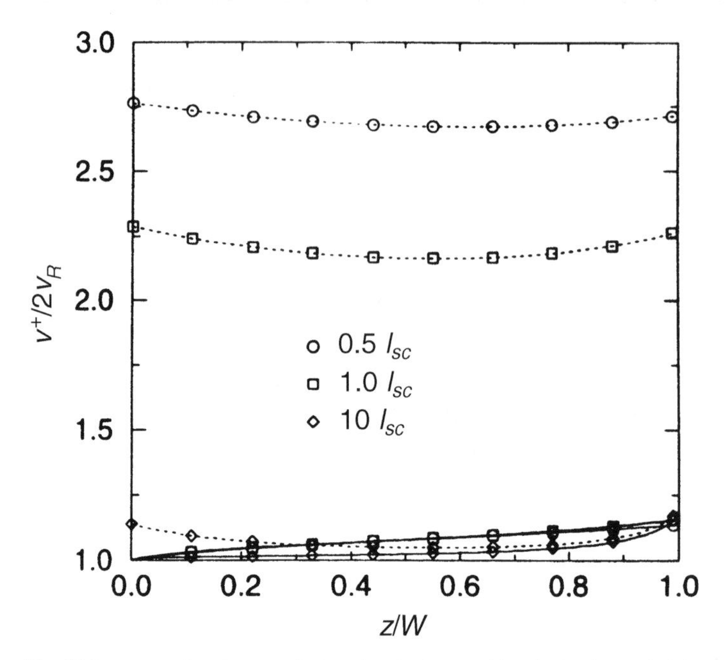

Fig. 18.6. Normalized, mean forward-velocity profiles for both graded-junction HBTs (full lines) and abrupt-junction HBTs (dashed lines) for three different base widths and the same bias conditions as in Figure 18.4 [4]. Note: $W \equiv W_B$. (A.R. St Denis and D.L. Pulfrey, *Journal of Applied Physics*, Vol. 84, pp. 4959–4965, 1998.)

JWKB approximation, is the same as for the injected component, but the backward tunneling current is orders of magnitude less, because of the huge difference between n_E and $n(0)$.

The mean, forward-velocity profiles for each type of junction are shown in Figure 18.6. For the graded-emitter HBT, the hemi-Maxwellian nature of the injected distribution fixes the boundary condition of $v^+(0) = 2v_R$, where $v_R = \left[k_B T/(2\pi m^*)\right]^{1/2}$ is the Richardson velocity. Scattering tends to drive any distribution towards a Maxwellian, as Figure 18.3 illustrates, so $v^+(z)$ remains close to $2v_R$ over the entire width of the base in the graded-emitter case. There is a slight increase towards the collector end of the base, where the forward-directed flow is not mitigated by scattering in the immediate vicinity of the absorbing collector [6,9].

The $v^+(z)$ profiles for the abrupt-junction HBT differ from those for the graded-emitter HBT, not only in magnitude, but also in the fact that they are thickness- and voltage-dependent. In a short-base, abrupt-junction device, $v^+(z)$ is high, in accordance with the highly focused, energetic nature of the tunneling contribution to the injected distribution. As the base thickens, there is more backscattering and, from this component of the distribution, which tends towards a hemi-Maxwellian, reflection into the forward direction occurs at the conduction-band spike. Thus, the mean velocity of the forward component falls and, at $W_B = 10 l_{sc}$, the situation is almost

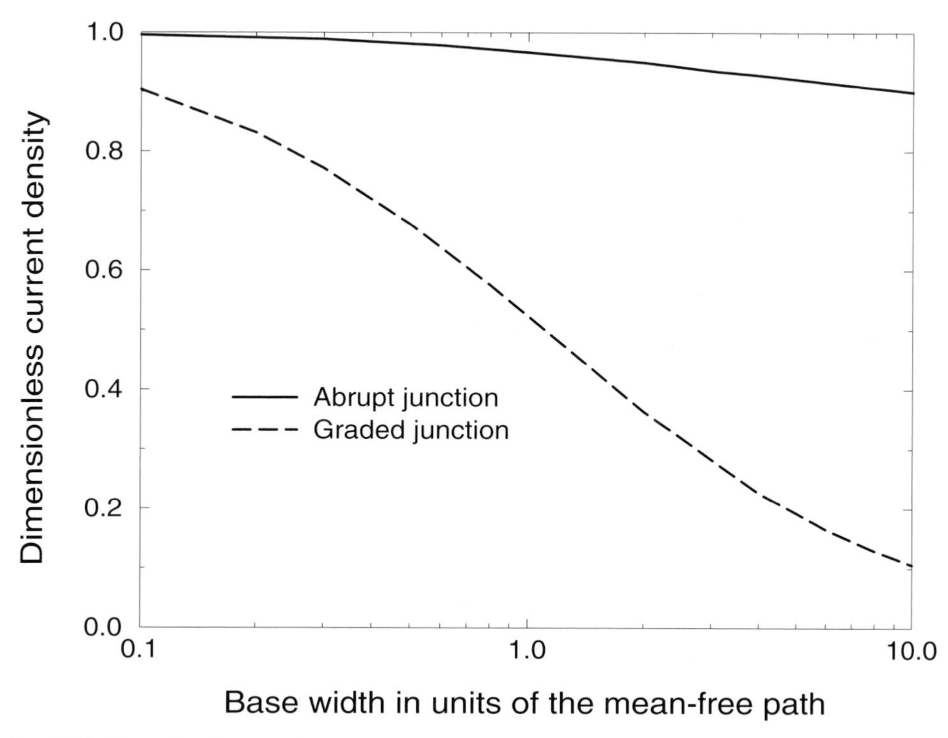

Fig. 18.7. Normalized current density as a function of base width for both graded- and abrupt-junction HBTs operating at $V_{BE} = 0.8V_{bi}$ and $V_{BC} = 0$ [10]. For both devices the current is normalized to its ballistic value. (D.L. Pulfrey, A.R. St Denis and M. Vaidyanathan, *Proceedings of the IEEE Conference on Optical and Microelectronic Materials and Devices*, IEEE, Piscataway, pp. 81–85, 1998.)

identical to that in a graded-emitter transistor, i.e., scattering and reflection essentially destroy the focused, high-energy character of the injected electron distribution.

Collector current

The collector current density as a function of base width for the two types of HBT is shown in Figure 18.7 [10]. The normalization in this case is to the ballistic current density, which is not the same for the two devices, i.e.,

$$\left.\begin{aligned} J_{bal,graded} &= -q\frac{n_E^*}{2}2v_R, \\ J_{bal,abrupt} &= -qn_{TTE}^+ v_{TTE}^+. \end{aligned}\right\} \qquad (18.10)$$

In the abrupt-junction case, n_{TTE}^+ is the concentration of injected carriers arising from tunneling and thermionic emission, and v_{TTE}^+ is their mean velocity. For the conditions cited in Figure 18.7, it is found that $n_{TTE}^+ v_{TTE}^+ = 17.1\, n_E^* v_R$.

The most surprising feature of Figure 18.7 is the near constancy of J_C in the abrupt-junction case, at least for base widths up to about $W_B = 10 l_{sc}$. As W_B increases, the backscattered component of the electron distribution increases, but due to specular reflection at the emitter–base junction, there is no contribution of this flux to the net current which, therefore, continues to be dominated by the injected component. It is only when $n(0)$ builds up to the extent that backward tunneling into the emitter becomes significant that J_C drops appreciably below its ballistic value.

18.3 Compact modeling of HBTs

18.3.1 Introduction

Microscopic models, such as the one described in the previous section, provide quantitative information on the details of carrier transport within a device. This knowledge comes at the expense of a significant investment, both in model development and in program execution time. For engineering device-design purposes, at least in the early iterations of the design cycle, models that are less time-intensive, and more easily mastered, are more likely to find widespread application. Particularly useful are compact models, in which insight into the key factors determining device performance is provided by a set of analytical expressions that relate the terminal behavior of a device to its composition and layout. Ideally, the simplifications inherent in the analytical expressions can be justified by appeal to the more rigorous results of a microscopic model. In this vein, compact models for J_C are presented here, and are contrasted with the results from the microscopic model considered above. With confidence in the expressions for J_C, it is then possible to proceed to compact models for the related high-frequency performance metrics, f_T and f_{max}, and also to a model suitable for assessing high-speed performance.

18.3.2 Compact models for the collector current

The path towards a compact model for J_C starts at the second moment of the BTE [11] or, alternatively, at Equation (9.16), which can be written as:

$$J_n = q\mu_n n\mathcal{E} + qD_n\frac{\partial n}{\partial z} + 2\mu_n n\frac{\partial u}{\partial z} - \tau_{sc}\frac{\partial J_n}{\partial t} \tag{18.11}$$

where τ_{sc} is an average time related to scattering processes, μ_n is the mobility, D_n is the diffusivity, and u is the average z-directed kinetic energy, defined by the following expressions:

$$\frac{1}{\tau_{sc}} = \frac{\int [v_z/\tau^\star(\mathbf{v})] f(z, \mathbf{v}, t)\, d\mathbf{v}}{\int v_z f(z, \mathbf{v}, t)\, d\mathbf{v}}, \tag{18.12}$$

$$\mu_n = \frac{q\tau_{sc}}{m^*}, \tag{18.13}$$

$$D_n = \left[\frac{2u}{q}\right]\mu_n, \tag{18.14}$$

$$u = \frac{\int [m^* v_z^2/2] f(z, \mathbf{v}, t)\, d\mathbf{v}}{\int f(z, \mathbf{v}, t)\, d\mathbf{v}}, \tag{18.15}$$

where, with respect to the variables used in Equations (18.1)–(18.3), $f(z, \mathbf{v}, t) \equiv f(z, k, \theta, t)$ and $\tau^\star(\mathbf{v}) \equiv \tau(k, \theta) = \tau(k)$ for isotropic scattering.

To proceed further, the term involving $\partial J/\partial t$ in (18.11) can be safely dropped in the static case. If the energy-gradient term involving $\partial u/\partial z$ could also be ignored, then the non-equilibrium DDE would result:

$$J_n = q\mu_n n\mathcal{E} + qD_n\frac{\partial n}{\partial z}. \tag{18.16}$$

The validity of dropping the energy-gradient term can be judged by studying Figure 18.8, which shows the concentration gradient (diffusion) and energy-gradient terms from Equation (18.11) for a field-free base. For both graded and abrupt devices the magnitudes of the two terms are similar in the region close to the collector. This is a consequence of the specified nature of the collector. Its absorbing property leads to an increase in the mean forward-going velocity, and its non-injecting property (in the normal, active mode of operation) leads to a decrease in the mean backward-going velocity. Thus, there is a strong energy gradient in this region which opposes the flow due to the concentration gradient. Apart from this, for graded-emitter HBTs it would appear from Figure 18.8(a) to be not unreasonable to build a compact model on consideration of the diffusion term alone. For the abrupt-junction device, it would appear that a similar construction should not be attempted with so much confidence.

If Equation (18.16) is judged to be appropriate, then a final simplification would follow if the velocity distribution function, $f(z, \mathbf{v}, t)$, could be considered to maintain a form close to its equilibrium Maxwellian value over the base region. Under these circumstances, Equation (18.16) would reduce to the equilibrium drift-diffusion equation (DDE):

$$J_n = q\mu_{n0} n\mathcal{E} + qD_{n0}\frac{\partial n}{\partial z}. \tag{18.17}$$

From the foregoing discussion and the results of the microscopic model it can be appreciated that Equation (18.17) may be applicable to graded-emitter HBTs, but not even Equation (18.16) is likely to be valid for abrupt-emitter HBTs [12].

Graded-emitter HBTs

It is *sine qua non* that high-frequency devices must have small dimensions, so any compact model for J_C must acknowledge the increasing importance of the boundary conditions as the base width shrinks. This can be elegantly done by invoking the one-flux method to represent the forward- and backward-going carrier flows at the

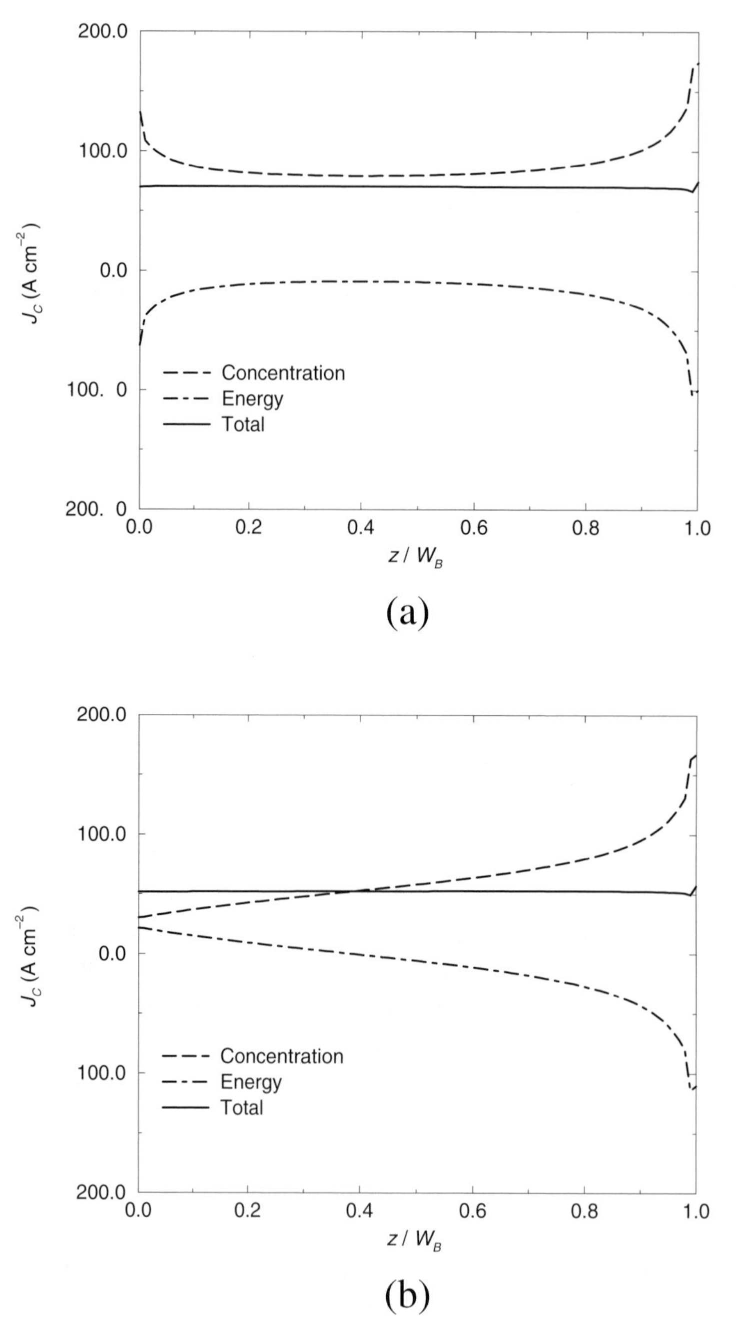

Fig. 18.8. Contributions to the collector current of the concentration-gradient term (dashed line) and the energy-gradient term (dashed-dotted line), as specified by the second and third terms, respectively, in Equation (18.11). The simulations are for $W_B = 1l_{sc}$ and $J_C \approx 50\,\mathrm{A\,cm^{-2}}$: (a) graded-junction HBT; (b) abrupt-junction HBT. (Data courtesy of A. St Denis.)

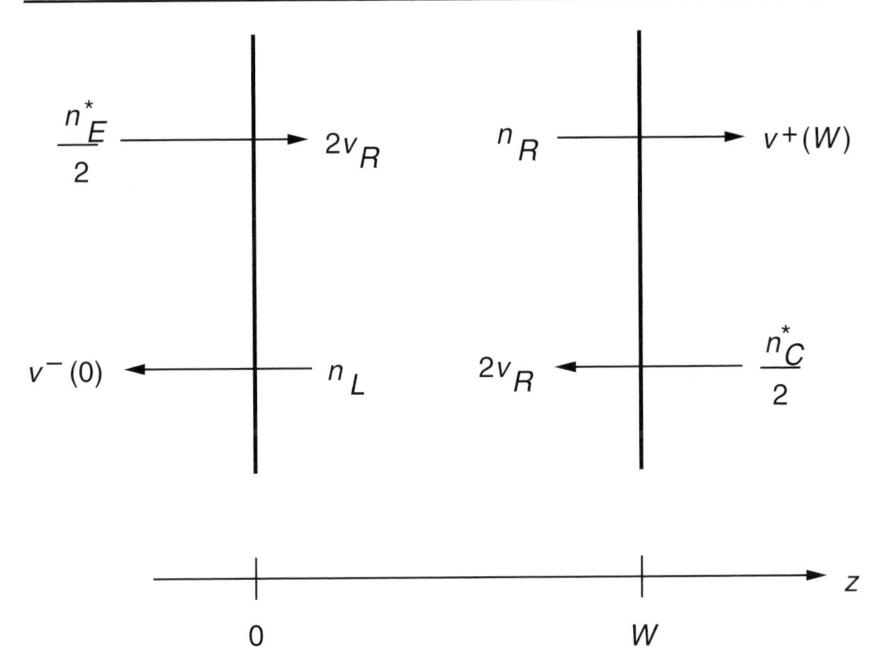

Fig. 18.9. Incoming and outgoing fluxes at the junctions to the base of a graded-emitter HBT [6]. Here, $v^-(0) = 2v_R$, as is appropriate for a hemi-Maxwellian distribution. Note: $W \equiv W_B$. (A.R. St Denis and D.L. Pulfrey, *Solid State Electronics*, Vol. 38, pp. 1431–1436, 1995.)

interfaces to the base [13], as depicted in Figure 18.9. The one-flux model is, in effect, a one-speed solution to the near-equilibrium BTE because it treats all fluxes as hemi-Maxwellians with a velocity of $2v_R$. This is a fair representation of the situation in the base of a graded-emitter device, as the microscopic results of Figure 18.6 indicate. The use of a constant velocity means that the energy-gradient term in the BTE (Equation 18.11) can be ignored, and the use of the equilibrium value for this velocity means that the one-flux method is equivalent to solving the equilibrium DDE with appropriate boundary conditions [14]. The relevant boundary conditions were proposed by Hansen [15]. In the derivation that follows, the exit velocity $v^+(W_B)$ is not explicitly set to $2v_R$ in order that the influence of a real, non-absorbing collector may be considered later, i.e.,

$$v^+(W_B) = v_{coll}. \tag{18.18}$$

It is implicit that the collection velocity v_{coll} does not differ from $2v_R$ to an extent that the conditions for application of the equilibrium DDE are violated. Thus, with reference to Figure 18.9, the boundary conditions are:

$$\begin{aligned} J(0) &= -q\left(\frac{n^*_E}{2} - n_L\right)2v_R, \\ J(W_B) &= -q\left(n_R v_{coll} - \frac{n^*_C}{2}2v_R\right) \end{aligned} \tag{18.19}$$

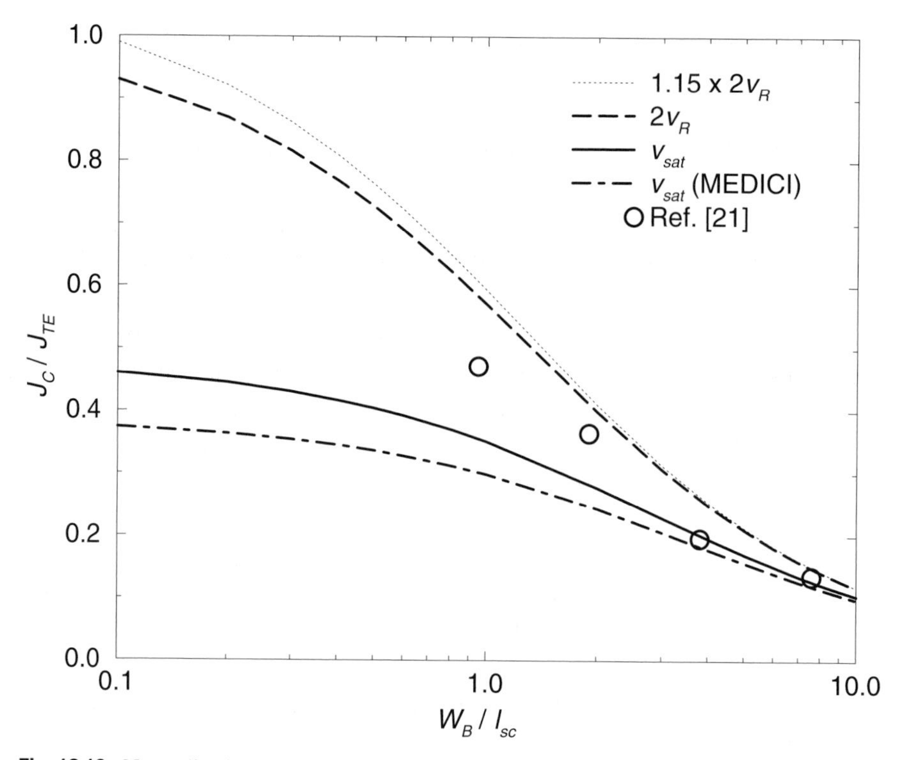

Fig. 18.10. Normalized collector current as a function of normalized base width for a graded-emitter HBT, from Equation (18.21) for different values of v_{coll}. The curves are for operating conditions of $V_{BE} = 0.8V_{bi}$ and $V_{BC} = 0$. The circles are experimental data, and are plotted using $l_{sc} = 52.5$ nm [21]. These data have been scaled to coincide with the predicted curves at the long-base width limit.

and

$$n(0) = n_E^*/2 + n_L,$$
$$n(W_B) = n_C^*/2 + n_R. \tag{18.20}$$

For operation in the normal, active mode, $n_C^* \approx 0$, so combining the expressions for J with the field-free, equilibrium version of the DDE (Equation 18.17), and ignoring recombination in the base, gives:

$$J_C = \frac{-qn_E^*}{\left(\frac{W_B}{D_{n0}} + \frac{1}{2v_R} + \frac{1}{v_{coll}}\right)}. \tag{18.21}$$

The form of Equation (18.21), with each of the denominator terms being a reciprocal effective velocity, is characteristic of serial-flow transport situations [16,17]. It is useful in identifying the bottleneck to charge flow [18] which, in this case, can be the base ($v_{eff} = D_{n0}/W_B$), or either of the two junctions ($v_{eff} = 2v_R$ for the emitter–base, or $v_{eff} = v_{coll}$ for the base–collector). Results are shown in Figure 18.10 for various values of the collection, or exit, velocity, v_{coll}.

When $v_{coll} = 2v_R$, the agreement with the prediction from the microscopic model (Figure 18.7) is excellent. Similar agreement for the case of Si homojunction bipolar transistors has been observed and commented upon previously [6,10,15]. However, as pointed out in [10], this agreement should not be construed as evidence that Equation (18.17) is rigorously valid on a microscopic level. The assumptions made about the point value of D_n, the insignificance of the energy-gradient term and the departure of v_{coll} from $2v_R$, all give cause for concern as the base width shrinks. If one of these assumptions is removed, e.g., by making $v_{coll} = 1.15 \times 2v_R$, as suggested by the results of the microscopic model, then, as Figure 18.10 indicates, Equation (18.21) actually gives poorer agreement with the results from the microscopic analysis. This suggests that the assumptions made in the analytical model are somehow compensatory [6]. Nevertheless, Equation (18.21) captures sufficient of the physics and gives adequate agreement with more sophisticated models to be viewed as a useful compact model.

Giving further consideration to v_{coll}, the assumption made in the microscopic model of an absorbing collector may need modification because, in real devices, phenomena such as velocity saturation and velocity overshoot could determine the effective value of the collection velocity. In homojunction devices, this velocity is often taken to be v_{sat}, the high-field saturation velocity [19,pp. 236–237]. In HBTs, because of the high doping density in the base, bandgap-narrowing phenomena are likely to induce a heterojunction at the base–collector interface (see Figure 18.1), even if there is no change in material composition, and this may further reduce v_{coll} [20]. Some idea of the importance of these collection-velocity-limiting factors can be obtained by using $v_{coll} = v_{sat}$ in Equation (18.21), and also displaying some experimental data from GaAs homojunction devices [21], which behave in the same way as the graded-emitter devices being considered here. As Figure 18.10 indicates, setting $v_{coll} = v_{sat}$ does significantly reduce the current, but, judging from the experimental data shown, a more realistic value of v_{coll} would appear to lie closer to the absorbing limit of $2v_R$.

Because of the widespread use of commercial, DDE-based simulators, such as MEDICI*, it is instructive to demonstrate their capabilities for predicting the current in short-base devices. In the heterojunction mode, MEDICI employs current balancing at the interface but, in essence, uses Maxwellian distributions throughout, so the velocity of n_E^* electrons injected into the base, for example, is v_R. This leads to the factor of $1/2v_R$ in Equation (18.21) being doubled. For the case of $v_{coll} = v_{sat}$, this has only a slight effect on the predicted current, as demonstrated by the lowest curve in Figure 18.10.

Abrupt-junction HBTs

The results from the microscopic model for J_C in abrupt-junction HBTs, as shown in Figure 18.7, indicate that a compact model should focus on describing the tunnel

* Available from Avant! Corp., www.avanticorp.com

and thermionic-emission currents, as the base width has little influence on the charge flow. Accordingly, it is reasonable to start with the JWKB approximation and consider additional simplifications to produce an analytical expression for the interfacial current. As noted in the discussion of Figure 18.3, the tunneling flux reaches a maximum at some energy, which turns out to be about $0.8E_{pk}$ [8]. By making a second order Taylor series expansion about this point, the integral for the tunneling current can be greatly simplified. Further, if the limits to the integration are allowed to extend to $\pm\infty$, rather than being limited by the bounds of the tunneling barrier, a particularly convenient expression materializes [8]:

$$\begin{aligned} J_{TTE} &= J_{TUN} + J_{TE} \\ &= J_0 C \exp\left[-E'_{pk} \tanh(U'_{oo})/(U'_{oo})\right] + J_0 \exp(-E'_{pk}), \end{aligned} \tag{18.22}$$

where

$$J_0 = -qn_E v_R,$$

$$C = \left[\frac{4\pi E'_{pk} \sinh(U'_{oo})(U'_{oo})}{\cosh^3(U'_{oo})}\right]^{1/2},$$

$$U'_{oo} = \frac{\hbar q}{2k_B T}\left(\frac{N_E}{m^*_E \epsilon_E}\right)^{1/2},$$

where N_E, m^*_E and ϵ_E refer to the emitter parameters of doping density, electron effective mass, and permittivity, respectively, and E'_{pk} is the peak barrier height normalized to $k_B T$.

J_{TTE} from Equation (18.22) agrees almost exactly with the prediction from the microscopic model, as shown in Figure 18.11. Note that the normalization in this figure is not to the ballistic current, as is the case in Figure 18.7, but, instead, to the thermionic-emission current:

$$J_{TE} = J_0 \exp(-E'_{pk}) = -qn^*_E v_R. \tag{18.23}$$

Practical, high-frequency HBTs are likely to have $W_B \approx 1 l_{sc}$, so the comparison in Figure 18.11 suggests that the compact model of Equation (18.22) should prove useful in the prediction of the collector current in these devices.

The remaining curve on Figure 18.11 is from MEDICI simulations. In this simulator, for abrupt-junction devices, the interfacial current resulting from the balancing of currents computed from the JWKB approximation is coupled to drift-diffusion currents in the bulk regions of the device [23,34]. Analytically, the allowance for tunneling, and the presence of a conduction-band spike, can be represented by the addition of two symbols, γ and δ, respectively, to Equation (18.21):

$$J_C = \frac{-qn^*_E/\delta}{\left(\dfrac{W_B}{D_{n0}} + \dfrac{1}{\gamma\delta v_R} + \dfrac{1}{v_{coll}}\right)} \tag{18.24}$$

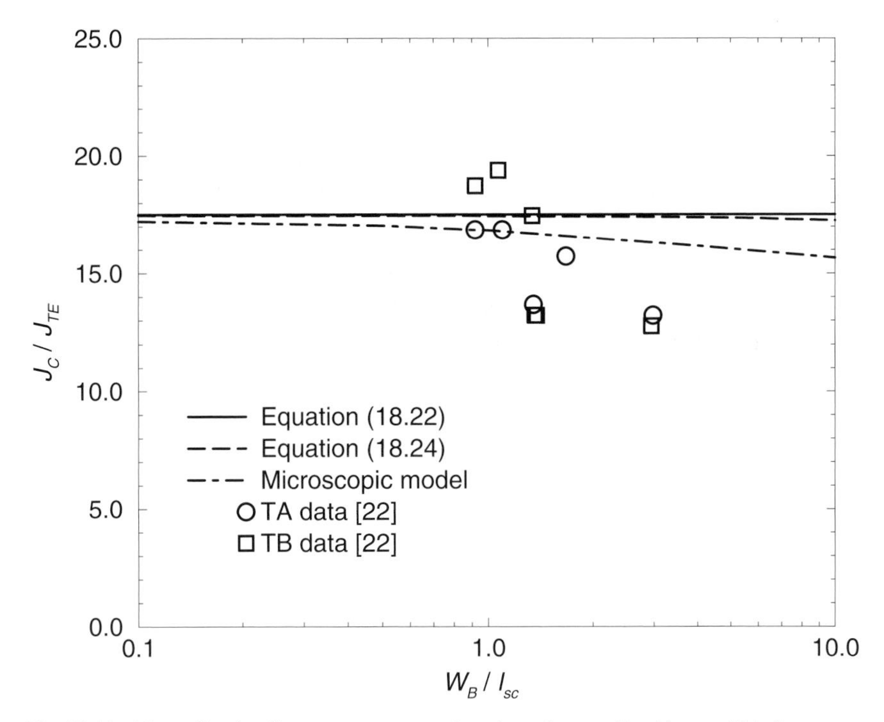

Fig. 18.11. Normalized collector current as a function of normalized base width for an abrupt-emitter HBT. The curves are for operating conditions of $V_{BE} = 0.8V_{bi}$ and $V_{BC} = 0$. The data points are experimental results [22], and are plotted using the same mean free path length as in Figure 18.10. The circles refer to high-gain devices (TA series) and the squares to normal-gain devices (TB series). The data have been extracted from information given in [22], and correspond to operation at the same bias conditions as used for the model calculations. These data have been scaled so that the current for the thinnest high-gain device coincides with the current predicted by the microscopic model.

where $\gamma = J_{TTE}/J_{TE}$ is the ratio of the total current to the thermionic-emission current and $\delta = \exp(-\Delta E_n/k_B T)$, with ΔE_n being the barrier height shown on Figure 18.1. Typically, γ does not exceed 10, but δ can be as small as 10^{-4}, so $\gamma\delta v_R \ll v_{coll}$ and, as $W_B \to 0$, $J_C \to J_{TTE}$, as desired. Even though Equation (18.24) has a thickness-dependent term, it turns out that the effective base transport velocity, D_{n0}/W_B, is orders of magnitude higher than the effective injection velocity, $\gamma\delta v_R$, so this expression predicts a current that is essentially constant, with the same value as that given by the compact model of Equation (18.22)*.

Some results from experimental, abrupt-junction, $Al_xGa_{1-x}As$/GaAs devices, with $x = 0.25$ [22], are also displayed in Figure 18.11. The ratio of the highest value of J_C/J_{TE} to the lowest value is 1.3 for the high-gain devices (TA series), and 1.5 for the TB devices.

* Note that Equation (18.24) is derived assuming full-Maxwellian distributions, consistent with the treatment in MEDICI. If a hemi-Maxwellian flux approach is used, consistent with the treatment used to derive Equation (18.21), then Equation (18.24) is modified slightly (see Problem 18.3).

Over the same thickness range ($W/l_{sc} \approx$ 1–3), the corresponding ratio for the graded-emitter devices is 1.9 (see Figure 18.10). While this value is higher than those for the abrupt-junction devices, clearly more experimental data are needed before any definite corroboration can be claimed for the theory that abrupt-junction devices exhibit less thickness dependence of J_C than graded-emitter devices.

18.3.3 Compact models for f_T

The charge-control method

The frequency response of a transistor is related in some way to the rate at which the charge distribution within the device responds to a change in applied bias. One such relation, from basic charge-control theory [25], links the total emitter-to-collector, signal-delay time, τ_{EC}, to the changes in regional charge, ΔQ_i, that are induced by a small change in collector current, ΔI_C, with the collector–emitter voltage, V_{CE}, held constant:

$$\tau_{EC} = \frac{1}{2\pi f_T} = \sum_i \tau_i = \sum_i \left.\frac{\Delta Q_i}{\Delta I_C}\right|_{V_{CE}}. \tag{18.25}$$

With appropriately chosen boundaries, Q can refer to the charge of either the electrons or the holes contained within the specified region [20]. Before identifying convenient boundaries for the regions, i, it is instructive to show that Equation (18.25) is equivalent to the familiar result from the small-signal, hybrid-π, equivalent circuit. To do this involves a bit of calculus:

$$Q = Q(V_{BE}, V_{CB})$$

$$\Delta Q = \left.\frac{\partial Q}{\partial V_{BE}}\right|_{V_{CB}} \Delta V_{BE} + \left.\frac{\partial Q}{\partial V_{CB}}\right|_{V_{BE}} \Delta V_{CB}$$

$$\Delta Q = \left.\frac{\partial Q}{\partial V_{BE}}\right|_{V_{CB}} \Delta V_{BE} + \left.\frac{\partial Q}{\partial V_{BC}}\right|_{V_{BE}} \Delta V_{BE}$$

$$\begin{aligned}\left.\frac{\Delta Q}{\Delta I_C}\right|_{V_{CE}} &= \left.\frac{\partial Q}{\partial V_{BE}}\right|_{V_{CB}} \left.\frac{\Delta V_{BE}}{\Delta I_C}\right|_{V_{CE}} + \left.\frac{\partial Q}{\partial V_{BC}}\right|_{V_{BE}} \left.\frac{\Delta V_{BE}}{\Delta I_C}\right|_{V_{CE}} \\ &= C_\pi \frac{1}{g_m} + C_\mu \frac{1}{g_m},\end{aligned}$$

where C_π and C_μ are the capacitances associated with charge changes due to changes in V_{BE} and V_{BC}, respectively, and g_m is the transconductance. This is the same as the result obtained from the hybrid-π equivalent circuit when parasitic resistances are neglected [19, p. 175]. If these resistances need to be considered, as they are in the full hybrid-π circuit shown in Figure 18.12, then they can be factored into the charge-control derivation by distinguishing between terminal voltages and junction voltages*.

* See Problem 18.5.

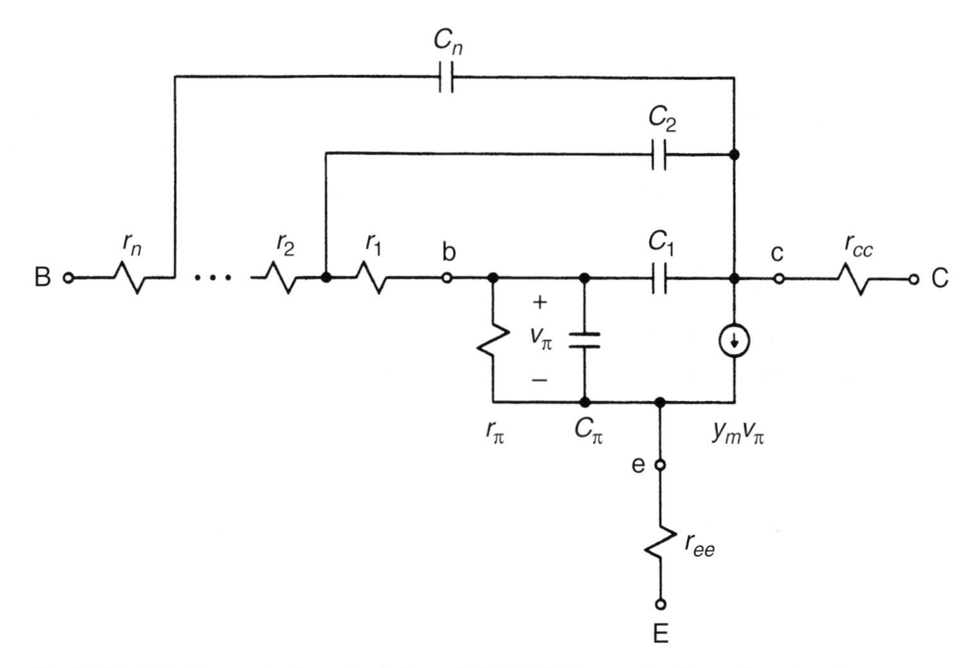

Fig. 18.12. Hybrid-π equivalent circuit for an HBT [35]. The total collector–base junction capacitance is given by $C_{jc} = \sum_{i=1}^{n} C_i$, and, for operation in the active mode, is equal to C_μ, as introduced following Equation (18.25). (M. Vaidyanathan and D.L. Pulfrey, *IEEE Transactions on Electron Devices*, Vol. 46, pp. 301–309, 1999.)

A convenient separation of the device into 'quasi-neutral' and 'space-charge regions' is shown in Figure 18.13. Based on this partition, the following regional delay times may be defined:

$$\tau_E = \left.\frac{\Delta Q_E}{\Delta I_C}\right|_{V_{CE}}, \tag{18.26}$$

$$\tau_B = \left.\frac{\Delta Q_B}{\Delta I_C}\right|_{V_{CE}}, \tag{18.27}$$

$$\tau_C = \left.\frac{\Delta Q_C}{\Delta I_C}\right|_{V_{CE}}, \tag{18.28}$$

where τ_E accounts for the signal delay through the quasi-neutral emitter and the emitter–base space-charge region, τ_B represents the delay of the quasi-neutral base region, and τ_C represents the delay through the collector–base space-charge region and the quasi-neutral collector.

To compute the changes in stored charge, access is needed to some simulator that can represent the entire device, from emitter terminal to collector terminal. In the section on compact models for J_C, it was shown that a numerical simulator, such as MEDICI, can provide adequate results for the collector current, even in short-base

Emitter | Base | Collector | Subcollector

τ_E τ_B τ_C

0 → z

(a)

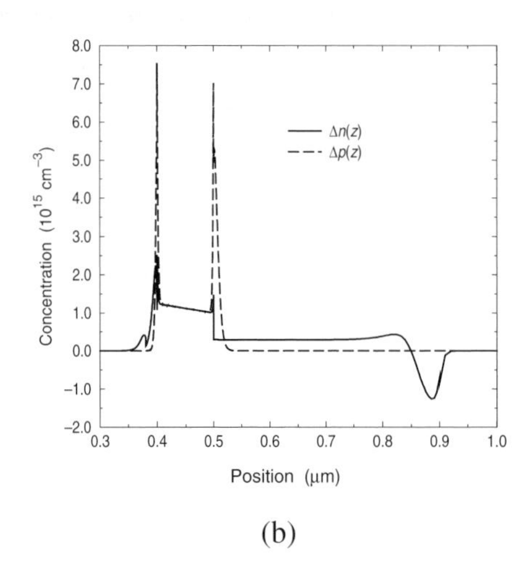

(b)

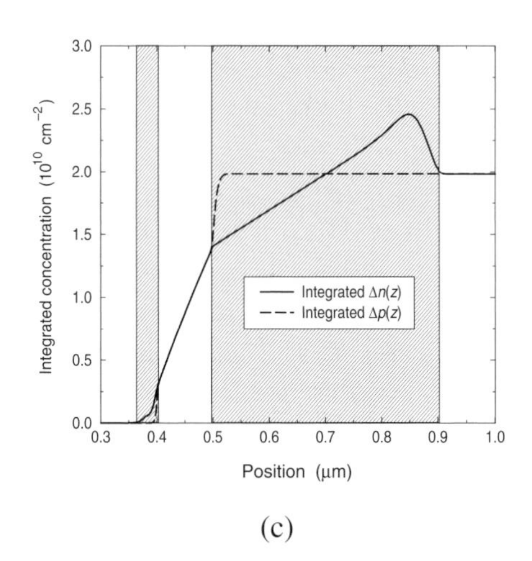

(c)

Fig. 18.13. (a) Conceptual partition of an HBT into quasi-neutral and space-charge regions (shaded). (b) Sample plots of the local charge changes and (c) corresponding plots of the integrated charge changes, versus position z for $J_C = 3 \times 10^4$ A cm^{-2}. For (b) and (c) the MEDICI simulations were for a graded-emitter device with the metallurgical base extending from 0.4 to 0.5 μm and a subcollector commencing at 0.9 μm [20]. (D.L. Pulfrey, S. Fathpour, A.R. St Denis, M. Vaidyanathan, W.A. Hagley and R.K. Surridge, *Journal of Vacuum Science and Technology A*, Vol. 18, No. 2, pp. 775–779, 2000.)

devices. It is convenient, therefore, to use this simulator to determine the changes in regional stored charge needed to evaluate the regional delay times. The changes in stored charge can be found simply as the integrated change in electron *or* hole concentration in each region: $\Delta Q_i \equiv q \int_i \Delta n(z)\, dz = q \int_i \Delta p(z)\, dz$, where $\Delta n(z)$ and $\Delta p(z)$ represent the concentration changes brought about by the change in current ΔI_C. The values of τ_E, τ_B and τ_C so calculated can then serve as the benchmarks against which compact expressions for the regional delay times can be judged.

Extraction of the regional delay times from the numerical simulations is straightforward, provided the edges of the quasi-neutral base are taken to demarcate the three regions of the device associated with the delay times τ_E, τ_B and τ_C, as shown in Figure 18.13(a). In a quasi-neutral region the changes in electron and hole concentration must be equal at each point (see Figure 18.13(b)): $\Delta p(z) = \Delta n(z)$ and, furthermore, on integrating entirely across a quasi-neutral region, the total change in stored electron and hole concentrations must be equal (see Figure 18.13(c)): $\int_{QN} \Delta p(z)\, dz = \int_{QN} \Delta n(z)\, dz$. This region is sharply defined in the base of HBTs because of the very high doping density that exists in this region*. Numerically, equality of the electron and hole quantities involved can be defined to exist if the difference between the quantities is a small fraction of either quantity taken alone.

Unfortunately, it is not feasible with this scheme to subdivide τ_E into portions that can be associated with the quasi-neutral emitter and the emitter space-charge layer. This is because such a boundary is blurred by the large number of free charges present during forward-bias operation. Thus, if the numerical condition used to define equality of the electron and hole quantities involved is changed by a small amount, the resulting space-charge-region edge in the emitter changes by an unacceptable amount.

Regional signal-delay expressions

Traditionally in bipolar transistor analysis and design, and also commonly in HBT studies [26], the following expressions are used to estimate the regional signal-delay times:

$$\tau_E = \frac{\epsilon A}{W_{BE} g_m}, \tag{18.29}$$

$$\tau_B = \frac{W_B^2}{2D_{n0}} + \frac{W_B}{v_{coll}}, \tag{18.30}$$

$$\tau_C = \frac{\epsilon A}{W_{BC}} \left(\frac{1}{g_m} + \frac{W_E}{A\sigma_E} + \frac{W_C}{A\sigma_C} \right) + \frac{W_{BC}}{2v_{sat}}, \tag{18.31}$$

where the new symbols are as follows: ϵ is the permittivity; A is the cross-sectional device area; σ_E and σ_C are the conductivities of the emitter and collector materials; W_E and W_C are the quasi-neutral widths of the emitter and collector; W_{BE} and W_{BC} are the space-charge region widths at the emitter–base and collector–base junctions, respectively. The expression for τ_E accounts only for the charging and discharging of the emitter–base junction capacitance via the dynamic resistance, $1/g_m$; the effects of stored free charge in the neutral emitter and in the emitter–base space-charge region are ignored. The expression for τ_B assumes a finite electron velocity v_{coll} at the collector side of the quasi-neutral base. The expression for τ_C accounts for the effects of charging and discharging the collector–base junction capacitance via the dynamic resistance and the parasitic emitter and collector resistances; the delay due to stored free charge within the collector–base space-charge region is also included, with the factor of $\frac{1}{2}$ arising specifically from the simplifying assumption of a uniform electron

* The onset of the space-charge regions at either end of the base can also be identified by the relatively large change in majority-carrier concentration, as required to charge and discharge the majority-carrier, space-charge capacitance.

velocity [19, pp. 238–240]. Equations (18.29) and (18.31) are most easily applied when the region widths are estimated by invoking the depletion approximation.

To test the suitability of the above compact expressions for estimating the regional delay times in $Al_xGa_{1-x}As/GaAs$ HBTs, a comparison is made below with results from MEDICI, using the relevant, resident models for mobility, recombination, Fermi–Dirac statistics and bandgap narrowing. The devices studied here are n-$Al_{0.3}Ga_{0.7}As$/p-GaAs HBTs with metallurgical emitter and base widths of 4000 and 1000 Å, respectively, and associated doping densities of $6 \times 10^{17}\,\mathrm{cm}^{-3}$ and $3 \times 10^{19}\,\mathrm{cm}^{-3}$. The collector and subcollector have widths of 3000 and 4000 Å, respectively, with doping levels of $4 \times 10^{16}\,\mathrm{cm}^{-3}$ and $4 \times 10^{18}\,\mathrm{cm}^{-3}$. Two devices are considered: an abrupt-junction device and a graded device with 200 Å of linear grading in the emitter. Unless indicated otherwise, the results presented are for $V_{CE} = 2\,\mathrm{V}$ and $J_C \equiv I_C/A = 1 \times 10^4\,\mathrm{A/cm}^2$, which represents a reasonable operating point prior to the onset of high-current effects.

Emitter delay

Figure 18.14(a) shows values of τ_E versus J_C for the graded device. In employing Equation (18.29) to obtain the analytical predictions, the terminal parameter g_m was set equal to that employed in the MEDICI simulation, and W_{BE} was estimated from the depletion approximation. As shown, τ_E is significantly underestimated by Equation (18.29), with the error increasing at high bias. In general, the failure of Equation (18.29) can be attributed to two factors, namely: (i) the breakdown of the depletion approximation due to the presence of large numbers of free carriers in the emitter–base, space-charge region, as noted earlier; (ii) the presence of significant stored charge in the quasi-neutral emitter, which is ignored in writing Equation (18.29). Figure 18.14(b), which shows $\int_0^z \Delta p(z)\,dz$ in the metallurgical emitter, illustrates that stored emitter charge is unimportant in the graded device, but is very significant in the abrupt device. This is because the hole barrier is actually lower in the abrupt-junction case, due to the larger value of V_{BE} required to produce a given collector current.

Attempts have been made to improve the accuracy of the analytical expression for τ_E by including terms to account for the above factors [27,28]. However, the suggested terms are either unwieldy, limited in their bias range of applicability, or require knowledge of poorly defined effective junction velocities.

The best that can be offered at present is to note from Figure 18.14 that, for graded-junction devices, where stored charge in the quasi-neutral emitter can be neglected, τ_E from Equation (18.29) can be made to match the simulation results by multiplying by a correction factor of value around 2 [20].

Practically, the emitter delay is usually the least likely to be of concern because it can be reduced, without adversely affecting the other delay components, by operating at current densities at least twice as high as used in this comparison. Such currents can be obtained at a low enough forward bias to avoid neutral-emitter storage problems,

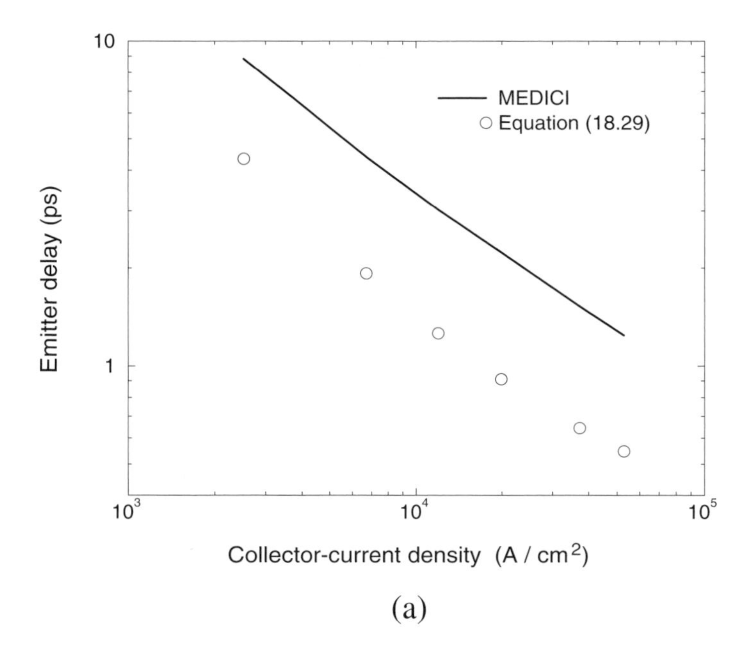

(a)

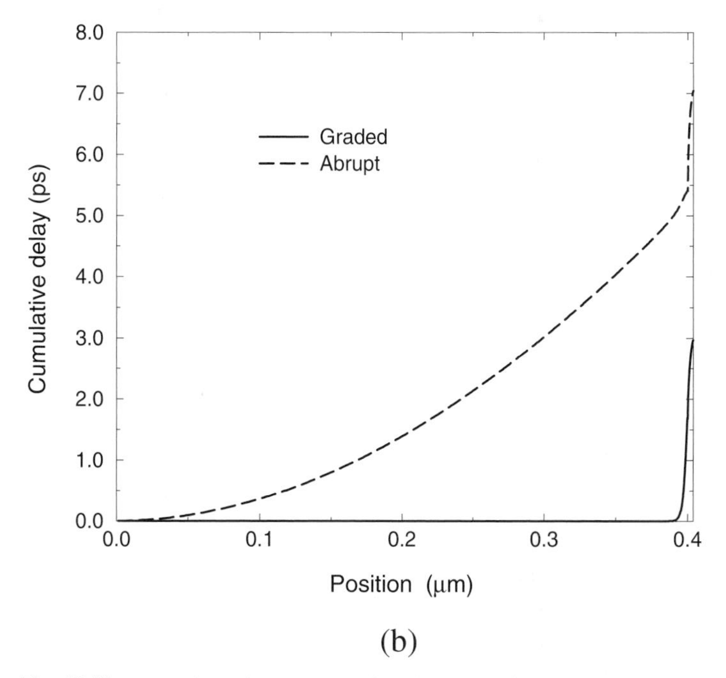

(b)

Fig. 18.14. (a) Emitter delay τ_E as a function of J_C for the graded-emitter HBT. (b) Profiles of cumulative delay in the τ_E region, for both graded and abrupt devices [20]. (D.L. Pulfrey, S. Fathpour, A.R. St Denis, M.Vaidyanathan, W.A. Hagley and R.K. Surridge, *Journal of Vacuum Science and Technology A*, Vol. 18, No. 2, pp. 775–779, 2000.)

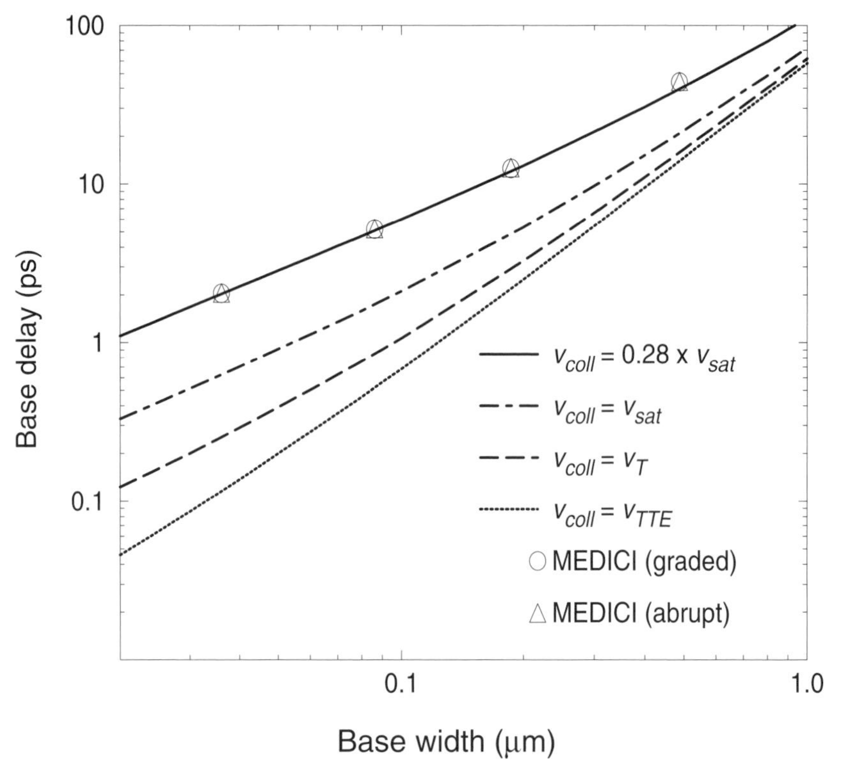

Fig. 18.15. Base delay τ_B as a function of quasi-neutral base width W_B [20]. The curves are from Equation (18.30) for various values of v_{coll}. Note that $v_T \equiv 2v_R$. (D.L. Pulfrey, S. Fathpour, A.R. St Denis, M.Vaidyanathan, W.A. Hagley and R.K. Surridge, *Journal of Vacuum Science and Technology A*, Vol. 18, No. 2, pp. 775-779, 2000.)

even in abrupt-junction GaAs HBTs, by using, for example, an emitter of InGaP, which has a favorably high bandgap and low conduction-band offset.

Base delay

In Figure 18.10 the importance of the collection velocity, v_{coll}, in determining the collector current in short-base, graded-junction HBTs is apparent. The lower the velocity, the greater the charge needed to carry a given current and, consequently, the larger the stored charge in the base. This effect is embodied in Equation (18.30) by the term W_B/v_{coll}, which arises analytically from the supposed trapezoidal shape of the minority-carrier charge profile. Figure 18.5 indicates that this is not a bad description of the charge profile, even in a short-base, abrupt-junction device. The predictions of Equation (18.30) for τ_B are shown in Figure 18.15. The MEDICI simulations predict $v_{coll} = 0.28v_{sat}$, and when this value is used in Equation (18.30), the agreement between the analytical and numerical results is excellent. What may be surprising is that MEDICI gives the same result for both graded-emitter and abrupt-emitter HBTs. This is because in this type of DDE simulator, at least when the energy-balance option

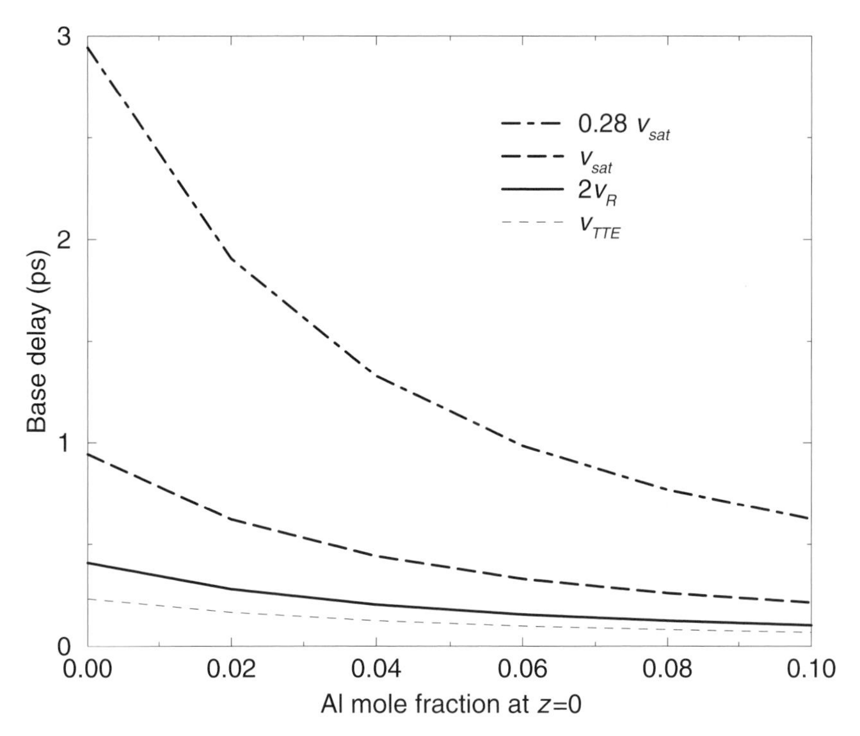

Fig. 18.16. Base delay as a function of Al mole fraction at the emitter end of the base for a linearly graded-base, abrupt-junction HBT. The results are from Equation (18.32) for various values of v_{coll}. $W_B = 1l_{sc}$, $D_n = 75\ \text{cm}^2\,\text{s}^{-1}$ [21], $V_{BE} = 0.8V_{bi}$.

is not operative, the full height and width of the induced heterojunction barrier at the base–collector interface are 'felt' by the flux exiting the base. This serves, in our examples, to reduce v_{coll} to the pessimistic value of $0.28v_{sat}$. In reality, the forward-going flux in a short-base device will be less affected by the presence of a base–collector barrier because its velocity is not v_R, as implied by the full-Maxwellian description of charge concentrations in a classical DDE simulator, but is, as we have seen from the microscopic model, either $2v_R$ in a graded device, or v_{TTE}^+ in an abrupt device. In practical HBTs, therefore, τ_B can be expected to lie closer to the limit given by Equation (18.30) when employing either of these velocities, rather than when using v_{coll} as predicted by MEDICI.

With short bases and potentially high electron velocities, τ_B is no longer the major contributor to the overall delay in modern bipolar transistors. It can be further reduced by employing compositional grading in the base, in which case a convenient expression for τ_B is*:

$$\tau_B = \frac{W_B^2}{bD_{n0}}\left[1 - \frac{[1-\exp(-b)]}{b}\right] + \frac{W_B}{v_{coll}}\frac{[1-\exp(-b)]}{b}, \tag{18.32}$$

* See Problem 18.7.

where $b = \Delta E_g/kT$ and ΔE_g is the bandgap change in the base due to linear compositional grading. The results in Figure 18.16 show how useful a graded base could be in mitigating the effect of a low collection velocity.

Collector delay

Figure 18.17 shows values of τ_C versus J_C for the graded device (similar results are obtained for the abrupt-junction device); results from MEDICI are presented along with the predictions of Equation (18.31). In employing Equation (18.31), the terminal parameter g_m was set equal to that employed in the MEDICI simulation, and the value of W_E was set equal to the metallurgical emitter width. The value of W_{BC} was computed by employing the usual, one-sided, depletion approximation in two different circumstances. In the first instance, the space-charge region on the collector side of the junction was assumed to be fully depleted, giving a charge density of magnitude qN_C, where N_C is the collector doping density. In the second instance, the magnitude of the charge density was taken to be $q(N_C - n_c)$, where n_c accounts for the finite electron concentration required to support the collector current, $J_C = -qn_c v_{sat}$. The results emphasize the importance of accounting for this mobile charge in the space-charge region, even at current-density magnitudes below the onset of the Kirk effect [29], which occurs in these devices when $|J_C| > qN_C v_{sat} \approx 4 \times 10^4$ A/cm^2.

Further subtle effects in the space-charge region can occur due to: (i) the dependence of the electron velocity on both the electric field and the junction self-heating [30]; (ii) the sensitivity of the velocity profile to the electron space-charge density [30,31]. The former effect can increase τ_C with increasing collector voltage, while the latter can decrease the delay at high currents. For a fully depleted collector these effects can be incorporated in an improved expression for the transit-related term in Equation (18.31) [30]:

$$\tau_{Cd} = [\kappa_0 + \kappa_2(\Delta T)]\frac{W_{BC}}{2} + \kappa_1\frac{V_{bi} - V_{BC}}{2} + \kappa_1\frac{q(N_C - 2\bar{n})W_{BC}^2}{12\epsilon}, \qquad (18.33)$$

where $\bar{n}$ is the average electron density in the space-charge region, ΔT is the rise in junction temperature, and the coefficients κ come from a linear fit to the inverse velocity versus field relation for GaAs at high field strengths.

As τ_C is presently the most significant of the delay times in high-performance HBTs, there is no shortage of collector designs to reduce this portion of the signal delay. Exploiting velocity overshoot, by engineering a low-field region in the part of the space-charge region proximal to the base, can be worthwhile [32]. From calculations using a phenomenological model, it has been suggested that this improvement can be simply quantified in GaAs junctions by changing the factor of $\frac{1}{2}$ in Equation (18.31) to $\frac{1}{3}$ [33].

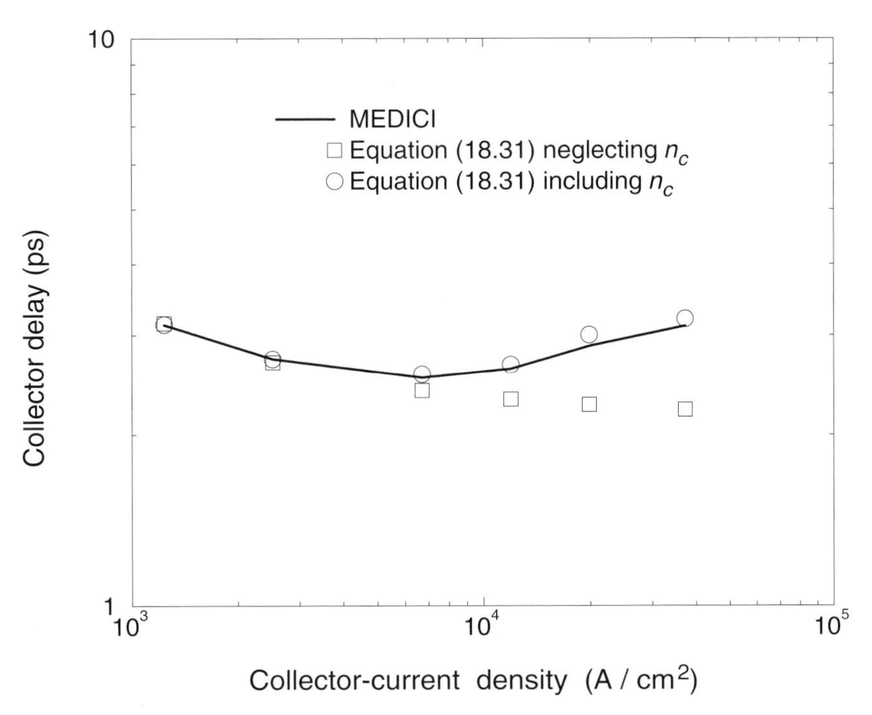

Fig. 18.17. Collector delay as a function of J_C for the graded-emitter device [20]. (D.L. Pulfrey, S. Fathpour, A.R. St Denis, M.Vaidyanathan, W.A. Hagley and R.K. Surridge, *Journal of Vacuum Science and Technology A*, Vol. 18, No. 2, pp. 775–779, 2000.)

18.3.4 Compact models for f_{max}

Base–collector network

The cut-off frequency, f_T, relates to the current gain of a transistor. While this is an important property of a device, of more practical significance is the ability of a transistor to produce a power gain at high frequencies. The appropriate figure of merit is f_{max}, which has long been described by the compact expression [19, p. 177]:

$$f_{max} = \left(\frac{f_T}{8\pi RC} \right)^{1/2}, \tag{18.34}$$

where RC is some base–collector time constant, as discussed below. f_{max} refers to the frequency at which the power gain becomes unity, when determined by extrapolation at -20 dB/decade from lower frequencies. It is relevant to employ an extrapolated figure of merit because, in reality, electronic instrumentation is presently incapable of directly measuring the high values of f_{max} which are characteristic of high-performance HBTs.

In early transistors it was permissible to view the base resistance, r_b, and the collector–base junction capacitance, C_{jc}, as lumped elements, i.e., $RC = r_b C_{jc}$. More recently, the distributed nature of the base–collector RC network has been accounted for by choosing an effective value for this time constant, namely: $RC = (r_b C_{jc})_{eff}$ [34]. In the equivalent circuit of Figure 18.12, for example, r_b and C_{jc} are each shown as being distributed into n components. The case of $n = 3$ for a conventional, mesa-style HBT is worth considering briefly as the three components relate directly to obvious features of the physical structure. With reference to Figure 18.18, the three features are the intrinsic, extrinsic and contact regions of the device, and are represented by (r_{bi}, C_{jci}), (r_{bx}, C_{jcx}) and (r_{bc}, C_{jcc}), respectively. It can be shown* that, at high frequencies, the appropriate base–collector network is given by the circuit in Figure 18.18(b), where r_{cv} is the purely vertical part of the base contact resistance [35]. In this network, the three components of r_b and C_{jc} to substitute into Figure 18.12 are clearly identifiable.

Extrapolated f_{max}

If Equation (18.34) is to remain useful for modern, high-performance HBTs, the RC time constant in this compact expression must be further modified. Modification is necessary because of violation of one of the assumptions on which the original derivation of Equation (18.34) is based, namely: that the collector and emitter resistances, r_{cc} and r_{ee}, and the dynamic resistance, $1/g_m$, are negligible compared to the base resistance. This assumption is invalid for high-performance HBTs due to the employment of very highly doped base regions.

To find the appropriate form for RC in order that Equation (18.34) can be applied to high-performance HBTs, it is necessary to start with an equivalent circuit, such as in Figure 18.12, and move towards the small-signal parameters on which Equation (18.34) is based [36], systematically dropping terms of relatively minor importance. The approach has been carefully documented [35,37], and leads to compact expressions for RC, which can then be substituted into Equation (18.34) to compute f_{max}. When f_{max} is based on the extrapolation of Mason's unilateral gain, U, to 0 dB [38], the relevant expression for RC is:

$$RC = (RC)^U_{\mathit{eff}} = (r_b C_{jc})^U_{\mathit{eff}} + (2\pi f_T r_{cc} C_{jc})\left(r_{ee} + \frac{1}{g_m}\right) C_{jc}, \tag{18.35}$$

where $(r_b C_{jc})^U_{\mathit{eff}}$ accounts for the distributed nature of the base resistance and the collector capacitance [34]. This expression has been tested against results from the SPICE simulation of the circuit in Figure 18.12 for the case of a high-performance device in which $r_{ee} = 6.2\ \Omega$ and $r_{cc} = 16\ \Omega$ [35]. Excellent agreement was demonstrated. By contrast, if the new terms in Equation (18.35) are neglected, $(RC)^U_{\mathit{eff}}$ is underestimated by about 30%.

It can be appreciated from Equation (18.35) that the collector resistance must be minimized if extremely high values of f_{max} are to be obtained. It is noteworthy in this regard that the world-record device described at the end of the following chapter has negligible r_{cc}, due primarily to its employment of a Schottky-barrier collector contact.

* See Problem 18.8.

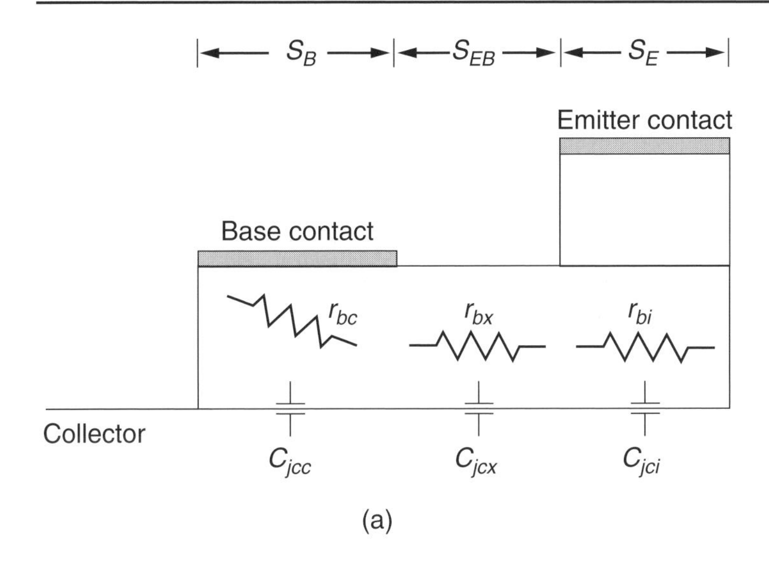

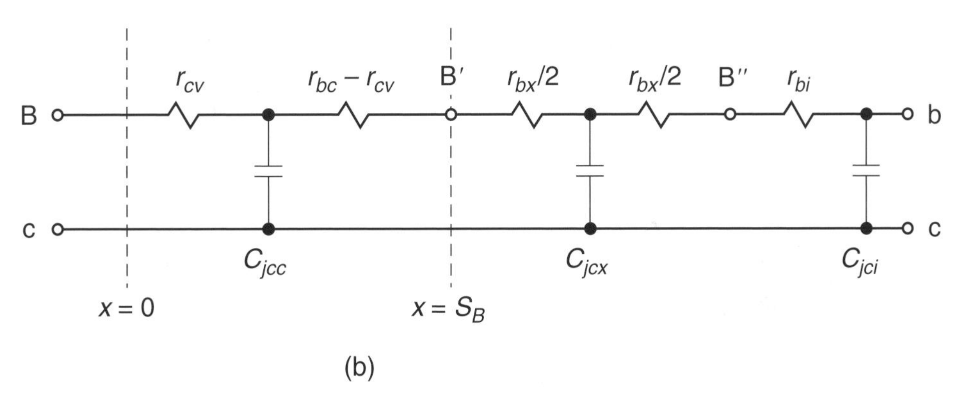

Fig. 18.18. (a) Schematic cross-section of a mesa-type HBT, identifying the components of base resistance and collector capacitance. (b) Reduced equivalent circuit for the base–collector network [35]. (M. Vaidyanathan and D.L. Pulfrey, *IEEE Transactions on Electron Devices*, Vol. 46, pp. 301–309, 1999.)

18.3.5 Compact model for large-signal analysis

In the previous two sections we considered the small-signal modeling of HBTs, such as would be appropriate for high-frequency amplifiers, for example. In this section we briefly examine large-signal modeling, which has relevance for applications in high-speed switching, for example.

An Ebers–Moll model

The well-known Ebers–Moll representation of a three-terminal bipolar device is a convenient starting point for the formulation of a large-signal model for HBTs. In fact, the traditional model for BJTs applies without modification to graded-emitter HBTs, at least in those cases where there is no significant conduction-band spike due to bandgap narrowing (see Figure 18.1). This is because the electronic collector current, J_C, has the same ideal Boltzmann dependence ($\exp(qV/k_BT)$) as the hole currents considered in the intrinsic model. This follows from Equation (18.21), where

$$n_E^* = n_E \exp(-E_{pk}/k_BT) = n_E \exp(-qV_{bi}/k_BT)\exp(qV_{BE}/k_BT). \qquad (18.36)$$

On the other hand, in an abrupt-junction HBT, the presence of a large conduction-band spike at the emitter–base junction causes the electron- and hole-injection currents to have different voltage dependences. In an equivalent circuit, such as the one shown in Figure 18.19, this phenomenon can be represented by using ideal diodes for the hole currents (IPC and IPE), a current source based on non-ideal diodes for the collector electron current (ICT) and non-ideal diodes for the quasi-neutral base current (INC + INE). Details of this circuit are given elsewhere [2,40]: here we concentrate on the representation of ICT in an abrupt-junction HBT.

Although quasi-neutral base recombination is an important contributor to the base current in some HBTs [22], it can usually be ignored in computing J_C, as we have done throughout this chapter. Under these circumstances, and for short-base transistors, Equation (18.24) yields:

$$J_C \approx -\gamma q n_E^* v_R, \qquad (18.37)$$

which is then precisely equal to $J_{TUN} + J_{TE}$, as given by Equation (18.22). There are two ways in which the conduction-band spike contributes to the non-ideal nature of these components of J_C. Firstly, because E_{pk} is now determined by the potential energy difference across only the emitter side of the junction, rather than across the entire junction, the thermionic emission component of the current is not dependent on the full value of V_{BE}. Instead, elaborating on Equation (18.23), it is given by [8]:

$$\begin{aligned} J_{TE} &= J_0 \exp(-E'_{pk}) \\ &= J_0 \exp(-qN_{rat}V_{bi}/k_BT)\exp(-qN_{rat}V_{BE}/n_2k_BT), \end{aligned} \qquad (18.38)$$

where $N_{rat} = \epsilon_B N_B/(\epsilon_B N_B + \epsilon_E N_E)$, and the new diode ideality factor is $n_2 = 1/N_{rat}$. In reality, because of the very high base doping density in HBTs, $n_2 \approx 1.0$, so this deviation from ideality is barely significant. However, the second factor, which concerns the tunneling current, is significant. The deviation from ideality of J_{TUN} is due to not only the splitting of the potential drops at the junction, but also to the fact that the tunneling current is carried by electrons of energy less than E_{pk}. These effects are embodied in Equation (18.22) which, when written out in more detail, leads to another diode current, namely [8]:

$$J_{TUN} = J_0 C^* \exp[(-qN_{rat}V_{bi}/k_BT)\tanh(U'_{oo})/U'_{oo}]\exp(qV_{BE}/n_1k_BT), \qquad (18.39)$$

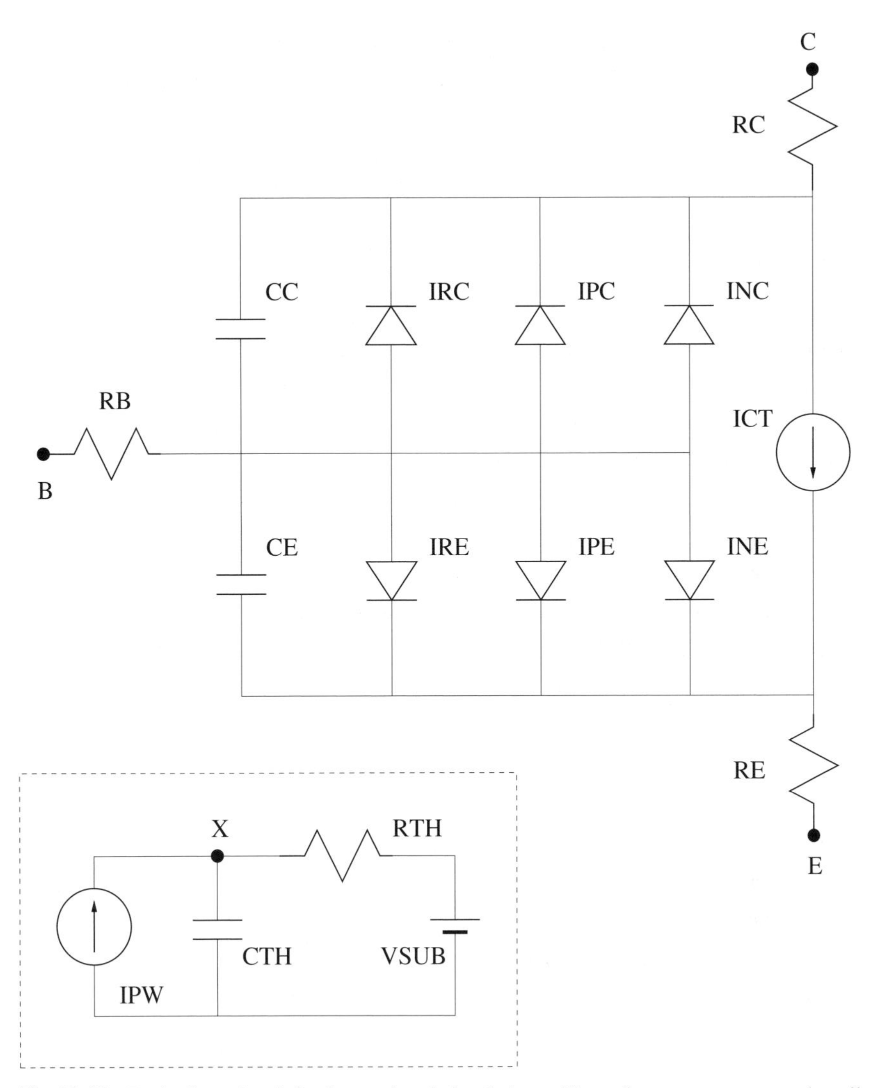

Fig. 18.19. Equivalent circuit for large-signal simulations. The only components not described in the text are the diodes IRC and IRE: these represent space-charge-region recombination currents [39]. (S. Searles and D.L. Pulfrey, *IEEE Transactions on Electron Devices*, Vol. 41, pp. 476–483, 1994.)

where C^* is C from Equation (18.22) evaluated at the desired V_{BE}, and the diode ideality factor is $n_1 = U'_{oo}/N_{rat}\tanh(U'_{oo})$. Thus, when using Figure 18.19 for an abrupt-junction HBT, ICT ($= J_C \cdot$ Area) can be represented by a current source controlled by the voltage across two parallel diodes, with characteristics described by Equations (18.38) and (18.39).

Regarding the resistive components RB, RE and RC in Figure 18.19, these are the parasitic resistances equivalent to r_b, r_{ee} and r_{cc}, respectively, introduced previously in the small-signal equivalent circuit of Figure 18.12. Similarly, the capacitances CE and CC are equivalent to C_π and C_μ, respectively. In a SPICE implementation of the large-signal equivalent circuit, the storage component of CE would be represented by (ICT · TF/V_t), where V_t is the thermal voltage and TF is the forward transit time. A reasonable value for the latter parameter can be obtained by adding the base and collector transit times, as given by Equations (18.32) and (18.33).

Thermal considerations

In practical n–p–n HBTs the barrier height presented to electrons attempting to enter the base is either E_{pk}, in the case of thermionic emission, or somewhat less than this if tunneling is the dominant flow mechanism (see Equation (18.22)). In either case, this barrier is less than the corresponding barrier height, say E_{val}, which inhibits holes from being back-injected into the emitter. Thus, the emitter injection efficiency is approximately $\propto \exp(E_{val} - E_{pk}/k_B T)$. Because there is very little quasi-neutral-base recombination in high-performance HBTs, it follows that the common-emitter current gain, β, will fall with temperature. This can present a problem in devices built upon substrates of poor thermal conductivity, such as GaAs. Transient local heating can occur when the collector current is switched to high densities, and this would lead to a negative differential output conductance, which is usually not desirable.

A simple way of accounting for thermal effects is to add the electrical equivalent of a thermal network to the main electrical equivalent circuit discussed in the previous subsection. The subcircuit shown in Figure 18.19 comprises a current source IPW $\equiv I_C V_{CE}$(W), a resistor RTH representing the thermal resistance (°C/W), a capacitor such that CTH · RTH is equivalent to the thermal time constant (s), and a DC voltage source VSUB which represents the substrate temperature (°C) [41]. Thus, the voltage at node X of the subcircuit corresponds to the temperature of the device. This temperature can then be used to recompute the temperature-dependent model-parameter values for the elements in the main electrical equivalent circuit. In this manner, some thermoelectric feedback is included in the large-signal simulation.

18.4 Conclusion

From the material presented in this chapter on the modeling of short-base, high-performance HBTs, the following observations can be made:

1. Microscopic modeling is able to show that, over most of the base region of a graded-emitter device, the electron distribution function maintains a near-equilibrium form, and the concentration-gradient term in the BTE dominates over the energy-gradient term. These factors suggest that the equilibrium DDE, with appropriate boundary conditions, may adequately describe base transport in these devices. For abrupt-emitter HBTs this is not the case.
2. Guided by the microscopic results, compact models for the collector current in both

abrupt and graded devices can be developed. The different dependences of J_C on base width for these two types of device can be largely reproduced by the compact models.

3. Compact models for the signal delay time can adequately represent the contributions to f_T from delays in the base and collector regions of HBTs.
4. A compact model for f_{max} can be derived which takes into account the effects of emitter and collector resistances, as demanded in the modeling of modern HBTs exhibiting low values of base resistance.
5. It is possible to incorporate the compact model for J_C into a simple Ebers–Moll large-signal equivalent circuit which can aid in the evaluation of the high-speed switching performance of HBTs.

Acknowledgement

Over the last 10 years it has been my privilege to work on HBT modeling issues at UBC with a group of very talented graduate students: S. Ho, O.-S. Ang, A. Laser, H. Zhou, B. Ghodsian, J. Feng, S. Searles, M. Vaidyanathan, A. St Denis and S. Fathpour. My present understanding of the subject is due, in large part, to the endeavors of these gentlemen. I sincerely thank them all. I particularly acknowledge Drs St Denis, Vaidyanathan and Searles, whose work has contributed directly to the material in this chapter.

18.5 Bibliography

[1] H. Kroemer, 'Theory of a wide-gap emitter for transistors', *Proceedings of the IRE*, Vol. 45, pp. 1535–1538, 1957.

[2] D.L. Pulfrey, 'Heterojunction bipolar transistor', *Wiley Encyclopedia of Electrical and Electronics Engineering*, J.G. Webster, Ed., John Wiley & Sons Inc., New York, Vol. 8, pp. 690–706, 1999.

[3] A.R. St Denis, 'A one-dimensional solution of the Boltzmann transport equation with application to the compact modeling of modern bipolar transistors', PhD thesis, University of British Columbia, 1999.

[4] A.R. St Denis and D.L. Pulfrey, 'Quasiballistic transport in GaAs-based heterojunction and homojunction bipolar transistors', *Journal of Applied Physics*, Vol. 84, pp. 4959–4965, 1998.

[5] A.A. Grinberg and S. Luryi, 'Diffusion in a short base', *Solid State Electronics*, Vol. 35, pp. 1299–1309, 1992.

[6] A.R. St Denis and D.L. Pulfrey, 'An analytical expression for the current in short-base transistors', *Solid State Electronics*, Vol. 38, pp. 1431–1436, 1995.

[7] E.S. Harmon, M.L. Lovejoy, M.R. Melloch, M.S. Lundstrom, T.J. de Lyon and J.M. Woodall, 'Experimental observation of minority carrier mobility enhancement in degenerately doped p-type GaAs', *Applied Physics Letters*, Vol. 63, pp. 536–538, 1993.

[8] S. Searles, D.L. Pulfrey and T.C. Kleckner, 'Analytical expressions for the tunnel current at abrupt semiconductor-semiconductor heterojunctions', *IEEE Transactions on Electron Devices*, Vol. 44, pp. 1851–1856, 1997.

[9] M.A. Stettler and M.S. Lundstrom, 'A microscopic study of transport in thin base silicon bipolar transistors', *IEEE Transactions on Electron Devices*, Vol. 41, pp. 1027–1033, 1994.

[10] D.L. Pulfrey, A.R. St Denis and M. Vaidyanathan, 'Compact modeling of high-frequency, small-dimension bipolar transistors', *Proceedings of the IEEE Conference on Optical and Microelectronic Materials and Devices*, Perth, Australia, IEEE, Piscataway, pp. 81–85, 1998.

[11] M.S. Lundstrom, *Fundamentals of Carrier Transport*, Vol. X, *Modular Series on Solid State Devices*, Addison-Wesley, Reading, MA, 1990.

[12] F. Assad, K. Banoo and M.S. Lundstrom, 'The drift-diffusion equaton revisited', *Solid State Electronics*, Vol. 42, pp. 283–295, 1998.

[13] S. Tanaka and M.S. Lundstrom, 'A compact HBT device model based on a one-flux treatment of carrier transport', *Solid State Electronics*, Vol. 37, pp. 401–410, 1994.

[14] M. Vaidyanathan and D.L. Pulfrey, 'An appraisal of the one-flux method for treating carrier transport in modern semiconductor devices', *Solid State Electronics*, Vol. 39, pp. 827–832, 1996.

[15] O. Hansen, 'Diffusion in a short base', *Solid State Electronics*, Vol. 37, pp. 1663–1669, 1994.

[16] A. Marty, G. Rey and J.P. Bailbe, 'Electrical behavior of an *npn* GaAlAs/GaAs heterojunction bipolar transistor', *Solid State Electronics*, Vol. 22, pp. 549–557, 1979.

[17] M.A. Stettler and M.S. Lundstrom, 'A detailed investigation of heterojunction transport using a rigorous solution to the Boltzmann Transport Equation', *IEEE Transactions on Electron Devices*, Vol. 41, pp. 592–600, 1994.

[18] A.R. St Denis, D.L. Pulfrey and A. Marty, 'Reciprocity in heterojunction bipolar transistors', *Solid State Electronics*, Vol. 35, pp. 1633–1637, 1992.

[19] D.J. Roulston, *Bipolar Semiconductor Devices*, McGraw-Hill, New York, 1990.

[20] D.L. Pulfrey, S. Fathpour, A.R. St Denis, M.Vaidyanathan, W.A. Hagley and R.K. Surridge, 'Applicability of the traditional compact expressions for estimating the regional signal-delay times of HBTs', *Journal of Vacuum Science and Technology A*, Vol. 18, No. 2, pp. 775–779, 2000.

[21] E.S. Harmon, M.R. Melloch, M.S. Lundstrom and F. Cardone, 'Thermal velocity limits to diffusive electron transport in thin-base np^+n GaAs bipolar transistors', *Applied Physics Letters*, Vol. 64, pp. 205–207, 1994.

[22] R.E. Wesler, N. Pan, D.-P. Vu, P.J. Zampardi and B.T. McDermott, 'Role of neutral base recombination in high gain AlGaAs/GaAs HBTs', *IEEE Transactions on Electron Devices*, Vol. 46, pp. 1599–1607, 1999.

[23] A.A. Grinberg, M.S. Shur, R.J. Fischer and H. Morkoç, 'An investigation of the effect of graded layers and tunneling on the performance of AlGaAs/GaAs heterojunction bipolar transistors', *IEEE Transactions on Electron Devices*, Vol. 31, pp. 1758–1765, 1984.

[24] K. Yang, J.R. East and G.I. Haddad, 'Numerical modeling of abrupt heterojunctions using a thermionic-field emission boundary condition', *Solid State Electronics*, Vol. 36, pp. 321–330, 1993.

[25] H.K. Gummel and H.C. Poon, 'An integral charge control model of bipolar transistors', *Bell System Technical Journal*, Vol. 49, pp. 827–852, 1970.

[26] W. Liu, *Fundamentals of III–V Devices*, p. 265, John Wiley & Sons, New York, 1999.

[27] J.J. Liou, F.A. Lindholm and D.C. Malocha, 'Forward-voltage capacitance of heterojunction space-charge regions', *Journal of Applied Physics*, Vol. 63, pp. 5015–5022, 1988.

[28] G.-B. Gao, J.-I. Chyi, J. Chen and H. Morkoç, 'Emitter region delay time of AlGaAs/GaAs heterojunction bipolar transistors', *Solid State Electronics*, Vol. 33, pp. 389–390, 1990.

[29] J.C.T. Kirk, 'A theory of transistor cut-off frequency (f_T) falloff at high current densities', *IRE Transactions on Electron Devices*, Vol. 9, pp. 164–174, 1962.

[30] L.H. Camnitz and N. Moll, 'An analysis of the cut-off frequency behavior of microwave heterostructure transistors', S. Tiwari, Ed., IEEE Press, New York, pp. 21–45, 1993.

[31] Y. Betser and D. Ritter, 'Reduction of the base–collector capacitance in InP/GaInAs heterojunction bipolar transistors due to electron velocity modulation', *IEEE Transactions on Electron Devices*, Vol. 46, pp. 628–633, 1999.

[32] C.M. Maziar, M.E. Klausmeier-Brown and M.S. Lundstrom, 'A proposed structure for collector transit-time reduction in AlGaAs/GaAs bipolar transistors', *IEEE Electron Device Letters*, Vol. EDL-7, pp. 483–485, 1986.

[33] H. Zhou and D.L. Pulfrey, 'Computation of transit and signal delay times for the collector depletion region of GaAs-based HBTs', *Solid State Electronics*, Vol. 35, pp. 113–115, 1992.

[34] M. Vaidyanathan and D.J. Roulston, 'Effective base–collector time constants for calculating the maximum oscillation frequency of bipolar transistors', *Solid State Electronics*, Vol. 38, pp. 509–516, 1995.

[35] M. Vaidyanathan and D.L. Pulfrey, 'Extrapolated f_{max} of heterojunction bipolar transistors', *IEEE Transactions on Electron Devices*, Vol. 46, pp. 301–309, 1999.

[36] R.L. Pritchard, *Electrical Characteristics of Transistors*, McGraw-Hill, New York, Ch. 8, 1967.

[37] M. Vaidyanathan, 'Compact models for the high-frequency characteristics of modern bipolar transistors', PhD thesis, University of British Columbia, 1998.

[38] S.J. Mason, 'Power gain in feedback amplifier', *Transactions of the IRE*, Vol. CT-1, pp. 20–25, 1954.

[39] S. Searles and D.L. Pulfrey, 'An analysis of space-charge-region recombination in HBT's', *IEEE Transactions on Electron Devices*, Vol. 41, pp. 476–483, 1994.

[40] J.J.X. Feng, D.L. Pulfrey, J. Sitch and R.K. Surridge, 'A physics-based HBT SPICE model for large-signal applications', *IEEE Transactions on Electron Devices*, Vol. 42, pp. 8–14, 1995.

[41] P. Roblin, personal communication, January 2000.

18.6 Problems

18.1 Derive Equation (18.24) for the collector current in an abrupt-junction HBT by balancing the diffusion current in the base against both the net current at the emitter–base junction and the exit current $(-qn(W_B)v_{coll})$ at the base–collector junction.

18.2 Is it of any consequence that the equilibrium diffusivity is used in Equation (18.24), rather than some effective diffusivity that would account for not only the non-Maxwellian nature of the distribution (see Figure 18.3), but also for the presence of an appreciable charge flow due to changes in the electron kinetic energy (see Figure 18.8)?

18.3 Equation (18.24) can be improved if a hemi-Maxwellian flux approach is used in the derivation of J_C, in the same manner as described in the development of Equation (18.21). The key to the derivation is a proper accounting for the flux that is reflected back into the base from the energy barrier at the emitter–base junction (see Figure 18.2).
Show that the above considerations lead to:

$$J_C = -q\frac{n_E^*}{2}(1-\gamma+2/\delta) \Big/ \left[\frac{W_B}{D_{n0}} + \frac{(2-\gamma\delta)}{\gamma\delta 2 v_R} + \frac{1}{v_{coll}}\right].$$

Note that this equation reduces to Equation (18.21) in the case of a homojunction, and that it reduces to Equation (18.24) if $\gamma\delta \ll 2$, as will usually be the case for a heterojunction when an appreciable band spike is present.

18.4 In Figure 18.15 it is shown that the presence of a barrier at the base–collector junction, through its influence on v_{coll}, has a large effect on τ_B. However, from Equation (18.24), it can be shown that this barrier barely affects the collector current density in a short-base, abrupt-emitter HBT. Is there a contradiction here? Explain.

18.5 Amend the derivation of $(\Delta Q/\Delta I_C)|_{V_{CE}}$ at the beginning of Section 18.3 so that the effects of parasitic resistance in the emitter and collector are taken into account.
Confirm that the result obtained is identical to the expression for $1/(2\pi f_T)$ resulting from the hybrid-π equivalent circuit in Figure 18.12, for the case of a lumped representation of the collector capacitance.

18.6 Establish the conditions under which the differential charge formulaton for the emitter–collector delay, $\Delta Q/\Delta I_C$, is equivalent to the integral formulation, Q/I_C.
Are these conditions met for an abrupt-junction HBT?

18.7 Derive Equation (18.32) for the delay in the linearly graded base region of an abrupt-junction HBT.

18.8 Starting from the schematic cross-section of an HBT shown in Figure 18.18, draw a general equivalent circuit for the base–collector RC network, and show that it reduces to the circuit given in Figure 18.18(b).

19 Practical high-frequency HBTs

Nicolas Moll

The test of all knowledge is experiment. Experiment is the sole judge of scientific 'truth'.

The Feynman Lectures on Physics, RICHARD FEYNMAN

19.1 Introduction

In the previous chapter, we laid the foundation for quantitative understanding of the performance of heterojunction bipolar transistors (HBTs) with a particular emphasis on f_T and f_{max}. In particular, the connection of these quantities to microscopic physics such as carrier transport was described, as was their connection to higher-level parametric descriptions of the transistor, as embodied in an equivalent circuit.

In this chapter, we will discuss the practical constraints imposed by real material systems, epitaxial growth techniques and fabrication processes. The effect of these constraints on f_T and f_{max} will be studied using the results of Chapter 18. The overall goal of the chapter is to develop physical insight into three areas: (i) the key elements which distinguish material technologies that are suitable for HBTs, and the relationship between the choice of material system and device performance; (ii) the interplay of process development issues and device performance; (iii) other factors than transport, doping behavior, and device geometry (such as reliability) that further constrain device performance. In particular, we will cover the history and evolution of material systems, from AlGaAs/GaAs to InP/GaAsSb, which have been used for HBTs, and we will describe a representative fabrication sequence. Armed with this background knowledge, we will apply some of the theoretical results of Chapter 18 to a state-of-the-art production HBT, and examine the prospects for improving its performance. We will also look at some examples of other problems that arise when a device is scaled and operated for maximum f_T and f_{max}. These include the effect of a large emitter periphery-to-area ratio on current-gain, thermal problems, particularly for HBTs with multiple emitter fingers, and constraints on the collector current density that arise from reliability considerations.

19.2 Material choices for HBTs

19.2.1 History and evolution

The basic concepts underlying the heterojunction bipolar transistor (HBT) were outlined in 1948 by William Schockley [1] as a specific claim in his patent on the transistor. About ten years later, Herbert Kroemer published the first journal article that described the remarkable advantages that an HBT holds over a conventional bipolar transistor [2]. Another decade passed before a meaningful experimental realization of an HBT was reported on [3]. The growing experimental interest in HBTs was fueled in large part by the growing experimental abilities of the semiconductor research community in III–V epitaxy. III–V semiconductors were important in the early practical development of HBTs for at least four key reasons.

First, it turns out that heterojunctions in many III–V material systems, and in particular in the AlGaAs/GaAs material system, can be grown with a very low density of interface states. This is particularly a key feature for HBTs, because too many interface states at the emitter–base junction could lead to a large recombination at this interface, and low current gain. Figure 19.1 shows schematically the various charged carrier fluxes which may flow in a single heterojunction n–p–n HBT (an HBT with a heterojunction only at the emitter–base). From that figure, we see that base current may arise from a number of mechanisms, including bulk recombination in the neutral base, space-charge recombination, recombination at an electrostatic potential saddle-point where the emitter–base space-charge region meets the emitter mesa, recombination at the base surface and reverse injection of holes into the emitter:

$$I_b = I_{rec} + I_{scr} + I_{sp} + I_{surf} + I_{inj}. \tag{19.1}$$

The space-charge recombination current will be strongly influenced by interface states that appear in the forbidden gap, and which may originate because of impurities, strain, local bonding defects or other crystalline imperfections. These interface states turn out to be low in density for the material system considered here, partly out of good fortune, and partly for fundamental reasons. If the two materials composing the heterojunction could not be well matched in lattice constant, the resulting dangling bonds, vacancies and dislocations would likely lead to a high concentration of interface states. But in fact it is possible to independently adjust the lattice parameter and bandgap of III–V alloys. This is illustrated in Figure 19.2, which shows bandgap energy versus lattice parameter for a variety of III–V semiconductors. Some study of this figure will show the reader that, once a particular substrate is chosen, thereby fixing the target lattice parameter, there is a variety of alloys, and hence of bandgaps, which can at least be considered for lattice-matched growth on that substrate. The nearly vertical tie-line between GaAs and AlAs, for example, means that on GaAs substrates, nearly any

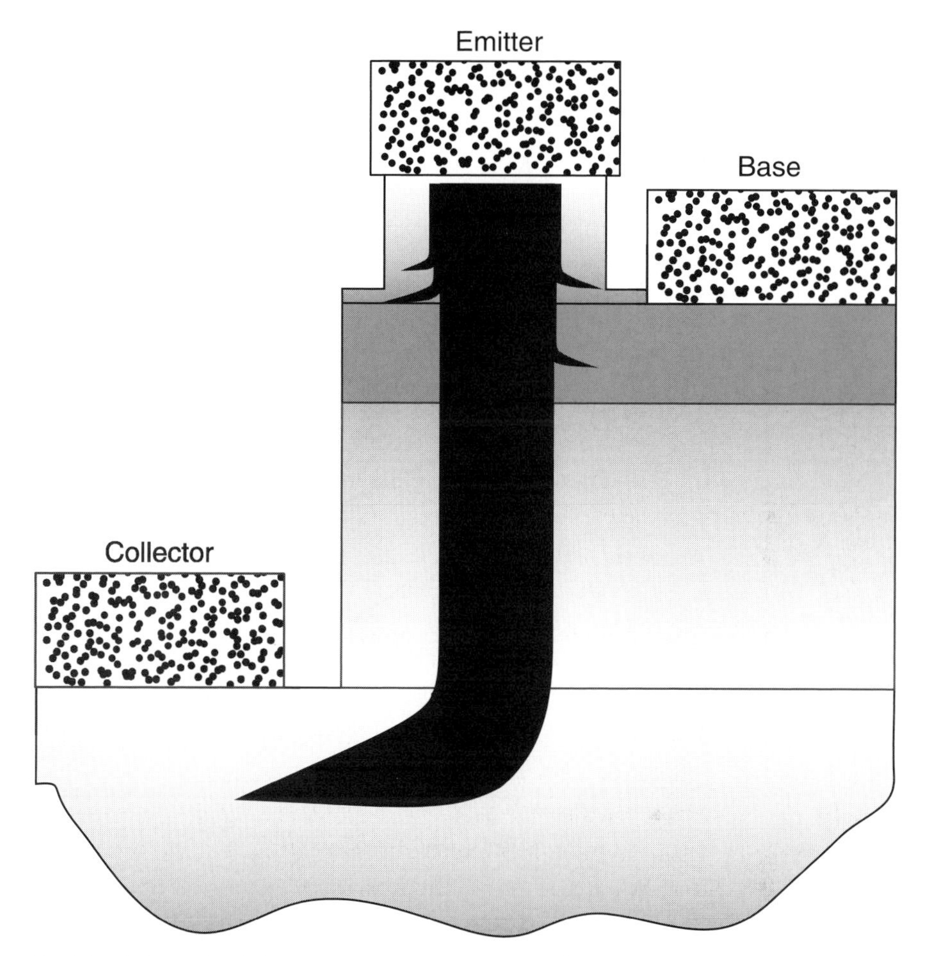

Fig. 19.1. Schematic of electron flow in an n–p–n HBT. Electron fluxes that end inside the device, in the bulk emitter, emitter–base space-charge region, base or saddle-point, will be matched by hole flux. In most real devices, nearly all of the electrons are collected, and only one or two of the other charge paths will be important.

alloy of the form $Ga_{(1-x)}Al_xAs$ will be close enough to being lattice matched, that it can be considered as a potential semiconductor for use in an HBT fabricated on a GaAs substrate. Conversely, only one composition of the $In_xGa_{(1-x)}P$ system is so suited. At this writing, GaAs and InP are the practical commercially available substrates. It is left as an exercise to work out the approximate alloy compositions of ternary alloys that can be lattice-matched to InP. We have also mentioned in passing one feature of the AlGaAs alloy system that made it such an interesting starting point for early heterojunction work: lattice matching is automatic. That is to say, it is not sensitive to composition. In the early days of III–V epitaxy, this was an advantage not to be taken lightly.

Second, the transport properties of GaAs are significantly better than the transport properties of silicon. It has roughly 10 times the electron mobility, comparable hole

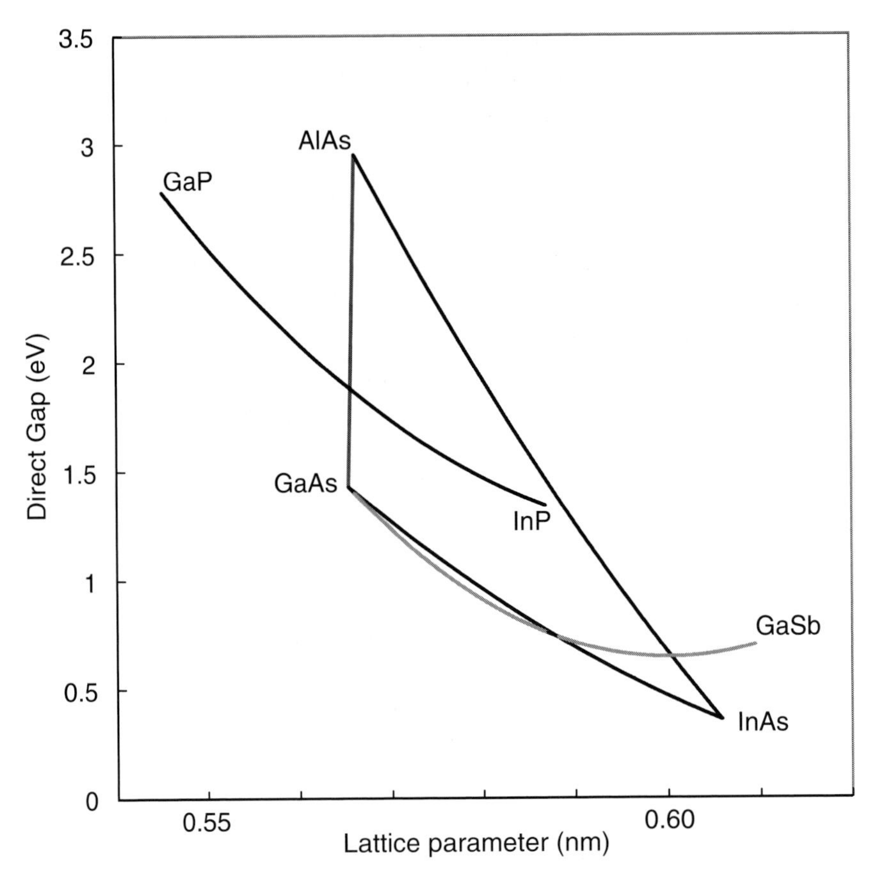

Fig. 19.2. The direct bandgap energy versus lattice parameter for various ternary III–V alloys. Note that, since GaP and AlAs are indirect gap semiconductors, alloys close to them will also be indirect, with a smaller bandgap than is shown.

mobility, and a higher effective saturation velocity. This leads, for reasons that we shall see later, to significantly better potential performance for III–V HBTs than for existing Si bipolar junction transistors. Without this performance advantage, the technological driving force behind work on HBTs would have been too weak to proceed on a comparatively complex and immature technology.

Third, if the devices can be realized at all, they can be easily realized on a semiinsulating substrate. This is a key advantage, already well demonstrated by early work in GaAs FET integrated circuits (ICs), for a technology that is being driven by high-speed applications because of the ease with which matching elements can be incorporated into circuits [4].

Fourth, Molecular beam epitaxy of III–V compounds is relatively easy. The advent of MBE, which allowed structures to be grown with practical doping levels, desirable band offsets and reasonably low background defect levels, was clearly the enabling technology that lead to early reports of good HBT microwave performance [5].

All of these factors put together led to rapid development of an HBT with an AlGaAs emitter and GaAs base, collector and substrate, and much pioneering work in understanding of device physics issues, growth issues, and fabrication techniques. However, the range of theoretically possible systems is much larger than just this one, and other material systems have proved to offer their own advantages for high-speed devices. Since the chapter is organized more around physical effects than material systems, we will only mention these material systems here, and cover the details in the relevant section. Before we do that, we should observe that throughout this chapter we will focus on n–p–n transistors. While interesting work has been done on p–n–p HBTs, including some that shows promise for high f_{max} devices, only n–p–n devices are characterized by the high f_T that is an essential figure-of-merit for high-speed devices. The reason for this is the poor hole transport, reflected in the poor low-field mobility, for practically every III–V material system, as well as for silicon. This in turn leads to a long base delay; additionally p–n–p transistors present daunting bandgap engineering problems in some material systems.

The closest variation to the AlGaAs emitter GaAs base HBT is the InGaP emitter GaAs base HBT which features a band lineup suitable for an abrupt emitter–base junction, lower I_{scr}, and improved reliability. There is also a family of HBTs grown on InP substrates with lattice-matched InGaAs bases. The emitter material in these devices can be either InP or AlInAs. These devices operate at substantially lower emitter–base bias than GaAs base devices, which is a significant advantage in some power-sensitive applications. They have other performance advantages as well, but at the expense of less technological maturity. The 'single' HBT (SHBT) device by definition has an emitter–base heterojunction and a collector–base homojunction, and hence an InGaAs collector drift region, which leads to poor high-power characteristics. The double HBT (DHBT) typically has an InP collector drift region, which should lead to excellent high-power characteristics, while retaining all the other performance advantages of InGaAs base SHBTs. However, they do present a thorny collector–base junction design problem.

An interesting and different HBT technology that started to show significant promise around the same time as the InGaAs base HBTs is the SiGe base HBT. In this device the base layer is significantly strained, and the strain plays a beneficial role in determining the band lineup at the heterojunctions. The result is a device with negligible conduction-band discontinuity, and many other desirable characteristics that derive from the deep understanding of how to do silicon processing. They do not quite match the best III–V devices for high-speed applications, however, because of their intrinsically poorer transport properties. Their growth and fabrication technology is quite different from that for the III–V devices, and we will not go into much further detail on this topic.

Another interesting and different HBT technology is the AlGaN/GaN SHBT [6]. The material technology for these devices is difficult and still embryonic. They present two main attractions that both derive from the large bandgap of the nitrides: the potential for very-high-temperature operation, and the potential for very-high-voltage operation because of the high avalanche field expected in a GaN collector drift region. This comes at the expense of poor low-field transport and p-type doping properties. These disadvantages are least important for high-speed high-voltage applications, where the device delay is dominated by transit through the collector drift region.

The last material-based variation worth mentioning at the time of writing is the GaAsSb/InP DHBT. Good results for these devices have been obtained only since 1998, in spite of early work that pointed out the potential of this interesting material [7]. GaAsSb has been described as in some sense an alternative to InGaAs as a base material in InP DHBTs. That is, it has a comparable bandgap, and retains the advantage of a low emitter–base bias voltage. It also presents an assortment of tradeoffs both desirable and problematic. The most interesting feature of the InP/GaAsSb heterojunction is its type II band lineup. Although GaAsSb has a substantially smaller bandgap than InP, its conduction-band energy is slightly higher than that of InP, by an amount on the order of 0.1 eV. This energy is both small enough, and of the right sign, that HBT layers can be grown with abrupt heterojunctions without major negative consequences for the emitter–base junction, while presenting a theoretically ideal collector–base junction which is completely free of the design issues related to electron pile-up that plague InGaAs/InP DHBTs. Also, GaAsSb can easily be doped with carbon, a desirable p-type dopant, to high concentrations, and excellent p ohmic contacts can be obtained. However, the hole mobility in this material is quite poor, and devices with high f_{max} have yet to be demonstrated.

19.2.2 Growth techniques

There are two main techniques that are used for the epitaxial growth of HBT layer structures. These are MBE and organometallic vapor phase epitaxy, or OMVPE. MBE started out as an ultra-high vacuum evaporation technique from elemental sources such as metallic gallium, aluminum, and arsenic, onto a heated substrate. The flux from each source is precisely controlled by adjusting the source temperature, and by opening and closing shutters, so that beams of the elemental constituents of the desired semiconductor, as well as of dopant atoms, impinge simultaneously on the substrate. Typical growth rates are on the order of 1 μm per hour so that exquisite control over layer thickness and dopant concentration can easily be obtained. In its modern form MBE still reflects the basic approach demonstrated in early work. However, sources for both the semiconductor constituents and dopants have evolved considerably. Many group V elements are now supplied by hydride gases such as arsine. Control over their flux is thus effected by mass flow controllers and is more precise than ever. The p-type dopant of choice has become CBr_4 which is also supplied in gas form through a mass flow controller. The precise choice of growth conditions represents a tradeoff between various aspects of material quality such as morphology, background defect concentration, dopant activation and the desired material composition. As MBE hardware has evolved, along with the understanding of detailed growth mechanisms, these tradeoffs have actually become less difficult.

Today the device designer can obtain practically any epitaxial layer structure imaginable from a knowledgeable and well-equipped crystal grower. Of course, the laws of physics still apply, so that a transistor made with highly strained base material which is doped to 10^{20} carbon per cubic centimeter will perform poorly. Some heterojunctions will be easier to grow than others; abrupt ones are easily obtained by opening and closing the appropriate source shutters. Smooth but rapidly graded ones are more difficult because of the thermal time constants of the sources. Morphology is good, but not perfect. The main problem is a small density of oval defects, on the order of 10–100 per square centimeter, which are thought to be related to the use of elemental group III sources.

In OMVPE, the source atoms that will become epitaxial semiconductor layers are supplied as organometallic vapors such as trimethyl gallium. The substrate is heated to a high enough temperature that the source molecules, as well as dopant molecules, are cracked (thermally broken apart) when they impinge on the substrate. As a growth technology, OMVPE scales to large numbers of substrates quite gracefully. In volume production it offers considerable economic advantages over MBE, and some technical advantages as well as some technical drawbacks. The general consensus is that it is a superior growth technique for the production of optical devices such as lasers; this is due to a slightly lower non-radiative recombination rate than in comparable MBE material. Since composition is controlled by the gas flow rates through mass flow controllers, linear grading or more complicated grading can easily be obtained in OMVPE. Atomically abrupt interfaces are more difficult. Its suitability for optical devices also makes it an attractive growth technology for other minority-carrier devices such as HBTs; low non-radiative recombination is important both in the neutral base, and in the emitter–base and collector–base depletion regions. Growth from the vapor phase can result in nearly perfect surface morphology, and HBT ICs based on OMVPE layers can be made with very high manufacturing yield.

There are two main difficulties with this technology. Some source gases can be adsorbed, and then later reemitted, from the reactor walls. This 'memory effect,' when present, makes it difficult to effectively turn off certain dopants such as Mg for p-type doping, and Te, Se, and S for n-type doping [8, 9, 10]. The higher growth temperature, relative to MBE, can also result in a more difficult tradeoff of the growth conditions. In particular, the p-type dopant may either diffuse if magnesium or zinc is used, or fail to incorporate if carbon (the dopant of choice if diffusion is to be minimized) is used. Carbon doping turns out to be fairly easy in GaAs and GaAsSb, but less so in InGaAs.

19.3 Processing techniques and device design

19.3.1 Introduction

From the simplest point of view, a bipolar transistor is fundamentally a one-dimensional device. As illustrated schematically in Figure 19.1, the main current flow is along a single path perpendicular to the emitter base and collector base junctions. This point of view leaves out issues related to contacts; since in practice contacts have to be included, the realization of this one-dimensional device has to be two-dimensional. The starting point for the device designer contemplating the development of an HBT technology is to understand those two-dimensional effects, and then to minimize them as much as possible. In other words, the task is to make the device function as if it were one-dimensional. This amounts to minimizing R_b, minimizing R_c, minimizing periphery-related recombination, and minimizing the ratio of the extrinsic collector–base capacitance to the intrinsic collector–base capacitance. The final outcome of this piece of design work is the definition of the lateral device geometry. The three most important ingredients that go into the design are base spreading resistance considerations, fabrication technology considerations and thermal considerations. Neither the base spreading resistance issues nor the thermal issues are different, from a fundamental point of view, from the issues that arise in silicon bipolar device design. However, the quantitative outcome for the spreading resistance is quite different. Likewise, fabrication of III–V devices tends to be quite different from their silicon counterparts. So we will start by outlining the fabrication technology, then move on to the problems of spreading resistance and other issues.

19.3.2 III–V processing technology

With rare exceptions, III–V HBTs are mesa-style transistors. In a typical fabrication sequence the starting point is a complete set of epitaxial layers grown on a semi-insulating substrate. As can be inferred from Figure 19.1, the finished device will be an emitter-up device so that the epitaxial layers consist, from top to bottom, of a heavily doped InGaAs contact layer, the wide bandgap emitter material, the base layer, the collector drift layer and a highly conductive subcollector layer. A typical fabrication process, loosely based on [11], might begin with lithographic definition of the isolation pattern, and implant isolation through the subcollector layer. This step defines the extent of the subcollector and of part of the collector base junction. This is followed by the formation of a dummy photoresist and silicon nitride emitter. This same lithography step also leaves photoresist outside the outer boundary of the base contact metal pattern. The epitaxial layers are now etched away by dry etching down to the base layer, the base contact metal is evaporated and the excess lifted. Note that, particularly for small devices, with the emitter sizes on the order of a micrometer, clever choice of the isolation geometry can significantly reduce the parasitic collector base capacitance. This is illustrated in Figure 19.3(a) and (b), where we see that much of the material underneath the base contact metal has been rendered semi-insulating by the isolation implant. Next the nitride in the field, but not the dummy emitter, is removed and the epitaxial layers are etched with a wet etchant to finish defining the emitter. This etch step typically stops just short of the metallurgical base leaving behind a so-called emitter ledge; the ledge is required for devices with gallium arsenide bases, for reasons that we will discuss later. It may be unnecessary in other material systems. After emitter mesa formation, the side walls of the emitter mesa are protected with a dielectric such as silicon oxynitride. This protects the area between the emitter and the base contact, and another etching step is performed to etch away the material outside the base contact metal. A second etching step is performed to etch away the material outside the base contact metal. This last etching step exposes the subcollector layer. Finally, the dummy emitter is removed, and a photoresist mask is patterned to define the n-type ohmic contact metal to the emitter and subcollector. That metal is then evaporated, and the excess is lifted away. At this point, the transistor itself is complete. All that remains to be done is the formation of any additional dielectric layers and metal layers. In a typical IC technology, resistors and capacitors will also be added. All this is done in a manner similar to silicon IC processing and gallium arsenide FET IC processing.

19.4 Further discussion of f_T, f_{max}

19.4.1 Origin and distribution of delay times

In our discussion so far, we have outlined the general set of material and processing choices faced by the device designer. We now turn to a more detailed discussion of the effect of these choices on microwave performance, as measured by f_T and f_{max}. These figures-of-merit are carefully discussed in Section 18.3 from a theoretical point of view. What we would now like to do is to connect that discussion with some of the limitations imposed by practical material choices and fabrication technology. As a starting point, let us consider an InGaP emitter HBT process in production at Agilent Technologies [12]. Typical device parameters for a 2×4 μm emitter transistor realized with that process are shown in Table 19.1.

Note that the expression for collector delay shown in Equation (18.31) can be rewritten as

$$\tau_C = C_{BC}\left(\frac{1}{g_m} + R_E + R_C\right) + \frac{W_{BC}}{2v_{sat}}. \tag{19.2}$$

These devices have a typical total delay of 2.15 ps, at 6×10^4 A/cm^2 and V_{ce} of 1.5 V.

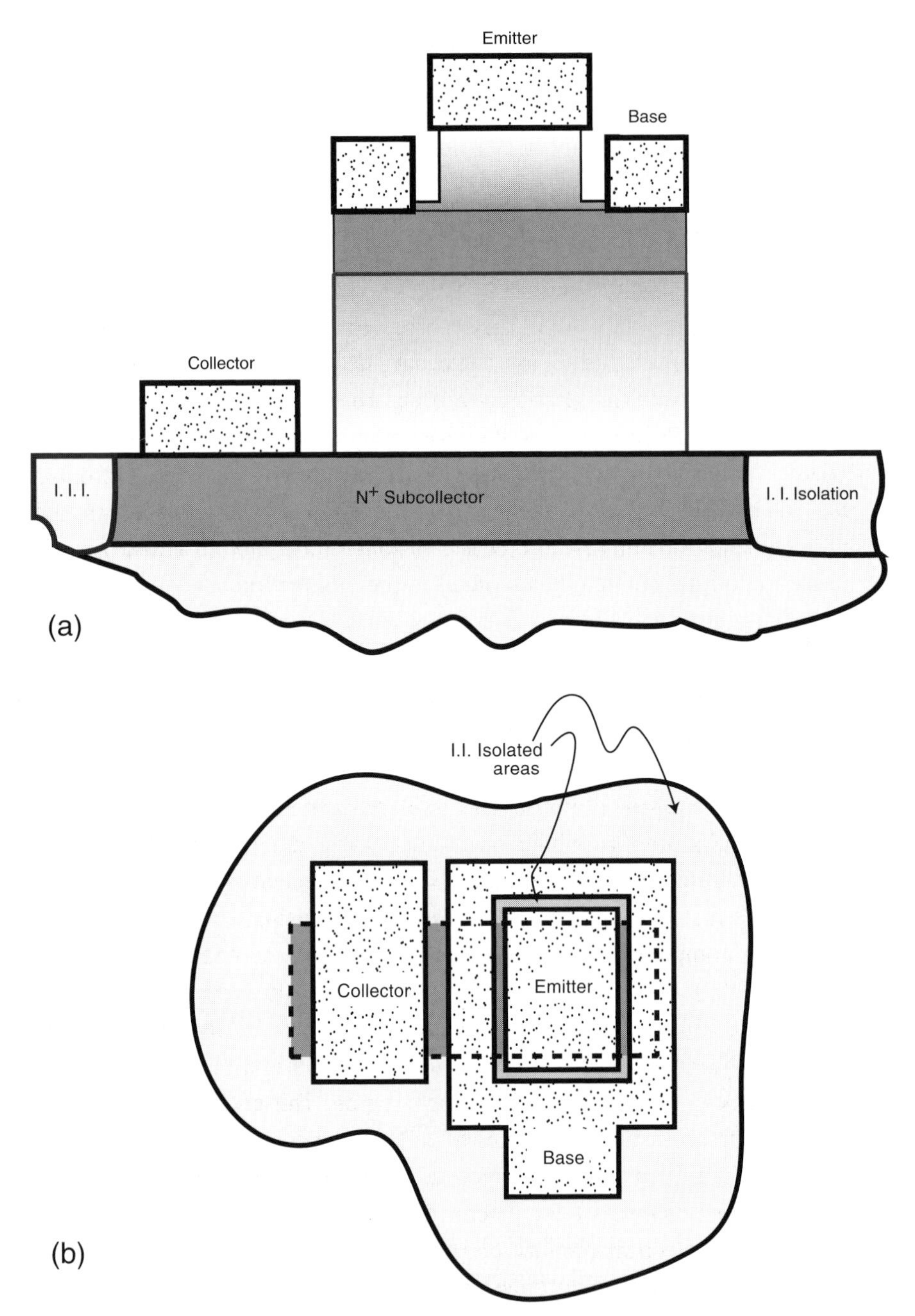

Fig. 19.3. Cross-sectional and top views of a mesa and ion-implantion-isolated HBT. The cross-section shown in (a) is taken along a horizontal cut of (b). In (b) the broken line indicates the boundary of implant isolation, the lightest gray indicates implant isolated subcollector, and the medium gray indicates implant isolated base–collector–subcollector layers. The darkest gray indicates unimplanted active device area. Note the use of base-layer isolation, seen in (b), to reduce the collector–base area.

Table 19.1. *Typical parameters for a 2 × 4 μm emitter InGaP/GaAs abrupt single-heterojunction bipolar transistor at V_{ce} = 1.5 V and I_c = 4.8 mA.*

f_T	f_{max}	β	$R_E(\Omega)$	$R_B(\Omega)$	$R_C(\Omega)$	$C_{BC}(fF)$
74	83	132	7.2	43	9.0	14.2

Equation (19.2) shows that 0.31 ps can be attributed to collector charging (the first term). The emitter charging time can be estimated from the slope of the delay versus $1/J_c$ as 0.4 ps. Note that the effect of C_{BC} must be accounted for in this procedure. As discussed in Section 18.3.3, the base delay depends on the exit velocity into the collector. From Figure 18.15, we get an estimated base transit time of 0.4–0.8 ps for an 800 Å base, depending on the exact choice of exit velocity. This leaves 0.65–1.05 ps of delay that can be attributed to collector transit time through a 4000 Å collector.

19.4.2 Improvement of delay times

In analyzing these delays, our ultimate objective is to shorten them as much as possible, preferably without degrading any other device characteristics. In turn, such a reduction will depend either on a different material choice, or on changes in the device structure and concomitant improvements in process technology.

The first observation, which practically leaps out, is that collector delay is the single largest contribution to the total delay. A second observation, however, must go with this first one: the collector delay that we estimated from our analysis of the other delay components is far less than the delay calculated by blind application of Equation (19.2). That calculation would lead to a collector transit delay of 2.5 ps for an optimistic v_{sat} of 8×10^6 cm/s – more than the total measured delay at a finite current density! In general, this discrepancy must have its origin in a significant variation in electron velocity through the collector drift region. This could be due to non-stationary transport in the collector region (i.e. velocity overshoot) which we expect always to be present in III–V semiconductors, and which leads to high electron velocity close to where the electron enters the collector drift region. It could also be due to the static dependence of electron velocity on the electric field [13]. Most likely both effects play a role; the difference between the delay calculated from the high-field velocity and the delay estimated from actual measurements is really too large to be due only to the variation of static electron velocity with field. This points to the importance of the non-stationary transport properties of the collector material as a key factor in the choice of a material system. Monte Carlo results [14] show that, theoretically, an InP collector should show substantially more overshoot than a GaAs one, at comparable bias. InAs and InGaAs lattice-matched to InP show even more, but at very low collector bias.

Examination of Equation (19.2) also shows that there is less benefit than might be thought to be gained from shortening of the collector drift region. While the contribution of the transit term can obviously be reduced, the charging delay will actually increase as the drift region is shortened. As shown in Problem 19.2, control of this time is critical if the collector length is to be scaled.

The collector charging time, base delay and emitter charging time are all comparable in the device we are considering here. So in terms of performance gain, there is no *a priori* advantage in focusing on any one of them. In other words, in the absence of any other knowledge, we might expect that halving any of these delays could be equally problematic. Let us then examine in more detail what would be required to reduce each of these delays by a significant amount. The resistive contribution to the collector charging time arises from the sum of the dynamic emitter resistance, parasitic emitter contact and body resistance, and the series collector resistance. In the case considered here, this amounts to about 22 Ω. None of these contributions can be reduced easily. The dynamic emitter resistance cannot easily be altered. Operation at a higher current density would have a substantial negative impact on reliability, both through weak current activation of the device's failure mechanism, and because of the higher junction temperature that comes with greater power density. The parasitic emitter resistance is less than specific contact resistances that are typically described in the literature as excellent. Substantial improvement in this parameter would be difficult at best. The parasitic collector resistance can only be reduced by changes in device layout, or by decreasing the sheet resistance of the subcollector layer either through increased doping or with a thicker layer. Changes in the device layout would require, in this case, scaling of the lithographically defined dimensions. As shown in Problem 19.3, R_C responds well to scaling. The practical difficulties of doing this in a manufacturing environment should not be underestimated, however. Reduction of the subcollector sheet resistance by increasing its doping is not necessarily possible, due to limitations of the growth process, or to reliability considerations [15]. Increasing its thickness would require changes to the isolation process that carry with them serious manufacturability issues, either in terms of required implantation energy, or device topography. Even if R_C could be reduced by 50%, the overall impact on the collector charging time would be small – on the order of 20%.

The capacitive contribution to the collector charging time is a different story. If we think of the problem as reducing the collector area while keeping the emitter area constant, it becomes clear that either clever layout tricks or scaling may have benefits. Two examples of layout tricks are: use of an isolation implant that overlaps part of the base contact, and use of a single base contact finger between two emitter mesas. The first approach is, of course, process-dependent. The second requires scaling of the lithography process to keep the base spreading resistance equivalent. Straightforward scaling of the base lithography will produce comparable results. Reduction of the base lithography CD (critical dimension) would reduce C_{BC}, typically, by 30%. This will translate directly into a comparable reduction in collector charging time, of about 0.1 ps.

Earlier in this section, we estimated the base delay to be between 0.4 and 0.8 ps, based on Figure 18.15. The spread in this estimate depends on the effective exit velocity for injected electrons as they pass from the base into the collector space-charge region; this velocity has a direct and substantial impact on the total stored minority-carrier charge in the base. As pointed out in the preceding chapter, microscopic modeling tends to favor the view that the exit velocity is high. However, it is not possible to be certain that all of the necessary physics has been put into even elaborate microscopic models, without verification through comparison with experiment. In the particular case of base delay time in short bases, convincing experimental evidence about the exit velocity does not exist. We can then only say that several tenths of a picosecond might be shaved from the delay by improving the carrier dynamics at the end of the neutral base, or that no improvement can be had because the carrier dynamics are already excellent. The surest way to reduce base delay is by scaling the thickness of the base (the base width). Even if this results in an only marginal increase in the carrier velocity across the base, it will still reduce the stored base charge, and the transit time, in proportion to the scaling factor. From Figure 18.15 in the preceding chapter, we see that reducing the base width from 800 to 400 Å will reduce the base delay by 0.2–0.4 ps, with the likely improvement, according to microscopic models, being on the low end. However, scaling the base width still entails other technological considerations. The processing technology used for emitter mesa and base contact formation has to be consistent with maintaining good process yield and reliability with the thinner base layer. And the lithography has to be scaled by the same factor to maintain a roughly equivalent base resistance. This scaling can only be avoided if the base sheet resistance can be maintained by increasing the doping. Typically, this is not possible, because the base doping will already be at some maximum value set by other considerations. Such considerations include the onset of Auger recombination, reliability issues or dopant solubility.

To summarize then, even dramatic changes in the scaling of the 2 ps transistor that we have been considering here, say by a factor of 2, lead to rather smaller improvements in f_T. We could expect to gain about 0.1 ps in collector charging time, or 0.2–0.4 ps in base transit time. Note, however, that the lithography scaling which we must do when scaling the base width excludes scaling to reduce τ_{cc}. Halving the collector length will also produce an improvement of the same order, but at the expense of operating voltage. Thus to dramatically improve the f_T of an HBT technology that has been optimally designed, scaling alone will not suffice. Major changes, such as adoption of a different material system, dramatic improvement of the safe operating current density, or a totally different approach to device fabrication that greatly reduces C_{BC}/C_{BE}, must be made.

The other figure-of-merit, f_{max}, can be described as arising from a time constant that is the geometric mean of the total forward delay, τ_{EC}, and another time-constant $4(r_b C_c)$ that is related to charging of the collector capacitance through the base resistance. As pointed out in the preceding chapter, this problem is actually more complex than the charging of a single capacitor through a single resistor, and computation of the effective value of $(r_b C_c)$ is equally complex. We can set aside these complexities in this discussion because we are more interested in the overall scaling behavior than in precise predictions. Using, then, the inaccurate but simple expression

$$\tau_{max} = 2\left(r_b C_c \tau_{EC}\right)^{1/2} \tag{19.3}$$

we note that f_{max} is quite insensitive to changes in the collector drift length: the increase in τ_{EC} roughly balances the decrease in C_c that accompanies an increase in the collector length. On the other hand, it is quite sensitive to changes in r_b. If we consider the effect of lithography scaling on the four components of r_b shown in Figure 18.18(b), we observe that r_{cv} is constant, but the other three components scale. Although r_b does not quite decrease proportionally when the lithography shrinks, C_c remains essentially constant, so shrinking is an excellent way to improve f_{max} without degrading f_T. However, use of a low-contact-resistance base contact is a key element of this strategy.

There is a more radical approach than straightforward (but not easy) device scaling to the problem of reducing HBT delay times. As we saw above, both τ_C and τ_{max} are sensitive to the collector capacitance, which is determined mainly by the extrinsic collector area. As shown in Equation (18.35), τ_{max} also contains a term that depends on the series collector resistance. In 1995, Rodwell's group at UCSB demonstrated a method of minimizing the collector capacitance by nearly eliminating the extrinsic collector–base junction, and of eliminating the series collector resistance by replacing the subcollector with a Schottky contact [16]. This was accomplished by an ambitious and clever process flow [17] in which a thick dielectric layer is deposited over the transistor after the emitter and base mesas, and metal traces to contact these layers, have been formed in the usual manner. Gold thermal vias are then formed through the dielectric to the emitter mesa, and a carrier is bonded to the emitter side of the wafer. Finally, the substrate is etched away, a Schottky contact that is scarcely larger than the active device is formed on the back of the collector drift region, and the collector material outside this contact is partially etched away. The resulting structure is shown in Figure 19.4.

Because this structure eliminates the problematic scaling issues associated with poor scaling of C_{cb} and counterproductive scaling of r_b, extremely high-speed results have been demonstrated using it. As the device has evolved, mostly by scaling the collector, its performance has steadily improved. The cut-off frequency f_T has gone from 127 to 252 GHz, and f_{max} from 277 to 820 GHz, for 0.6–0.7 μm emitters as the collector lithography has shrunk to 0.6 μm [17, 18]. A delay analysis similar to the one above

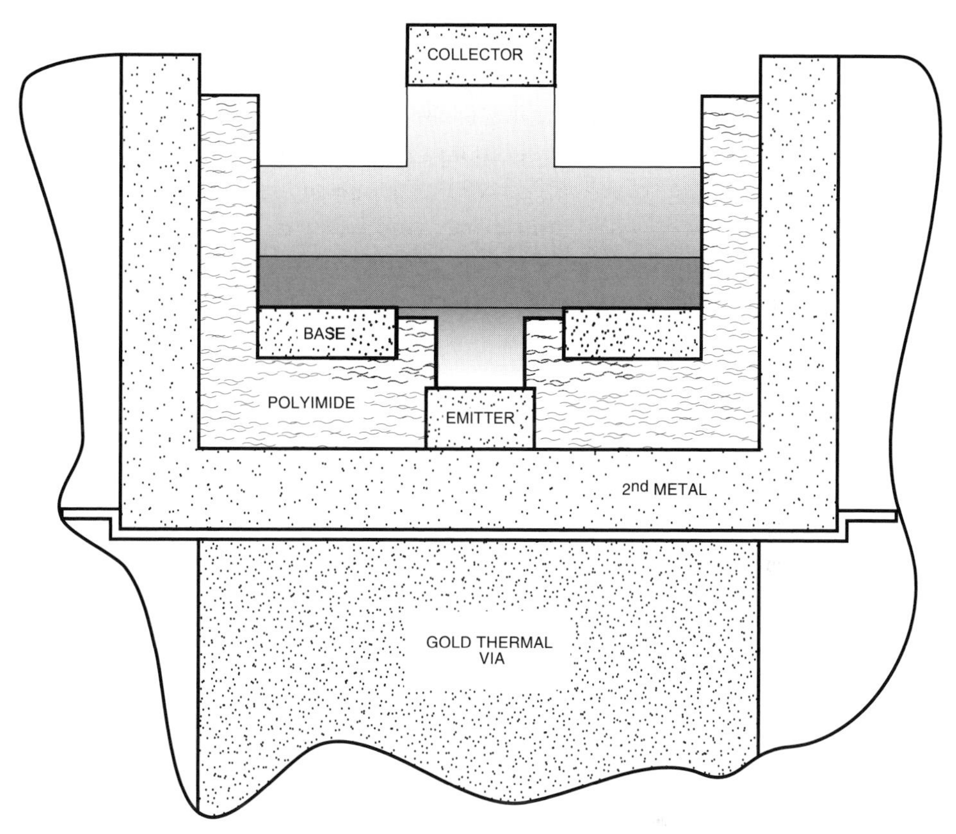

Fig. 19.4. Cross-section of a finished transferred-substrate HBT. The emitter is insulated from the thermal via by a thin layer of Si_3N_4. Note the small extrinsic collector area.

shows that $\tau_B + \tau_C$ is 0.41 ps for a device with a 40 nm InGaAs base and 200 nm InGaAs collector. Further reduction of parasitic emitter resistance and the use of InP for the collector seem likely to lead to f_T well over 400 GHz.

19.5 III–V surfaces and the emitter base saddle-point

One of the distinguishing features of most III–V semiconductor systems is their poor surface properties, compared to silicon. This comes about because they typically have a large number of surface states, which pin the Fermi level at the surface somewhere in the forbidden gap. These surface states also enhance the surface recombination velocity. In the case of gallium arsenide, which is the oldest and best studied III–V semiconductor, controversy over the exact physics which gives rise to these surface states has raged for years [19]. However, it is generally agreed that in n-type material, the Fermi level is pinned close to mid-gap. In p-type material it is pinned 0.4–0.5 eV above the valence band [20]. Early in the development of the AlGaAs emitter HBT it was realized that this surface pinning leads to an unusual shape for the electrostatic potential at the mesa surface of the emitter base junction [21]. Figure 19.5 shows the conduction-band energy near the bottom of the emitter mesa, at roughly 1 V forward bias. This plot reveals the existence of a saddle point in the potential energy, inside the emitter base space-charge region, close to the intersection of the emitter mesa surface and the base. Transistor operation is based on thermal injection of carriers perpendicularly over the emitter–base potential barrier into the neutral base. As shown in the figure, this saddle-point provides a potential path for electron injection laterally into the surface depletion region of the extrinsic base. Electrons that are injected into this region are then swept directly to the surface where they can recombine with holes through the surface states. If, as is the typically the case, the surface recombination velocity is large, the local current density for this lateral injection will be orders of magnitude larger than the normal injection current. All this lateral current appears as a periphery-related base current, which then reduces the current gain in devices with a large perimeter-to-area ratio, that is to say small devices. As shown in [21], the current gain of a device with a perimeter-to-area ratio of 10^4 cm^{-1} will not be much over 10. Thus in order to scale devices for high-frequency performance, this problem must be overcome.

There are two ways to accomplish this. Either the recombination velocity for laterally injected electrons must be reduced, or the saddle-point potential must be increased so that fewer electrons are injected. The use of a ledge – a thin layer of depleted emitter material in the extrinsic base region – accomplishes both of these objectives and has been shown to be effective in eliminating the periphery-associated recombination current in the AlGaAs/GaAs and InGaP/GaAs material systems [22, 23]. Use of a different material system can also reduce or eliminate this problem; for example InGaAs/InP HBTs can be fabricated with negligible periphery current and no ledge [24, 25]. This comes about through more favorable pinning potentials and lower surface recombination velocity.

19.6 Thermal considerations

High-frequency transistors must be operated at high current density in order to minimize the emitter charging delay, and the portion of the collector charging delay associated with $1/g_m$. In applications where output power is a consideration, the collector voltage may also be high. As a consequence, the operating power density of these devices can easily exceed 10^5 W/cm^2, and thermal effects play an important role in their operation. Typical thermal resistances for HBTs on GaAs substrates are on the order of 2×10^3 K/W for 10-μm-long emitters, or 3×10^3 K/W for 5-μm-long emitters, on very thin (75 μm) substrates with 25 °C heat sinking [26]. In realistic applications, heat-sink temperatures can easily be 50–75 °C. The thermal conductivity

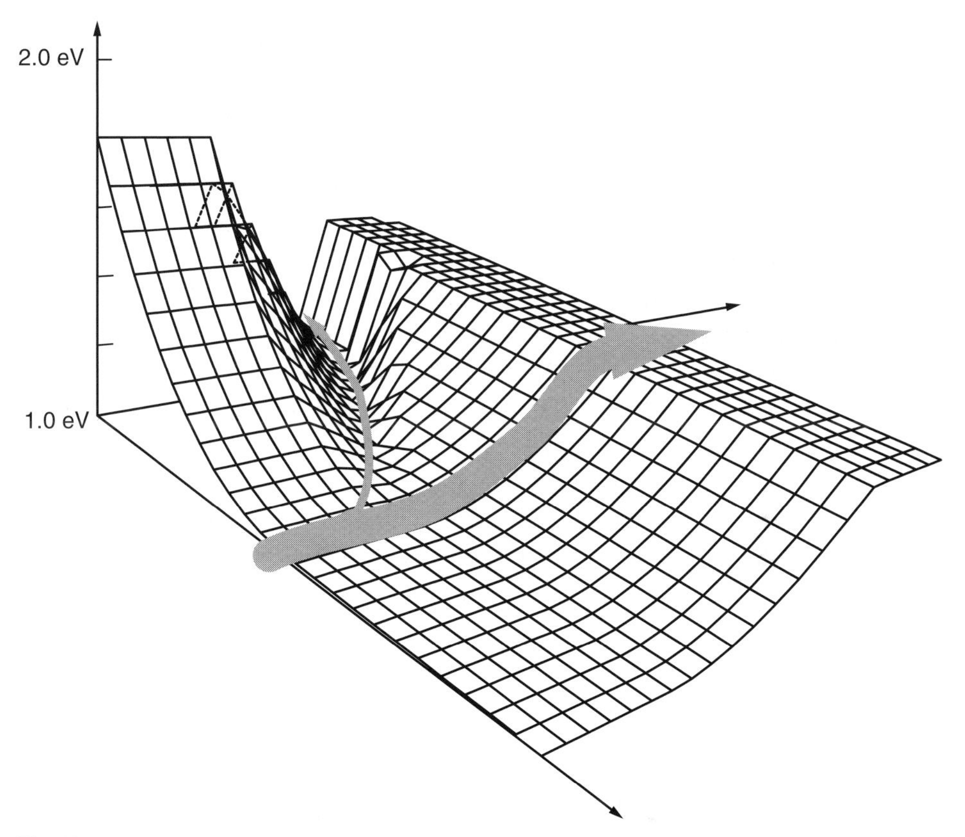

Fig. 19.5. Perspective view of the conduction-band energy of a graded-junction AlGaAs/GaAs HBT with no ledge. The main current path is indicated by the thick arrow. Note the low potential of the saddle point, which leads to the parasitic electron injection as shown.

of GaAs is actually temperature dependent; it has the general form [27]

$$K = A/T^{1.2}. \tag{19.4}$$

A depends on the doping concentration, falling from 544 W $K^{0.2}$ cm^{-1} for very pure material to 357 W $K^{0.2}$ cm^{-1} for heavily doped material. Figure 19.6 compares the thermal conductivity of undoped GaAs, InP [28], and InGaAs as a function of temperature. That figure shows that we should generally expect substantially lower thermal resistances for HBTs grown on InP; this is indeed the case. Devices on thick (375 μm) substrates have been shown to have a thermal resistance of 1.8×10^3 K/W for 10-μm-long emitters [29]. One caveat is worth pointing out regarding InP-based HBTs. Often these devices will incorporate a layer of heavily doped InGaAs in the subcollector layer, to facilitate formation of the collector contact and to enhance the subcollector sheet conductance. InGaAs lattice-matched to InP is thought to have an extremely low thermal conductivity – 0.05 W $K^{-1}cm^{-1}$ – compared to InP [29]. This figure, however, is for the lattice conductivity [30]. The combination of low

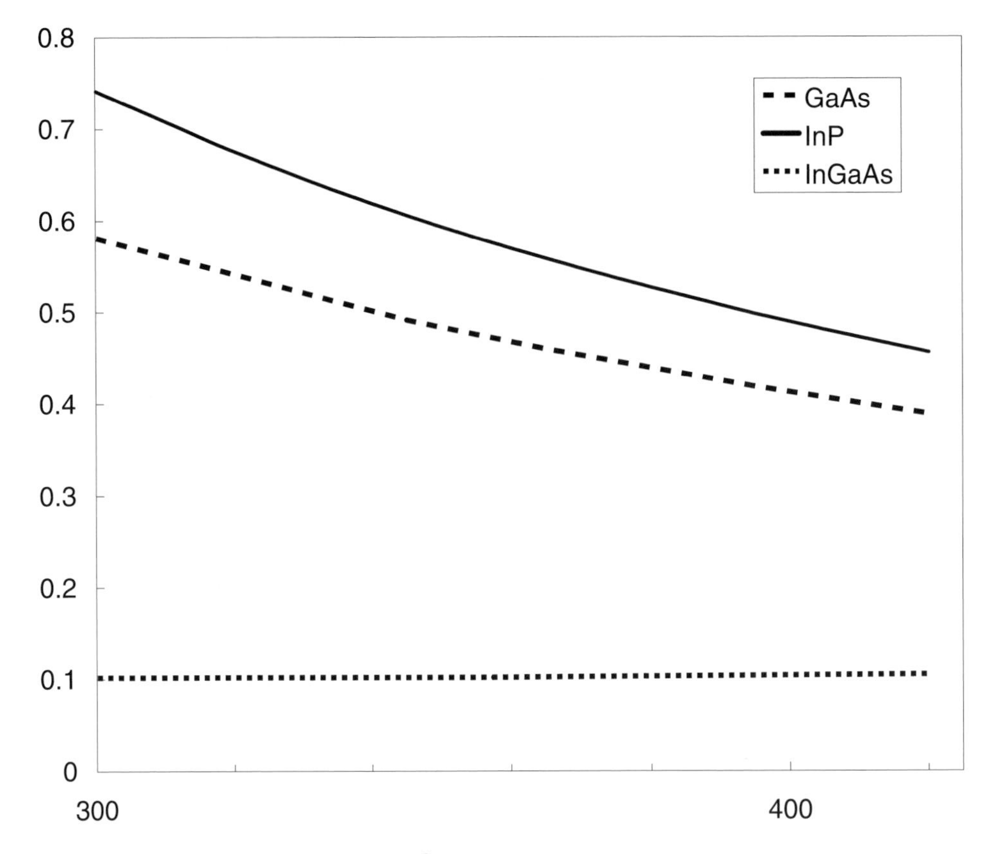

Fig. 19.6. Thermal conductivity, in W/cm^2 K, of three common collector materials, vs. temperature.

lattice thermal conductivity and high electron mobility leads to a situation where the electronic component of the thermal conductivity is significant. For strongly degenerate semiconductors [31],

$$k_e = \pi^2 k_b^2 \sigma T / q^2, \tag{19.5}$$

which has a room temperature value of 0.05 W K^{-1} cm^{-1} for 2×10^{19} cm^{-3} n-type material. Nonetheless, positioned as it is directly below the collector drift region where most of the power is dissipated, such a layer has the maximum deleterious effect possible. Heat flow through layered structures with layer thickness on the order of the lateral device dimensions is fairly complex, and numerical modeling is the surest way of accurately solving these problems. Qualitatively, use of InP instead of GaAs, avoidance of InGaAs layers and scaling of devices to long, skinny emitters, is desirable.

Junction temperature is only one of the thermal design problems presented by HBTs. Multi-emitter-finger devices can exhibit a variety of thermal runaway problems, depending on the details of the temperature coefficients associated with a particular device, and its thermal resistance. The origin of any thermal runaway problem is the temperature dependence of the injection current. We saw in the previous chapter that the exact form of J_C depends on whether the emitter–base junction is abrupt, or graded. However, from (18.21) and (18.24) we see that it always has the form

$$J_C = -qn_E^* v_{eff}/\delta, \tag{19.6a}$$

where

$$\delta = \exp\left(-\Delta E_n/k_B T\right) \tag{19.6b}$$

$$n_E^* = n_E \exp\left(-E_{pk}/k_B T\right) \tag{19.6c}$$

and ΔE_n is the height of the conduction-band spike (hence, 0 in the graded case). The temperature dependence of v_{eff} is small compared to that of n_E^*. Then at fixed voltage across the emitter–base junction, i.e., ignoring resistive drops,

$$\frac{1}{J_C}\frac{\partial J_C}{\partial T} \approx \frac{1}{T}\frac{E_{pk} - \Delta E_n}{k_B T}. \tag{19.7}$$

When the emitter–base junction heats up, the collector current density increases substantially. In a typical circuit application no transistor will be biased at constant v_{be}; in fact the collector bias current will be held more or less constant. But in a multi-emitter device, each emitter finger is at substantially the same v_{be}. If one emitter finger starts to heat up, it can rob current from the other fingers without affecting the total collector current. In so doing, its temperature will rise further, and it will draw even more current. Whether this process in fact runs away depends on the thermal resistance of each finger, the thermal coupling between fingers, the parasitic emitter resistance associated with each emitter finger and the bias point. Given knowledge of the thermal resistance emitter resistance, and electrical temperature coefficients, a set of non-linear equations can be solved to determine a critical current as a function of V_{CE} [29, 32]. Operation above this current can lead to thermal instability. However, since the interfinger thermal coupling is typically left out of these analyses, transistors with very tight thermal coupling between emitter fingers can be successfully operated above the critical current. If the transistor does enter the thermally unstable regime, one emitter finger may hog most of the emitter current; i.e., it will find itself carrying current intended to be spread out among several fingers, at a junction temperature and current density well beyond intended design limits. At the least, this will result in rapid degradation, or even burn-out, of the affected device. It may also result in pathological electrical behavior. For example AlGaAs/GaAs HBTs have a negative temperature coefficient for current gain, so when runaway current-hogging occurs, it is accompanied by a substantial decrease in β. This phenomenon is called gain collapse, and its signature is a marked collapse of the common emitter family curves beyond the region where thermal instability occurs. A similar phenomenon has been observed in InP-based HBTs, although the β of these devices is essentially temperature-independent [29]. However, the devices studied in [29] do have a *current*-dependent β, so they still exhibit a collapse phenomenon. A multi-finger transistor that had perfectly temperature- and current-independent electrical properties would be free of obvious electrical misbehavior like current collapse. Current-hogging will still occur with the wrong combination of bias point, thermal characteristics and emitter resistance. In turn, early failure or device burn-out at less than the expected power dissipation will result.

19.7 Reliability issues

19.7.1 Introduction

Compound semiconductor HBTs present some very interesting problems in reliability physics. The manifestations of reliability problems are disparate, ranging from a significant increase in emitter–base voltage, to a significant increase in emitter–base leakage, to increases in collector–base mesa side wall leakage, to name just a few. Still, most of these problems can be tied together as a single theme with two unifying elements. The first element is the involvement of defect chemistry. In other words, the appearance or movement of some lattice imperfection (and in this sense even intentionally introduced dopants are defects) either in the bulk semiconductor or at the semiconductor surface takes place. These changes in turn lead to an irreversible change in the electrical characteristics of the device over the course of operation. This is the definition of a reliability problem. The second element is that the defect chemistry is driven by quanta of energy deposited into the lattice by non-radiative recombination. To be sure, not all reliability problems fit into this schema; it is always possible to create a device that will fail, for example, through a purely thermally driven process by making a poor choice of metallization. But the theme holds up well enough that the beryllium diffusion problem is a useful paradigm.

19.7.2 The beryllium diffusion problem

Historically, the p-type dopant with the most predictable and most material-independent incorporation properties for MBE-grown material has been beryllium, so this has been the earliest and most popular choice of dopant. As it turns out, the beryllium atom can be displaced from its substitutional gallium site, where it acts as an acceptor, fairly easily. Once out, the interstitial atom is a fast diffuser [33]. So the diffusivity of beryllium is determined by the rate at which it is displaced from its substitutional site, and the rate at which interstitial beryllium can be returned to a substitutional site. One example of the defect chemistry that would govern this behavior is the reaction equation

$$\mathrm{Ga}_{\mathrm{Be}}^{-} + 2\mathrm{p}^{+} \leftrightarrow \mathrm{I}_{\mathrm{Be}}^{0} + \mathrm{V}_{\mathrm{Ga}}^{+}, \tag{19.8}$$

which describes the displacement of a beryllium atom from a substitutional gallium site to an interstitial site, leaving behind a gallium vacancy. Two holes are consumed in the process, to conserve charge. Some energy must be supplied to break the Be–As bonds. Before proceeding further, we should note that the precise equations which govern these processes remain controversial [34]. There is not full agreement on what the charge states of interstitial and vacancy defects really are; but the choice of charge state can have a profound effect on how well a given model based on a particular set of defect reactions fits the experimental data. The true charge states are slowly yielding to numerical calculation [35, 36], but much work still needs to be done in this area, and the numerical results have to be adopted by other workers. Equation (19.8) should thus be viewed as an example, rather than as a necessarily precise description. Nonetheless, it leads to qualitatively correct predictions about various aspects of Be diffusion. It leads to the equilibrium constant

$$K = \frac{[\mathrm{I}_{\mathrm{Be}}^{0}][\mathrm{V}_{\mathrm{Ga}}^{+}]}{\mathrm{p}^2[\mathrm{Ga}_{\mathrm{Be}}^{-}]} \tag{19.9}$$

whose form shows that production of interstitial beryllium, and hence the diffusivity of beryllium, is enhanced by high hole concentration. This makes HBTs particularly susceptible to beryllium diffusion problems because of the scaling considerations which always favor high base doping. Conversely, diffusion can be reduced by having a high gallium vacancy concentration; this can at least be achieved during MBE growth by using a high arsenic flux [33].

There is more to beryllium, however, than diffusion during high-temperature exposure over the course of epitaxial growth [33, 37]. It was eventually noticed that the emitter–base voltage of AlGaAs/GaAs HBTs exhibited a significant and irreversible increase after high-current device operation. An example of the effect is shown in Figure 19.7. This behavior is due to beryllium diffusion into the emitter, at junction temperatures too cold to cause purely thermally activated diffusion. Soon after the first reports [38, 39], Uematsu and Wada showed compelling evidence of the same effect in tunnel diodes, which had the advantage that the effective diffusion coefficient for beryllium could be determined directly from the device characteristics [40]. Their data can be reanalyzed to show that the diffusion coefficient has the form:

$$D_{\mathrm{Be}}(j) = 0.6 \times 10^{-11} \left(\frac{J}{1000\ \mathrm{A/cm^2}} \right)^{3.9} \exp\left(-0.59q/kT\right)\ \mathrm{cm^2/s} \tag{19.10}$$

for current densities between 1000 and 2000 A/cm^2. The activation energy for purely thermal diffusion is 1.8 ± 0.23 eV. The 0.59 eV activation energy for current-activated diffusion shows that the physics of the two processes are very different. The obvious difference is that current flow is accompanied by non-radiative recombination events. Each of these events delivers energy to the crystal lattice in a packet that is less than the bandgap energy, but of the same order of magnitude. Ordinarily, the defects

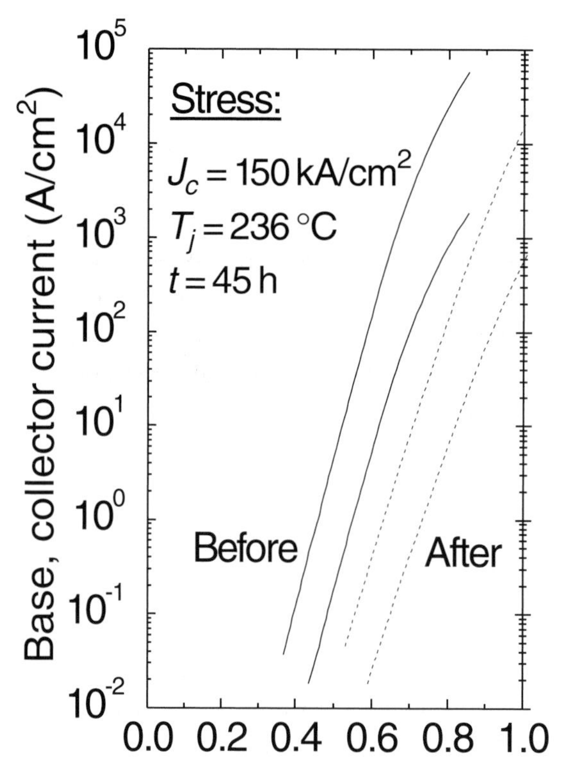

Fig. 19.7. Gummel plots for and AlInAs/InGaAs HBT with a 4×10^{19} beryllium-doped base, before and after current stress at elevated temperature. The plot shows the shift in v_{be} characteristic of beryllium diffusion. (Courtesy of S. Bahl.)

responsible for non-radiative recombination are roughly mid-gap defects, so the energy delivered will be roughly half the bandgap. The energetics of this process are probed in more depth in Problem 19.4, where we see that non-radiative recombination can be argued to deliver energy in quanta of *at least* half the bandgap (e.g. 0.7–1.4 eV for GaAs). This energy will be delivered to a single defect, which can thereby be raised to a highly excited state. If we think of the defect as comprising either an atom or a vacancy, and the surrounding atoms that it is bonded to, then the idea that the defect will undergo some profound change such as migration, or even dissociation into daughter defects, is plausible. The most likely explanation for the decrease in activation energy then is that the recombination energy results in the creation of a new kind of defect that is not otherwise present. In turn, this new defect enables a different diffusion mechanism with a lower activation energy, which therefore dominates over the process governing purely thermal diffusion. For example, if thermal Be diffusion is indeed completely described by the interstitial-substitutional model and the defect chemistry of Equation (19.8), the activation energy would correspond to the energy required to remove the substitutional beryllium from a gallium lattice site.

19.7.3 Beryllium diffusion solutions

A common speculation is that recombination-enhanced generation of interstitial Ga defects is at the root of the current-induced diffusion problem [40]. Interstitial Ga is expected to exchange sites with some of the substitutional Be. In fact, there is not consensus that this hypothesis is correct in a detailed sense [41], but there is growing agreement on the idea that defect creation through recombination, as well as the interaction of these defects with beryllium, plays a key role in causing the beryllium diffusion [42]. Even if it is not a perfect representation of the details, the general picture outlined here is accurate enough to be useful in reducing the beryllium diffusion problem, and thus improving the reliability. There are essentially three elements to this picture. First, the physical process of recombination-current-induced defect generation plays an essential role. Second, the defects so generated cause the production of interstitial beryllium through some defect reaction. Once this happens, the interstitial beryllium then diffuses rapidly. Third, since beryllium is a fast interstitial diffuser which is susceptible to displacement from the lattice, it is also particularly susceptible to this class of reliability problem. This suggests several means which might be used to improve HBT reliability: reduction of the non-radiative recombination current, reduction of the emitter–base junction material's susceptibility to defect production, or outright elimination of the beryllium.

The first approach is generally not fruitful since most non-radiative recombination paths will have been wrung out of a technology early in its development, because of their impact on current gain. The second approach has turned out to be quite viable. Streit *et al.* demonstrated one of the earliest improvements in beryllium-based HBT reliability, by carefully controlling the MBE conditions during layer growth [41]. By adjusting those conditions to increase the concentration of gallium vacancies grown into the base, they were able to demonstrate an extrapolated mean time to failure (MTTF) of $> 10^8$ hours at a 125 °C junction temperature, and a collector current density of 3 kA/cm^2. While this particular extrapolation is open to criticism [42, 43], and the strong dependence of beryllium diffusion on current density suggests that these devices will not be reliable at the bias conditions required for high-frequency performance, the result still demonstrates that attention to the defect problem improves the reliability. The basis of the improvement is the large number of grown-in Ga vacancies; these vacancies act as a sink for any interstitial Ga that might be created during current stress, as well as any interstitial beryllium that might subsequently be created. An even greater improvement can be obtained by changing material systems. By using an InAlAs emitter and InGaAs base, the amount of beryllium diffusion into the emitter can be reduced considerably compared to that in an AlGaAs/GaAs HBT [44]. In fact, devices with MTTF in excess of 10^6 hours at a 125 °C junction temperature, and a collector current density of 70 kA/cm^2, can easily be obtained [45, 46]. In these devices, most of the non-radiative recombination takes place in

InGaAs, whose small bandgap leads to a much smaller release of energy by each recombination event.

The most straightforward solution to the beryllium problem has been to replace it with carbon [47]. This completely eliminates diffusion problems, but a new assortment of problems appears. These range from increased emitter–base leakage due to trap formation in that junction's space-charge region, to changes in the effective base doping when large amounts of grown-in hydrogen are driven from the base [43], to creation of carbon precipitates at the junction [48]. All of these mechanisms exhibit strong dependence on the bias current, suggesting that recombination-induced defect reactions continue to play an essential role in the degradation mechanism. Moreover, the use of an InGaP emitter, which dramatically reduces space-charge-region recombination current (because of the large, abrupt valence-band discontinuity) raises the current density at which these various failure mechanisms become a problem [43, 48, 49]. Carbon-doped InGaP devices have an MTTF $> 10^6$ hours, at a 125 °C junction temperature and a collector current density of 60 kA/cm^2. At the slightly higher current density required for operation at maximum f_T, the reliability will not be as good, but quite adequate for many applications. As material systems evolve and further progress is made in this area, the details of the failure mechanisms will certainly change, but it seems likely that the fundamentals – non-radiative recombination driving microscopic changes in the location of atoms – will not.

19.8 Conclusion

We have examined some of the key problems in realizing practical high-speed HBTs. The following general features have emerged from that examination.

1. Successful material systems for III–V HBTs exhibit four important properties: (i) a low density of heterointerface states; (ii) significantly better transport than in Si; (iii) semiinsulating substrate; (iv) relatively easy epitaxial growth, with nearly atomic scale accuracy.
2. Scaling of vertical device dimensions to improve f_T centers around reduction of the total collector delay; however, this is difficult to do because any decrease in the transit delay is accompanied by an increase in the collector charging time.
3. Velocity overshoot plays a significant role in collector transport.
4. Scaling of the emitter size can be constrained by the problem of periphery-dependent base current, depending on the details of fabrication technology and material properties.
5. Careful thermal design is essential for multi-emitter transistors.
6. Reliability considerations place an upper limit on the practical operating current density, through the combination of thermal effects and recombination-enhanced

defect reactions. In the examples considered, the base dopant plays a central role in the important defect reactions, and hence in failure. The role of the base dopant may not be central in all material systems, but the general theme of defect reactions driven by large quanta of energy delivered to the lattice will be.

Acknowledgement
This chapter represents the accumulated knowledge of nearly two decades of work on III–V heterojunctions, and the contributions of many talented and stimulating colleagues at Hewlett-Packard Labs and Agilent Technologies. The thought in this chapter is theirs as much as mine. Many thanks to Alice Fischer-Colbrie, Dan Mars, Virginia Robbins, Denny Houng, Jeff Miller, Karen Seaward, Mark Hueschen, Steve Kofol, Arlene Wakita, Hans Rohdin, Sandeep Bahl, Stretch Camnitz, Tom Low, Steve Lester, and Ying-Lan Chang. Thanks also to Sandip Tiwari of Cornell University for encouragement when most needed, and many stimulating discussions.

19.9 Bibliography

[1] US Patent 2569347

[2] H. Kroemer, 'Theory of a wide-gap emitter for transistors', *Proceedings of the IRE*, Vol. 45, pp. 1535–1537, 1957.

[3] W. Dumke, J. Woodall and V. Rideout, 'GaAs–GaAlAs heterojunction transistor for high frequency operation', *Solid State Electronics*, Vol. 15, pp. 1339–1343, 1972.

[4] C. Liechti, 'GaAs IC technology–impact on the semiconductor industry', *Technical Digest of the International Electron Devices Meeting*, IEEE, Piscataway, pp. 13–18, 1984.

[5] P. Asbeck *et al.*, '(Ga,Al)As/GaAs bipolar transistors for digital integrated circuits', *IEEE Electron Device Letters*, Vol. 3, pp. 403–404, 1982.

[6] L.S. McCarthy, P. Kozodoy, M. Rodwell, S. DenBaars and U. Mishra, 'AlGaN/GaN heterojunction bipolar transistor', *IEEE Electron Device Letters*, pp. 277–279, 1999.

[7] C. Bolognesi *et al.*, 'Low-offset n–p–n InP–GaAsSb–InP double heterojunction bipolar transistors with abrupt interfaces and ballistically launched collector electrons', *56th Annual Device Research Conference*, IEEE, Piscataway, pp. 30–31, 1998.

[8] H. Tews, R. Neumann, R. Treichler and P. Zwicknagl, 'Mg-doped base layers for GaAs–GaAlAs heterobipolar transistors', *ITG-Fachberichte*, Vol. 112, pp. 67–72, 1990.

[9] Y.M. Houng and T.S. Low, 'The doping of GaAs and $Al_xGa_{1-x}As$ using diethyltellurium in low pressure OMVPE', *Journal of Crystal Growth*, Vol. 77, pp. 272–80, 1986.

[10] A.R. Clawson and C.M. Hanson, 'MOCVD grown Si-doped n^+ InP layers for the subcollector region in HBTs', *Proceedings of the 1994 6th International Conference on Indium Phosphide and Related Materials*, IEEE, Piscataway, pp. 114–117, 1994.

[11] T. Lester *et al.*, 'A manufacturable process for HBT circuits', *Proceedings of the 20th International Symposium on Gallium Arsenide and Related Compounds*, Institute of Physics, pp. 449–454, 1993.

[12] T. Low *et al.*, 'InGaP HBT technology for RF and microwave instrumentation', *Solid State Electronics*, Vol. 43, pp. 1437–1444, 1999.

[13] L. Camnitz and N. Moll, 'An analysis of the cutoff-frequency behavior of microwave heterostructure bipolar transistors', in *Compound Semiconductor Transistors*, S. Tiwari, Ed., pp. 21–45, IEEE Press, Piscataway, 1993.

[14] S. Tiwari, M. Fischetti and S. Laux, 'Overshoot in transient and steady-state in GaAs, InP, GaInAs, and InAs bipolar transistors', *Technical Digest of the International Electron Devices Meeting*, IEEE, Piscataway, pp. 435–438, 1990.

[15] T. Low and D. Mars, 'Space-charge recombination in N-AlGaAs/p$^+$–GaAs heterojunction diodes', *Applied Physics Letters*, Vol. 55, pp. 2423–2425, 1989.

[16] U. Battacharya *et al.*, 'Transferred substrate Schottky-collector heterojunction bipolar transistors: First results and scaling laws for high f_{max}', *IEEE Electron Devices*, Vol. 16, pp. 357–359, 1995.

[17] B. Agarwal *et al.*, 'A 277-GHz f_{max} transferred-substrate heterojunction bipolar transistor', *IEEE Electron Device Letters*, Vol. 18, pp. 228–232, 1997.

[18] M. Rodwell *et al.*, 'Transferred-substrate heterojunction bipolar transistor integrated circuit technology', *Proceedings of the 11th International Conference on Indium Phosphide and Related Materials*, IEEE, Piscataway, pp. 169–174, 1999.

[19] W. Spicer and A. Green, 'Reaching consensus and closure on key questions, a history of success, and failure of GaAs surfaces and interfaces at the Proceedings of the Physics and Chemistry of Semiconductor Interfaces', *Journal of Vacuum Science and Technology B*, Vol. 11, pp. 1347–1353, 1993.

[20] V. Emiliani *et al.* 'Interaction mechanisms of near-surface quantum wells with oxidized and H-passivated AlGaAs surfaces', *Journal of Applied Physics*, Vol. 75, pp. 5114–5122, 1994.

[21] S. Tiwari and D. Frank, 'Analysis of the operation of GaAlAs/GaAs HBT's', *IEEE Transactions on Electron Devices*, Vol. 36, pp. 2105–2121, 1989.

[22] V.M. Andreev *et al.*, 'Reduction of surface recombination currents in AlGaAs/GaAs p–n junctions', *Pis'ma v Zhurnal Tekhnicheskoi Fizika*, Vol. 13, pp. 1481–1485, 1987.

[23] M.T. Fresina, O. Hartmann, G. Stillman, 'Selective self-aligned emitter ledge formation for heterojunction bipolar', *IEEE Electron Device Letters*, Vol. 17, pp. 555–556, 1996.

[24] R. Driad *et al.*, 'Passivation of InGaAs surfaces and InGaAs/InP heterojunction bipolar transistors by sulfur treatment', *Applied Physics Letters*, Vol. 73, pp. 665–667, 1998.

[25] T. Kikawa *et al.*, 'Passivation of InP-based heterostructure bipolar transistors in relation to surface fermi level', *Japanese Journal of Applied Physics*, Vol. 38, pp. 1195–1199, 1999.

[26] J. Higgins, 'Thermal properties of power HBTs', *IEEE Transactions on Electron Devices*, Vol. 40, pp. 2171–2177, 1993.

[27] J. Brice, 'Thermal conductivity of GaAs', in *Properties of Gallium Arsenide*, p. 1.9, The Institution of Electrical Engineers, London, 1986.

[28] J. Brice, 'Thermal conductivity of InP', in *Properties of Indium Phosphide*, p. 1.9, The Institution of Electrical Engineers, London, 1991.

[29] W. Lui *et al.*, 'Thermal properties and thermal instabilities of InP-based heterojunction bipolar transistors', *IEEE Transactions on Electron Devices*, Vol. 43, pp. 388–395, 1996.

[30] W. Nakwaski, 'Thermal conductivity of binary, ternary, and quaternary III–V compounds', *Journal of Applied Physics*, Vol. 64, pp. 159–166, 1988.

[31] R. Smith, *Wave Mechanics of Crystalline Solids*, pp. 358–362, Chapman and Hall, London, 1969.

[32] K. Lu and C. Snowden, 'Analysis of thermal instability in multi-finger power AlGaAs/GaAs HBTs', *IEEE Transactions on Electron Devices*, Vol. 43, pp. 1799–1805, 1996.

[33] J. Miller, D. Collins and N. Moll, 'Control of Be diffusion in molecular beam epitaxy GaAs', *Applied Physics Letters*, Vol. 46, pp. 960–962, 1985.

[34] R.M. Cohen, 'Application of the charged point-defect model to diffusion and interdiffusion in GaAs', *Journal of Applied Physics*, Vol. 67, pp. 7268–7273, 1990.

[35] S. Zhang and J. Northrup, 'Chemical potential dependence of defect formation energies in GaAs: application to Ga self-diffusion', *Physical Review Letters*, Vol. 67, p2339–2342, 1991.

[36] D. Chadi, 'Self-interstitial bonding configurations in GaAs and Si', *Physical Review B*, Vol. 46, pp. 9400–9408, 1992.

[37] D. Miller and P. Asbeck, 'Be redistribution during growth of GaAs and AlGaAs by molecular beam epitaxy', *Journal of Applied Physics*, Vol. 57, pp. 1816–1822, 1985.

[38] M. Hafizi *et al.*, 'Reliability analysis of GaAs/AlGaAs HBTs under forward current/temperature stress', *Technical Digest of the GaAs IC Symposium*, IEEE, Piscataway, pp. 329–332, 1990.

[39] O. Nakajima *et al.*, 'Current induced degradation of Be-doped AlGaAs/GaAs HBTs and its suppression by Zn diffusion into extrinsic base layer', *Technical Digest of the International Electron Devices Meeting*, IEEE, Piscataway, pp. 673–676, 1990.

[40] M. Uematsu and K. Wada, 'Recombination enhanced impurity diffusion in Be-doped GaAs', *Applied Physics Letters*, Vol. 58, pp. 2015–2017, 1991.

[41] D. Streit *et al.*, 'High-reliability GaAs-AlGaAs HBTs by MBE with Be base doping and InGaAs emitter contacts', *IEEE Electron Device Letters*, Vol. 12, pp. 471–473, 1991.

[42] T. Henderson, 'The GaAs heterojunction bipolar transistor: an electron device with optical device reliability', *Microelectronics Reliability*, Vol. 36, pp. 1819–1886, 1996.

[43] T. Henderson, 'Physics of degradation in GaAs-based heterojunction bipolar transistors', *Microelectronics Reliability*, Vol. 39, pp. 1033–1042, 1999.

[44] S. Tanaka *et al.*, 'Characterization of current-induced degradation in Be-doped HBTs based in GaAs and InP', *IEEE Transactions on Electron Devices*, Vol. 40, pp. 1194–1201, 1993.

[45] M. Hafizi, R. Metzger and W. Stanchina, 'Stability of beryllium doped compositionally graded and abrupt AlInAs/GaInAs heterojunction bipolar transistors', *Applied Physics Letters*, Vol. 63, pp. 93–95, 1993.

[46] S. Bahl *et al.*, 'Be diffusion in InGaAs/InP heterojunction bipolar transistors' submitted to *IEEE Electron Device Letters*, 2000.

[47] G. Wang *et al.*, 'High-performance MOCVD-grown AlGaAs/GaAs heterojunction bipolar transistors with carbon-doped base', *IEEE Electron Device Letters*, Vol. 12, pp. 347–349, 1991.

[48] O. Ueda *et al.*, 'Current status of reliability of InGaP/GaAs HBTs', *Solid State Electronics*, Vol. 41, pp. 1605–1610, 1997.

Table 19.2.

GaAs	0.565 32
AlAs	0.566 11
InP	0.586 8
InAs	0.605 8
GaSb	0.609 4

[49] T. Low *et al.*, 'Migration from an AlGaAs to an InGaP emitter HBT IC process for improved reliability', *IEEE GaAs IC Symposium Technical Digest*, pp. 153–156, 1998.

19.10 Problems

19.1 The lattice parameters, in nanometers, of InP, AlAs, InAs, GaAs, and GaSb are shown in Table 19.2. All materials have a cubic zinc-blende crystal structure.

(a) Assuming that lattice parameter is a linear function of composition, work out the compositions of the four ternary alloys (alloys of the form $A_{1-x}B_xC$ which can be formed from these materials, and lattice-matched to InP.

(b) Which alloys might be most suitable as an emitter material, and which as a base? Explain.

19.2 Assume that the collector drift region is fully depleted, and that the capacitance is given by the appropriate parallel-plate capacitance. Calculate $\partial\tau_C/\partial W_{BC}$ vs W_{BC}, with A_c/A_e as a parameter. Scale the values from Table 19.1 for R_E.

19.3 Assume that the subcollector is a 1 μm thick layer containing 2×10^{19} electrons/cm^3, with a mobility of 1000 cm^2/V s. If the collector contact has a specific contact resistance of 3×10^{-7} Ω cm^2, estimate R_C for a one-sided contact to a 2 × 4-μm emitter device. Calculate and plot R_C as the emitter width is scaled, keeping the emitter area constant.

19.4 In a semiconductor under forward injection conditions, i.e. $pn \gg n_i^2$, non-radiative recombination takes place through a process where a trap level alternately captures an electron and a hole.

(a) Draw an energy diagram that shows the energy dissipated in each capture event.

(b) Plot the energy that must be dissipated by the trap, as a function of the energetic position of the trap within the bandgap. Assume that the trap energy is the same for the occupied and unoccupied trap.

Index

Italic numbers denote pages with tables and figures.